Quantum Mechanics

Lecture notes

Quantum Mechanics

Lecture notes

Konstantin K Likharev

IOP Publishing, Bristol, UK

ISBN   978-0-7503-1410-7 (ebook)
ISBN   978-0-7503-1411-4 (print)
ISBN   978-0-7503-1412-1 (mobi)

DOI   10.1088/2053-2563/aaf3a3

Version: 20190501

IOP Expanding Physics
ISSN 2053-2563 (online)
ISSN 2054-7315 (print)

British Library Cataloguing-in-Publication Data: A catalogue record for this book is available from the British Library.

Published by IOP Publishing, wholly owned by The Institute of Physics, London

IOP Publishing, Temple Circus, Temple Way, Bristol, BS1 6HG, UK

US Office: IOP Publishing, Inc., 190 North Independence Mall West, Suite 601, Philadelphia, PA 19106, USA

Contents

Preface to the EAP Series   ix

Preface to *Quantum Mechanics: Lecture notes*   xii

Acknowledgments   xiii

Notation   xiv

1   Introduction   **1-1**

1.1   Experimental motivations   1-1

1.2   Wave mechanics postulates   1-8

1.3   Postulates' discussion   1-13

1.4   Continuity equation   1-15

1.5   Eigenstates and eigenvalues   1-18

1.6   Time evolution   1-21

1.7   Spatial dependence   1-23

1.8   Dimensionality reduction   1-27

1.9   Problems   1-30

     References   1-32

2   1D wave mechanics   **2-1**

2.1   Basic relations   2-1

2.2   Free particle: wave packets   2-4

2.3   Particle reflection and tunneling   2-12

2.4   Motion in soft potentials   2-20

2.5   Resonant tunneling, and metastable states   2-30

2.6   Localized state coupling, and quantum oscillations   2-40

2.7   Periodic systems: energy bands and gaps   2-49

2.8   Periodic systems: particle dynamics   2-63

2.9   Harmonic oscillator: brute force approach   2-73

2.10 Problems   2-79

     References   2-87

3   Higher dimensionality effects   **3-1**

3.1   Quantum interference and the AB effect   3-1

3.2   Landau levels and quantum Hall effect   3-12

3.3   Scattering and diffraction   3-17

3.4 Energy bands in higher dimensions   3-27
3.5 Axially-symmetric systems   3-35
3.6 Spherically-symmetric systems: brute force approach   3-42
3.7 Atoms   3-51
3.8 Spherically-symmetric scatterers   3-61
3.9 Problems   3-65
    References   3-72

4   Bra–ket formalism   **4-1**

4.1 Motivation   4-1
4.2 States, state vectors, and linear operators   4-4
4.3 State basis and matrix representation   4-10
4.4 Change of basis, and matrix diagonalization   4-16
4.5 Observables: expectation values and uncertainties   4-27
4.6 Quantum dynamics: three pictures   4-33
4.7 Coordinate and momentum representations   4-45
4.8 Problems   4-55

5   Some exactly solvable problems   **5-1**

5.1 Two-level systems   5-1
5.2 The Ehrenfest theorem   5-8
5.3 The Feynman path integral   5-11
5.4 Revisiting harmonic oscillator   5-18
5.5 Glauber states and squeezed states   5-25
5.6 Revisiting spherically-symmetric systems   5-35
5.7 Spin and its addition to orbital angular momentum   5-40
5.8 Problems   5-47
    References   5-53

6   Perturbative approaches   **6-1**

6.1 Eigenproblems   6-1
6.2 The Stark effect   6-7
6.3 Fine structure of atomic levels   6-10
6.4 The Zeeman effect   6-16
6.5 Time-dependent perturbations   6-20

6.6   Quantum-mechanical golden rule   6-27
6.7   Golden rule for step-like perturbations   6-34
6.8   Problems   6-38
      References   6-42

7   Open quantum systems   **7-1**

7.1   Open systems, and the density matrix   7-1
7.2   Coordinate representation, and the Wigner function   7-8
7.3   Open system dynamics: dephasing   7-16
7.4   Fluctuation–dissipation theorem   7-22
7.5   The Heisenberg–Langevin approach   7-35
7.6   Density matrix approach   7-39
7.7   Problems   7-58
      References   7-60

8   Multiparticle systems   **8-1**

8.1   Distinguishable and indistinguishable particles   8-1
8.2   Singlets, triplets, and the exchange interaction   8-5
8.3   Multiparticle systems   8-14
8.4   Perturbative approaches   8-27
8.5   Quantum computation and cryptography   8-33
8.6   Problems   8-55
      References   8-60

9   Introduction to relativistic quantum mechanics   **9-1**

9.1   Electromagnetic field quantization   9-1
9.2   Photon absorption and counting   9-8
9.3   Photon emission: spontaneous and stimulated   9-15
9.4   Cavity QED   9-20
9.5   The Klein–Gordon and relativistic Schrödinger equations   9-25
9.6   Dirac's theory   9-29
9.7   Low-energy limit   9-33
9.8   Problems   9-39
      References   9-41

10   Making sense of quantum mechanics   **10-1**

10.1 Quantum measurements   10-1
10.2 QND measurements   10-7
10.3 Hidden variables and local reality   10-12
10.4 Interpretations of quantum mechanics   10-18
        Reference   10-19

Appendices

A   Selected mathematical formulas   **A-1**

B   Selected physical constants   **B-1**

Bibliography   **13-1**

Preface to the EAP Series

Essential Advanced Physics

Essential Advanced Physics (EAP) is a series of lecture notes and problems with solutions, consisting of the following four parts[1]:

- *Part CM*: *Classical Mechanics* (a one-semester course),
- *Part EM*: *Classical Electrodynamics* (two semesters),
- *Part QM*: *Quantum Mechanics* (two semesters), and
- *Part SM*: *Statistical Mechanics* (one semester).

Each part includes two volumes: *Lecture Notes* and *Problems with Solutions*, and an additional file *Test Problems with Solutions*.

Distinguishing features of this series—in brief

- condensed lecture notes (~250 pp per semester)—much shorter than most textbooks
- emphasis on simple explanations of the main notions and phenomena of physics
- a focus on problem solution; extensive sets of problems with detailed model solutions
- additional files with test problems, freely available to qualified university instructors
- extensive cross-referencing between all parts of the series, which share style and notation

Level and prerequisites

The goal of this series is to bring the reader to a general physics knowledge level necessary for professional work in the field, regardless on whether the work is theoretical or experimental, fundamental or applied. From the formal point of view, this level (augmented by a few special topic courses in a particular field of concentration, and of course by an extensive thesis research experience) satisfies the typical PhD degree requirements. Selected parts of the series may be also valuable for graduate students and researchers of other disciplines, including astronomy, chemistry, mechanical engineering, electrical, computer and electronic engineering, and material science.

The entry level is a notch lower than that expected from a physics graduate from an average US college. In addition to physics, the series assumes the reader's familiarity with basic calculus and vector algebra, to such an extent that the meaning of the formulas listed in appendix A, 'Selected mathematical formulas' (reproduced at the end of each volume), is absolutely clear.

---

[1] Note that the (very ambiguous) term *mechanics* is used in these titles in its broadest sense. The acronym *EM* stems from another popular name for classical electrodynamics courses: *Electricity and Magnetism*.

Origins and motivation

The series is a by-product of the so-called 'core physics courses' I taught at Stony Brook University from 1991 to 2013. My main effort was to assist the development of students' problem-solving skills, rather than their idle memorization of formulas. (With a certain exaggeration, my lectures were not much more than introductions to problem solution.) The focus on this main objective, under the rigid time restrictions imposed by the SBU curriculum, had some negatives. First, the list of covered theoretical methods had to be limited to those necessary for the solution of the problems I had time to discuss. Second, I had no time to cover some core fields of physics—most painfully general relativity[2] and quantum field theory, beyond a few quantum electrodynamics elements at the end of *Part QM*.

The main motivation for putting my lecture notes and problems on paper, and their distribution to students, was my desperation to find textbooks and problem collections I could use, with a clear conscience, for my purposes. The available graduate textbooks, including the famous *Theoretical Physics* series by Landau and Lifshitz, did not match the minimalistic goal of my courses, mostly because they are far too long, and using them would mean hopping from one topic to another, picking up a chapter here and a section there, at a high risk of losing the necessary background material and logical connections between the course components—and the students' interest with them. In addition, many textbooks lack even brief discussions of several traditional and modern topics that I believe are necessary parts of every professional physicist's education[3].

On the problem side, most available collections are not based on particular textbooks, and the problem solutions in them either do not refer to any background material at all, or refer to the included short sets of formulas, which can hardly be used for systematic learning. Also, the solutions are frequently too short to be useful, and lack discussions of the results' physics.

Style

In an effort to comply with the Occam's Razor principle[4], and beat Malek's law[5], I have made every effort to make the discussion of each topic as clear as the time/space (and my ability :-) permitted, and as simple as the subject allowed. This effort has resulted in rather succinct lecture notes, which may be thoroughly read by a student during the semester. Despite this briefness, the introduction of every new

---

[2] For an introduction to this subject, I can recommend either a brief review by S Carroll, *Spacetime and Geometry* (2003, New York: Addison-Wesley) or a longer text by A Zee, *Einstein Gravity in a Nutshell* (2013, Princeton University Press).

[3] To list just a few: the statics and dynamics of elastic and fluid continua, the basics of physical kinetics, turbulence and deterministic chaos, the physics of computation, the energy relaxation and dephasing in open quantum systems, the reduced/RWA equations in classical and quantum mechanics, the physics of electrons and holes in semiconductors, optical fiber electrodynamics, macroscopic quantum effects in Bose–Einstein condensates, Bloch oscillations and Landau–Zener tunneling, cavity quantum electrodynamics, and density functional theory (DFT). All these topics are discussed, if only briefly, in my lecture notes.

[4] *Entia non sunt multiplicanda praeter necessitate*—Latin for 'Do not use more entities than necessary'.

[5] 'Any simple idea will be worded in the most complicated way'.

physical notion/effect and of every novel theoretical approach is always accompanied by an application example or two.

The additional exercises/problems listed at the end of each chapter were carefully selected[6], so that their solutions could better illustrate and enhance the lecture material. In formal classes, these problems may be used for homework, while individual learners are strongly encouraged to solve as many of them as practically possible. The few problems that require either longer calculations, or more creative approaches (or both), are marked by asterisks.

In contrast with the lecture notes, the model solutions of the problems (published in a separate volume for each part of the series) are more detailed than in most collections. In some instances they describe several alternative approaches to the problem, and frequently include discussions of the results' physics, thus augmenting the lecture notes. Additional files with sets of shorter problems (also with model solutions) more suitable for tests/exams, are available for qualified university instructors from the publisher, free of charge.

Disclaimer and encouragement

The prospective reader/instructor has to recognize the limited scope of this series (hence the qualifier *Essential* in its title), and in particular the lack of discussion of several techniques used in current theoretical physics research. On the other hand, I believe that the series gives a reasonable introduction to the *hard core* of physics—which many other sciences lack. With this hard core knowledge, today's student will always feel at home in physics, even in the often-unavoidable situations when research topics have to be changed at a career midpoint (when learning from scratch is terribly difficult—believe me :-). In addition, I have made every attempt to reveal the remarkable logic with which the basic notions and ideas of physics subfields merge into a wonderful single construct.

Most students I taught liked using my materials, so I fancy they may be useful to others as well—hence this publication, for which all texts have been carefully reviewed.

---

[6] Many of the problems are original, but it would be silly to avoid some old good problem ideas, with long-lost authorship, which wander from one textbook/collection to another one without references. The assignments and model solutions of all such problems have been re-worked carefully to fit my lecture material and style.

Preface to *Quantum Mechanics: Lecture notes*

The structure of this course is more or less traditional, with most attention paid to the non-relativistic quantum mechanics, and only chapter 9 reviewing the relativistic effects–first in electrodynamics, and then for particles with a non-zero rest mass.

One deviation from the tradition is that, due to the counter-intuitive character of quantum mechanics, I have found it necessary to start the course from a short discussion, in the beginning of chapter 1, of the experimental facts that, by the 1920s, has necessitated its development.

However, the feature that distinguishes this course most strongly from many modern textbooks on quantum mechanics is that the discussion of Dirac's bra-ket formalism is postponed until chapter 4, i.e. until after the discussion of numerous wave-mechanical effects in one- and multi-dimensional systems, respectively, in chapters 2 and 3. One reason for that decision was the author's serious adherence (declared in the general *Preface to the EAP Series*) to the Occam Razor principle, in particular to using only the simplest theoretical tools possible for discussions of particular physical phenomena. Another motivation was to discuss the most important quantum effects, including the energy band theory, without the heavy artillery of the bra-ket formalism, to make the discussion more accessible to the potential readership from the electrical engineering and material science communities. Finally, I believe that it is useful for the reader to see how the inconveniences and pitfalls of the wave mechanics approach gradually accumulate, thus justifying the eventual introduction of a more general formalism.

Another distinguishing feature of the course is its large attention to the notions of *dephasing* (alternatively called 'decoherence') and *energy relaxation*–the effects whose description needs to go beyond the usual idealization of a closed (Hamiltonian) quantum system. A clear understanding of these effects is necessary for any educated discussion of the conceptual issues of quantum measurements, and also of the recent numerous experiments with macroscopic-scale quantum systems (such as mechanical and electromagnetic resonators, superconductor qubits, etc), because of a substantial coupling of such systems to their environment. As a result, I felt compelled to give, in chapter 7, a discussion of open quantum systems, which is more typically reserved for statistical mechanics courses.

One more not-very-traditional topic, quantum computation and cryptography, is discussed at the end of chapter 8. Since this is a hot research field, with many aspects still actively debated, the style of its discussion is closer to that of a (brief) research review than to a textbook.

Finally, two related, still-controversial topics, quantum measurements and interpretations of quantum mechanics, are also so special that I have found it natural to place their discussion into a separate, albeit short, chapter 10 at the very end of the course.

Acknowledgments

I am extremely grateful to my faculty colleagues and other readers of the preliminary (circa 2013) version of this series, who provided feedback on certain sections; here they are listed in alphabetical order[7]: A Abanov, P Allen, D Averin, S Berkovich, P-T de Boer, M Fernandez-Serra, R F Hernandez, A Korotkov, V Semenov, F Sheldon, and X Wang. (Obviously, these kind people are not responsible for any remaining deficiencies.)

A large part of my scientific background and experience, reflected in these materials, came from my education, and then research, in the Department of Physics of Moscow State University from 1960 to 1990. The Department of Physics and Astronomy of Stony Brook University provided a comfortable and friendly environment for my work during the following 25+ years.

Last but not least, I would like to thank my wife Lioudmila for all her love, care, and patience—without these, this writing project would have been impossible.

I know very well that my materials are still far from perfection. In particular, my choice of covered topics (always very subjective) may certainly be questioned. Also, it is almost certain that despite all my efforts, not all typos have been weeded out. This is why all remarks (however candid) and suggestions from readers will be greatly appreciated. All significant contributions will be gratefully acknowledged in future editions.

Konstantin K Likharev
Stony Brook, NY

---

[7] I am very sorry for not keeping proper records from the beginning of my lectures at Stony Brook, so I cannot list all the numerous students and TAs who have kindly attracted my attention to typos in earlier versions of these notes. Needless to say, I am very grateful to all of them as well.

Notation

Abbreviations	Fonts	Symbols
c.c. complex conjugate	F, $\not{F}$ scalar variables[8]	$\cdot$ time differentiation operator (d/dt)
h.c. Hermitian conjugate	$\mathbf{F}$, $\vec{\not{F}}$ vector variables	∇ spatial differentiation vector (*del*)
	$\hat{F}$, $\hat{\not{F}}$ scalar operators	$\approx$ approximately equal to
	$\hat{\mathbf{F}}$, $\hat{\vec{\not{F}}}$ vector operators	$\sim$ of the same order as
	F matrix	$\propto$ proportional to
	$F_{jj'}$ matrix element	$\equiv$ equal to by definition (or evidently)
		$\cdot$ scalar ('dot-') product
		$\times$ vector ('cross-') product
		$-$ time averaging
		$\langle\ \rangle$ statistical averaging
		[,] commutator
		{ , } anticommutator

Prime signs

The prime signs (', ", etc) are used to distinguish similar variables or indices (such as j and j' in the matrix element above), rather than to denote derivatives.

Parts of the series

Part CM: Classical Mechanics   *Part EM: Classical Electrodynamics*
Part QM: Quantum Mechanics   *Part SM: Statistical Mechanics*

Appendices

Appendix A: Selected mathematical formulas
Appendix B: Selected physical constants

Formulas

The abbreviation Eq. may mean any displayed formula: either the equality, or inequality, or equation, etc.

---

[8] The same letter, typeset in different fonts, typically denotes different variables.

IOP Publishing

Quantum Mechanics

Lecture notes

Konstantin K Likharev

Chapter 1

Introduction

This introductory chapter briefly reviews the major experimental motivations for quantum mechanics, and then discusses its simplest formalism—the Schrödinger's wave mechanics. Much of this material (perhaps besides the last section) may be found in undergraduate textbooks[1], so that the discussion is rather brief, and focused on the most important conceptual issues.

1.1 Experimental motivations

By the beginning of the 1900s, physics (which by that time included what we now call non-relativistic classical mechanics, classical statistics and thermodynamics, and classical electrodynamics including the geometric and wave optics) looked an almost completed discipline, with most human-scale phenomena reasonably explained, and just a couple of mysterious 'dark clouds'[2] on the horizon. However, rapid technological progress and the resulting development of more refined scientific instruments have led to a fast multiplication of observed phenomena that could not be explained on the classical basis. Let me list the most consequential of those experimental findings.

(i) *The blackbody radiation measurements*, pioneered by G Kirchhoff in 1859, have shown that in the thermal equilibrium, the power of electromagnetic radiation by a fully absorbing ('black') surface, per unit frequency interval, drops exponentially at high frequencies. This is not what could be expected from the combination of the classical electrodynamics and statistics, which predicted an infinite growth of the radiation density with frequency. Indeed, the classical electrodynamics shows[3] that

---

[1] See, for example, [1].

[2] This famous expression was used in a 1900 talk by Lord Kelvin (born W Thomson) in reference to the blackbody radiation measurements and the results of the Michelson-Morley experiments, i.e. the precursors of the quantum mechanics and the relativity theory.

[3] See, e.g. *Part EM* section 7.8, in particular Eq. (7.211).

electromagnetic field modes evolve in time as harmonic oscillators, and that the number dN of these modes in a large free-space volume $V \gg \lambda^3$, in a small frequency interval $d\omega \ll \omega$ near some frequency ω, is

$$dN = 2V \frac{d^3k}{(2\pi)^3} = 2V \frac{4\pi k^2 dk}{(2\pi)^3} = V \frac{\omega^2}{\pi^2 c^3} d\omega, \tag{1.1}$$

where $c \approx 3 \times 10^8$ m s^{-1} is the free-space speed of light, $k = \omega/c$ the free-space wave number, and $\lambda = 2\pi/k$ is the radiation wavelength. On the other hand, the classical statistics[4] predicts that in the thermal equilibrium at temperature T, the average energy E of each 1D harmonic oscillator should be equal to $k_B T$, where k_B is the Boltzmann constant[5].

Combining these two results, we readily get the so-called *Rayleigh–Jeans formula* for the average electromagnetic wave energy per unit volume:

$$u \equiv \frac{1}{V}\frac{dE}{d\omega} = \frac{k_B T}{V}\frac{dN}{d\omega} = \frac{\omega^2}{\pi^2 c^3} k_B T, \tag{1.2}$$

that diverges at $\omega \to \infty$. On the other hand, the blackbody radiation measurements, improved by O Lummer and E Pringsheim, and also by H Rubens and F Kurlbaum to reach a 1% scale accuracy, were compatible with the phenomenological law suggested in 1900 by Max Planck:

$$u = \frac{\omega^2}{\pi^2 c^3} \frac{\hbar\omega}{\exp\{\hbar\omega/k_B T\} - 1}. \tag{1.3a}$$

This law may be reconciled with the fundamental equation (1.1) if the following replacement is made for the average energy of each field oscillator:

$$k_B T \to \frac{\hbar\omega}{\exp(\hbar\omega/k_B T) - 1}, \tag{1.3b}$$

with a constant factor

$$\hbar \approx 1.055 \times 10^{-34} \text{J s}, \tag{1.4}$$

now called the *Planck's constant*[6]. At low frequencies ($\hbar\omega \ll k_B T$), the denominator in Eq. (1.3) may be approximated as $\hbar\omega/k_B T$, so that the average energy (1.3b) tends to its classical value $k_B T$, and the Planck law (1.3a) reduces to the Rayleigh–Jeans formula (1.2). However, at higher frequencies ($\hbar\omega \gg k_B T$), Eq. (1.3) describes the experimentally observed rapid decrease of the radiation density—see figure 1.1.

---

[4] See, e.g. *Part SM* section 2.2.
[5] In the SI units, used through this series, $k_B \approx 1.38 \times 10^{-23}$ J K^{-1}—see appendix B for more exact value.
[6] Max Planck himself wrote $\hbar\omega$ as $h\nu$, where $\nu = \omega/2\pi$ is the 'cyclic' frequency (the number of periods per second), so that in early texts on quantum mechanics the term 'Planck's constant' referred to $h \equiv 2\pi\hbar$, while $\hbar$ was called 'the Dirac constant' for a while. I will use the contemporary terminology, and abstain from using the 'old Planck's constant' h at all, in order to avoid confusion.

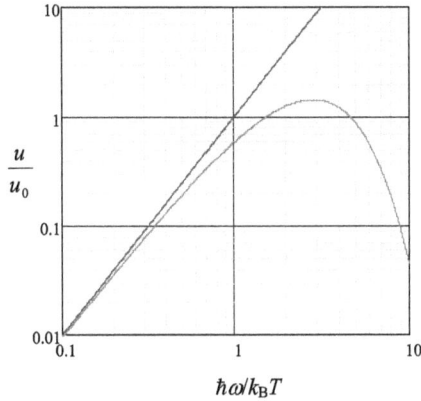

Figure 1.1. The blackbody radiation density u, expressed in units of $u_0 \equiv (k_B T)^3/\pi^2 \hbar^2 c^3$, as a function of frequency, according to: the Rayleigh–Jeans formula (blue line) and the Planck's law (red line).

(ii) *The photoelectric effect*, discovered in 1887 by H Hertz, shows a sharp lower bound on the frequency of the incident light that may kick electrons out from metallic surfaces, regardless of the light intensity. Albert Einstein, in one of his three famous 1905 papers, noticed that this threshold ω_{min} could be readily explained assuming that light consisted of certain particles (now called *photons*) with energy

$$E = \hbar\omega, \tag{1.5}$$

with the same Planck's constant that participates in Eq. (1.3).[7] Indeed, with this assumption, at the photon absorption by the surface, its energy $E = \hbar\omega$ is divided between a fixed energy U_0 (nowadays called the *workfunction*) of electron binding inside the metal, and the excess kinetic energy $m_e v^2/2 > 0$ of the freed electron—see figure 1.2. In this picture, the frequency threshold finds a natural explanation as $\omega_{min} = U_0/\hbar$.[8] Moreover, as was shown by S Bose in 1924, Eq. (1.5) readily explains[9] the Planck's law (1.3).

(iii) *The discrete frequency spectra* of the electromagnetic radiation by excited atomic gases, known since the 1600s, could not be explained by classical physics. (Applied to the planetary model of atoms, proposed by E Rutherford, classical electrodynamics predicts the collapse of electrons on nuclei in $\sim 10^{-10}$ s, due to electric dipole radiation of electromagnetic waves[10].) Especially challenging was the observation by J Balmer (in 1885) that the radiation frequencies of simple atoms may be well described by simple formulas. For example, for the lightest atom, the

---

[7] As a reminder, A Einstein received his only Nobel Prize (in 1922) for exactly this work, which essentially jump-started quantum mechanics, rather than for his relativity theory.

[8] For most metals, U_0 is between 4 and 5 electron volts (eV), so that the threshold corresponds to $\lambda_{max} = 2\pi c/\omega_{min} = 2\pi c/(U_0/\hbar) \approx 300$ nm—approximately at the border between the visible light and the ultraviolet radiation.

[9] See, e.g. *Part SM* section 2.5.

[10] See, e.g. *Part EM* section 8.2.

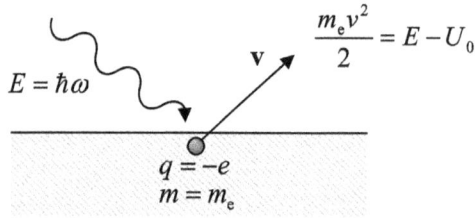

Figure 1.2. The Einstein's explanation of the photoelectric effect's frequency threshold.

hydrogen, all radiation frequencies may be numbered with just two positive integers n and n':

$$\omega_{n,n'} = \omega_0\left(\frac{1}{n^2} - \frac{1}{n'^2}\right), \tag{1.6}$$

with $\omega_0 \equiv \omega_{1,\infty} \approx 2.07 \times 10^{16}\,\text{s}^{-1}$. This observation, and the experimental value of ω_0, have found their first explanation in the famous 1913 theory by Niels Bohr, which was a phenomenological precursor for quantum mechanics. In this theory, $\omega_{n,n'}$ was interpreted as the frequency of a photon that obeys the Einstein's formula (1.5), with its energy $E_{n,n'} = \hbar\omega_{n,n'}$ being the difference between two *quantized* (discrete) energy levels of the atom (figure 1.3):

$$E_{n,n'} = E_{n'} - E_n > 0. \tag{1.7}$$

Bohr showed that Eq. (1.6) may be obtained from Eq. (1.7) and the non-relativistic[11] classical mechanics, augmented with just one additional postulate, equivalent to the assumption that the angular momentum $L = m_e vr$ of the electron moving on a circular trajectory of radius r about the hydrogen's nuclei (i.e. the proton, assumed to stay at rest because of its much higher mass), is *quantized* as

$$L = \hbar n, \tag{1.8}$$

where $\hbar$ is again the same Planck's constant (1.4), and n is an integer. (In Bohr's theory, n could not be equal to zero, though in the genuine quantum mechanics, it can.)

Indeed, it is sufficient to solve Eq. (1.8), $m_e vr = \hbar n$, together with the equation

$$m_e\frac{v^2}{r} = \frac{e^2}{4\pi\varepsilon_0 r^2}, \tag{1.9}$$

which expresses Newton's 2nd law for the electron rotating in the Coulomb field of the nucleus, for the electron's velocity v and the radius r. (Here $e \approx 1.6 \times 10^{-19}\,\text{C}$ is the fundamental electric charge, and $m_e \approx 0.91 \times 10^{-30}\,\text{kg}$ is the electron's rest mass.) The result for r is

---

[11] The non-relativistic approach to the problem is justified *a posteriori* by the fact the resulting energy scale E_H, given by Eq. (1.13), is much smaller than electron's rest energy, $m_e c^2 \approx 0.5$ MeV.

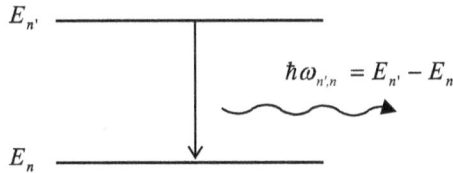

Figure 1.3. The electromagnetic radiation of a system at a result of transition between its quantized energy levels.

$$r = n^2 r_{\mathrm{B}}, \qquad \text{where } r_{\mathrm{B}} \equiv \frac{\hbar^2/m_e}{e^2/4\pi\varepsilon_0} \approx 0.0529 \text{ nm.} \tag{1.10}$$

The constant r_{B}, called the *Bohr radius*, is the most important spatial scale of phenomena in atomic, molecular and condensed matter physics—as well as in all chemistry and biochemistry.

Now plugging these results into the non-relativistic expression for the full electron's energy (with the free electron's rest energy taken for reference),

$$E = \frac{m_e v^2}{2} - \frac{e^2}{4\pi\varepsilon_0 r}, \tag{1.11}$$

we get the following simple expression for the energy levels (which, together with Eqs. (1.5) and (1.7), immediately gives Eq. (1.6) for the radiation frequencies):

$$E_n = -\frac{E_{\mathrm{H}}}{2n^2} < 0, \tag{1.12}$$

where E_{H} is called the so-called *Hartree energy constant* (or just the 'Hartree energy')[12]

$$E_{\mathrm{H}} \equiv \frac{(e^2/4\pi\varepsilon_0)^2}{\hbar^2/m_e} \approx 4.360 \times 10^{-18} \text{ J} \approx 27.21 \text{eV.} \tag{1.13a}$$

(Note the useful relations, which follow from Eqs. (1.10) and (1.13a):

$$E_{\mathrm{H}} = \frac{e^2}{4\pi\varepsilon_0 r_{\mathrm{B}}} = \frac{\hbar^2}{m_e r_{\mathrm{B}}^2}, \qquad \text{i.e. } r_{\mathrm{B}} = \frac{e^2/4\pi\varepsilon_0}{E_{\mathrm{H}}} = \left(\frac{\hbar^2/m_e}{E_{\mathrm{H}}}\right)^{1/2}; \tag{1.13b}$$

the first of them shows, in particular, that r_{B} is the distance at which the coefficient-free scales of the electron's potential and kinetic energies are equal.)

Note also that Eq. (1.8), in the form $pr = \hbar n$, where $p = m_e v$ is the electron momentum's magnitude, may be rewritten as the condition than an integer number (n) of wavelengths λ of certain (then hypothetic) waves[13] fits the circular orbit's

---

[12] Unfortunately, another name, the 'Rydberg constant', is sometimes used for either this energy unit or its half, $E_{\mathrm{H}}/2 \approx 13.6$ eV. To add to the confusion, the same term 'Rydberg constant' is used in some sub-fields of physics for the reciprocal free-space wavelength ($1/\lambda_0 = \omega_0/2\pi c$) corresponding to the frequency $\omega_0 = E_{\mathrm{H}}/2\hbar$.
[13] This fact was first noticed and discussed in 1924 by L de Broglie (in his PhD thesis!), so that instead of wavefunctions, especially of free particles, we are still frequently speaking of the *de Broglie waves*.

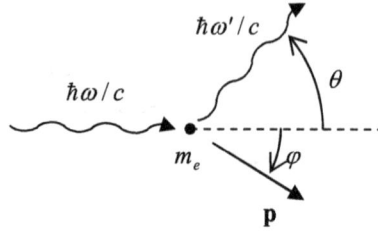

Figure 1.4. The Compton effect.

perimeter: $2\pi r \equiv 2\pi\hbar n/p = n\lambda$. Dividing both parts of the last relation by n, we see that for this statement to be true, the wave number $k \equiv 2\pi/\lambda$ of the de Broglie waves should be proportional to the electron's momentum $p = mv$:

$$p = \hbar k, \tag{1.14}$$

again with the same Planck's constant as in Eq. (1.5).

(iv) *The Compton effect*[14] is the reduction of frequency of x-rays at their scattering on free (or nearly-free) electrons—see figure 1.4. The effect may be explained assuming that the x-ray photon also has a momentum that obeys the vector-generalized version of Eq. (1.14):

$$\mathbf{p}_{\text{photon}} = \hbar\mathbf{k} = \frac{\hbar\omega}{c}\mathbf{n}, \tag{1.15}$$

where $\mathbf{k}$ is the wavevector (whose magnitude is equal to the wave number k, and direction coincides with the unit vector, $\mathbf{n}$, directed along the wave propagation[15]), and that the momenta magnitudes of both the photon and the electron are related to their energies E by the classical relativistic formula[16]

$$E^2 = (cp)^2 + (mc^2)^2. \tag{1.16}$$

(For a photon, the rest energy is zero, and this relation is reduced to Eq. (1.5): $E = cp = \hbar k = \hbar\omega$.) Indeed, a straightforward solution of the following system of three equations,

$$\hbar\omega + m_e c^2 = \hbar\omega' + [(cp)^2 + (m_e c^2)^2]^{1/2}, \tag{1.17}$$

$$\frac{\hbar\omega}{c} = \frac{\hbar\omega'}{c}\cos\theta + p\cos\varphi, \tag{1.18}$$

$$0 = \frac{\hbar\omega'}{c}\sin\theta - p\sin\varphi, \tag{1.19}$$

---

[14] This effect was observed (in 1922) and explained a year later by A Compton.
[15] See, e.g. *Part EM* section 7.1.
[16] See, e.g. *Part EM* section 9.3, in particular Eq. (9.78).

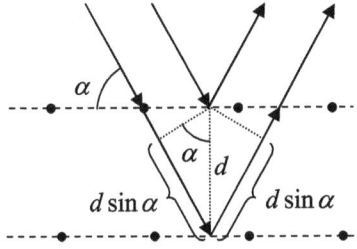

Figure 1.5. The De Broglie wave interference at electron scattering from a crystal lattice.

(which describe, respectively, the conservation of the full energy of the system, and of the two relevant Cartesian components of its full momentum, at the scattering—see figure 1.4), yields the following result,

$$\frac{1}{\hbar\omega'} = \frac{1}{\hbar\omega} + \frac{1}{m_e c^2}(1 - \cos\theta), \qquad (1.20a)$$

which is traditionally represented as the relation between the initial and final values of the photon's wavelength $\lambda = 2\pi/k = 2\pi/(\omega/c)$:

$$\lambda' = \lambda + \frac{2\pi\hbar}{m_e c}(1 - \cos\theta) \equiv \lambda + \lambda_C(1 - \cos\theta), \quad \text{with } \lambda_C \equiv \frac{2\pi\hbar}{m_e c}, \qquad (1.20b)$$

and is in agreement with experiment[17].

(v) *De Broglie wave diffraction.* In 1927, following the suggestion by W Elassger (who was excited by the de Broglie's conjecture of 'matter waves'), C Davisson and L Germer, and independently G Thomson succeeded in observing the diffraction of electrons on solid crystals (figure 1.5). Specifically, they have found that the intensity of the elastic reflection of electrons from a crystal increases sharply when the angle α between the incident beam of electrons and the crystal's atomic planes, separated by distance d, satisfies the following relation:

$$2d \sin\alpha = n\lambda, \qquad (1.21)$$

where $\lambda = 2\pi/k = 2\pi\hbar/p$ is the de Broglie wavelength of the electrons, and n is an integer. As figure 1.5 shows, this is just the well-known condition[18] that the path

---

[17] The constant λ_C, which participates in this relation, is close to 2.46×10^{-12} m, and is called the *Compton wavelength* of the electron. This term is somewhat misleading: as the reader can see from Eqs. (1.17)–(1.19), no wave in the Compton problem has such a wavelength—either before or after the scattering.

[18] See, e.g. *Part EM* section 8.4, in particular figure 8.9 and Eq. (8.82). Frequently, Eq. (1.21) is called the *Bragg condition*, due to the pioneering experiments by W Bragg with x-ray scattering from crystals (that started in 1912).

difference $\Delta l = 2d \sin \alpha$ between the de Broglie waves reflected from two adjacent crystal planes coincides with an integer number of λ, i.e. of the constructive interference of the waves[19].

To summarize, all the listed effects may be explained starting from two very simple (and similarly looking) formulas: Eq. (1.5) (at that stage, for photons only), and Eq. (1.15) for both photons and electrons—both relations involving the same Planck's constant. This might give an impression of sufficient experimental evidence to declare the light consisting of discrete particles (photons), and, conversely, electrons being some 'matter waves' rather than particles. However, by that time (the mid-1920s), physics had accumulated overwhelming evidence of *wave* properties of light, such as interference and diffraction[20]. In addition, there was also strong evidence for lumped-particle ('corpuscular') behavior of electrons. It is sufficient to mention the famous oil-drop experiments by R Millikan and H Fletcher (1909–13) in which only single (and whole!) electrons could be added to an oil drop, changing its total electric charge by multiples of electron's charge ($-e$)—and never its fraction. It was apparently impossible to reconcile these observations with a purely wave picture, in which an electron and hence its charge need to be spread over the wave's extension, so that its arbitrary part could be cut out using an appropriate experimental setup.

Thus the founding fathers of quantum mechanics faced a formidable task of reconciling the wave and corpuscular properties of electrons and photons—and other particles. The decisive breakthrough in that task has been achieved in 1926 by Ervin Schrödinger and Max Born, who formulated what is now known either formally as the *Schrödinger picture of non-relativistic quantum mechanics of the orbital motion*[21] *in the coordinate representation* (this term will be explained later in the course), or informally just as the *wave mechanics*. I will now formulate the main postulates of this theory.

1.2 Wave mechanics postulates

Let us consider a spinless[22], non-relativistic point-like particle, whose classical dynamics may be described by a certain Hamiltonian function $H(\mathbf{r}, \mathbf{p}, t)$,[23] where $\mathbf{r}$ is the particle's radius-vector and $\mathbf{p}$ is its momentum[24]. Wave mechanics of such

---

[19] Later, spectacular experiments with diffraction and interference of heavier particles (with much smaller de Broglie wavelength), e.g. neutrons and even C_{60} molecules, have also been performed—see, e.g. a review [2] and a later publication [3]. Nowadays, such interference of heavy particles is used, for example, for ultrasensitive measurements of gravity—see, e.g. a popular review [4], and recent advanced experiments [5].

[20] See, e.g. *Part EM* section 8.4.

[21] The *orbital motion* is the historic (and very unfortunate) term used for *any* motion of the particle as a whole.

[22] Actually, in wave mechanics, the spin of the described particle has not to be equal zero. Rather, it is assumed that the spin's *effects* on the orbital motion of the particle are negligible.

[23] As a reminder, for many systems (including those whose kinetic energy is a quadratic-homogeneous function of generalized velocities, like $mv^2/2$), H coincides with the total energy E—see, e.g. *Part CM* section 2.3.

[24] Note that this restriction is very important. In particular, it excludes from our current discussion the particles whose interaction with their environment is *irreversible*, for example it provides a friction leading to particle energy's decay. Such 'open' systems need a more general description, which will be discussed in chapter 7.

Hamiltonian particles may be based on the following set of postulates[25] that are comfortingly elegant—though their final justification is given only by the agreement of all their corollaries with experiment.

(i) *Wavefunction and probability.* Such variables as **r** or **p** cannot always be measured exactly, even at 'perfect conditions' when all external uncertainties, including measurement instrument imperfection, macroscopic uncertainties of the initial state preparation, and unintended particle interactions with its environment, have been removed[26]. Moreover, **r** and **p** of the same particle can *never* be measured exactly simultaneously. Instead, even the *most detailed* description of the particle's state, allowed by Nature[27], is given by a certain complex function $\Psi(\mathbf{r}, t)$, called the *wavefunction* (or 'wave function'), which generally enables only *probabilistic* predictions of the measured values of **r**, **p**, and other directly measurable variables —in quantum mechanics, usually called *observables*.

Specifically, the probability dW of finding a particle inside an elementary volume $dV \equiv d^3r$ is proportional to this volume, and hence may be characterized by a volume-independent *probability density* $w \equiv dW/d^3r$, which in turn is related to the wavefunction as

$$w = |\Psi(\mathbf{r}, t)|^2 \equiv \Psi^*(\mathbf{r}, t)\Psi(\mathbf{r}, t), \qquad (1.22a)$$

where the sign * denotes the usual complex conjugation. As a result, the total probability of finding the particle somewhere inside a volume V may be calculated as

$$W = \int_V w\, d^3r = \int_V \Psi^*\Psi\, d^3r. \qquad (1.22b)$$

In particular, if the volume V contains the particle *definitely* (i.e. with the 100% probability, $W = 1$), Eq. (1.22b) is reduced to the so-called *normalization condition*

$$\int_V \Psi^*\Psi\, d^3r = 1. \qquad (1.22c)$$

(ii) *Observables and operators.* With each observable A, quantum mechanics associates a certain *linear operator* $\hat{A}$, such that, in the perfect conditions mentioned above, the average measured value (also called the *expectation value*) of A is expressed as[28]

---

[25] Generally, quantum mechanics, as any theory, may be built on different sets of postulates ('axioms') leading to the same conclusions. In this text, I will not try to beat down the number of postulates to the absolute possible minimum, not only because this would require longer argumentation, but chiefly because such attempts typically result in making certain implicit assumptions hidden from the reader—a practice as common as it is regrettable.

[26] I will imply such perfect conditions further on, until the discussion of particle's interaction with environment in chapter 7 and beyond.

[27] This is one more important caveat. As will be discussed in detail in chapter 7, in many cases even Hamiltonian systems cannot be described by certain wavefunctions, and allow only a more general (and less precise) description, e.g. by the *density matrix*.

[28] This key measurement postulate is sometimes called the *Born rule*, though sometimes this term is used for the (less general) Eqs. (1.22).

$$\langle A \rangle = \int_V \Psi^* \hat{A} \Psi d^3 r, \tag{1.23}$$

where $\langle \ldots \rangle$ means the statistical average, i.e. the result of averaging the measurement results over a large *ensemble* (set) of macroscopically similar experiments, and Ψ is the normalized wavefunction, which obeys Eq. (1.22c). Note immediately that for Eqs. (1.22) and (1.23) to be compatible, the *identity* ('unit') *operator* defined by the relation

$$\hat{I} \Psi = \Psi, \tag{1.24}$$

has to be associated with a particular type of measurement, namely with the particle's detection.

(iii) *The Hamiltonian and the Schrödinger equation.* Another particular operator, the *Hamiltonian* $\hat{H}$, whose observable is the particle's energy E, also plays in wave mechanics a very special role, because it participates in the *Schrödinger equation*,

$$i\hbar \frac{\partial \Psi}{\partial t} = \hat{H} \Psi, \tag{1.25}$$

that determines the wavefunction's dynamics, i.e. its time evolution.

(iv) *The radius-vector and momentum operators.* In the wave mechanics, i.e. in the coordinate representation, the (vector-) operator of particle's radius-vector $\mathbf{r}$ just multiples the wavefunction by this vector, while the operator of particle's momentum[29] is proportional to the spatial derivative:

$$\hat{\mathbf{r}} = \mathbf{r}, \qquad \hat{\mathbf{p}} = -i\hbar \nabla, \tag{1.26a}$$

where ∇ is the *del* (or 'nabla') vector operator[30]. Thus in the Cartesian coordinates,

$$\hat{\mathbf{r}} = \mathbf{r} = \{x, \ y, \ z\}, \qquad \hat{\mathbf{p}} = -i\hbar \left\{ \frac{\partial}{\partial x}, \ \frac{\partial}{\partial y}, \ \frac{\partial}{\partial z} \right\}. \tag{1.26b}$$

(v) *The correspondence principle.* In the limit when quantum effects are insignificant, e.g. when the characteristic scale of action[31] (i.e. the product of the relevant energy and time scales of the problem) is much larger than Planck's constant $\hbar$, all wave mechanics results have to tend to those given by classical mechanics. Mathematically, this correspondence is achieved by duplicating the classical relations between various observables by similar relations between the corresponding operators. For example, for a free particle, the Hamiltonian (which in this particular case corresponds to the kinetic energy $T = p^2/2m$ alone) has the form

---

[29] For an electrically charged particle in magnetic field, this relation is valid for its *canonical* momentum—see section 3.1 below.
[30] See, e.g. sections 8–10 of the *Selected Mathematical Formulas* appendix (appendix A). Note that according to these formulas, the del operator follows all the geometric rules of the usual vectors. This is, by definition, true for other vector operators of quantum mechanics—to be discussed below.
[31] See, e.g. *Part CM* section 10.3.

$$\hat{H} = \hat{T} = \frac{\hat{p}^2}{2m} = -\frac{\hbar^2}{2m}\nabla^2. \tag{1.27}$$

Now, even before a deeper discussion of the postulates' physics (offered in the next section), we may immediately see that they indeed provide a formal way toward the resolution of the apparent contradiction between the wave and corpuscular properties of particles. For a free particle, the Schrödinger equation (1.25), with the substitution of Eq. (1.27), takes the form

$$i\hbar\frac{\partial\Psi}{\partial t} = -\frac{\hbar^2}{2m}\nabla^2\Psi, \tag{1.28}$$

whose particular, but most important solution is a plane, single-frequency ('monochromatic') traveling wave[32],

$$\Psi(\mathbf{r}, t) = ae^{i(\mathbf{k}\cdot\mathbf{r} - \omega t)}, \tag{1.29}$$

where a, $\mathbf{k}$ and ω are constants. Indeed, plugging Eq. (1.29) into Eq. (1.28), we immediately see that the plane wave, with an arbitrary amplitude a, is indeed a solution of this Schrödinger equation, provided a specific *dispersion relation* between the wavevector $\mathbf{k}$ and the frequency ω:

$$\hbar\omega = \frac{(\hbar k)^2}{2m}. \tag{1.30}$$

The constant a may be calculated, for example, assuming that the wave (1.29) is extended over a certain volume V, while beyond it, $\Psi = 0$. Then from the normalization condition (1.22c) and Eq. (1.29), we get[33]

$$|a|^2 \, V = 1. \tag{1.31}$$

Now we can use Eqs. (1.23), (1.26) and (1.27) to calculate the expectation values of the particle's momentum $\mathbf{p}$ and energy E (which, for a free particle, coincides with its Hamiltonian function H). The result is

$$\langle\mathbf{p}\rangle = \hbar\mathbf{k}, \quad \langle E\rangle = \langle H\rangle = \frac{(\hbar k)^2}{2m}; \tag{1.32}$$

according to Eq. (1.30), the last equality may be rewritten as $\langle E\rangle = \hbar\omega$.

Next, Eq. (1.23) enables us to calculate not only the average (in the math speak, the *first moment*) of an observable, but also its higher moments, notably the *second moment* (in physics, usually called either the *variance* or *dispersion*):

$$\langle\tilde{A}^2\rangle \equiv \langle(A - \langle A\rangle)^2\rangle = \langle A^2\rangle - \langle A\rangle^2, \tag{1.33}$$

and hence its root mean square (rms) fluctuation,

---

[32] See, e.g. *Part CM* section 6.4 and/or *Part EM* section 7.1.
[33] For infinite space ($V \to \infty$), Eq. (1.31) yields $a \to 0$, i.e. wavefunction (1.29) vanishes. This formal problem may be readily resolved considering sufficiently long wave packets—see section 2.2 below.

$$\delta A \equiv \langle \tilde{A}^2 \rangle^{1/2}, \tag{1.34}$$

that characterizes the scale of deviations $\tilde{A} \equiv A - \langle A \rangle$ of measurement results from the average, i.e. the *uncertainty* of the observable A. In the particular case when the uncertainty δA equals zero, every measurement of the observable A will give the value $\langle A \rangle$; such a state is said to have a *definite value* of the variable. For example, in application to the wavefunction (1.29), these relations yield $\delta E = 0$, $\delta \mathbf{p} = 0$. This means that in the plane-wave, monochromatic state (1.29), the energy and momentum of the particle have definite values, so that the statistical average signs in Eqs. (1.32) might be removed. Thus, these relations are reduced to the experimentally-inferred Eqs. (1.5) and (1.15)—though the relation of $\mathbf{k}$ and ω to experimental observations still has to be clarified.

Hence the wave mechanics postulates may indeed explain the observed wave properties of non-relativistic particles. (For photons, we would need a relativistic formalism—see chapter 9 below.) On the other hand, due to the linearity of the Schrödinger equation (1.25), any sum of its solutions is also a solution—the so-called *linear superposition principle*. For a free particle, this means that any set of plane waves (1.29) is also a solution of this equation. Such sets, with close values of $\mathbf{k}$ and hence $\mathbf{p} = \hbar \mathbf{k}$ (and, according to Eq. (1.30), of ω as well), may be used to describe spatially localized 'pulses', called *wave packets*—see figure 1.6. In section 2.1, I will prove (or rather reproduce H Weyl's proof) that the wave packet's extension δx in any direction (say, x) is related to the width δk_x of the distribution of the corresponding component of its wave vector as $\delta x \delta k_x \geqslant \frac{1}{2}$, and hence, according to Eq. (1.15), to the width δp_x of the momentum component distribution as

$$\delta x \cdot \delta p_x \geqslant \frac{\hbar}{2}. \tag{1.35}$$

This is the famous *Heisenberg's uncertainty principle*, which quantifies the first postulate's point that the coordinate and the momentum cannot be defined exactly simultaneously. However, since Planck's constant, $\hbar \sim 10^{-34}$ J s, is extremely small on the human scale of things, it still allows for a particle's localization in a very small volume even if the momentum spread in the wave packet is also small on that scale.

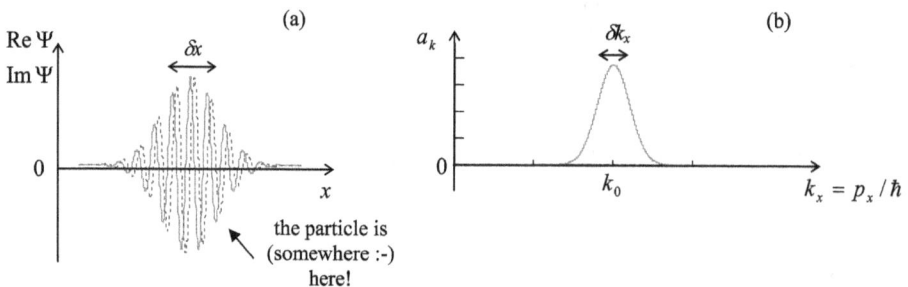

Figure 1.6. (a) A snapshot of a typical wave packet propagating along axis x, and (b) the corresponding distribution of the wave numbers k_x, i.e. the momenta p_x.

For example, according to Eq. (1.35), a 0.1% spread of momentum of a 1 keV electron ($p \sim 1.7 \times 10^{-24}$ kg m s^{-1}) allows its wave packet to be as small as $\sim 3 \times 10^{-10}$ m. (For a heavier particle such as a proton, the packet would be even tighter.) As a result, wave packets may be used to describe the particles that are quite point-like from the macroscopic point of view.

In a nutshell, this is the main idea of the wave mechanics, and the first part of this course (chapters 1–3) will be essentially a discussion of various effects described by this approach. During this discussion, however, we will not only evidence wave mechanics' many triumphs within its applicability domain, but also gradually accumulate evidence for its handicaps, which will force an eventual transfer to a more general formalism—to be discussed in chapter 4 and beyond.

1.3 Postulates' discussion

The wave mechanics' postulates listed in the previous section (hopefully, familiar to reader from his or her undergraduate studies) may look very simple. However, the physics of these axioms is very deep, leading to some counter-intuitive conclusions, and their in-depth discussion requires solutions of several key problems of wave mechanics. This is why in this section I will give only an initial, admittedly superficial discussion of the postulates, and will be repeatedly returning to the conceptual foundations of quantum mechanics throughout the course, especially in chapter 10.

First of all, the fundamental uncertainty of observables, which is in the core of the first postulate, is very foreign to the basic ideas of classical mechanics, and historically has made the quantum mechanics so hard to swallow for many star physicists, notably including A Einstein—despite his 1905 work, which essentially launched the whole field! However, this fact has been confirmed by numerous experiments, and (more importantly) there has not been a single confirmed experiment which would contradict this postulate, so that quantum mechanics was long ago promoted from a theoretical hypothesis to the rank of a reliable scientific theory.

One more remark in this context is that Eq. (1.25) itself is *deterministic*, i.e. conceptually enables an *exact* calculation of the wavefunction's distribution in space at any instant t, provided that its initial distribution, and the particle's Hamiltonian, are known exactly. Note that in the classical statistical mechanics, the probability density distribution $w(\mathbf{r}, t)$ may be also calculated from deterministic differential equations, for example the Liouville equation[34]. The quantum-mechanical description differs from that situation in two important aspects. First, in the perfect conditions outlined above (the exact initial state preparation and the best possible measurements), the Liouville equation reduces to the 2nd Newton law of classical mechanics, i.e. the statistical uncertainty disappears. In quantum mechanics this is not true: the quantum uncertainly, such as described by Eq. (1.35), persists even in this limit. Second, the wavefunction $\Psi(\mathbf{r}, t)$ gives more information than just $w(\mathbf{r}, t)$, because besides the modulus of Ψ, involved in Eq. (1.22), this complex function also

---

[34] See, e.g. *Part SM* section 6.1.

has the *phase* $\varphi \equiv \arg \Psi$, which may affect some observables, describing, in particular, the interference of the de Broglie waves.

Next, it is very important to understand that the relation between the quantum mechanics and experiment, given by the second postulate, necessarily involves another key notion: that of the corresponding *statistical ensemble*. This ensemble may be defined as a set of many experiments carried out at apparently (*macroscopically*) similar conditions, including the initial conditions, which nevertheless may lead to different measurement results (*outcomes*). Indeed, the probability of a certain (*n*th) outcome of an experiment may be only defined for a certain ensemble, as the limit

$$W_n \equiv \lim_{M \to \infty} \frac{M_n}{M}, \qquad \text{with } M \equiv \sum_{n=1}^{N} M_n, \qquad (1.36)$$

where M is the total number of experiments, M_n is the number of outcomes of the *n*th type, and N is the number of different outcomes.

Note that a particular choice of an ensemble may affect probabilities W_n very significantly. For example, if we pull out playing cards at random from a standard pack of 52 different cards of 4 suits, the probability W_n of getting a certain card (e.g. the queen of spades) is 1/52. However, if the cards of a certain suit (say, hearts) had been taken out from the pack in advance, the probability of getting the queen of spades is higher, 1/39. It is important that we would also get the last number for the probability even if we had used the full 52 card pack, but by some reason discarded results of all experiments giving us any rank of hearts. Hence, the ensemble definition (or its *re-definition* in the middle of the game) may change outcome probabilities.

In quantum wave mechanics, with its fundamental relation (1.22) between w and Ψ, this means not only the outcome probabilities, but the wavefunction itself also may depend on the statistical ensemble we are using, i.e. not only on the preparation of the system and the experimental setup, but also on the subset of outcomes taken into account. The sometimes accounted attribution of the wavefunction to a single experiment, both before and after the measurement, may lead to very unphysical interpretations of the results, including a wavefunction's evolution not described by the Schrödinger equation (the so-called *wave packet reduction*), subluminal action on distance, etc. Later in the course we will see that minding the fundamentally statistical nature of quantum mechanics, and in particular the dependence of wavefunctions on the statistical ensembles' definition (or re-definition), readily resolves some, though not all, paradoxes of quantum measurements.

Note, however, that the standard quantum mechanics, as discussed in chapters 1–6 of this course, is limited to statistical ensembles with *the least uncertainty* of the considered systems, i.e. with best possible knowledge about their state[35]. This condition requires, first, the least uncertain initial preparation of the system, and

---

[35] The reader should not be surprised by the use of the notion of 'knowledge' (or 'information') in this context. Indeed, due to the statistical character of experiment outcomes, quantum mechanics (or at least its relation to experiment) is intimately related to the information theory. In contrast to much of classical physics, which may be discussed without any reference to information, in quantum mechanics, as in classical statistical physics, such abstraction is possible only in some very special (and not the most interesting) cases.

second, its total isolation from the rest of the world, or at least from its disordered part (the 'environment'), in the course of its evolution in time. Only such ensembles may be described by certain wavefunctions. A detailed discussion of more general ensembles, which are necessary if these conditions are not satisfied, will be given in chapters 7, 8, and 10.

Finally, regarding Eq. (1.23), a better feeling of this definition may be obtained by its comparison with the general definition of the expectation value (i.e. the statistical average) in the probability theory. Namely, let each of N possible outcomes in a set of M experiments give a certain value A_n of observable A; then

$$\langle A \rangle \equiv \lim_{M \to \infty} \frac{1}{M} \sum_{n=1}^{N} A_n M_n = \sum_{n=1}^{N} A_n W_n. \tag{1.37}$$

Taking into account Eq. (1.22), which relates W and Ψ, the structures of Eq. (1.23) and the final form of Eq. (1.37) are similar. Their exact relation will be further discussed in section 4.1.

1.4 Continuity equation

The wave mechanics postulates survive one more sanity check: they satisfy the natural requirement that the particle does not appear or vanish in the course of the quantum evolution[36]. Indeed, let us use Eq. (1.22) to calculate the rate of change of the probability W to find a particle within a certain volume V:

$$\frac{dW}{dt} = \frac{d}{dt} \int_V \Psi \Psi^* d^3 r. \tag{1.38}$$

Assuming for simplicity that the boundaries of the volume V do not move, it is sufficient to carry out the partial differentiation of the product $\Psi \Psi^*$ inside the integral. Using the Schrödinger equation (1.25), together with its complex conjugate,

$$-i\hbar \frac{\partial \Psi^*}{\partial t} = (\hat{H}\Psi)^*, \tag{1.39}$$

we readily get

$$\frac{dW}{dt} = \int_V \frac{\partial}{\partial t}(\Psi\Psi^*)\, d^3r \equiv \int_V \left(\Psi^* \frac{\partial \Psi}{\partial t} + \Psi \frac{\partial \Psi^*}{\partial t} \right) d^3r$$

$$= \frac{1}{i\hbar} \int_V [\Psi^*(\hat{H}\Psi) - \Psi(\hat{H}\Psi)^*]\, d^3r. \tag{1.40}$$

Let the particle move in a field of external forces (not necessarily constant in time), so that its classical Hamiltonian function H is the sum of the particle's kinetic

---

[36] Note that this requirement may be violated in the relativistic quantum theory—see chapter 9.

energy $T = p^2/2m$ and its potential energy $U(\mathbf{r}, t)$.[37] According to the correspondence principle, and Eq. (1.27), the Hamiltonian operator may be represented as the sum[38],

$$\hat{H} = \hat{T} + \hat{U} = \frac{\hat{p}^2}{2m} + U(\mathbf{r}, t) = -\frac{\hbar^2}{2m}\nabla^2 + U(\mathbf{r}, t). \qquad (1.41)$$

At this stage we should notice that this operator, when acting on a real function, returns a real function[39]. Hence, the result of its action on an arbitrary complex function $\Psi = a + ib$ (where a and b are real) is

$$\hat{H}\Psi = \hat{H}(a + ib) = \hat{H}a + i\hat{H}b, \qquad (1.42)$$

where $\hat{H}a$ and $\hat{H}b$ are also real, while

$$(\hat{H}\Psi)^* = (\hat{H}a + i\hat{H}b)^* = \hat{H}a - i\hat{H}b = \hat{H}(a - ib) = \hat{H}\Psi^*. \qquad (1.43)$$

This means that Eq. (1.40) may be rewritten as

$$\frac{dW}{dt} = \frac{1}{i\hbar}\int_V [\Psi^*\hat{H}\Psi - \Psi\hat{H}\Psi^*]\, d^3r$$
$$= -\frac{\hbar^2}{2m}\frac{1}{i\hbar}\int_V [\Psi^*\nabla^2\Psi - \Psi\nabla^2\Psi^*]\, d^3r \qquad (1.44)$$

Now, let us use general rules of vector calculus[40] to write the following identity:

$$\nabla \cdot (\Psi^*\nabla\Psi - \Psi\nabla\Psi^*) = \Psi^*\nabla^2\Psi - \Psi\nabla^2\Psi^*, \qquad (1.45)$$

A comparison of Eqs. (1.44) and (1.45) shows that we may write

$$\frac{dW}{dt} = -\int_V (\nabla \cdot \mathbf{j})\, d^3r, \qquad (1.46)$$

where the vector $\mathbf{j}$ is defined as

$$\mathbf{j} \equiv \frac{i\hbar}{2m}(\Psi\nabla\Psi^* - \text{c.c.}) \equiv \frac{\hbar}{m}\text{Im}(\Psi^*\nabla\Psi), \qquad (1.47)$$

where c.c. means the complex conjugate of the previous expression—in this case, $(\Psi\nabla\Psi^*)^*$, i.e. $\Psi^*\nabla\Psi$. Now using the well-known divergence theorem[41], Eq. (1.46) may be rewritten as the *continuity equation*

---

[37] As a reminder, such description is valid not only for conservative forces (in that case U has to be time-independent), but also for any force $\mathbf{F}(\mathbf{r}, t)$ that may be expressed via the gradient of $U(\mathbf{r}, t)$—see, e.g. *Part CM* chapters 2 and 10. (A good example when such a description is *impossible* is given by the magnetic component of the Lorentz force—see, e.g. *Part EM* section 9.7, and also section 3.1 below.)

[38] Historically, this was the main step made (in 1926) by E Schrödinger on the background of L de Broglie's idea. The probabilistic interpretation of the wavefunction was put forward, almost simultaneously, by M Born.

[39] In chapter 4, we will discuss a more general family of *Hermitian operators*, which have this property.

[40] See, e.g. Eq. (A.11.4a), combined with the del operator's definition $\nabla^2 \equiv \nabla \cdot \nabla$.

[41] See, e.g. Eq. (A.12.2).

$$\frac{dW}{dt} + I = 0, \quad \text{with } I \equiv \int_S j_n d^2r, \qquad (1.48)$$

where j_n is the component of the vector $\mathbf{j}$ along the outwardly directed normal to the closed surface S that limits volume V, i.e. the scalar product $\mathbf{j} \cdot \mathbf{n}$, where $\mathbf{n}$ is the unit vector along this normal.

Eqs (1.47) and (1.48) show that if the wavefunction on the surface vanishes, the total probability W of finding the particle within the volume does not change, providing the required sanity check. In the general case, Eq. (1.48) says that dW/dt equals the flux I of the vector $\mathbf{j}$ through the surface, with the minus sign. It is clear that this vector may be interpreted as the *probability current density*—and I, as the total *probability current* through the surface S. This interpretation may be further supported by rewriting Eq. (1.47) for the wavefunction represented in the polar form $\Psi = ae^{i\varphi}$, with real a and φ:

$$\mathbf{j} = a^2 \frac{\hbar}{m} \nabla \phi. \qquad (1.49)$$

Note that for a real wavefunction, or even for a wavefunction with an arbitrary but space-constant phase φ, the probability current density vanishes. In contrast, for the traveling wave (1.29), with a constant probability density $w = a^2$, Eq. (1.49) yields a non-zero (and physically very transparent) result:

$$\mathbf{j} = w\frac{\hbar}{m}\mathbf{k} = w\frac{\mathbf{p}}{m} = w\mathbf{v}, \qquad (1.50)$$

where $\mathbf{v} = \mathbf{p}/m$ is particle's velocity. If multiplied by the particle's mass m, the probability density w turns into the (average) mass density ρ, and the probability current density—into the *mass flux density* $\rho\mathbf{v}$. Similarly, if multiplied by the total electric charge q of the particle, with w turning into the *charge density* σ, $\mathbf{j}$ becomes the electric current density. As the reader (hopefully :-) knows, both currents satisfy classical continuity equations similar to Eq. (1.48)[42].

Finally, let us recast the continuity equation, rewriting Eq. (1.46) as

$$\int_V \left(\frac{\partial w}{\partial t} + \nabla \cdot \mathbf{j}\right) d^3r = 0. \qquad (1.51)$$

Now we may argue that this equality may be true for any choice of volume V only if the expression under the integral vanishes everywhere, i.e. if

$$\frac{\partial w}{\partial t} + \nabla \cdot \mathbf{j} = 0. \qquad (1.52)$$

This differential form of the continuity equation may be more convenient than its integral form (1.48).

---

[42] See, e.g. respectively, *Part CM* section 8.3 and *Part EM* section 4.1.

1.5 Eigenstates and eigenvalues

Now let us discuss the most important corollaries of wave mechanics' *linearity*. First of all, it uses only *linear operators*. This term means that the operators must obey the following two rules[43]:

$$(\hat{A}_1 + \hat{A}_2)\,\Psi = \hat{A}_1\Psi + \hat{A}_2\Psi, \tag{1.53}$$

$$\hat{A}(c_1\Psi_1 + c_2\Psi_2) = \hat{A}(c_1\Psi_1) + \hat{A}(c_2\Psi_2) = c_1\hat{A}\Psi_1 + c_2\hat{A}\Psi_2, \tag{1.54}$$

where Ψ_n are arbitrary wavefunctions, while c_n are arbitrary constants (in quantum mechanics, frequently called *c-numbers*, to distinguish them from operators and wavefunctions). The most important examples of linear operators are given by:

(i) the multiplication by a function, such as for the operator $\hat{\mathbf{r}}$ given by Eq. (1.26), and

(ii) the spatial or temporal differentiation of the wavefunction, such as in Eqs. (1.25)–(1.27).

Next, it is of key importance that the Schrödinger equation (1.25) is also linear. (We have already used this fact when we discussed wave packets in the last section.) This means that if each of several functions Ψ_n are (particular) solutions of Eq. (1.25) with a certain Hamiltonian, then their arbitrary linear combination

$$\Psi = \sum_n c_n \Psi_n \tag{1.55}$$

is also a solution of the same equation[44].

Let us use this linearity to accomplish an apparently impossible feat: immediately find the *general* solution of the Schrödinger equation for the most important case when system's Hamiltonian does not depend on time explicitly—for example, like in Eq. (1.41) with time-independent potential energy $U = U(\mathbf{r})$, when the Schrödinger equation has the form

$$i\hbar\frac{\partial\Psi}{\partial t} = -\frac{\hbar^2}{2m}\nabla^2\Psi + U(\mathbf{r})\Psi. \tag{1.56}$$

First of all, let us prove that the following product,

$$\Psi_n = a_n(t)\psi_n(\mathbf{r}), \tag{1.57}$$

---

[43] By the way, if any equality involving operators is valid for an arbitrary wavefunction, the latter is frequently dropped from notation, resulting in an *operator equality*. In particular, Eq. (1.53) may be readily used to prove that the operators are *commutative*: $\hat{A}_2 + \hat{A}_1 = \hat{A}_1 + \hat{A}_2$, and *associative*: $(\hat{A}_1 + \hat{A}_2) + \hat{A}_3 = \hat{A}_1 + (\hat{A}_2 + \hat{A}_3)$.

[44] At first glance, it may seem strange that the *linear* Schrödinger equation correctly describes quantum properties of systems whose classical dynamics is described by *nonlinear* equations of motion (e.g. an anharmonic oscillator —see, e.g. *Part CM* section 5.2). Note, however, that statistical equations of classical dynamics (see, e.g. *Part SM* chapters 5 and 6) also have this property, so it is not specific to quantum mechanics.

qualifies as a (particular) solution of such an equation. Indeed, plugging Eq. (1.56) into Eq. (1.25) with any time-independent Hamiltonian, using the fact that in this case

$$\hat{H} a_n(t)\psi_n(\mathbf{r}) = a_n(t)\hat{H}\psi_n(\mathbf{r}), \tag{1.58}$$

and dividing both parts of the equation by $a_n\psi_n$, we get

$$\frac{i\hbar}{a_n}\frac{da_n}{dt} = \frac{\hat{H}\psi_n}{\psi_n}. \tag{1.59}$$

The left-hand side of this equation may depend only on time, while the right hand one depends only on coordinates. These facts may be only reconciled if we assume that each of these parts is equal to (the same) constant of the dimension of energy, which I will denote as E_n.[45] As a result, we are getting two separate equations for the temporal and spatial parts of the wavefunction:

$$\hat{H}\psi_n = E_n\psi_n, \tag{1.60}$$

$$i\hbar\frac{da_n}{dt} = E_n a_n. \tag{1.61a}$$

The latter of these equations, rewritten in the form

$$\frac{da_n}{a_n} = -i\frac{E_n}{\hbar}dt, \tag{1.61b}$$

is readily integrable, giving

$$\ln a_n = -i\omega_n t + \text{const}, \quad \text{so that } a_n = \text{const} \times \exp\{-i\omega_n t\},$$
$$\text{with } \omega_n \equiv \frac{E_n}{\hbar}. \tag{1.62}$$

Now plugging Eqs. (1.57) and (1.62) into Eq. (1.22), we see that in the quantum state described by Eqs. (1.57)–(1.62), the probability w of finding the particle at a certain location does not depend on time:

$$w \equiv \psi_n^*(\mathbf{r})\psi_n(\mathbf{r}) = w(\mathbf{r}). \tag{1.63}$$

With the same substitution, Eq. (1.23) shows that the expectation value of any operator that does not depend on time explicitly is also time-independent:

$$\langle A \rangle \equiv \int \psi_n^*(\mathbf{r})\hat{A}\psi_n(\mathbf{r})d^3r = \text{const.} \tag{1.64}$$

---

[45] This argumentation, leading to *variable separation*, is very common in mathematical physics—see, e.g. its discussion in *Part CM* section 6.5 and *Part EM* section 2.5 and beyond.

Due to this property, the states described by Eqs. (1.57)–(1.62), are called *stationary*; they are fully defined by the possible solutions (called *eigenfunctions*[46]) of the *stationary* (or 'time-independent') *Schrödinger* equation (1.60).[47]

Note that for the time-independent Hamiltonian (1.41), the stationary Schrödinger equation (1.60),

$$-\frac{\hbar^2}{2m}\nabla^2\psi_n + U(\mathbf{r})\psi_n = E_n\psi_n, \qquad (1.65)$$

is a linear, homogeneous differential equation for the function ψ_n, with *a priory* unknown parameter E_n. Such equations fall into the mathematical category of *eigenproblems*, in which the eigenfunctions ψ_n and *eigenvalues* E_n should be found simultaneously, i.e. self-consistently[48]. Mathematics tells us that for the such equations with space-confined eigenfunctions ψ_n, tending to zero at $r \to \infty$, the spectrum of eigenvalues is *discrete*. It also proves that the eigenfunctions corresponding to different eigenvalues are *orthogonal*, i.e. that space integrals of the products $\psi_n\psi_{n'}^*$ vanish for all pairs with $n \neq n'$. Due to the Schrödinger equation's linearity, each of these functions may be multiplied by a proper constant coefficient to make their set *orthonormal*:

$$\int \psi_n^*\psi_n d^3r = \delta_{n,n'} \equiv \begin{cases} 1, & \text{if } n = n', \\ 0, & \text{if } n \neq n'. \end{cases} \qquad (1.66)$$

Moreover, the eigenfunctions $\psi_n(\mathbf{r})$ form a *full set*, meaning that an arbitrary function $\psi(\mathbf{r})$, in particular the actual wavefunction Ψ of the system in the initial moment of its evolution (which I will take for $t = 0$, with a few exceptions), may be represented as a unique expansion over the eigenfunction set[49]:

$$\Psi(\mathbf{r}, 0) = \sum_n c_n\psi_n(\mathbf{r}). \qquad (1.67)$$

The expansion coefficients c_n may be readily found by multiplying both parts of Eq. (1.67) by $\psi_{n'}^*$, integrating the result over the space, and using Eq. (1.66). The result is

$$c_n = \int \psi_n^*(\mathbf{r})\Psi(\mathbf{r}, 0)d^3r. \qquad (1.68)$$

Now let us consider the following wavefunction

$$\Psi(\mathbf{r}, t) = \sum_n c_n a_k(t)\psi_k(\mathbf{r}) = \sum_n c_n \exp\left\{-i\frac{E_n}{\hbar}t\right\}\psi_n(\mathbf{r}). \qquad (1.69)$$

---

[46] From the German root *eigen*, meaning 'particular' or 'characteristic'.
[47] For contrast, the full *Schrödinger* equation (1.25) is frequently called *time-dependent* or *non-stationary*.
[48] Eigenvalues of energy are frequently called *eigenenergies*, and it is often said that eigenfunction ψ_n and eigenenergy E_n together characterize the nth *stationary eigenstate* of the system.
[49] If the reader has any doubt in these properties of linear, homogeneous differential equations, I may recommend reviewing section 9.3 of the wonderful handbook by G Korn and T Korn, listed in section A.16 (ii).

Since each term of the sum has the form (1.57) and satisfies the Schrödinger equation, so does the sum as the whole. Moreover, if the coefficients c_n are derived in accordance with Eq. (1.68), then the solution (1.69) satisfies the initial conditions as well. At this moment we can use one more bit of help from mathematicians, who tell us that the linear, partial differential equation of type (1.65), with fixed initial conditions, may have only one (*unique*) solution. This means that in our case of motion in a time-independent potential Hamiltonian, Eq. (1.69) gives the *general* solution of the Schrödinger equation (1.65).

So, we have succeeded in our apparently over-ambitious goal. Now let us stop this mad mathematical dash for a minute, and discuss this key result.

1.6 Time evolution

For the time-dependent factor, $a_n(t)$, of each component state (1.57) of the general solution (1.69), our procedure gave a very simple and universal result (1.62), describing a linear change of the phase $\varphi_n \equiv \arg(a_n)$ of this complex function in time, with the constant rate

$$\frac{d\varphi_n}{dt} = -\omega_n = -\frac{E_n}{\hbar}, \tag{1.70}$$

so that the real and imaginary parts of a_n oscillate sinusoidally with this frequency. The relation (1.70) coincides with the Einstein's conjecture (1.5) for photons, but could these oscillations of the wavefunctions represent a physical reality? Indeed, for photons, described by Eq. (1.5), E may be (and as we will see in chapter 9, is) the actual, well-defined energy of one photon, and ω is the frequency of the radiation so quantized. However, for non-relativistic particles, described by wave mechanics, the potential energy U, and hence the full energy E, are defined to an arbitrary constant, because we may measure them from an arbitrary reference level. How can such a change of the energy reference level (which may be made just in our mind) alter the frequency of oscillations of a variable?

According to Eqs. (1.22) and (1.23), this time evolution of a wavefunction does not affect the particle's probability distribution, or even any observable (including the energy E, provided that it is always referred to the same origin as U), in any stationary state. However, as will be proved later in the course using the combination of Einstein's formula (1.5) with Bohr's assumption (1.7),

$$\hbar\omega_{nn'} = E_{n'} - E_n, \tag{1.71}$$

the *difference* of the eigenfrequencies ω_n (evidently, independent on the energy reference) of two eigenstates is absolutely physical, because it determines the measurable frequency of the electromagnetic radiation (or possibly a wave of a different physical nature) emitted or absorbed at the quantum transition between the states.

As one more example, consider two similar, independent particles 1 and 2, each in the same (say, the lowest, *ground*) eigenstate, but with the potential energies (and

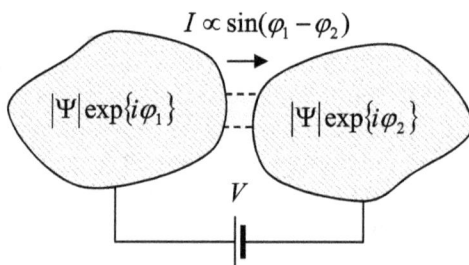

Figure 1.7. The Josephson effect in a weak link between two bulk superconductor electrodes.

hence the ground state energies $E_{1,2}$) different by a constant $\Delta U \equiv U_1 - U_2$. Then, according to Eq. (1.70), the difference $\varphi \equiv \varphi_1 - \varphi_2$ of their wavefunction phases evolves in time with a reference-independent rate

$$\frac{d\varphi}{dt} = -\frac{\Delta U}{\hbar}. \tag{1.72}$$

Certain measurement instruments, weakly coupled to each particle, may allow an observation of this evolution, while keeping the particle's quantum dynamics virtually unperturbed, i.e. Eq. (1.70) intact. Perhaps the most dramatic measurement of this type is possible using the *Josephson effect* in weak links between two superconductors—see figure 1.7.

As a brief reminder[50], superconductivity may be explained by a specific coupling between conduction electrons in solids, that leads, at low temperatures, to the formation of the so-called *Cooper pairs*. Such pairs, each consisting of two electrons with opposite spins and momenta, behave as Bose particles, and form a coherent *Bose–Einstein condensate*[51]. Most properties of such a condensate may be described by a single, common wavefunction Ψ, evolving in time just as that of a free particle, with the effective potential energy $U = q\phi = -2e\phi$, where ϕ is the electrochemical potential[52], and $q = -2e$ is the electric charge of a Cooper pair. As a result, for the system shown in figure 1.7, in which an externally applied voltage V fixes the difference $\phi_1 - \phi_2$ between the electrochemical potentials of two bulk super-conductors, Eq. (1.72) takes the form

$$\frac{d\varphi}{dt} = \frac{2e}{\hbar}V, \tag{1.73}$$

where $V = \phi_1 - \phi_2$ is the applied voltage. If the link between the superconductors is weak enough, the electric current I of the Cooper pairs (called the *supercurrent*) through the link may be approximately described by the following simple relation[53],

---

[50] For a more detailed discussion, including the derivation of Eq. (1.74), see, e.g. *Part EM* section 6.5.

[51] A detailed discussion of the Bose–Einstein condensation may be found, e.g. in *Part SM* section 3.4.

[52] For more on this notion see, e.g. *Part SM* section 6.3.

[53] In some cases, the function $I(\varphi)$ may somewhat deviate from Eq. (1.74), but these deviations do not affect its fundamental 2π-periodicity. As a result, no corrections to the fundamental relations (1.75)–(1.76) have ever been found (yet :-).

$$I = I_c \sin \varphi, \qquad (1.74)$$

where I_c is some constant, dependent on the weak link's strength. Now combining Eqs. (1.73) and (1.74), we see that if the applied voltage V is constant in time, the current oscillates sinusoidally, with the so-called *Josephson frequency*

$$\omega_J \equiv \frac{2e}{\hbar} V, \qquad (1.75)$$

as high as ~484 MHz per each microvolt of applied dc voltage. This effect may be readily observed experimentally: though its direct detection is a bit tricky, it is easy to observe the *phase locking* (synchronization)[54] of the Josephson oscillations by an external microwave signal of frequency ω. Such phase locking results in the relation $\omega_J = n\omega$ fulfilled within certain current intervals, and hence in the formation, on the weak-link's dc I–V curve, of virtually vertical current steps at dc voltages

$$V_n = n\frac{\hbar\omega}{2e}, \qquad (1.76)$$

where n is an integer[55]. Since frequencies may be stabilized and measured with very high precision, this effect is being used in highly accurate standards of dc voltage.

1.7 Spatial dependence

In contrast to the simple and universal time dependence (1.62) of the stationary states, the spatial distributions of their wavefunction $\psi_n(\mathbf{r})$ need to be calculated from the problem-specific stationary Schrödinger equation (1.65). The solution of this equation for various particular cases is a major focus of the next two chapters. For now, let us consider just the simplest example, which nevertheless will be the basis for our discussion of more complex problems, namely a particle confined inside a rectangular hard-wall box. Such confinement may be described by the following potential energy[56]:

$$U(\mathbf{r}) = \begin{cases} 0, & \text{for } 0 < x < a_x, \ 0 < y < a_y, \text{ and } 0 < z < a_z, \\ +\infty, & \text{otherwise.} \end{cases} \qquad (1.77)$$

The only way to keep the product $U(\mathbf{r})\psi_n$ in Eq. (1.65) finite outside the box, is to have $\psi = 0$ in these regions. Also, the function have to be continuous everywhere, to avoid the divergence of its Laplace operator—which would give an unphysical

---

[54] For the discussion of this general effect, see, e.g. *Part CM* section 5.4.

[55] If ω is not too high, the size of these current steps may be calculated from Eqs. (1.73) and (1.74). Let me leave this task for the reader's exercise.

[56] Another common name for such potentials, especially of lower dimensionality, is the *potential well*, in our current case with a flat 'bottom', and infinitely high 'walls'. Note that sometimes, very unfortunately, such potential profiles are called 'quantum wells'. (This term seems to imply that the particle's confinement in such a well is a phenomenon specific for quantum mechanics. However, as we will repeatedly see in this course, the opposite is true: quantum effects do as much as they can to overcome the particle's confinement in a potential well, letting it partly penetrate the 'classically forbidden' regions beyond the well's walls.)

divergence of its kinetic energy (1.27). Hence, we may solve the stationary Schrödinger equation (1.60) just inside the box, i.e. with $U = 0$, so that it takes a simple form[57]

$$-\frac{\hbar^2}{2m}\nabla^2\psi_n = E_n\psi_n, \tag{1.78a}$$

with zero boundary conditions on all the walls. For our particular geometry, it is natural to express the Laplace operator in the Cartesian coordinates $\{x, y, z\}$ aligned with the box sides, with the origin at one of the corners of its rectangular $a_x \times a_y \times a_z$ volume, so that we get the following boundary problem:

$$-\frac{\hbar^2}{2m}\left(\frac{\partial^2}{\partial x^2} + \frac{\partial^2}{\partial y^2} + \frac{\partial^2}{\partial z^2}\right)\psi_n = E_n\psi_n,$$

for $0 < x < a_x$, $0 < y < a_y$, and $0 < z < a_z$, \hfill (1.78b)

with $\psi_n = 0$ for: $x = 0$ and a_x; $y = 0$ and a_y; $z = 0$ and a_z.

This problem may be readily solved using the same variable separation method as in section 1.5, now to separate the Cartesian spatial variables from each other, by looking for a partial solution of Eq. (1.78) in the form

$$\psi(\mathbf{r}) = X(x)Y(y)Z(z). \tag{1.79}$$

(It is convenient to postpone taking care of the proper indices for a minute.) Plugging this expression into Eq. (1.78b) and dividing all terms by the product XYZ, we get

$$-\frac{\hbar^2}{2m}\frac{1}{X}\frac{d^2X}{dx^2} - \frac{\hbar^2}{2m}\frac{1}{Y}\frac{d^2Y}{dy^2} - \frac{\hbar^2}{2m}\frac{1}{Z}\frac{d^2Z}{dz^2} = E. \tag{1.80}$$

Now let us repeat the standard argumentation of the variable separation method: since each term on the left-hand side of this equation may be only a function of the corresponding argument, the equality is possible only if each of them is a constant—in our case, with the dimensionality of energy. Calling these constants E_x, etc, we get three similar 1D equations

$$-\frac{\hbar^2}{2m}\frac{1}{X}\frac{d^2X}{dx^2} = E_x, \qquad -\frac{\hbar^2}{2m}\frac{1}{Y}\frac{d^2Y}{dy^2} = E_y, \qquad -\frac{\hbar^2}{2m}\frac{1}{Z}\frac{d^2Z}{dx^2} = E_z, \tag{1.81}$$

with Eq. (1.80) turning into the following energy-matching condition:

$$E_x + E_y + E_z = E. \tag{1.82}$$

---

[57] Rewritten as $\nabla^2 f + k^2 f = 0$, this is just the *Helmholtz equation*, which describes waves of any nature (with the wave vector **k**) in a uniform, isotropic, linear medium—see, e.g. *Part EM* sections 7.5–7.9 and 8.5.

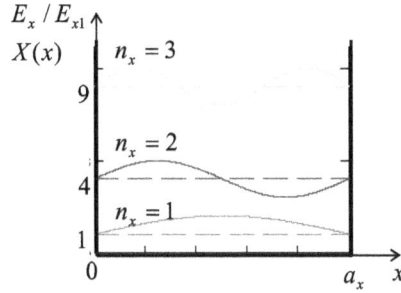

Figure 1.8. The lowest eigenfunctions (solid lines) and eigenvalues (dashed lines) of Eq. (1.83) for a potential well of length a_x. Solid black lines show the effective potential energy profile for the 1D eigenproblem (1.83).

All three ordinary differential equation (1.81), and their solutions, are similar. For example, for $X(x)$, we have a 1D Helmholtz equation

$$\frac{d^2 X}{dx^2} + k_x^2 X = 0, \quad \text{with } k_x^2 \equiv \frac{2mE_x}{\hbar^2}, \tag{1.83}$$

and simple boundary conditions: $X(0) = X(a_x) = 0$. Let me hope that the reader knows how to solve this well-known 1D boundary problem—describing, for example, the usual mechanical waves on a guitar string. The problem allows an infinite number of sinusoidal standing-wave eigenfunctions[58],

$$X \propto \sin k_x x, \quad \text{with } k_x = \frac{\pi n_x}{a_x},$$

$$\text{so that } X = \left(\frac{2}{a_x}\right)^{1/2} \sin \frac{\pi n_x x}{a_x}, \quad \text{with } n_x = 1, 2, \ldots , \tag{1.84}$$

corresponding to the eigenvalues $k_x = \pi n_x / a_x$, and hence the following eigenenergies:

$$E_x = \frac{\hbar^2}{2m} k_x^2 = \frac{\pi^2 \hbar^2}{2ma_x^2} n_x^2 \equiv E_{x1} n_x^2. \tag{1.85}$$

Figure 1.8 shows these simple results, using a somewhat odd but very graphic and hence common way, where the eigenenergy values (frequently called the *energy levels*) are used as horizontal axes for plotting the eigenfunctions, despite their completely different dimensionality.

Due to the similarity of all Eqs. (1.81), $Y(y)$ and $Z(z)$ are absolutely similar functions of their arguments, and may also be numbered by integers (say, n_y and n_z) independent of n_x, so that the spectrum of values of the total energy (1.82) is

$$E_{n_x, n_y, n_z} = \frac{\pi^2 \hbar^2}{2m} \left(\frac{n_x^2}{a_x^2} + \frac{n_y^2}{a_y^2} + \frac{n_z^2}{a_z^2} \right). \tag{1.86}$$

---

[58] The front coefficient in the last expression for X ensures the (ortho)normality condition (1.66).

Thus, in this 3D problem, the role of index n in the general Eq. (1.69) is played by a set of 3 independent integers $\{n_x, n_y, n_z\}$. In quantum mechanics, such integers play a key role, and thus have a special name, the *quantum numbers*. Using them, the general solution of our simple problem may be represented as the sum

$$\Psi(\mathbf{r}, t) = \sum_{n_x, n_y, n_z=1}^{\infty} c_{n_x,n_y,n_z} \sin \frac{\pi n_x x}{a_x} \sin \frac{\pi n_y y}{a_y} \sin \frac{\pi n_z z}{a_z} \exp\left\{-i\frac{E_{n_x,n_y,n_z}}{\hbar}t\right\}, \quad (1.87)$$

with the front coefficients that may be readily calculated from the initial wave-function $\Psi(\mathbf{r}, 0)$, using Eq. (1.68)—again with the replacement $n \to \{n_x, n_y, n_z\}$. This simplest problem is a good illustration of typical results the wave mechanics gives for spatially-confined motion, including the discrete energy spectrum, and (in this case, evidently) orthogonal eigenfunctions. Perhaps most importantly, its solution shows that the lowest value of the particle's kinetic energy, reached in the so-called *ground state* (in our case, the state with $n_x = n_y = n_z = 1$) is above zero.

An example of the opposite case of a *continuous spectrum* for *unconfined motion* of a free particle is given by the plane waves (1.29). With the account of relations $E = \hbar\omega$ and $\mathbf{p} = \hbar\mathbf{k}$, this wavefunction may be viewed as the product of the time-dependent factor (1.62) by the eigenfunction,

$$\psi_{\mathbf{k}} = a_{\mathbf{k}} \exp\{i\mathbf{k} \cdot \mathbf{r}\}, \quad (1.88)$$

which is the solution of the stationary Schrödinger equation (1.78a) if it is valid in the whole space[59]. The reader should not be worried too much by the fact that the fundamental solution (1.86) in free space is a *traveling* wave (having, in particular, a nonvanishing value of the probability current $\mathbf{j}$), while those inside a quantum box are *standing* waves, with $\mathbf{j} = 0$, even though the free space may be legitimately considered as the ultimate limit of a quantum box with volume $V = a_x \times a_y \times a_z \to \infty$. Indeed, due to the linearity of wave mechanics, two traveling-wave solutions (1.88) with equal and opposite values of the momentum (and hence with the same energy) may be readily combined to give a standing-wave solution, for example, $\exp\{i\mathbf{k} \cdot \mathbf{r}\} + \exp\{-i\mathbf{k} \cdot \mathbf{r}\} = 2\cos(\mathbf{k}\cdot\mathbf{r})$, with the net current $\mathbf{j} = 0$.[60] Thus, depending on convenience for solution of a particular problem, we can represent the general solution as a sum of either traveling-wave or standing-wave eigenfunctions.

Since in the unlimited free space there are no boundary conditions to satisfy, the Cartesian components of the wave vector $\mathbf{k}$ in Eq. (1.88) can take any real values. (This is why it is more convenient to label these wavefunctions, and the corresponding eigenenergies,

---

[59] In some systems (e.g. a particle interacting with a potential well of a finite depth), a discrete energy spectrum within a certain interval of energies may coexist with a continuous spectrum in a complementary interval. However, the conceptual philosophy of eigenfunctions and eigenvalues remains the same in this case as well.
[60] This is, of course, the general property of waves of any physical nature, propagating in a linear medium—see, e.g. *Part CM* section 6.5 and/or *Part EM* section 7.3.

$$E_{\mathbf{k}} = \frac{\hbar^2 k^2}{2m} \geqslant 0, \tag{1.89}$$

with their wave vector $\mathbf{k}$ rather than an integer index.) However, one aspect of continuous-spectrum systems requires a bit more math caution: the summation (1.69) should be replaced by the integration over a continuous index or indices—in our current case, three Cartesian components of the vector $\mathbf{k}$. The main rule of such replacement may be readily extracted from Eq. (1.84): according to this relation, for standing-wave solutions, the eigenvalues of k_x are *equidistant*, i.e. separated by equal intervals $\Delta k_x = \pi/a_x$ (with the similar relations for other two Cartesian components of vector $\mathbf{k}$). Hence the number of different eigenvalues of the standing wave vector $\mathbf{k}$ (with k_x, k_y, $k_z \geqslant 0$), within a volume $d^3 k \gg 1/V$ of the $\mathbf{k}$ space is $dN = d^3 k / (\Delta k_x \Delta k_x \Delta k_x) = (V/\pi^3) d^3 k$. Since in the continuum it is more convenient to work with traveling waves (1.88), we should take into account that, as was just discussed, there are two different traveling wave numbers (say, $+k_x$ and $-k_x$) corresponding to each standing wave vector's $k_x > 0$. Hence the same number of physically different states corresponds to a $2^3 = 8$ fold larger $\mathbf{k}$ space (which now is infinite in all directions) or, equivalently, to an 8-fold smaller number of states per unit volume $d^3 k$:

$$dN = \frac{V}{(2\pi)^3} d^3 k. \tag{1.90}$$

For $dN \gg 1$, this expression is independent on the boundary conditions, and is frequently represented as the following *summation rule*

$$\lim_{k^3 V \to \infty} \sum_{\mathbf{k}} f(\mathbf{k}) = \int f(\mathbf{k}) dN = \frac{V}{(2\pi)^3} \int f(\mathbf{k}) d^3 k, \tag{1.91}$$

where $f(\mathbf{k})$ is an arbitrary function of $\mathbf{k}$. Note that if the same wave vector $\mathbf{k}$ corresponds to several internal quantum states (such as spin—see chapter 4), the right-hand side of Eq. (1.91) requires its multiplication by the corresponding *degeneracy factor*.

1.8 Dimensionality reduction

To conclude this introductory chapter, let me discuss the conditions when the spatial dimensionality of a wave mechanics problem may be reduced[61]. Naively, one may think that if the particle's potential energy depends on just one spatial coordinate, say $U = U(x, t)$, then its wavefunction has to be one-dimensional as well: $\psi = \psi(x, t)$. Our discussion of the particular case $U = \text{const}$ in the previous section shows that this assumption is wrong. Indeed, though this potential is just a special case of the potential $U(x, t)$, most of its eigenfunctions, given by Eqs. (1.87) or (1.88), do depend on other two coordinates. This is why the solutions $\psi(x, t)$ of the 1D Schrödinger equation

---

[61] Many textbooks on quantum mechanics jump to the formal solution of 1D problems without such discussion, and most of my beginning graduate students did not understand that in realistic physical systems, such dimensionality restriction is adequate only under very specific conditions.

$$i\hbar\frac{\partial\Psi}{\partial t} = -\frac{\hbar^2}{2m}\frac{\partial^2}{\partial x^2}\Psi + U(x,\,t)\Psi, \tag{1.92}$$

which follows from Eq. (1.65) by assuming $\partial\Psi/\partial y = \partial\Psi/\partial z = 0$, are insufficient to form the general solution of Eq. (1.65) for this case.

This fact is easy to understand physically for the simplest case of a stationary 1D potential: $U = U(x)$. The absence of the y- and z-dependence of the potential energy U may be interpreted as a potential well which is flat in two directions, y and z. Replicating the arguments of the previous section for this case, we see that the eigenfunctions of a particle in such a well have the form

$$\psi(\mathbf{r}) = X(x)\exp\{i(k_y y + k_z z)\}, \tag{1.93}$$

where $X(x)$ are the eigenfunction of the following stationary 1D Schrödinger equation:

$$-\frac{\hbar^2}{2m}\frac{d^2 X}{dx^2} + U_{\text{ef}}(x)X = EX, \tag{1.94}$$

where $U_{\text{ef}}(x)$ is not the full potential energy of the particle, as would follow from Eq. (1.92), but rather its effective value including the kinetic energy of lateral motion:

$$U_{\text{ef}} \equiv U + (E_y + E_z) = U + \frac{\hbar^2}{2m}(k_y^2 + k_z^2). \tag{1.95}$$

In plain English, the particle's partial wavefunction $X(x)$, and its full energy, depends of its transverse momenta, which have continuous spectrum—see the discussion of Eq. (1.89). This means that Eq. (1.92) is adequate only if the condition $k_y = k_z = 0$ is somehow enforced, and in most physical problems, it is not. For example, if a de Broglie (or any other) plane wave $\Psi(x,\,t)$ is incident on a potential step, it would be reflected exactly back, i.e. with $k_y = k_z = 0$, only if the wall's surface is a perfect plane and exactly normal to the axis x. Any imperfection (and there are so many of them in real physical systems -:) may cause excitation of waves with nonvanishing values of k_y and k_z, due the continuous character of the functions $E_y(k_y)$ and $E_z(k_z)$.[62]

There is essentially one, perhaps counter-intuitive way to make the 1D solutions 'robust' to small perturbations: that is to provide a *rigid lateral confinement*[63] in two other directions. As the simplest example, consider a narrow *quantum wire* (figure 1.9a), provided by the potential

---

[62] This problem is not specific for quantum mechanics. The classical motion of a particle in a 1D potential may be also unstable with respect to lateral perturbations, especially is the potential is time-dependent, i.e. capable of exciting low-energy lateral modes.

[63] The term 'quantum confinement', sometimes used to describe this phenomenon, is as unfortunate as the 'quantum well', because of the same reason: the confinement is a purely classical effect, and as we will repeatedly see in this course, the quantum mechanical effects reduce, rather than enable it.

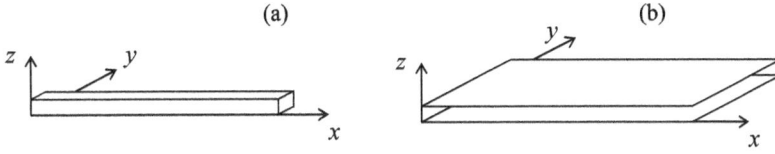

Figure 1.9. Partial confinement in: (a) two dimensions, and (b) one dimensions.

$$U(\mathbf{r}) = \begin{cases} U(x), & \text{for } 0 < y < a_y, \text{ and } 0 < z < a_z, \\ +\infty, & \text{otherwize.} \end{cases} \tag{1.96}$$

Performing the standard variable separation (1.79), we see that the corresponding stationary Schrödinger equation is satisfied if the partial wavefunction $X(x)$ obeys Eqs. (1.94)–(1.95), but now with a discrete energy spectrum in the transverse directions:

$$U_{\text{ef}} = U + \frac{\pi^2 \hbar^2}{2m} \left(\frac{n_y^2}{a_y^2} + \frac{n_z^2}{a_z^2} \right). \tag{1.97}$$

If the lateral confinement is tight, $a_y, a_z \to 0$, then there is a large energy gap,

$$\Delta U \sim \frac{\pi^2 \hbar^2}{2m a_{y,z}^2}, \tag{1.98}$$

between the ground-state energy of the lateral motion (with $n_y = n_z = 1$) and that for all its excited states. As a result, if the particle is initially placed into the lateral ground state, and its energy E is much smaller than ΔU, it would stay in this state, i.e. it may be described by a 1D Schrödinger equation similar to Eq. (1.92)—even in the time-dependent case, if the characteristic frequency of energy variations is much smaller than $\Delta U/\hbar$. Absolutely similarly, the strong lateral confinement in just one dimension (say, z, see figure 1.9b) enables systems with a robust 2D evolution of the particle's wavefunction.

The tight lateral confinement may ensure the dimensionality reduction even if the potential well is not exactly rectangular in the lateral direction(s), as described by Eq. (1.96), but is described by some x- and t-independent profile, if it still provides a sufficiently large energy gap ΔU. For example, many 2D quantum phenomena, such as the quantum Hall effect[64], have been studied experimentally using electrons confined at semiconductor heterojunctions (e.g. epitaxial interfaces GaAs/Al$_x$Ga$_{1-x}$As), where the potential well in the direction perpendicular to the interface has a nearly triangular shape, and provides the energy gap ΔU of the order of 10^{-2} eV.[65] This splitting energy corresponds to $k_B T$ with $T \sim 100$ K, so that careful experimentation at liquid helium temperatures (4 K and below) may keep the electrons performing purely 2D motion in the 'lowest subband' ($n_z = 1$).

---

[64] To be discussed in section 3.2.
[65] See, e.g. [6].

Finally, note that in systems with a reduced dimensionality, Eq. (1.90) for the number of states at large $\mathbf{k}$ (i.e. for an essentially free particle motion) should be replaced accordingly: in a 2D system of area $A \gg 1/k^2$,

$$dN = \frac{A}{(2\pi)^2}d^2k, \tag{1.99}$$

while in a 1D system of length $l \gg 1/k$,

$$dN = \frac{l}{2\pi}dk, \tag{1.100}$$

with the corresponding changes of the summation rule (1.91). This change has important implications for the density of states on the energy scale, dN/dE: it is straightforward (and hence left for the reader) to use Eqs. (1.90), (1.99), and (1.100) to show that for free 3D particles the density increases with E (proportionally to $E^{1/2}$), for free 2D particles it does not depend on energy at all, while for free 1D particles it scales as $E^{-1/2}$, i.e. decreases with energy.

1.9 Problems

Problem 1.1. The actual postulate made by N Bohr in his original 1913 paper was not directly Eq. (1.8), but an assumption that at quantum leaps between adjacent large (quasiclassical) orbits with $n \gg 1$, the hydrogen atom either emits or absorbs energy $\Delta E = \hbar\omega$, where ω is its classical radiation frequency—according to classical electrodynamics, equal to the angular velocity of electron's rotation[66]. Prove that this postulate is indeed compatible with Eqs. (1.7) and (1.8).

Problem 1.2. Use Eq. (1.53) to prove that the linear operators of quantum mechanics are commutative: $\hat{A}_2 + \hat{A}_1 = \hat{A}_1 + \hat{A}_2$, and associative: $(\hat{A}_1 + \hat{A}_2) + \hat{A}_3 = \hat{A}_1 + (\hat{A}_2 + \hat{A}_3)$.

Problem 1.3. Prove that for any time-independent Hamiltonian operator $\hat{H}$ and two arbitrary complex functions $f(\mathbf{r})$ and $g(\mathbf{r})$,

$$\int f(\mathbf{r})\hat{H}g(\mathbf{r})\,d^3r = \int \hat{H}f(\mathbf{r})g(\mathbf{r})\,d^3r.$$

Problem 1.4. Prove that the Schrödinger equation (1.25) with the Hamiltonian operator given by Eq. (1.41), is Galilean form-invariant, provided that the wavefunction is transformed as

$$\Psi'(\mathbf{r}', t') = \Psi(\mathbf{r}, t)\exp\left\{-i\frac{m\mathbf{v}\cdot\mathbf{r}}{\hbar} + i\frac{mv^2t}{2\hbar}\right\},$$

where the prime sign denotes the variables measured in the reference frame 0′ that moves, without rotation, with a constant velocity $\mathbf{v}$ relatively to the 'lab' frame 0. Give a physical interpretation of this transformation.

---

[66] See, e.g. *Part EM* section 8.2.

Problem 1.5.* Prove the so-called *Hellmann–Feynman theorem*[67]:

$$\frac{\partial E_n}{\partial \lambda} = \left\langle \frac{\partial H}{\partial \lambda} \right\rangle_n,$$

where λ is some *c*-number parameter, on which the time-independent Hamiltonian $\hat{H}$, and hence its eigenenergies E_n, depend.

Problem 1.6.* Use Eqs. (1.73) and (1.74) to analyze the effect of phase locking of Josephson oscillations on the dc current flowing through a weak link between two superconductors (frequently called the *Josephson junction*), assuming that an external microwave source applies to the junction a sinusoidal ac voltage with frequency ω and amplitude A.

Problem 1.7. Calculate $\langle x \rangle$, $\langle p_x \rangle$, δx, and δp_x for the eigenstate $\{n_x, n_y, n_z\}$ of a particle in a rectangular, hard-wall box, described by Eq. (1.77), and compare the product $\delta x \delta p_x$ with the Heisenberg's uncertainty relation.

Problem 1.8. Looking at the lower (red) line in figure 1.8, it seems plausible that the 1D ground-state function (1.84) of the simple potential well (1.77) may be well approximated with an inverted quadratic parabola:

$$X_{\text{trial}}(x) = C \, x \, (a_x - x),$$

where C is a normalization constant. Explore how good this approximation is.

Problem 1.9. A particle, placed in a hard-wall, rectangular box with sides a_x, a_y, and a_z, is in its ground state. Calculate the average force acting on each face of the box. Can the forces be characterized by a certain pressure?

Problem 1.10. A 1D quantum particle was initially in the ground state of a very deep, rectangular potential well of width a:

$$U(x) = \begin{cases} 0, & \text{for } -a/2 < x < +a/2, \\ +\infty, & \text{otherwise.} \end{cases}$$

At some instant, the well's width is abruptly increased to a new value $a' > a$, leaving the potential symmetric with respect to the point $x = 0$, and then left constant. Calculate the probability that after the change, the particle is still in the ground state of the system.

Problem 1.11. At $t = 0$, a 1D particle of mass m is placed into a hard-wall, flat-bottom potential well

$$U(x) = \begin{cases} 0, & \text{for } 0 < x < a, \\ +\infty, & \text{otherwise,} \end{cases}$$

---

[67] Despite the theorem's common name, H Hellmann (in 1937) and R Feynman (in 1939) were not the first ones in the long list of physicists who have (apparently, independently) discovered this fact. Indeed, it may be traced back at least to a 1922 paper by W Pauli, and was carefully proved by P Güttinger in 1931.

in a 50/50 linear superposition of the lowest (ground) state and the first excited state. Calculate:

(i) the normalized wavefunction $\Psi(x, t)$ for arbitrary time $t \geqslant 0$, and
(ii) the time evolution of the expectation value $\langle x \rangle$ of the particle's coordinate.

Problem 1.12. Calculate the potential profiles $U(x)$ for that the following wavefunctions,

(i) $\Psi = c \exp\{-ax^2 - ibt\}$, and
(ii) $\Psi = c \exp\{-a|x| - ibt\}$,

(with real coefficients $a > 0$ and b), satisfy the 1D Schrödinger equation for a particle with mass m. For each case, calculate $\langle x \rangle$, $\langle p_x \rangle$, δx, and δp_x, and compare the product $\delta x \delta p_x$ with the Heisenberg's uncertainty relation.

Problem 1.13. A 1D particle of mass m, moving in the field of a stationary potential $U(x)$, has the following eigenfunction

$$\psi(x) = \frac{C}{\cosh \kappa x},$$

where C is the normalization constant, and κ is a real constant. Calculate the function $U(x)$ and the state's eigenenergy E.

Problem 1.14. Calculate the density dN/dE of traveling-wave states in large rectangular potential wells of various dimensions: $d = 1$, 2, and 3.

Problem 1.15.* Use the finite-difference method with steps $a/2$ and $a/3$ to find as many eigenenergies as possible for a 1D particle in the infinitely deep, hard-wall 1D potential well of width a. Compare the results with each other, and with the exact formula[68].

References

[1] Griffith D 2016 *Quantum Mechanics* 2nd ed (Cambridge: Cambridge University Press)
[2] Zeilinger A *et al* 1988 *Rev. Mod. Phys.* **60** 1067
[3] Nairz O *et al* 2003 *Am. J. Phys.* **71** 319
[4] Arndt M 2014 *Phys. Today* **67** 30
[5] Hamilton P *et al* 2015 *Phys. Rev. Lett.* **114** 100–405
[6] Harrison P 2010 *Quantum Wells, Wires, and Dots* 3rd ed (Wiley)

---

[68] You may like to start from reading about the finite-difference method—see, e.g. *Part CM* section 8.5 or *Part EM* section 2.11.

IOP Publishing

Quantum Mechanics
Lecture notes
Konstantin K Likharev

Chapter 2

1D wave mechanics

The simplest, 1D version of wave mechanics enables a quantitative discussion of many important quantum-mechanical effects. The order of their discussion in this chapter is dictated mostly by the mathematical convenience—going from the simplest potential profiles to more complex ones, so that we may build up on the previous results. However, I would advise the reader to focus more not on the math, but rather on the variety of the non-classical physical phenomena it describes, ranging from the particle penetration to classically-forbidden regions, to quantum-mechanical tunneling, to the metastable state decay, to covalent bonding and quantum oscillations, to energy bands and gaps.

2.1 Basic relations

As was discussed at the end of chapter 1, in several cases (most importantly, at strong confinement within the $[y, z]$ plane), the general (3D) Schrödinger equation may be reduced to the 1D equation, similar to Eq. (1.92):

$$i\hbar \frac{\partial \Psi(x, t)}{\partial t} = -\frac{\hbar^2}{2m} \frac{\partial^2 \Psi(x, t)}{\partial x^2} + U(x, t)\Psi(x, t). \qquad (2.1)$$

It is important, however, to remember that according to the discussion in section 1.8, $U(x, t)$ in this equation is generally an effective potential energy, which may include the energy of the lateral motion, while $\Psi(x, t)$ may be just one factor in the complete wavefunction $\Psi(x, t)\chi(y, z)$. If the transverse factor $\chi(y, z)$ is normalized to 1, then the integration of Eq. (1.22a) over the 3D space within a segment $[x_1, x_2]$ gives the following probability to find the particle on this segment:

$$W(t) \equiv \int_{x_1}^{x_2} \Psi(x, t)\Psi^*(x, t)dx. \qquad (2.2)$$

doi:10.1088/2053-2563/aaf3a3ch2

If the particle under analysis is definitely somewhere inside the system, the normalization of its 1D wavefunction $\Psi(x, t)$ is provided by extending the integral (2.2) to the whole axis x:

$$\int_{-\infty}^{+\infty} w(x, t)dx = 1, \quad \text{where } w(x, t) \equiv \Psi(x, t)\Psi^*(x, t). \tag{2.3}$$

A similar integration of Eq. (1.23) shows that the expectation value of any operator depending only on the coordinate x (and possibly time), may be expressed as

$$\langle A \rangle(t) = \int_{-\infty}^{+\infty} \Psi^*(x, t)\hat{A}\,\Psi(x, t)dx. \tag{2.4}$$

It is also useful to introduce the notion of the *probability current* along the x-axis (a *scalar*):

$$I(x, t) \equiv \int j_x dy dz = \frac{\hbar}{m}\text{Im}\left(\Psi^*\frac{\partial}{\partial x}\Psi\right) = \frac{\hbar}{m}|\Psi(x, t)|^2\,\frac{\partial\varphi}{\partial x}, \tag{2.5}$$

where j_x is the x-component of the current density *vector* $\mathbf{j}(\mathbf{r}, t)$. Then the continuity equation (1.48) for the segment $[x_1, x_2]$ takes the form

$$\frac{dW}{dt} + I(x_2) - I(x_1) = 0. \tag{2.6}$$

The above formulas are the basis for the analysis of 1D problems of wave mechanics, but before proceeding to particular cases, let me deliver on my earlier promise to prove that the Heisenberg's uncertainty relation (1.35) is indeed valid for any wavefunction $\Psi(x, t)$. For that, let us consider an evidently positive (or at least non-negative) integral

$$J(\lambda) \equiv \int_{-\infty}^{+\infty}\left|x\Psi + \lambda\frac{\partial\Psi}{\partial x}\right|^2 dx \geqslant 0, \tag{2.7}$$

where λ is an arbitrary real constant, and assume that at the at $x \to \pm\infty$ the wavefunction vanishes, together with its first derivative—as we will see below, a very common case. Then the left-hand side of Eq. (2.7) may be recast as

$$\int_{-\infty}^{+\infty}\left|x\Psi + \lambda\frac{\partial\Psi}{\partial x}\right|^2 dx \equiv \int_{-\infty}^{+\infty}\left(x\Psi + \lambda\frac{\partial\Psi}{\partial x}\right)\left(x\Psi + \lambda\frac{\partial\Psi}{\partial x}\right)^* dx$$

$$= \int_{-\infty}^{+\infty} x^2\Psi\Psi^*dx + \lambda\int_{-\infty}^{+\infty} x\left(\Psi\frac{\partial\Psi^*}{\partial x} + \frac{\partial\Psi}{\partial x}\Psi^*\right)dx \tag{2.8}$$

$$+ \lambda^2\int_{-\infty}^{+\infty}\frac{\partial\Psi}{\partial x}\frac{\partial\Psi^*}{\partial x}dx.$$

According to Eq. (2.4), the first term in the last form of Eq. (2.8) is just $\langle x^2 \rangle$, while the second and the third integrals may be worked out by parts:

$$\int_{-\infty}^{+\infty} x\left(\Psi\frac{\partial\Psi^*}{\partial x} + \frac{\partial\Psi}{\partial x}\Psi^*\right)dx = \int_{-\infty}^{+\infty} x\frac{\partial}{\partial x}(\Psi\Psi^*)\,dx$$

$$= \int_{x=-\infty}^{x=+\infty} xd(\Psi\Psi^*) \tag{2.9}$$

$$= \Psi\Psi^*x\,\Big|\,\begin{matrix}x=+\infty\\x=-\infty\end{matrix} - \int_{-\infty}^{+\infty}\Psi\Psi^*dx = -1,$$

$$\int_{-\infty}^{+\infty}\frac{\partial\Psi}{\partial x}\frac{\partial\Psi^*}{\partial x}dx = \int_{x=-\infty}^{x=+\infty}\frac{\partial\Psi}{\partial x}d\Psi^* = \frac{\partial\Psi}{\partial x}\Psi^*\,\Big|\,\begin{matrix}x=+\infty\\x=-\infty\end{matrix} - \int_{-\infty}^{+\infty}\Psi^*\frac{\partial^2\Psi}{\partial x^2}dx$$

$$= \frac{1}{\hbar^2}\int_{-\infty}^{+\infty}\Psi^*\hat{p}_x^2\Psi dx = \frac{\langle p_x^2\rangle}{\hbar^2}. \tag{2.10}$$

As a result, Eqs. (2.7) takes the following form:

$$J(\lambda) = \langle x^2\rangle - \lambda + \lambda^2\frac{\langle p_x^2\rangle}{\hbar^2} \geqslant 0, \quad \text{i.e. } \lambda^2 + a\lambda + b \geqslant 0,$$

$$\text{with } a \equiv -\frac{\hbar^2}{\langle p_x^2\rangle}, \quad b \equiv \frac{\hbar^2\langle x^2\rangle}{\langle p_x^2\rangle}. \tag{2.11}$$

This inequality should be valid for any real λ, so that the corresponding quadratic equation, $\lambda^2 + a\lambda + b = 0$, can have either one (degenerate) real root—or no real roots at all. This is only possible if its determinant, $\text{Det} = a^2 - 4b$, is non-positive, leading to the following requirement:

$$\langle x^2\rangle\langle p_x^2\rangle \geqslant \frac{\hbar^2}{4}. \tag{2.12}$$

In particular, if $\langle x\rangle = 0$ and $\langle p_x\rangle = 0$,[1] then according to Eq. (1.33), Eq. (2.12) takes the form

$$\langle\tilde{x}^2\rangle\langle\tilde{p}_x^2\rangle \geqslant \frac{\hbar^2}{4}, \tag{2.13}$$

which, according to the definition (1.34) of the rms uncertainties, is equivalent to Eq. (1.35).

Now let us notice that the Heisenberg's uncertainty relation looks very similar to the *commutation relation* between the corresponding operators:

---

[1] Eq. (2.13) may be proved even if $\langle x\rangle$ and $\langle p_x\rangle$ are not equal to zero, by making the following replacements, $x \to x - \langle x\rangle$, $\partial/\partial x \to \partial/\partial x + i\langle p\rangle/\hbar$, in Eq. (2.7), and then repeating all the calculations—which in this case become somewhat bulky. In chapter 4, equipped with the bra-ket formalism, we will derive a more general uncertainty relation, which includes the Heisenberg's relation (2.13) as a particular case, in a more efficient way.

$$\left[\hat{x}, \hat{p}_x\right]\Psi \equiv (\hat{x}\hat{p}_x - \hat{p}_x\hat{x})\Psi = x\left(-i\hbar\frac{\partial\Psi}{\partial x}\right) - \left(-i\hbar\frac{\partial}{\partial x}\right)(x\Psi) = i\hbar\Psi. \qquad (2.14a)$$

Since this relation is valid for any wavefunction $\Psi(x, t)$, we may represent it as an operator equality:

$$\left[\hat{x}, \hat{p}_x\right] = i\hbar \neq 0. \qquad (2.14b)$$

In section 4.5 we will see that the relation between Eqs. (2.13) and (2.14) is just a particular case of a general relation between the expectation values of non-commuting operators, and their commutators.

2.2 Free particle: wave packets

Let us start our discussion of particular problems with the free 1D motion, i.e. with $U(x, t) = 0$. From Eq. (1.29), it is evident that in the 1D case, a similar 'fundamental' (i.e. a particular but the most important) solution of the Schrödinger equation (2.1) is a sinusoidal ('monochromatic') wave

$$\Psi_0(x, t) = \text{const} \times \exp\{i(k_0 x - \omega_0 t)\}. \qquad (2.15)$$

According to Eqs. (1.32), it corresponds to a particle with an exactly defined momentum[2] $p_0 = \hbar k_0$ and energy $E_0 = \hbar\omega_0 = \hbar^2 k_0^2/2m$. However, for this wavefunction, the product $\Psi^*\Psi$ does not depend on either x or t, so that the particle is completely delocalized, i.e. its probability is spread all over axis x, at all times.

In order to describe a space-localized state, let us form, at the initial moment of time ($t = 0$), a wave packet of the type shown in figure 1.6, by multiplying the sinusoidal waveform (2.15) by some smooth *envelope function* $A(x)$. As the most important particular example, consider the *Gaussian wave packet*

$$\Psi(x, 0) = A(x)e^{ik_0 x}, \quad \text{with } A(x) = \frac{1}{(2\pi)^{1/4}(\delta x)^{1/2}}\exp\left\{-\frac{x^2}{(2\delta x)^2}\right\}. \qquad (2.16)$$

(By the way, figure 1.6 shows exactly such a packet.) The pre-exponential factor in this envelope function has been selected in the way to have the initial probability density,

$$w(x, 0) = \Psi^*(x, 0)\Psi(x, 0) = A^*(x)A(x) = \frac{1}{(2\pi)^{1/2}\delta x}\exp\left\{-\frac{x^2}{2(\delta x)^2}\right\}, \qquad (2.17)$$

normalized as in Eq. (2.3), for any parameters δx and k_0.[3]

---

[2] From this point on, and to the end of this chapter, I will drop the index x in the notation for the x-component of the vectors **k** and **p**.

[3] This fact may be readily proven using the well-known integral of the Gaussian function (2.17), in infinite limits—see, e.g. Eq. (A.36b). It is also straightforward to use Eq. (A.36c) to prove that for the wave packet (2.16), the parameter δx is indeed the rms uncertainty (1.34) of the coordinate x, thus justifying its notation.

In order to explore the evolution of this wave packet in time, we could try to solve Eq. (2.1) with the initial condition (2.16) directly, but in the spirit of the discussion in section 1.5, it is easier to proceed differently. Let us first represent the initial wavefunction (2.16) as a sum (1.67) of the eigenfunctions $\psi_k(x)$ of the corresponding stationary 1D Schrödinger equation (1.60), in our current case

$$-\frac{\hbar^2}{2m}\frac{d^2\psi_k}{dx^2} = E_k\psi_k, \quad \text{with } E_k \equiv \frac{\hbar^2k^2}{2m}, \tag{2.18}$$

which are simply monochromatic waves,

$$\psi_k = a_k e^{ikx}. \tag{2.19}$$

Since (as was discussed in section 1.7) at the unconstrained motion the spectrum of possible wave numbers k is continuous, the sum (1.67) should be replaced with an integral[4]:

$$\Psi(x, 0) = \int a_k\psi_k(x)dp = \int a_k e^{ikx}dk. \tag{2.20}$$

Now let us notice that from the point of view of mathematics, Eq. (2.20) is just the usual Fourier transform from the variable k to the 'conjugate' variable x, and we can use the well-known formula of the reciprocal Fourier transform to calculate

$$a_k = \frac{1}{2\pi}\int \Psi(x, 0)e^{-ikx}dx$$
$$= \frac{1}{2\pi}\frac{1}{(2\pi)^{1/4}(\delta x)^{1/2}}\int \exp\left\{-\frac{x^2}{(2\delta x)^2} - i\tilde{k}x\right\}dx, \quad \text{where } \tilde{k} \equiv k - k_0, \tag{2.21}$$

This *Gaussian integral* may be worked out by the following standard method. Let us complement the exponent to the full square of a linear combination of x and k, adding a compensating term independent of x:

$$-\frac{x^2}{(2\delta x)^2} - i\tilde{k}x \equiv -\frac{1}{(2\delta x)^2}[x + 2i(\delta x)^2\tilde{k}]^2 - \tilde{k}^2(\delta x)^2. \tag{2.22}$$

Since the integration in the right-hand side of Eq. (2.21) should be performed at constant $\tilde{k}$, in the infinite limits of x, its result would not change if we replace dx by $dx' \equiv d[x + 2i(\delta x)^2\,\tilde{k}].$[5] As a result, we get,

$$a_k = \frac{1}{2\pi}\frac{1}{(2\pi)^{1/4}(\delta x)^{1/2}}\exp\{-\tilde{k}^2(\delta x)^2\}\int \exp\left\{-\frac{x'^2}{(2\delta x)^2}\right\}dx'$$
$$= \left(\frac{1}{2\pi}\right)^{1/2}\frac{1}{(2\pi)^{1/4}(\delta k)^{1/2}}\exp\left\{-\frac{\tilde{k}^2}{(2\delta k)^2}\right\}, \tag{2.23}$$

---

[4] For the notation brevity, from this point on the infinite limit signs will be dropped in all 1D integrals.
[5] The fact that the argument shift is imaginary is not important, because the function under the integral is analytical, and tends to zero at Re $x' \to \pm\infty$.

so that a_k also has a Gaussian distribution, now along the k-axis, centered to the value k_0 (figure 1.6b), with the constant δk defined as

$$\delta k \equiv 1/2\delta x. \tag{2.24}$$

Thus we may represent the initial wave packet (2.16) as

$$\Psi(x, 0) = \left(\frac{1}{2\pi}\right)^{1/2} \frac{1}{(2\pi)^{1/4}(\delta k)^{1/2}} \int \exp\left\{-\frac{(k - k_0)^2}{(2\delta k)^2}\right\} e^{ikx} dk. \tag{2.25}$$

From the comparison of this formula with Eq. (2.16), it is evident that the rms uncertainty of the wave number k in this packet is indeed equal to δk defined by Eq. (2.24), thus justifying the notation. The comparison of that relation with Eq. (1.35) shows that the Gaussian packet represents the ultimate case in which the product $\delta x \delta p = \delta x(\hbar \delta k)$ has the lowest possible value ($\hbar/2$); for any other envelope's shape the uncertainty product may only be larger.

We could of course get the same result for δk from Eq. (2.16) using the definitions (1.23), (1.33), and (1.34); the real advantage of Eq. (2.24) is that it can be readily generalized to $t > 0$. Indeed, we already know that the time evolution of the wavefunction is given by Eq. (1.69), for our case[6]

$$\Psi(x, t) = \left(\frac{1}{2\pi}\right)^{1/2} \frac{1}{(2\pi)^{1/4}(\delta k)^{1/2}} \int \exp\left\{-\frac{(k - k_0)^2}{(2\delta k)^2}\right\} e^{ikx} \exp\left\{-i\frac{\hbar k^2}{2m}t\right\} dk. \tag{2.26}$$

Figure 2.1 shows several snapshots of the real part of the wavefunction (2.26), for a particular case $\delta k = 0.1\, k_0$.

The plots clearly show the following effects:

(i) the wave packet as a whole (as characterized by its envelope) moves along the x axis with a certain *group velocity* v_{gr},

(ii) the 'carrier' quasi-sinusoidal wave inside the packet moves with a different, *phase velocity* v_{ph}, which may be defined as the velocity of the spatial points where the wave's phase $\varphi(x, t) \equiv \arg \Psi$ takes a certain fixed value (say, $\varphi = \pi/2$, where Re Ψ vanishes), and

(iii) the wave packet's spatial width gradually increases with time—the packet *spreads*.

All these effects are common for waves of any physical nature[7]. Indeed, let us consider a 1D wave packet of the type (2.26), but more general:

$$\Psi(x, t) = \int a_k e^{i(kx-\omega t)} dk, \tag{2.27}$$

---

[6] Note that this packet is equivalent to Eq. (2.16) and hence is properly normalized to 1—see Eq. (2.3). Hence the wave packet introduction offers a natural solution to the problem of infinite wave normalization, which was mentioned in section 1.2.

[7] See, e.g. brief discussions in *Part CM* section 6.3 and *Part EM* section 7.2.

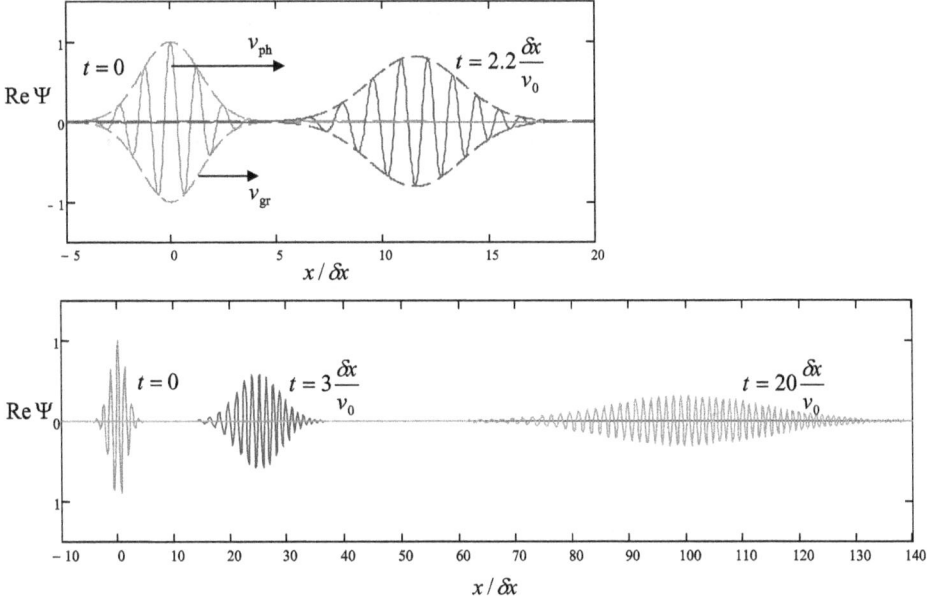

Figure 2.1. The time evolution of a 1D wave packet on (a) smaller and (b) larger time scales. The dashed lines show the packet's envelope, i.e. $\pm |\Psi|$.

let it propagate in a medium with an arbitrary (but smooth!) dispersion relation $\omega(k)$, and assume that the wave number distribution a_k is narrow: $\delta k \ll \langle k \rangle \equiv k_0$—see figure 1.6b. Then we may expand the function $\omega(k)$ into the Taylor series near the central wave number k_0, and keep only three of its leading terms:

$$\omega(k) \approx \omega_0 + \frac{d\omega}{dk}\tilde{k} + \frac{1}{2}\frac{d^2\omega}{dk^2}\tilde{k}^2, \quad \text{where } \tilde{k} \equiv k - k_0, \ \omega_0 \equiv \omega(k_0), \quad (2.28)$$

where both derivatives have to be evaluated at the point $k = k_0$. In this approximation[8], the expression in the parentheses on the right-hand side of Eq. (2.27) may be rewritten as

$$kx - \omega(k)t \approx k_0 x + \tilde{k}x - \left(\omega_0 + \frac{d\omega}{dk}\tilde{k} + \frac{1}{2}\frac{d^2\omega}{dk^2}\tilde{k}^2\right)t$$

$$\equiv (k_0 x - \omega_0 t) + \tilde{k}\left(x - \frac{d\omega}{dk}t\right) - \frac{1}{2}\frac{d^2\omega}{dk^2}\tilde{k}^2 t, \quad (2.29)$$

so that Eq. (2.27) is reduced to the following integral

$$\Psi(x, t) = e^{i(k_0 x - \omega_0 t)} \int a_k \exp\left\{i\left[\tilde{k}\left(x - \frac{d\omega}{dk}t\right) - \frac{1}{2}\frac{d^2\omega}{dk^2}\tilde{k}^2 t\right]\right\}d\tilde{k}. \quad (2.30)$$

---

[8] By the way, in the particular case of de Broglie waves, described by the dispersion relation (1.30), Eq. (2.28) is exact, because $\omega = E/\hbar$ is a quadratic function of $k = p/\hbar$, and all higher derivatives of ω over k vanish for any k_0.

First, let us neglect the last term in square brackets (which is much smaller than the first term if the dispersion relation is smooth enough and/or the time interval t is sufficiently small), and compare the result with the initial form of the wave packet (2.27)

$$\Psi(x, 0) = \int a_k e^{ikx} dk = A(x) e^{ik_0 x}, \quad \text{with } A(x) \equiv \int a_k e^{i\tilde{k}x} d\tilde{k}. \tag{2.31}$$

The comparison shows that Eq. (2.30) is reduced to

$$\Psi(x, t) = A(x - v_{gr}t) e^{ik_0(x - v_{ph}t)}, \tag{2.32}$$

where v_{gr} and v_{ph} are two constants with the dimension of velocity:

$$v_{gr} \equiv \left. \frac{d\omega}{dk} \right|_{k=k_0}, \quad v_{ph} \equiv \left. \frac{\omega}{k} \right|_{k=k_0}. \tag{2.33a}$$

Clearly, this general result describes the effects (i) and (ii) listed above. For the particular case of de Broglie waves, whose dispersion law is given by Eq. (1.30), we get

$$v_{gr} \equiv \left. \frac{d\omega}{dk} \right|_{k=k_0} = \frac{\hbar k_0}{m} \equiv v_0, \quad v_{ph} \equiv \left. \frac{\omega}{k} \right|_{k=k_0} = \frac{\hbar k_0}{2m} = \frac{v_{gr}}{2}. \tag{2.33b}$$

We see that (very fortunately) the velocity of the wave packet's envelope equals to that of the classical particle moving by inertia, in accordance with the correspondence principle.

Next, the last term in the square brackets of Eq. (2.30) describes the effect (iii), the wave packet's spread. It may be readily evaluated if the packet (2.27) is initially Gaussian, as in our example (2.25):

$$a_k = \text{const} \times \exp\left\{ -\frac{\tilde{k}^2}{(2\delta k)^2} \right\}. \tag{2.34}$$

In this case the integral (2.30) is Gaussian, and may be worked out exactly as the integral (2.21), i.e. by representing the merged exponents under the integral as a full square of a linear combination of x and k:

$$-\frac{\tilde{k}^2}{(2\delta k)^2} + i\tilde{k}(x - v_{gr}t) - \frac{i}{2}\frac{d^2\omega}{dk^2}\tilde{k}^2 t$$
$$\equiv -\Delta(t)\left(\tilde{k} + i\frac{x - v_{gr}t}{2\Delta(t)} \right)^2 - \frac{(x - v_{gr}t)^2}{4\Delta(t)} + ik_0 x - \frac{i}{2}\frac{d^2\omega}{dk^2}k_0^2 t, \tag{2.35}$$

where I have introduced the following complex function of time:

$$\Delta(t) \equiv \frac{1}{4(\delta k)^2} + \frac{i}{2}\frac{d^2\omega}{dk^2}t = (\delta x)^2 + \frac{i}{2}\frac{d^2\omega}{dk^2}t, \tag{2.36}$$

and, at the second step, used Eq. (2.24). Now integrating over $\tilde{k}$, we get

$$\Psi(x,\,t) \propto \exp\left\{-\frac{(x-v_{\text{gr}}t)^2}{4\Delta(t)} + i\left(k_0 x - \frac{1}{2}\frac{d^2\omega}{dk^2}k_0^2 t\right)\right\}. \tag{2.37}$$

The imaginary part of the ratio $1/\Delta(t)$ in this exponent gives just an additional contribution to the wave's phase, and does not affect the resulting probability distribution

$$w(x,\,t) = \Psi^*\Psi \propto \exp\left\{-\frac{(x-v_{\text{gr}}t)^2}{2}\operatorname{Re}\frac{1}{\Delta(t)}\right\}. \tag{2.38}$$

This is again a Gaussian distribution over axis x, centered to point $\langle x \rangle = v_{\text{gr}}t$, with the rms width

$$(\delta x')^2 \equiv \left\{\operatorname{Re}\left[\frac{1}{\Delta(t)}\right]\right\}^{-1} = (\delta x)^2 + \left(\frac{1}{2}\frac{d^2\omega}{dk^2}t\right)^2\frac{1}{(\delta x)^2}. \tag{2.39a}$$

In the particular case of de Broglie waves, $d^2\omega/dk^2 = \hbar/m$, so that

$$(\delta x')^2 = (\delta x)^2 + \left(\frac{\hbar t}{2m}\right)^2\frac{1}{(\delta x)^2}. \tag{2.39b}$$

The physics of the packet spreading is very simple: if $d^2\omega/dk^2 \neq 0$, the group velocity $d\omega/dk$ of each small group dk of the monochromatic components of the wave is different, resulting in the gradual (eventually, linear) accumulation of the differences of the distances traveled by the groups. The most curious feature of Eq. (2.39) is that the packet width at $t > 0$ depends on its initial width $\delta x'(0) = \delta x$ in a non-monotonic way, tending to infinity at both $\delta x \to 0$ and $\delta x \to \infty$. Because of that, for a given time t, there is an optimal value of δx that minimizes $\delta x'$:

$$(\delta x')_{\min} = \sqrt{2}\,(\delta x)_{\text{opt}} = \left(\frac{\hbar t}{m}\right)^{1/2}. \tag{2.40}$$

This expression may be used for spreading effect estimates. Due to the smallness of the Planck constant $\hbar$ on the human scale of things, for macroscopic bodies this effect is extremely small even for very long time intervals; however, for light particles it may be very noticeable: for an electron ($m = m_e \approx 10^{-30}$ kg), and $t = 1$ s, Eq. (2.40) yields $(\delta x')_{\min} \sim 1$ cm.

Note also that for any $t \neq 0$, the wave packet retains its Gaussian envelope, but the ultimate relation (2.24) is *not* satisfied, $\delta x'\delta p > \hbar/2$—due to a gradually accumulated phase shift between the component monochromatic waves. The last remark on this topic: in quantum mechanics, the wave packet spreading is *not* a ubiquitous effect! For example, in chapter 5 we will see that in a quantum oscillator, the spatial width of a Gaussian packet (for that system, called the *Glauber state* of the oscillator) does not grow monotonically but rather either stays constant or oscillates in time.

Now let us briefly discuss the case when the initial wave packet is not Gaussian, but is described by an arbitrary initial wavefunction. In order to make the forthcoming result more aesthetically appealing, it is beneficial to generalize out calculations to an arbitrary initial time t_0; it is evident that if U does not depend on time explicitly, it is sufficient to replace t with $(t - t_0)$ in all the above formulas. With this replacement, Eq. (2.27) becomes

$$\Psi(x, t) = \int a_k e^{i[kx - \omega(t-t_0)]} dk, \tag{2.41}$$

and the reciprocal transform (2.21) reads

$$a_k = \frac{1}{2\pi} \int \Psi(x, t_0) e^{-ikx} dx. \tag{2.42}$$

If we want to express these two formulas with one relation, i.e. plug Eq. (2.42) into Eq. (2.41), we should give the integration variable x some other name, e.g. x_0. (Such notation is appropriate, because this variable describes the coordinate argument in the initial wave packet.) The result is

$$\Psi(x, t) = \frac{1}{2\pi} \int dk \int dx_0 \Psi(x_0, t_0) e^{i[k(x-x_0)-\omega(t-t_0)]}. \tag{2.43}$$

Changing the order of integration, this expression may be rewritten in the following general form:

$$\Psi(x, t) = \int G(x, t; x_0, t_0) \, \Psi(x_0, t_0) dx_0, \tag{2.44}$$

where the function G, usually called *kernel* in mathematics, in quantum mechanics is called the *propagator*[9]. The physical sense of the propagator may be understood by considering the following special initial conditions[10]:

$$\Psi(x_0, t_0) = \delta(x_0 - x'), \tag{2.45}$$

where x' is a certain point within the domain of particle's motion. In this particular case, Eq. (2.44) evidently gives

$$\Psi(x, t) = G(x, t; x', t_0). \tag{2.46}$$

Hence, the propagator, considered as a function of its arguments x and t only, is just the wavefunction of the particle, at the δ-functional initial conditions (2.45). Thus just as Eq. (2.41) may be understood as a mathematical expression of the linear superposition principle in the momentum (i.e. reciprocal) space domain, Eq. (2.44) is an expression of this principle in the direct space domain: the system's 'response' $\Psi(x, t)$ to an arbitrary initial condition $\Psi(x_0, t_0)$ is just a sum of its responses to

[9] Its standard notation by letter G stems from the fact that the propagator is essentially the spatial-temporal *Green's function* of Eq. (2.18), defined very similarly to Green's functions of other ordinary and partial differential equations describing various physics systems—see, e.g. *Part CM* section 5.1 and/or *Part EM* sections 2.7 and 7.3.
[10] Note that such initial condition is *not* equivalent to a δ-functional initial probability density (2.2).

elementary spatial 'slices' of this initial function, with the propagator $G(x, t; x_0, t_0)$ representing the weight of each slice in the final sum.

According to Eqs. (2.43) and (2.44), in the particular case of a free particle the propagator is equal to

$$G(x, t; x_0, t_0) = \frac{1}{2\pi} \int e^{i[k(x-x_0)-\omega(t-t_0)]} dk, \tag{2.47}$$

Calculating this integral, one should remember that here ω is not a constant but a function of k, given by the dispersion relation for the particular waves. In particular, for the de Broglie waves, with $\hbar\omega = \hbar^2 k^2/2m$,

$$G(x, t; x_0, t_0) \equiv \frac{1}{2\pi} \int \exp\left\{i\left[k(x - x_0) - \frac{\hbar k^2}{2m}(t - t_0)\right]\right\} dk. \tag{2.48}$$

This is a Gaussian integral again, and may be readily calculated just as done (twice) above, by completing the exponent to the full square. The result is

$$G(x, t; x_0, t_0) = \left(\frac{m}{2\pi i \hbar(t - t_0)}\right)^{1/2} \exp\left\{-\frac{m(x - x_0)^2}{2i\hbar(t - t_0)}\right\}. \tag{2.49}$$

Please note the following features of this complex function (plotted in figure 2.2):

(i) It depends only on the differences $(x - x_0)$ and $(t - t_0)$. This is natural, because the free-particle propagation problem is uniform (*translation-invariant*) both in space and time.

(ii) The function's shape does not depend on its arguments—they just rescale the same function: its snapshot (figure 2.2), if plotted as a function of un-normalized x, just becomes broader and lower with time. It is curious that the spatial broadening scales as $(t - t_0)^{1/2}$—just as at the classical diffusion, as a result of a deep mathematical analogy between quantum mechanics and classical statistics—to be discussed further in chapter 7.

(iii) In accordance with the uncertainty relation, the ultimately compressed wave packet (2.45) has an infinite width of momentum distribution, and the quasi-sinusoidal tails of the free-particle propagator, clearly visible in figure 2.2, are the results of the free propagation of the fastest (highest-momentum) components of that distribution, in both directions from the packet center.

In the following sections, I will mostly focus on monochromatic wavefunctions (that, for unconfined motion, may be interpreted as wave packets of a very large spatial width δx), and only rarely revisit the wave packet discussion. My best excuse is the linear superposition principle, i.e. our conceptual ability to restore the general solution from that of monochromatic waves of all possible energies. However, the reader should not forget that, as the above discussion has illustrated, mathematically such restoration is not always trivial.

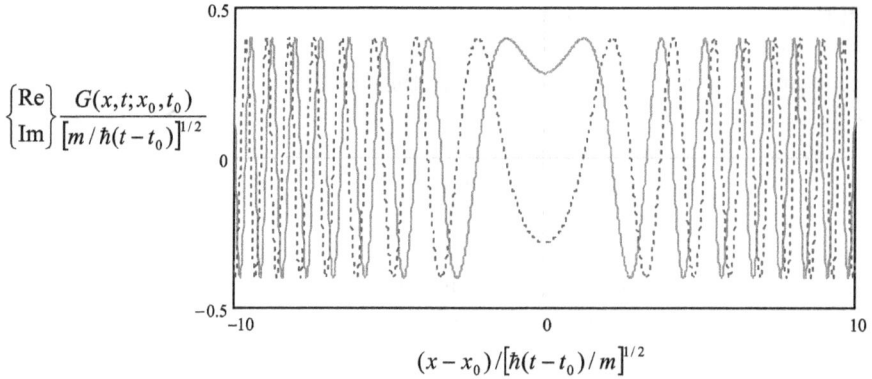

Figure 2.2. The real (solid line) and imaginary (dashed line) parts of the 1D free particle's propagator.

2.3 Particle reflection and tunneling

Now, let us proceed to the cases when a 1D particle moves in various potential profiles $U(x)$ that are constant in time. Conceptually, the simplest of such profiles is a potential step—see figure 2.3.

As I am sure the reader knows, in the classical mechanics the particle's kinetic energy $p^2/2m$ cannot be negative, so if the particle is incident on such a step (in figure 2.3, from the left), it can only travel through the *classically accessible* region, where its (conserved) full energy,

$$E = \frac{p^2}{2m} + U(x), \tag{2.50}$$

is larger than the local value $U(x)$. Let the initial velocity $v = p/m$ be positive, i.e. directed toward the step. Before it has reached the *classical turning point* x_c, defined by equation

$$U(x_c) = E, \tag{2.51}$$

the particle's kinetic energy $p^2/2m$ is positive, so that it continues to move in the initial direction. On the other hand, the particle cannot penetrate that *classically forbidden region* $x > x_c$, because there its kinetic energy would be negative. Hence when the particle reaches the point $x = x_c$, its velocity has to change sign, i.e. the particle is reflected back from the classical turning point.

In order to see what the wave mechanics says about this situation, let us start from the simplest, sharp potential step shown with the bold black line in figure 2.4:

$$U(x) = U_0\theta(x) \equiv \begin{cases} 0, & \text{at } x < 0, \\ U_0, & \text{at } 0 < x. \end{cases} \tag{2.52}$$

For this choice, and any energy within the interval $0 < E < U_0$, the classical turning point is $x_c = 0$.

Let us represent an incident particle with a wave packet so long that the spread $\delta k \sim 1/\delta x$ of its wave number spectrum, and hence the energy uncertainty

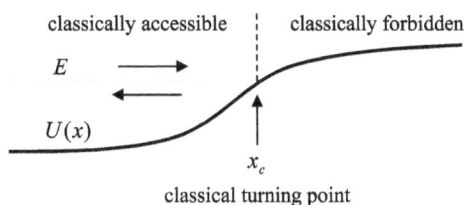

classically accessible : classically forbidden

E

$U(x)$

x_c

classical turning point

Figure 2.3. Classical 1D motion in a potential profile $U(x)$.

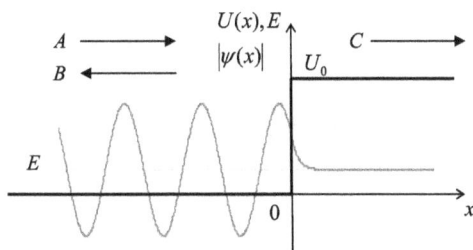

Figure 2.4. The reflection of a monochromatic wave from a potential step $U_0 > E$. (This particular wavefunction's shape is for $U_0 = 5E$.) The wavefunction is plotted with the same schematic vertical offset by E as those in figure 1.8.

$\delta E = \hbar\delta\omega = \hbar(d\omega/dk)\delta k$ is negligible in comparison with its average value $E < U_0$, as well as with $(U_0 - E)$. In this case, E may be considered as a given constant, the time dependence of the wavefunction is given by Eq. (1.62), and we can calculate its spatial factor $\psi(x)$ from the 1D version of the stationary Schrödinger equation (1.65):[11]

$$-\frac{\hbar^2}{2m}\frac{d^2\psi}{dx^2} + U(x)\psi = E\psi. \tag{2.53}$$

At $x < 0$, i.e. at $U = 0$, the equation is reduced to the Helmholtz equation (1.78), and may be satisfied with either of two traveling waves, proportional to $\exp\{+ikx\}$ and $\exp\{-ikx\}$ correspondingly, with k satisfying the dispersion equation (1.30):

$$k^2 \equiv \frac{2mE}{\hbar^2}. \tag{2.54}$$

Thus the general solution of Eq. (2.53) in this region may be represented as

$$\psi_-(x) = Ae^{+ikx} + Be^{-ikx}. \tag{2.55}$$

The second term on the right-hand side of Eq. (2.55) evidently describes an (infinitely long) wave packet traveling to the left, arising because of the particle's reflection from the potential step. If $B = -A$, this solution is reduced to Eq. (1.84) for the

---

[11] Note that this is *not* the eigenproblem like the one we have solved in section 1.4 for a potential well. Indeed, now the energy E is considered given—e.g. by the initial conditions that launch a long wave packet upon the potential step—in figure 2.4, from the left.

potential well with infinitely high walls, but for our current case of a finite step height U_0, the relation between the coefficients B and A may be different.

To show this, let us solve Eq. (2.53) for $x > 0$, where $U = U_0 > E$. In this region the equation may be rewritten as

$$\frac{d^2\psi_+}{dx^2} = \kappa^2\psi_+, \tag{2.56}$$

where κ is a real constant defined by the relation similar to Eq. (2.54):

$$\kappa^2 \equiv \frac{2m(U_0 - E)}{\hbar^2} > 0. \tag{2.57}$$

The general solution of Eq. (2.56) is the sum of $\exp\{+\kappa x\}$ and $\exp\{-\kappa x\}$, with arbitrary coefficients. However, in our case the wavefunction should be finite at $x \to \infty$, so only the latter exponent is acceptable:

$$\psi_+(x) = Ce^{-\kappa x}. \tag{2.58}$$

Such penetration of the wavefunction into the classically forbidden region, and hence a non-zero probability to find the particle there, is one of the most fascinating predictions of quantum mechanics, and has been repeatedly observed in experiment, e.g. via tunneling experiments—see the next section[12]. From Eq. (2.58), it is evident that the constant κ, defined by Eqs. (2.57), may be interpreted as the reciprocal penetration depth. Even for the lightest particles this depth is usually very small. Indeed, for $E \ll U_0$ that equation yields

$$\delta \equiv \frac{1}{\kappa}\bigg|_{E=0} = \frac{\hbar}{(2mU_0)^{1/2}}. \tag{2.59}$$

For example, for a conduction electron in a typical metal, that runs, at its surface, into a sharp potential step whose height equals metal's workfunction $U_0 \approx 5$ eV (see the discussion of the photoelectric effect in section 1.1), δ is close to 0.1 nm, i.e. is close to a typical size of an atom. For heavier elementary particles (e.g. protons) the penetration depth is correspondingly lower, and for macroscopic bodies it is hardly measurable.

Returning to Eqs. (2.55) and (2.58), we still should relate the coefficients B and C to the amplitude A of the incident wave, using the boundary conditions at $x = 0$. Since E is a finite constant, and $U(x)$ is a finite function, Eq. (2.53) says that $d^2\psi/dx^2$ should be finite as well. This means that the first derivative should be continuous:

$$\lim_{\varepsilon \to 0}\left(\frac{d\psi}{dx}\bigg|_{x=+\varepsilon} - \frac{d\psi}{dx}\bigg|_{x=-\varepsilon}\right) = \lim_{\varepsilon \to 0}\int_{-\varepsilon}^{+\varepsilon}\frac{d^2\psi}{dx^2}dx$$

$$= \frac{2m}{\hbar^2}\lim_{\varepsilon \to 0}\int_{-\varepsilon}^{+\varepsilon}[U(x) - E]\psi\,dx = 0. \tag{2.60}$$

---

[12] Note that this effect is pertinent to waves of any type, including the electromagnetic waves propagating in a medium—see, e.g. *Part EM* sections 7.3–7.7.

Repeating this calculation for the wavefunction $\psi(x)$ itself, we see that it also should be continuous at all points, including the border point $x = 0$, so that the boundary conditions in our problem are

$$\psi_-(0) = \psi_+(0), \qquad \frac{d\psi_-}{dx}(0) = \frac{d\psi_+}{dx}(0). \tag{2.61}$$

Plugging Eqs. (2.55) and (2.58) into Eqs. (2.61), we get a system of two linear equations

$$A + B = C, \quad ikA - ikB = -\kappa C, \tag{2.62}$$

whose (easy) solution enables us to express B and C via A:

$$B = A\frac{k - i\kappa}{k + i\kappa}, \quad C = A\frac{2k}{k + i\kappa}. \tag{2.63}$$

We immediately see that the numerator and denominator in the first of these fractions have equal moduli, so that $|B| = |A|$. This means that, as we could expect, a particle with energy $E < U_0$ is totally reflected from the step—just as in classical mechanics. As a result, at $x < 0$ our solution (2.55) may be represented by a standing wave

$$\psi_- = 2iAe^{i\theta} \sin(kx - \theta), \quad \text{with} \quad \theta \equiv \tan^{-1}\frac{k}{\kappa}. \tag{2.64}$$

Note that the shift $\Delta x \equiv \theta/k = (\tan^{-1} k/\kappa)/k$ of the standing wave to the right, due to the partial penetration of the wavefunction under the potential step, is commensurate with, but generally not equal to the penetration depth $\delta \equiv 1/\kappa$. The red line in figure 2.4 shows the full behavior of the wavefunction, for a particular case $E = U_0/5$, at which $k/\kappa \equiv [E/(U_0 - E)]^{1/2} = 1/2$.

According to Eq. (2.59), as the particle's energy E is increased to approach U_0, the penetration depth $1/\kappa$ diverges. This raises an important issue: what happens at $E > U_0$, i.e. if there is no classically forbidden region in the problem? Again, in the classical mechanics the incident particle would continue to move to the right, though with a reduced velocity, corresponding to the new kinetic energy $E - U_0$, so there would be no reflection. In quantum mechanics, however, the situation is different. In order to analyze it, it is not necessary to re-solve the whole problem; it is sufficient to note that all our calculations, and hence Eqs. (2.63) are still valid if we take[13]

$$\kappa = -ik', \quad \text{with } k'^2 \equiv \frac{2m(E - U_0)}{\hbar^2} > 0. \tag{2.65}$$

---

[13] Our earlier discarding of the particular solution $\exp\{\kappa x\}$, now becoming $\exp\{-ik'x\}$, is still valid, but now on a different grounds: this term would describe a wave packet incident on the potential step from the right, and this is not the problem under our consideration.

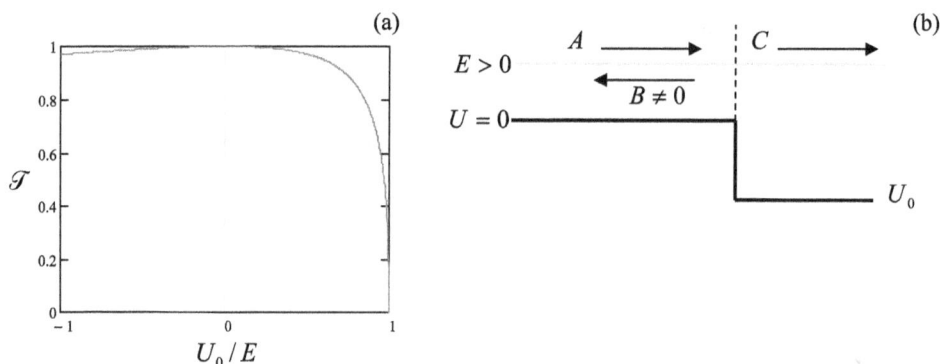

Figure 2.5. (a) The transparency of a potential step with $U_0 < E$ as a function of its height, according to Eq. (2.75), and (b) the potential profile at $U_0 < 0$.

With this replacement, Eq. (2.63) becomes[14]

$$B = A\frac{k - k'}{k + k'}, \quad C = A\frac{2k}{k + k'}. \tag{2.66}$$

The most important result of this change is that now the particle's reflection is *not* total: $|B| < |A|$. In order to evaluate this effect qualitatively, it is more fair to use not the B/A or C/A ratios, but rather that of the probability currents (2.5) carried by the de Broglie waves traveling to the right, with amplitudes C and A, in the corresponding regions (respectively, $x > 0$ and $x < 0$):

$$\mathscr{T} \equiv \frac{I_C}{I_A} = \frac{k'|C|^2}{k|A|^2} = \frac{4kk'}{(k + k')^2} \equiv \frac{4[E(E - U_0)]^{1/2}}{[E^{1/2} + (E - U_0)^{1/2}]^2}. \tag{2.67}$$

(The parameter $\mathscr{T}$ so defined is called the *transparency* of the system, in our current case of the potential step.) The result given by Eq. (2.67) is plotted in figure 2.5a as a function of the ratio U_0/E. Note its most important features:

(i) At $U_0 = 0$, the transparency is full, $\mathscr{T} = 1$—naturally, because there is no step at all.

(ii) At $U_0 \rightarrow E$, the transparency tends to zero, giving a proper connection with the case $E < U_0$.

(iii) Nothing in our solution's procedure prevents us from using Eq. (2.67) even for $U_0 < 0$, i.e. for the *step-down* (or 'cliff') potential profile—see figure 2.5b. Very counter-intuitively, the particle is (partly) reflected even from such a cliff, and the transmission diminishes (though rather slowly) at $U_0 \rightarrow -\infty$.

The most important conceptual conclusion of this analysis is that the quantum particle is *partly reflected* from a potential step with $U_0 < E$, in the sense that there is

---

[14] These formulas are completely similar to those describing the partial reflection of classical waves from a sharp interface between two uniform media, at normal incidence (see, e.g. *Part CM* section 6.4 and *Part EM* section 7.4), with the effective impedance Z of de Broglie waves being proportional to their wave number k.

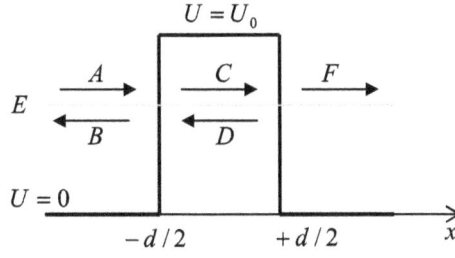

Figure 2.6. A rectangular potential barrier, and the partial waves taken into account at its analysis.

a nonvanishing probability $\mathcal{T} < 1$ to find it passed over the step, while there is also some probability $(1 - \mathcal{T})$ to have it reflected.

The last property is exhibited, but for *any* relation between E and U_0, by another simple potential profile $U(x)$, the famous *potential* (or 'tunnel') *barrier*. Figure 2.6 shows its simple, 'rectangular' version:

$$U(x) = \begin{cases} 0, & \text{for } x < -d/2, \\ U_0, & \text{for } -d/2 < x < +d/2, \\ 0, & \text{for } +d/2 < x. \end{cases} \tag{2.68}$$

In order to analyze this problem, it is sufficient to look for the solution to the Schrödinger equation in the form (2.55) at $x \leqslant -d/2$. At $x > +d/2$, i.e. behind the barrier, we may use the arguments presented above (no wave source on the right!) to keep just one traveling wave, now with the same wave number:

$$\psi_+(x) = Fe^{ikx}. \tag{2.69}$$

However, under the barrier, i.e. at $-d/2 \leqslant x \leqslant +d/2$, we should generally keep both exponential terms,

$$\psi_b(x) = Ce^{-\kappa x} + De^{+\kappa x}, \tag{2.70}$$

because our previous argument, used in the potential step problem's solution, is no longer valid. (Here k and κ are still defined, respectively, by Eqs. (2.54) and (2.57).) In order to express the coefficients B, C, D, and F via the amplitude A of the incident wave, we need to plug these solutions into the boundary conditions similar to Eqs. (2.61), but now at two boundary points, $x = \pm d/2$.

Solving the resulting system of four linear equations, we get four ratios B/A, C/A, etc; in particular,

$$\frac{F}{A} = \left[\cosh \kappa d + \frac{i}{2} \left(\frac{\kappa}{k} - \frac{k}{\kappa} \right) \sinh \kappa d \right]^{-1} e^{-ikd}, \tag{2.71a}$$

and hence the barrier's transparency

$$\mathcal{T} \equiv \left| \frac{F}{A} \right|^2 = \left[\cosh^2 \kappa d + \left(\frac{\kappa^2 - k^2}{2\kappa k} \right)^2 \sinh^2 \kappa d \right]^{-1}. \tag{2.71b}$$

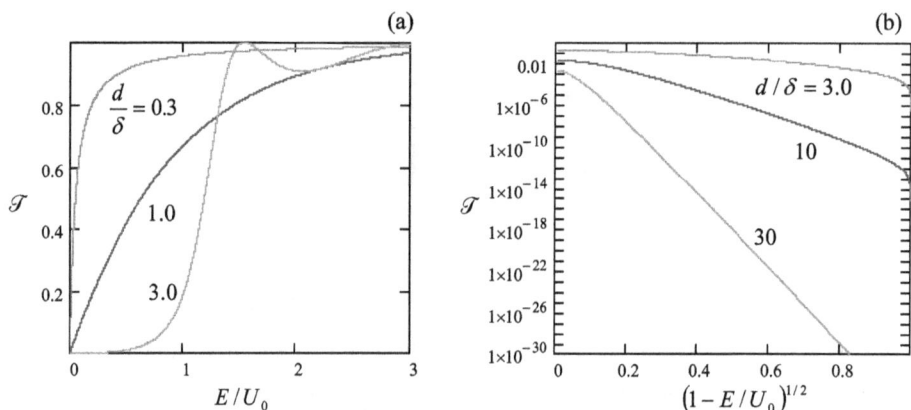

Figure 2.7. The transparency of the rectangular potential barrier as a function of the particle's energy E.

So, quantum mechanics indeed allows particles with energies $E < U_0$ to pass 'through' the potential barrier—see figure 2.6. This is the famous effect of *quantum-mechanical tunneling*. Figure 2.7a shows the barrier transparency as a function of the particle energy E, for several characteristic values of its thickness d, or rather of the ratio d/δ, with the δ is defined by Eq. (2.59).

The plots show that generally, the transparency grows gradually with the particle's energy. This growth is natural, because the penetration constant κ decreases with the growth of E, i.e. the wavefunction penetrates more and more into the barrier, so that more and more of it is 'picked up' at the second interface ($x = +d/2$) and transferred into the wave $F\exp\{ikx\}$ propagating behind the barrier.

Now let consider an important limit of a very thin and high rectangular barrier, $d \ll \delta$, $E \ll U_0$ (giving $k \ll \kappa \ll 1/d$). In this limit, Eq. (2.71) yields

$$\mathcal{T} \equiv \left|\frac{F}{A}\right|^2 \to \frac{1}{|1 + i\alpha|^2} = \frac{1}{1 + \alpha^2}, \quad \text{where}$$

$$\alpha \equiv \frac{1}{2}\left(\frac{\kappa^2 - k^2}{\kappa k}\right)\kappa d \approx \frac{1}{2}\frac{\kappa^2 d}{k} \approx \frac{m}{\hbar^2 k}U_0 d, \tag{2.72}$$

The last product, $U_0 d$, is just the 'area'

$$\mathcal{W} \equiv \int_{U(x)>E} U(x)dx \tag{2.73}$$

of the barrier. This fact implies that the very simple result (2.72) may be correct for a barrier of any shape, provided that it is sufficiently thin and high. In order to prove this, let us consider the tunneling problem for a very thin barrier with κd, $kd \ll 1$ approximating it with the Dirac's δ-function (figure 2.8):

$$U(x) = \mathcal{W}\delta(x), \tag{2.74}$$

where the parameter $\mathcal{W}$ evidently satisfies Eq. (2.73).

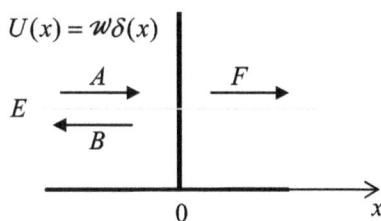

Figure 2.8. A delta-functional potential barrier.

The solutions of the tunneling problem at all points but $x = 0$ still may be taken in the form of Eqs. (2.55) and (2.69), so we only need to analyze the boundary conditions in that point. However, due to the special character of the δ-function, we should be careful here. Indeed, instead of Eq. (2.60) we now get

$$\lim_{\varepsilon \to 0}\left(\frac{d\psi}{dx}\bigg|_{x=+\varepsilon} - \frac{d\psi}{dx}\bigg|_{x=-\varepsilon}\right) = \lim_{\varepsilon \to 0}\int_{-\varepsilon}^{+\varepsilon}\frac{d^2\psi}{dx^2}dx$$

$$= \lim_{\varepsilon \to 0}\frac{2m}{\hbar^2}\int_{-\varepsilon}^{+\varepsilon}[U(x) - E]\psi\, dx \qquad (2.75)$$

$$= \frac{2m}{\hbar^2}\mathcal{W}\psi(0).$$

According to this relation, at a finite $\mathcal{W}$, the derivatives $d\psi/dx$ are also finite, so that the wavefunction itself is still continuous:

$$\lim_{\varepsilon \to 0}(\psi|_{x=+\varepsilon} - \psi|_{x=+\varepsilon}) = \lim_{\varepsilon \to 0}\int_{-\varepsilon}^{+\varepsilon}\frac{d\psi}{dx}\, dx = 0. \qquad (2.76)$$

Using these two boundary conditions, we readily get the following system of two linear equations,

$$A + B = F, \quad ikF - (ikA - ikB) = \frac{2m\mathcal{W}}{\hbar^2}F, \qquad (2.77)$$

whose solution yields

$$\frac{B}{A} = \frac{-i\alpha}{1 + i\alpha}, \quad \frac{F}{A} = \frac{1}{1 + i\alpha}, \quad \text{where } \alpha \equiv \frac{m\mathcal{W}}{\hbar^2 k}. \qquad (2.78)$$

(This expression for α is compatible with that in Eq. (2.72).) For the barrier transparency $\mathcal{T} \equiv |F/A|^2$, this result again gives the first of Eqs. (2.72), which is therefore indeed general for such thin barriers. That formula may be recast to give the following simple expression (valid only for $E \ll U_{\max}$):

$$\mathcal{T} = \frac{1}{1 + \alpha^2} \equiv \frac{E}{E + E_0}, \quad \text{where } E_0 \equiv \frac{m\mathcal{W}^2}{2\hbar^2}, \qquad (2.79)$$

which shows that as energy becomes larger than the constant E_0, the transparency approaches 1.

Now proceeding to another important limit of thick barriers ($d \gg \delta$), Eq. (2.71) shows that in this case, the transparency is dominated by what is called the *tunnel exponent*,

$$\mathscr{T} = \left(\frac{4k\kappa}{k^2 + \kappa^2}\right)^2 e^{-2\kappa d}, \tag{2.80}$$

which may be clearly seen as the straight segments in semi-log plots (figure 2.7b) of T as a function of the combination $(1 - E/U_0)^{1/2}$, which is proportional to κ—see Eq. (2.57). This exponential dependence on the barrier thickness is the most important factor for various applications of the quantum-mechanical tunneling—from the field emission of electrons to vacuum[15] to the scanning tunneling microscopy[16]. Note also very substantial negative implications of the effect for the electronic technology progress, most importantly imposing limits on the so-called *Dennard scaling* of field-effect transistors in semiconductor integrated circuits (and hence on the well-known *Moore's law*), due to the increase of tunneling both through the gate oxide and along the channel of the transistors, from source to drain[17].

Finally, one more feature visible in figure 2.7a (for case $d = 3\delta$) are the oscillations of the transparency as a function of energy, at $E > U_0$, with $\mathscr{T} = 1$, i.e. the reflection completely vanishing, at some points[18]. This is our first glimpse at one more interesting quantum effect: *resonant tunneling*. This effect will be discussed in more detail in section 2.5 below, using another potential profile, where it is more clearly pronounced.

2.4 Motion in soft potentials

Before moving on to exploring other quantum-mechanical effects, let us see how the results discussed in the previous section are modified in the opposite limit of the so-called *soft* (also called 'smooth') potential profiles, like the one sketched in figure 2.3.[19] The most efficient analytical tool to study this limit is the so-called *WKB* (or 'JWKB', or 'quasiclassical') *approximation* developed by H Jeffrey, G Wentzel, A Kramers, and L Brillouin in 1925–27. In order to derive its 1D version, let us rewrite the Schrödinger equation (2.53) in a simpler form

$$\frac{d^2\psi}{dx^2} + k^2(x)\psi = 0, \tag{2.81}$$

---

[15] See, e.g. [1].
[16] See, e.g. [2].
[17] See, e.g. [3], and references therein. (A brief discussion of the field-effect transistors may be found in *Part SM* section 6.4.)
[18] Let me mention in passing the curious case of the potential well $U(x) = -(\hbar^2/2m)\nu(\nu+1)/\cosh^2(x/a)$, with any positive integer ν and any real a, which is reflection-free ($\mathscr{T} = 1$) for the incident de Broglie wave of *any* energy E, and hence for any incident wave packet. Unfortunately, a proof of this fact would require more time/space than I can afford. (Note that it was first discussed in a 1930 paper by P Epstein, prior to the 1933 publication by G Pöschl and E Teller, which is responsible for the common name of this *Pöschl–Teller potential*.)
[19] Quantitative conditions of the 'softness' will be formulated later in this section.

where the local wave number $k(x)$ is defined similarly to Eq. (2.65),

$$k^2(x) \equiv \frac{2m[E - U(x)]}{\hbar^2};$$ (2.82)

besides that now it may be a function of x. We already know that for $k(x) = $ const, the fundamental solutions of this equation are $A\exp\{+ikx\}$ and $B\exp\{-ikx\}$, which may be represented in a single form

$$\psi(x) = e^{i\Phi(x)},$$ (2.83)

where $\Phi(x)$ is a complex function, in these two simplest cases equal, respectively, to $(kx - i\ln A)$ and $(-kx - i\ln B)$. This is why we may try use Eq. (2.83) to look for solution of Eq. (2.81) even in the general case, $k(x) \neq$ const. Differentiating Eq. (2.83) twice, we get

$$\frac{d\psi}{dx} = i\frac{d\Phi}{dx}e^{i\Phi}, \quad \frac{d^2\psi}{dx^2} = \left[i\frac{d^2\Phi}{dx^2} - \left(\frac{d\Phi}{dx}\right)^2\right]e^{i\Phi}.$$ (2.84)

Plugging the last expression into Eq. (2.81) and requiring the factor before $\exp\{i\Phi(x)\}$ to vanish, we get

$$i\frac{d^2\Phi}{dx^2} - \left(\frac{d\Phi}{dx}\right)^2 + k^2(x) = 0.$$ (2.85)

This is still an exact, general equality. At first sight, it looks worse than the initial Eq. (2.81), because Eq. (2.85) is nonlinear. However, it is ready for simplification in the limit when the potential profile is very soft, $dU/dx \to 0$. Indeed, for a uniform potential, $d^2\Phi/dx^2 = 0$. Hence, in the so-called 0th approximation, $\Phi(x) \to \Phi_0(x)$, we may try to keep that result, so that Eq. (2.85) is reduced to

$$\left(\frac{d\Phi_0}{dx}\right)^2 = k^2(x), \quad \text{i.e.} \quad \frac{d\Phi_0}{dx} = \pm k(x), \quad \Phi_0(x) = \pm\int^x k(x')dx',$$ (2.86)

so that its general solution is a linear superposition of two functions (2.83), with Φ replaced with Φ_0:

$$\psi_0(x) = A\exp\left\{+i\int^x k(x')dx'\right\} + B\exp\left\{-i\int^x k(x')dx'\right\},$$ (2.87)

where the lower limits of integration affect only the constants A and B. The physical sense of this result is simple: it is a sum of the forward- and back-propagating de Broglie waves, with the coordinate-dependent local wave number $k(x)$ that self-adjusts to the potential profile.

Let me emphasize the non-trivial nature of this approximation[20]. First, any attempt to address the problem with the standard perturbation approach (say, $\psi = \psi_0 + \psi_1 + ...$, with ψ_n proportional to the nth power of some small parameter, see chapter 6) would fail for most potentials, because as Eq. (2.86) shows, even a slight but persisting deviation of $U(x)$ from a constant leads to a gradual accumulation of the phase Φ_0, impossible to describe by any small perturbation of ψ. Second, the dropping of term $d^2\Phi/dx^2$ in Eq. (2.85) is not too easy to justify. Indeed, since we are committed to the 'soft potential limit' $dU/dx \rightarrow 0$, we should be ready to assume the characteristic length a of the spatial variation of Φ to be large, and neglect the terms that are the smallest ones in the limit $a \rightarrow \infty$. However, both first terms in Eq. (2.85) are apparently of the same order in a, namely $O(a^{-2})$; why have we neglected just one of them?

The price we have paid for such a 'sloppy' treatment is substantial: Eq. (2.87) does *not* satisfy the fundamental property of the Schrödinger equation, the probability current conservation. Indeed, since Eq. (2.81) describes a fixed-energy (stationary) spatial part of the general Schrödinger equation, its probability density $w = \Psi\Psi^* = \psi\psi^*$, and should not depend on time. Hence, according to Eq. (2.6), we should have $I(x) = $ const. However, this is not true for any component of Eq. (2.87); for example for the first, forward-propagating component on its right-hand side, Eq. (2.5) yields

$$I_0(x) = \frac{\hbar}{m} |A|^2 k(x), \qquad (2.88)$$

evidently not a constant if $k(x) \neq$ const. The brilliance of the WKB theory is that the problem may be fixed without revising the 0th approximation, just by amending it. Indeed, let us explore the next, 1st approximation instead:

$$\Phi(x) \rightarrow \Phi_{WKB}(x) \equiv \Phi_0(x) + \Phi_1(x), \qquad (2.89)$$

where Φ_0 still obeys Eq. (2.86), while Φ_1 describes a correction to the 0th approximation, which is small in the following sense[21]:

$$\left| \frac{d\Phi_1}{dx} \right| \ll \left| \frac{d\Phi_0}{dx} \right| = k(x). \qquad (2.90)$$

Plugging Eq. (2.89) into Eq. (2.85), with the account of the definition (2.86), we get

$$i\left(\frac{d^2\Phi_0}{dx^2} + \frac{d^2\Phi_0}{dx^2} \right) - \frac{d\Phi_1}{dx}\left(2\frac{d\Phi_0}{dx} + \frac{d\Phi_1}{dx} \right) = 0. \qquad (2.91)$$

---

[20] Philosophically, this space-domain method is very close to the time-domain *van der Pol method* in classical mechanics, and the very similar *rotating wave approximation* (RWA) in quantum mechanics—see, e.g. *Part CM* sections 5.2–5.5, and also sections 6.5, 7.6, 9.2, and 9.4 of this course.
[21] For certainty, I will use the discretion given by Eq. (2.82) to define $k(x)$ as the *positive* root of its right-hand side.

Using the condition (2.90), we may neglect $d^2\Phi_1/dx^2$ in comparison with $d^2\Phi_0/dx^2$ inside the first parentheses, and $d\Phi_1/dx$ in comparison with $2d\Phi_0/dx$ inside the second parenthesis. As a result, we get the following (still approximate) result:

$$\frac{d\Phi_1}{dx} = \frac{i}{2}\frac{d^2\Phi_0}{dx^2}\bigg/\frac{d\Phi_0}{dx} = \frac{i}{2}\frac{d}{dx}\left(\ln\frac{d\Phi_0}{dx}\right) = \frac{i}{2}\frac{d}{dx}[\ln k(x)] = i\frac{d}{dx}[\ln k^{1/2}(x)], \quad (2.92)$$

$$i\Phi|_{\mathrm{WKB}} \equiv i\Phi_0 + i\Phi_1 = \pm i\int^x k(x')dx' + \ln\frac{1}{k^{1/2}(x)}, \quad (2.93)$$

$$\psi_{\mathrm{WKB}}(x) = \frac{a}{k^{1/2}(x)}\exp\left\{i\int^x k(x')dx'\right\}$$
$$+ \frac{b}{k^{1/2}(x)}\exp\left\{-i\int^x k(x')dx'\right\}, \quad \text{for } k^2(x) > 0. \quad (2.94)$$

(Again, the lower integration limit is arbitrary, because its choice may be incorporated into the complex constants a and b.) This modification of the 0th approximation (2.87) overcomes the problem of current continuity; for example, for the forward-propagating wave, Eq. (2.5) gives

$$I_{\mathrm{WKB}}(x) = \frac{\hbar}{m}|a|^2 = \text{const.} \quad (2.95)$$

Physically, the factor $k^{1/2}$ in the denominator of the WKB wavefunction's pre-exponent is easy to understand. The smaller the local group velocity (2.33) of the wave packet, $v_{\mathrm{gr}}(x) = \hbar k(x)/m$, the 'easier' (more probable) it should be to find the particle within a certain interval dx. This is exactly the result that the WKB approximation gives: $w(x) = \psi\psi^* \propto 1/k(x) \propto 1/v_{\mathrm{gr}}$. Another value of the 1st approximation is a clarification of the WKB theory's validity condition: it is given by Eq. (2.90). Plugging into this relation the first form of Eq. (2.92), and estimating $|d^2\Phi_0/dx^2|$ as $|d\Phi_0/dx|/a$, where a is the spatial scale of a substantial change of $|d\Phi_0/dx| = k(x)$, we may write the condition as

$$ka \gg 1. \quad (2.96)$$

In plain English, this means that the region where $U(x)$, and hence $k(x)$, change substantially should contain many de Broglie wavelengths $\lambda = 2\pi/k$.

So far I have implied that $k^2(x) \propto E - U(x)$ is positive, i.e. particle moves in the classically accessible region. Now let us extend the WKB approximation to the situation where the difference $E - U(x)$ may change sign, for example to the reflection problem sketched in figure 2.3. Just as we did for the sharp potential step, we first need to find the appropriate solution in the classically forbidden region, in this case for $x > x_c$. For that, there is again no need to redo our calculations, because they are still valid if we, just as in the sharp-step problem, take $k(x) = i\kappa(x)$, where

$$\kappa^2(x) \equiv \frac{2m[U(x) - E]}{\hbar^2} > 0, \quad \text{for } x > x_c, \tag{2.97}$$

and keep just one of two possible solutions (with $\kappa > 0$), in analogy with Eq. (2.58). The result is

$$\psi_{\text{WKB}}(x) = \frac{c}{\kappa^{1/2}(x)} \exp\left\{-\int^x \kappa(x')dx'\right\}, \quad \text{for } k^2 < 0, \text{ i.e. } \kappa^2 > 0, \tag{2.98}$$

with the lower limit at some point with $\kappa^2 > 0$ as well. This is a really wonderful formula! It describes the quantum-mechanical penetration of the particle into the classically forbidden region, and provides a natural generalization of Eq. (2.58)—leaving intact, of course, our estimates of the depth $\delta \sim 1/\kappa$ of such penetration.

Now we have to do what we have done for the sharp-step problem in section 2.2: use the boundary conditions in the interface point $x = x_c$ to relate the constants a, b, and c. However, now this operation is a tad more complex, because both WKB functions (2.94) and (2.98) diverge, albeit weakly, at the classical turning point, where both $k(x)$ and $\kappa(x)$ tend to zero. This *connection problem* may be solved, however, in the following way[22]. Let us use our commitment of the potential 'softness', assuming that it allows us to keep just two leading terms in the Taylor expansion of the function $U(x)$ at the point x_c:

$$U(x) \approx U(x_c) + \frac{dU}{dx}\bigg|_{x=x_c} (x - x_c) \equiv E + \frac{dU}{dx}\bigg|_{x=x_c} (x - x_c). \tag{2.99}$$

Using this truncated expansion, and introducing the following dimensionless variable for the coordinate's deviation from the classical turning point,

$$\zeta \equiv \frac{x - x_c}{x_0}, \quad \text{with } x_0 \equiv \left[\frac{\hbar^2}{2m \, (dU/dx)_{x=x_c}}\right]^{1/3}, \tag{2.100}$$

we reduce the Schrödinger equation (2.53) to the so-called *Airy equation*

$$\frac{d^2\psi}{d\zeta^2} - \zeta \psi = 0. \tag{2.101}$$

This simple linear, ordinary, homogenous differential equation of the second order has been very well studied. Its general solution may be represented as a linear combination of two fundamental solutions, called the *Airy functions* Ai(ζ) and Bi(ζ), shown in figure 2.9a.[23]

---

[22] An alternative way to solve the connection problem, without involving the Airy functions but using an analytical extension of WKB formulas to the plane of complex argument, may be found, e.g. in section 47 of textbook [4].

[23] Note the following (exact) integral formulas,

$$\text{Ai}(\zeta) = \frac{1}{\pi}\int_0^\infty \cos\left(\frac{\xi^3}{3} + \zeta\xi\right)d\xi, \quad \text{Bi}(\zeta) = \frac{1}{\pi}\int_0^\infty \left[\exp\left\{-\frac{\xi^3}{3} + \zeta\xi\right\} + \sin\left(\frac{\xi^3}{3} + \zeta\xi\right)\right]d\xi,$$

frequently more convenient for practical calculations of the Airy functions than the differential equation (2.101).

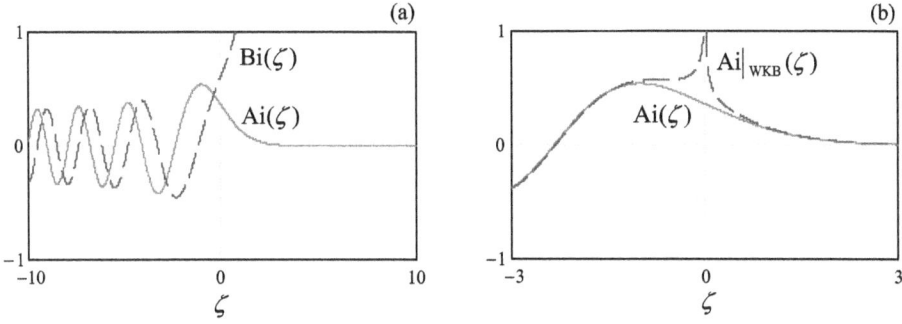

Figure 2.9. (a) The Airy functions Ai and Bi, and (b) the WKB approximation for the function Ai(ζ).

The latter function diverges at $\zeta \to \infty$, and thus is not suitable for our current problem (figure 2.3), while the former function has the following asymptotic behaviors at $|\zeta| \gg 1$:

$$\text{Ai}(\zeta) \to \frac{1}{\pi^{1/2} |\zeta|^{1/4}} \times \begin{cases} \dfrac{1}{2} \exp\left\{ -\dfrac{2}{3}\zeta^{3/2} \right\}, & \text{for } \zeta \to +\infty, \\[2ex] \sin\left\{ \dfrac{2}{3}(-\zeta)^{3/2} + \dfrac{\pi}{4} \right\}, & \text{for } \zeta \to -\infty. \end{cases} \quad (2.102)$$

Now let us apply the WKB approximation to the Airy equation (2.101). Taking the classical turning point ($\zeta = 0$) for the lower limit, for $\zeta > 0$ we get

$$\kappa^2(\zeta) = \zeta, \quad \kappa(\zeta) = \zeta^{1/2}, \quad \int_0^\zeta \kappa(\zeta')d\zeta' = \frac{2}{3}\zeta^{3/2}, \quad (2.103)$$

i.e. exactly the exponent in the top line of Eq. (2.102). Making a similar calculation for $\zeta < 0$, with the natural assumption $|b| = |a|$ (the full reflection from the potential step), we arrive at the following result:

$$\text{Ai}_{\text{WKB}}(\zeta) = \frac{1}{|\zeta|^{1/4}} \times \begin{cases} c' \exp\left\{ -\dfrac{2}{3}\zeta^{3/2} \right\}, & \text{for } \zeta > 0, \\[2ex] a' \sin\left\{ \dfrac{2}{3}(-\zeta)^{3/2} + \varphi \right\}, & \text{for } \zeta < 0. \end{cases} \quad (2.104)$$

This approximation differs from the exact solution at small values of ζ, i.e. close to the classical turning point—see figure 2.9b. However, at $|\zeta| \gg 1$, Eqs. (2.104) describe the Airy function exactly, provided that

$$\varphi = \frac{\pi}{4}, \quad c' = \frac{a'}{2}. \quad (2.105)$$

These *connection formulas* may be used to rewrite Eq. (2.104) as the following,

$$\text{Ai}_{\text{WKB}}(\zeta) = \frac{a'}{2\,|\zeta|^{1/4}}$$

$$\times \begin{cases} \exp\left\{-\dfrac{2}{3}\zeta^{3/2}\right\}, & \text{for } \zeta > 0, \quad (2.106) \\[2mm] \dfrac{1}{i}\left[\exp\left\{+i\dfrac{2}{3}\zeta^{3/2} + i\dfrac{\pi}{4}\right\} - \exp\left\{-i\dfrac{2}{3}\zeta^{3/2} - i\dfrac{\pi}{4}\right\}\right], & \text{for } \zeta < 0, \end{cases}$$

and hence are described by two simple mnemonic rules:

(i) If the classical turning point is taken for lower limit in the WKB integrals in the classically allowed and the classically forbidden regions, then the moduli of the quasi-amplitudes of the exponents are equal.

(ii) Reflecting from a 'soft' potential step, the wavefunction acquires an additional phase shift $\Delta\varphi = \pi/2$, if compared with the reflection from a 'hard', infinite potential wall at $x = x_c$ (for which, according to Eq. (2.63) with $\kappa = 0$, we have $B = -A$).

In order for the connection formulas (2.105)–(2.106) to be valid, deviations from the linear approximation (2.99) of the potential profile should be relatively small within the region where the WKB approximation differs from for the exact Airy equation: $|\zeta| \sim 1$, i.e. $|x - x_c| \sim x_0$. The deviations may be estimated using the next term of the Taylor expansion, $(d^2U/d^2x)(x - x_c)^2/2$. As a result, the condition of validity of the connection formulas (i.e. of the 'softness' of the reflecting potential profile) may be expressed as $|d^2U/d^2x|x_0 \ll |dU/dx|$ at $x \approx x_c$ (meaning the $\sim x_0$–wide vicinity of the point x_c). With the account of Eq. (2.100) for x_0, this condition becomes

$$\left|\frac{d^2U}{dx^2}\right|^3_{x\approx x_c} \ll \frac{2m}{\hbar^2}\left(\frac{dU}{dx}\right)^4_{x\approx x_c}, \qquad (2.107)$$

As an example of a very useful application of the WKB approximation, let us use the rule (ii) to calculate the energy spectrum of 1D particle in a soft 1D potential well (figure 2.10).

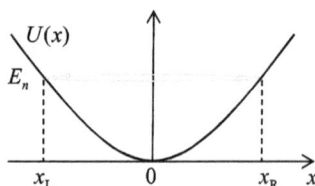

Figure 2.10. The WKB treatment of an eigenstate of a particle in a soft 1D potential well.

As was discussed in section 1.7, we may consider the standing wave describing an eigenfunction ψ_n (corresponding to an eigenenergy E_n) as a traveling de Broglie wave going back and forth between the walls, being sequentially reflected by each of them. Let us apply the WKB approximation to such a traveling wave. First, according to Eq. (2.94), propagating from the left classical turning point x_L to the right point x_R, it acquires the phase change

$$\Delta\varphi_\rightarrow = \int_{x_L}^{x_R} k(x)dx. \tag{2.108}$$

At the reflection from the soft wall at x_R, according to the connection formulas (2.105)–(2.106), the wave acquires an additional shift $\pi/2$. Now, traveling back from x_R to x_L, the wave gets a shift similar to one given by Eq. (2.108): $\Delta\varphi_\leftarrow = \Delta\varphi_\rightarrow$. Finally, at the reflection from x_L it gets one more $\pi/2$. Summing up all these contributions at the wave's roundtrip, we may write the self-consistency condition (that the wavefunction 'catches its own tail with its teeth') in the form

$$\Delta\varphi_{total} \equiv \Delta\varphi_\rightarrow + \frac{\pi}{2} + \Delta\varphi_\leftarrow + \frac{\pi}{2}$$
$$\equiv 2\int_{x_L}^{x_R} k(x)dx + \pi = 2\pi n, \quad \text{with } n = 1, 2, \ldots \tag{2.109}$$

Rewriting this result in the terms of the particle's momentum $p(x) = \hbar k(x)$, we arrive at the famous 1D *Bohr–Sommerfeld quantization rule*

$$\oint_C p(x)dx = 2\pi\hbar\left(n - \frac{1}{2}\right), \tag{2.110}$$

where the closed path C means the full period of classical motion[24].

Let us see what this quantization rule gives for the very important particular case of a quadratic potential profile of a harmonic oscillator of frequency ω_0. In this case,

$$U(x) = \frac{m}{2}\omega_0^2 x^2, \tag{2.111}$$

and the classical turning points (where $U(x) = E$) are the roots of a simple equation

$$\frac{m}{2}\omega_0^2 x_c^2 = E_n, \quad \text{so that } x_R = \omega_0^{-1}\left(\frac{2E_n}{m}\right)^{1/2} > 0, \quad x_L = -x_R < 0. \tag{2.112}$$

Due to potential's symmetry, the integration required by Eq. (2.110) is also simple:

$$\int_{x_L}^{x_R} p(x)dx = \int_{x_L}^{x_R} \{2m[E_n - U(x)]\}^{1/2}dx \equiv (2mE_n)^{1/2}2\int_0^{x_R}\left(1 - \frac{x^2}{x_R^2}\right)^{1/2}dx$$

$$\equiv (2mE_n)^{1/2}2x_R\int_0^1 (1 - \xi^2)^{1/2}d\xi = (2mE_n)^{1/2}2x_R\frac{\pi}{4} \equiv \frac{\pi E_n}{\omega_0}, \tag{2.113}$$

---

[24] Note that at the motion in more than one dimension, a closed classical trajectory may have no turning points. In this case, the constant 1/2 in the parentheses of Eq. (2.110), arising from the turns, should be dropped. The simplest example is the circular motion of the electron about the proton in Bohr's picture of the hydrogen atom, for which the so-modified quantization (2.110) condition takes the form (1.8). (A similar relation for the radial motion of a particle is sometimes called the *Sommerfeld–Wilson quantization rule*.)

so that Eq. (2.110) yields

$$E_n = \hbar\omega_0\left(n' + \frac{1}{2}\right), \quad \text{with } n' \equiv n - 1 = 0, 1, 2, \ldots. \tag{2.114}$$

In order to estimate the validity of this result, we have to check the condition (2.96) at all points of the classically allowed region, and Eq. (2.107) at the turning points. A straightforward calculation shows that both conditions are valid only for $n \gg 1$. However, we will see in section 2.9 below that Eq. (2.114) is actually *exactly correct for all energy levels*—thanks to special properties of potential profile (2.111).

Now let us apply the mnemonic rule (i) to examine particle's penetration into the classically forbidden region of a potential step of a height $U_0 > E$. For this case, the rule, i.e. the second of Eqs. (2.105), yields the following relation of the quasi-amplitudes in Eqs. (2.94) and (2.98): $|c| = |a|/2$. If we now naively applied this relation to the sharp step sketched in figure 2.4, forgetting that it does not satisfy Eq. (2.107), we would get the following relation of the full amplitudes, defined by Eqs. (2.55) and (2.58):

$$\left|\frac{C}{\kappa}\right| = \frac{1}{2}\left|\frac{A}{k}\right|. \qquad \text{(WRONG!)} \tag{2.115}$$

This result differs from the correct Eq. (2.63), and hence we may expect that the WKB approximation's prediction for more complex potentials, most importantly for tunneling through a soft potential barrier (figure 2.11) should be also different from the exact result (2.71) for the rectangular barrier shown in figure 2.6.

In order to analyze tunneling through such a soft barrier, we need (just as in the case of a rectangular barrier) to take into consideration five partial 'waves', but now they should be taken in the WKB form[25]:

$$\psi_{\text{WKB}} = \begin{cases} \dfrac{a}{k^{1/2}(x)}\exp\left\{i\displaystyle\int^x k(x')dx'\right\} \\ \quad + \dfrac{b}{k^{1/2}(x)}\exp\left\{-i\displaystyle\int^x k(x')dx'\right\}, & \text{for } x < x_c, \\ \dfrac{c}{\kappa^{1/2}(x)}\exp\left\{-\displaystyle\int^x \kappa(x')dx'\right\} \\ \quad + \dfrac{d}{\kappa^{1/2}(x)}\exp\left\{\displaystyle\int^x \kappa(x')dx'\right\}, & \text{for } x_c < x < x_c', \\ \dfrac{f}{k^{1/2}(x)}\exp\left\{i\displaystyle\int^x k(x')dx'\right\}, & \text{for } x_c' < x, \end{cases} \tag{2.116}$$

where the lower limits of integrals are arbitrary (each within the corresponding range of x). Since on the right of the left classical point we have two exponents rather than

---

[25] Sorry, but the same letter d is used here for the barrier thickness (defined in this case as the classically forbidden region length, $x_c' - x_c$), and the constant in one of the wave amplitudes—see Eq. (2.116). Let me hope that the difference between these uses is absolutely evident from the context.

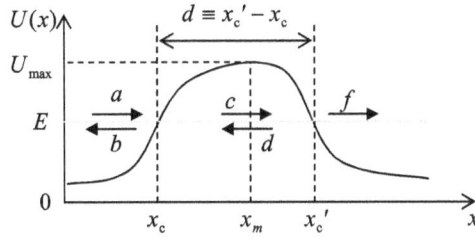

Figure 2.11. Tunneling through a soft 1D potential barrier of an arbitrary shape.

one, and on the right of the second point, one traveling wave rather than two, the connection formulas (2.105) have to be generalized, using asymptotic formulas not only for Ai(ζ), but also for the second Airy function, Bi(ζ). The analysis, absolutely similar to that carried out above (though naturally a bit more bulky)[26], gives a remarkably simple result:

$$\mathscr{T}_{\text{WKB}} \equiv \left| \frac{f}{a} \right|^2 = \exp\left\{ -2 \int_{x_c}^{x_c'} \kappa(x)dx \right\}$$

$$\equiv \exp\left\{ -\frac{2}{\hbar} \int_{x_c}^{x_c'} (2m[U(x) - E])^{1/2} dx \right\},$$

(2.117)

with the pre-exponential factor equal to 1—the fact which might be readily expected from the mnemonic rule (i) of the connection formulas.

This formula is broadly used in applied quantum mechanics, despite the approximate character of its pre-exponential coefficient for insufficiently soft barriers that do not satisfy Eq. (2.107). For example, Eq. (2.80) shows that for a thick rectangular barrier, the WKB approximation (2.117) underestimates $\mathscr{T}$ by a factor of $[4k\kappa/(k^2 + \kappa^2)]^2$—equal, for example, to 4, if $k = \kappa$, i.e. $U_0 = 2E$. However, on the appropriate logarithmic scale (see figure 2.7b), such a factor, about half an order of magnitude, still is a small correction.

Note also that when E approaches the barrier top U_{max} (figure 2.11), the points x_c and x_c' merge, so that according to Eq. (2.117), $\mathscr{T}_{WKB} \to 1$, i.e. the particle reflection vanishes at $E = U_{\text{max}}$. So, the WKB approximation does not describe the effect of the over-barrier reflection at $E > U_{\text{max}}$. (This fact could be noticed already from Eq. (2.95): in the absence of the classical turning points, the WKB probability current is constant for any barrier profile.) This conclusion is incorrect even for apparently smooth barriers where one could naively expect the WKB approximation to work perfectly. Indeed, near the point $x = x_m$ where the potential reaches maximum (i.e. $U(x_m) = U_{\text{max}}$), we may always approximate a smooth function $U(x)$ with the quadratic term of the Taylor expansion, i.e. with an inverted parabola:

$$U(x) \approx U_{\text{max}} - \frac{m\omega_0^2(x - x_m)^2}{2}.$$

(2.118)

---

[26] For the most important case $\mathscr{T}_{WKB} \ll 1$, Eq. (2.117) may be simply derived from Eqs. (2.105)–(2.106)—an exercise left for the reader.

Calculating the derivatives dU/dx and d^2U/dx^2 of this function and plugging them into condition (2.107), we see that the WKB approximation is only valid if $|U_{max} - E| \gg \hbar\omega_0$. Just for the reader's reference, an exact analysis of tunneling through the barrier (2.118) gives the following *Kemble formula*[27]:

$$\mathscr{T} = \frac{1}{1 + \exp\{-2\pi(E - U_{max})/\hbar\omega_0\}}, \qquad (2.119)$$

valid for any sign of the difference $(E - U_{max})$. This formula describes a gradual approach of $\mathscr{T}$ to 1, i.e. a *gradual* reduction of reflection at particle energy's increase, with $\mathscr{T} = 1/2$ (rather than 1) at $E = U_{max}$.

The last remark of this section: philosophically, the WKB approximation opens a straight way toward an alternative formulation of quantum mechanics, based on the *Feynman path integral*. However, I will postpone its discussion until a more compact notation has been introduced in chapter 4.

2.5 Resonant tunneling, and metastable states

Now let us move to other, conceptually different quantum effects, taking place in more elaborate potential profiles. The piecewise-constant and smooth-potential models of $U(x)$ are not too convenient for their quantitative description, because they both require 'stitching' local solutions in each classical turning point, which may lead to very cumbersome calculations. However, we may get a very good insight into the physics phenomena in such profiles using their approximation by sets of Dirac's delta-functions.

A further help in studying such profiles is provided by the notions of the *scattering* and *transfer matrices*, very useful for other cases as well. Consider an arbitrary but finite-length potential 'bump' (more formally called a *scatterer*), localized somewhere between points x_1 and x_2, on the flat potential background, say $U = 0$ (figure 2.12).

From section 2.2, we know that the general solutions of the stationary Schrödinger equation, with a certain energy E, outside the interval $[x_1, x_2]$ are sets

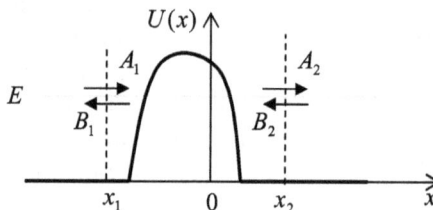

Figure 2.12. A single 1D scatterer.

---

[27] It was derived (in a more general form, valid for an arbitrary soft potential barrier) by E Kemble in 1935. In some communities it is known as the 'Hill–Wheeler formula', after D Hill and J Wheeler's 1953 paper where it was spelled out for the quadratic profile (2.118). Note that mathematically Eq. (2.119) is similar to the Fermi distribution in statistical physics, with an effective temperature $T_{ef} = \hbar\omega_0/2\pi k_B$. This similarity has some interesting implications for the statistics of Fermi gas tunneling, and hence for the electron tunneling in solids.

of two sinusoidal waves, traveling in the opposite directions. Let us represent them in the form

$$\psi_j = A_j e^{ik(x-x_j)} + B_j e^{-ik(x-x_j)}, \qquad (2.120)$$

where the index j (for now) equals either 1 or 2, and $(\hbar k)^2/2m = E$. Note that each of the two wave pairs (2.129) has, in this notation, its own reference point x_j, because this is very convenient for what follows.

As we have already discussed, if the de Broglie wave/particle is incident from the left, the solution of the linear Schrödinger equation within the scatterer range ($x_1 < x < x_2$) can provide only linear expressions for the transmitted (A_2) and reflected (B_1) wave amplitudes via the incident wave amplitude A_1:

$$A_2 = S_{21}A_1, \quad B_1 = S_{11}A_1, \qquad (2.121)$$

where S_{11} and S_{21} are certain (generally, complex) coefficients. In this case, $B_2 = 0$. Alternatively, if a wave, with amplitude B_2, is incident on the scatterer from the right, it also can induce a transmitted wave (B_1) and reflected wave (A_2), with amplitudes

$$B_1 = S_{12}B_2, \quad A_2 = S_{22}B_2, \qquad (2.122)$$

where coefficients S_{22} and S_{12} are generally different from S_{11} and S_{21}. Now we can use the linear superposition principle to argue that if the waves A_1 and B_2 are simultaneously incident on the scatterer (say, because the wave B_2 has been partly reflected back by some other scatterer located at $x > x_2$), the resulting *scattered wave* amplitudes A_2 and B_1 are just the sums of their values for separate incident waves:

$$\begin{aligned} B_1 &= S_{11}A_1 + S_{12}B_2, \\ A_2 &= S_{21}A_1 + S_{22}B_2. \end{aligned} \qquad (2.123)$$

These linear relations may be conveniently represented using the so-called *scattering matrix* S:

$$\begin{pmatrix} B_1 \\ A_2 \end{pmatrix} = S \begin{pmatrix} A_1 \\ B_2 \end{pmatrix}, \quad \text{with} \quad S \equiv \begin{pmatrix} S_{11} & S_{12} \\ S_{21} & S_{22} \end{pmatrix}. \qquad (2.124)$$

Scattering matrices, duly generalized, are an important tool for the analysis of wave scattering in more than one dimension; for 1D problems, however, another matrix is more convenient to represent the same linear relations (2.123). Indeed, let us solve this system for A_2 and B_2. The result is

$$\begin{aligned} A_2 &= T_{11}A_1 + T_{12}B_1, \\ B_2 &= T_{21}A_1 + T_{22}B_1, \end{aligned} \quad \text{i.e.} \quad \begin{pmatrix} A_2 \\ B_2 \end{pmatrix} = T \begin{pmatrix} A_1 \\ B_1 \end{pmatrix}, \qquad (2.125)$$

where T is the *transfer matrix*, with the following elements:

$$T_{11} = S_{21} - \frac{S_{11}S_{22}}{S_{12}}, \quad T_{12} = \frac{S_{22}}{S_{12}}, \quad T_{21} = -\frac{S_{11}}{S_{21}}, \quad T_{22} = \frac{1}{S_{12}}. \qquad (2.126)$$

The matrices S and T have some universal properties, valid for an arbitrary (but time-independent) scatterer; they may be readily found from the probability current conservation and the time-reversal symmetry of the Schrödinger equation. Let me leave finding these relations for the reader's exercise. The results show, in particular, that the scattering matrix may be rewritten in the following form:

$$S = e^{i\theta}\begin{pmatrix} re^{i\varphi} & t \\ t & -re^{-i\varphi} \end{pmatrix}, \qquad (2.127a)$$

where four real parameters r, t, θ, and φ satisfy a universal relation

$$r^2 + t^2 = 1, \qquad (2.127b)$$

so that only three of these parameters are independent. As a result of this symmetry, T_{11} may be also represented in a simpler form, similar to T_{22}: $T_{11} = \exp\{i\theta\}/t = 1/S_{12}^* = 1/S_{21}^*$. The last form allows a ready expression of the scatterer's transparency via just one coefficient of the transfer matrix:

$$\mathscr{T} \equiv \left| \frac{A_2}{A_1} \right|^2_{B_2=0} = |S_{21}|^2 = |T_{11}|^{-2}. \qquad (2.128)$$

In our current context, the most important property of the 1D transfer matrices is that in order to find the total transfer matrix T of a system consisting of several (say, N) sequential arbitrary scatterers (figure 2.13), it is sufficient to multiply their matrices. Indeed, extending the definition (2.125) to other points x_j ($j = 1, 2, ..., N + 1$), we can write

$$\begin{pmatrix} A_2 \\ B_2 \end{pmatrix} = T_1 \begin{pmatrix} A_1 \\ B_1 \end{pmatrix}, \quad \begin{pmatrix} A_3 \\ B_3 \end{pmatrix} = T_2 \begin{pmatrix} A_2 \\ B_2 \end{pmatrix} = T_2 T_1 \begin{pmatrix} A_1 \\ B_1 \end{pmatrix}, \quad \text{etc.} \qquad (2.129)$$

(where the matrix indices correspond to the scatterers' order on axis x), so that

$$\begin{pmatrix} A_{N+1} \\ B_{N+1} \end{pmatrix} = T_N T_{N-1} ... T_1 \begin{pmatrix} A_1 \\ B_1 \end{pmatrix}. \qquad (2.130)$$

But we can also define the *total transfer matrix* similarly to Eq. (2.125), i.e. as

$$\begin{pmatrix} A_{N+1} \\ B_{N+1} \end{pmatrix} \equiv T \begin{pmatrix} A_1 \\ B_1 \end{pmatrix}, \qquad (2.131)$$

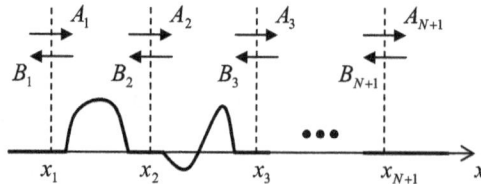

Figure 2.13. A sequence of several 1D scatterers.

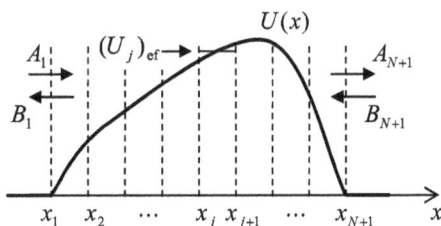

Figure 2.14. The transfer matrix approach to a long potential barrier with an arbitrary potential profile.

so that comparing Eqs. (2.130) and (2.131) we get

$$T = T_N T_{N-1} \ldots T_1. \tag{2.132}$$

This formula is valid even if the flat-potential gaps between component scatterers are shrunk to zero, so that it may be applied to a scatterer with an arbitrary profile $U(x)$, by fragmenting its length into many small segments $\Delta x = x_{j+1} - x_j$, and treating each fragment as a rectangular barrier of the average height $(U_j)_{ef} = [U(x_{j+1}) - U(x_j)]/2$ —see figure 2.14. Since very efficient numerical algorithms are readily available for fast multiplication of matrices (especially as small as 2×2 in our case), this approach is broadly used in practice for the computation of transparency of potential barriers with complicated profiles $U(x)$. (Computationally, this procedure is much more efficient then the direct numerical solution of the stationary Schrödinger equation.)

In order to use this approach for several particular, conceptually important systems, let us calculate the transfer matrices for a few elementary scatterers, starting from the delta-functional barrier located at $x = 0$—see figure 2.8. Taking $x_1 = x_2 = 0$, we can merely change the notation of the wave amplitudes in Eq. (2.78) to get

$$S_{11} = \frac{-i\alpha}{1 + i\alpha}, \quad S_{21} = \frac{1}{1 + i\alpha}. \tag{2.133}$$

An absolutely similar analysis of the wave incidence from the left yields

$$S_{22} = \frac{-i\alpha}{1 + i\alpha}, \quad S_{12} = \frac{1}{1 + i\alpha}, \tag{2.134}$$

and using Eqs. (2.126), we get

$$T_\alpha = \begin{pmatrix} 1 - i\alpha & -i\alpha \\ i\alpha & 1 + i\alpha \end{pmatrix}. \tag{2.135}$$

As a sanity check, Eq. (2.128), applied to this result, immediately brings us back to Eq. (2.79).

The next example may seem strange at first glance: what if there is no scatterer at all between points x_1 and x_2? If the points coincide, the answer is indeed trivial and can be obtained, e.g. from Eq. (2.135) by taking $\mathcal{W} = 0$, i.e. $\alpha = 0$:

$$T_0 = \begin{pmatrix} 1 & 0 \\ 0 & 1 \end{pmatrix} \equiv I \tag{2.136}$$

—the so-called *identity matrix*. However, we are free to choose the reference points $x_{1,2}$ participating in Eq. (2.120) as we wish. For example, what if $x_2 - x_1 = a$? Let us first take the forward-propagating wave alone: $B_2 = 0$ (and hence $B_1 = 0$); then

$$\psi_2 = \psi_1 = A_1 e^{ik(x-x_1)} \equiv A_1 e^{ik(x_2-x_1)} e^{ik(x-x_2)}. \tag{2.137}$$

The comparison of this expression with the definition (2.120) for $j = 2$ shows that $A_2 = A_1 \exp\{ik(x_2 - x_1)\} = A_1 \exp\{ika\}$, i.e. $T_{11} = \exp\{ika\}$. Repeating the calculation for the back-propagating wave, we see that $T_{22} = \exp\{-ika\}$, and since the space interval provides no particle reflection, we finally get

$$T_a = \begin{pmatrix} e^{ika} & 0 \\ 0 & e^{-ika} \end{pmatrix}, \tag{2.138}$$

independently of a common shift of points x_1 and x_2. At $a = 0$, we naturally recover the special case (2.136).

Now let us use these results to analyze the *double-barrier system* shown in figure 2.15. We could of course calculate its properties as before, writing down explicit expressions for all four traveling waves shown by arrows in figure 2.15, and then using the boundary conditions (2.124) and (2.125) at each of points $x_{1,2}$ to get a system of four linear equations, and then solving it for four amplitude ratios.

However, the transfer matrix approach simplifies the calculations, because we may immediately use Eqs. (2.132), (2.135), and (2.138) to write

$$T = T_a T_a T_a = \begin{pmatrix} 1 - i\alpha & -i\alpha \\ i\alpha & 1 + i\alpha \end{pmatrix} \begin{pmatrix} e^{ika} & 0 \\ 0 & e^{-ika} \end{pmatrix} \begin{pmatrix} 1 - i\alpha & -i\alpha \\ i\alpha & 1 + i\alpha \end{pmatrix}. \tag{2.139}$$

Let me hope that the reader remembers the 'row by column' rule of the multiplication of square matrices[28]; using it for the last two matrices, we may reduce Eq. (2.139) to

$$T = \begin{pmatrix} 1 - i\alpha & -i\alpha \\ i\alpha & 1 + i\alpha \end{pmatrix} \begin{pmatrix} (1 - i\alpha)e^{ika} & -i\alpha e^{ika} \\ i\alpha e^{-ika} & (1 + i\alpha)e^{-ika} \end{pmatrix}. \tag{2.140}$$

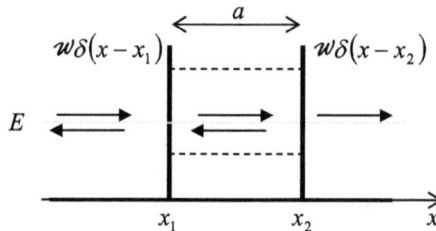

Figure 2.15. The double-barrier system. The dashed lines show (schematically) the quasi-levels of the metastable-state energies.

[28] In the analytical form: $(AB)_{jj'} = \sum_{j''=1}^{N} A_{jj''} B_{j''j'}$, where N is the matrix rank (in our current case, $N = 2$).

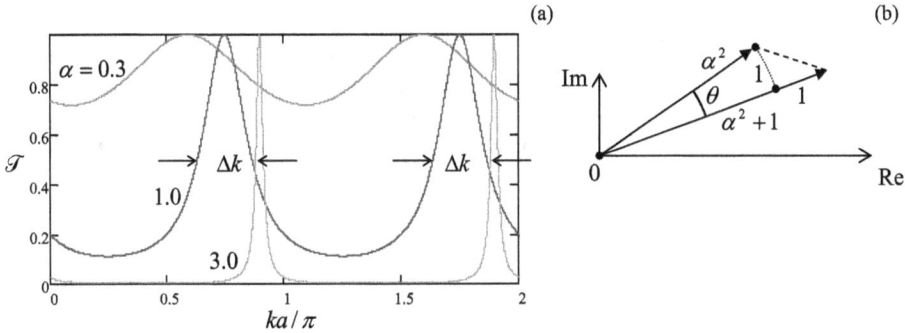

Figure 2.16. Resonant tunneling through a potential well with delta-functional walls: (a) the system's transparency as a function of ka, and (b) calculating the resonance's FWHM at $\alpha \gg 1$.

Now there is no need to calculate all elements of the full product T, because, according to Eq. (2.128), for the calculation of barrier's transparency $\mathscr{T}$ we need only one of its elements, T_{11}:

$$\mathscr{T} = \frac{1}{|T_{11}|^2} = \frac{1}{|\alpha^2 e^{-ika} + (1 - i\alpha)^2 e^{ika}|^2}.$$ (2.141)

This result is somewhat similar to that following from Eq. (2.71) for $E > U_0$: the transparency is a π-periodic function of the product ka, reaching its maximum ($\mathscr{T} = 1$) at some point of each period—see figure 2.16a. However, Eq. (2.141) is different in that for $\alpha \gg 1$, the resonance peaks of transparency are very narrow, reaching their maxima at $ka \approx k_n a \equiv n\pi$, with $n = 1, 2, \ldots$.

The physics of this *resonant tunneling* effect[29] is the so-called *constructive interference*, absolutely similar to that of electromagnetic waves (for example, light) in a *Fabry–Perot resonator* formed by two parallel semi-transparent mirrors[30]. Namely, the incident de Broglie wave may either tunnel through the two barriers, or undertake, on its way, several sequential reflections from these semi-transparent walls. At $k = k_n$, i.e. at $2ka = 2k_n a = 2\pi n$, the phase differences between all these partial waves are multiples of 2π, so that they add up in phase—'constructively'. Note that the same constructive interference of numerous reflections from the walls may be used to interpret the standing-wave eigenfunctions (1.84), so that the resonant tunneling at $\alpha \gg 1$ may be also considered as a result of the incident wave's resonance induction of such a standing wave, with a very large amplitude, in the space between the barriers, with the transmitted wave's amplitude proportionately increased.

As a result of this resonance, the maximum transparency of the system is *perfect* ($\mathscr{T}_{max} = 1$) even at $\alpha \to \infty$, i.e. in the case of a very *low* transparency of each of two

---

[29] In older literature, it is sometimes called the *Townsend* (or 'Ramsauer–Townsend') *effect*. However, it is more common to use that term only for the similar effect at 3D scattering—to be discussed in chapter 3.

[30] See, e.g. *Part EM* section 7.9. Note that despite the abundance of resonance functions similar to Eq. (2.141) in classical physics (see, e.g. *Part CM* section 5.1), some texts on quantum mechanics use for them the term *Breit–Wigner distribution* (or the 'Breit–Wigner function'), specific for this field.

component barriers. Indeed, the denominator in Eq. (2.141) may be interpreted as the squared length of the difference between two 2D vectors, one of length α^2, and another of length $|(1 - i\alpha)^2| = 1 + \alpha^2$, with the angle $\theta = 2ka + \text{const}$ between them—see figure 2.16b. At the resonance, the vectors are aligned, and their difference is smallest (equal to 1), so that $\mathscr{T}_{\max} = 1$. (This result is exact only if the two barriers are exactly equal.)

The same vector diagram may be used to calculate the so-called FWHM, a common acronym for the *Full Width* [of the resonance curve at its] *Half-Maximum*. By definition, this is the difference $\Delta k = k_+ - k_-$ between such two values of k, on the opposite slopes of the same resonance, at that $\mathscr{T} = \mathscr{T}_{\max}/2$—see the arrows in figure 2.16a. Let the vectors in figure 2.16b, drawn for $\alpha \gg 1$, be misaligned by a small angle $\theta \sim 1/\alpha^2 \ll 1$, so that the length of the difference vector (of the order of $\alpha^2\theta \sim 1$) is still much smaller than the length of each vector. In order to double its length squared, and hence to reduce $\mathscr{T}$ by a factor of 2 in comparison with its maximum value 1, the arc $\alpha^2\theta$ between the vectors should also become equal to ± 1, i.e. $\alpha^2(2k_{\pm}a + \text{const}) = \pm 1$. Subtracting these two equalities from each other, we finally get

$$\Delta k \equiv k_+ - k_- = \frac{1}{a\alpha^2} \ll k_{\pm}. \qquad (2.142)$$

Now let us use the simple system shown in figure 2.15 to discuss an issue of large conceptual significance. For that, consider what would happen if at some initial moment (say, $t = 0$) we have placed a 1D quantum particle inside the double-barrier well with $\alpha \gg 1$, and left it there alone, without any incident wave. To simplify the analysis, let us assume that the initial state of the particle coincides with one of the stationary states of the infinite-wall well of the same size—see Eq. (1.84):

$$\Psi(x, 0) = \psi_n(x) = \left(\frac{2}{a}\right)^{1/2} \sin\left[k_n(x - x_1)\right], \quad \text{where } k_n = \frac{\pi n}{a}. \qquad (2.143)$$

At $\alpha \to \infty$, this is just an eigenstate of the system, and from our analysis in section 1.5 we know the time evolution of its wavefunction:

$$\Psi(x, t) = \psi_n(x)\exp\{-i\omega_n t\}$$
$$= \left(\frac{2}{a}\right)^{1/2} \sin\left[k_n(x - x_1)\right]\exp\{-i\omega_n t\}, \quad \text{with } \omega_n = \frac{E_n}{\hbar} = \frac{\hbar k_n^2}{2m}, \qquad (2.144)$$

telling us that the particle remains in the well at all times with constant probability $W(t) = W(0) = 1$.

However, if the parameter α is large but finite, the de Broglie wave should slowly 'leak out' from the well, so that $W(t)$ would slowly decrease. Such a state is called *metastable*. Let us derive the law of its time evolution, assuming that the slow leakage, with a characteristic time $\tau \gg 1/\omega_n$, does not affect the instant wave

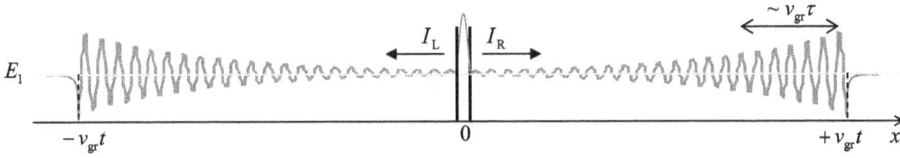

Figure 2.17. The metastable state's decay in the simple model of a 1D potential well formed by two low-transparent walls—schematically.

distribution inside the well, besides the gradual, slow reduction of W.[31] Then we can generalize Eq. (2.144) as

$$\Psi(x,\, t) = \left(\frac{2W}{a}\right)^{1/2} \sin\left[k_n(x - x_1)\right]\exp\{-i\omega_n t\}$$
$$\equiv A \exp\{i\,(k_n x - \omega_n t)\} + B\{-i\,(k_n x + \omega_n t)\}, \tag{2.145}$$

making the probability of finding the particle in the well equal to $W \leqslant 1$. This function is the sum of two traveling waves, with equal magnitudes of their amplitudes and probability currents

$$|A| = |B| = \left(\frac{W}{2a}\right)^{1/2}, \quad I_A = \frac{\hbar}{m}|A|^2 k_n = \frac{\hbar}{m}\frac{W}{2a}\frac{\pi n}{a}, \quad I_B = -I_A. \tag{2.146}$$

But we already know from Eq. (2.79) that at $\alpha \gg 1$, the delta-functional wall's transparency $\mathscr{T}$ equals $1/\alpha^2$, so that the wave carrying current I_A, incident on the right wall from the inside, induces an outcoming wave outside of the well (figure 2.17) with the following probability current:

$$I_R = \mathscr{T}I_A = \frac{1}{\alpha^2}I_A = \frac{1}{\alpha^2}\frac{\pi n\hbar W}{2ma^2}. \tag{2.147}$$

Absolutely similarly,

$$I_L = \frac{1}{\alpha^2}I_B = -I_R. \tag{2.148}$$

Now we may combine the 1D version (2.6) of the probability conservation law for the well's interior,

$$\frac{dW}{dt} + I_R - I_L = 0, \tag{2.149}$$

with Eqs. (2.147) to write

$$\frac{dW}{dt} = -\frac{1}{\alpha^2}\frac{\pi n\hbar}{ma^2}W. \tag{2.150}$$

---

[31] This virtually evident assumption finds its formal justification in the perturbation theory to be discussed in chapter 6.

This is just the standard differential equation,

$$\frac{dW}{dt} = -\frac{1}{\tau}W, \tag{2.151}$$

of the exponential decay, $W(t) = W(0)\exp\{-t/\tau\}$, where the constant τ, in our case equal to

$$\tau = \frac{ma^2}{\pi n\hbar}\alpha^2, \tag{2.152}$$

is called the *metastable state's lifetime*. Using Eq. (2.33b) for the de Broglie waves' group velocity, for our particular wave vector giving $v_{gr} = \hbar k_n/m = \pi n\hbar/ma$, Eq. (2.159) may be rewritten as a more general form,

$$\tau = \frac{t_a}{\mathcal{T}}, \tag{2.153}$$

where in our case the *attempt time* t_a is equal to a/v_{gr}, and $\mathcal{T} = 1/\alpha^2$. Eq. (2.153), which is valid for a broad class of similar metastable systems[32], may be interpreted in the following semi-classical way. The particle travels back and forth between the confining potential barriers, with the time interval t_a between the adjacent moments of incidence, each time making an attempt to leak through the wall, with a success probability equal to $\mathcal{T}$, so the reduction of W per each incidence is $\Delta W = -W\mathcal{T}$, in the limit $\alpha \gg 1$ (i.e. $\mathcal{T} \ll 1$) immediately leading to the differential equation (2.151) with the lifetime (2.153).

Another useful look at Eq. (2.152) may be taken by returning to the resonant tunneling problem in the same system, and expressing the resonance width (2.142) in terms of the incident particle's energy:

$$\Delta E = \Delta\left(\frac{\hbar^2 k^2}{2m}\right) \approx \frac{\hbar^2 k_n}{m}\Delta k = \frac{\hbar^2 k_n}{m}\frac{1}{a\alpha^2} = \frac{\pi n\hbar^2}{ma^2\alpha^2}. \tag{2.154}$$

Comparing Eqs. (2.152) and (2.154), we get a remarkably simple, parameter-independent formula[33]

$$\Delta E \cdot \tau = \hbar. \tag{2.155}$$

This *energy-time uncertainty relation* is certainly *more* general than our simple model; for example, it is valid for the lifetime and resonance tunneling width of any metastable state in the potential profile of any shape. This seems very natural,

---

[32] Essentially the only requirement is to have the attempt time Δt_A to be much longer than the effective time (the *instanton time*, see section 5.3 below) of tunneling through the barrier. In the delta-functional approximation for the barrier, the latter time is equal to zero, so that this requirement is always fulfilled.

[33] Note that the metastable state's decay (2.151) may be formally obtained from the basic Schrödinger equation (1.61) by adding an imaginary part, equal to $(-\Delta E/2)$, to its eigenenergy E_n. Indeed, in this case Eq. (1.62) becomes $a_n(t) = \text{const} \times \exp\{-i(E_n - i\Delta E/2)t/\hbar\} \equiv \text{const} \times \exp\{-iE_n t/\hbar\} \times \exp\{-\Delta Et/2\hbar\} = \text{const} \times \exp\{-iE_n t/\hbar\} \times \exp\{-t/2\tau\}$, so that $W(t) \propto |a_n(t)|^2 \propto \exp\{-t/\tau\}$. Such formalism (which hides the physical origin of the state's decay) may be convenient for some calculations, but misleading in other cases, and I will not use it in this course.

since because of the energy identification with frequency, $E = \hbar\omega$, typical for quantum mechanics, Eq. (2.155) may be rewritten as $\Delta\omega\cdot\tau = 1$ and seems to follow directly from the Fourier transform in time, just as the Heisenberg's uncertainty relation (1.35) follows from the Fourier transform in space. In some cases, these two relations are indeed interchangeable; for example, Eq. (2.24) for the Gaussian wave packet width may be rewritten as $\delta E \cdot \Delta t = \hbar$, where $\delta E = \hbar(d\omega/dk)\delta k = \hbar v_{\mathrm{gr}}\delta k$ is the rms spread of energies of monochromatic components of the packet, while $\Delta t \equiv \delta x/v_{\mathrm{gr}}$ is the time scale of the packet passage through a fixed observation point x.

However, Eq. (2.155) is much *less* general than the Heisenberg's uncertainty relation (1.35). Indeed, in the non-relativistic quantum mechanics, the Cartesian coordinates of a particle, the Cartesian components of its momentum, and the energy E are regular observables, represented by operators. In contrast, the time is treated as a *c*-number argument, and is *not* represented by an operator, so that Eq. (2.155) cannot be derived in such general assumptions as Eq. (1.35). Thus the time–energy uncertainty relation should be used with caution. Unfortunately, not everybody is so careful. One can find, for example, wrong claims that due to this relation, the energy dissipated by any system performing an elementary (single-bit) calculation during a time interval Δt has to be larger than $\hbar/\Delta t$.[34] Another incorrect statement is that the energy of a system cannot be measured, during a time interval Δt, with an accuracy better than $\hbar/\Delta t$.[35]

Now that we have a quantitative mathematical description of the metastable state's decay (valid, again, only if $\alpha \gg 1$, i.e. if $\tau \gg t_{\mathrm{a}}$), we may use it for discussion for two important conceptual issues of quantum mechanics. First, this is one of the simplest examples of systems that may be considered, from two different points of view, as either Hamiltonian (and hence *time-reversible*), or open (and hence *irreversible*). Indeed, from the former point of view, the system is certainly described by a time-independent Hamiltonian of the type (1.41), with the potential energy

$$U(x) = \mathcal{W}[\delta(x - x_1) + \delta(x - x_2)] \qquad (2.156)$$

(at $\mathcal{W} > 0$, evidently describing the profile shown in figure 2.15, and used to obtain the picture sketched in figure 2.17). In this picture, the wavefunction's time evolution, described by the Schrödinger equation (2.1) is reversible, the total probability of finding the particle *somewhere* on the axis x remains equal to 1, and the full system's energy, calculated from Eq. (1.23),

$$\langle E \rangle = \int_{-\infty}^{+\infty} \Psi^*(x, t)\, \hat{H}\, \Psi(x, t)\, d^3x, \qquad (2.157)$$

---

[34] On this issue, I dare to refer the reader to my own old work [5], which presented a constructive proof (for a particular system) that at *reversible computation* (the notion introduced in 1973 by C Bennett—see, e.g. *Part SM* section 2.3), the energy dissipation may be lower than this apparent 'quantum limit'.

[35] See, e.g. a discussion of this issue in the monograph [6].

remains constant and completely definite ($\delta E = 0$). On the other hand, since the 'emitted' wave packets would never return to the potential well[36], and it makes sense to look at the well region alone. For such a truncated, *open* system (for which the open ends beyond the interval $[x_1, x_2]$ serve as its environment), the probability W of finding the particle in it, and hence its energy $\langle E \rangle = WE_n$, decay exponentially in accordance with Eq. (2.151)—the decay equation typical for irreversible systems. We will return to the discussion of dynamics of such open quantum systems in chapter 7.

Second, the same model enables a preliminary discussion of one important aspect of quantum measurements. As Eq. (2.151) and figure 2.17 show, at $t \gg \tau$, the well becomes virtually empty ($W \approx 0$), and the whole probability is localized in two clearly separated wave packets with equal amplitudes, moving from each other with the speed v_{gr}, each 'carrying the particle away' with a probability of 50%. Now assume that an experiment has detected the particle on the left side of the well. Though the formalisms suitable for a quantitative analysis of the detection process will not be discussed until chapter 10, due to the wide separation $\Delta x = 2v_{gr}t \gg 2v_{gr}\tau$ of the packets, we may safely assume that such detection may be done without any actual physical effect on the counterpart wave packet[37]. But if we know that the particle has been found on the left side, there is no chance to find it on the right side. If we attributed the full wavefunction to all time stages of this particular experiment, this situation might be rather confusing. Indeed, this would mean that the wave-function at the right packet's location should instantly turn into zero—the so-called *wave packet reduction*—a hypothetical, irreversible process that cannot be described by the Schrödinger equation for this system, even including the particle detectors.

However, if (as was already discussed in section 1.3) we attribute the wave-function to a certain statistical ensemble of similar experiments, there is no need to involve such artificial notion. While the two-packet picture we have calculated (figure 2.17) describes the full ensemble of experiments with all systems prepared in the initial state (2.143), i.e. does not depend on the particle detection results, the 'reduced packet' picture (with no wave packet on the right of the well) describes only a sub-ensemble of such experiments, in which the particles have been detected on the left side. As was discussed on classical examples in section 1.3, for such re-defined ensemble the probability distribution, and hence the wavefunction, may be rather different. I will return to this important discussion in section 10.1.

2.6 Localized state coupling, and quantum oscillations

Now let us discuss one more effect specific for quantum mechanics. Its mathematical description may be simplified using a model consisting of two very short and deep

---

[36] For more realistic 2D and 3D systems, this statement is true even if the system as a whole is confined inside some closed volume, much larger than the potential well housing the metastable states. Indeed, if the walls providing such confinement are even slightly uneven, the emitted wave packets will be reflected from them, but would never return back exactly to the well. (See *Part SM* section 2.1 for more detailed discussion of this issue.)

[37] This argument is especially convincing if the particle's detection time is much shorter than the time $t_c = 2v_{gr}t/c$, where c is the speed of light in vacuum, i.e. the maximum velocity of any information transfer.

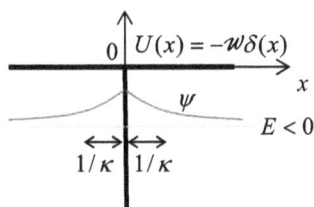

Figure 2.18. A delta-functional potential well and its localized eigenstate (schematically).

potential wells. But first, let us analyze the properties of a single well of this type (figure 2.18), which may be modeled similarly to the short and high potential barrier—see Eq. (2.74), but with a negative 'area':

$$U(x) = -\mathscr{W}\delta(x), \quad \text{with } \mathscr{W} > 0. \tag{2.158}$$

In contrast to its tunnel-barrier counterpart (2.74), such potential sustains a stationary state with a negative eigenenergy $E < 0$, and a *localized* eigenfunction ψ, with $|\psi| \to 0$ at $x \to \pm\infty$. Indeed, at $x \neq 0$, $U(x) = 0$, so the 1D Schrödinger equation of the system is reduced to the Helmholtz equation (1.83), whose localized solutions with $E < 0$ are single exponents, vanishing at large distances[38]:

$$\psi(x) = \psi_0(x) \equiv \begin{cases} Ae^{-\kappa x}, & \text{for } x > 0, \\ Ae^{+\kappa x}, & \text{for } x < 0, \end{cases} \quad \text{with } \frac{\hbar^2 \kappa^2}{2m} = -E, \quad \kappa > 0. \tag{2.159}$$

(The coefficients before the exponents have been selected equal to satisfy the boundary condition (2.76) of the wavefunction's continuity at $x = 0$.) Plugging Eq. (2.159) into the second boundary condition, given by Eq. (2.75), but now with the negative sign before $\mathscr{W}$, we get

$$(-\kappa A) - (+\kappa A) = -\frac{2m\mathscr{W}}{\hbar^2}A, \tag{2.160}$$

in which the common factor $A \neq 0$ may be cancelled. The remaining equation[39], has one (and just one) solution for any $\mathscr{W} > 0$:

$$\kappa = \kappa_0 \equiv \frac{m\mathscr{W}}{\hbar^2}, \tag{2.161}$$

and hence the system has only one (ground) localized state, with the following eigenenergy[40]:

---

[38] See Eqs. (2.56)–(2.58), with $U_0 = 0$.

[39] Such algebraic equations for linear differential equations are frequently called *characteristic*.

[40] Note that this E_0 is equal, by magnitude, to the constant E_0 that participates in Eq. (2.79). Note also that this result was actually already obtained, 'backwards', in the solution of problem 1.12(ii), but that solution did not address the issue whether the calculated potential (2.158) could sustain any other localized eigenstates.

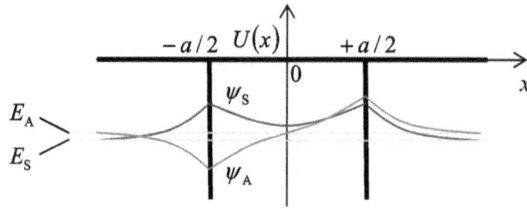

Figure 2.19. A system of two coupled potential wells, and its localized eigenstates (schematically).

$$E = E_0 \equiv -\frac{\hbar^2 \kappa_0^2}{2m} = -\frac{m \mathcal{w}^2}{2\hbar^2}. \tag{2.162}$$

Now we are ready to analyze localized states of the two-well potential shown in figure 2.19:

$$U(x) = -\mathcal{w}\left[\delta\left(x - \frac{a}{2}\right) + \delta\left(x + \frac{a}{2}\right)\right], \quad \text{with } \mathcal{w} > 0. \tag{2.163}$$

Here we may still use the single-exponent solutions, similar to Eq. (2.159), for the wavefunction outside the interval $[-a/2, +a/2]$, but inside the interval we need to take into account both possible exponents:

$$\psi = C_+ e^{\kappa x} + C_- e^{-\kappa x} \equiv C_A \sinh \kappa x + C_S \cosh \kappa x, \quad \text{for } -\frac{a}{2} \leqslant x \leqslant +\frac{a}{2}, \tag{2.164}$$

with the parameter κ defined as in Eq. (2.159). The last of these equivalent expressions is more convenient, because due to the symmetry of the potential (2.163) with respect to the central point $x = 0$, the system's eigenfunctions should be either symmetric (even) or antisymmetric (odd) functions of x (see figure 2.19), so that they may be analyzed separately, and only for one half of the system, say $x \geqslant 0$.

For the *asymmetric* eigenfunction, Eqs. (2.159) and (2.164) yield

$$\psi_A \equiv C_A \times \begin{cases} \sinh \kappa x, & \text{for } x \leqslant \dfrac{a}{2}, \\[2mm] \sinh \dfrac{\kappa a}{2} \exp\left\{-\kappa\left(x - \dfrac{a}{2}\right)\right\}, & \text{for } x \geqslant \dfrac{a}{2}, \end{cases} \tag{2.165}$$

where the front coefficient in the lower line has been selected to satisfy the condition (2.76) of the wavefunction's continuity at $x = a/2$ (and hence at $x = -a/2$). What remains is to satisfy the condition (2.75), with a negative sign before $\mathcal{w}$, for the derivative's jump at that point. This condition yields the following characteristic equation:

$$\sinh \frac{\kappa a}{2} + \cosh \frac{\kappa a}{2} = \frac{2m\mathcal{w}}{\hbar^2 \kappa} \sinh \frac{\kappa a}{2}, \quad \text{i.e. } 1 + \coth \frac{\kappa a}{2} = 2\frac{(\kappa_0 a)}{(\kappa a)}, \tag{2.166}$$

where κ_0, given by Eq. (2.161), is the value of κ for a single well, i.e. the reciprocal spatial width of its localized eigenfunction—see figure 2.18.

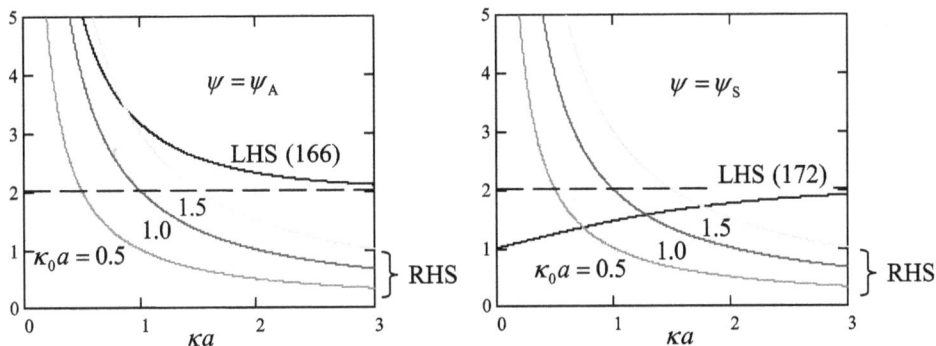

Figure 2.20. Graphical solutions of the characteristic equations of the two-well system, for: (a) the asymmetric eigenstate (2.165), and (b) the symmetric eigenstate (2.171).

Figure 2.20a shows both sides of Eq. (2.166) as functions of the dimensionless product κa, for several values of the parameter $\kappa_0 a$, i.e. of the normalized distance between the two wells. The plots show, first of all, that as the parameter $\kappa_0 a$ is decreased, the LHS and RHS lines cross (i.e. Eq. (2.166) has a solution) at lower and lower values of κa. At $\kappa a \ll 1$, the left-hand side of the last form of this equation may be approximated as $2/\kappa a$. Comparing this expression with the right-hand side, we see that this transcendental equation has a solution (i.e. the system has a localized asymmetric state) only if $\kappa_0 a > 1$, i.e. if the distance a between the two narrow potential wells is larger than the following value,

$$a_{\min} = \frac{1}{\kappa_0} \equiv \frac{\hbar^2}{m \mathcal{w}}, \qquad (2.167)$$

which is the characteristic spread of the wavefunction in a single well—see figure 2.18. (At $a \to a_{\min}$, $\kappa a \to 0$, meaning that the state becomes unlocalized.)

In the opposite limit of large distances between the potential wells, i.e. $\kappa_0 a \gg 1$, Eq. (2.166) shows that $\kappa a \gg 1$ as well, so that its left-hand side may be approximated as $2(1 + e^{-\kappa a})$, and the equation yields

$$\kappa \approx \kappa_0 \left(1 - \exp\{-\kappa_0 a\}\right) \approx \kappa_0. \qquad (2.168)$$

This result means that the eigenfunction is an asymmetric sum of two virtually unperturbed wavefunctions (2.159) of each partial potential well:

$$\psi_A(x) \approx \frac{1}{\sqrt{2}}[\psi_R(x) - \psi_L(x)], \quad \text{where}$$

$$\psi_R(x) \equiv \psi_0\left(x - \frac{a}{2}\right), \quad \psi_L(x) \equiv \psi_0\left(x + \frac{a}{2}\right), \qquad (2.169)$$

and the front coefficient is selected in such a way that if the eigenfunction ψ_0 of each well, given by Eqs. (2.159), is normalized, so is ψ_A. Plugging the middle (more exact) form of Eq. (2.168) into the last of Eq. (2.159), we can see that in this limit the asymmetric state's energy is only slightly higher than the eigenenergy E_0 of a single well, given by Eq. (2.162):

$$E_A \approx E_0(1 - 2\exp\{-\kappa_0 a\}) \equiv E_0 + \delta, \quad \text{where } \delta \equiv \frac{m\mathcal{W}^2}{\hbar^2}\exp\{-\kappa_0 a\} > 0. \quad (2.170)$$

The *symmetric* eigenfunction has a form reminding Eq. (2.165), but still different from it:

$$\psi = \psi_S \equiv C_S \times \begin{cases} \cosh \kappa x, & \text{for } x \leqslant \dfrac{a}{2}, \\[2mm] \cosh \dfrac{\kappa a}{2}\exp\left\{-\kappa\left(x - \dfrac{a}{2}\right)\right\}, & \text{for } x \geqslant \dfrac{a}{2}, \end{cases} \quad (2.171)$$

giving a characteristic equation similar in structure to Eq. (2.166), but with a different left-hand side:

$$1 + \tanh\frac{\kappa a}{2} = 2\frac{(\kappa_0 a)}{(\kappa a)}. \quad (2.172)$$

Figure 2.20b shows both sides of this equation for several values of the parameter $\kappa_0 a$. It is evident that in contrast to Eq. (2.166), Eq. (2.172) has a unique solution (and hence the system has a localized symmetric eigenstate) for *any* value of the parameter $\kappa_0 a$, i.e. for any distance between the partial wells. In the limit of very close wells (i.e. their strong coupling), $\kappa_0 a \ll 1$, we get $\kappa a \ll 1$, $\tanh(\kappa a/2) \to 0$, and Eq. (2.172) yields $\kappa \to 2\kappa_0$, leading to a four-fold increase of the eigenenergy 's magnitude in comparison with that of the single well:

$$E_S \approx 4E_0 \equiv -\frac{m(2\mathcal{W})^2}{2\hbar^2}, \quad \text{for } \kappa_0 a \ll 1. \quad (2.173)$$

The physical meaning of this result is very simple: two very close potential wells act (on the symmetric eigenfunction only!) together, so that their 'areas' $\mathcal{W} = \int U(x)dx$ just add up.

In the opposite, weak coupling limit, i.e. $\kappa_0 a \gg 1$, $\kappa a \gg 1$ as well, the left-hand side of Eq. (2.172) may be approximated as $2(1 - e^{-\kappa a})$, and the equation yields

$$\kappa \approx \kappa_0 (1 + \exp\{-\kappa_0 a\}) \approx \kappa_0. \quad (2.174)$$

In this limit, the eigenfunction is a symmetric combination of two virtually unperturbed wavefunctions (2.159) of each partial potential well:

$$\psi_S(x) \approx \frac{1}{\sqrt{2}}[\psi_R(x) + \psi_L(x)], \quad (2.175)$$

and the eigenenergy is also close to the energy E_0 of a partial well, but is slightly lower:

$$E_S \approx E_0(1 + 2\exp\{-\kappa_0 a\}) \equiv E_0 - \delta, \quad \text{so that } E_A - E_S = 2\delta, \quad (2.176)$$

where δ is again given by the last of Eqs. (2.170).

So, the eigenenergy of the symmetric state is always lower than that of the asymmetric state. The physics of this effect (which remains qualitatively the same in more complex two-component systems, most importantly in diatomic molecules such as H_2) is evident from the sketch of the wavefunctions ψ_A and ψ_S, given by Eqs. (2.165) and (2.171), in figure 2.19. In the antisymmetric mode, the wavefunction has to vanish in the center of the system, so that each of its halves is squeezed to one half of the system's spatial extension. Such a squeeze increases the function's gradient, and hence its kinetic energy (1.27), and hence its total energy. Conversely, in the symmetric mode the wavefunction effectively spreads into the counterpart well. As a result, it changes in space more slowly, and hence its kinetic energy is also lower.

Even more importantly, the symmetric state's energy decreases as the distance a is decreased, corresponding to the effective attraction of the partial wells. This is a good toy model of the strongest (and most important) type of atomic cohesion—the *covalent* (or 'chemical') *bonding*[41]. In the simplest case of the H_2 molecule, each of two electrons of the system[42] reduces its kinetic energy by spreading its wavefunction around both hydrogen nuclei (protons), rather than being confined near one of them—as it had to be in a single atom. The resulting bonding is very strong: in chemical units, 429 kJ mol^{-1}, i.e. 18.6 eV per molecule. Perhaps counter-intuitively, the covalent bonding is even stronger than the strongest classical (*ionic*) bonding due to electron transfer between atoms, leading to the Coulomb attraction of the resulting ions. (For example, the atomic cohesion in the NaCl molecule is just 3.28 eV.)

Now let us analyze dynamic properties of our model system (figure 2.19), because two weakly coupled potential wells are our first sample of the very important class of *two-level systems*[43]. It is easiest to do in the weak-coupling limit $\kappa_0 a \gg 1$, when the simple results (2.168)–(2.170) and (2.174)–(2.176) are quantitatively valid. In particular, Eqs. (2.169) and (2.175) allow us to represent the quasi-localized states of the particle in each partial well as linear combinations of its two eigenstates:

$$\psi_R(x) = \frac{1}{\sqrt{2}}[\psi_S(x) + \psi_A(x)], \quad \psi_L(x) = \frac{1}{\sqrt{2}}[\psi_S(x) - \psi_A(x)]. \tag{2.177}$$

Let us perform the following thought ('gedanken') experiment: place the particle, at $t = 0$, into one of the quasi-localized states, say $\psi_R(x)$, and leave the system alone to evolve, so that

$$\Psi(x, 0) = \psi_R(x) = \frac{1}{\sqrt{2}}[\psi_S(x) + \psi_A(x)]. \tag{2.178}$$

---

[41] Historically, the development of the quantum theory of such bonding in the H_2 molecule (by W Heitler and F London in 1927) was the breakthrough decisive for the acceptance of the then-emerging quantum mechanics by the community of chemists.

[42] Due to the opposite spins of these electrons, the Pauli principle allows them to be in the same orbital ground state—see chapter 8.

[43] As we will see later in chapter 4, these properties are similar to those of spin-1/2 particles; hence two-level systems are frequently called the *spin-1/2-like* systems.

According to the general solution (1.69) of the time-independent Schrödinger equation, the time dynamics of this wavefunction may be obtained by just multiplying each eigenfunction by the corresponding complex-exponential time factor:

$$\Psi(x,\, t) = \frac{1}{\sqrt{2}}\left[\psi_S(x)\exp\left\{-i\frac{E_S}{\hbar}t\right\} + \psi_A(x)\exp\left\{-i\frac{E_A}{\hbar}t\right\}\right].$$ (2.179)

From here, using Eqs. (2.170) and (2.176), and then Eqs. (2.169) and (2.175) again, we get

$$\Psi(x,\, t) = \frac{1}{\sqrt{2}}\left(\psi_S(x)\exp\left\{\frac{i\delta t}{\hbar}\right\} + \psi_A(x)\exp\left\{-\frac{i\delta t}{\hbar}\right\}\right)\exp\left\{-\frac{iE_0 t}{\hbar}\right\}$$

$$\equiv \left(\psi_R(x)\cos\frac{\delta t}{\hbar} + i\psi_L(x)\sin\frac{\delta t}{\hbar}\right)\exp\left\{-i\frac{E_0 t}{\hbar}\right\}.$$ (2.180)

This result implies, in particular, that the probabilities W_R and W_L of finding the particle, respectively, in the right and left wells change with time as

$$W_R = \cos^2\frac{\delta t}{\hbar}, \qquad W_L = \sin^2\frac{\delta t}{\hbar},$$ (2.181)

mercifully leaving the total probability constant: $W_R + W_L = 1$. (If our calculation had not passed this sanity check, we would be in big trouble.)

This is the famous effect of periodic *quantum oscillations*[44] of the particle's wavefunction between two similar, coupled subsystems, with the frequency

$$\omega = \frac{2\delta}{\hbar} \equiv \frac{E_A - E_S}{\hbar}.$$ (2.182)

In its last form, this result does not depend on the assumption of weak coupling, though the simple form (2.181) of the oscillations, with its 100% probability variations, does. (Indeed, at strong coupling of two subsystems, the very notion of the quasi-localized states ψ_R and ψ_L is ambiguous.) Qualitatively, this effect may be interpreted as follows: the particle, placed into one of the potential wells, tries to escape from it via tunneling through the potential barrier separating the wells. (In our particular system, shown in figure 2.17, the barrier is formed by the spatial segment of length a, which has the potential energy, $U = 0$, higher that the eigenstate energy $-E_0$.) However, in the two-well system the particle can only escape into the adjacent well. After the tunneling into that counterpart well, the particle tries to escape from it, and hence comes back, etc—very much as a classical 1D oscillator, initially deflected from its equilibrium position, at negligible damping.

Some care is required with using such interpretation for quantitative conclusions. In particular, let us compare the period $\mathcal{T} \equiv 2\pi/\omega$ of the oscillations (2.181) with the

---

[44] Sometimes they are called the *Bloch oscillations*, but more commonly the last term is reserved for a related, but different effect in spatially-periodic systems—to be discussed in section 2.8 below.

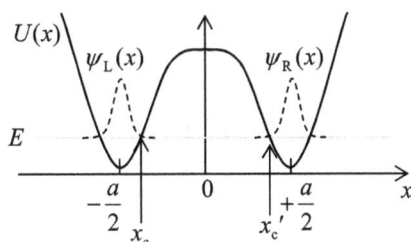

Figure 2.21. Weak coupling between two similar, soft potential wells.

metastable state's lifetime discussed in the previous section. For our particular model we may use the second of Eqs. (2.170) to write

$$\omega = \frac{4\,|E_0|}{\hbar}\exp\{-\kappa_0 a\},$$

$$\mathcal{T} = \frac{\pi\hbar}{\delta} = \frac{\pi\hbar}{2\,|E_0|}\exp\{\kappa_0 a\} = \frac{t_a}{2}\exp\{\kappa_0 a\}, \quad \text{for } \kappa_0 a \gg 1, \tag{2.183}$$

where $t_a \equiv 2\pi/\omega_0 \equiv 2\pi\hbar/|E_0|$ is the effective attempt time. On the other hand, according to Eq. (2.80), the transparency $\mathcal{T}$ of our potential barrier, in this limit, scales as $\exp\{-2\kappa_0 a\}$,[45] so that according to the general relation (2.153), the lifetime τ is of the order of $t_a \exp\{2\kappa_0 a\} \gg \mathcal{T}$. This is a rather counterintuitive result: the speed of a particle tunneling into a similar adjacent well is much higher than that, through a similar barrier, to the free space!

In order to show that this important result is not an artifact of our delta-functional model of the potential barrier, and also compare $\mathcal{T}$ and τ more directly, let us analyze the quantum oscillations between two weakly coupled wells, now assuming that the (symmetric) potential profile $U(x)$ is more soft (figure 2.21), so that all its eigenfunctions ψ_S and ψ_A are at least differentiable at all points[46]. If the barrier's transparency is low, the quasi-localized wavefunctions $\psi_R(x)$ and $\psi_L(x) = \psi_R(-x)$ and their eigenenergies may be found approximately by solving the Schrödinger equations in one of the wells, neglecting the tunneling through the barrier, but the calculation of δ requires a little bit more care. Let us write the stationary Schrödinger equations for the symmetric and antisymmetric solutions in the form

$$[E_A - U(x)]\,\psi_A = -\frac{\hbar^2}{2m}\frac{d^2\psi_A}{dx^2}, \quad [E_S - U(x)]\,\psi_S = -\frac{\hbar^2}{2m}\frac{d^2\psi_S}{dx^2}, \tag{2.184}$$

---

[45] It is hard to use Eq. (2.80) for a more exact evaluation of $\mathcal{T}$ in our current system, with its infinitely deep potential wells, because the meaning of the wave number k is not quite clear. However this is not too important, because in the limit $\kappa_0 a \gg 1$, the tunneling exponent makes the dominating contribution into the transparency—see, again, figure 2.7b.

[46] Such smooth well may have more than one quasi-localized stationary state, so that the proper index n is implied in all remaining formulas of this section.

multiply the former equation by ψ_S, and the latter one by ψ_A, subtract them from each other, and then integrate the result from 0 to ∞. The result is

$$(E_A - E_S) \int_0^\infty \psi_S \psi_A dx = \frac{\hbar^2}{2m} \int_0^\infty \left(\frac{d^2\psi_S}{dx^2} \psi_A - \frac{d^2\psi_A}{dx^2} \psi_S \right) dx. \qquad (2.185)$$

If $U(x)$, and hence $d^2\psi_{A,S}/dx^2$, are finite for all x, we may integrate the right-hand side by parts to get

$$(E_A - E_S) \int_0^\infty \psi_S \psi_A dx = \frac{\hbar^2}{2m} \left(\frac{d\psi_S}{dx} \psi_A - \frac{d\psi_A}{dx} \psi_S \right)_0^\infty. \qquad (2.186)$$

So far, this result is exact (provided that the derivatives participating in it are finite at each point); for *weakly* coupled wells, it may be further simplified. Indeed, in this case, the left-hand side of Eq. (2.186) may be approximated as

$$(E_A - E_S) \int_0^\infty \psi_S \psi_A dx \approx \frac{E_A - E_S}{2} \equiv \delta, \qquad (2.187)$$

because this integral is dominated by the vicinity of point a, where the second terms in each of Eqs. (2.169) and (2.175) are negligible, and the integral is equal to 1/2, assuming the proper normalization of the function $\psi_R(x)$. On the right-hand side of Eq. (2.186), the substitution at $x = \infty$ vanishes (due to the wavefunction's decay in the classically forbidden region), and so does the first term at $x = 0$, because for the antisymmetric solution, $\psi_A(0) = 0$. As a result, the energy half-split δ may be expressed in any of the following, equivalent forms:

$$\delta = \frac{\hbar^2}{2m} \psi_S(0) \frac{d\psi_A}{dx}(0) = \frac{\hbar^2}{m} \psi_R(0) \frac{d\psi_R}{dx}(0) = -\frac{\hbar^2}{m} \psi_L(0) \frac{d\psi_L}{dx}(0). \qquad (2.188)$$

It is straightforward (and hence left for the reader's exercise) to show that within the limits of the WKB approximation's validity, Eq. (2.188) may be reduced to

$$\delta = \frac{\hbar}{t_a} \exp\left\{ -\int_{x_c}^{x_c'} \kappa(x') dx' \right\}, \quad \text{so that } \mathcal{T} \equiv \frac{\pi\hbar}{\delta} = \frac{t_a}{2} \exp\left\{ \int_{x_c}^{x_c'} \kappa(x') dx' \right\}, \qquad (2.189)$$

where t_a is the time period of the classical motion of the particle, with the energy $E \approx E_A \approx E_S$, inside each well, the function $\kappa(x)$ is defined by Eq. (2.82), and x_c and x_c' are the classical turning points limiting the potential barrier at the level E of the particle's eigenenergy—see figure 2.21. The result (2.189) is evidently a natural generalization of Eq. (2.183), so that the strong relation between the times of particle tunneling into the continuum of states and into a discrete eigenstate, is indeed not specific for the delta-functional model. We will return to this fact, in its more general form, in the end of chapter 6.

2.7 Periodic systems: energy bands and gaps

Let us now proceed to the discussion of one of the most important issues of wave mechanics: a particle's motion through a periodic system. As a precursor to this discussion, let us calculate the transparency of the potential profile shown in figure 2.22 (frequently called the *Dirac comb*): a sequence of N similar, equidistant delta-functional potential barriers, separated by $(N - 1)$ potential-free intervals a.

According to Eq. (2.132), its transfer matrix is the following product

$$\mathrm{T} = \underbrace{\mathrm{T}_\alpha \mathrm{T}_a \mathrm{T}_\alpha \ldots \mathrm{T}_a \mathrm{T}_\alpha}_{(N-1)+N\,\text{terms}}, \tag{2.190}$$

with the component matrices given by Eqs. (2.135) and (2.138), and the barrier height parameter α defined by the last of Eqs. (2.78). Remarkably, this multiplication may be carried out analytically[47], giving

$$\mathscr{T} \equiv |T_{11}|^{-2} = \left[(\cos Nqa)^2 + \left(\frac{\sin ka - \alpha \cos ka}{\sin qa} \sin Nqa \right)^2 \right]^{-1}, \tag{2.191a}$$

where q is a new parameter, with the wave number dimensionality, defined by the following relation:

$$\cos qa \equiv \cos ka + \alpha \sin ka. \tag{2.191b}$$

For $N = 1$, Eqs. (2.191) immediately yield our old result (2.79), while for $N = 2$ they may be readily reduced to Eq. (2.141)—see figure 2.16a. Figure 2.20 shows its predictions for two larger numbers N, and several values of the dimensionless parameter α. Let us start the discussion of the plots from the case $N = 3$, when three barriers limit two coupled potential wells between them. The comparison of figures 2.23a and 2.16a shows that the transmission patterns, and their dependence on the parameter α, are very similar, besides that in the coupled-well system each resonant tunneling peak splits into two, with the ka-difference between them scaling as $1/\alpha$. From the discussion in the last section, we may now readily interpret this result: each

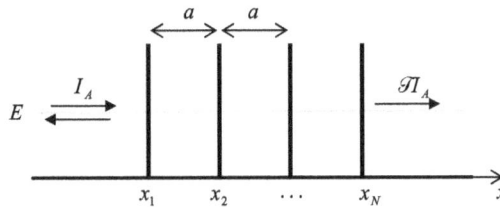

Figure 2.22. Tunneling through a system of N similar, equidistant barriers, i.e. $(N - 1)$ similar coupled potential wells.

---

[47] This formula will be easier to prove after we have discussed the properties of Pauli matrices in chapter 4.

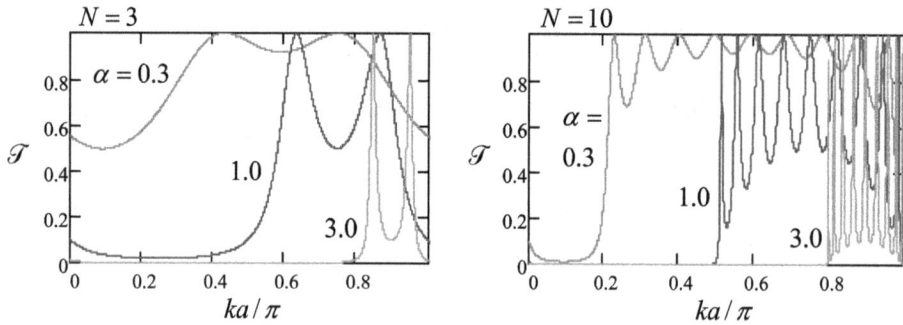

Figure 2.23. The transparency of the Dirac comb (figure 2.22), as a function of the product ka. Since the function $\mathscr{T}(ka)$ is π-periodic (just like it is for $N = 2$, see figure 2.16a), only one period is shown.

pair of resonance peaks of transparency corresponds to the alignment of the incident particle's energy E with the pair of energy levels E_A, E_S of the symmetric and antisymmetric states of the system. However, in contrast to the system shown in figure 2.19, these states are metastable, because the particle may leak out from these states just as it could in the system studied in section 2.5—see figure 2.15 and its discussion. As a result, each of the resonant peaks has a non-zero energy width ΔE, obeying Eq. (2.155).

A further increase of N (see, e.g. figure 2.23b) results in the increase of the number of resonant peaks per period to $(N - 1)$, and at $N \to \infty$ the peaks merge into the so-called *allowed energy bands* (frequently called just the 'energy bands') with a transparency $\mathscr{T} \sim 1$, separated from similar bands in the adjacent periods of function $\mathscr{T}(ka)$ by *energy gaps*[48] where $\mathscr{T} \to 0$. Notice the following important features of the pattern:

(i) at $N \to \infty$, the band/gap edges become sharp for any α, and tend to fixed positions (determined by α but independent of N);

(ii) the larger is the interwell coupling ($\alpha \to 0$), the broader are the allowed energy bands and the narrower are the gaps between them.

Our previous discussion of the resonant tunneling gives us an evident clue for a semi-quantitative interpretation of this pattern: if $(N - 1)$ potential wells are weakly coupled by tunneling through the potential barriers separating them, the system's energy spectrum consists of groups of $(N - 1)$ metastable energy levels, each group close to one of unperturbed eigenenergies of the well. (According to Eq. (1.84), for our current example, shown in figure 2.22, with its rectangular potential wells, these eigenenergies correspond to $k_n a = \pi n$.) Now let us recall that in the case $N = 2$, analyzed in the previous section, the eigenfunctions (2.169) and (2.175) differed only by the phase shift $\Delta\varphi$ between their localized components $\psi_R(x)$ and $\psi_L(x)$, with $\Delta\varphi = 0$ for one of them (ψ_S) and $\Delta\varphi = \pi$ for its counterpart. Hence it is natural to

---

[48] In the solid state (especially semiconductor) physics and electronics, the term *bandgaps* is more common.

expect that for other N as well, each metastable energy level corresponds to an eigenfunction that is the set of similar localized functions in each potential well, but with certain phase shifts $\Delta\varphi$ between them.

Moreover, we may expect that at $N \to \infty$, i.e. for *periodic structures*[49], with

$$U(x + a) = U(x), \qquad (2.192)$$

when the system does not have the ends which could affect its properties, the phase shift $\Delta\varphi$ between the localized wavefunctions in all couples of adjacent potential wells should be the same, i.e.

$$\psi(x + a) = \psi(x)e^{i\Delta\varphi} \qquad (2.193a)$$

for all x.[50] This equality is the (1D version of the) much-celebrated *Bloch theorem*[51]. Mathematical rigor aside[52], it is a virtually evident fact, because the particle's density $w(x) = \psi^*(x)\psi(x)$, that has to be periodic in this a-periodic system, may be so only $\Delta\varphi$ is constant. For what follows, it is more convenient to represent the real constant $\Delta\varphi$ in the form qa, so that the Bloch theorem takes the form

$$\psi(x + a) = \psi(x)e^{iqa}. \qquad (2.193b)$$

The physical sense of the parameter q will be discussed in detail below, but we may immediately notice that according to Eq. (2.193b), an addition of $(2\pi/a)$ to this parameter yields the same wavefunction; hence all observables have to be $(2\pi/a)$-periodic functions of q.[53]

Now let us use the Bloch theorem to calculate the eigenfunctions and eigenenergies for the infinite version of the system shown in figure 2.22, i.e. for an infinite set of delta-functional potential barriers—see figure 2.24. To start, let us consider two points separated by one period a: one of them, x_j, just left of the position of one of the barriers, and another one, x_{j+1}, just left of the following barrier. The eigenfunctions at each of the points may be represented as linear superpositions of two simple waves $\exp\{\pm ikx\}$, and the amplitudes of their components should be related by a 2×2

---

[49] This is a reasonable 1D model for solid state crystals, whose samples may feature up to $\sim 10^9$ similar atoms or molecules in each direction of the crystal lattice.

[50] A reasonably fair classical image of $\Delta\varphi$ is the geometric angle between similar objects—e.g. similar paper clips—attached at equal distances to a long, uniform rubber band. If the band's ends are twisted, the twist is equally distributed between the structure's periods, representing the constancy of $\Delta\varphi$. I am ashamed to confess that, due to lack of time, this was the only 'lecture demonstration' in my QM courses.

[51] Named after F Bloch who applied this concept to the wave mechanics in 1929, i.e. very soon after its formulation. Note, however, in mathematics, an equivalent statement, called the *Floquet theorem*, has been known since at least 1883.

[52] I will recover this rigor in two steps. Later in this section, we will see that the function obeying Eq. (2.193) is indeed *a* solution of the Schrödinger equation. However, to save time/space, it will be better for us to postpone the proof that *any* eigenfunction of the equation, with periodic boundary conditions, obeys the Bloch theorem, until chapter 4. As a partial reward for this delay, that proof will be valid for an arbitrary spatial dimensionality.

[53] The product $\hbar q$, which has the dimensionality of momentum, is called either the *quasi-momentum* or (especially in solid state physics) the 'crystal momentum' of the particle. Informally, it is very convenient (and common) to use the name 'quasi-momentum' for the bare q as well, despite its evidently different dimensionality.

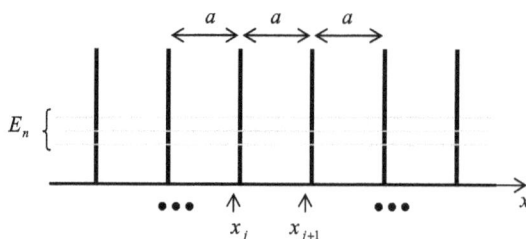

Figure 2.24. The simplest periodic potential: the infinite Dirac comb.

transfer matrix T of the potential fragment separating them. According to Eq. (2.132), this matrix may be found as the product of the matrix (2.135) of one delta-functional barrier, and the matrix (2.138) of one zero-potential interval a:

$$\begin{pmatrix} A_{j+1} \\ B_{j+1} \end{pmatrix} = T_a T_\alpha \begin{pmatrix} A_j \\ B_j \end{pmatrix} = \begin{pmatrix} e^{ika} & 0 \\ 0 & e^{-ika} \end{pmatrix} \begin{pmatrix} 1 - i\alpha & -i\alpha \\ i\alpha & 1 + i\alpha \end{pmatrix} \begin{pmatrix} A_j \\ B_j \end{pmatrix}. \tag{2.194}$$

However, according to the Bloch theorem (2.193b), the component amplitudes should be also related as

$$\begin{pmatrix} A_{j+1} \\ B_{j+1} \end{pmatrix} = e^{iqa} \begin{pmatrix} A_j \\ B_j \end{pmatrix} = \begin{pmatrix} e^{iqa} & 0 \\ 0 & e^{iqa} \end{pmatrix} \begin{pmatrix} A_j \\ B_j \end{pmatrix}. \tag{2.195}$$

The condition of self-consistency of these two equations leads to the following characteristic equation:

$$\left| \begin{pmatrix} e^{ika} & 0 \\ 0 & e^{-ika} \end{pmatrix} \begin{pmatrix} 1 - i\alpha & -i\alpha \\ i\alpha & 1 + i\alpha \end{pmatrix} - \begin{pmatrix} e^{iqa} & 0 \\ 0 & e^{iqa} \end{pmatrix} \right| = 0. \tag{2.196}$$

In section 2.5, we have already calculated the matrix product participating in this equation—see Eq. (2.140). Using it, we see that Eq. (2.196) is reduced to the same simple Eq. (2.191b) that has jumped at us from the solution of the somewhat different (resonant tunneling) problem.

Let us explore that simple result in detail. First of all, the right-hand side of Eq. (2.191b) is a sinusoidal function of ka, with the amplitude $(1 + \alpha^2)^{1/2}$—see figure 2.25, while its left-hand side is a sinusoidal function of qa with the unit amplitude. As a result, within each half-period $\Delta(ka) = \pi$ of the right-hand side, there is an interval where the characteristic equation does not have a real solution for q. These intervals correspond to the energy gaps clearly visible in figure 2.23, while the complementary intervals of ka, where a real solution for q exists, correspond to the allowed energy bands. In contrast, the parameter q can take *any* real values, so it is more convenient to plot the eigenenergy $E = \hbar^2 k^2/2m$ as the function of q (or, even more conveniently, qa) rather than ka.[54] While doing that, we need to recall that the parameter α,

---

[54] A more important reason for taking q as the argument is that for a general potential $U(x)$, the particle's momentum $\hbar k$ is not a constant of motion, while (according to the Bloch theorem), the quasi-momentum $\hbar q$ is.

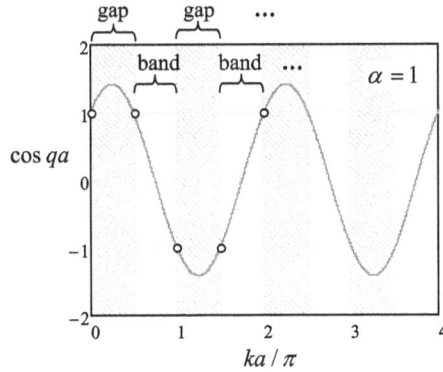

Figure 2.25. The graphical representation of the characteristic Eq. (2.191b) for a fixed value of the parameter α. The ranges of ka that yield |cos qa| < 1, correspond to allowed energy bands, while those with |cos qa| > 1, correspond to energy gaps between them.

defined by the last of Eqs. (2.78), depends on the wave vector k as well, so that if we vary q (and hence k), it is better to characterize the structure by another, k-independent dimensionless parameter, for example

$$\beta \equiv (ka)\,\alpha \equiv \frac{\mathcal{W}}{\hbar^2/ma}, \tag{2.197}$$

so that our characteristic Eq. (2.191b) becomes

$$\cos qa \equiv \cos ka + \beta \frac{\sin ka}{ka}. \tag{2.198}$$

Figure 2.26 shows the plots of k and E, following from Eq. (2.198), as functions of the dimensionless product qa, for a particular, moderate value of the parameter β. The band structure of the energy spectrum is apparent. Another evident feature is the 2π-periodicity of the pattern in the argument qa, which we have already predicted from the general Bloch theorem arguments. (Due to this periodicity, the complete band/gap pattern may be studied, for example, on just one interval $-\pi \leqslant qa \leqslant +\pi$, called the 1st *Brillouin zone*—the so-called *reduced zone picture*. For some applications, however, it is more convenient to use the *extended zone picture* with $-\infty \leqslant qa \leqslant +\infty$—see, e.g. the next section.)

However, maybe the most *important* fact, clearly visible in figure 2.26, is that there is an infinite number of energy bands, with different energies $E_n(q)$ for the same value of q. Mathematically, it is evident from Eq. (2.198)—see also figure 2.25. Indeed, for each value of qa there is a solution ka to this equation on each half-period $\Delta(ka) = \pi$—see also panel (a) in figure 2.26. Each of such solutions gives a specific value of the particle energy $E = \hbar^2 k^2/2m$. A continuous set of similar solutions for various qa forms a particular energy band.

Since the energy band picture is one of the most practically important results of quantum mechanics, it is imperative to understand its physics. It is natural to interpret this physics in different ways in two opposite potential strength limits. In

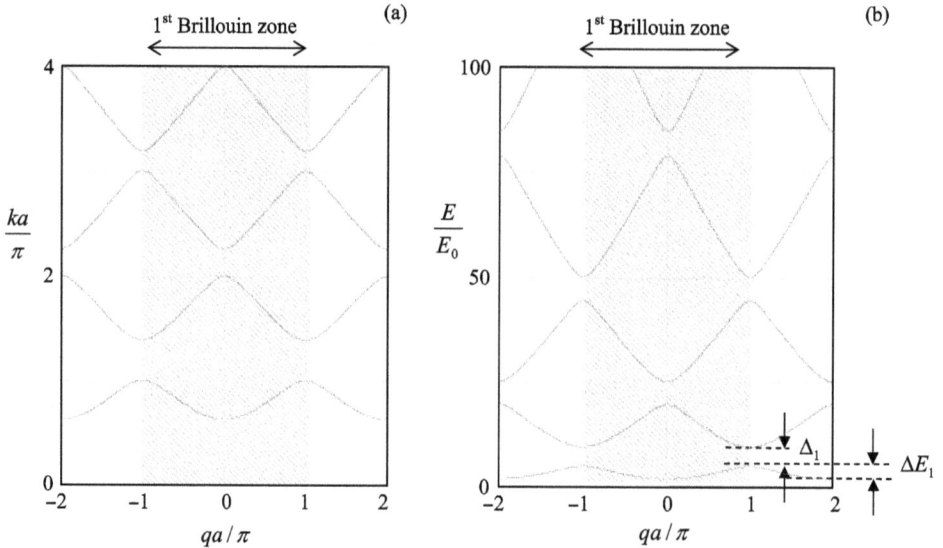

Figure 2.26. (a) The 'real' momentum k of a particle in an infinite Dirac comb (figure 2.24), and (b) its energy $E = \hbar^2 k^2/2m$ (in units of $E_0 \equiv \hbar^2/2ma^2$), as functions of the quasi-momentum q, for a particular value ($\beta = 3$) of the dimensionless potential parameter β. Arrows in the lower right corner of the panel (b) illustrate the definition of the energy bands (ΔE_n) and energy gaps (Δ_n).

parallel, we will use this discussion to obtain simpler expressions for the energy band/gap structure, in each limit. An important advantage of this approach is that both analyses may be carried out for an arbitrary periodic potential $U(x)$, rather than for the particular model shown in figure 2.24, and used to obtain the patterns shown in figures 2.25 and 2.26.

(i) *Tight-binding approximation.* This approximation is sound when the eigenenergy E_n of the states quasi-localized at the energy profile minima is much lower than the height of the potential barriers separating them—see figure 2.27. As should be clear from our discussion in section 2.6, essentially the only role of coupling between these states (via tunneling through the potential barriers separating the minima) is to establish a certain phase shift $\Delta\varphi \equiv qa$ between the adjacent quasi-localized wavefunctions $u_n(x - x_j)$ and $u_n(x - x_{j+1})$.

To describe this effect quantitatively, let us first return to the problem of two coupled wells considered in section 2.6, and recast the result (2.180), with restored eigenstate index n, as

$$\Psi_n(x, t) = [a_R(t)\psi_R(x) + a_L(t)\psi_L(x)]\exp\left\{-i\frac{E_n}{\hbar}t\right\}, \tag{2.199}$$

where the probability amplitudes a_R and a_L oscillate sinusoidally in time:

$$a_R(t) = \cos\frac{\delta_n}{\hbar}t, \quad a_L(t) = i\sin\frac{\delta_n}{\hbar}t. \tag{2.200}$$

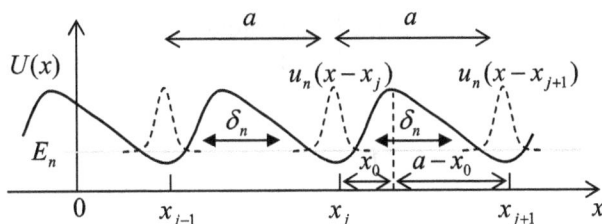

Figure 2.27. The tight binding approximation (schematically).

This evolution satisfies the following system of two equations, whose structure is similar to Eq. (1.61):

$$i\hbar\dot{a}_R = -\delta_n a_L, \quad i\hbar\dot{a}_L = -\delta_n a_R. \tag{2.201}$$

These equations may be readily generalized to the case of many similar coupled wells. In this case, instead of Eq. (2.199), we evidently should write

$$\Psi_n(x, t) = \left[\sum_j a_j(t)u_n(x - x_j)\right]\exp\left\{-i\frac{E_n}{\hbar}t\right\}, \tag{2.202}$$

where E_n are the eigenenergies, and u_n the eigenfunctions of each well. In the tight binding limit, only the adjacent wells are coupled, so that instead of Eq. (2.201) we should write an infinite system of similar equations

$$i\hbar\dot{a}_j = -\delta_n a_{j-1} - \delta_n a_{j+1}, \tag{2.203}$$

for each well number j, where parameters δ_n describe the coupling between two adjacent potential wells. Repeating the calculation outlined in the end of the last section for our new situation, for a smooth potential we may get an expression essentially similar to the last form of Eq. (2.188):

$$\delta_n = \frac{\hbar^2}{m}u_n(x_0)\frac{du_n}{dx}(a - x_0), \tag{2.204}$$

where x_0 is the distance between the well bottom and the middle of the potential barrier on the right of it—see figure 2.27. The only substantial new feature of this expression in comparison with Eq. (2.188) is that the sign of δ_n alternates with the level number n: $\delta_1 > 0$, $\delta_2 < 0$, $\delta_3 > 0$, etc. Indeed, the number of zeros (and hence, 'wiggles') of eigenfunctions $u_n(x)$ of any potential well increases as n—see, e.g. figure 1.8,[55] so that the difference of the exponential tails of the functions, sneaking under the left and right barriers limiting the well also alternates with n.

The infinite system of ordinary differential equations (2.203) allows one to explore a large range of important problems (such as the spread of the wavefunction that was initially localized in one well, etc), but our task right now is just to find its stationary states, i.e. the solutions proportional to $\exp\{-i(\varepsilon_n/\hbar)t\}$, where ε_n is a still

---

[55] Below, we will see several other examples of this behavior. This alternation rule is also described by the Bohr–Sommerfeld quantization condition (2.110).

unknown, q-dependent addition to the background energy E_n of the nth energy level. In order to satisfy the Bloch theorem (2.193) as well, such a solution should have the form

$$a_j(t) = a \exp\left\{ iqx_j - i\frac{\varepsilon_n}{\hbar}t + \text{const} \right\}. \tag{2.205}$$

Plugging this solution into Eq. (2.203) and canceling the common exponent, we get

$$E = E_n + \varepsilon_n = E_n - \delta_n(e^{-iqa} + e^{iqa}) \equiv E_n - 2\delta_n \cos qa, \tag{2.206}$$

so that in this approximation, the energy band width ΔE_n (see figure 2.26b) equals $4|\delta_n|$.

The relation (2.206), whose validity is restricted to $|\delta_n| \ll E_n$, describes the lowest energy bands plotted in figure 2.26b reasonably well. (For larger β, the agreement would be even better.) So, this calculation explains what the energy bands really are—in the tight-binding limit they are best interpreted as isolated well's energy levels E_n, broadened into the bands by the interwell interaction. Also, this result gives a clear proof that the energy band extremes correspond to $qa = 2\pi l$ and $qa = 2\pi(l + 1/2)$, with integer l. Finally, the sign alteration of the coupling coefficient δ_n (2.204) explains why the energy maxima of one band are aligned, on the qa axis, with energy minima of the adjacent bands—see figure 2.26.

(ii) *Weak-potential limit.* Amazingly, the energy band structure is also compatible with a completely different physical picture that may be developed in the opposite limit. Let the particle energy E be so high that the periodic potential $U(x)$ may be treated as a small perturbation. Naively, in this limit we could expect a slightly and smoothly deformed parabolic dispersion relation $E = \hbar^2 k^2/2m$. However, if we are plotting energy as a function of q rather than k, we need to add $2\pi l/a$, with an arbitrary integer l, to the argument. Let us show this by expanding all variables into the spatial Fourier series. For a periodic potential energy $U(x)$ such an expansion is straightforward[56]:

$$U(x) = \sum_{l''} U_{l''} \exp\left\{ -i\frac{2\pi x}{a}l'' \right\}, \tag{2.207}$$

where the summation is over all integers l'', from $-\infty$ to $+\infty$. However, for the wavefunction we should show a due respect to the Bloch theorem (2.193). To understand how to proceed, let us define another function

$$u(x) \equiv \psi(x)e^{-iqx}, \tag{2.208}$$

and study its periodicity:

$$u(x + a) = \psi(x + a)e^{-iq(x+a)} = \psi(x)e^{-iqx} = u(x). \tag{2.209}$$

We see that the new function is a-periodic, and hence we can use Eqs. (2.208) and (2.209) to rewrite the Bloch theorem as

---

[56] The benefits of such unusual choice of the summation index (l'' instead of, say, l) will be clear in a few lines.

$$\psi(x) = u(x)e^{iqx}, \quad \text{with } u(x + a) = u(x).\tag{2.210}$$

Now it is safe to expand the periodic function $u(x)$ exactly as $U(x)$:

$$u(x) = \sum_{l'} u_{l'} \exp\left\{-i\frac{2\pi x}{a}l'\right\},\tag{2.211}$$

so that, according to the Bloch theorem in the form (2.210),

$$\psi(x) = e^{iqx} \sum_{l'} u_{l'} \exp\left\{-i\frac{2\pi x}{a}l'\right\} = \sum_{l'} u_{l'} \exp\left\{i\left(q - \frac{2\pi}{a}l'\right)x\right\}.\tag{2.212}$$

The only nontrivial part of plugging this expression into the stationary Schrödinger equation (2.53) is the calculation of the product term, using the expansions (2.207) and (2.211):

$$U(x)\psi = \sum_{l',l''} U_{l''} u_{l'} \exp\left\{i\left(q - \frac{2\pi x}{a}(l' + l'')\right)\right\}.\tag{2.213}$$

At fixed l', we may change the summation over l'' to that over $l \equiv l' + l''$ (so that $l'' \equiv l - l'$), and write:

$$U(x)\psi = \sum_{l} \exp\left\{i\left(q - \frac{2\pi x}{a}l\right)\right\}\sum_{l'} u_{l'} U_{l-l'}.\tag{2.214}$$

Now plugging Eqs. (2.212) (with the summation index l' replaced with l) and (2.214) into the stationary Schrödinger equation (2.61), and requiring the coefficients of each spatial exponent to match, we get an infinite system of linear equations for u_l:[57]

$$\sum_{l'} U_{l-l'} u_{l'} = \left[E - \frac{\hbar^2}{2m}\left(q - \frac{2\pi}{a}l\right)^2\right]u_l.\tag{2.215}$$

So far, this system of equations is an equivalent alternative to the initial Schrödinger equation, for any potential's strength[58]. In the weak-potential limit, i.e. if all the Fourier coefficients U_n are small[59], we can complete all the calculation

---

[57] Note that by this calculation we have essentially proved that the Bloch wavefunction (2.210) is indeed *a* solution of the Schrödinger equation, provided that the quasi-momentum q is selected in a way to make the system of linear equation (2.215) compatible, i.e. is a solution of its characteristic equation—see, e.g. Eq. (2.223) below.

[58] By the way, the system is very efficient for fast numerical solutions of the Schrödinger equation for any periodic profile $U(x)$, even though in systems with large U_n it may require taking into account a large number of harmonics u_l.

[59] Besides, possibly, a constant potential U_0, which, as was discussed in chapter 1, may be taken for the energy reference. As a result, in the following calculations I will take $U_0 = 0$.

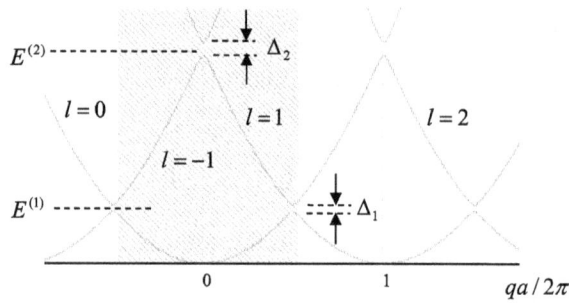

Figure 2.28. The energy band/gap picture in the weak potential limit ($\Delta_n \ll E^{(n)}$), with the shading showing the 1st Brillouin zone.

analytically[60]. Indeed, in the so-called 0th approximation we can ignore *all* U_n, so that in order to have at least one u_l different from 0, Eq. (2.215) requires that

$$E \to E_l \equiv \frac{\hbar^2}{2m}\left(q - \frac{2\pi l}{a}\right)^2. \tag{2.216}$$

(u_l itself should be obtained from the normalization condition). This result means that the dispersion relation $E(q)$ has an infinite number of similar quadratic branches numbered by integer l—see figure 2.28.

On any branch, the eigenfunction has just one Fourier coefficient, i.e. represents a monochromatic traveling wave

$$\psi_l \to u_l e^{ikx} = u_l \exp\left\{i\left(q - \frac{2\pi l}{a}\right)x\right\}. \tag{2.217}$$

This fact allows us to rewrite Eq. (2.215) in a more transparent form

$$\sum_{l'\neq l} U_{l'-l}u_{l'} = (E - E_l)u_l, \tag{2.218}$$

which may be formally solved for u_l:

$$u_l = \frac{1}{E - E_l}\sum_{l'\neq l}U_{l'-l}u_{l'}. \tag{2.219}$$

This formula shows that if the Fourier coefficients U_n are nonvanishing but small, the wavefunctions do acquire other Fourier components (besides the main one, with the index corresponding to the branch number), but these additions are all small, besides narrow regions near the points $E_l = E_{l'}$ where two branches (2.216) of the dispersion relation $E(q)$, with some specific numbers l and l', cross. According to Eq. (2.216), this happens when

---

[60] This method is so powerful that its multi-dimensional version is not much more complex than the 1D version described here—see, e.g. section 3.2 in the classical textbook [7].

$$\left(q - \frac{2\pi}{a}l\right) \approx -\left(q - \frac{2\pi}{a}l'\right), \tag{2.220}$$

i.e. at $q \approx q_m \equiv \pi m/a$ (with the integer $m \equiv l + l'$)[61] corresponding to

$$E_l \approx E_{l'} \approx \frac{\hbar^2}{2ma^2}[\pi(l + l') - 2\pi l]^2 = \frac{\pi^2\hbar^2}{2ma^2}n^2 \equiv E^{(n)}, \tag{2.221}$$

with integer $n \equiv l - l'$. (Eq. (2.221) shows that the index n is just the number of the branch crossing on the energy scale—see figure 2.28.) In such a region, E has to be close to both E_l and $E_{l'}$, so that the denominator in just one of the infinite number of terms in Eq. (2.219) is very small, making the term substantial despite the smallness of U_n. Hence we can take into account only one term in each of the sums (written for l and l'):

$$U_{-n}u_{l'} = (E - E_l)u_l,$$
$$U_nu_l = (E - E_{l'})u_{l'}. \tag{2.222}$$

Taking into account that for any real function $U(x)$ the Fourier coefficients in series (2.207) have to be related as $U_{-n} = U_n^*$, Eq. (2.222) yields the following simple characteristic equation

$$\begin{vmatrix} E - E_l & -U_n^* \\ -U_n & E - E_{l'} \end{vmatrix} = 0, \tag{2.223}$$

with the solution

$$E_{\pm} = E_{\text{ave}} \pm \left[\left(\frac{E_l - E_{l'}}{2}\right)^2 + U_nU_n^*\right]^{1/2}, \quad \text{with } E_{\text{ave}} \equiv \frac{E_l + E_{l'}}{2} = E^{(n)}. \tag{2.224}$$

According to Eq. (2.216), close to the branch crossing point $q_m = \pi(l + l')/a$, the fraction participating in this result may be approximated as[62]

$$\frac{E_l - E_{l'}}{2} \approx \gamma\,\tilde{q},$$

$$\gamma \equiv \left.\frac{dE_l}{dq}\right|_{q=q_m} = \frac{\pi\hbar^2 n}{ma} = \frac{2aE^{(n)}}{\pi n}, \quad \text{and} \quad \tilde{q} \equiv q - q_m, \tag{2.225}$$

while the parameters $E_{\text{ave}} = E^{(n)}$ and $U_nU_n^* = \hbar U_n\hbar^2$ do not depend on $\tilde{q}$, i.e. on the distance from the central point q_m. This is why Eq. (2.224) may be plotted as the famous *level anticrossing* (also called 'avoided crossing', or 'intended crossing', or 'non-crossing') *diagram* (figure 2.29), with the energy gap width Δ_n equal to $2|U_n|$, i.e. just double the magnitude of the nth Fourier harmonic of the periodic potential

---

[61] Let me hope that the difference between this new integer and the particle's mass, both called m, is absolutely clear from the context.

[62] Physically, $\beta/\hbar = \hbar(n\pi/a)m = \hbar k^{(n)}/m$ is just the velocity of a free classical particle with energy $E^{(n)}$.

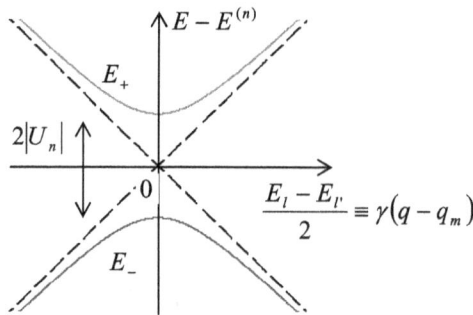

Figure 2.29. The level anticrossing diagram.

$U(x)$. Such anticrossings are also clearly visible in figure 2.28, which shows the result of the exact solution of Eq. (2.198) for $\beta = 0.5$.[63]

We will run into the anticrossing diagram again and again in the course, notably of the discussion of spin-1/2 and other two-level systems. It is also repeatedly met in classical mechanics, for example at the calculation of eigenfrequencies of coupled oscillators[64,65]. In our current case of the weak potential limit of the band theory, the diagram describes the interaction of two sinusoidal de Broglie waves (2.216), with oppositely directed wave vectors, l and $-l'$, via the $(l - l')$th (i.e. the nth) Fourier harmonic of the potential profile $U(x)$.[66] This effect exists also in the classical wave theory, and is known as the *Bragg reflection*, describing, for example, the 1D case of the X-wave reflection by a crystal lattice (see, e.g. figure 1.5) in the limit of weak interaction between the incident wave and each atom.

The anticrossing diagram shows that, rather counter-intuitively, even a weak periodic potential changes the topology of the initially-parabolic dispersion relation radically, connecting its different branches, and thus creating the energy gaps. Let me hope that the reader has enjoyed the elegant description of this effect, discussed above, as well as one more illustration of the wonderful ability of physics to give completely different interpretations, and approximate quantitative approaches to the same effect in the opposite limits.

So, we have explained analytically two asymptotic trends of the particular band structure shown in figure 2.26. Now we may wonder how general the rest of this structure is, i.e. how much does it depend on the peculiar properties of the delta-function model (figure 2.24). For that, let us represent the detailed band pattern, such as that shown in figure 2.26b (plotted for a particular value of the parameter β,

---

[63] From that figure, it is also clear that in the weak potential limit, the width ΔE_n of the nth energy band is just $E^{(n)} - E^{(n-1)}$—see Eq. (2.221). Note that this is exactly the distance between the adjacent energy levels of the simplest 1D potential well of an infinite depth—cf Eq. (1.85).

[64] See, e.g. *Part CM* section 6.1 and in particular figure 6.2.

[65] Actually, we could obtain this diagram earlier in this section, for the system of two weakly coupled potential wells (figure 2.21), if we assumed the wells to be slightly dissimilar.

[66] In the language of the de Broglie wave scattering, to be discussed in section 3.3, Eq. (2.220) may be interpreted as the condition at that each of these partial waves, scattered on the nth Fourier harmonic of the potential profile, constructively interferes with its counterpart, leading to a strong enhancement of their interaction.

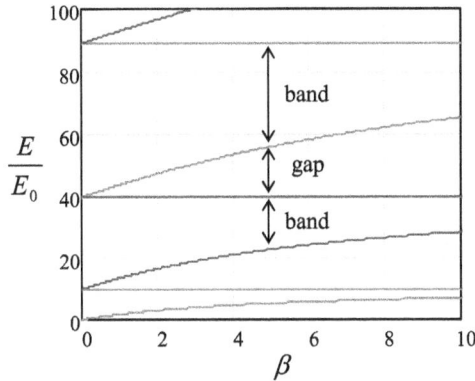

Figure 2.30. Characteristic curves of the Schrödinger equation for the infinite Dirac comb (figure 2.24).

characterizing the potential barrier strength) in a more condensed form, which would allow us to place the results for a range of β on a single comprehensible plot. The way to do this should be clear from figure 2.26b: since the dependence of energy on the quasi-momentum in each energy band is not too eventful, we may plot just the highest and smallest values of the particle's energy $E = \hbar^2 k^2 / 2m$, as functions of $\beta \equiv ma\mathcal{W}/\hbar^2$—see figure 2.30. (As figure 2.26b shows, they may be obtained from Eq. (2.198) with $qa = 0$ and $qa = \pi$.)

These plots (in mathematics, frequently called the *characteristic curves*) show, first of all, that at small β, all energy gaps are equal and proportional to this parameter, and hence to $\mathcal{W}$. This feature is in a full agreement with the main conclusion (2.224) of our analysis of the weak-potential limit, because for this potential (figure 2.24),

$$U(x) = \mathcal{W} \sum_{j=-\infty}^{+\infty} \delta\,(x - ja + \text{const}), \qquad (2.226)$$

all Fourier harmonic amplitudes, defined by Eq. (2.207), are equal by magnitude: $|U_l| = \mathcal{W}/a$. As β is further increased, the gaps grow and the allowed energy bands shrink, but rather slowly. This is also natural, because, as Eq. (2.79) shows, the transparency $\mathcal{T}$ of the delta-functional barriers, separating the quasi-localized states (and hence the coupling parameters $\delta_n \propto \mathcal{T}^{1/2}$) decrease with $\mathcal{W} \propto \beta$ very gradually.

These features may be compared with similar curves for more realistic and relatively simple periodic functions $U(x)$, for example the sinusoidal potential $U(x) = A\cos(2\pi x/a)$—see figure 2.31a. For this potential, the stationary Schrödinger equation (2.53) takes the following form:

$$-\frac{\hbar^2}{2m}\frac{d^2\psi}{dx^2} + A\cos\frac{2\pi x}{a}\psi = E\psi. \qquad (2.227)$$

By the introduction of dimensionless variables

$$\xi \equiv \frac{\pi x}{a}, \quad \alpha \equiv \frac{E}{E^{(1)}}, \quad 2\beta \equiv \frac{A}{E^{(1)}}, \qquad (2.228)$$

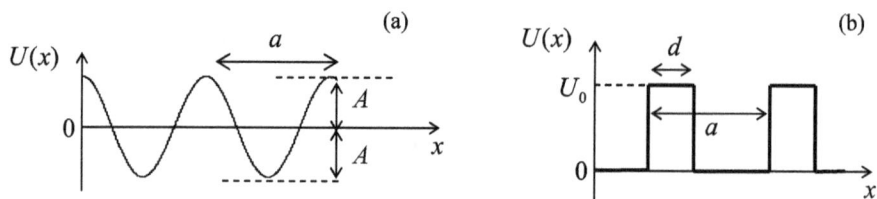

Figure 2.31. Two other simple periodic potential profiles: (a) the sinusoidal ('Mathieu') potential and (b) the Kronig–Penney potential.

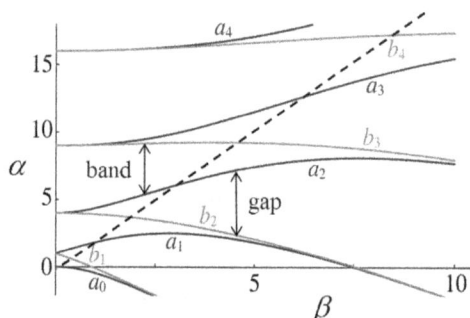

Figure 2.32. Characteristic curves of the Mathieu equation. The dashed line corresponds to the equality $\alpha = 2\beta$, i.e. $E = A \equiv U_{max}$, separating the regions of under-barrier tunneling and over-barrier motion. Adapted from figure 28.2.1 at http://dlmf.nist.gov. (Contribution by US Government, not subject to copyright.)

where $E^{(1)}$ is defined by Eq. (2.221), Eq. (2.227) is reduced to the canonical form of the well-studied *Mathieu equation*[67]

$$\frac{d^2\psi}{d\xi^2} + (\alpha - 2\beta \cos 2\xi)\,\psi = 0. \tag{2.229}$$

(Note that this definition of β is quantitatively different from that for the Dirac comb (2.226), but in both cases this parameter is proportional to the amplitude of the potential modulation.)

Figure 2.32 shows the characteristic curves of this equation. We see that now at small β the first energy gap grows much faster than the higher ones: $\Delta_n \propto \beta^n$. This feature is in accord with the weak-coupling result $\Delta_1 = 2|U_1|$, which is valid only in the linear approximation in U_n, because for the Mathieu potential, $U_l = A(\delta_{l,+1} + \delta_{l,-1})/2$. Another clearly visible feature is the exponentially fast shrinkage of the allowed energy bands at $2\beta > \alpha$ (in figure 2.32, on the right from the dashed line), i.e. at $E < A$. It may be readily explained by our tight-binding approximation result (2.206): as soon as the eigenenergy drops significantly below the potential maximum $U_{max} = A$ (see figure 2.31a), the quantum states in the adjacent potential wells are

[67] This equation, first studied in the 1860s by É Mathieu in the context of a rather practical problem of vibrating elliptical drumheads (!), has many other important applications in physics and engineering, notably including the parametric excitation of oscillations—see, e.g. *Part CM* section 5.5.

Quantum Mechanics: Lecture notes

connected only by tunneling through the high separating potential barriers, so that the coupling amplitudes δ_n become exponentially small—see, e.g. Eq. (2.189).

Another simple periodic profile is the *Kronig–Penney potential*, shown in figure 2.31b, which allows one to get relatively simple analytical expressions for the characteristic curves. Its advantage is a more realistic law of the decrease of the Fourier harmonics U_l at $l \gg 1$, and hence of the energy gaps in the weak-potential limit:

$$\Delta_n \to 2 \, |U_n| \propto \frac{U_0}{n}, \quad \text{at} \quad E \sim E^{(n)} \gg U_0. \tag{2.230}$$

Leaving a detailed analysis of the Kronig–Penney potential for the reader's exercise, let me conclude this section by addressing the effect of the potential modulation on the number of eigenstates in 1D systems of a large but finite length $l \gg a$, k^{-1}. Surprisingly, the Bloch theorem makes the analysis of this problem elementary, for arbitrary $U(x)$. Indeed, let us assume that l is comprised of an integer number of periods a, and its ends are described by the similar boundary conditions—both assumptions evidently inconsequential for $l \gg a$. Then, according to Eq. (2.210), the boundary conditions impose, on the quasi-momentum q, exactly the same quantization condition as we had for k for a free 1D motion. Hence, instead of Eq. (1.100) we can write

$$dN = \frac{l}{2\pi} dq, \tag{2.231}$$

with the corresponding change of the summation rule:

$$\sum_q f(q) \to \frac{l}{2\pi} \int f(q) dk. \tag{2.232}$$

Hence, the density of states in 1D *q-space*, $dN/dq = l/2\pi$, does not depend on the potential profile at all! Note, however, that the profile does affect the density of states *on the energy axis*, dN/dE. As an extreme example, on the bottom and at the top of each energy band we have $dE/dq \to 0$, and hence

$$\frac{dN}{dE} = \frac{dN}{dq} \Big/ \frac{dE}{dq} = \frac{l}{2\pi} \Big/ \frac{dE}{dq} \to \infty. \tag{2.233}$$

This effect of state concentration at the band/gap edges (which survives in higher spatial dimensionalities as well) has important implications for the operation of several electronic and optical devices, in particular semiconductor lasers.

2.8 Periodic systems: particle dynamics

The band structure of the energy spectrum has profound implications not only on the density of states, but also on the dynamics of particles in periodic potentials. Indeed, let us consider the simplest case of a wave packet composed of the Bloch functions (2.210), all belonging to the same (say, nth) energy band. Similarly to Eq. (2.27) for the a free particle, we can describe such a packet as

$$\Psi(x,\,t) = \int a_q u_q(x) e^{i[qx - \omega(q)t]} dq, \tag{2.234}$$

where the a-periodic functions $u(x)$, defined by Eq. (2.208), are now indexed to emphasize their dependence on the quasi-momentum, and $\omega(q) \equiv E_n(q)/\hbar$ is the function of q describing the shape of the corresponding energy band—see, e.g. figure 2.26b or figure 2.28. If the packet is narrow in the q-space, i.e. the width δq of the distribution a_q is much smaller than all the characteristic q-scales of the dispersion relation $\omega(q)$, in particular π/a, we may simplify Eq. (2.234) exactly as was done in section 2.2 for a free particle, despite the presence of the periodic factors $u_q(x)$ under the integral. In the linear approximation of the Taylor expansion, we again get Eq. (2.32), but now with

$$v_{\mathrm{gr}} = \left.\frac{d\omega}{dq}\right|_{q=q_0}, \quad \text{and} \quad v_{\mathrm{ph}} = \left.\frac{\omega}{q}\right|_{q=q_0}, \tag{2.235}$$

where q_0 is the central point of the quasi-momentum distribution. Despite the formal similarity with Eq. (2.33) for the free particle, this result is much more eventful; for example, as evident from the dispersion relation's topology (see figures 2.26a, 2.28), the group velocity vanishes not only at $q = 0$, but at all values of q that are multiples of (π/a), at the bottom and on the top of each energy band. Even more intriguing is that the group velocity's sign changes periodically with q.

This group velocity alternation leads to fascinating, counter-intuitive phenomena if a particle in a periodic potential is the subject of an additional external force $F(t)$. (For electrons in a crystal, this may be, for example, the Lorentz force of the applied electric field.) Let the force be relatively weak, so that the product Fa (i.e. the scale of the energy increment from the additional force per one lattice period) is much smaller than both relevant energy scales of the dispersion relation $E(q)$—see figure 2.26b:

$$Fa \ll \Delta E_n, \, \Delta_n. \tag{2.236}$$

This strong relation allows one to neglect the force-induced interband transitions, so that the wave packet (2.234) includes the Bloch eigenfunctions belonging to only one (initial) energy band at all times. For the time evolution of its center q_0, theory yields[68] an extremely simple equation of motion

$$\dot{q}_0 = \frac{1}{\hbar} F(t). \tag{2.237}$$

This equation is physically very transparent: it is essentially the 2nd Newton law for the time evolution of the quasi-momentum $\hbar q$ under the effect of the additional force $F(t)$ only, excluding the periodic force $-\partial U(x)/\partial x$ of the background potential $U(x)$.

---

[68] The proof of Eq. (2.237) is not difficult, but becomes more compact in the bra-ket formalism, to be discussed in chapter 4. This is why I recommend to the reader its proof as an exercise after reading that chapter. For a generalization of this theory to the case of essential interband transitions see, e.g. section 55 in [8].

This is very natural, because $\hbar q$ is essentially the particle's momentum averaged over the potential's period, and the periodic force effect drops out at such an averaging.

Despite the simplicity of Eq. (2.237), the results of its solution may be highly nontrivial. First, let us use Eqs. (2.235) and (2.237) to find the instant *group acceleration* of the particle (i.e. the acceleration of its wave packet's envelope):

$$
\begin{aligned}
a_{gr} &\equiv \frac{dv_{gr}}{dt} \equiv \frac{d}{dt}\frac{d\omega(q_0)}{dq_0} \\
&\equiv \frac{d}{dq_0}\frac{d\omega(q_0)}{dq_0}\frac{dq_0}{dt} = \frac{d^2\omega(q_0)}{dq_0^2}\frac{dq_0}{dt} = \frac{1}{\hbar}\frac{d^2\omega}{dq^2}\bigg|_{q=q_0} F(t).
\end{aligned}
\tag{2.238}
$$

This means that the second derivative of the dispersion $\omega(q)$ relation (specific for each energy band) plays the role of the effective reciprocal mass of the particle at this particular value of q_0:

$$
m_{ef} = \frac{\hbar}{d^2\omega/dq^2} \equiv \frac{\hbar^2}{d^2E_n/dq^2}.
\tag{2.239}
$$

For the particular case of a free particle, for which Eq. (2.216) is exact, this expression is reduced to the original (and constant) mass m, but generally the effective mass depends on the wave packet's momentum. According to Eq. (2.239), at the bottom of any energy band, m_{ef} is always positive, but depends on the strength of a particle's interaction with the periodic potential. In particular, according to Eq. (2.206), in the tight-binding limit, the effective mass is very large:

$$
|m_{ef}|_{q=(\pi/a)n} = \frac{\hbar^2}{2\delta_n a^2} \equiv m\frac{E^{(1)}}{\pi^2\delta_n} \gg m.
\tag{2.240}
$$

Conversely, in the weak potential limit, the effective mass is close to m at most points of each energy band, but at the edges of the (narrow) bandgaps it is much smaller. Indeed, expanding Eq. (2.224) in the Taylor series near point $q = q_m$, we get

$$
E_\pm|_{E\approx E^{(n)}} - E_{ave} \approx \pm|U_n| \pm \frac{1}{2\,|U_n|}\left(\frac{dE_l}{dq}\right)^2_{q=q_m}\tilde{q}^2 = \pm|U_n| \pm \frac{\gamma^2}{2\,|U_n|}\tilde{q}^2, \tag{2.241}
$$

where γ and $\tilde{q}$ are defined by Eq. (2.225), so that

$$
|m_{ef}|_{q=q_m} = |U_n|\frac{\hbar^2}{\gamma^2} \equiv m\frac{|U_n|}{2E^{(n)}} \ll m.
\tag{2.242}
$$

The effective mass effects in real atomic crystals may be very significant. For example, the charge carriers in silicon have $m_{ef} \approx 0.19\, m_e$ in the lowest, normally-empty energy band (traditionally called the *conduction band*), and $m_{ef} \approx 0.98\, m_e$ in the adjacent lower, normally-filled *valence band*. In some semiconducting compounds the conduction-band electron mass may be even smaller—down to $0.0145\, m_e$ in InSb!

The absolute value of the effective mass is not the most surprising effect. The more fascinating corollary of Eq. (2.239) is that on the top of each energy band the effective mass is *negative*—please revisit figures 2.26b, 2.28, and 2.29 again. This means that the particle (or more strictly its wave packet's envelope) is accelerated in the direction *opposite* to the force. This is exactly what electronics engineers, working with electrons in semiconductors, call *holes*, characterizing them by the *positive mass* $|m_e|$, but compensating this sign change by taking their charge e positive. If the particle stays in close vicinity to the energy band's top (say, due to frequent scattering effects, typical for the semiconductors used in engineering practice), such double sign flip does not lead to an error in calculations of a hole's dynamics, because the Lorentz force is proportional to the particle's charge, so that the particle's acceleration a_{gr} is proportional to the charge-to-mass ratio[69].

However, at some phenomena such simple representation is unacceptable[70]. For example, let us form a narrow wave packet at the bottom of the lowest energy band[71], and then exert on it a constant force $F > 0$—say, due to a constant external electric field directed along axis x. According to Eq. (2.237), this force would lead to a linear growth of q_0 in time, so that in the quasi-momentum space, the packet's center would slide, with a constant speed, along the q axis—see figure 2.33a. Close to the energy band's bottom, this motion would correspond to a positive effective mass (possibly, somewhat different than the genuine particle's mass m), and hence be close to the free particle's acceleration. However, as soon as q_0 has reached the inflection point, where $d^2E_1/dq^2 = 0$, the effective mass, and hence its acceleration (2.238) change signs to negative, i.e. the packet starts to slow down (in the direct space), while still moving ahead in the quasi-momentum space. Finally, at the energy band's top the particle stops at certain x_{max}, while continuing to move in the q-space.

Now we have two alternative ways to look at the further time evolution of the wave packet along the quasi-momentum axis. From the extended zone picture (which is the simplest for this analysis, see figure 2.33a),[72] we may say that the particle crosses the 1st Brillouin zone boundary and continues to go forward in q, i.e. down the lowest energy band. According to Eq. (2.235), this region (up to the next energy minimum at $qa = 2\pi$) corresponds to a negative group velocity. After q_0 has reached that minimum, the whole process repeats again (and again, and again).

---

[69] More discussion of this issue may be found in *Part SM* section 6.4.

[70] The balance of this section describes effects which are not discussed in most quantum mechanics textbooks. Though, in my opinion, every educated physicist should be aware of them, some readers may skip them at the first reading, jumping directly to the next section 2.9.

[71] Physical intuition tells us (and the theory of open systems, to be discussed in chapter 7, confirms) that this may be readily done, for example, by weakly coupling the system to a relatively low-temperature environment, and letting it to relax to the lowest possible energy.

[72] This phenomenon may be also discussed from the point of view of the reduced zone picture, but then it requires the introduction of instant jumps between the Brillouin zone boundary points (see the dashed red line in figure 2.33) that correspond to physically equivalent states of the particle. Evidently, for the description of this particular phenomenon, this language is more artificial.

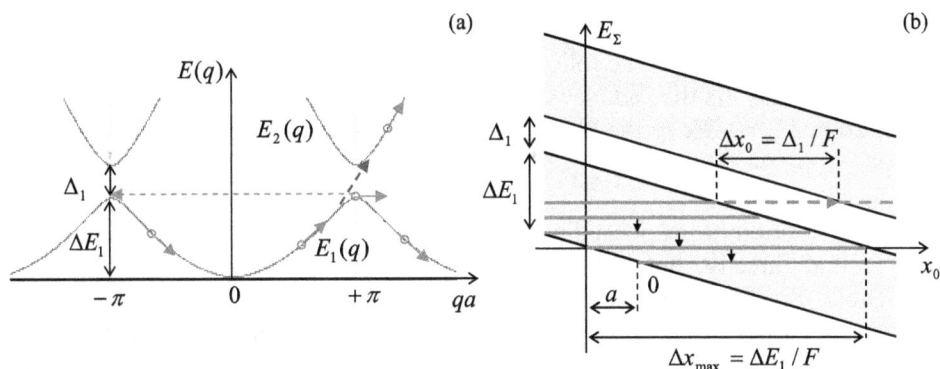

Figure 2.33. The Bloch oscillations (red lines) and the Landau–Zener tunneling (blue arrows) represented in: (a) the reciprocal space of q, and (b) the direct space. On panel (b), the tilted gray strips show the allowed energy bands, while the bold red line.

These are the famous *Bloch oscillations*—the effect which was predicted, by the same F Bloch, as early as in 1929, but evaded experimental observation until the 1980s (see below) due to the strong scattering effects in solid-state crystals. The time period of the oscillations may be readily found from Eq. (2.237):

$$\Delta t_B = \frac{\Delta q}{dq/dt} = \frac{2\pi/a}{F/\hbar} = \frac{2\pi\hbar}{Fa},\tag{2.243}$$

so that their frequency is expressed by a very simple formula

$$\omega_B \equiv \frac{2\pi}{\Delta t_B} = \frac{Fa}{\hbar},\tag{2.244}$$

and hence is independent of any peculiarities of the energy band/gap structure.

The direct-space motion of the wave packet's center $x_0(t)$ during the Bloch oscillation process may be analyzed by integrating the first of Eqs. (2.235) over some time interval Δt, again using Eq. (2.237):

$$\begin{aligned}\Delta x_0(t) &\equiv \int_0^{\Delta t} v_{gr}dt \\ &= \int_0^{\Delta t}\frac{d\omega(q_0)}{dq_0}dt = \int_0^{\Delta t}\frac{d\omega(q_0)}{dq_0/dt} = \frac{\hbar}{F}\int_{t=0}^{t=\Delta t}d\omega = \frac{\hbar}{F}\Delta\omega(q_0).\end{aligned}\tag{2.245}$$

If the interval Δt is equal to the Bloch oscillation period Δt_B (2.243), the initial and final values of $E(q_0) = \hbar\omega(q_0)$ are equal, giving $\Delta x_0 = 0$: in the end of the period, the wave packet returns to its initial position in space. However, if we carry this integration only from the smallest to the largest values of $\omega(q_0)$, i.e. the adjacent points where the group velocity vanishes, we get the oscillation swing

$$\Delta x_{max} = \frac{\hbar}{F}(\omega_{max} - \omega_{min}) \equiv \frac{\Delta E_1}{F}.\tag{2.246}$$

This simple result may be interpreted using an alternative energy diagram (figure 2.33b), which results from the following arguments. The additional force F may be described not only via the 2nd Newton law's version (2.237), but, alternatively, by its contribution $U_F = -Fx$ to the Gibbs potential energy[73]

$$U_\Sigma(x) = U(x) - Fx \qquad (2.247)$$

The direct solution of the Schrödinger equation (2.61) with such a potential may be hard to find directly, but if the force is sufficiently weak, as we are assuming throughout this discussion, the second term in Eq. (2.247) may be considered as a constant on the scale of $a \ll \Delta x_{max}$. In this case, our quantum-mechanical treatment of the periodic potential $U(x)$ is still almost correct, but with an energy shift depending on the position x_0 of the packet's center. In this approximation, the total energy of the wave packet is

$$E_\Sigma = E(q_0) - Fx_0. \qquad (2.248)$$

In a plot of such energy as a function of x_0 (figure 2.33b), the energy dependence on q_0 is hidden, but as was discussed above it is rather uneventful, and may be well characterized by the position of band-gap edges on the energy axis[74]. In this representation, the Bloch oscillations keep the full energy E_Σ constant, i.e. follow a horizontal line in figure 2.33b, limited by the classical turning points corresponding to the bottom and the top of the allowed energy band. The distance Δx_{max} between these points is evidently given by Eq. (2.246).

Besides this second look at the oscillation swing result, the total energy diagram shown in figure 2.33b enables one more remarkable result. Let a wave packet be so narrow in the momentum space that $\delta x \sim 1/\delta q \gg \Delta x_{max}$; then the horizontal line segment in figure 2.33b represents the spatial extension of the eigenfunction of the Schrödinger equation with the potential (2.247). But this equation is exactly invariant with respect to the following simultaneous translation of the coordinate and the energy:

$$x \rightarrow x + a, \quad E \rightarrow E - Fa. \qquad (2.249)$$

This means that it is satisfied by an infinite set of similar solutions, each corresponding to one of the horizontal red lines shown in figure 2.33b. This is the famous *Wannier–Stark ladder*[75], with the step height

$$\Delta E_{WS} = Fa. \qquad (2.250)$$

---

[73] Physically, this is just the relevant part of the potential energy of the total system comprised of our particle (in the periodic potential) and the source of the force F—see, e.g. *Part CM* section 1.4.

[74] In semiconductor physics and engineering, such plots are called the *bandedge diagrams*, and are the virtually unavoidable components of any discussion/publication. In this series, a few examples of such diagrams may be found in *Part SM* section 6.4.

[75] It was first discussed in detail by G Wannier in his 1959 monograph, while the name of J Stark is associated with virtually any effect of electric field on atomic systems, after he had discovered the first of them in 1913.

The importance of this alternative representation of the Bloch oscillations is due to the following fact. In most experimental realizations, the power of the electromagnetic radiation with the frequency (2.244), that may be extracted from the oscillations of a charged particle, is very low, so that their direct detection represents a hard problem[76]. However, let us apply to a Bloch oscillator an additional ac field at frequency $\omega \sim \omega_B$. As these frequencies are brought close together, the external signal should synchronize ('phase lock') the Bloch oscillations[77], resulting in certain changes of time-independent observables—for example, a resonant change of absorption of the external radiation. Now let us notice that the combination of Eqs. (2.244) and (2.250) yield the following simple relation:

$$\Delta E_{WS} = \hbar \omega_B. \tag{2.251}$$

This means that the phase locking at $\omega \approx \omega_B$ allows for an alternative (but equivalent) interpretation—as the result of ac-field-induced quantum transitions[78] between the steps of the Wannier–Stark ladder. (Again, such occasions when two very different languages may be used for alternative interpretations of the same phenomenon is one of the most beautiful features of physics.)

This phase-locking effect has been used for first experimental confirmations of the Bloch oscillation theory[79]. For this purpose, the natural periodic structures, solid state crystals, are inconvenient due to their very small period $a \sim 10^{-10}$ m. Indeed, according to Eq. (2.244), such structures require very high forces F (and hence very high electric fields $\mathscr{E} = F/e$) to bring ω_B to an experimentally convenient range. This problem has been overcome using artificial periodic structures (*superlattices*) of certain semiconductor compounds, such as $Ga_{1-x}Al_xAs$ with various degrees x of the gallium-to-aluminum atom replacement, whose layers may be grown over each other epitaxially, i.e. with very few crystal structure violations. These superlattices, with periods $a \sim 10$ nm, have enabled a clear observation of the resonance at $\omega \approx \omega_B$, and hence the measurement of the Bloch oscillation frequency, in particular its proportionality to the applied dc electric field, predicted by Eq. (2.244).

Very soon after this discovery, the Bloch oscillations were observed[80] in small Josephson junctions, where they result from the quantum dynamics of the Josephson phase difference φ in a 2π-periodic potential profile, created by the junction. A straightforward translation of Eq. (2.244) to this case (left for the reader's exercise) shows that the frequency of such Bloch oscillations is

$$\omega_B = \frac{\pi \bar{I}}{2e}, \quad \text{i.e.} \, f_B \equiv \frac{\omega_B}{2\pi} = \frac{\bar{I}}{2e}, \tag{2.252}$$

---

[76] In systems with many independent particles (such as electrons in semiconductors), the detection problem is exacerbated by the phase incoherence of the Bloch oscillations performed by each particle. This drawback is absent in atomic Bose–Einstein condensates whose Bloch oscillations (in a periodic potential created by standing optical waves) were eventually observed by M Ben Dahan *et al* [9].

[77] A simple analysis of phase locking of a classical oscillator may be found, e.g. in *Part CM* section 5.4. (See also the brief discussion of the phase locking of the Josephson oscillations in the end of section 1.6 of this volume.)

[78] A qualitative theory of such transitions will be discussed in section 6.6 and then in chapter 7.

[79] E Mendez *et al* [10].

[80] D Haviland *et al* [11].

where $\bar{I}$ is the dc current passed through the junction—the effect not to be confused with the 'classical' Josephson oscillations with frequency (1.75). It is curious that Eq. (2.252) may be legitimately interpreted as a result of a periodic transfer, through the Josephson junction, of discrete Cooper pairs (of charge $-2e$), between two coherent Bose–Einstein condensates in the superconducting electrodes of the junction[81].

The Bloch oscillation discussion above was based on the premise that the wave packet of the particle stays within one (say, the lowest) energy band. However, just one look at figure 2.28 shows that this assumption becomes unrealistic if the energy gap separating this band from the next one becomes very small, $\Delta_1 \to 0$. Indeed, in the weak-potential approximation, which is adequate in this limit, $|U_1| \to 0$, the two dispersion curve branches (2.216) cross without any interaction, so that if our particle (meaning its wave packet) is driven to approach that point, it should continue to move up in energy—see the dashed blue arrow in figure 2.33a. Similarly, in the real-space representation shown in figure 2.33b, it is intuitively clear that at $\Delta_1 \to 0$, the particle residing at one of the steps of the Wannier–Stark ladder should be able to somehow overcome the vanishing spatial gap $\Delta x_0 = \Delta_1/F$ and to leak into the next band—see the horizontal dashed blue arrow on that panel.

This process, called the *Landau–Zener* (or 'interband', or 'band-to-band') *tunneling*[82] is indeed possible. In order to analyze it, let us first take $F = 0$, and consider what happens if a quantum particle, described by an x-long (i.e. E-narrow) wave packet, is incident from the free space upon a periodic structure of a large but finite length $l = Na \gg a$—see, e.g. figure 2.22. If the packet's energy E is within one of the energy bands, it may evidently propagate through the structure (though may be partly reflected from its ends). The corresponding quasi-momentum may be found by solving the dispersion relation for q; for example, in the weak-potential limit, Eq. (2.224), valid near the gap, yields

$$q = q_m + \tilde{q}, \quad \text{with}$$

$$\tilde{q} = \pm \frac{1}{\gamma} [\tilde{E}^2 - |U_n|^2]^{1/2}, \quad \text{for } |U_n|^2 \leqslant \tilde{E}^2, \quad \text{where } \tilde{E} \equiv E_{\pm} - E^{(n)}, \quad (2.253)$$

and $\gamma = 2aE^{(n)}/\pi n$—see the second of Eqs. (2.225).

Now, if the energy E is inside one of the energy gaps Δ_n, the wave packet's propagation in an infinite periodic lattice is impossible, so that it is completely reflected back from it. However, our analysis of the potential step problem in section 2.3 implies that the packet's wavefunction should still have an exponential tail protruding into the structure and decaying on some length δ—see Eq. (2.58) and figure 2.4. Indeed, a review of the calculation leading to Eq. (2.253) shows that it remains valid for energies within the gap as well, if the quasi-momentum is understood as a purely imaginary number:

[81] See, e.g. D Averin *et al* [12]. This effect is qualitatively similar to the transfer of single electrons, with the similar frequency $f = I/e$, in tunnel junctions between normal (non-superconducting) metals—see, e.g. *Part EM* section 2.9 and references therein.

[82] It was predicted independently by L Landau and C Zener in 1932.

$$\tilde{q} \to \pm i\kappa, \quad \text{where} \quad \kappa \equiv \frac{1}{\gamma}[|U_n|^2 - \tilde{E}^2]^{1/2}, \quad \text{for } \tilde{E}^2 \leqslant |U_n|^2. \tag{2.254}$$

With this contribution, the Bloch solution (2.193b) indeed describes an exponential decay of the wavefunction at length $\delta \sim 1/\kappa$.

Returning to the effects of weak force F in the real-space approach, represented by Eq. (2.248) and illustrated in figure 2.33b, we may recast Eq. (2.254) as

$$\kappa \to \kappa(x) = \frac{1}{\gamma}[|U_n|^2 - (F\tilde{x})^2]^{1/2}, \tag{2.255}$$

where $\tilde{x}$ is the particle's (i.e. the wave packet center's) deviation from the mid-gap point. Thus the gap creates a potential barrier of a finite width $\Delta x_0 = 2|U_n|/F$, through which the wave packet may tunnel with a non-zero probability. As we already know, in the WKB approximation (in our case requiring $\kappa\Delta x_0 \gg 1$) this probability is just the potential barrier's transparency $\mathscr{T}$, which may be calculated from Eq. (2.117):

$$-\ln\mathscr{T} = 2\int_{\kappa(x)^2>0} \kappa(x)dx$$
$$= \frac{2}{\gamma}\int_{-x_c}^{x_c}[|U_n|^2 - (F\tilde{x})^2]^{1/2}d\tilde{x} = \frac{2|U_n|}{\gamma}2x_c\int_0^1(1-\xi^2)^{1/2}d\xi. \tag{2.256}$$

where $\pm x_c \equiv \pm\Delta x_0/2 = \pm|U_n|/F$ are the classical turning points. Working out this simple integral (which may be viewed upon as the quarter of the unit circle's area, and hence equal to $\pi/4$), we get

$$\mathscr{T} = \exp\left\{-\frac{\pi|U_n|^2}{\gamma F}\right\}. \tag{2.257}$$

This famous result was obtained by Landau and Zener in a more complex way, whose advantage is a constructive proof that Eq. (2.257) is valid for an arbitrary relation between γF and $|U_n|^2$, i.e. arbitrary $\mathscr{T}$, while our simple derivation was limited to the WKB approximation, valid only at $\mathscr{T} \ll 1$.[83] Returning to Eqs. (2.225) and (2.237), we can rewrite the product γF, participating in Eq. (2.257), as

$$\gamma F = \frac{1}{2}\left|\frac{d(E_l - E_{l'})}{dq_0}\right|_{E_l=E_{l'}=E^{(n)}} \hbar\frac{dq_0}{dt} = \frac{\hbar}{2}\left|\frac{d(E_l - E_{l'})}{dt}\right|_{E_l=E_{l'}=E^{(n)}} \equiv \frac{\hbar u}{2}, \tag{2.258}$$

where u has the meaning of the 'speed' of the energy level crossing in the absence of the gap. Hence, Eq. (2.257) may be rewritten in the form

$$\mathscr{T} = \exp\left\{-\frac{2\pi|U_n|^2}{\hbar u}\right\}, \tag{2.259}$$

---

[83] Note that Eq. (2.257) is limited to the hyperbolic dispersion relation, i.e. (in the band theory) to the weak-potential limit. In chapter 6, it will be derived using a different method, based on the so-called *Golden Rule* of quantum mechanics.

which is more physically transparent. Indeed, the fraction $2|U_n|/u = \Delta_n u$ gives the time scale Δt of the energy's crossing the gap region, and according to the Fourier transform, its reciprocal, $\omega_{\max} \sim 1/\Delta t$ gives the upper cutoff of the frequencies involved in the Bloch oscillation process. Hence Eq. (2.259) means that

$$- \ln \mathscr{T} \approx \frac{\Delta_n}{\hbar \omega_{\max}}. \tag{2.260}$$

This formula allows us to interpret the Landau–Zener tunneling as the system's excitation across the energy gap Δ_n, by the maximum energy quantum $\hbar\omega_{\max}$ available from the Bloch oscillation process. This interpretation remains valid even in the opposite, tight-binding limit, in which, according to Eqs. (2.206) and (2.237), the Bloch oscillations are purely sinusoidal, so that the Landau–Zener tunneling is completely suppressed at $\hbar\omega_B < \Delta_1$.

The interband tunneling is an important ingredient of several physical phenomena and even some practical electron devices, for example the *tunneling* (or 'Esaki') *diodes*. This simple device is just a junction of two semiconductor electrodes, one of them so strongly *n*-doped by electron donors that the additional electrons form a degenerate Fermi gas at the bottom of the conduction band. Similarly, the counterpart semiconductor electrode is *p*-doped so strongly that the Fermi level in the valence band is shifted below the band edge (figure 2.34).[84]

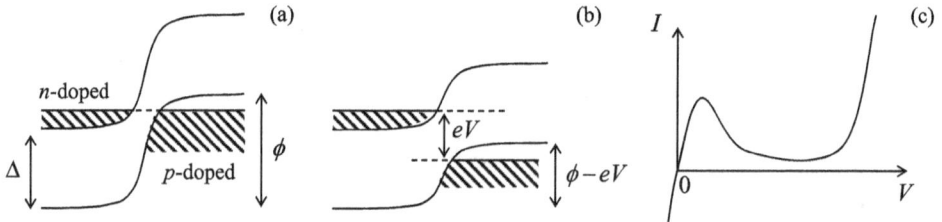

Figure 2.34. The tunneling ('Esaki') diode: (a) the band edge diagram of the device at zero bias; (b) the same diagram at a modest positive bias $eV \sim \Delta/2$, and (c) the *I–V* curve (schematically). Dashed lines show the Fermi level positions.

In thermal equilibrium, and in the absence of an external voltage bias, the Fermi levels self-align, leading to the build-up of the *contact potential difference* ϕ/e, with ϕ somewhat larger than the energy bandgap Δ—see figure 2.34a. This potential difference creates an internal electric field that tilts the energy bands (just as the external field did in figure 2.33b), and leads to the formation of the so-called *depletion layer*, in which the Fermi level is located within the energy gap and hence there are no charge carriers ready to move. In usual *p–n* junctions, this layer is broad and prevents any current at applied voltages V lower than $\sim\Delta/e$. In contrast, in a tunneling diode the depletion layer is so thin (below $\sim$10 nm) that the interband tunneling is possible and provides a substantial Ohmic current at small applied voltages—see figure 2.34c. However, at a substantial positive bias, $eV \sim \Delta/2$, the conduction band becomes

---

[84] Here I have to rely on the reader's knowledge of elementary properties of semiconductors; they will be discussed in more detail in *Part SM* section 6.4.

aligned with the middle of the energy gap in the p-doped electrode, and electrons cannot tunnel there. Similarly, there are no electrons in the n-doped semiconductor to tunnel into the available states just above the Fermi level in the p-doped electrode—see figure 2.34b. As a result, at such voltages the current drops significantly, to grow again only when eV exceeds $\sim\Delta$, enabling electron motion within each energy band. Thus the tunnel junction's I–V curve has a part with a *negative differential resistance* ($dV/dI < 0$)—see figure 2.34c. This effect, equivalent in its effect to a negative kinematic friction in mechanics, may be used for the amplification of analog signals, including the self-excitation of electronic oscillators (i.e. ac signal generation)[85], and the signal swing restoration in digital electronics.

2.9 Harmonic oscillator: brute force approach

To complete our review of the basic 1D wave mechanics, we have to consider the famous harmonic oscillator, i.e. a 1D particle moving in the quadratic-parabolic potential (2.111), so that the stationary Schrödinger equation (2.53) is

$$-\frac{\hbar^2}{2m}\frac{d^2\psi}{dx^2} + \frac{m\omega_0^2 x^2}{2}\psi = E\psi. \tag{2.261}$$

Conceptually, on the background of the fascinating quantum effects discussed in the previous sections, this is not a very interesting system: Eq. (2.261) is a standard 1D eigenproblem, resulting in a discrete energy spectrum E_n, with smooth eigenfunctions $\psi_n(x)$ vanishing at $x \to \pm\infty$ (because the potential energy tends to infinity there)[86]. However, as we will repeatedly see later in the course, the problem's solutions have an enormous range of applications, so we have to know their basic properties.

The direct analytical solution of the problem is not very simple (see below), so let us start with trying some indirect approaches to it. First, as was discussed in section 2.4, the WKB-approximation-based Bohr–Sommerfeld quantization rule (2.110), applied to this potential, yields the eigenenergy spectrum (2.114). With the common quantum number convention, this result is

$$E_n = \hbar\omega_0\left(n + \frac{1}{2}\right), \quad \text{with } n = 0, 1, 2, \ldots, \tag{2.262}$$

so that (in contrast to the 1D rectangular potential well) the ground-state energy corresponds to $n = 0$. However, as was discussed in the end of section 2.4, for the quadratic potential (2.111) the WKB approximation's conditions are strictly satisfied only at $E_n \gg \hbar\omega_0$, so that so far we can only trust Eq. (2.262) for high levels, with $n \gg 1$, rather than for the (most important) ground state.

Consequently, let us apply to Eq. (2.261) another approximate approach, called the *variational method*, whose simplest form is aimed exactly at characterizing the ground states. The method is based on the following observation. (Here I am

---

[85] See, e.g. *Part CM* section 5.4.
[86] The stationary states of the harmonic oscillator are sometimes called its *Fock states*, to distinguish them from other fundamental *Glauber states*, which will be discussed in section 5.5.

presenting its 1D wave mechanics form, though the method is much more general.) Let ψ_n be the exact, full and orthonormal set of stationary wavefunctions of the system under study, and E_n the set of corresponding energy levels, satisfying Eq. (1.60):

$$\hat{H}\psi_n = E_n\psi_n. \tag{2.263}$$

Then we may use this set for the unique expansion of an arbitrary *trial wavefunction* ψ_{trial}:

$$\psi_{\text{trial}} = \sum_n \alpha_n\psi_n, \quad \text{so that} \quad \psi_\alpha^* = \sum_n \alpha_n^*\psi_n^*, \tag{2.264}$$

where α_n are some (generally, complex) coefficients. Let us require the trial function to be normalized, using the condition (1.66) of orthonormality of the eigenfunctions ψ_n:

$$\int \psi_{\text{trial}}^*\psi_{\text{trial}}d^3x \equiv \sum_{n,n'} \int \alpha_n^*\psi_n^*\alpha_n\psi_n d^3x$$

$$\equiv \sum_{n,n'} \alpha_{n'}\alpha_n^* \int \psi_n^*\psi_n d^3x \equiv \sum_{n,n'} \alpha_{n'}\alpha_n^*\delta_{n,\,n'} \equiv \sum_n W_n = 1, \tag{2.265}$$

where each of the coefficients W_n, defined as

$$W_n \equiv \alpha_n^*\alpha_n \equiv |\alpha_n|^2 \geqslant 0, \tag{2.266}$$

may be interpreted as the probability for the particle, in the trial state, to be found in the nth stationary state. Now let us use Eq. (1.23) for the absolutely similar calculation of the expectation value of the system's Hamiltonian in the trial state[87]:

$$\langle H \rangle_{\text{trial}} = \int \psi_{\text{trial}}^*\hat{H}\psi_{\text{trial}}d^3x \equiv \sum_{n,n'} \int \alpha_n^*\psi_n^*\hat{H}\alpha_n\psi_n d^3x$$

$$\equiv \sum_{n,n'} \alpha_{n'}\alpha_n^*E_{n'} \int \psi_n^*\psi_n d^3x \tag{2.267}$$

$$\equiv \sum_{n,n'} \alpha_{n'}\alpha_n^*E_n\delta_{n,\,n'} \equiv \sum_n W_nE_n.$$

Since the exact ground state energy E_g is, by definition, the lowest one of the set E_n, i.e. $E_n \geqslant E_g$, Eqs. (2.265) and (2.267) yield the following inequality:

$$\langle H \rangle_{\text{trial}} \geqslant \sum_n W_nE_g \equiv E_g\sum_n W_n = E_g. \tag{2.268}$$

Thus, the genuine ground state energy of the system is always lower then (or equal to) its energy in any trial state. Hence, if we make several attempts with reasonably

---

[87] It is easy (and hence left for the reader) to show that the uncertainty δH in *any* state of a Hamiltonian system, including the trial state (2.264), vanishes, so that the $\langle H \rangle_{\text{trial}}$ may be interpreted as the definite energy of the state. For our current goals, however, this fact is not important.

selected trial states, we may expect the lowest of the results to approximate the genuine ground state energy reasonably well. Even more conveniently, if we select some reasonable class of trial wavefunctions dependent on a free parameter λ, then we may use the necessary condition of the minimum,

$$\frac{\partial \langle H \rangle_{\text{trial}}}{\partial \lambda} = 0, \qquad (2.269)$$

to find the closest of them to the genuine ground state. Even better results may be obtained using trial wavefunctions dependent on several parameters. Note, however, that the variational method does not tell us how exactly the trial function should be selected, or how close its final result is to the genuine ground-state function. In this sense, this method has 'uncontrollable accuracy', and differs from both the WKB approximation and the perturbation methods (to be discussed in chapter 6), for which we have certain accuracy criteria. Because of this drawback, the variational method is typically used as the last resort—though sometimes (as in the example below) it works remarkably well[88].

Let us apply this method to the harmonic oscillator. Since the potential (2.111) is symmetric with respect to point $x = 0$, and continuous at all points (so that, according to Eq. (2.261), $d^2\psi/dx^2$ has to be continuous as well), the most natural selection of the ground-state trial function is the Gaussian function

$$\psi_{\text{trial}}(x) = C \exp\{-\lambda x^2\}, \qquad (2.270)$$

with some real $\lambda > 0$. The normalization coefficient C may be immediately found either from the standard Gaussian integration of $|\psi_{\text{trial}}|^2$, or just from the comparison of this expression with Eq. (2.16), in which $\lambda = 1/4(\delta x)^2$, i.e. $\delta x = 1/2\lambda^{1/2}$, giving $|C|^2 = (2\lambda/\pi)^{1/2}$. Now the expectation value of the particle's Hamiltonian,

$$\hat{H} = \frac{\hat{p}^2}{2m} + U(x) = -\frac{\hbar^2}{2m}\frac{d^2}{dx^2} + \frac{m\omega_0^2 x^2}{2}, \qquad (2.271)$$

in the trial state, may be calculated as

$$\begin{aligned}
\langle H \rangle_{\text{trial}} &\equiv \int_{-\infty}^{+\infty} \psi_{\text{trial}}^* \left(-\frac{\hbar^2}{2m}\frac{d^2}{dx^2} + \frac{m\omega_0^2 x^2}{2} \right) \psi_{\text{trial}} dx \\
&= \left(\frac{2\lambda}{\pi}\right)^{1/2} \left[\frac{\hbar^2\lambda}{m} \int_0^\infty \exp\{-2\lambda x^2\} dx \right. \\
&\quad \left. + \left(\frac{m\omega_0^2}{2} - \frac{2\hbar^2\lambda^2}{m}\right) \int_0^\infty x^2 \exp\{-2\lambda x^2\} dx \right].
\end{aligned} \qquad (2.272)$$

Both involved integrals are of the same well-known Gaussian type[89], giving

---

[88] Note that the variational method may be used also to estimate the first excited state (or even a few lowest excited states) of the system, by requiring the new trial function to be orthogonal to the previously calculated eigenfunctions of the lower-energy states, though the method's error typically grows with the state number.
[89] See, e.g. Eqs. (A.36b) and (A.36c).

$$\langle H \rangle_{\text{trial}} = \frac{\hbar^2}{2m}\lambda + \frac{m\omega_0^2}{8\lambda}. \tag{2.273}$$

As a function of λ, this expression has a single minimum at the value λ_{opt} that may be found from the requirement (2.269), giving $\lambda_{\text{opt}} = m\omega_0/2\hbar$. The resulting minimum of $\langle H \rangle_{\text{trial}}$ is *exactly* equal to ground-state energy following from Eq. (2.262),

$$E_0 = \frac{\hbar\omega_0}{2}. \tag{2.274}$$

Such a coincidence of the WKB and the variational-method results is very unusual. It implies (though does not strictly prove) that Eq. (2.274) is exact. As a minimum, this coincidence gives a strong motivation to plug the trial wavefunction (2.270), with $\lambda = \lambda_{\text{opt}}$, i.e.

$$\psi_0 = \left(\frac{m\omega_0}{\pi\hbar}\right)^{1/4} \exp\left\{-\frac{m\omega_0 x^2}{2\hbar}\right\}, \tag{2.275}$$

and the energy (2.274), into the Schrödinger equation (2.261). Such substitution[90] shows that the equation is indeed exactly satisfied.

According to Eq. (2.275), the characteristic scale of the wavefunction's spatial spread[91] is

$$x_0 \equiv \left(\frac{\hbar}{m\omega_0}\right)^{1/2}. \tag{2.276}$$

Due to the importance of this scale, let us give its crude estimates for several representative systems:

(i) For atom-bound electrons in solids and fluids, $m \sim 10^{-30}$ kg, and $\omega_0 \sim 10^{15}$ s^{-1}, giving $x_0 \sim 0.3$ nm, of the order of the typical inter-atomic distances in condensed matter. As a result, classical mechanics is not valid at all for the analysis of their motion.

(ii) For atomic nuclei in solids, $m \approx 10^{-24}$–10^{-26} kg, and $\omega_0 \sim 10^{13}$ s^{-1}, giving $x_0 \sim 0.01$–0.1 nm, i.e. somewhat smaller than inter-atomic distances. Because of that, the methods based on classical mechanics (e.g. molecular dynamics) are approximately valid for the analysis of atomic motion, though they may miss some fine effects exhibited by lighter atoms—e.g. the so-called *quantum diffusion* of hydrogen atoms, due to their tunneling through the energy barriers of the potential profile created by other atoms.

(iii) Recently, the progress of patterning technologies has enabled the fabrication of high-quality *micromechanical* oscillators consisting of zillions of atoms. For example, the oscillator used in one of the pioneering experiments in this field[92] was a ~1 μm thick membrane with a 60 μm diameter, and had $m \sim 2 \times 10^{-14}$ kg and

---

[90] Actually, this is a twist on one of the tasks of problem 1.12.
[91] Quantitatively, as was already mentioned in section 2.1, $x_0 = \sqrt{2}\delta x = \langle 2x^2 \rangle^{1/2}$.
[92] A O'Connell *et al* [13].

$\omega_0 \sim 3 \times 10^{10}$ s^{-1}, so that $x_0 \sim 4 \times 10^{-16}$ m. It is remarkable that despite such extreme smallness of x_0 (much smaller than not only any atom, but even any atomic nucleus!), quantum states of such oscillators may be manipulated and measured, using their coupling to electromagnetic (in particular, optical) resonant cavities[93].

Returning to the Schrödinger equation (2.261), in order to analyze its higher eigenstates, we need some help from mathematics. Let us recast this equation into a dimensionless form by introducing the dimensionless variable $\xi \equiv x/x_0$. This gives

$$-\frac{d^2\psi}{d\xi^2} + \xi^2\psi = \varepsilon\psi, \tag{2.277}$$

where $\varepsilon \equiv 2E/\hbar\omega_0 = E/E_0$. In this notation, the ground-state wavefunction (2.275) is proportional to $\exp\{-\xi^2/2\}$. Using this clue, let us look for the solutions to Eq. (2.277) in the form

$$\psi = C\exp\left\{-\frac{\xi^2}{2}\right\}H(\xi), \tag{2.278}$$

where $H(\xi)$ is a new function. With this substitution, Eq. (2.277) yields

$$\frac{d^2H}{d\xi^2} - 2\xi\frac{dH}{d\xi} + (\varepsilon - 1)H = 0. \tag{2.279}$$

It is evident that $H = $ const and $\varepsilon = 1$ is one of its solutions, describing the ground-state eigenfunction (2.275) and energy (2.274), but what are the other eigenstates and eigenvalues? Fortunately, the linear differential equation (2.274) was studied in detail in the mid-1800s by C Hermite who has shown that all its eigenvalues are given by the set

$$\varepsilon_n - 1 = 2n, \quad \text{with} \quad n = 0, 1, 2, \ldots, \tag{2.280}$$

so that Eq. (2.262) is indeed exact for any n.[94] The eigenfunction of Eq. (2.279), corresponding to the eigenvalue ε_n, is a polynomial (called the *Hermite polynomial*) of degree n, which may be most conveniently calculated using the following explicit formula:

$$H_n = (-1)^n \exp\{\xi^2\}\frac{d^n}{d\xi^n}\exp\{-\xi^2\}. \tag{2.281}$$

It is easy to use this formula to spell out several lowest-degree polynomials—see figure 2.35a:

$$H_0 = 1, \quad H_1 = 2\xi, \quad H_2 = 4\xi^2 - 2, \quad H_3 = 8\xi^3 - 12\xi,$$
$$H_4 = 16\xi^4 - 48\xi^2 + 12, \ldots \tag{2.282}$$

---

[93] For a recent review of such experiments, see M Aspelmeyer *et al* [14].
[94] Perhaps the most important property of this energy spectrum is that it is equidistant: $E_{n+1} - E_n = \hbar\omega_0 = $ const.

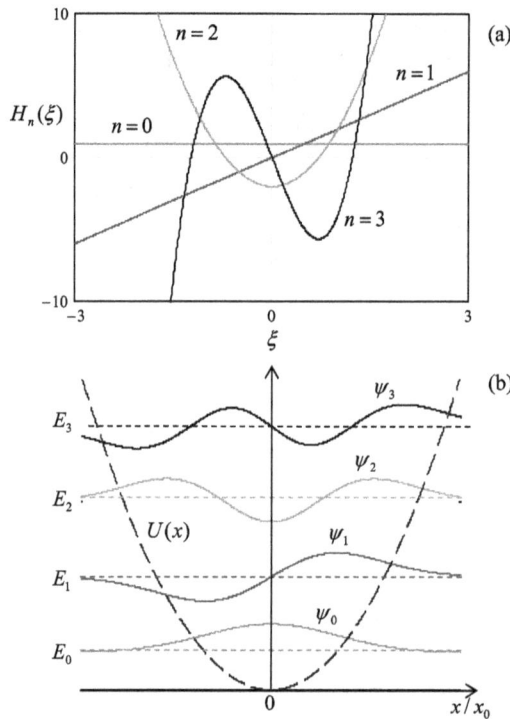

Figure 2.35. (a) A few lowest Hermite polynomials and (b) the corresponding eigenenergies (horizontal dashed lines) and eigenfunctions (solid lines) of the harmonic oscillator. The black dashed curve shows the potential profile $U(x)$, drawn on the same scale as the energies E_n, so that its crossings with the energy levels correspond to the classical turning points.

The properties of these polynomials, most important for applications, are as follows:

(i) the function $H_n(\xi)$ has exactly n zeros (crosses the ξ-axis exactly n times); as a result, the 'parity' (symmetry–antisymmetry) of these functions alternates with n, and

(ii) the polynomials are mutually orthonormal in the following sense:

$$\int_{-\infty}^{+\infty} H_n(\xi)H_{n'}(\xi)\exp\{-\xi^2\}d\xi = \pi^{1/2}2^n n! \delta_{n,n'}. \tag{2.283}$$

Using the last property, we may calculate, from Eq. (2.278), the normalized eigenfunctions $\psi_n(x)$ of the harmonic oscillator—see figure 2.35b:

$$\psi_n(x) = \frac{1}{(2^n n!)^{1/2}\pi^{1/4}x_0^{1/2}}\exp\left\{-\frac{x^2}{2x_0^2}\right\}H_n\left(\frac{x}{x_0}\right). \tag{2.284}$$

It is very instructive to compare these eigenfunctions with those of a 1D rectangular potential well, with its ultimately-hard walls—see figure 1.8. Let us list their similar features:

(i) The wavefunctions oscillate in the classically-allowed regions with $E_n > U(x)$, while dropping exponentially beyond the boundaries of that region. (For the rectangular well with infinite walls, the latter regions are infinitesimally narrow.)

(ii) Each step up of the energy level ladder increases the number of the oscillation half-waves (and hence the number of its zeros), by one[95].

Here are the major features specific for the soft (e.g. the quadratic—parabolic) confinement:

(i) The spatial spread of the wavefunction grows with n, following the gradual widening of the classically allowed region.

(ii) Correspondingly, E_n exhibits a slower growth than the $E_n \propto n^2$ law given by Eq. (1.85), because the gradual reduction of the spatial confinement moderates the growth of the kinetic energy.

Unfortunately, the 'brute-force' approach to the harmonic oscillator problem, discussed above, is not too appealing. First, the proof of Eq. (2.281) is rather longish—so I do not have time/space for it. More importantly, it is hard to use Eq. (2.284) for the calculation of the expectation values of observables, and the so-called *matrix elements* of the system—as we will see in chapter 4, virtually the only numbers important for applications. Finally, it is also almost evident that there has to be some straightforward math leading to any formula as simple as Eq. (2.262) for E_n. Indeed, there is a much more efficient, operator-based approach to this problem; it will be described in section 5.4.

2.10 Problems

Problem 2.1. The initial wave packet of a free 1D particle is described by Eq. (2.20) of the lecture notes:

$$\Psi(x, 0) = \int a_k e^{ikx} dk.$$

(i) Obtain a compact expression for the expectation value $\langle p \rangle$ of the particle's momentum. Does $\langle p \rangle$ depend on time?
(ii) Calculate $\langle p \rangle$ for the case when the function $|a_k|^2$ is symmetric with respect to some value k_0.

Problem 2.2. Calculate the function a_k, defined by Eq. (2.20), for the wave packet with a rectangular spatial envelope:

---

[95] In mathematics, a slightly more general statement, valid for a broader class of ordinary linear differential equations, is frequently called the *Sturm oscillation theorem*, and is a part of the *Sturm–Liouville theory* of such equations—see, e.g. chapter 10 in the handbook by G Arfken *et al*, recommended in section A.16.

$$\Psi(x, 0) = \begin{cases} C \exp\{ik_0 x\}, & \text{for } -a/2 \leqslant x \leqslant +a/2, \\ 0, & \text{otherwise.} \end{cases}$$

Analyze the result in the limit $k_0 a \to \infty$.

Problem 2.3. Prove Eq. (2.49) for the 1D propagator of a free quantum particle, starting from Eq. (2.48).

Problem 2.4. Express the 1D propagator, defined by Eq. (2.44), via the eigenfunctions and eigenenergies of a particle moving in an arbitrary stationary potential $U(x)$.

Problem 2.5. Calculate the change of the wavefunction of a 1D particle, resulting from a short pulse of an external classical force that may be approximated by the delta-function[96]:

$$F(t) = P\delta(t).$$

Problem 2.6. Calculate the transparency $\mathscr{T}$ of the rectangular potential barrier,

$$U(x) = \begin{cases} 0, & \text{for } x < -d/2, \\ U_0, & \text{for } -d/2 < x < +d/2, \\ 0, & \text{for } d/2 < x, \end{cases}$$

for a particle of energy $E > U_0$. Analyze and interpret the result, taking into account that U_0 may be either positive or negative. (In the latter case, we are speaking about the particle's passage over a rectangular potential well of a finite depth $|U_0|$.)

Problem 2.7. Prove Eq. (2.117) for the case $\mathscr{T}_{WKB} \ll 1$, using the connection formulas (2.104).

Problem 2.8. Spell out the stationary wavefunctions of a harmonic oscillator in the WKB approximation, and use them to calculate the expectation values $\langle x^2 \rangle$ and $\langle x^4 \rangle$ for the eigenstate number $n \gg 1$.

Problem 2.9. Use the WKB approximation to express the expectation value of the kinetic energy of a 1D particle confined in a soft potential well, in its nth stationary state, via the derivative dE_n/dn, for $n \gg 1$.

Problem 2.10. Use the WKB approximation to calculate the transparency $\mathscr{T}$ of the following triangular potential barrier:

$$U(x) = \begin{cases} 0, & \text{for } x < 0, \\ U_0 - Fx, & \text{for } x > 0, \end{cases}$$

with $F, U_0 > 0$, as a function of the incident particle's energy E.

---

[96] The constant P is called the force's *impulse*. (In higher dimensionalities, it is a vector—just as the force is.)

Hint: Be careful with the sharp potential step at $x = 0$.

Problem 2.11.* Prove that the element symmetry of the 1D scattering matrix S, describing an arbitrary time-independent scatterer, allows its representation in the form (2.127).

Problem 2.12. Prove the universal relations between elements of the 1D transfer matrix T of a stationary (but otherwise arbitrary) scatterer, mentioned in section 2.5.

Problem 2.13. A 1D particle had been localized in a very narrow and deep potential well, with the 'area' $\int U(x)dx$ equal to $-\mathcal{W}$, where $\mathcal{W} > 0$. Then (say, at $t = 0$) the well's bottom is suddenly lifted up, so that the particle becomes free to move. Calculate the probability density, $w(k)$, to find the particle in a state with the wave number k at $t > 0$, and the total final energy of the system.

Problem 2.14. Calculate the lifetime of the metastable localized state of a 1D particle in the potential

$$U(x) = -\mathcal{W}\delta(x) - Fx, \quad \text{with } \mathcal{W} > 0,$$

using the WKB approximation. Formulate the condition of validity of the result.

Problem 2.15. Calculate the energy levels and the corresponding eigenfunctions of a 1D particle placed into a flat-bottom potential well of width $2a$, with infinitely hard walls, and a transparent, short potential barrier in the middle—see figure below. Discuss the dynamics of the particle in the limit $\mathcal{W} \to \infty$.

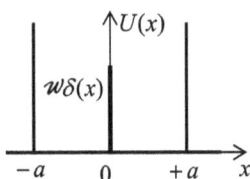

Problem 2.16.* Consider a symmetric system of two potential wells of the type shown in figure 2.21, but with $U(0) = U(\pm\infty) = 0$—see figure below. What is the sign of the well interaction force due to sharing a quantum particle of mass m, for the cases when the particle is in:

(i) a symmetric localized eigenstate, with $\psi_S(-x) = \psi_S(x)$?
(ii) an asymmetric localized eigenstate, with $\psi_A(-x) = -\psi_A(x)$?

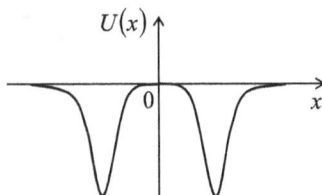

Use an alternative approach to confirm your result for the particular case of delta-functional wells.

Problem 2.17. Derive and analyze the characteristic equation for the localized eigenstates of a 1D particle in a rectangular well of a finite depth (see figure below):

$$U(x) = \begin{cases} -U_0, & \text{for } |x| \leqslant a/2, \\ 0, & \text{otherwise.} \end{cases}$$

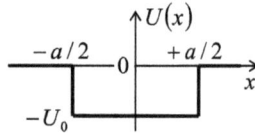

In particular, calculate the number of localized states as a function of well's width a, and explore the limit $U_0 \ll \hbar^2/2ma^2$.

Problem 2.18. Calculate the energy of a 1D particle localized in a potential well of an arbitrary shape $U(x)$, provided that its width a is finite, and the average depth is very small:

$$|\bar{U}| \ll \frac{\hbar^2}{2ma^2}, \quad \text{where } \bar{U} \equiv \frac{1}{a} \int_{\text{well}} U(x)dx.$$

Problem 2.19. A particle of mass m is moving in a field with the following potential:

$$U(x) = U_0(x) + \mathcal{W}\delta(x),$$

where $U_0(x)$ is a smooth, symmetric function with $U_0(0) = 0$, growing monotonically at $x \to \pm\infty$.

(i) Use the WKB approximation to derive the characteristic equation for the particle's energy spectrum, and
(ii) semi-quantitatively describe the spectrum structure evolution at the increase of $|\mathcal{W}|$, for both signs of this parameter.

Make both results more specific for the quadratic-parabolic potential (2.111): $U_0(x) = m\omega_0^2 x^2/2$.

Problem 2.20. Prove Eq. (2.189), starting from Eq. (2.188).

Problem 2.21. For the problem discussed in the beginning of section 2.7, i.e. the 1D particle's motion in an infinite Dirac comb potential shown in figure 2.24,

$$U(x) = w \sum_{j=-\infty}^{+\infty} \delta(x - ja), \quad \text{with } w > 0,$$

(where j takes integer values), write explicit expressions for the eigenfunctions at the very bottom and at the very top of the lowest energy band. Sketch both functions.

Problem 2.22. A 1D particle of mass m moves in an infinite periodic system of very narrow and deep potential wells that may be described by delta-functions:

$$U(x) = w \sum_{j=-\infty}^{+\infty} \delta(x - ja), \quad \text{with } w < 0.$$

(i) Sketch the energy band structure of the system for very small and very large values of the potential well's 'area' $|w|$, and
(ii) calculate explicitly the ground state energy of the system in these two limits.

Problem 2.23. For the system discussed in the previous problem, write explicit expressions for the eigenfunctions of the system, corresponding to:

(i) the bottom of the lowest energy band,
(ii) the top of that band, and
(iii) the bottom of each higher energy band.

Sketch these functions.

*Problem 2.24.** The 1D 'crystal', analyzed in the last two problems, now extends only to $x > 0$, with a sharp step to a flat potential plateau at $x < 0$:

$$U(x) = \begin{cases} w \sum_{j=1}^{+\infty} \delta(x - ja), & \text{with } w < 0, \ \text{for } x > 0, \\ U_0 > 0, & \text{for } x < 0. \end{cases}$$

Prove that the system can have a set of the so-called *Tamm states*, localized near the 'surface' $x = 0$, and calculate their energies in the limit when U_0 is very large but finite. (Quantify this condition.)[97]

Problem 2.25. Calculate the whole transfer matrix of the rectangular potential barrier, specified by Eq. (2.68), for particle energies both below and above U_0.

Problem 2.26. Use results of the previous problem to calculate the transfer matrix of one period of the periodic Kronig-Penney potential shown in figure 2.31b.

---

[97] In applications to electrons in solid-state crystals, the delta-functional potential wells model the attractive potentials of atomic nuclei, while U_0 represents the workfunction, i.e. the energy necessary for the extraction of an electron from the crystal to the free space—see, e.g. section 1.1(ii), and also *Part EM* section 2.6 and *Part SM* section 6.3.

Problem 2.27. Using the results of the previous problem, derive the characteristic equations for a particle's motion in the periodic Kronig–Penney potential, for both $E < U_0$ and $E > U_0$. Try to bring the equations to a form similar to that obtained in section 2.5 for the delta-functional barriers—see Eq. (2.198). Use the equations to formulate the conditions of applicability of the tight-binding and weak-potential approximations, in terms of the system's parameters, and the particle's energy E.

Problem 2.28. For the Kronig–Penney potential, use the tight-binding approximation to calculate the widths of the allowed energy bands. Compare the results with those of the previous problem (in the corresponding limit).

Problem 2.29. For the same Kronig–Penney potential, use the weak-potential limit formulas to calculate the energy gap widths. Again, compare the results with those of problem 27, in the corresponding limit.

Problem 2.30. 1D periodic chains of atoms may exhibit what is called the *Peierls instability*, leading to the *Peierls transition* to a phase in which atoms are slightly displaced by $\Delta x_j = (-1)^j \Delta x$, with $\Delta x \ll a$, where j is the atom's number in the chain, and a is its initial period. These displacements lead to the alternation of the coupling amplitudes δ_n (see Eq. (2.204)) between some values δ_n^+ and δ_n^-. Use the tight-binding approximation to calculate the resulting change of the nth energy band, and discuss the result.

Problem 2.31.* Use Eqs. (1.73) and (1.74) of the lecture notes to derive Eq. (2.252), and discuss the relation between these Bloch oscillations and the Josephson oscillations of frequency (1.75).

Problem 2.32. A 1D particle of mass m is placed into the following triangular potential well:

$$U(x) = \begin{cases} +\infty, & \text{for } x < 0, \\ Fx, & \text{for } x > 0, \end{cases} \quad \text{with } F > 0.$$

  (i) Calculate its energy spectrum using the WKB approximation.
 (ii) Estimate the ground state energy using the variational method, with two different trial functions.
(iii) Calculate the three lowest energy levels, and also for the 10th level, with at least 0.1% accuracy, from the exact solution of the problem.
(iv) Compare and discuss the results.

Hint: The values of the first zeros of the Airy function, necessary for task (iii), may be found in many math handbooks, for example, in table 10.13 of the collection edited by Abramowitz and Stegun—see section A.16(i).

Problem 2.33. Use the variational method to estimate the ground state energy E_g of a particle in the following potential well:

$$U(x) = -U_0 \exp\{-\alpha x^2\}, \quad \text{with } \alpha > 0, \text{ and } U_0 > 0.$$

Spell out the results in the limits of small and large U_0, and give their interpretation.

Problem 2.34. For a 1D particle of mass m, placed into a potential well with the following profile,

$$U(x) = ax^{2s}, \quad \text{with } a > 0, \text{ and } s > 0,$$

(i) calculate its energy spectrum using the WKB approximation, and
(ii) estimate the ground state energy using the variational method.

Compare the ground state energy results for the parameter s equal to 1, 2, 3, and 100.

Problem 2.35. Use the variational method to estimate the 1st excited state of the 1D harmonic oscillator.

Problem 2.36. Assuming the quantum effects to be small, calculate the lower part of the energy spectrum of the following system: a small bead of mass m, free to move without friction along a ring of radius R, which is rotated about its vertical diameter with a constant angular velocity ω—see figure below[98]. Formulate a quantitative condition of validity of your results.

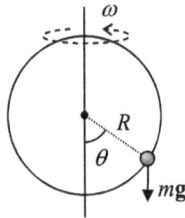

Problem 2.37. A 1D harmonic oscillator, with mass m and frequency ω_0, had been in its ground state; then an additional force F was suddenly applied, and then retained constant in time. Calculate the probability of the oscillator staying in its ground state.

Problem 2.38. A 1D particle of mass m has been placed into a quadratic potential well (2.111),

---

[98] This system was used as the analytical mechanics 'testbed problem' in *Part CM* of this series, and the reader is welcome to use any relations derived there—but remember that they pertain to the classical mechanics domain!

$$U(x) = \frac{m\omega_0^2 x^2}{2},$$

and allowed to relax into the ground state. At $t = 0$, the well was fast accelerated to move with velocity v, without changing its profile, so that at $t \geqslant 0$ the above formula for U is valid with the replacement $x \to x' \equiv x - vt$. Calculate the probability for the system to still be in the ground state at $t > 0$.

Problem 2.39. A 1D harmonic oscillator had initially been in its ground state. At a certain moment of time, its spring constant κ is abruptly increased, so that its frequency $\omega_0 = (\kappa/m)^{1/2}$ is increased by a factor of α, and then is kept constant at the new value. Calculate the probability that after the change, the oscillator is still in its ground state.

Problem 2.40. A 1D particle is placed into the following potential well:

$$U(x) = \begin{cases} +\infty, & \text{for } x < 0, \\ m\omega_0^2 x^2/2, & \text{for } x \geqslant 0. \end{cases}$$

(i) Find its eigenfunctions and eigenenergies.
(ii) This system had been let to relax into its ground state, and then the potential wall at $x < 0$ was rapidly removed, so that the system was instantly turned into the usual harmonic oscillator (with the same m and ω_0). Find the probability for the oscillator to be in its ground state.

Problem 2.41. Prove the following formula for the propagator of the 1D harmonic oscillator:

$$G(x, t; x_0, t_0) = \left(\frac{m\omega_0}{2\pi i\hbar \sin[\omega_0(t - t_0)]} \right)^{1/2}$$
$$\times \exp\left\{ \frac{im\omega_0}{2\hbar \sin[\omega_0(t - t_0)]}\left[\left(x^2 + x_0^2\right)\cos[\omega_0(t - t_0)] - 2xx_0 \right] \right\}.$$

Discuss the relation between this formula and the propagator of a free 1D particle.

Problem 2.42. In the context of the Sturm oscillation theorem mentioned in section 2.9, prove that the number of eigenfunction zeros of a particle confined in an arbitrary but finite potential well, always increases with the corresponding eigenenergy.

Hint: You may like to use the suitably modified Eq. (2.186).

Problem 2.43.* Use the WKB approximation to calculate the lifetime of the metastable ground state of a 1D particle of mass m in the 'pocket' of the potential profile

$$U(x) = \frac{m\omega_0^2}{2}x^2 - \alpha x^3.$$

Contemplate the significance of this problem.

References

[1] Fursey G 2005 *Field Emission in Vacuum Microelectronics* (New York: Kluwer)
[2] Binning G and Rohrer H 1982 *Helv. Phys. Acta* **55** 726
[3] Sverdlov V *et al* 2003 *IEEE Trans. Electron Devices* **50** 1926
[4] Landau L and Lifshitz E 1977 *Quantum Mechanics, Non-Relativistic Theory* 3rd ed (Pergamon)
[5] Likharev K 1982 *Int. J. Theor. Phys.* **21** 311
[6] Braginsky V and Khalili F 1992 *Quantum Measurement* (Cambridge: Cambridge University Press)
[7] Ziman J 1979 *Principles of the Theory of Solids* 2nd ed (Cambridge: Cambridge University Press)
[8] Lifshitz E and Pitaevskii L 1980 *Statistical Physics, Part 2* (Pergamon)
[9] Ben Dahan M *et al* 1996 *Phys. Rev. Lett.* **76** 4508
[10] Mendez E *et al* 1998 *Phys. Rev. Lett.* **60** 2426
[11] Havliand D *et al* 1991 *Z. Phys.* B **85** 339
[12] Averin D *et al* 1985 *Sov. Phys. JETP* **61** 407
[13] O'Connell A *et al* 2010 *Nature* **464** 697
[14] Aspelmeyer M *et al* 2014 *Rev. Mod. Phys.* **86** 1391

IOP Publishing

Quantum Mechanics
Lecture notes
Konstantin K Likharev

Chapter 3

Higher dimensionality effects

The extension of the description of the basic quantum-mechanical effects, discussed in the previous chapter, to multi-dimensional systems is mostly straightforward. As a result, this chapter is focused on the phenomena (such as the AB effect and the Landau levels) that cannot take place in one dimension due to topological reasons, and also on a few key 3D problems (such as the Born approximation in the scattering theory, and the axially- and spherically-symmetric systems) whose solutions are important for numerous applications.

3.1 Quantum interference and the AB effect

In the past two chapters, we have already discussed some effects of the de Broglie wave interference. For example, the standing waves inside a potential well, or even on the top of a potential barrier, may be considered as a result of the constructive interference of the incident and reflected waves. However, there are some remarkable new effects made possible by the spatial separation of such waves, and such separation requires a higher (either 2D or 3D) dimensionality. A good example of wave separation is provided by the Young-type experiment (figure 3.1) in which particles, emitted by the same source, are passed through two narrow holes (or slits) is an otherwise opaque partition.

If the particles do not interact (which is always true if the emission rate is sufficiently low), the average rate of particle counting by the detector is proportional to the probability density $w(\mathbf{r}, t) = \Psi(\mathbf{r}, t) \Psi^*(\mathbf{r}, t)$ to find a single particle at the detector's location $\mathbf{r}$, where $\Psi(\mathbf{r}, t)$ is the solution of the single-particle Schrödinger equation (1.25) for the system. Let us describe this experiment for the case when the incident particles may be represented by virtually-monochromatic waves of energy E (e.g. very $\mathbf{r}$-long wave packets), so that their wavefunction may be taken in the form

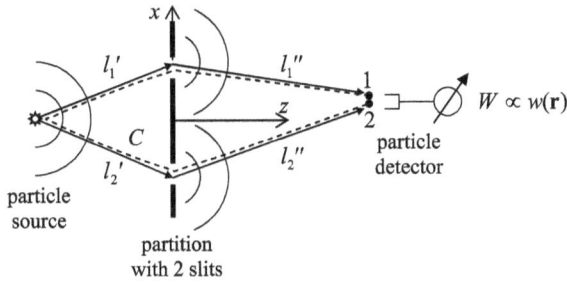

Figure 3.1. The scheme of the 'two-slit' (Young-type) interference experiment.

given by Eqs. (1.57) and (1.62): $\Psi(\mathbf{r},\ t) = \psi(\mathbf{r})\exp\{-iEt/\hbar\}$. In this case, in the free-space parts of the system, where $U(\mathbf{r}) = 0$, $\psi(\mathbf{r})$ satisfies the stationary Schrödinger equation (1.78a):

$$-\frac{\hbar^2}{2m}\nabla^2\psi = E\psi. \qquad (3.1a)$$

With the standard definition $k \equiv (2mE)^{1/2}/\hbar$, it may be rewritten as the 3D Helmholtz equation:

$$\nabla^2\psi + k^2\psi = 0. \qquad (3.1b)$$

The opaque parts of the partition may be well described as classically forbidden regions, so if their size scale a is much larger than the wavefunction penetration depth δ, described by Eq. (2.59), we may use on their surface S the same boundary conditions as for the well's walls of infinite height:

$$\psi|_S = 0. \qquad (3.2)$$

Eqs. (3.1) and (3.2) describe the standard boundary problem of the theory of propagation of scalar waves of any nature. For an arbitrary geometry, this problem does not have a simple analytical solution. However, for a conceptual discussion of the wave interference we may use certain natural assumptions that will allow us to find its particular, approximate solution.

First, let us discuss the wave emission, into free space, by a small-size, isotropic source located at the origin. Naturally, the emitted wave should be spherically-symmetric: $\psi(\mathbf{r}) = \psi(r)$. Using the well-known expression for the Laplace operator in spherical coordinates[1], we may reduce Eq. (3.1) to the following ordinary differential equation:

$$\frac{1}{r^2}\frac{d}{dr}\left(r^2\frac{d\psi}{dr}\right) + k^2\psi = 0. \qquad (3.3)$$

---

[1] See, e.g. Eq. (A.61).

Let us introduce a new function, $f(r) \equiv r\psi(r)$. Plugging the reciprocal relation $\psi = f/r$ into Eq. (3.3), we see that it is reduced to the 1D wave equation,

$$\frac{d^2f}{dr^2} + k^2f = 0, \tag{3.4}$$

whose solutions were discussed in detail in section 2.2. For a fixed k, the general solution of Eq. (3.4) is

$$f = f_+ e^{ikr} + f_- e^{-ikr}, \tag{3.5}$$

so that the full wavefunction

$$\psi(\mathbf{r}) = \frac{f_+}{r} e^{ikr} + \frac{f_-}{r} e^{-ikr}, \qquad \text{i.e. } \Psi(\mathbf{r},\, t) = \frac{f_+}{r} e^{i(kr-\omega t)} + \frac{f_-}{r} e^{-i(kr+\omega t)},$$
$$\text{with } \omega \equiv \frac{E}{\hbar} = \frac{\hbar k^2}{2m}. \tag{3.6}$$

If the source is located at point $\mathbf{r}' \neq 0$, the obvious generalization of Eq. (3.6)

$$\Psi(\mathbf{r},\, t) = \frac{f_+}{R} e^{i(kR-\omega t)} + \frac{f_-}{R} e^{-i(kR+\omega t)}, \qquad \text{with } R \equiv |\,\mathbf{R}\,|, \quad \mathbf{R} \equiv \mathbf{r} - \mathbf{r}'. \tag{3.7}$$

 The first term of this solution describes a spherically-symmetric wave propagating from the source outward, and the second one, a wave converging onto the source point $\mathbf{r}'$ from large distances. Though the latter solution is possible in some very special circumstances (say, when the outgoing wave is reflected back from a spherical shell), for our problem only the outgoing waves are relevant, so that we may keep only the first term (proportional to f_+) in Eq. (3.7). Note that the factor R in the denominator (that was absent in the 1D geometry) has a simple physical sense: it provides the independence of the full probability current $I = 4\pi R^2 j(R)$, with $j(R) \propto k\Psi\Psi^* \propto 1/R^2$, of the distance R between the observation point and the source.

 Now let us assume that the partition's geometry is not too complicated—for example, it is planar as shown in figure 3.1, or nearly-planar, and consider the region of the particle detector location far behind the partition (at $z \gg 1/k$), and at a relatively small angle to it: $|x| \ll z$. Then it should be physically clear that the spherical waves (3.7) emitted by each point inside the slit cannot be perturbed too much by the opaque parts of the partition, and their only role is the restriction of the set of such emitting points by the area of the slits. Hence, an approximate solution of the boundary problem is given by the following *Huygens principle*: the wave behind the partition looks as if it was the sum of contributions (3.7) of point sources located in the slits, with each source's strength f_+ proportional to the amplitude of the wave arriving at this pseudo-source from the real source—see figure 3.1. This principle finds its confirmation in the strict wave theory, which shows[2] that with our

---

[2] For a proof of Eq. (3.8), see, e.g. *Part EM* section 8.5.

assumptions, the solution of the boundary problem (3.1)–(3.2) may be represented as the following *Kirchhoff integral*:

$$\psi(\mathbf{r}) = c \int_{\text{slits}} \frac{\psi(\mathbf{r}')}{R} e^{ikR} d^2r', \quad \text{with } c = \frac{k}{2\pi i}. \qquad (3.8)$$

If the source is also far from the partition, its wave front is almost parallel to the slit plane, and if the slits are not too broad, we can take $\psi(\mathbf{r}')$ constant ($\psi_{1,2}$) at each slit, so that Eq. (3.8) is reduced to

$$\psi(\mathbf{r}) = a''_1 \exp\{ikl''_1\} + a''_2 \exp\{ikl''_2\}, \quad \text{with } a''_{1,2} = \frac{cA_{1,2}}{l''_{1,2}}\psi_{1,2}, \qquad (3.9)$$

where $A_{1,2}$ are the slit areas, and $l''_{1,2}$ are the distances from the slits to the detector. The wavefunctions on the slits be calculated approximately[3] by applying the same Eq. (3.7) to the space *before* the slits: $\psi_{1,2} \approx (f_+/l'_{1,2})\exp\{ikl'_{1,2}\}$, where $l'_{1,2}$ are the distances from source to the slits—see figure 3.1. As a result, Eq. (3.9) may be rewritten as

$$\psi(\mathbf{r}) = a_1 \exp\{ikl_1\} + a_2 \exp\{ikl_2\}, \quad \text{with } l_{1,2} \equiv l'_{1,2} + l''_{1,2};$$
$$a_{1,2} \equiv \frac{c f_+ A_{1,2}}{l'_{1,2}l''_{1,2}}. \qquad (3.10)$$

(As figure 3.1 shows, each of $l_{1,2}$ is the full length of the classical path of the particle from the source, through the corresponding slit, and further to the observation point $\mathbf{r}$.)

According to Eq. (3.10), the resulting rate of particle counting at point $\mathbf{r}$ is proportional to

$$w(\mathbf{r}) = \psi(\mathbf{r})\psi^*(\mathbf{r}) = |a_1|^2 + |a_2|^2 + 2|a_1a_2| \cos\varphi_{12}, \qquad (3.11)$$

where

$$\varphi_{12} \equiv k(l_2 - l_1) \qquad (3.12)$$

is the difference of the total wave phase accumulations along each of two alternative trajectories. The last expression may be evidently generalized as

$$\varphi_{12} = \oint_C \mathbf{k} \cdot d\mathbf{r}, \qquad (3.13)$$

with integration along the virtually closed contour C (see the dashed line in figure 3.1), i.e. from the point 1, in the positive (i.e. counterclockwise) direction to the point 2.

---

[3] A possible (and reasonable) concern about the application of Eq. (3.7) to the field in the slits is that it ignores the effect of opaque parts of the partition. However, as we know from chapter 2, the main role of the classically forbidden region is providing the reflection of the incident wave toward its source (i.e. to the left in figure 3.1). As a result, the contribution of this reflection to the field inside the slits is insignificant if $A_{1,2} \gg \lambda^2$, and even in the opposite case provides just some rescaling of the amplitudes $a_{1,2}$, which is unimportant for our conceptual discussion.

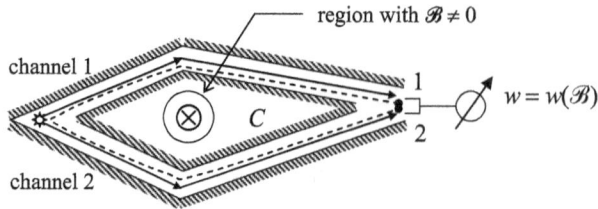

Figure 3.2. The AB effect.

(From our discussion of the 1D WKB approximation, we may expect such general-ization to be valid even if k changes, sufficiently slowly, along the paths.)

Our result (3.11) shows that the counting rate oscillates as a function of the difference $(l_2 - l_1)$, which in turn changes with the detector's position, giving the famous interference pattern, with the amplitude proportional to the product $|a_1a_2|$, and hence vanishing if any of the slits is closed. For a wave theory, this is a well-known result[4], but for particle physics, it was (and still is :-) rather shocking. Indeed, our analysis is also valid for a very low particle emission/detection rate, so that there is no way to interpret the pattern other than resulting from a particle's interference *with itself*, or rather the interference of its de Broglie waves passing through each of two slits[5]. Nowadays, such interference is reliably observed not only for electrons, but for such heavy particles as atoms and molecules; moreover, atomic interfer-ometers are used as ultra-sensitive instruments for measurements of the gravity field, rotation velocity, and tilt[6].

Let us now discuss a very interesting effect of magnetic field on the quantum interference. In order to make the discussion simpler, let us consider a slightly different version of the two-slit experiment, in which each of the two alternative paths is constricted to a narrow channel using a partial confinement—see figure 3.2. (In this arrangement, moving the particle detector without changing channels' geometry, and hence local values of k may be more problematic experimentally, so let us think about its position **r** as fixed.) In this case, because of the effect of the walls providing the path confinement, we cannot use Eqs. (3.10) for the amplitudes $a_{1,2}$. However, from the discussions in sections 1.6 and 2.2, it should be clear that the first of the expressions (3.10) remains valid, though maybe with a value of k specific for each channel.

In this geometry, we can now apply some local magnetic field $\mathscr{B}$, say normal to the plane of particle motion, whose lines would pierce, but not touch the contour

---

[4] See, e.g. a detailed discussion in *Part EM* section 8.4.

[5] Here I have to mention the fascinating experiments (first performed in 1987 by C Hong *et al* with photons, and recently, in 2015, by R Lopes *et al*, with non-relativistic particles—helium atoms) on the interference of de Broglie waves of *independent* but *identical* particles, in the same internal quantum state and virtually the same values of E and k. These experiments raise the important issue of the particle *indistinguishability*, which will be discussed in section 8.1.

[6] See, e.g. the review paper [1].

C drawn along the particle propagation channels—see the dashed line in figure 3.2. In classical electrodynamics[7], the external magnetic field's effect on a particle with electric charge q is described by the *Lorentz force*

$$\mathbf{F}_{\mathscr{B}} = q\mathbf{v} \times \mathscr{B}, \qquad (3.14)$$

where $\mathscr{B}$ is the field value at the point of its particle's location, so that for the experiment shown in figure 3.2, $\mathbf{F}_{\mathscr{B}} = 0$, and the field would not affect the particle motion at all. In quantum mechanics, this is not so, and the field does affect the probability density w, even if $\mathscr{B} = 0$ at all points where the wavefunction $\psi(\mathbf{r})$ is not equal to zero.

In order to describe this surprising effect, let us first develop a general framework for account of effects of electromagnetic fields on a quantum particle, which will also give us some important by-product results. In order to do that, we need to calculate the Hamiltonian operator of a charged particle in the field. For an electrostatic field, this task is easy. Indeed, from classical electrodynamics we know that this field may be represented as a gradient of its electrostatic potential ϕ,

$$\mathscr{E} = -\nabla\phi(\mathbf{r}), \qquad (3.15)$$

so that the force exerted by the field on a particle with electric charge q,

$$\mathbf{F}_{\mathscr{E}} = q\mathscr{E}, \qquad (3.16)$$

may be described by adding the field-induced potential energy,

$$U(\mathbf{r}) = q\phi(\mathbf{r}), \qquad (3.17)$$

to other (possible) components of the full potential energy of the particle. As was already discussed in section 4.1, such a potential energy may be included into the particle's Hamiltonian operator just by adding it to the kinetic energy operator—see Eq. (1.41).

However, the magnetic field's effect is peculiar: since its Lorentz force (3.14) cannot do any work on a classical particle:

$$d\mathscr{W}_{\mathscr{B}} \equiv \mathbf{F}_{\mathscr{B}} \cdot d\mathbf{r} = \mathbf{F}_{\mathscr{B}} \cdot \mathbf{v}dt = q(\mathbf{v} \times \mathscr{B}) \cdot \mathbf{v}\, dt = 0, \qquad (3.18)$$

the field cannot be represented by any potential energy, so it may not be immediately clear how to account for it in the Hamiltonian. The crucial help comes from the analytical-mechanics approach to classical electrodynamics[8]: in the non-relativistic limit, the Hamiltonian function of a particle in electromagnetic field looks superficially like that in the electric field only:

$$H = \frac{mv^2}{2} + U = \frac{p^2}{2m} + q\phi; \qquad (3.19)$$

however, the momentum $\mathbf{p} \equiv m\mathbf{v}$ that participates in this expression is now the difference

---

[7] See, e.g. *Part EM* section 5.1. Note that Eq. (3.14), as well as all other formulas of this course, are in SI units.
[8] See, e.g. *Part EM* section 9.7, in particular Eq. (9.196).

$$\mathbf{p} = \mathbf{P} - q\mathbf{A}. \tag{3.20}$$

Here $\mathbf{A}$ is the vector-potential, defined by the well-known relations for the electric and magnetic field[9]:

$$\mathscr{E} = -\nabla\phi - \frac{\partial\mathbf{A}}{\partial t}, \qquad \mathscr{B} = \nabla \times \mathbf{A}, \tag{3.21}$$

while $\mathbf{P}$ is the *canonical momentum* whose Cartesian components may be calculated (in classics) from the Lagrangian function L using the standard formula of analytical mechanics,

$$P_j \equiv \frac{\partial L}{\partial v_j}. \tag{3.22}$$

To emphasize the difference between the two momenta, $\mathbf{p} = m\mathbf{v}$ is frequently called the *kinematic momentum* (or '$m\mathbf{v}$-momentum'). The distinction between $\mathbf{p}$ and $\mathbf{P} = \mathbf{p} + q\mathbf{A}$ becomes more clear if we notice that the vector-potential is not *gauge-invariant*: according to the second of Eqs. (3.21), at the so-called *gauge transformation*

$$\mathbf{A} \to \mathbf{A} + \nabla\chi, \tag{3.23}$$

with an arbitrary single-valued scalar *gauge function* $\chi = \chi(\mathbf{r}, t)$, the magnetic field does not change. Moreover, according to the first of Eqs. (3.21), if we make the simultaneous replacement

$$\phi \to \phi - \frac{\partial\chi}{\partial t}, \tag{3.24}$$

the gauge transformation does not affect the electric field either. With that, the gauge function's choice does not affect the classical particle's equation of motion, and hence the velocity $\mathbf{v}$ and momentum $\mathbf{p}$. Hence, the kinematic momentum is gauge-invariant, while $\mathbf{P}$ is not, because according to Eqs. (3.20) and (3.23), the introduction of χ changes it by $q\nabla\chi$.

Now the standard way of transfer to the quantum mechanics is to treat the canonical rather than kinematic momentum according to the correspondence postulate discussed in section 1.2. This means that in the wave mechanics, the operator of this variable is still given by Eq. (1.26):[10]

$$\hat{\mathbf{P}} = -i\hbar\nabla. \tag{3.25}$$

Hence the Hamiltonian operator corresponding to the classical function (3.19) is

$$\hat{H} = \frac{1}{2m}(-i\hbar\nabla - q\mathbf{A})^2 + q\phi \equiv -\frac{\hbar^2}{2m}\left(\nabla - \frac{iq}{\hbar}\mathbf{A}\right)^2 + q\phi, \tag{3.26}$$

---

[9] See, e.g. *Part EM* section 6.1, in particular Eqs. (6.7).
[10] The validity of this choice is clear from the fact that if the *kinetic* momentum was described by this differential operator, the Hamiltonian operator corresponding to the classical Hamiltonian function (3.19), and the corresponding Schrödinger equation would not describe the magnetic field effects at all.

so that the stationary Schrödinger equation (1.60) of a particle moving in an electromagnetic field (but otherwise free) is

$$-\frac{\hbar^2}{2m}\left(\nabla - \frac{iq}{\hbar}\mathbf{A}\right)^2\psi + q\phi\psi = E\psi,\tag{3.27}$$

We may now repeat all the calculations of section 1.4 for the case $\mathbf{A} \neq 0$, and readily get the following generalized expression for the probability current density:

$$\mathbf{j} = \frac{\hbar}{2im}\left[\psi^*\left(\nabla - \frac{iq}{\hbar}\mathbf{A}\right)\psi - \text{c.c}\right] \equiv \frac{1}{2m}[\psi^*\hat{\mathbf{p}}\psi - \text{c.c.}] \equiv \frac{\hbar}{m}|\psi|^2\left(\nabla\varphi - \frac{q}{\hbar}\mathbf{A}\right).\tag{3.28}$$

We see that the current density is gauge-invariant (as required for any observable) only if the wavefunction's phase φ changes as

$$\varphi \to \varphi + \frac{q}{\hbar}\chi.\tag{3.29}$$

This may be a point of conceptual concern: since the quantum interference is described by the spatial dependence of phase φ, can the observed interference pattern depend on the gauge function's choice (which would not make sense)? Fortunately, this is not true, because the spatial phase *difference* between two interfering paths, participating in Eq. (3.12), is gauge-transformed as

$$\varphi_{12} \to \varphi_{12} + \frac{q}{\hbar}(\chi_2 - \chi_1).\tag{3.30}$$

But χ has to be a single-valued function of coordinates, hence in the limit when the points 1 and 2 coincide, $\chi_1 = \chi_2$, so that $\Delta\varphi$, and hence the interference pattern are gauge-invariant.

However, the difference φ *may be* affected by the magnetic field, even if it is localized outside the channels in which the particle propagates. Indeed, in this case the field cannot not affect the particle's velocity $\mathbf{v}$ and the probability current density $\mathbf{j}$:

$$\mathbf{j}(\mathbf{r})|_{\mathscr{B}\neq0} = \mathbf{j}(\mathbf{r})|_{\mathscr{B}=0},\tag{3.31}$$

so that the last form of Eq. (3.28) yields

$$\nabla\varphi(\mathbf{r})|_{\mathscr{B}\neq0} = \nabla\varphi(\mathbf{r})|_{\mathscr{B}=0} + \frac{q}{\hbar}\mathbf{A}.\tag{3.32}$$

Integrating this equation along the contour C (figure 3.2), for the phase difference between points 1 and 2 we get

$$\varphi_{12}|_{\mathscr{B}\neq0} = \varphi_{12}|_{\mathscr{B}=0} + \frac{q}{\hbar}\oint_C \mathbf{A} \cdot d\mathbf{r},\tag{3.33}$$

where the integral should be taken along the same contour C as before (in figure 3.2, from point 1, counterclockwise along the dashed line to point 2). But from the classical electrodynamics we know[11] that as the points 1 and 2 tend to each other, i.e.

---

[11] See, e.g. *Part EM* section 5.3.

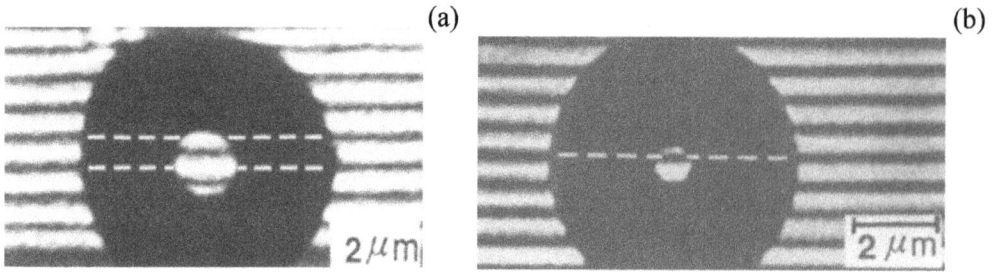

Figure 3.3. Typical results of a two-paths interference experiment by A Tonomura *et al* [2], showing the AB effect for electrons well shielded from the applied magnetic field. In this particular experimental geometry, the AB effect produces a relative shift of the interference patterns inside and outside the dark ring. (a) $\Phi = \Phi_0'/2$, (b) $\Phi = \Phi_0'$. Copyright 1986 by the American Physical Society.

the contour C becomes closed, the last integral is just the magnetic flux $\Phi \equiv \int \mathscr{B}_n d^2r$ through any smooth surface limited by the contour, so that Eq. (3.33) may be rewritten as

$$\varphi_{12}|_{\mathscr{B}\neq0} = \varphi_{12}|_{\mathscr{B}=0} + \frac{q}{\hbar}\Phi. \tag{3.34a}$$

In terms of the interference pattern, this means a shift of interference fringes, proportional to the magnetic flux (figure 3.3). This phenomenon is usually called the 'Aharonov–Bohm' (or just the *AB*) *effect*[12]. For particles with a single elementary charge, $q = \pm e$, this result is frequently represented as

$$\varphi_{12}|_{\mathscr{B}\neq0} = \varphi_{12}|_{\mathscr{B}=0} \pm 2\pi\frac{\Phi}{\Phi_0'}, \tag{3.34b}$$

where the fundamental constant $\Phi_0' \equiv 2\pi\hbar/e \approx 4.14 \times 10^{-15}$ Wb has the meaning of the magnetic flux necessary to change φ_{12} by 2π, i.e. to shift the interference pattern (3.11) by one period, and is called the *normal magnetic flux quantum*—'normal' because of the reasons we will soon discuss.

The AB effect may be 'almost explained' classically, in terms of Faraday's electromagnetic induction. Indeed, a change $\Delta\Phi$ of magnetic flux in time causes a vortex-like electric field $\Delta\mathscr{E}$ around it. That field is *not* restricted to the magnetic field's location, i.e. may reach the particle's trajectories. The field's magnitude (or rather of its integral along the contour C) may be readily calculated by integration of the first of Eqs. (3.21):

---

[12] I prefer the latter, less personable name, because the effect had been actually predicted by W Ehrenberg and R Siday in 1949, before it was rediscovered (also theoretically) by Y Aharonov and D Bohm in 1959. To be fair to Aharonov and Bohm, it was their work that triggered a wave of interest to the phenomenon, resulting in its first experimental observation by R Chambers in 1960 and several other groups soon after that. Later, the experiments were improved, using ferromagnetic cores and/or superconducting shielding to provide better separation between the electron trajectories and the applied magnetic field—as in the work whose results are shown in figure 3.3.

$$\Delta V \equiv \oint_C \Delta \mathscr{E} \cdot d\mathbf{r} = -\frac{d\Phi}{dt}, \tag{3.35}$$

I hope that in this expression the reader readily recognizes the integral ('under-graduate') form of Faraday's induction law[13]. Now let us assume that the variable separation described in section 1.5 may be applied to the end points 1 and 2 of a particle's alternative trajectories as two independent systems[14], and that the magnetic flux' change by certain amount $\Delta \Phi$ does not change the spatial parts ψ_j of wavefunctions of these systems. Then the change (3.35) leads to the change of the potential energy difference $\Delta U = q \Delta V$ between the two points, and repeating the arguments that were used in section 1.6 at the discussion of the Josephson effect, we may rewrite Eq. (1.72) as

$$\frac{d\varphi_{12}}{dt} = -\frac{\Delta U}{\hbar} = -\frac{q}{\hbar} \Delta V = \frac{q}{\hbar} \frac{d\Phi}{dt}. \tag{3.36}$$

Integrating this relation over the time of magnetic field's change, we get

$$\Delta \varphi_{12} = \frac{q}{\hbar} \Delta \Phi, \tag{3.37}$$

—superficially, the same result as given by Eq. (3.34).

However, this interpretation of the AB effect is limited. Indeed, it requires the particle to be in the system (on the way from the source to the detector) during the flux change, i.e. when the induced electric field $\mathscr{E}$ may affect its dynamics. Conversely, Eq. (3.34) predicts that the interference pattern would shift even if the field change has been made when there was no particle in the system, and hence field $\mathscr{E}$ could not be felt by it. Experiment confirms the latter conclusion. Hence, there is *something* in the space where a particle propagates (i.e. outside of the magnetic field region), which transfers the information about even the static magnetic field to the particle. The standard interpretation of this surprising fact is as follows: the vector-potential $\mathbf{A}$ is not just a convenient mathematical tool, but a physical reality (just as its electric counterpart ϕ), despite the large freedom of choice we have in prescribing specific spatial and temporal dependences of these potentials without affecting any observable—see Eqs. (3.23) and (3.24).

To conclude this section, let me briefly discuss the very interesting form taken by the AB effect in superconductivity. In this case, our results require two changes. The first one is simple: since superconductivity may be interpreted as the Bose–Einstein condensate of Cooper pairs with electric charge $q = -2e$, Φ_0' has to be replaced by the so-called *superconducting flux quantum*[15].

---

[13] See, e.g. *Part EM* section 6.1.

[14] This assumption may seem a little bit of a stretch, but the resulting relation (3.37) may be indeed proven for a rather realistic model, though that would take more time and space that I can afford.

[15] One more bad, but common term: a metallic wire can (super)conduct, but a quantum hardly can!

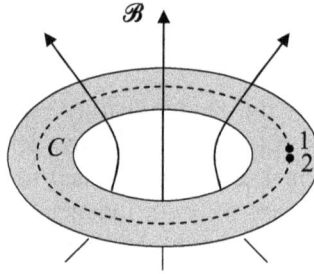

Figure 3.4. The magnetic flux quantization in a superconducting loop (schematically).

$$\Phi_0 \equiv \frac{\pi\hbar}{e} \approx 2.07 \times 10^{-15}\,\mathrm{Wb} = 2.07 \times 10^{-7}\mathrm{Gs} \cdot \mathrm{cm}^2. \tag{3.38}$$

Second, since the pairs are Bose particles and are all condensed in the same quantum state, described by the same wavefunction, the total electric current density, proportional to the probability current density j, may be extremely large—in real superconducting materials, up to $\sim 10^{12}\,\mathrm{A}\,\mathrm{m}^{-2}$. In these conditions, one cannot neglect the contribution of that current into the magnetic field and hence into its flux Φ, which (according to the Lenz rule of the Faraday induction law) tries to compensate the changes in external flux. In order to see possible results of this contribution, let us consider a closed superconducting loop (figure 3.4). Due to the Meissner effect (which is just another version of the flux' self-compensation), the current and magnetic field penetrate into a superconductor by only a small distance (called the *London penetration depth*) $\delta_{\mathrm{L}} \sim 10^{-7}\,\mathrm{m}$.[16] If the loop is made of a superconducting 'wire' that is considerably thicker than δ_{L}, we can draw a contour deep inside the wire, at which the current density is negligible. According to Eq. (3.28), everywhere at the contour,

$$\nabla\varphi - \frac{q}{\hbar}\mathbf{A} = 0. \tag{3.39}$$

Integrating this equation along the contour as before (in figure 3.4, from some point 1 to the virtually coinciding point 2), we need to have the phase difference φ_{12} equal to $2\pi n$, because the wavefunction $\psi \propto \exp\{i\varphi\}$ in the initial and final points 1 and 2 should be 'essentially' the same, i.e. produce the same observables. As a result, we get

$$\Phi \equiv \oint_C \mathbf{A} \cdot d\mathbf{r} = \frac{\hbar}{q}2\pi n \equiv n\Phi_0. \tag{3.40}$$

This is the famous *flux quantization effect*[17], which justifies the term 'magnetic flux quantum' for the constant Φ_0 given by Eq. (3.38).

Unfortunately, in this course I have no space/time here to discuss very interesting effects of 'partial flux quantization', that arise when a superconductor loop is closed

---

[16] For more detail, see *Part EM* section 6.4.

[17] It was predicted in 1949 by F London and experimentally discovered (independently and virtually simultaneously) in 1961 by two experimental groups: B Deaver and W Fairbank, and R Doll and M Näbauer.

by a Josephson junction, forming the so-called *Superconductor QUantum Interference Device*—'SQUID'. Such devices are used, in particular, for super-sensitive magnetometry and ultrafast, low-power computing[18].

3.2 Landau levels and quantum Hall effect

In the last section, we have used the Schrödinger equation (3.27) for an analysis of static magnetic field effects in 'almost-1D', circular geometries shown in figures 3.1, 3.2, and 3.4. However, this equation describes very interesting effects in fully-higher-dimensions as well, especially in the 2D case. Let us consider a quantum particle free moving in the [x, y] plane only (say, due do its strong confinement in the perpendicular direction z—see the discussion in section 1.8). Taking the confinement energy for the reference, we may reduce Eq. (3.27) to the similar equation, but with the Laplace operator acting only in the directions x and y:

$$-\frac{\hbar^2}{2m}\left(\mathbf{n}_x\frac{\partial}{\partial x} + \mathbf{n}_y\frac{\partial}{\partial y} - i\frac{q}{\hbar}\mathbf{A}\right)^2\psi = E\psi. \tag{3.41}$$

Let us find its solutions for the simplest case when the applied static magnetic field is uniform and perpendicular to the motion plane:

$$\mathscr{B} = \mathscr{B}\,\mathbf{n}_z. \tag{3.42}$$

According to the second of Eq. (3.21), this relation imposes the following restriction on the choice of vector-potential:

$$\mathscr{B} = \frac{\partial A_y}{\partial x} - \frac{\partial A_x}{\partial y}, \tag{3.43}$$

but the gauge transformations still give us a lot of freedom in its choice. The 'natural' axially-symmetric form, $A = \mathbf{n}_\varphi\rho\mathscr{B}/2$, where $\rho = (x^2 + y^2)^{1/2}$ is the distance from some z-axis, leads to cumbersome math. In 1930, L Landau realized that the energy spectrum of Eq. (3.41) may be obtained by making a much simpler, though counter-intuitive choice:

$$A_x = 0, \quad A_y = \mathscr{B}(x - x_0), \tag{3.44}$$

(with arbitrary x_0), which evidently satisfies Eq. (3.43), though ignores the physical equivalence of the x and y directions.

Now, expanding the eigenfunction into the Fourier integral in the y-direction:

$$\psi(x, y) = \int X_k(x)\exp\{ik(y - y_0)\}dk, \tag{3.45}$$

---

[18] A brief review of these effects, and recommendations for further reading may be found in *Part EM* section 6.5.

we see that for each component of this integral, Eq. (3.41) yields a specific equation

$$-\frac{\hbar^2}{2m}\left\{\mathbf{n}_x\frac{d}{dx} + i\mathbf{n}_y\left[k - \frac{q}{\hbar}\mathscr{B}\,(x - x_0)\right]\right\}^2 X_k = EX_k. \tag{3.46}$$

Since the two vectors inside the curly brackets are mutually perpendicular, its square has no cross-terms, so that Eq. (3.46) reduces to

$$-\frac{\hbar^2}{2m}\frac{d^2}{dx^2}X_k + \frac{q^2}{2m}\mathscr{B}^2(x - x_0')^2 X_k = EX_k, \quad \text{where } x_0' \equiv x_0 + \frac{\hbar k}{q\mathscr{B}}. \tag{3.47}$$

But this 1D Schrödinger equation is identical to Eq. (2.261) for a 1D harmonic oscillator[19], with the center at the point x_0', and the frequency ω_0 equal to

$$\omega_c = \frac{|q\mathscr{B}|}{m}. \tag{3.48}$$

In the last expression, it is easy to recognize the *cyclotron frequency* of the classical particle's rotation in the magnetic field. (It may be readily obtained using the 2nd Newton law for a circular orbit of radius r,

$$m\frac{v^2}{r} = F_{\mathscr{B}} \equiv qv\mathscr{B}, \tag{3.49}$$

and noting that the resulting ratio $v/r = |q\mathscr{B}|/m$ is just the radius-independent angular velocity ω_c of the particle's rotation.) Hence, the energy spectrum for each Fourier component of the expansion (3.45) is the same:

$$E_n = \hbar\omega_c\left(n + \frac{1}{2}\right), \tag{3.50}$$

independent of either x_0, or y_0, or k.

This is a good example of a highly *degenerate* system: for each eigenvalue E_n, there are many different eigenfunctions that differ by the positions $\{x_0, y_0\}$ of their center on axis x, and the rate k of their phase change along axis y. They may be used to assemble a large variety of linear combinations, including 2D wave packets whose centers move along classical circular orbits with some radius r determined by initial conditions. Note, however, that this radius cannot be smaller than the so-called *Landau radius*,

$$r_L \equiv \left(\frac{\hbar}{m\omega_c}\right)^{1/2} \equiv \left(\frac{\hbar}{|q\mathscr{B}|}\right)^{1/2}, \tag{3.51}$$

which characterizes the minimum size of the wave packet, and follows from Eq. (2.276) after the replacement $\omega_0 \to \omega_c$. This radius is remarkably independent

---

[19] This result may become a bit less puzzling if we recall that at the classical circular cyclotron motion of a particle, each of its Cartesian coordinates, including x, performs sinusoidal oscillations with the frequency (3.48), just as a 1D harmonic oscillator with this frequency.

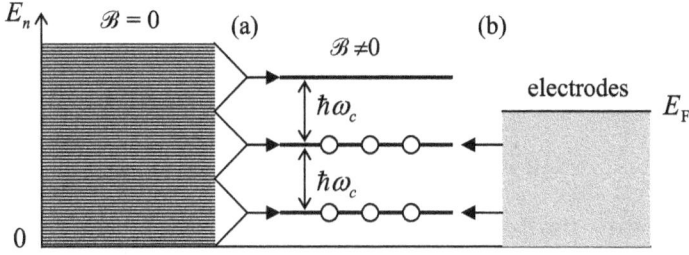

Figure 3.5. (a) The 'condensation' of 2D states on Landau levels, and (b) filling the levels with electrons at the quantum Hall effect.

on the particle's mass, and may be interpreted in the following way: the scale $\mathscr{B}A_{min}$ of the applied magnetic field's flux through the effective area $A_{min} = 2\pi r_L{}^2$ of the smallest wave packet is just one normal flux quantum $\Phi_0' \equiv 2\pi\hbar/|q|$.

A detailed analysis of such wave packets (for which we would not have time in this course) shows that the magnetic field does not change the *average* density dN_2/dE of different 2D states on the energy scale, following from Eq. (1.99), but just 'assembles' them on the Landau levels (see figure 3.5a), so that the number of different orbital states on each Landau level (per unit area) is

$$
\begin{aligned}
n_L &\equiv \frac{N_2}{A} = \frac{1}{A}\frac{dN_2}{dE}\Big|_{\mathscr{B}\neq 0}\Delta E \equiv \frac{1}{A}\frac{dN_2}{d^2k}\Big|_{\mathscr{B}=0}\frac{d^2k}{dk}\frac{1}{dE/dk}\Delta E \\
&= \frac{1}{A}\frac{A}{(2\pi)^2}2\pi k\frac{1}{\hbar^2 k/m} \quad \hbar\omega_c \equiv \frac{|q\mathscr{B}|}{2\pi\hbar}.
\end{aligned}
\tag{3.52}
$$

This expression may again be interpreted in terms of magnetic flux quanta: $n_L\Phi_0' = \mathscr{B}$, i.e. there is one particular state on each Landau level per each normal flux quantum.

The most famous application of the Landau levels picture is the explanation of the *quantum Hall effect*[20]. It is usually observed in the 'Hall bar' geometry sketched in figure 3.6, where electric current I is passed through a rectangular conducting sample placed into magnetic field $\mathscr{B}$ perpendicular to the sample's plane. The classical analysis of the effect is based on the notion of the Lorentz force (3.14). As the magnetic field is turned on, this force starts to divert the effective charge carriers (electrons or holes) from their straight motion from one external electrode to another, bending them to the insulated edges of the bar (in figure 3.6, parallel to the x-axis). Here the carriers accumulate, generating a gradually increasing electric field $\mathscr{E}$, until its force (3.16) exactly balances the Lorentz force (3.14):

$$
q\mathscr{E}_y = qv_x\mathscr{B},
\tag{3.53}
$$

where v_x is the drift velocity of the carriers along the bar (figure 3.6), providing the sustained balance condition $\mathscr{E}_y/v_x = \mathscr{B}$ at each point of the sample.

---

[20] It was first observed in 1980 by a group led by K von Klitzing, while the classical limit (3.54) of the effect was first observed by E Hall in 1879.

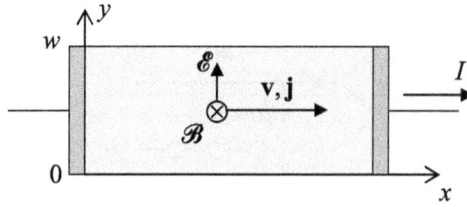

Figure 3.6. The Hall bar geometry. Darker rectangles show external (3D) electrodes.

With n_2 carriers per unit area, in a sample of width w, this condition yields the following classical expression for the so-called *Hall resistance* R_H:

$$R_H \equiv \frac{V_y}{I_x} = \frac{\mathscr{E}_y w}{q n_2 v_x w} = \frac{\mathscr{B}}{q n_2}. \qquad (3.54)$$

This formula is broadly used in practice for the measurement of the 2D density n_2 of the charge carriers, and of the carrier type—electrons with $q = -e < 0$, or holes with the effective charge $q = +e > 0$.

However, in experiments with high-quality (low-defect) 2D well structures, at very low, sub-kelvin temperatures[21] and high magnetic fields, the linear growth of R_H with $\mathscr{B}$, described by Eq. (3.54), is interrupted by virtually horizontal plateaus (figure 3.7) with constant values

$$R_H = \frac{1}{i} R_K, \qquad (3.55)$$

where i (only in this context, following tradition!) is a real integer, and the so-called *Klitzing constant*,

$$R_K = \frac{2\pi\hbar}{e^2} \approx 25.812\ 807\ 557\ k\Omega, \qquad (3.56)$$

is reproduced with an extremely high accuracy (up to $\sim 10^{-9}$) from experiment to experiment and from sample to sample. Such stability is a very rare exception in solid state physics where most results are noticeably dependent on the particular material and the particular sample under study.

This effect may be explained using the Landau level picture. The 2D sample is typically in weak contact with 3D electrodes whose conductivity electrons, at low temperatures, fill all states with energies below a certain Fermi energy E_F—see figure 3.5b. According to Eqs. (3.48) and (3.50), as $\mathscr{B}$ is increased, the spacing $\hbar\omega_c$ between the Landau levels increases proportionately, so that fewer and fewer of these levels are below E_F (and hence all their states are filled in equilibrium), and within broad

---

[21] In some crystals, such as *the graphene* (virtually perfect 2D sheets of carbon atoms, to be discussed in section 3.4 below), the effect may be more stable to thermal fluctuations, due to their topological properties, so that it may be observed even at room temperature—see, e.g. [3]. Also note that in some thin ferromagnetic layers, the quantum Hall effects may be observed in the absence of external magnetic field—see, e.g. the recent publication [4] and references therein.

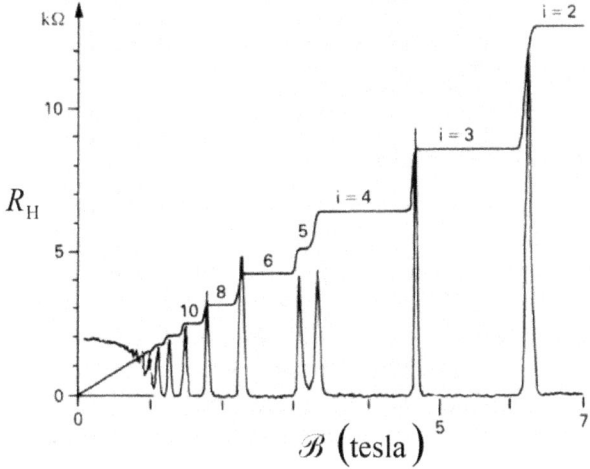

Figure 3.7. A typical record of the integer quantum Hall effect. The lower trace (with sharp peaks) shows the diagonal component, V_x/I_x, of the resistance tensor. (Adapted from https://www.nobelprize.org/nobel_prizes/physics/laureates/1998/press.html.)

ranges of the field variation, the number i of the filled levels is constant. (In figure 3.5b, $i = 2$.) So, plugging $n_2 = in_L$ and $q = -e$ into Eq. (3.54), and using Eq. (3.52) for n_L, we get

$$R_H = \frac{1}{i}\frac{\mathscr{B}}{qn_L} = \frac{1}{i}\frac{2\pi\hbar}{e^2},\qquad(3.57)$$

i.e. exactly the experimental result (3.55)–(3.56).

Admittedly, this oversimplified explanation of the quantum Hall effect does not take into account at least two important factors:

(i) the nonuniformity of the background potential $U(x, y)$ in realistic Hall bar samples, and the role of the quasi-1D *edge channels* this nonuniformity produces[22]; and

(ii) the Coulomb interaction of the electrons, in high-quality samples leading to the formation of R_H plateaus with not only integer, but also fractional values of i (1/3, 2/5, 3/7, etc)[23].

---

[22] Such quasi-1D regions, with the width of the order of r_L, form along the lines were the Landau levels cross the Fermi surface, and are actually responsible for all the electron transfer at the quantum Hall effect, giving the most famous example of the so-called *topological insulators*. The particle motion along these channels is effectively one-dimensional; because of this, it cannot be affected by modest unintentional nonuniformities of the potential $U(x, y)$. This fact is responsible for the extraordinary accuracy of Eqs. (3.55)–(3.56).

[23] This *fractional quantum Hall effect* was discovered in 1982 by D Tsui, H Stormer, and A Gossard. In contrast, the effect described by Eq. (3.55) with an integer i (figure 3.7) is now called the *integer quantum Hall effect*.

Unfortunately, a thorough discussion of these very interesting features is well beyond the framework of this course[24,25].

3.3 Scattering and diffraction

The second class of quantum effects that become more rich in multi-dimensional spaces is typically referred to as either *diffraction* or *scattering*—depending on the context. In classical physics, these two terms are used to describe very different effects. The term 'diffraction' is used for the interference of the *waves* re-emitted by many elementary components of an extended object, under the effect of a single incident *wave*. (The term 'interference' is typically reserved for the wave re-emission by a few components, such as two slits in the Young experiment[26].)

On the other case, the term 'scattering' is used in classical mechanics to describe the result of the interaction of a beam of incident *particles* with such an extended object, called the *scatterer*[27]. Figure 3.8 shows the general scattering situation.

Most commonly, the detector of the scattered particles is located at a large distance $r \gg a$ from the scatterer. In this case, the main observable independent of r is the *flux* (the number per unit time) of particles scattered in a certain direction, i.e. their flux per unit solid angle. Since this flux is proportional to the incident flux of particles per unit area, the efficiency of scattering in a particular direction may be characterized by the ratio of these two fluxes. This ratio has is called the *differential cross-section* of the scatterer:

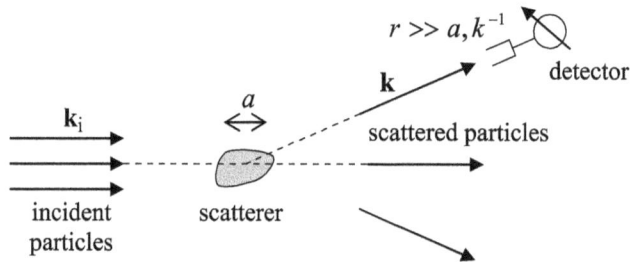

Figure 3.8. Scattering (schematically).

---

[24] For a comprehensive discussion of these effects I can recommend, e.g. either the monograph [5], or the review [6]. (See also the later publications cited above.)

[25] Note also that the quantum Hall effect is sometimes discussed in terms of the so-called *Berry phase*, one of the *geometric phases*—the notion apparently pioneered by S Pancharatnam in 1956. However, in the 'usual' quantum Hall effect the Berry phase equals zero, and I believe that this concept should be saved for topologically more involved systems. Unfortunately, I will have no time/space for a discussion of such systems in this course, and have to refer the interested reader to special literature—see, e.g. either the key papers collected by A Shapere and F Wilczek [7], or the monograph by A Bohm [8].

[26] See, e.g. the discussion of diffraction and interference of electromagnetic waves in *Part EM* sections 8.3–8.8.

[27] In the context of classical waves, the term 'scattering' is typically reserved for wave interaction with disordered sets of small objects—see, e.g. *Part EM* section 8.3.

$$\frac{d\sigma}{d\Omega} \equiv \frac{\text{flux of scatterd particles per unit solid angle}}{\text{flux of incident particles per unit area}}. \tag{3.58}$$

Such terminology and notation stem from the fact that the integral of $d\sigma/d\Omega$ over all scattering angles,

$$\sigma \equiv \oint \frac{d\sigma}{d\Omega} d\Omega = \frac{\text{total flux of scattered particles}}{\text{incident flux per per unit area}}, \tag{3.59}$$

evidently having the dimensionality of area, has a simple interpretation as the *full cross-section* of scattering. For the simplest case when a solid object scatters all classical particles hitting its surface, but does not affect the particles flying by it, σ is just the geometrical cross-section of the scatterer, as visible from the direction of the incident particles. In classical mechanics, we first calculate the particle's scattering angle as a function of its *impact parameter b*, and then average the result over all values of b, considered random[28].

In quantum mechanics, due to the particle/wave duality, a relatively broad, parallel beam of incident particles of the same energy E may be fairly represented with a plane de Broglie wave (1.88):

$$\psi_i = |\psi_i| \exp\{i\mathbf{k}_i \cdot \mathbf{r}\}, \tag{3.60}$$

with the free-space wave number $k_i = k = (2mE)^{1/2}/\hbar$. As a result, the particle scattering becomes a synonym of the de Broglie wave diffraction, and (somewhat counter-intuitively) the description of the effect becomes simpler, excluding the notion of the impact parameter. Indeed, the wave (3.60) corresponds to a constant probability current density (1.49):

$$\mathbf{j}_i = |\psi_i|^2 \frac{\hbar}{m}\mathbf{k}_i, \tag{3.61}$$

which is exactly the flux of incident particles per unit area that is used in the denominator of Eq. (3.58), while the numerator of that fraction may be simply expressed via the probability current density $\mathbf{j}_s$ of the scattered de Broglie waves:

$$\frac{d\sigma}{d\Omega} = \frac{j_s r^2}{j_i}, \quad \text{at } r \gg a. \tag{3.62}$$

Hence our task is reduced to the calculation of $\mathbf{j}_s$ at sufficiently large distances from the scatterer. For that, let us rewrite the stationary Schrödinger equation for the *elastic* scattering problem (when the energy E of the scattered particles is the same as that of the incident particles) in the form

$$(E - \hat{H}_0)\, \psi = U(\mathbf{r})\psi, \quad \text{with } \hat{H}_0 \equiv -\frac{\hbar^2}{2m}\nabla^2, \quad \text{and } E = \frac{\hbar^2 k^2}{2m}, \tag{3.63}$$

---

[28] See, e.g. *Part CM* section 3.5.

where the potential energy $U(\mathbf{r})$ describes the effect of scatterer. Looking for the solution of Eq. (3.62) in the natural form

$$\psi = \psi_i + \psi_s, \tag{3.64}$$

where ψ_i is the incident wave (3.60), and ψ_s has the sense of the scattered wave, and taking into account that former wave satisfies the free-space Schrödinger equation

$$\hat{H}_0 \psi_i = E \psi_i, \tag{3.65}$$

we may reduce Eq. (3.63) to either of the following equivalent forms:

$$(E - \hat{H}_0)\ \psi_s = U(\mathbf{r})(\psi_i + \psi_s), \qquad (\nabla^2 + k^2)\ \psi_s = \frac{2m}{\hbar^2} U(\mathbf{r})\ \psi. \tag{3.66}$$

For applications, an integral form of this equation is more convenient. To derive it, we may look at the second of Eqs. (3.66) as a linear, inhomogeneous differential equation for the function ψ_s, thinking of its right-hand side as a known 'source'. The solution of such an equation obeys the linear superposition principle, i.e. we may represent it as the sum of the waves outcoming from all elementary volumes d^3r' of the scatterer. Mathematically, this sum may be expressed as either

$$\psi_s(\mathbf{r}) = \frac{2m}{\hbar^2} \int U(\mathbf{r}')\psi(\mathbf{r}')G(\mathbf{r},\ \mathbf{r}')d^3r', \tag{3.67a}$$

or, equivalently, as[29]

$$\psi(\mathbf{r}) = \psi_i(\mathbf{r}) + \frac{2m}{\hbar^2} \int U(\mathbf{r}')\psi(\mathbf{r}')G(\mathbf{r},\ \mathbf{r}')d^3r', \tag{3.67b}$$

where $G(\mathbf{r},\ \mathbf{r}')$ is the *spatial Green's function*, defined as such an elementary, spherically-symmetric response of the 3D Helmholtz equation to a point source, i.e. the outward-propagating solution of the following equation[30]

$$(\nabla^2 + k^2)\ G = \delta(\mathbf{r} - \mathbf{r}'). \tag{3.68}$$

But we already know such a solution of this equation—see Eq. (3.7) and its discussion:

$$G(\mathbf{r},\ \mathbf{r}') = \frac{f_+}{R}e^{ikR}, \quad \text{where } \mathbf{R} \equiv \mathbf{r} - \mathbf{r}', \tag{3.69}$$

so that we need just to calculate the coefficient f_+ for Eq. (3.68). This can be done in several ways, for example by noticing that at $r \ll k^{-1}$, the second term on the left-

---

[29] This relation is sometimes called the *Lipmann–Schwinger equation,* though more frequently this term is reserved for either its operator form, or the resulting equation for the spatial Fourier components of ψ and ψ_i.
[30] Please notice both the similarity and difference between this Green's function and the propagator discussed in section 2.1. In both cases, we use the linear superposition principle to solve wave equations, but while Eq. (3.67) gives the solution of the *inhomogeneous* equation (3.66), Eq. (2.44) does that for a *homogeneous* Schrödinger equation. In the latter case the elementary wave sources are the elementary parts of the initial wavefunction, rather than of the equation's right-hand side as in our current problem.

hand side of Eq. (3.68) is negligible, so that it is reduced to the well-known Poisson equation with a delta-functional right-hand side, which describes, for example, the electrostatic potential induced by a point electric charge. Either recalling the Coulomb law, or applying the Gauss theorem[31], we readily get the asymptote

$$G \to -\frac{1}{4\pi R}, \quad \text{at } kR \ll 1, \qquad (3.70)$$

which is compatible with Eq. (3.69) only if $f_+ = -1/4\pi$, i.e. if

$$G(\mathbf{r}, \mathbf{r}') = -\frac{1}{4\pi R} e^{ikR}. \qquad (3.71)$$

Plugging this result into Eq. (3.67a), we get the following formal solution of Eq. (3.66):

$$\psi_s(\mathbf{r}) = -\frac{m}{2\pi\hbar^2} \int U(\mathbf{r}') \frac{\psi(\mathbf{r}')}{R} e^{ikR} d^3r'. \qquad (3.72)$$

Note that if the function $U(\mathbf{r})$ is smooth, the singularity in the denominator is integrable (i.e. not dangerous); indeed, the contribution of the sphere with some radius $\mathcal{R} \to 0$, with the center in point $\mathbf{r}'$, into this integral scales as

$$\int_{R<\mathcal{R}} \frac{d^3R}{R} \equiv 4\pi \int_0^{\mathcal{R}} \frac{R^2 dR}{R} \equiv 4\pi \int_0^{\mathcal{R}} R dR = 2\pi \mathcal{R}^2 \to 0. \qquad (3.73)$$

So far, our result (3.72) is exact, but its apparent simplicity is deceiving, because the wavefunction ψ on its right-hand side generally includes not only the incident wave ψ_i, but also the scattered wave ψ_s—see Eq. (3.64). The most straightforward, and most common simplification of this problem, called the *Born approximation*[32], is possible if the scattering potential $U(\mathbf{r})$ is in some sense small. (We will derive the quantitative condition of this smallness in a minute.) Since at $U(\mathbf{r}) = 0$ the scattering wave ψ_s has to disappear, at small but non-zero $U(\mathbf{r})$, $|\psi_s|$ has to be much smaller than $|\psi_i|$. In this case, on the right-hand side of Eq. (3.73) we may ignore ψ_s in comparison with ψ_i, getting

$$\psi_s(\mathbf{r}) = -\frac{m}{2\pi\hbar^2} |\psi_i| \int U(\mathbf{r}') \frac{\exp\{i\mathbf{k}_i \cdot \mathbf{r}'\}}{R} e^{ikR} d^3r'. \qquad (3.74)$$

Actually, Eq. (3.74) gives us even *more* than we wanted: it evaluates the scattered wave at *any* point, including those within of the scattering object, while in order to

---

[31] See, e.g. *Part EM* section 1.2.

[32] Named after M Born, who was the first to apply this approximation in quantum mechanics. However, the basic idea of this approach had been developed much earlier (in 1881) by Lord Rayleigh in the context of electromagnetic wave scattering—see, e.g. *Part EM* section 8.3. Note that the contents of that section repeat some aspects of our current discussion—perhaps regrettably but unavoidably so, because the Born approximation is a centerpiece of the theory of scattering/diffraction for both the electromagnetic waves and the de Broglie waves. Hence I felt I had to cover it in this volume for the benefit of readers who skipped the *EM* part of my series.

spell out Eq. (3.62), we only need to find the wave *far* from the scatterer, at $r \to \infty$. However, before going to that limit, we can use this general formula to find the quantitative criterion of the Born approximation's validity. For that, let us estimate the magnitude of the right-hand side of this equation for a scatterer of a linear size $\sim a$, and the potential magnitude's scale U_0, in the following two limits:

(i) If $ka \ll 1$, then inside the scatterer (i.e. at distances $r' \sim a$), both the $\exp\{i\mathbf{k}\cdot\mathbf{r}'\}$ and the second exponent under the integral in Eq. (3.74) change little, so that a crude estimate of the solution's magnitude is

$$| \psi_s | \sim \frac{m}{2\pi\hbar^2} | \psi_i | \, U_0 a^2. \tag{3.75}$$

(ii) In the opposite limit $ka \gg 1$, the function under the integral is nearly periodic in one of the spatial directions (that of the wave propagation), so that the integral accumulates only on distances of the order of the de Broglie wavelength, $\sim k^{-1}$, and the integral is correspondingly smaller:

$$| \psi_s | \sim \frac{m}{2\pi\hbar^2} | \psi_i | \, U_0 \frac{a^2}{ka}. \tag{3.76}$$

These relations allow us to spell out the Born approximation condition, $|\psi_s| \ll |\psi_i|$, as

$$U_0 \ll \frac{\hbar^2}{ma^2} \max[ka, \; 1]. \tag{3.77}$$

In the first factor on the right-hand side, we may readily recognize the scale of the kinetic (quantum-confinement) energy E_a of the particle inside a potential well of size $\sim a$, so that the Born approximation is valid essentially if the potential energy of a particle's interaction with the scatterer is smaller than E_a. Note, however, that the estimates (3.75) and (3.76) are not valid in some special situations when the effects of scattering *accumulate* in some direction. This is frequently the case for small angles θ of scattering by extended objects (when $ka \gg 1$, but $ka\theta \lesssim 1$).

Now let us proceed to large distances $r \gg r' \sim a$, and simplify Eq. (3.74) using an approximation similar to the dipole expansion in electrodynamics[33]. Namely, in the denominator's R, we can merely ignore r' in comparison with r, but the exponents require more care, because even if $r' \sim a \ll r$, the product $kr' \sim ka$ may still be larger than 1. In the first approximation in r', we can take (figure 3.9a):

$$R \equiv | \mathbf{r} - \mathbf{r}'| \approx r - \mathbf{n}_r \cdot \mathbf{r}', \tag{3.78}$$

and since the directions of the vectors $\mathbf{k}$ and $\mathbf{r}$ coincide, i.e. $\mathbf{k} = k\mathbf{n}_r$, we get

$$kR \approx kr - \mathbf{k} \cdot \mathbf{r}', \quad \text{and} \quad e^{ikR} \approx e^{ikr}e^{-i\mathbf{k}\cdot\mathbf{r}'}. \tag{3.79}$$

---

[33] See, e.g. *Part EM* section 8.2.

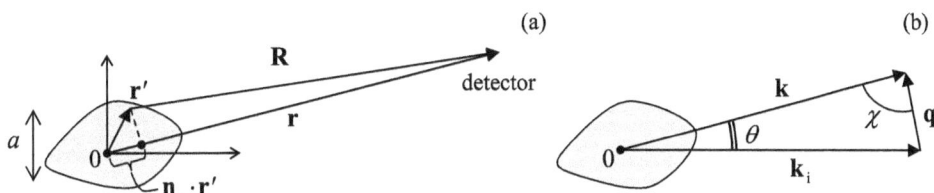

Figure 3.9. (a) The long-range expansion of R, and (b) the definitions of $\mathbf{q}$, χ, and θ.

With this replacement, Eq. (3.74) yields

$$\psi_s(\mathbf{r}) = -\frac{m}{2\pi\hbar^2}\frac{|\psi_i|}{r}e^{ikr}\int U(\mathbf{r}')\exp\{-i(\mathbf{k}-\mathbf{k}_i)\cdot\mathbf{r}'\}d^3r'. \tag{3.80}$$

This relation is a particular case of a more general formula[34]

$$\psi_s = |\psi_i|\frac{f(\mathbf{k},\mathbf{k}_i)}{r}e^{ikr}, \tag{3.81}$$

where $f(\mathbf{k},\mathbf{k}_i)$ is called the *scattering function*[35]. The physical sense of this function becomes clear from the calculation of the corresponding probability current density $\mathbf{j}_s$. For that, generally we need to use Eq. (1.47) with the gradient operator having all spherical-coordinate components[36]. However, at $kr \gg 1$, the main contribution to $\nabla\psi_s$, proportional to $k \gg 1/r$, is provided by differentiating the factor e^{ikr}, which changes in the common direction of vectors $\mathbf{r}$ and $\mathbf{k}$, so that

$$\nabla\psi_s \approx \mathbf{n}_r\frac{\partial}{\partial r}\psi_s \approx \mathbf{k}\psi_s, \quad \text{at } kr \gg 1, \tag{3.82}$$

and Eq. (1.47) yields

$$\mathbf{j}_s(\theta) \approx \frac{\hbar}{m}|\psi_i|^2\frac{|f(\mathbf{k},\mathbf{k}_i)|^2}{r^2}\mathbf{k}. \tag{3.83}$$

Plugging this expression into Eq. (3.62), for the differential cross-section we get simply

$$\frac{d\sigma}{d\Omega} = |f(\mathbf{k},\mathbf{k}_i)|^2, \tag{3.84}$$

---

[34] It is easy to prove that this form is an asymptotic form of *any* solution ψ_s of the scattering problem (even that beyond the Born approximation) at sufficiently large distances $r \gg a$, k^{-1}.

[35] Note that the function f has the dimension of length, and does not account for the incident wave. This is why sometimes a dimensionless function, $S = 1 + 2ikf$, is used instead. This function S is called the *scattering matrix*, because it may be considered a natural generalization of the 1D matrix S defined by Eq. (2.124), to higher dimensionality.

[36] See, e.g. Eq. (A.66).

while the total cross-section is

$$\sigma = \oint |f(\mathbf{k}, \mathbf{k}_i)|^2 \, d\Omega, \tag{3.85}$$

so that the scattering function $f(\mathbf{k}, \mathbf{k}_i)$ gives us everything we need—and in fact more, because the function also contains information about the phase of the scattered wave.

According to Eq. (3.80), in the Born approximation the scattering function is reduced to the so-called *Born integral*

$$f(\mathbf{k}, \mathbf{k}_i) = -\frac{m}{2\pi\hbar^2} \int U(\mathbf{r}) e^{-i\mathbf{q}\cdot\mathbf{r}} d^3r, \tag{3.86}$$

where for the notation simplicity, $\mathbf{r}'$ is replaced with $\mathbf{r}$, and the following *scattering vector* is introduced:

$$\mathbf{q} \equiv \mathbf{k} - \mathbf{k}_i, \tag{3.87}$$

with the length $q = 2k \sin(\theta/2)$, where θ is the *scattering angle* between the vectors $\mathbf{k}$ and $\mathbf{k}_i$—see figure 3.9b. For the differential cross-section, Eqs. (3.84) and (3.86) yield[37]

$$\frac{d\sigma}{d\Omega} = \left(\frac{m}{2\pi\hbar^2}\right)^2 \left| \int U(\mathbf{r}) e^{-i\mathbf{q}\cdot\mathbf{r}} d^3r \right|^2. \tag{3.88}$$

This is the main result of this section; it may be further simplified for spherically-symmetric scatterers, with

$$U(\mathbf{r}) = U(r). \tag{3.89}$$

In this case, it is convenient to represent the exponent in the Born integral as $\exp\{-iqr\cos\chi\}$, where χ is the angle between the vectors $\mathbf{k}$ (i.e. the direction $\mathbf{n}_r$ toward the detector) and $\mathbf{q}$ (rather than the incident wave vector $\mathbf{k}_i$!)—see figure 3.9b. Now, for a fixed $\mathbf{q}$, we can take this vector's direction for the polar axis of a spherical coordinate system, and reduce Eq. (3.86) to a 1D integral:

$$
\begin{aligned}
f(\mathbf{k}, \mathbf{k}_i) &= -\frac{m}{2\pi\hbar^2} \int_0^\infty r^2 dr\, U(r) \int_0^{2\pi} d\varphi \int_0^\pi \sin\chi d\chi \, \exp\{-iqr\cos\chi\} \\
&= -\frac{m}{2\pi\hbar^2} \int_0^\infty r^2 dr\, U(r) \; 2\pi \frac{2\sin qr}{qr} \\
&\equiv -\frac{2m}{\hbar^2 q} \int_0^\infty U(r) \sin(qr)\, r dr.
\end{aligned}
\tag{3.90}
$$

---

[37] Note that according to Eq. (3.88), in the Born approximation the scattering intensity does not depend on the sign of the potential U, and also that scattering in a certain direction is completely determined by a specific Fourier component of the function $U(\mathbf{r})$, namely by its harmonic with the wave vector equal to the scattering vector $\mathbf{q}$.

As a simple example, let us use the Born approximation to analyze scattering on the following spherically-symmetric potential:

$$U(\mathbf{r}) = U_0 \exp\left\{-\frac{r^2}{2a^2}\right\}. \tag{3.91}$$

In this particular case, it is better to avoid temptation to exploit the spherical symmetry by using Eq. (3.90), and instead use the generic Eq. (3.88), falling into a product of three similar Cartesian factors:

$$f(\mathbf{k}, \mathbf{k_i}) = -\frac{mU_0}{2\pi\hbar^2}I_xI_yI_z, \quad \text{with } I_x \equiv \int_{-\infty}^{+\infty} \exp\left\{-\left(\frac{x^2}{2a^2} + iq_xx\right)\right\}dx, \tag{3.92}$$

and similar integrals for I_y and I_z. From chapter 2, we already know that the Gaussian integrals like I_x may be readily worked out by complementing the exponent to the full square, in our current case giving

$$I_x = (2\pi)^{1/2}a \exp\left\{-\frac{q_x^2a^2}{2}\right\}, \text{ etc.,} \tag{3.93}$$

$$\frac{d\sigma}{d\Omega} = \left(\frac{mU_0}{2\pi\hbar^2}I_xI_yI_z\right)^2 = 2\pi a^2\left(\frac{mU_0a^2}{\hbar^2}\right)^2 e^{-q^2a^2}.$$

Now, the total cross-section σ is an integral of $d\sigma/d\Omega$ over all directions of vector $\mathbf{k}$. Since in our case the scattering intensity does not depend on the azimuthal angle φ, the only nontrivial integration is over the scattering angle θ—see figure 3.9b:

$$\sigma \equiv \oint \frac{d\sigma}{d\Omega}d\Omega = 2\pi \int_0^\pi \frac{d\sigma}{d\Omega}\sin\theta d\theta$$

$$= 4\pi^2a^2\left(\frac{mU_0a^2}{\hbar^2}\right)^2 \int_0^\pi \exp\left\{-\left(2k\sin\frac{\theta}{2}\right)^2 a^2\right\}\sin\theta d\theta$$

$$\equiv 4\pi^2a^2\left(\frac{mU_0a^2}{\hbar^2}\right)^2 \int_{\theta=0}^{\theta=\pi} \exp\{-2k^2a^2(1-\cos\theta)\}\, d(1-\cos\theta) \tag{3.94}$$

$$= \frac{2\pi^2}{k^2}\left(\frac{mU_0a^2}{\hbar^2}\right)^2 (1-e^{-4k^2a^2}).$$

Let us analyze these results. In the low-energy limit, $ka \ll 1$ (and hence $qa \ll 1$ for any scattering angle), the scattered wave is virtually isotropic: $d\sigma/d\Omega \approx \text{const}$—a very typical feature of a scalar-wave scattering[38] by small objects, in any approximation. Note that according to Eq. (3.77), the Born expression for σ, following from Eq. (3.94) in this limit,

---

[38] Note that this is only true for scalar (e.g. de Broglie) waves, and different for vector ones, in particular the electromagnetic waves, where the intensity of the dipole radiation, and hence the scattering by small objects vanishes in the direction of the incident field's polarization—see, e.g. *Part EM* Eqs. (8.26) and (8.139).

$$\sigma = 8\pi^2 a^2 \left(\frac{mU_0a^2}{\hbar^2}\right)^2, \tag{3.95}$$

is only valid if σ is much smaller than the scale a^2 of the physical cross-section of the scatterer. In the opposite, high-energy limit $ka \gg 1$, the scattering is dominated by small angles $\theta \approx q/k \sim 1/ka \sim \lambda/a$:

$$\frac{d\sigma}{d\Omega} \approx 2\pi a^2 \left(\frac{mU_0a^2}{\hbar^2}\right)^2 \exp\{-k^2a^2\theta^2\}. \tag{3.96}$$

This is, again, very typical for the diffraction. Note, however, that due to the smooth character of the Gaussian potential (3.91), the diffraction pattern (3.98) exhibits no oscillations of $d\sigma/d\Omega$ as a function of the diffraction angle θ.

Such oscillations naturally appear for scatterers with sharp borders. Indeed, let us consider a uniform spherical scatterer, described by the potential

$$U(\mathbf{r}) = \begin{cases} U_0, & \text{for } r < R, \\ 0, & \text{otherwise.} \end{cases} \tag{3.97}$$

For this case, an easy integration by parts of Eq. (3.90) yields:

$$f(\mathbf{k}, \mathbf{k}_i) = \frac{2mU_0}{\hbar^2 q^3}(qR \cos qR - \sin qR), \quad \text{so that}$$

$$\frac{d\sigma}{d\Omega} = \left(\frac{2mU_0}{\hbar^2 q^3}\right)^2 (qR \cos qR - \sin qR)^2. \tag{3.98}$$

According to this result, the scattered wave's intensity drops very fast with q, so that one needs a semi-log plot (such as shown in figure 3.10) to make visible the

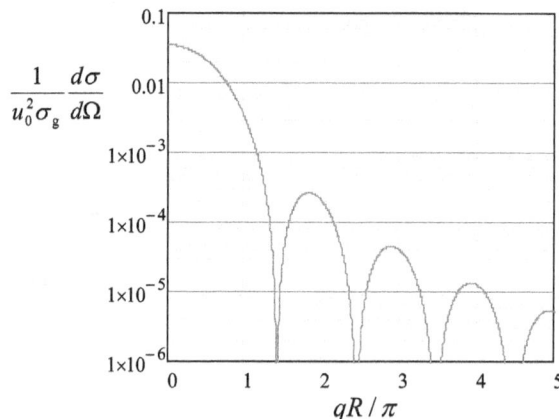

Figure 3.10. The differential cross-section of the Born scattering of a particle by a 'hard' (sharp-border) sphere (3.97), normalized to its geometric cross-section $\sigma_g \equiv \pi R^2$ and the square of the potential's magnitude parameter $u_0 \equiv U_0/(\hbar^2/2mR^2)$, as a function of the normalized magnitude of the scattering vector $\mathbf{q}$.

diffraction fringes[39], with the nth destructive interference (zero-intensity) point tending to $qR = \pi(n + 1/2)$ at $n \to \infty$. Since, as figure 3.9b shows, q may only change from 0 to $2k$, these intensity minima are only observable at sufficiently large values of the parameter kR, when they correspond to real values of the scattering angle θ. (At $kR \gg 1$, approximately kR/π of these minima, i.e. 'dark rings' of low scattering probability, are observable.) Conversely, at $kR \ll 1$ all allowed values of qR are much smaller than 1, and in this limit, the differential cross-section does not depend on qR, i.e. the scattering by the sphere (as by *any* object in this limit) is isotropic.

This example shows that in quantum mechanics the notions of particle scattering and diffraction are essentially inseparable.

The Born approximation, while being very simple and used more often than any other scattering theory, is not without substantial shortcomings, as becomes clear from the following example. It is not too difficult to prove the following general *optical theorem*, valid for an arbitrary scatterer:

$$\operatorname{Im} f(\mathbf{k}_i, \mathbf{k}_i) = \frac{k}{4\pi}\sigma. \tag{3.99}$$

However, Eq. (3.86) shows that in the Born approximation, the function f is purely real at $q = 0$ (i.e. for $\mathbf{k} = \mathbf{k}_i$), and hence cannot satisfy the optical theorem. Even more evidently, it cannot describe such a simple effect as a dark shadow ($\psi \approx 0$) cast by a virtually opaque object (say, with $U \gg E$). There are several ways to improve the Born approximation, while still sticking to the general idea of an approximate treatment of U.

(i) Instead of the main assumption $\psi_s \propto U_0$, we may use a complete perturbation series:

$$\psi_s = \psi_1 + \psi_2 + \ldots \tag{3.100}$$

with $\psi_n \propto U_0{}^n$, and find successive approximations ψ_n one by one. In the 1st approximation we of course return to the Born formula, but already the 2nd approximation yields

$$\operatorname{Im} f_2(\mathbf{k}_i, \mathbf{k}_i) = \frac{k}{4\pi}\sigma_1, \tag{3.101}$$

where σ_1 is the full cross-section calculated in the 1st approximation, so that the optical theorem (3.99) is 'almost' satisfied.

(ii) As was mentioned above, the Born approximation does not work very well for the objects elongated along the direction (say, x) of the initial wave vector $\mathbf{k}_i$. This deficiency may be corrected by the so-called *eikonal*[40] approximation, which replaces the plane-wave representation (3.60) of the incident wave by a WKB-like exponent, though still in the first nonvanishing approximation in $U \to 0$:

---

[39] Their physics is very similar to that of the Fraunhofer diffraction on a 1D scatterer—see, e.g. *Part EM* section 8.4.

[40] From the Greek word εικον, meaning 'image'. In our current context, this term is purely historic.

$$\exp\{ik_0 x\} \to \exp\left\{i\int_0^x k(\mathbf{r}')dx'\right\}$$

$$\equiv \exp\left\{i\int_0^x \frac{\{2m[E - U(\mathbf{r}')]\}^{1/2}}{\hbar}dx'\right\} \tag{3.102}$$

$$\approx \exp\left\{i\left[k_0 x - \frac{m}{\hbar^2 k_0}\int_0^x U(\mathbf{r}')dx'\right]\right\}.$$

This approximation's results satisfy the optical theorem (3.99) already in the 1st approximation in U.

Another way toward quantitative results in the theory of scattering, beyond the Born approximation, may be pursued for spherically-symmetric potentials (3.89); I will discuss it in section 3.8, after a general discussion of the particle motion in such potentials in section 3.7.

3.4 Energy bands in higher dimensions

In section 2.7, we have discussed the 1D band theory for potential profiles $U(x)$ that obey the periodicity condition (2.192). For what follows, let us notice that that condition may be rewritten as

$$U(x + X) = U(x), \tag{3.103}$$

where $X = \tau a$, with τ being an arbitrary integer. One may say that the set of points X forms a periodic 1D *lattice* in the direct ($\mathbf{r}$-) space. We have also seen that each Bloch state (i.e. each eigenstate of the Schrödinger equation for such periodic potential) is characterized by the quasi-momentum $\hbar q$, and its energy does not change if q is changed by a multiple of $2\pi/a$. Hence if we form, in the reciprocal ($\mathbf{q}$-) space, a 1D lattice of points $Q = lb$, with $b \equiv 2\pi/a$ and integer l, any pair of points from these two mutually reciprocal lattices satisfies the following rule:

$$\exp\{iQX\} = \exp\left\{il\frac{2\pi}{a}\tau a\right\} \equiv e^{2\pi i \tau l} = 1. \tag{3.104}$$

In this form, the results of section 2.7 may be readily extended to d-dimensional periodic potentials whose translational symmetry obeys the following natural generalization of Eq. (3.103):

$$U(\mathbf{r} + \mathbf{R}) = U(\mathbf{r}), \tag{3.105}$$

where the points $\mathbf{R}$, which may be numbered by d integers τ_j, form the so-called *Bravais lattice*[41]:

---

[41] Named after A Bravais, the crystallographer who introduced this notion in 1850.

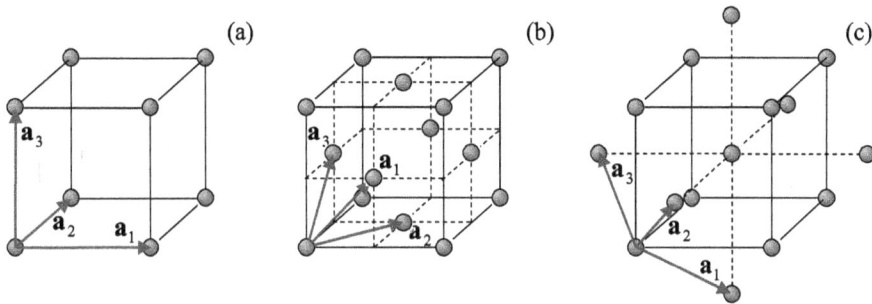

Figure 3.11. The simplest (and most common) 3D Bravais lattices: (a) simple cubic, (b) face-centered cubic (fcc), and (c) body-centered cubic (bcc), and possible choices of their primitive vector sets (blue arrows).

$$\mathbf{R} = \sum_{j=1}^{d} \tau_j \mathbf{a}_j, \qquad (3.106)$$

with d *primitive vectors* $\mathbf{a}_j$. The simplest example of a 3D Bravais lattice are given by the *simple cubic lattice* (figure 3.11a), which may be described by a system of mutually perpendicular primitive vectors $\mathbf{a}_j$ of equal length. However, not in any lattice these vectors are perpendicular; for example figures 3.11b,c show possible sets of the primitive vectors describing, respectively, the *face-centered cubic lattice* (fcc) and the *body-centered cubic lattice* (bcc). In 3D, the science of crystallography, based on the group theory, distinguishes, by their symmetry properties, 14 Bravais lattices grouped into 7 different *lattice systems*[42].

Note, however, not all highly symmetric sets of points form Bravais lattices. As probably the most striking example, the nodes of a very simple 2D *honeycomb lattice* (figure 3.12a)[43] cannot be described by a Bravais lattice—while the 2D hexagonal lattice, shown in figure 3.12b, can. The most prominent 3D case of such a lattice is the diamond structure (figure 3.12c), which describes, in particular, silicon[44]. In cases like these, the band theory is much facilitated by the fact that the Bravais lattices using some point assemblies (called *primitive unit cells*) may describe these point systems[45]. For example, figure 3.12a shows a possible choice of the primitive vectors for the honeycomb lattice, with the primitive unit cell formed by any two adjacent points of the original lattice (say, within the dashed ovals on that panel).

---

[42] A very clear, well illustrated introduction to the Bravais lattices is given in chapters 4 and 7 of the famous textbook [9].

[43] This structure describes, for example, the now-famous *graphene*—isolated monolayer sheets of carbon atoms arranged in a honeycomb lattice with the interatomic distance of 0.142 nm.

[44] This diamond structure may be best understood as an overlap of two fcc lattices of side a, mutually shifted by the vector $\{1, 1, 1\} \times a/4$, so that the distances between each point of the combined lattice and its 4 nearest neighbors (see the solid gray lines in figure 3.12c) are all equal.

[45] A harder case is presented by *quasicrystals* (whose idea may be traced back to medieval Islamic tilings, but was discovered in natural crystals, by D Shechtman *et al*, only in 1984), which obey a high (say, five-fold) rotational symmetry, but cannot be described by a Bravais lattice with *any* finite primitive unit cell. For a popular review of quasicrystals see, for example, [10].

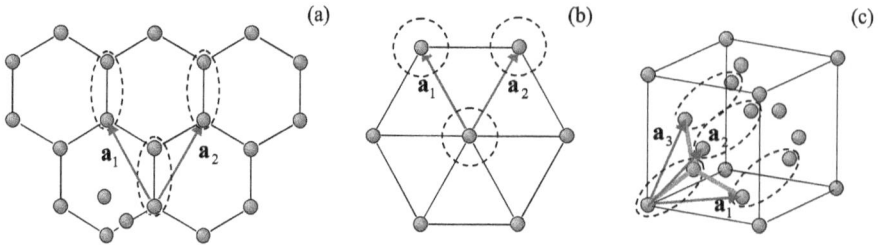

Figure 3.12. Some important periodic structures that require two-point primitive cells for their Bravais lattice presentation: (a) 2D honeycomb lattice and (c) 3D diamond lattice, and their primitive vectors. For a contrast, panel (b) shows the 2D hexagonal structure which forms a Bravais lattice with a single-point primitive cell.

Similarly, the diamond lattice may be described as an fcc Bravais lattice with two-point primitive unit cell—see figure 3.12c.

Now we are ready for the following generalization of the 1D Bloch theorem, given by Eqs. (2.193) and (2.210), to higher dimensions: any eigenfunction of the Schrödinger equation describing particle's motion in the infinite periodic potential (3.105) may be represented either as

$$\psi(\mathbf{r} + \mathbf{R}) = \psi(\mathbf{r})e^{i\mathbf{q}\cdot\mathbf{R}}, \tag{3.107}$$

or as

$$\psi(\mathbf{r}) = u(\mathbf{r})e^{i\mathbf{q}\cdot\mathbf{r}}, \quad \text{with } u(\mathbf{r} + \mathbf{R}) = u(\mathbf{r}), \tag{3.108}$$

where the quasi-momentum $\hbar\mathbf{q}$ is again a constant of motion, but now it is a vector. The key notion of the band theory in d dimensions is the *reciprocal lattice* in the wave-vector ($\mathbf{q}$) space, formed as

$$\mathbf{Q} = \sum_{j=1}^{d} l_j \mathbf{b}_j, \tag{3.109}$$

with integer l_j, and vectors $\mathbf{b}_j$ selected in such way that the following natural generalization of Eq. (3.104) is valid for any pair of points of the direct and reciprocal lattices:

$$e^{i\mathbf{Q}\cdot\mathbf{R}} = 1. \tag{3.110}$$

One way to describe the physical sense of the lattice $\mathbf{Q}$ is to say that according to Eqs. (3.80) and/or (3.86), it gives the set of the vectors $\mathbf{k} - \mathbf{k}_i$ for which the interference of the waves scattered by all Bravais lattice points is constructive, and hence strongly enhanced[46]. Another way to look at the reciprocal lattice follows

---

[46] This is why the notion of the $\mathbf{Q}$-lattice is also the main starting point of x-ray diffraction studies of crystals. Indeed, it allows rewriting the well-known Bragg condition for diffraction peaks in an extremely simple form $\mathbf{k} = \mathbf{k}_i + \mathbf{Q}$, where $\mathbf{k}_i$ and $\mathbf{k}$ are the wave vectors of the, respectively, incident and diffracted waves—see, e.g. *Part EM* section 8.4 (where is was more convenient to use the notation $\mathbf{k}_0$ for $\mathbf{k}_i$).

from the first formulation of the Bloch theorem, given by Eq. (3.107): if we add to the quasi-momentum $\mathbf{q}$ of a particle any vector $\mathbf{Q}$ of the reciprocal lattice, the wavefunction does not change. This means, in particular, that all information about the system's eigenfunctions is contained in just one elementary cell of the reciprocal space $\mathbf{q}$. Its most frequent choice, called the 1st *Brillouin zone*, is the set of all points $\mathbf{q}$ that are closer to the origin than to any other point of the lattice $\mathbf{Q}$. (Evidently, the 1st Brillouin zone in one dimension, discussed in section 2.7, falls under this definition—see, e.g. figures 2.26 and 2.28.)

It is easy to see that the primitive vectors $\mathbf{b}_j$ of the reciprocal lattice may be constructed as

$$\mathbf{b}_1 = 2\pi \frac{\mathbf{a}_2 \times \mathbf{a}_3}{\mathbf{a}_1 \cdot (\mathbf{a}_2 \times \mathbf{a}_3)}, \quad \mathbf{b}_2 = 2\pi \frac{\mathbf{a}_3 \times \mathbf{a}_1}{\mathbf{a}_1 \cdot (\mathbf{a}_2 \times \mathbf{a}_3)}, \quad \mathbf{b}_3 = 2\pi \frac{\mathbf{a}_1 \times \mathbf{a}_2}{\mathbf{a}_1 \cdot (\mathbf{a}_2 \times \mathbf{a}_3)}. \quad (3.111)$$

Indeed, from the 'operand rotation rule' of the vector algebra[47] it is evident that $\mathbf{a}_j \cdot \mathbf{b}_{j'} = 2\pi\delta_{jj'}$. Hence, with the account of Eq. (3.109), the exponent on the left-hand side of Eq. (3.110) is reduced to

$$e^{i\mathbf{Q}\cdot\mathbf{R}} = \exp\{2\pi i(l_1\tau_1 + l_2\tau_2 + l_3\tau_3)\}. \quad (3.112)$$

Since all l_j and τ_j are integers, the expression in the parentheses is also an integer, so that the exponent indeed equals 1, thus satisfying the definition of the reciprocal lattice given by Eq. (3.110).

As the simplest example, let us return to the simple cubic lattice of period a (figure 3.11a), oriented in space so that

$$\mathbf{a}_1 = a\mathbf{n}_x, \quad \mathbf{a}_2 = a\mathbf{n}_y, \quad \mathbf{a}_3 = a\mathbf{n}_z, \quad (3.113)$$

According to Eq. (3.111), its reciprocal lattice is also cubic:

$$\mathbf{Q} = \frac{2\pi}{a}(l_x\mathbf{n}_x + l_y\mathbf{n}_y + l_z\mathbf{n}_z), \quad (3.114)$$

so that the 1st Brillouin zone is a cube with the side $b = 2\pi/a$.

Almost equally simple calculations show that the reciprocal lattice of fcc is bcc, and vice versa. Figure 3.13 shows the resulting 1st Brillouin zone of the fcc lattice.

The notion of the reciprocal lattice makes the multi-dimensional band theory not much more complex than that in 1D, especially for numerical calculations, at least for the single-point Bravais lattices. Indeed, repeating all the steps that have led us to Eq. (2.215), but now with a d-dimensional Fourier expansion of functions $U(\mathbf{r})$ and $u_l(\mathbf{r})$, we readily get its generalization:

$$\sum_{l' \neq l} U_{l'-l}u_{l'} = (E - E_l)u_l, \quad (3.115)$$

---

[47] See, e.g. Eq. (A.48).

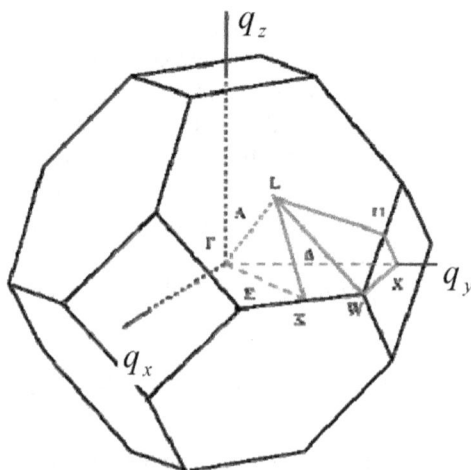

Figure 3.13. The 1st Brillouin zone of the fcc lattice, and the traditional notation of its main directions. Adapted from http://en.wikipedia.org/wiki/Band_structure, as a public domain material.

where **l** is now a *d*-dimensional vector of the integer indices l_j. The summation in Eq. (3.115) should be carried over all essential components of this vector (i.e. over all relevant nodes of the reciprocal lattice), so writing a corresponding computer code requires a bit more care than in 1D. However, this is just a homogeneous system of linear equations, and numerous routines of finding its eigenvalues E are readily available from both public sources and commercial software packages.

What is indeed more complex than in 1D is the representation (and hence the comprehension), of the calculation results and experimental data. Typically, the representation is limited to plotting the Bloch state eigenenergy as a function of components of the vector q along certain special directions the reciprocal space of quasi-momentum (see, e.g. the red lines in figure 3.13), typically plotted on a single panel. Figure 3.14 shows perhaps the most famous (and certainly the most practically important) of such plots, the band structure of electrons in silicon. The dashed horizontal lines mark the so-called *indirect gap* of the width ~1.12 eV between the 'valence' (nominally occupied) and the next 'conduction' (nominally unoccupied) energy bands.

In order to understand the reason for such complexity, let us see how would we start to calculate such a picture in the weak-potential approximation, for the simplest case of a 2D square lattice—which is a subset of the cubic lattice (3.106), with $\tau_3 = 0$. Its 1st Brillouin zone is of course also a square, of the area $(2\pi/a)^2$. Let us draw the lines of the constant energy of a free particle ($U = 0$) in this zone. Repeating the arguments of section 2.7 (see especially figure 2.28 and its discussion), we conclude that Eq. (2.216) should be now generalized as follows,

$$E = \frac{\hbar^2 k^2}{2m} = \frac{\hbar^2}{2m}\left[\left(q_x - \frac{2\pi\,l_x}{a}\right)^2 + \left(q_y - \frac{2\pi\,l_y}{a}\right)^2\right], \qquad (3.116)$$

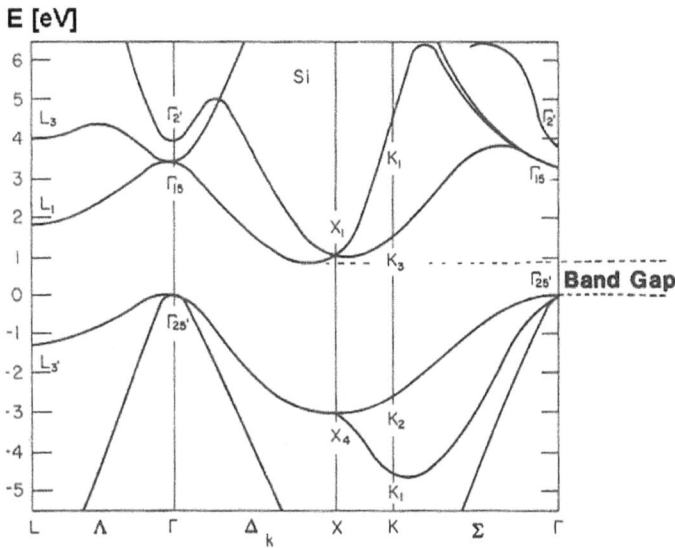

Figure 3.14. Band structure of silicon, along the special directions shown in figure 3.13. (Adapted from http://www.tf.uni-kiel.de/matwis/amat/semi_en/.)

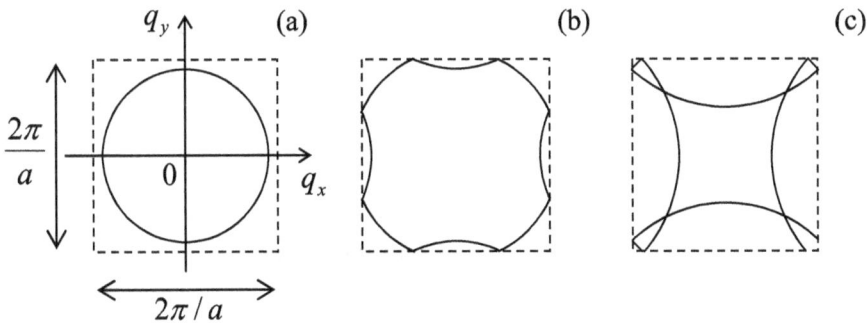

Figure 3.15. The lines of constant energy E of a free particle, within the 1st Brillouin zone of a square Bravais lattice, for: (a) $E/E_1 \approx 0.95$, (b) $E/E_1 \approx 1.05$; and (c) $E/E_1 \approx 2.05$, where $E_1 \equiv \pi^2 \hbar^2 / 2ma^2$.

with all possible integers l_x and l_y. Considering this result only within the 1st Brillouin zone, we see that as the particle's energy E grows, the lines of equal energy, for the lowest energy band, evolve as shown in figure 3.15. Just like in 1D, the weak-potential effects are only important at the Brillouin zone boundaries, and may be crudely considered as the appearance of narrow energy gaps, but one can see that the band structure in **q**-space is complex enough even without these effects—and becomes even more involved at higher E.

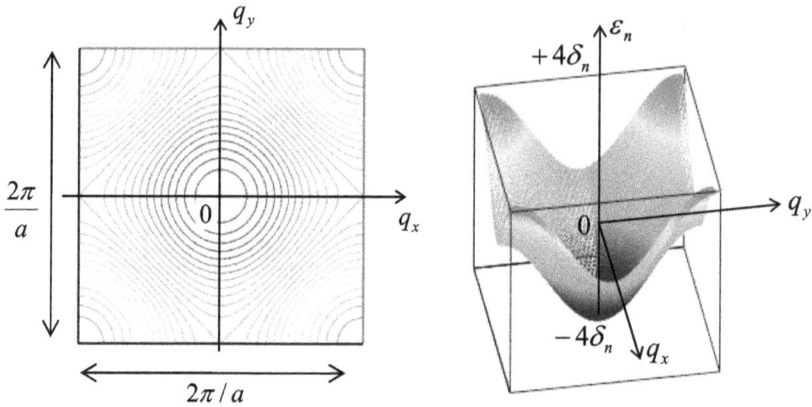

Figure 3.16. The allowed band energy $\varepsilon_n \equiv E - E_n$ for a square 2D lattice, in the tight-binding approximation.

The tight-binding approximation is usually easier to follow. For example, for the same square 2D lattice, we may repeat the arguments that have led us to Eq. (2.203), to write[48]

$$i\hbar \dot{a}_{0,0} = -\delta_n(a_{-1,0} + a_{+1,0} + a_{0,+1} + a_{0,-1}), \tag{3.117}$$

where the indices correspond to the deviations of the integers τ_x and τ_y from an arbitrarily selected minimum of the potential energy—and hence of the wave-function's 'hump', quasi-localized at this minimum. Now, looking for the stationary solution of these equations, that would obey the Bloch theorem (3.107), instead of Eq. (2.206) we get

$$\begin{aligned} E &= E_n + \varepsilon_n \\ &= E_n - \delta_n(e^{iq_x a} + e^{-iq_x a} + e^{iq_y a} + e^{-iq_y a}) \\ &\equiv E_n - 2\delta_n(\cos q_x a + \cos q_y a). \end{aligned} \tag{3.118}$$

Figure 3.16 shows this result, within the 1st Brillouin zone, in two forms: as the color-coded lines of equal energy, and as a 3D plot (also enhanced by color). It is evident that the plots of this function along different lines on the **q**-plane, for example along one of axes (say, q_x) and along a diagonal of the 1st Brillouin zone (say, with $q_x = q_y$) give different curves $E(q)$, qualitatively similar to those of silicon (figure 3.14). However, the latter structure is further complicated by the fact that the primitive cell of its Bravais lattice contains more than two atoms—see figure 3.12c and its discussion. In this case, even the tight-binding picture becomes more complex. Indeed, even if the atoms at different positions of the primitive unit cell are similar (as they are, for example, in both graphene and silicon), and hence the potential well shape near those points and the corresponding local wavefunctions $u(\mathbf{r})$ are similar as

---

[48] Actually, using the same values of δ_n in both directions (x and y) implies some sort of symmetry of the quasi-localized states. For example, the s-states of axially-symmetric potentials (see the next section) always have such a symmetry.

well, the Bloch theorem (which only pertains to Bravais lattices!) does not forbid them to have different complex probability amplitudes $a(t)$ whose time evolution should be described by a specific differential equation.

As the simplest example, in order to describe the honeycomb lattice shown in figure 3.12a, we have to prescribe different probability amplitudes to the 'top' and 'bottom' points of its primitive cell—say, α and β, correspondingly. Since each of these points is surrounded (and hence weakly interacts) with three neighbors of the opposite type, instead of Eq. (3.117) we have to write two equations

$$i\hbar\dot{\alpha} = -\delta_n \sum_{j=1}^{3} \beta_j, \qquad i\hbar\dot{\beta} = -\delta_n \sum_{j'=1}^{3} \alpha_{j'}, \qquad (3.119)$$

where each summation is over three next-neighbor points. (In these two sums, I am using different summation indices just to emphasize that these directions are different for the 'top' and 'bottom' points of the primitive cell—see figure 3.12a.) Now using the Bloch theorem (3.107) in the form similar to Eq. (2.205), we get two coupled systems of linear algebraic equations:

$$(E - E_n)\alpha = -\delta_n \beta \sum_{j=1}^{3} e^{i\mathbf{q}\cdot\mathbf{r}_j}, \qquad (E - E_n)\beta = -\delta_n \alpha \sum_{j'=1}^{3} e^{i\mathbf{q}\cdot\mathbf{r}'_{j'}}, \qquad (3.120)$$

where $\mathbf{r}_j$ and $\mathbf{r}'_{j'}$ are the next-neighbor positions, as seen from the top and bottom points, respectively. Writing the condition of consistency of this system of homogeneous linear equations, we get two equal and opposite values for energy correction for each value of $\mathbf{q}$:

$$E_{\pm} = E_n \pm \delta_n \Sigma^{1/2}, \quad \text{where } \Sigma \equiv \sum_{j,j'=1}^{3} e^{i\mathbf{q}\cdot\left(\mathbf{r}_j + \mathbf{r}'_{j'}\right)}. \qquad (3.121)$$

According to Eq. (3.120), these two energy bands correspond to the phase shifts (on the top of the regular Bloch shift $\mathbf{q}\cdot\Delta\mathbf{r}$) of either 0 or π between the adjacent quasi-localized wavefunctions $u(\mathbf{r})$.

The most interesting corollary of such energy symmetry, augmented by the honeycomb lattice's symmetry, is that for certain values $\mathbf{q}_D$ of the vector $\mathbf{q}$ (that turn out to be in each of six corners of the honeycomb-shaped 1st Brillouin zone), the double sum Σ vanishes, i.e. the two band surfaces $E_{\pm}(\mathbf{q})$ touch each other. As a result, in vicinities of these so-called *Dirac points*[49] the dispersion relation is linear:

$$E_{\pm}|_{\mathbf{q}\approx\mathbf{q}_D} \approx E_n \pm \hbar v_n |\tilde{\mathbf{q}}|, \quad \text{where } \tilde{\mathbf{q}} \equiv \mathbf{q} - \mathbf{q}_D, \qquad (3.122)$$

with $v_n \propto \delta_n$ being a constant with the dimension of velocity (for graphene, close to 10^6 m s^{-1}). Such a linear dispersion relation ensures several interesting transport properties of graphene, in particular of the quantum Hall effect in it (as was already

---

[49] This term is based on a (rather indirect) analogy with the Dirac theory of relativistic quantum mechanics, to be discussed in chapter 9 below.

mentioned in section 2). For their more detailed discussion, I have to refer the reader to special literature[50].

3.5 Axially-symmetric systems

I cannot conclude this chapter (and hence our review of wave mechanics) without addressing the exact solutions of the stationary Schrödinger equation[51] possible in the cases of highly symmetric functions $U(\mathbf{r})$. Such solutions are very important, in particular, for atomic and nuclear physics, and will be used, in particular, in the later chapters of this course.

In some rare cases such symmetries may be exploited by the separation of variables in Cartesian coordinates. The most famous example is the d-dimensional harmonic oscillator—a particle moving inside the potential

$$U = \frac{m\omega_0^2}{2}\sum_{j=1}^{d} r_j^2. \tag{3.123}$$

Separating the variables exactly as we did in section 1.7 for the rectangular hard-wall box (1.77), for each degree of freedom we get the Schrödinger equation (2.261) of a 1D oscillator, whose eigenfunctions are given by Eq. (2.284), and the energy spectrum is described by Eq. (2.162). As a result, the total energy spectrum may be indexed by a vector $\mathbf{n} = \{n_1, n_2, ..., n_d\}$ of d independent integer quantum numbers:

$$E_{\mathbf{n}} = \hbar\omega_0\left(\sum_{j=1}^{d} n_j + \frac{d}{2}\right), \tag{3.124}$$

each ranging from 0 to ∞. Note that every energy level of this system, with the only exception of the ground state,

---

[50] See, e.g. reviews [11] and [12]. Note that the transport properties of graphene are determined by coupling of $2p$-state electrons of its carbon atoms (see sections 3.6 and 3.7 below), whose wavefunctions are proportional to $\exp\{\pm i\varphi\}$ rather than being axially-symmetric as implied by Eqs. (3.120). However, due to the lattice symmetry this fact does not affect the above dispersion relation $E(\mathbf{q})$.

[51] This is my only chance to mention, in passing, that the eigenfunctions $\psi_n(\mathbf{r})$ of any such problem do not feature the instabilities typical for the *deterministic chaos* effects of classical mechanics—see, e.g. *Part CM* chapter 9. (This is why the term *quantum mechanics of classically chaotic systems* is preferable to the occasionally used term 'quantum chaos'.) It is curious that at the initial stages of the time evolution of the wavefunctions of such systems, their certain correlation functions still grow exponentially, reminding the *Lyapunov exponents* λ of their classical chaotic dynamics. This growth stops at the so-call *Ehrefect times* $t_E \sim \lambda^{-1} \ln(S/\hbar)$, where S is the action scale of the problem—see, e.g. [13]. In the stationary quantum state, the most essential trace of the classical chaos in a system is an unusual statistics of its eigenvalues, in particular of the energy spectra. We will have a chance for a brief look at such statistics in chapter 5, but unfortunately, I will not have time/space to discuss this field in any detail. Perhaps the best available book for further reading is the monograph [14].

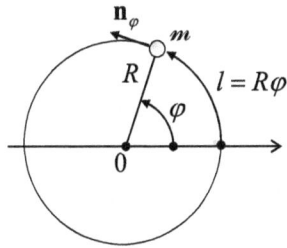

Figure 3.17. The planar rigid rotator.

$$\psi_{\rm g} = \prod_{j=1}^{d} \psi_0(r_j) = \frac{1}{\pi^{d/4}x_0^{d/2}} \exp\left\{-\frac{1}{2x_0^2}\sum_{j=1}^{d} r_j^2\right\}, \tag{3.125}$$

is *degenerate*: several different wavefunctions, each with its own different set of quantum numbers n_j, but the same value of their sum, have the same energy.

However, the harmonic oscillator problem is an exception: for other central- and spherically-symmetric problems the solution is made easier by using more appropriate curvilinear coordinates. Let us start with the simplest axially-symmetric problem: the so-called *planar rigid rotator* (or 'rotor'), i.e. a particle of mass m,[52] constrained to move along a plane, round circle of radius R (figure 3.17)[53].

The classical planar rotator may be described by just *one* degree of freedom, say the angle displacement φ (or equivalently the arc displacement $l \equiv R\varphi$) from some reference point, with the energy (and the Hamiltonian function) $H = p^2/2m$, where $\mathbf{p} \equiv m\mathbf{v} = m\mathbf{n}_\varphi(dl/dt)$, $\mathbf{n}_\varphi$ being the unit vector in the azimuthal direction—see figure 3.17. This function is similar to that of a free 1D particle (with the replacement $x \to l \equiv R\varphi$), and hence the rotator's quantum properties may be described by a similar Hamiltonian operator:

$$\hat{H} = \frac{\hat{p}^2}{2m}, \quad \text{with } \hat{\mathbf{p}} = -i\hbar\mathbf{n}_\varphi\frac{\partial}{\partial l} \equiv -i\frac{\hbar}{R}\mathbf{n}_\varphi\frac{\partial}{\partial \varphi}, \tag{3.126}$$

whose eigenfunctions have a similar structure:

$$\psi = Ce^{ikl} \equiv Ce^{ikR\varphi}. \tag{3.127}$$

The 'only' new feature is that in the rotator, all observables should be 2π-periodic functions of the angle φ. Hence, as we have already discussed in the context of the magnetic flux quantization (see figure 3.4 and its discussion), as the particle makes

---

[52] From this point on (until the chapter's end), I will use this exotic font for the particle's mass, in order to avoid any chance of its confusion with the impending 'magnetic' quantum number m, traditionally used in axially-symmetric problems.

[53] This is a reasonable model for the confinement of light atoms, notably hydrogen, in some organic compounds, but I am addressing this system mostly as the basis for the following, more complex problems.

one turn about the center, its wavefunction's phase $kR\varphi$ may only change by $2\pi m$, with an arbitrary integer m (from $-\infty$ to $+\infty$):

$$\psi_m(\varphi + 2\pi) = \psi_m(\varphi)e^{2\pi i m}. \tag{3.128}$$

With the eigenfunctions (3.127), this periodicity condition immediately gives $2\pi k R = 2\pi m$. Thus, the wavenumber k can take only quantized values $k_m = m/R$, so that the eigenfunctions should be indexed by this *magnetic* quantum number m:

$$\psi_m = C_m \exp\left\{im\frac{l}{R}\right\} \equiv C_m \exp\{im\varphi\}, \tag{3.129}$$

and the energy spectrum is discrete:

$$E_m = \frac{p_m^2}{2m} = \frac{\hbar^2 k_m^2}{2m} = \frac{\hbar^2 m^2}{2mR^2}. \tag{3.130}$$

This simple model allows an exact analysis of the external magnetic field effects on a confined motion of an electrically charged particle. Indeed, in the simplest case when this field is axially-symmetric (or just uniform) and directed normally to the rotator's plane, it does not violate the axial symmetry of the system. According to Eq. (3.26), in this case we have to generalize Eq. (3.126) as

$$\hat{H} = \frac{1}{2m}\left(-i\hbar\mathbf{n}_\varphi\frac{\partial}{\partial l} - q\mathbf{A}\right)^2 \equiv \frac{1}{2m}\left(-i\frac{\hbar}{R}\mathbf{n}_\varphi\frac{\partial}{\partial\varphi} - q\mathbf{A}\right)^2. \tag{3.131}$$

Here, in contrast to the Cartesian gauge choice (3.44), which was so instrumental for the solution of the Landau level problem, it is beneficial to take the vector-potential in the axially-symmetric form $\mathbf{A} = A(\rho)\mathbf{n}_\varphi$, where $\boldsymbol{\rho} \equiv \{x, y\}$ is the 2D radius-vector, with the magnitude $\rho = (x^2 + y^2)^{1/2}$. Using the well-known expression for the curl operator in the cylindrical coordinates[54], we can readily check that the requirement $\nabla \times \mathbf{A} = \mathscr{B}\mathbf{n}_z$, with $\mathscr{B} = $ const, is satisfied by the following function:

$$\mathbf{A} = \mathbf{n}_\varphi\frac{\mathscr{B}\rho}{2}. \tag{3.132}$$

For the planar rotator, $\rho = R = $ const, so that the stationary Schrödinger equation becomes

$$\frac{1}{2m}\left(-i\frac{\hbar}{R}\frac{\partial}{\partial\varphi} - q\frac{\mathscr{B}R}{2}\right)^2\psi_m = E_n\psi_m. \tag{3.133}$$

A little surprisingly, this equation is still satisfied with the eigenfunctions (3.127). Moreover, since the periodicity condition (3.128) is also unaffected by the applied

---

[54] See, e.g. Eq. (A.63).

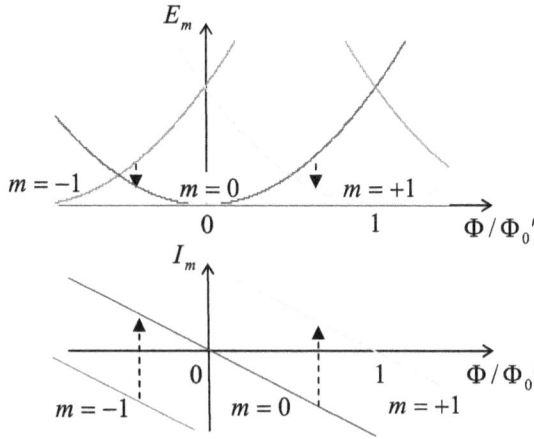

Figure 3.18. The magnetic field effect on a charged planar rotator. Dashed arrows show possible inelastic transitions between metastable and ground states, due to weak interaction with environment, as the external magnetic field is increased.

magnetic field, we return to the periodic eigenfunctions (3.129), independent of $\mathscr{B}$. However, the field does affect the system's eigenenergies:

$$
\begin{aligned}
E_m &= \frac{1}{\psi_m} \frac{1}{2m} \left(-i\frac{\hbar}{R} \frac{\partial}{\partial \varphi} - q\frac{\mathscr{B}R}{2} \right)^2 \psi_m \\
&= \frac{1}{2m} \left(\frac{\hbar m}{R} - q\frac{\mathscr{B}R}{2} \right)^2 \equiv \frac{\hbar^2}{2mR^2} \left(m - \frac{\Phi}{\Phi_0'} \right)^2,
\end{aligned}
\tag{3.134}
$$

where $\Phi \equiv \pi R^2 \mathscr{B}$ is the magnetic flux through the area limited by the particle's trajectory, and $\Phi_0' \equiv 2\pi\hbar/q$ is the 'normal' magnetic flux quantum we have already met in the AB effect's context—see Eq. (3.34) and its discussion. The field also changes the electric current of the particle in each eigenstate:

$$
I_m = q\frac{\hbar}{2imR} \left[\psi_m^* \left(\frac{\partial}{\partial \varphi} - \frac{iqR\mathscr{B}}{2\hbar} \right) \psi_m - \text{c.c.} \right] = q\frac{\hbar}{mR} |C_m|^2 \left(m - \frac{\Phi}{\Phi_0'} \right).
\tag{3.135}
$$

Normalizing the wavefunction (3.129) to have $W_m = 1$, we get $|C_m|^2 = 1/2\pi R$, so that Eq. (3.135) becomes

$$
I_m = \left(m - \frac{\Phi}{\Phi_0'} \right) I_0, \quad \text{with } I_0 \equiv \frac{\hbar q}{2\pi m R^2}.
\tag{3.136}
$$

The functions $E_m(\Phi)$ and $I_m(\Phi)$ are shown in figure 3.18. Note that since $\Phi_0' \propto 1/q$, for any sign of the particle's charge q, $dI_m/d\Phi < 0$. It is easy to verify that this means that the current is *diamagnetic*[55]: the field-induced current flows in such a direction

---

[55] This effect, whose qualitative features remain the same for all 2D or 3D localized states (see chapter 6 below), is frequently referred to as the *orbital diamagnetism*. In magnetic materials consisting of particles with uncompensated spins, this effect competes with an opposite effect, *spin paramagnetism*—see, e.g. *Part EM* section 5.5.

that its own magnetic field tries to compensate the external magnetic flux applied to the loop. This result may be interpreted as a different manifestation of the AB effect[56]. In contrast to the interference experiment that was discussed in section 3.1, in the situation shown in figure 3.17 the particle is not absorbed by the detector, but travels around the ring continuously. As a result, its wavefunction is *rigid*: due to the periodicity condition (3.128), the quantum number m is discrete, and the applied magnetic field cannot change the wavefunction gradually. In this sense, the system is similar to a superconducting loop—see figure 3.4 and its discussion. The difference between these systems is two-fold:

(i) For a single charged particle, in macroscopic systems with practicable values of q, R, and m, the scale I_0 of the induced current is very small. For example, for $m = m_e$, $q = -e$, and $R = 1\ \mu$m, Eq. (3.136) yields $I_0 \approx 3$ pA.[57] With the ring's inductance L of the order of $\mu_0 R$,[58] the contribution $\Phi_I = LI \sim \mu_0 R I_0 \sim 10^{-24}$ Wb of such a small current into the net magnetic flux Φ is negligible in comparison with $\Phi_0' \sim 10^{-15}$ Wb, so that the quantization of m does not lead to the magnetic flux quantization.

(ii) As soon as the magnetic field raises the eigenstate energy E_m above that of another eigenstate $E_{m'}$, the former state becomes metastable, and a weak interaction of the system with its environment (which is neglected in our simple model, but will be discussed in chapter 7) may induce a quantum transition of the system to the lower-energy state, thus reducing the diamagnetic current's magnitude—see the dashed lines in figure 3.18. The flux quantization in superconductors is much more robust to such perturbations[59].

Now let us return, once again, to the key Eq. (3.129), and see what it gives for one more important observable, the particle's angular momentum

$$\mathbf{L} \equiv \mathbf{r} \times \mathbf{p}, \qquad (3.137)$$

In this particular geometry, the vector $\mathbf{L}$ has just one component, normal to the rotator plane:

$$L_z = Rp. \qquad (3.138)$$

In classical mechanics, L_z of the rotator should be conserved (due to the absence of external torque), but it can take arbitrary values. In quantum mechanics, the

---

[56] It is straightforward to check that the last forms of Eqs. (3.134)—(3.136) remain valid even if the magnetic field is localized well inside rotator's ring, so that its lines do not touch the particle's trajectory.

[57] Such weak persistent, macroscopic diamagnetic currents in non-superconducting systems have been experimentally observed by measuring the weak magnetic field induced by the currents, in systems of a large number ($\sim 10^7$) of similar conducting rings—see, e.g. [15]. Due to the dephasing effects of electron scattering by phonons and other electrons (unaccounted for in our simple theory), the effect's observation requires submicron samples and millikelvin temperatures.

[58] See, e.g. *Part EM* section 5.3.

[59] Interrupting a superconducting ring with a weak link (Josephson junction), i.e. forming a SQUID, we may get a switching behavior similar to that shown with dashed arrows in figure 3.18—see, e.g. *Part EM* section 6.5.

situation changes: with $p = \hbar k$, our result $k_m = m/R$ for the mth eigenstate may be rewritten as

$$(L_z)_m = R\hbar k_m = \hbar m. \tag{3.139}$$

Thus, the angular momentum is quantized: it may be only a multiple of the Planck constant $\hbar$—confirming N Bohr's guess—see Eq. (1.8). As we will see in chapter 5, this result is very general (though it may be modified by spin effects), and the wavefunctions (3.129) may be interpreted as eigenfunctions of the angular momentum operator.

Let us see whether this quantization persists in more general, but still axial-symmetric systems. In order to implement the planar rotator in our 3D world, we needed to provide a rigid confinement of the particle both in the motion plane, and along the 2D radius ρ. Let us consider a more general situation when only the former confinement is strict, i.e. to the case when a 2D particle moves in an arbitrary centrally-symmetric potential

$$U(\boldsymbol{\rho}) = U(\rho). \tag{3.140}$$

Using the well-known expression for the 2D Laplace operator in polar coordinates[60], we may represent the 2D stationary Schrödinger equation in the form

$$-\frac{\hbar^2}{2m}\left[\frac{1}{\rho}\frac{\partial}{\partial\rho}\left(\rho\frac{\partial}{\partial\rho}\right) + \frac{1}{\rho^2}\frac{\partial^2}{\partial\varphi^2}\right]\psi + U(\rho)\psi = E\psi. \tag{3.141}$$

Separating the radial and angular variables as[61]

$$\psi = \mathcal{R}(\rho)\mathcal{F}(\varphi), \tag{3.142}$$

we get, after the division by ψ and the multiplication by ρ^2, the following equation:

$$-\frac{\hbar^2}{2m}\left[\frac{\rho}{\mathcal{R}}\frac{d}{d\rho}\left(\rho\frac{d\mathcal{R}}{d\rho}\right) + \frac{1}{\mathcal{F}}\frac{d^2\mathcal{F}}{d\varphi^2}\right] + \rho^2 U(\rho) = \rho^2 E. \tag{3.143}$$

It is clear that the fraction $(d^2\mathcal{F}/d\varphi^2)/\mathcal{F}$ should be a constant (because all other terms of the equation may be functions only of ρ), so that for the function $\mathcal{F}(\varphi)$ we get an ordinary differential equation,

$$\frac{d^2\mathcal{F}}{d\varphi^2} + \nu^2\mathcal{F} = 0, \tag{3.144}$$

where ν^2 is the variable separation constant. The fundamental solution of Eq. (3.144) is evidently $F \propto \exp\{\pm i\nu\varphi\}$. Now requiring, as we did for the planar rotator, the 2π periodicity of any observable, i.e.

---

[60] See, e.g. Eq. (A.61) with $\partial/\partial z = 0$.
[61] At this stage, I do not want to mark the particular solution (eigenfunction) ψ and corresponding eigenenergy E with any single index, because based on our experience in section 3.1, we already may suspect that in a 2D problem the role of this index will be played by *two* integers—two quantum numbers.

$$\mathcal{F}(\varphi + 2\pi) = \mathcal{F}(\varphi)e^{2\pi i m}, \tag{3.145}$$

where m is an integer, we see that the constant ν has to be equal to m, and get, for the angular factor, the same result as for the full wavefunction of the rotator—cf Eq. (3.129):

$$\mathcal{F}_m = C_m e^{im\varphi}, \quad \text{with } m = 0, \pm 1, \pm 2,... \tag{3.146}$$

Plugging the resulting relation $(d^2\mathcal{F}/d\varphi^2)/\mathcal{F} = -m^2$ into Eq. (3.143), we may rewrite it as

$$-\frac{\hbar^2}{2m}\left[\frac{1}{\rho\mathcal{R}}\frac{d}{d\rho}\left(\rho\frac{d\mathcal{R}}{d\rho}\right) - \frac{m^2}{\rho^2}\right] + U(\rho) = E. \tag{3.147}$$

The physical interpretation of this equation is that the full energy is a sum,

$$E = E_\rho + E_\varphi, \tag{3.148}$$

of the radial-motion part

$$E_\rho = -\frac{\hbar^2}{2m}\frac{1}{\rho}\frac{d}{d\rho}\left(\rho\frac{d\mathcal{R}}{d\rho}\right) + U(\rho). \tag{3.149}$$

and the angular-motion part

$$E_\varphi = \frac{\hbar^2 m^2}{2m\rho^2}. \tag{3.150}$$

Now let us notice that a similar separation exists in classical mechanics[62], because the total energy of a particle moving in a central field may be represented as

$$E = \frac{m}{2}v^2 + U(\rho) = \frac{m}{2}(\dot{\rho}^2 + \rho^2\dot{\varphi}^2) + U(\rho) \equiv E_\rho + E_\varphi, \tag{3.151}$$

$$\text{with } E_\rho \equiv \frac{p_\rho^2}{2m} + U(\rho), \quad \text{and} \quad E_\varphi \equiv \frac{m}{2}\rho^2\dot{\varphi}^2 \equiv \frac{p_\varphi^2}{2m} \equiv \frac{L_z^2}{2m\rho^2}. \tag{3.152}$$

The comparison of the latter relation with Eqs. (3.139) and (3.150) gives us grounds to suspect that the quantization rule $L_z = m\hbar$ may be valid not only for this 2D problem, but in 3D cases as well. In section 5.6, we will see that this is indeed the case.

Returning to Eq. (3.147), on the basis of our experience with 1D wave mechanics we may expect that this ordinary, linear, second-order differential equation should have (for a motion confined to a certain final region of its argument ρ), for any fixed m, a discrete energy spectrum described by some other integer quantum number

---

[62] See, e.g. *Part CM* section 3.5.

(say, n). This means that the eigenfunctions (3.142), and corresponding eigenenergies (3.148) should be indexed by two quantum numbers, m and n. Note, however, that since the radial function obeys Eq. (3.147), which already depends on m, its eigenfunction $\mathcal{R}(\rho)$ should carry both indices, so the variable separation is not so 'clean' as it was for the rectangular potential well. Normalizing the angular function $\mathcal{F}$ to the full circle, $\Delta\varphi = 2\pi$, we may rewrite Eq. (3.142) as

$$\psi_{m,n} = \mathcal{R}_{m,n}(\rho)\mathcal{F}_m(\varphi) = \frac{1}{(2\pi)^{1/2}}\mathcal{R}_{m,n}(\rho)e^{im\varphi}. \tag{3.153}$$

A good (and important) example of a solvable problem of this type is a free 2D particle whose motion is rigidly confined to a disk of radius R:

$$U(\rho) = \begin{cases} 0, & \text{for } 0 \leqslant \rho < R, \\ +\infty, & \text{for } R < \rho. \end{cases} \tag{3.154}$$

In this case, the solutions $\mathcal{R}_{m,n}(\rho)$ of Eq. (3.147) are proportional to the first-order Bessel functions $J_m(k_n\rho)$,[63] with the spectrum of possible values k_n following from the boundary condition $\mathcal{R}_{m,n}(R) = 0$. Let me leave the detailed analysis of this problem for the reader's exercise.

3.6 Spherically-symmetric systems: brute force approach

Now let us proceed to the mathematically more involved, but practically even more important case of the 3D motion in a spherically-symmetric potential

$$U(\mathbf{r}) = U(r). \tag{3.155}$$

Let me start, again, with solving the eigenproblem for a rigid rotator—now a *spherical rotator*, i.e. a particle confined to move on the spherical surface of radius R. It has two degrees of freedom, because a position on the spherical surface is completely described by two coordinates—say, the polar angle θ and the azimuthal angle φ. In this case, the kinetic energy we need to consider is limited to its angular part, so that in the Laplace operator in spherical coordinates[64] we may keep only those parts, with fixed $r = R$. Because of this, the stationary Schrödinger equation becomes

$$-\frac{\hbar^2}{2mR^2}\left[\frac{1}{\sin\theta}\frac{\partial}{\partial\theta}\left(\sin\theta\frac{\partial}{\partial\theta}\right) + \frac{1}{\sin^2\theta}\frac{\partial^2}{\partial\varphi^2}\right]\psi = E\psi. \tag{3.156}$$

(Again, I will attach indices to ψ and E in a minute.) With the usual variable separation assumption,

$$\psi = \Theta(\theta)\mathcal{F}(\varphi), \tag{3.157}$$

---

[63] A short summary of properties of these functions, including a few graphic plots and a useful table of values, may be found in *Part EM* section 2.7.

[64] See, e.g. Eq. (A.67).

this equation, with all terms multiplied by $\sin^2\theta/\Theta\,\mathcal{F}$, yields

$$-\frac{\hbar^2}{2mR^2}\left[\frac{\sin\theta}{\Theta}\frac{d}{d\theta}\left(\sin\theta\frac{d\Theta}{d\theta}\right)+\frac{1}{\mathcal{F}}\frac{d^2\mathcal{F}}{d^2\varphi}\right]=E\sin^2\theta. \tag{3.158}$$

Just as in Eq. (3.143), the fraction $(d^2\mathcal{F}/d\varphi^2)/\mathcal{F}$ may be a function of φ only, and hence has to be constant, giving Eq. (3.144) for it. So, with the same periodicity condition (3.145), the azimuthal functions are expressed by (3.146) again; in the normalized form,

$$\mathcal{F}_m(\varphi)=\frac{1}{(2\pi)^{1/2}}e^{im\varphi}. \tag{3.159}$$

With that, the fraction $(d^2\mathcal{F}/d\varphi^2)/\mathcal{F}$ in Eq. (3.158) equals $(-m^2)$, and after its multiplication by $\Theta/\sin^2\theta$, it is reduced to the following ordinary, linear differential equation for the polar eigenfunctions $\Theta(\theta)$:

$$-\frac{1}{\sin\theta}\frac{d}{d\theta}\left(\sin\theta\frac{d\Theta}{d\theta}\right)+\frac{m^2}{\sin^2\theta}\Theta=\varepsilon\Theta,\quad\text{with }\varepsilon\equiv E\left/\left(\frac{\hbar^2}{2mR^2}\right)\right., \tag{3.160}$$

It is common to recast it into an equation for a new function $P(\xi)\equiv\Theta(\theta)$, with $\xi\equiv\cos\theta$:

$$\frac{d}{d\xi}\left[(1-\xi^2)\frac{dP}{d\xi}\right]+\left[l(l+1)-\frac{m^2}{1-\xi^2}\right]P=0, \tag{3.161}$$

where a new notation for the normalized energy is introduced: $l(l+1)\equiv\varepsilon$. The motivation for such notation is that, according to a mathematical analysis[65], Eq. (3.161) with integer m has solutions only if parameter l is integer: $l=0,1,2,\dots,$ and only if that integer is not smaller than $|m|$, i.e. if

$$-l\leqslant m\leqslant+l. \tag{3.162}$$

This fact immediately gives the following spectrum of the spherical rotator's energy E—and, as we will see below, the angular part of the energy of any spherically-symmetric system:

$$E_l=\frac{\hbar^2 l(l+1)}{2mR^2}, \tag{3.163}$$

so that the only effect of the magnetic quantum number m here is imposing the restriction (3.162) on the non-negative integer l—the so-called *orbital quantum number*. This means, in particular, that each energy level (3.163) corresponds to $(2l+1)$ different values of m, i.e. is $(2l+1)$–degenerate.

---

[65] This analysis was first carried out by A-M Legendre (1752–1833). Just as a historic note: besides many original mathematical achievements, Dr Legendre authored a famous textbook, *Éléments de Géométrie,* which dominated teaching geometry through the 19th century.

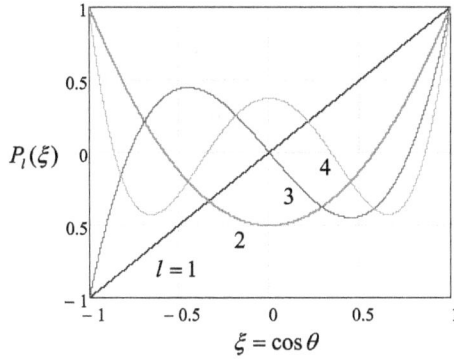

Figure 3.19. A few lowest Legendre polynomials.

To understand the physics of this degeneracy, we need to explore the corresponding eigenfunctions of Eq. (3.161). They are naturally numbered by two integers, m and l, and are called the *associated Legendre functions* P_l^m. (Note that m is an upper index, not a power!) For the particular, simplest case $m = 0$, these functions are the so-called *Legendre polynomials* $P_l(\xi) \equiv P_l^0(\xi)$, which may be defined as the solutions of the *Legendre equation*, following from Eq. (3.161) at $m = 0$:

$$\frac{d}{d\xi}\left[(1 - \xi^2)\frac{d}{d\xi}P\right] + l(l + 1)P = 0, \qquad (3.164)$$

but also may be calculated explicitly from the following *Rodrigues formula*[66]:

$$P_l(\xi) = \frac{1}{2^l l!}\frac{d^l}{d\xi^l}(\xi^2 - 1)^l, \quad l = 0, 1, 2, \ldots. \qquad (3.165)$$

Using this formula, it easy to spell out a few lowest Legendre polynomials:

$$P_0(\xi) = 1, \quad P_1(\xi) = \xi, \quad P_2(\xi) = \frac{1}{2}(3\xi^2 - 1), \quad P_3(\xi) = \frac{1}{2}(5\xi^3 - 3\xi), \ldots, \quad (3.166)$$

though such expressions become more and more bulky as l is increased. As these expressions (and figure 3.19) show, as the argument ξ is decreased, all these functions start in one point, $P_l(+1) = +1$, and end up either at the same point or at the opposite point: $P_l(-1) = (-1)^l$. On the way between these two end points, the lth polynomial crosses the horizontal axis exactly l times, i.e. has l roots[67]. Finally, it is easy to use the Rodrigues formula (3.165) and the integration by parts to show that on the segment $-1 \leqslant \xi \leqslant +1$, the Lagrange polynomials form a full orthogonal set of functions, with the following normalization rule:

---

[66] This wonderful formula may be readily proved by plugging it into Eq. (3.164), but was not so easy to discover! This was done independently by B O Rodrigues in 1816, J Ivory in 1824, and C Jacobi in 1827.
[67] In this behavior, we may readily recognize the 'standing wave' pattern typical for all 1D eigenproblems—cf figures 1.8 and 2.35, as well as the discussion of the Sturm oscillation theorem in the end of section 2.9.

$$\int_{-1}^{+1} P_l(\xi)P_{l'}(\xi)d\xi = \frac{2}{2l+1}\delta_{ll'}. \tag{3.167}$$

For $m > 0$, the associated Legendre functions (now *not* polynomials!), may be expressed via the Legendre polynomials (3.165) using the following formula[68]:

$$P_l^m(\xi) = (-1)^m(1-\xi^2)^{m/2}\frac{d^m}{d\xi^m}P_l(\xi), \tag{3.168}$$

while the functions with a negative magnetic quantum number may be found as

$$P_l^{-m}(\xi) = (-1)^m\frac{(l-m)!}{(l+m)!}P_l^m(\xi), \quad \text{for } m > 0. \tag{3.169}$$

On the segment $-1 \leqslant \xi \leqslant +1$, the associated Legendre functions with a fixed index m form a full orthogonal set, with the normalization relation,

$$\int_{-1}^{+1} P_l^m(\xi)P_{l'}^m(\xi)d\xi = \frac{2}{2l+1}\frac{(l+m)!}{(l-m)!}\delta_{ll'}, \tag{3.170}$$

which is evidently a generalization of Eq. (3.167) for arbitrary m.

Since the difference between the angles θ and φ is to much extent artificial (due to arbitrary direction of the polar axis), physicists prefer to use not the functions $\Theta(\theta) \propto P_l^m(\cos\theta)$ and $\mathscr{F}_m(\varphi) \propto e^{im\varphi}$ separately, but normalized products of the type (3.157), which are called the *spherical harmonics*:

$$Y_l^m(\theta, \varphi) \equiv \left[\frac{(2l+1)}{4\pi}\frac{(l-m)!}{(l+m)!}\right]^{1/2}P_l^m(\cos\theta)e^{im\varphi}. \tag{3.171}$$

The specific coefficient in Eq. (3.171) is chosen in a way to simplify the following two expressions: the relation of the spherical harmonics with opposite signs of the magnetic quantum number,

$$Y_l^{-m}(\theta, \varphi) = (-1)^m[Y_l^m(\theta, \varphi)]^*, \tag{3.172}$$

and the normalization relation

$$\oint_{4\pi} Y_l^m(\theta, \varphi)[Y_{l'}^{m'}(\theta, \varphi)]^*d\Omega = \delta_{ll'}\delta_{mm'}, \tag{3.173}$$

with the integration over the whole solid angle. The last formula shows that on a spherical surface, the spherical harmonics form an orthonormal set of functions. This set is also full, so that any function, defined on the surface, may be uniquely represented as a linear combination of Y_l^m.

---

[68] Note that some texts use different choices for the front factor (called the *Condon–Shortley phase*) in the functions P_l^m, which do not affect the final results for the spherical harmonics Y_l^m.

Despite a somewhat intimidating formulas given above, they yield rather simple expressions for the lowest spherical harmonics, which are most important for applications:

$$l = 0: \quad Y_0^0 = (1/4\pi)^{1/2}, \qquad (3.174)$$

$$l = 1: \quad \begin{cases} Y_1^{-1} = (3/8\pi)^{1/2} \sin\theta\, e^{-i\varphi}, \\ Y_1^0 = (3/4\pi)^{1/2} \cos\theta, \\ Y_1^1 = -(3/8\pi)^{1/2} \sin\theta\, e^{i\varphi}, \end{cases} \qquad (3.175)$$

$$l = 2: \quad \begin{cases} Y_2^{-2} = -(15/32\pi)^{1/2} \sin^2\theta\, e^{-2i\varphi}, \\ Y_2^{-1} = (15/8\pi)^{1/2} \sin\theta \cos\theta\, e^{-i\varphi}, \\ Y_2^0 = (3/16\pi)^{1/2} (3\cos^2\theta - 1), \quad \text{etc.} \\ Y_2^1 = -(15/8\pi)^{1/2} \sin\theta \cos\theta\, e^{i\varphi}, \\ Y_2^2 = (15/32\pi)^{1/2} \sin^2\theta\, e^{2i\varphi}, \end{cases} \qquad (3.176)$$

It is important to understand the general structure and symmetry of these functions. Since the spherical functions with $m \neq 0$ are complex, the most popular way of their graphical representation is to normalize their real and imaginary parts as[69]

$$Y_{lm} \equiv \sqrt{2}(-1)^m \times \begin{cases} \mathrm{Im}\left(Y_l^{|m|}\right) & \propto \sin m\varphi, \quad \text{for } m < 0, \\ \mathrm{Re}\left(Y_l^{|m|}\right) & \propto \cos m\varphi \quad \text{for } m > 0, \end{cases} \qquad (3.177)$$

(for $m = 0$, $Y_{l0} \equiv Y_l^0$), and then plot the magnitude of these real functions in the spherical coordinates as the distance from the origin, while using two colors to show their sign—see figure 3.20.

Let us start from the simplest case $l = 0$. According to Eq. (3.162), for this lowest orbital quantum number, there may be only one magnetic quantum number, $m = 0$. According to Eq. (3.174), the spherical harmonic corresponding to that state is just a constant, so that the wavefunction of this so-called s state[70], is uniformly distributed over the sphere. Since this function has no gradient in any angular direction, it is only natural that the angular kinetic energy (3.163) of the particle equals zero.

According to the same Eq. (3.162), for $l = 1$, there are three different p states, with $m = -1$, $m = 0$, and $m = +1$—see Eq. (3.175). As the second row of figure 3.20 shows,

---

[69] Such real functions Y_{lm}, which also form the full orthonormal set, and are frequently called the *real* (or 'tesseral') *spherical harmonics*, are more convenient than the complex harmonics Y_l^m for several applications, especially when the variables of interest are real by definition.

[70] The letter names for the states with various values of l stem from the history of optical spectroscopy—for example, the letter 's', used for the state with $l = 0$, originally denoted the 'sharp' optical line series, etc. The sequence of the letters is as follows: s, p, d, f, g, and then continuing in the alphabetical order.

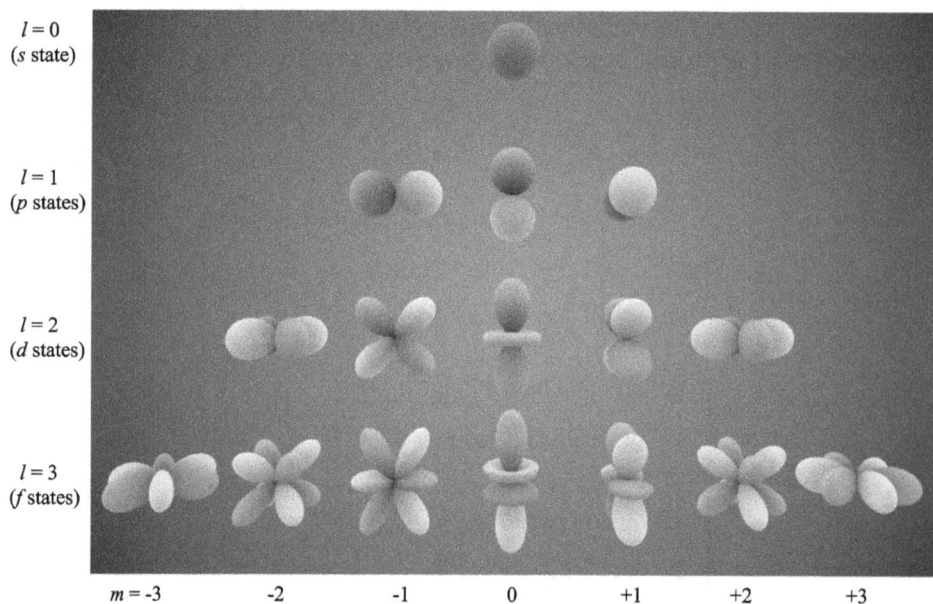

Figure 3.20. Radial plots of several lowest real spherical harmonics Y_{lm}. (Adapted from https://en.wikipedia. org/wiki/Spherical_harmonics under the CC BY-SA 3.0 license.)

these states are essentially identical in structure, and are just differently oriented in space, thus readily explaining the three-fold degeneracy of the kinetic energy (3.163). Such a simple explanation, however, is not valid for the five different *d states* ($l = 2$), shown in the third row of figure 3.20, as well as the states with higher l: despite their equal energies, they differ not only by their special orientation, but their structure as well. The states with $m = 0$ have a non-vanishing gradient only in the θ direction. In contrast, the states with the ultimate values of m ($\pm l$), change only monotonically (as $\sin^l\theta$) in the polar direction, while oscillating in the azimuthal direction. The states with intermediate values of m provide a gradual transition between these two extremes, oscillating in both directions, stronger and stronger in the azimuthal direction as $|m|$ is increased. Still, the magnetic quantum number, surprisingly, does not affect the energy for any l.

Another counter-intuitive feature of the spherical harmonics follows from the comparison of Eq. (3.163) with the second of the classical relations (3.152). These expressions coincide if we interpret the constant

$$L^2 \equiv \hbar^2 l(l+1), \tag{3.178}$$

as the value of the full angular momentum squared $L^2 = |\mathbf{L}|^2$ (including both its θ and φ components) in the eigenstate with the eigenfunction Y_l^m. On the other hand, the structure (3.159) of the azimuthal component $\mathscr{F}(\varphi)$ of the wavefunction is exactly the same as in 2D axially-symmetric problems, implying that Eq. (3.139) still gives correct values (in our notation, $L_z = m\hbar$) for the z-component of the angular momentum. If this is so, why for any state with $l > 0$, is $(L_z)^2 = m^2\hbar^2 \leqslant l^2\hbar^2$ less than

3-47

$L^2 = l(l + 1)\hbar^2$? In other words, what prevents the angular momentum vector being fully aligned with the axis z?

Besides the difficulty of answering this question using the above formulas, this analysis (though mathematically complete), is as intellectually unsatisfactory as the harmonic oscillator analysis in section 2.9. In particular, it does not explain the meaning of the extremely simple relations for the eigenvalues of the energy and the angular momentum coexisting with rather complicated eigenfunctions.

We will obtain natural answers to all these questions and concerns in section 5.6 below, and now proceed to the extension of our wave-mechanical analysis to the 3D motion in an *arbitrary* spherically-symmetric potential (3.155). In this case we have to use the full form of the Laplace operator in spherical coordinates[71]. The variable separation procedure is an evident generalization of what we have done before, with the particular solutions of the type

$$\psi = \mathcal{R}(\rho)\Theta(\theta)\mathcal{F}(\varphi), \tag{3.179}$$

whose substitution into the stationary Schrödinger equation yields

$$-\frac{\hbar^2}{2mr^2}\left[\frac{1}{\mathcal{R}}\frac{d}{dr}\left(r^2\frac{d\mathcal{R}}{dr}\right) + \frac{1}{\Theta}\frac{1}{\sin\theta}\frac{d}{d\theta}\left(\sin\theta\frac{d\Theta}{d\theta}\right) + \frac{1}{\sin^2\theta}\frac{1}{\mathcal{F}}\frac{d^2\mathcal{F}}{d\varphi^2}\right] + U(r) = E. \tag{3.180}$$

It is evident that the angular part of the left-hand side (the two last terms in the square brackets) separates from the radial part, and that for the former part we get Eq. (3.156) again, with the only change, $R \to r$. This change does not affect the fact that the eigenfunctions of that equation are still the spherical harmonics (3.171). This means that for the radial function $\mathcal{R}(r)$, Eq. (3.180) gives the following equation,

$$-\frac{\hbar^2}{2mr^2}\left[\frac{1}{\mathcal{R}}\frac{d}{dr}\left(r^2\frac{d\mathcal{R}}{dr}\right) - l(l + 1)\right] + U(r) = E. \tag{3.181}$$

Note that no information about the magnetic quantum number m has crept into the radial equation (besides setting the limitation (3.162) for the possible values of l), so that it includes only the latter, orbital quantum number.

Let us explore the radial equation for the simplest case when $U(r) = 0$, for example to solve the eigenproblem for the 3D motion of a particle free to move only inside the sphere of radius R—say, confined there by the potential[72]

$$U = \begin{cases} 0, & \text{for } 0 \leqslant r < R, \\ +\infty, & \text{for } R < r. \end{cases} \tag{3.182}$$

In this case, Eq. (3.181) is reduced to

$$-\frac{\hbar^2}{2mr^2}\left[\frac{1}{\mathcal{R}}\frac{d}{dr}\left(r^2\frac{d\mathcal{R}}{dr}\right) - l(l + 1)\right] = E. \tag{3.183}$$

---

[71] Again, see Eq. (A.67).

[72] This problem, besides giving the simplest example of quantization in spherically-symmetric systems, is also an important precursor for the discussion of scattering by spherically-symmetric potentials in section 3.8.

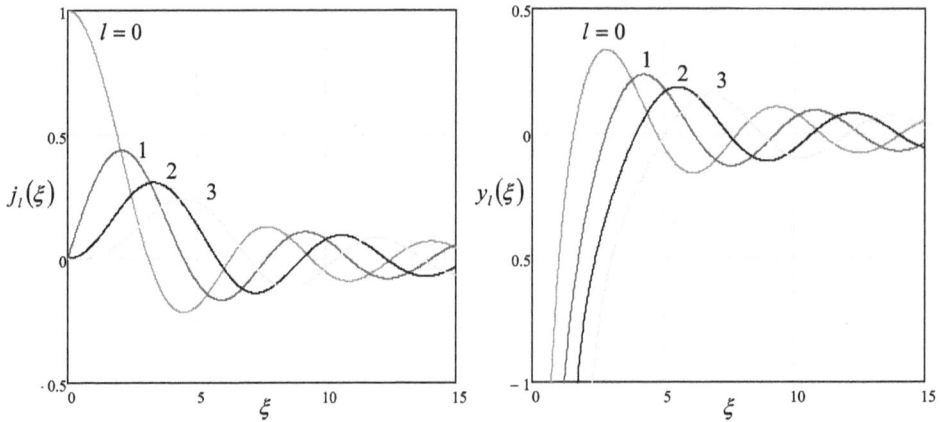

Figure 3.21. Several lowest-order spherical Bessel functions.

Multiplying both parts of this equality by $r^2 \mathcal{R}$, and introducing the dimensionless argument $\xi \equiv kr$, where k^2 is defined by the usual relation $\hbar^2 k^2 / 2m = E$, we obtain the canonical equation,

$$\xi^2 \frac{d^2 \mathcal{R}}{d\xi^2} + 2\xi \frac{d\mathcal{R}}{d\xi} + [\xi^2 - l(l+1)]\mathcal{R} = 0, \tag{3.184}$$

for the so-called *spherical Bessel functions* of the first and second kind, $j_l(\xi)$ and $y_l(\xi)$.[73] These functions are directly related to the Bessel functions of the semi-integer order[74],

$$j_l(\xi) = \left(\frac{\pi}{2\xi}\right)^{1/2} J_{l+\frac{1}{2}}(\xi), \qquad y_l(\xi) = \left(\frac{\pi}{2\xi}\right)^{1/2} Y_{l+\frac{1}{2}}(\xi), \tag{3.185}$$

but are actually much simpler than even the 'usual' Bessel functions, such as $J_n(\xi)$ and $Y_n(\xi)$ of an integer order n, because the former ones may be directly expressed via elementary functions:

$$j_0(\xi) = \frac{\sin \xi}{\xi}, \quad j_1(\xi) = \frac{\sin \xi}{\xi^2} - \frac{\cos \xi}{\xi},$$

$$j_2(\xi) = \left(\frac{3}{\xi^3} - \frac{1}{\xi}\right) \sin \xi - \frac{3}{\xi^2} \cos \xi, \dots,$$

$$y_0(\xi) = -\frac{\cos \xi}{\xi}, \quad y_1(\xi) = -\frac{\cos \xi}{\xi^2} - \frac{\sin \xi}{\xi}, \tag{3.186}$$

$$y_2(\xi) = -\left(\frac{3}{\xi^3} - \frac{1}{\xi}\right) \cos \xi - \frac{3}{\xi^2} \sin \xi, \dots,$$

---

[73] Alternatively, $y_l(\xi)$ are called the 'spherical Weber functions' or the 'spherical Neumann functions'.

[74] Note that the Bessel functions $J_\nu(\xi)$ and $Y_\nu(\xi)$ of *any* order ν obey the universal recurrent formulas and asymptotic formulas (discussed, e.g. in *Part EM* section 2.7), so that many properties of the functions $j_l(\xi)$ and $y_l(\xi)$ may be readily derived from these relations and Eq. (3.185).

A few of the lowest-order spherical harmonics are plotted in figure 3.21.

As these formulas and plots show, the functions $y_l(\xi)$ are diverging at $\xi \to 0$, and thus cannot be used in the solution of our current problem, so that we have to take

$$\mathcal{R}_l(r) = \text{const} \times j_l(kr). \tag{3.187}$$

Still, even for these functions, with the sole exception of the simplest function $j_0(\xi)$, the characteristic equation $j_l(kR) = 0$, resulting from the boundary condition $\mathcal{R}(R) = 0$, can be solved only numerically. However, the roots $\xi_{l,n}$ of the equations $j_l(\xi) = 0$, where the integer n (= 1, 2, 3,...) is the root's number, are tabulated in virtually any math handbook, and we may express the eigenvalues we are interested in,

$$k_{l,n} = \frac{\xi_{l,n}}{R}, \qquad E_{l,n} = \frac{\hbar^2 k_{l,n}^2}{2m} \equiv \frac{\hbar^2 \xi_{l,n}^2}{2mR^2}, \tag{3.188}$$

via these tabulated numbers. The table below lists several smallest roots, and the corresponding eigenenergies (normalized to their natural unit $E_0 \equiv \hbar^2/2mR^2$), in the order of their growth. It shows a very interesting effect: first the eigenenergies grow because of the increase of the orbital quantum number l, at the same (lowest) radial quantum number $n = 1$, due to the growth of the first roots of functions $j_l(\xi)$, but then suddenly the second root of $j_0(\xi)$ cuts into this orderly sequence, just to be followed by the first root of $j_3(\xi)$. With the growth of energy, the sequences of l and n become even more entangled.

l	n	$\xi_{l,n}$	$E_{l,n}/E_0 = (\xi_{l,n})^2$
0	1	$\pi \approx 3.1415$	$\pi^2 \approx 9.87$
1	1	4.493	20.19
2	1	5.763	33.21
0	2	$2\pi \approx 6.283$	$4\pi^2 \approx 39.48$
3	1	6.988	48.83

To complete the discussion of our current problem (3.182), note again that the energy levels, listed in the table above, are $(2l + 1)$-degenerate, because each of them corresponds to $(2l + 1)$ different eigenfunctions, each with a specific value of the magnetic quantum number m:

$$\psi_{n,l,m} = C_{l,n} j_l\left(\frac{\xi_{l,n} r}{R}\right) Y_l^m(\theta, \varphi), \quad \text{with } -l \leqslant m \leqslant +l. \tag{3.189}$$

3.7 Atoms

Let us proceed to the discussion of atoms, starting from the simplest, exactly solvable *Bohr atom problem*, i.e. that of a particle's motion in the so-called *attractive Coulomb potential*[75]

$$U(r) = -\frac{C}{r}, \quad \text{with } C > 0. \tag{3.190}$$

The natural scales of E and r in this problem are commonly defined by the following requirement of equality of the kinetic and potential energy magnitude scales (dropping all numerical coefficients):

$$E_0 \equiv \frac{\hbar^2}{mr_0^2} \equiv \frac{C}{r_0}, \tag{3.191}$$

similar to its particular case (1.13b). Solving these two equations, we get[76]

$$E_0 \equiv \frac{\hbar^2}{mr_0^2} \equiv m\left(\frac{C}{\hbar}\right)^2, \quad \text{and} \quad r_0 \equiv \frac{\hbar^2}{mC}. \tag{3.192}$$

In the normalized units $\varepsilon \equiv E/E_0$ and $\xi \equiv r/r_0$, Eq. (3.181), for our current case (3.190), looks relatively simple,

$$\frac{d^2\mathcal{R}}{d\xi^2} + \frac{2}{\xi}\frac{d\mathcal{R}}{d\xi} - l(l+1)\mathcal{R} + 2\left(\varepsilon + \frac{1}{\xi}\right)\mathcal{R} = 0, \tag{3.193}$$

but unfortunately its eigenfunctions may be called elementary only in the most generous meaning of the word. With the adequate normalization,

$$\int_0^\infty \mathcal{R}_{n,l}\mathcal{R}_{n',l}\, r^2 dr = \delta_{nn'}, \tag{3.194}$$

these (mutually orthogonal) functions may be represented as

$$\mathcal{R}_{n,l}(r) = -\left\{\left(\frac{2}{nr_0}\right)^3 \frac{(n-l-1)!}{2n[(n+l)!]^3}\right\}^{1/2} \exp\left\{-\frac{r}{nr_0}\right\}\left(\frac{2r}{nr_0}\right)^l L_{n-l-1}^{2l+1}\left(\frac{2r}{nr_0}\right). \tag{3.195}$$

Here $L_p^q(\xi)$ are the so-called *associated Laguerre* polynomials, which may be calculated as

---

[75] Historically, the solution of this problem in 1928, that reproduced the main result (1.12)–(1.13) of the 'old' quantum theory developed by N Bohr in 1912, without its phenomenological assumptions, was the decisive step toward the general acceptance of the Schrödinger's wave mechanics.

[76] For the most important case of the hydrogen atom, with $C = e^2/4\pi\varepsilon_0$, these scales are reduced, respectively, to the Bohr radius r_B (1.10) and the Hartree energy E_H (1.13a). Note also that according to Eq. (3.192), for the so-called *hydrogen-like atom* (actually, a positive ion), with $C = Z(e^2/4\pi\varepsilon_0)$, these two key parameters are rescaled as $r_0 = r_B/Z$ and $E_0 = Z^2 E_H$.

$$L_p^q(\xi) = (-1)^q \frac{d^q}{d\xi^q} L_{p+q}(\xi). \tag{3.196}$$

from the *simple Laguerre polynomials* $L_p(\xi) \equiv L_p^0(\xi)$.[77] In turn, the easiest way to obtain $L_p(\xi)$ is to use the following *Rodrigues formula*[78]:

$$L_p(\xi) = e^\xi \frac{d^p}{d\xi^p}(\xi^p e^{-\xi}). \tag{3.197}$$

Note that in contrast with the associated Legendre *functions* P_l^m, participating in the spherical harmonics, all L_p^q are just *polynomials*, and those with small indices p and q are indeed rather simple:

$$
\begin{array}{lll}
L_0^0(\xi) = 1, & L_1^0(\xi) = -\xi + 1, & L_2^0(\xi) = \xi^2 - 4\xi + 2, \\
L_0^1(\xi) = 1, & L_1^1(\xi) = -2\xi + 4, & L_2^1(\xi) = 3\xi^2 - 18\xi + 18, \\
L_0^2(\xi) = 2, & L_1^2(\xi) = -6\xi + 18, & L_2^2(\xi) = 12\xi^2 - 96\xi + 144, \ldots
\end{array}
\tag{3.198}
$$

Returning to Eq. (3.195), we see that the natural quantization of the radial equation (3.193) has brought us a new integer quantum number n. In order to understand its range, we should notice that according to Eq. (3.197), the highest power of terms in the polynomial L_{p+q} is $(p + q)$, and hence, according to Eq. (3.196), that of L_p^q is p, so that of the highest power in the polynomial participating in Eq. (3.195) is $(n - l - 1)$. Since the power cannot be negative (to avoid the unphysical divergence of wavefunctions at $r \to 0$), the radial quantum number n has to obey the restriction $n \geqslant l + 1$. Since l, as we already know, may take values $l = 0, 1, 2, \ldots$, n may only take the values

$$n = 1, 2, 3, \ldots \tag{3.199}$$

What makes this relation very important is the following, most surprising result of the theory: the eigenenergies corresponding to the wavefunctions (3.179), which are indexed with three quantum numbers:

$$\psi_{n,l,m} = \mathcal{R}_{n,l}(r) Y_l^m(\theta, \varphi), \tag{3.200}$$

depend only on one of them, n:

$$\varepsilon = \varepsilon_n = -\frac{1}{2n^2}, \quad \text{i.e. } E_n = -\frac{E_0}{2n^2} = -\frac{1}{2n^2} m\left(\frac{C}{\hbar}\right)^2. \tag{3.201}$$

i.e. agree with Bohr's formula (1.12). Because of this reason, n is usually called the *principal quantum number*, and the above relation between it and the 'more subordinate' orbital quantum number l is rewritten as

---

[77] In Eqs. (3.196) and (3.197), p and q are non-negative integers, with no relation whatsoever to the particle's momentum or electric charge. Sorry for this notation, but it is absolutely common, and can hardly result in any confusion.

[78] Named after the same B O Rodrigues, and belonging to the same class as his other famous result, Eq. (3.165) for the Legendre polynomials.

$$l \leqslant n - 1. \tag{3.202}$$

Together with the inequality (3.162), this gives us the following, very important hierarchy of the three quantum numbers involved in the Bohr atom problem:

$$1 \leqslant n \leqslant \infty \quad \Rightarrow \quad 0 \leqslant l \leqslant n - 1 \quad \Rightarrow \quad -l \leqslant m \leqslant + l. \tag{3.203}$$

Taking into account the $(2l + 1)$-degeneracy related to the magnetic number m, and using the well-known formula for the arithmetic progression[79], we see that each energy level (3.191) has the following *orbital degeneracy*:

$$g = \sum_{l=0}^{n-1}(2l + 1) \equiv 2\sum_{l=0}^{n-1}l + \sum_{l=0}^{n-1}1 = 2\frac{n(n-1)}{2} + n \equiv n^2. \tag{3.204}$$

Due to its importance for atoms, let us spell out the hierarchy (3.203) of a few lowest-energy states, using the traditional state notation, in which the value of n is followed by the letter that denotes the value of l:

$$n = 1: \quad l = 0 \quad \text{(one } 1s \text{ state)} \quad m = 0. \tag{3.205}$$

$$\begin{aligned} n = 2: \quad &l = 0 \text{ (one } 2s \text{ state)} \quad m = 0, \\ &l = 1 \text{ (three } 2p \text{ states)} \quad m = 0, \ \pm 1. \end{aligned} \tag{3.206}$$

$$\begin{aligned} n = 3: \quad &l = 0 \text{ (one } 3s \text{ state)} \quad m = 0, \\ &l = 1 \text{ (three } 3p \text{ states)} \quad m = 0, \ \pm 1, \\ &l = 2 \text{ (five } 3d \text{ states)} \quad m = 0, \ \pm 1, \ \pm 2. \end{aligned} \tag{3.207}$$

Figure 3.22 shows the plots of the radial functions (3.195) of the listed states. The most important of them is of course the ground ($1s$) state with $n = 1$ and hence $E = -E_0/2$. According to Eqs. (3.195) and (3.198), its radial function is just

$$\mathcal{R}_{1,0}(r) = \frac{2}{r_0^{3/2}}e^{-r/r_0}, \tag{3.208}$$

while its angular distribution is uniform—see Eq. (3.174). The gap between the ground energy and the energy $E = -E_0/8$ of the lowest excited states (with $n = 2$) in a hydrogen atom (in which $E_0 = E_H \approx 27.2$ eV) is as large as ~ 10 eV, so that their thermal excitation requires temperatures as high as $\sim 10^5$ K, and the overwhelming part of all hydrogen atoms in the visible Universe are in their ground state. Since atomic hydrogen makes up about 75% of 'normal' matter[80], we are very fortunate that such simple formulas as Eqs. (3.174) and (3.208) describe the atomic states prevalent in Mother Nature!

---

[79] See, e.g. Eq. (A.8a).
[80] Excluding the so-far hypothetical *dark matter* and *dark energy*.

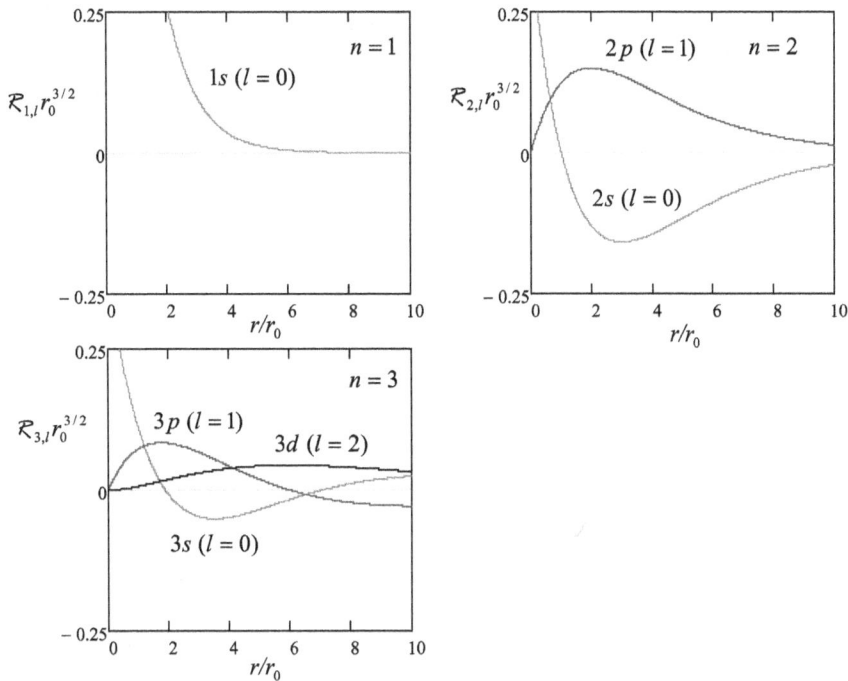

Figure 3.22. The lowest radial functions of the Bohr atom.

According to Eqs. (3.195) and (3.198), the radial functions of the lowest excited states, $2s$ (with $n = 2$ and $l = 0$), and $2p$ (with $n = 2$ and $l = 1$) are also not too complicated:

$$\mathcal{R}_{2,0}(r) = \frac{1}{(2r_0)^{3/2}}\left(2 - \frac{r}{r_0}\right)e^{-r/2r_0}, \quad \mathcal{R}_{2,1}(r) = \frac{1}{(2r_0)^{3/2}}\frac{r}{3^{1/2}r_0}e^{-r/2r_0}, \quad (3.209)$$

with the former of these states ($2s$) having a uniform angular distribution, and the three latter ($2p$) states, with different $m = 0, \pm 1$, having simple angular distributions, which differ only by their spatial orientation—see Eq. (3.175) and the second row of figure 3.20. The most important trend here, clearly visible from the comparison of the two top panels of figure 3.22 as well, is a larger radius of the decay exponent in the radial functions ($2r_0$ for $n = 2$ instead of r_0 for $n = 1$), and hence the larger radial extension of the states. This trend is confirmed by the following general formula[81]:

$$\langle r \rangle_{n,l} = \frac{r_0}{2}[3n^2 - l(l + 1)]. \quad (3.210)$$

The second important trend is that at a fixed n, the orbital quantum number l determines how fast the wavefunction changes with r near the origin, and how much it oscillates in the radial direction at larger values of r. For example, the $2s$

---

[81] Note that even at the largest value of l, equal to $(n - 1)$, the second term $l(l + 1)$ in Eq. (3.210) is equal to $(n^2 - n)$, and hence cannot over-compensate the first term $3n^2$.

eigenfunction $\mathcal{R}_{2,0}(r)$ is nonvanishing at $r = 0$, and 'makes one wiggle' (has one root) in the radial direction, while the eigenfunctions $2p$ equal zero at $r = 0$, and do not oscillate in the radial direction. Instead, those wavefunctions oscillate as the functions of an angle—see the second row of figure 3.20. The same trend is clearly visible for $n = 3$ (see the bottom panel of figure 3.22), and continues for the higher values of n.

The states with $l = l_{max} \equiv n - 1$ may be viewed as crude analogs of the circular motion of a particle in a plane whose orientation defines the quantum number m, with an almost fixed radius $r \approx r_0(n^2 \pm n)$. On the other hand, the best classical image of the s-state ($l = 0$) is the purely radial, spherically-symmetric motion of the particle to and from the attracting center. (The latter image is especially imperfect, because the motion needs to happen simultaneously in all radial directions.) The classical language becomes reasonable only for the highly degenerate *Rydberg states*, with $n \gg 1$, whose linear superpositions may be used to compose wave packets closely following the classical (circular or elliptic) trajectories of the particle—just as was discussed in section 2.2 for the free 1D motion.

Besides Eq. (3.210), mathematics gives us several other simple relations for the radial functions $\mathcal{R}_{n,l}$ (and, since the spherical harmonics are normalized to 1, for the eigenfunctions as the whole), including those that we will use later in the course[82]:

$$\left\langle \frac{1}{r} \right\rangle_{n,l} = \frac{1}{n^2 r_0}, \quad \left\langle \frac{1}{r^2} \right\rangle_{n,l} = \frac{1}{n^3(l + \frac{1}{2}) r_0^2},$$

$$\left\langle \frac{1}{r^3} \right\rangle_{n,l} = \frac{1}{n^3 l(l + \frac{1}{2})(l + 1) r_0^3}. \tag{3.211}$$

In particular, the first of these formulas means that for *any* eigenfunction $\psi_{n,l,m}$, with all its complicated radial and angular dependences, there is a simple relation between the potential and full energies:

$$\langle U \rangle_{n,l} = -C \left\langle \frac{1}{r} \right\rangle_{n,l} = -\frac{C}{n^2 r_0} = -\frac{E_0}{n^2} = 2E_n, \tag{3.212}$$

so that the average kinetic energy of the particle, $\langle T \rangle_{n,l} = E_n - \langle U \rangle_{n,l}$, is equal to $E_n - 2E_n = |E_n| > 0$.

As in the several previous cases we have met, the simple results (3.201), (3.210)–(3.212) are in a sharp contrast with the rather complicated expressions for the corresponding eigenfunctions. Historically this contrast gave an additional motivation for the development of more general approaches to quantum mechanics, that would replace, or at least complement our brute-force (wave-mechanics) analysis. A discussion of such an approach will be the main topic of the next chapter.

Rather strikingly, the above classification of the quantum numbers, together with very modest steals from the further theory, allows a semi-quantitative explanation of

---

[82] The first of these relations may be readily proved using the Heller–Feynman theorem (see chapter 1); this proof is left for the reader's exercise, after the more general form of this theorem has been proved in chapter 6. Note also that the last of Eqs. (3.211) diverges at $l = 0$, in particular in the ground state (with $n = 1$, $l = 0$).

1 H																	2 He
3 Li	4 Be	alkali metals		transition metals		metalloids						5 B	6 C	7 N	8 O	9 F	10 Ne
11 Na	12 Mg	alkali-earth metals		nonmetals		halogens						13 Al	14 Si	15 P	16 S	17 Cl	18 Ar
19 K	20 Ca	rare-earth metals		other metals		noble gases											
19 K	20 Ca	21 Sc	22 Ti	23 V	24 Cr	25 Mn	26 Fe	27 Co	28 Ni	29 Cu	30 Zn	31 Ga	32 Ge	33 As	34 Se	35 Br	36 Kr
37 Rb	38 Sr	39 Y	40 Zr	41 Nb	42 Mo	43 Tc	44 Ru	45 Rh	46 Pd	47 Ag	48 Cd	49 In	50 Sn	51 Sb	52 Te	53 I	54 Xe
55 Cs	56 Ba	57- 71	72 Hf	73 Ta	74 W	75 Re	76 Os	77 Ir	78 Pt	79 Au	80 Hg	81 Tl	82 Pb	83 Bi	84 Po	85 At	86 Rn
87 Fr	88 Ra	89- 102	104 Rf	105 Db	106 Sg	107 Bh	108 Hs	109 Mt	110 Ds	111 Rg	112 Cn	113 Uut	114 Fl	115 Uup	116 Lv	117 Uus	118 Uuo

Property legend:

Lanthanides:	57 La	58 Ce	59 Pr	60 Nd	61 Pm	62 Sm	63 Eu	64 Gd	65 Tb	66 Dy	67 Ho	68 Er	69 Tm	70 Yb	71 Lu
Actinides:	89 Ac	89 Ac	90 Th	91 Pa	92 U	93 Np	94 Pu	95 Am	96 Cm	97 Bk	98 Cf	99 Es	100 Fm	101 Md	102 Lr

Figure 3.23. The periodic table of elements, showing their atomic numbers and chemical symbols, as well as their color-coded basic physical/chemical properties at the so-called *ambient* (meaning usual laboratory) *conditions*.

the whole system of chemical elements. The 'only' two additions we need are the following facts, borrowed from the further chapters of this course:

(i) due to their unavoidable interaction with relatively low-temperature environments, atoms tend to relax into their lowest-energy state, and

(ii) due to the Pauli principle (valid for electrons as the Fermi particles), each *orbital eigenstate* discussed above may be occupied with two electrons with opposite spins.

Of course, atomic electrons do interact, so that their quantitative description requires quantum mechanics of multiparticle systems, which is rather complex. (Its main concepts will be discussed in chapter 8.) However, the lion's share of this interaction is reduced to simple electrostatic *screening*, i.e. the partial compensation of the electric charge of the atomic nucleus, as felt by a particular electron, by other electrons of the atom. This screening changes the qualitative results (such as the energy scale E_0) dramatically; however, the quantum number hierarchy, and hence their classification, is not affected.

The system of atoms is most often presented as the famous *periodic table of chemical elements*[83], whose simple version is shown in figure 3.23. (The table in figure 3.24 presents a sequential list of the elements and their electron configurations, following the convention already used in Eqs. (3.205)–(3.207), with the additional upper index showing the number of electrons with the indicated values of quantum numbers n and l.) The numbers in the table's cells, and in the first column of the list,

---

[83] Also called the *Mendeleev table*, after D Mendeleev who put forward the concept of the quasi-periodicity of chemical element properties as functions of Z phenomenologically in 1869. (The explanation of this periodicity had to wait for 60 more years until the quantum mechanics' formulation in the late 1920s.)

Atomic number	Atomic symbol	Electron states
Period 1		
1	H	$1s^1$
2	He	$1s^2$
Period 2		[He] shell, plus:
3	Li	$2s^1$
4	Be	$2s^2$
5	B	$2s^2 2p^1$
6	C	$2s^2 2p^2$
7	N	$2s^2 2p^3$
8	O	$2s^2 2p^4$
9	F	$2s^2 2p^5$
10	Ne	$2s^2 2p^6$
Period 3		[Ne] shell, plus:
11	Na	$3s^1$
12	Mg	$3s^2$
13	Al	$3s^2 3p^1$
14	Si	$3s^2 3p^2$
15	P	$3s^2 3p^3$
16	S	$3s^2 3p^4$
17	Cl	$3s^2 3p^5$
18	Ar	$3s^2 3p^6$
Period 4		[Ar] shell, plus:
19	K	$4s^1$
20	Ca	$4s^2$
21	Sc	$3d^1 4s^2$
22	Ti	$3d^2 4s^2$
23	V	$3d^3 4s^2$
24	Cr	$3d^5 4s^1$
25	Mn	$3d^5 4s^2$
26	Fe	$3d^6 4s^2$
27	Co	$3d^7 4s^2$
28	Ni	$3d^8 4s^2$
29	Cu	$3d^9 4s^1$
30	Zn	$3d^{10} 4s^2$
31	Ga	$3d^{10} 4s^2 4p^1$
32	Ge	$3d^{10} 4s^2 4p^2$
33	As	$3d^{10} 4s^2 4p^3$
34	Se	$3d^{10} 4s^2 4p^4$
35	Br	$3d^{10} 4s^2 4p^5$
36	Kr	$3d^{10} 4s^2 4p^6$
Period 5		[Kr] shell, plus:
37	Rb	$5s^1$
38	Sr	$5s^2$
39	Y	$4d^1 5s^2$
40	Zr	$4d^2 5s^2$
41	Nb	$4d^4 5s^1$
42	Mo	$4d^5 5s^1$
43	Tc	$4d^6 5s^1$
44	Ru	$4d^7 5s^1$
45	Rh	$4d^8 5s^1$
46	Pd	$4d^{10}$
47	Ag	$4d^{10} 5s^1$
48	Cd	$4d^{10} 5s^2$
49	In	$4d^{10} 5s^2 5p^1$
50	Sn	$4d^{10} 5s^2 5p^2$
51	Sb	$4d^{10} 5s^2 5p^3$
52	Te	$4d^{10} 5s^2 5p^4$
53	I	$4d^{10} 5s^2 5p^5$
54	Xe	$4d^{10} 5s^2 5p^6$
Period 6		[Xe] shell, plus:
55	Cs	$6s^1$
56	Ba	$6s^2$
57	La	$5d^1 6s^2$
58	Ce	$4f^1 5d^1 6s^2$
59	Pr	$4f^3 6s^2$
60	Nd	$4f^4 6s^2$
61	Pm	$4f^5 6s^2$
62	Sm	$4f^6 6s^2$
63	Eu	$4f^7 6s^2$
64	Gd	$4f^7 5d^1 6s^2$
65	Tb	$4f^9 6s^2$
66	Dy	$4f^{10} 6s^2$
67	Ho	$4f^{11} 6s^2$
68	Er	$4f^{12} 6s^2$
69	Tm	$4f^{13} 6s^2$
70	Yb	$4f^{14} 6s^2$
71	Lu	$4f^{14} 5d^1 6s^2$
72	Hf	$4f^{14} 5d^2 6s^2$
73	Ta	$4f^{14} 5d^3 6s^2$
74	W	$4f^{14} 5d^4 6s^2$
75	Re	$4f^{14} 5d^5 6s^2$
76	Os	$4f^{14} 5d^6 6s^2$
77	Ir	$4f^{14} 5d^7 6s^2$
78	Pt	$4f^{14} 5d^9 6s^1$
79	Au	$4f^{14} 5d^{10} 6s^1$
80	Hg	$4f^{14} 5d^{10} 6s^2$
81	Tl	$4f^{14} 5d^{10} 6s^2 6p^1$
82	Pb	$4f^{14} 5d^{10} 6s^2 6p^2$
83	Bi	$4f^{14} 5d^{10} 6s^2 6p^3$
84	Po	$4f^{14} 5d^{10} 6s^2 6p^4$
85	At	$4f^{14} 5d^{10} 6s^2 6p^5$
86	Rn	$4f^{14} 5d^{10} 6s^2 6p^6$
Period 7		[Rn] shell, plus:
87	Fr	$7s^1$
88	Ra	$7s^2$
89	Ac	$6d^1 7s^2$
90	Th	$6d^2 7s^2$
91	Pa	$5f^2 6d^1 7s^2$
92	U	$5f^3 6d^1 7s^2$
93	Np	$5f^4 6d^1 7s^2$
94	Pu	$5f^6 7s^2$
95	Am	$5f^7 7s^2$
96	Cm	$5f^7 6d^1 7s^2$
97	Bk	$5f^9 7s^2$
98	Cf	$5f^{10} 7s^2$
99	Es	$5f^{11} 7s^2$
100	Fm	$5f^{12} 7s^2$
101	Md	$5f^{13} 7s^2$
102	No	$5f^{14} 7s^2$
103	Lr	$5f^{14} 6d^1 7s^2$
104	Rf	$5f^{14} 6d^2 7s^2$
105	Db	$5f^{14} 6d^3 7s^2$
106	Sg	$5f^{14} 6d^4 7s^2$
107	Bh	$5f^{14} 6d^5 7s^2$
108	Hs	$5f^{14} 6d^6 7s^2$
109	Mt	$5f^{14} 6d^7 7s^2$
110	Ds	$5f^{14} 6d^8 7s^2$
111	Rg	$5f^{14} 6d^9 7s^2$
112	Cn	$5f^{14} 6d^{10} 7s^2$
113	Uut	$5f^{14} 6d^{10} 7s^2 7p^1$
114	Fl	$5f^{14} 6d^{10} 7s^2 7p^2$
115	Uup	$5f^{14} 6d^{10} 7s^2 7p^3$
116	Lv	$5f^{14} 6d^{10} 7s^2 7p^4$
117	Uus	$5f^{14} 6d^{10} 7s^2 7p^5$
118	Uuo	$5f^{14} 6d^{10} 7s^2 7p^6$

Figure 3.24. Atomic electron configurations. The upper index shows the number of electrons in the states with the indicated quantum numbers n (the first digit) and l (letter-coded as was discussed above).

are the *atomic numbers* Z, that physically is the number of protons in the particular atomic nucleus, and hence the number of electrons in an electrically-neutral atom.

The simplest atom, with $Z = 1$, is hydrogen (chemical symbol H)—the only atom for which each theory discussed above is quantitatively correct[84]. According to Eq. (3.191), the $1s$ ground state of its only electron corresponds to the quantum

---

[84] Besides very small *fine-structure* and *hyperfine-splitting* corrections—to be discussed, respectively, in chapters 6 and 8.

number values $n = 1$, $l = 0$, and $m = 0$—see Eq. (3.205). In most versions of the periodic table, the cell of H is placed in the top left corner.

In the next atom, helium (symbol He, $Z = 2$), the same orbital quantum state ($1s$) is filled with two electrons with different spins. As will be discussed in detail in chapter 8, electrons of the same atom are actually *indistinguishable*, and their quantum states are not independent, and may be *entangled*. These factors are important for several properties of helium atoms (and heavier elements as well); however, a bit counter-intuitively, for atom classification purposes, they are not crucial. Due to the twice higher electric charge of the nucleus of the helium atom, i.e. the twice higher value of the constant C in Eq. (3.190), resulting in a four-fold increase of the constant E_0 given by Eq. (3.192), the binding energy of each electron is crudely four times higher than that of the hydrogen atom—though the electron interaction decreases it by about 25%—see section 8.2. This is why taking one electron away (i.e. the *positive ionization* of the helium atom) requires a very high energy, ~ 23.4 eV, which is not available in the usual chemical reactions. On the other hand, a neural helium atom cannot bind one more electron (i.e. form a *negative ion*) either. As a result, the helium, and all other elements with fully completed electron *shells* (meaning the sets of states with eigenenergies well separated from higher energy levels) is a chemically inert *noble gas*, thus starting the whole right-most column of the periodic table, allocated for such elements.

The situation changes rather dramatically as we move to the next element, lithium (Li), with $Z = 3$ electrons. Two of them are still accommodated by the inner shell with $n = 1$ (listed in figure 3.24 as the *helium shell* [He]), but the third one has to reside in the next shell with $n = 2$, $l = 0$, and $m = 0$, i.e. in the $2s$ state. According to Eq. (3.201), the binding energy of this electron is much lower, especially if we take into account that according to Eqs. (3.210) and (3.211), the $1s$ electrons of the [He] shell are much closer to the nucleus and almost completely compensate two thirds of its electric charge $+3e$. As a result, the $2s$-state electron is approximately, but reasonably well described by Eq. (3.201) with $Z = 1$ and $n = 2$, giving a binding energy close to just 3.4 eV (experimentally, ~ 5.39 eV), so that a lithium atom can give out that electron rather easily—to either atoms of other elements to form chemical compounds, or into the common conduction band of the solid-state lithium; as a result, at the ambient conditions it is a typical *alkali metal*. The similarity of chemical properties of lithium and hydrogen, with the *chemical valence* of one[85], places Li as the starting element of the second *period* (row), with the first period limited to only H and He—see figure 3.23.

In the next element, beryllium (symbol Be, $Z = 4$), the $2s$ state ($n = 2$, $l = 0$, $m = 0$) picks up one more electron, with the opposite spin. Due to the higher electric charge of the nucleus, $Q = 4e$, with only half of it compensated by $1s$ electrons of the [He] shell, the binding energy of the $2s$ electrons is somewhat higher than that in lithium,

---

[85] The chemical valence (or 'valency') is a not very precise term, physically describing the number of an atom's electrons involved in a chemical reaction. For the same atom, especially with a large number of electrons in the outer, unfilled shell, this number may depend on the chemical compound formed. (For example, the valence of the atoms of iron is 2 in the *ferrous oxide*, FeO, and is 3 in the *ferric oxide*, Fe_2O_3.)

so that the ionization energy increases to ~9.32 eV. As a result, beryllium is also chemically active with the valence of two, but not as active as lithium, and is also metallic in its solid-state phase, but with a lower electric conductivity than lithium.

Moving in this way along the second row of the periodic table (from $Z = 3$ to $Z = 10$), we see a gradual filling of the rest of the total $2n^2 = 2 \times 2^2 = 8$ different electron states of the $n = 2$ shell (see Eq. (3.204), with the additional spin degeneracy factor of 2), including two $2s$ states with $m = 0$, and six $2p$ states with $m = 0, \pm 1,$[86] with a gradually growing ionization potential (up to ~21.6 eV in Ne with $Z = 10$), i.e. a growing reluctance to conduct electricity or form *positive* ions. However, the final elements of the row, such as oxygen (O, with $Z = 8$) and especially fluorine (F, with $Z = 9$) can readily pick up extra electrons to fill up their $2p$ states, i.e. form *negative* ions. As a result, these elements are chemically active, with the double valence for oxygen and the single valence for fluorine. However, the final element of this row, neon, has its $n = 2$ shell completely full, and cannot form a stable negative ion. This is why it is a noble gas, like helium. Traditionally, in the periodic table such elements are placed right under helium (figure 3.23), to emphasize the similarity of their chemical and physical properties. But this necessitates making an at least six-cell gap in the 1st row. (Actually, the gap is often made larger, to accommodate next rows—keep reading.)

Period 3, i.e. the 3rd row of the table, starts exactly like period 2, with sodium (Na, with $Z = 11$), also a chemically active alkali metal whose atom features 10 electrons filling the shells with $n = 1$ and $n = 2$ (in figure 3.24, collectively called the neon shell [Ne]), plus one electron in the $3s$ state ($n = 3, l = 0, m = 0$), which may be again reasonably well described by the hydrogen atom theory—see, e.g. the red curve on the last panel of figure 3.22. Continuing along this row, we could naively expect that, according to Eq. (3.204), and with the account of double spin degeneracy, this period of the table should have $2n^2 = 2 \times 3^2 = 18$ elements, with a gradual, sequential filling of two $3s$ states, then six $3p$ states, and then ten $3d$ states. However, here we run into a big surprise: after argon (Ar, with $Z = 18$), a relatively inert element with the ionization energy of ~15.7 eV due to the fully filled $3s$ and $3p$ shells, the next element, potassium (K, with $Z = 19$) is an alkali metal again!

The reason for that is the difference of the actual electron energies from those of the hydrogen atom, which is due mostly to inter-electron interactions, and gradually accumulates with the growth of Z. It may be semi-quantitatively understood from the results of section 3.6. In hydrogen-like atoms, the electron state energies do not depend on the quantum number l (as well as m)—see Eq. (3.201). However, the orbital quantum number does affect the wavefunction of an electron. As figure 3.22 shows, the larger l the less the probability for an electron to be close to the nucleus, where its positive charge is less compensated by other electrons. As a result of this effect (and also the relativistic corrections to be discussed in section 6.3), the electron's energy grows with l. Actually, this effect is visible even in the period 2 of the table: it manifests itself in the filling order—the p states after the s states.

---

[86] The specific order of filling of the states within each shell follow the so-called *Hund rules*—see section 8.3.

However, for potassium (K, with $Z = 19$) and calcium (Ca, with $Z = 20$), the energies of the $3d$ states become so high that the energies of two $4s$ states are lower, and they are filled first. As described by Eq. (3.210), and also by the first of Eqs. (3.211), the effect of the principal number n on the distance from the nucleus is much stronger than that of l, so that the $4s$ wavefunctions of K and Ca are relatively far from the nucleus, and determine the chemical valence (equal to 1 and 2, correspondingly) of these elements. The next atoms, from Sc ($Z = 21$) to Zn ($Z = 30$), with the gradually filled 'internal' $3d$ states, are the so-called *transition metals* whose (comparable) ionization energies and chemical properties are determined by the $4s$ electrons.

This fact is the origin of the difference between various forms of the 'periodic' table. In its most popular option, shown in figure 3.23, K is used to start the next period 4, and then a new period is started each time and only when the first electron with the next principal quantum number (n) appears[87]. This topology provides a very clear mapping on the chemical properties of the first element of each period (an alkali metal), as well as its last element (a noble gas). This also automatically means making gaps in all previous rows. Usually, this gap is made between the atoms with completely filled s states and with the first electron in a p state, because here the properties of the elements make a somewhat larger step. (For example, the step from Be to B makes the material an insulator, but the step from Mg to Al does not make a similar difference.) As a result, the elements of the same column have only *approximately* similar chemical valence and physical properties.

In order to accommodate longer lowest rows, such representation is inconvenient, because the whole table would be too broad. This is why the so-called *rare earths*, including *lanthanides* (with Z from 57 to 70, of the 6th row, with a gradual filling of the $4f$ and $5d$ states) and the *actinides* (Z from 89 to 103, of the 7th row, with a gradual filling of the $5f$ and $6d$ states), are usually represented as outlet rows—see figure 3.23. This is quite acceptable for the purposes of the standard chemistry, because the chemical properties of the elements within each such group are rather close.

To summarize this very short review[88], the 'periodic table of elements' is not periodic in the strict sense of the word. Nevertheless, it has had enormous historic significance for chemistry, as well as atomic and solid state physics, and is still very convenient for many purposes. For our course, the most important aspect of its discussion is the surprising possibility to describe, at least for classification purposes, such a complex multi-electron system as an atom as a set of quasi-independent electrons in certain quantum states indexed with the same quantum numbers n, l, and m as those of the hydrogen atom. This fact enables the use of various *perturbation theories*, which give more quantitative description of atomic properties. Some of these techniques will be reviewed in chapters 6 and 8.

---

[87] Another popular option is to return to the first column as soon as an atom has one electron in the s state (like for Cu, Ag, and Au, in addition to the alkali metals).

[88] For a bit more detailed (but still succinct) discussion of the valence and other chemical aspects of atomic structure, I can recommend chapter 5 of the very clear text by L Pauling, *General Chemistry*, Dover, 1988.

3.8 Spherically-symmetric scatterers

The machinery of the Legendre polynomials and the spherical Bessel functions, discussed in section 3.6, may be also used for an analysis of particle scattering by spherically-symmetric potentials (3.155), well beyond the Born approximation (section 3.3), provided that such a potential $U(r)$ is also localized, i.e. reduces sufficiently fast at $r \to \infty$. (The quantification of this condition is left for the reader's exercise.)

Indeed, directing the axis z along the direction of the incident plane de Broglie wave ψ_i, and taking its origin in the center of the scatterer, we may expect the scattered wave ψ_s to be (at least) axially symmetric, so that its expansion in the series over the spherical harmonics includes only the terms with $m = 0$. Hence, the solution (3.64) of the stationary Schrödinger equation (3.63) in this case may be represented as[89]

$$\psi = \psi_i + \psi_s = a_i\left[e^{ikz} + \sum_{l=0}^{\infty} \mathcal{R}_l(r)P_l(\cos\theta) \right], \qquad (3.213)$$

where $k \equiv (2mE)^{1/2}/\hbar$ is defined by the energy E of the incident particle, while the radial functions $\mathcal{R}_l(r)$ have to satisfy Eq. (3.181), and be finite at $r \to 0$. At large distances $r \gg R$, where R is the effective radius of the scatterer, the potential $U(r)$ is negligible, and Eq. (3.181) is reduced to Eq. (3.183). In contrast to its analysis in section 3.6, we should look for its solution using a linear superposition of the spherical Bessel functions of both kinds, $j_l(kr)$ and $y_l(kr)$, because Eq. (3.183) is now invalid at $r \to 0$, and our former argument for dropping the latter functions is no longer valid:

$$\mathcal{R}_l(r) = A_l j_l(kr) + B_l y_l(kr), \quad \text{at } r \gg R. \qquad (3.214)$$

Here A_l and B_l are some complex coefficients, determined by the scattering potential $U(r)$, i.e. by the solution of Eq. (3.181) at $r \sim R$.

As the explicit expressions (3.186) show, the spherical Bessel functions $j_l(\xi)$ and $y_l(\xi)$ represent *standing* de Broglie waves, with equal real amplitudes, so that their simple linear combinations (called the *spherical Hankel functions of the first and second kind*),

$$h_l^{(1)}(\xi) \equiv j_l(\xi) + iy_l(\xi), \quad \text{and} \quad h_l^{(2)}(\xi) \equiv j_l(\xi) - iy_l(\xi), \qquad (3.215)$$

represent *traveling* waves propagating, respectively, from the origin (i.e. from the center of the scatterer), and toward the origin, from infinity. In particular, at $\xi \gg 1, l$, i.e. at large distances $r \gg 1/k, l/k,$[90]

---

[89] The particular terms in the sum over l are frequently called the *partial waves*.
[90] For arbitrary l, this result may be confirmed using Eqs. (3.185) and the asymptotic formulas for the 'usual' Bessel functions—see, e.g. *Part EM* Eqs. (2.135) and (2.152), valid for an arbitrary (not necessarily integer) index n.

$$h_l^{(1)}(kr) \rightarrow \frac{(-i)^{l+1}}{kr}e^{ikr} \qquad h_l^{(2)}(kr) \rightarrow \frac{i^{l+1}}{kr}e^{-ikr}. \tag{3.216}$$

But using the same physical argument as in the beginning of section 3.1, we may conclude that in the case of a localized scatterer, there should be no latter waves at $r \gg R$; hence, we have to require the amplitude of the term proportional to $h_l^{(2)}$ to be zero. With the relations reciprocal to Eqs. (3.125),

$$j_l(\xi) = \frac{1}{2}\left[h_l^{(1)}(\xi) + h_l^{(2)}(\xi)\right], \qquad y_l(\xi) = \frac{1}{2i}\left[h_l^{(1)}(\xi) - h_l^{(2)}(\xi)\right], \tag{3.217}$$

which enable us to rewrite Eq. (3.214) as

$$
\begin{aligned}
\mathcal{R}_l(r) &= \frac{A_l}{2}\left[h_l^{(1)}(\xi) + h_l^{(2)}(\xi)\right] + \frac{B_l}{2i}\left[h_l^{(1)}(\xi) - h_l^{(2)}(\xi)\right] \\
&\equiv \left(\frac{A_l - iB_l}{2}\right)h_l^{(1)}(\xi) + \left(\frac{A_l + iB_l}{2}\right)h_l^{(2)}(\xi),
\end{aligned}
\tag{3.218}
$$

this means that the combination $(A_l + iB_l)$ has to be equal zero, so that $B_l = iA_l$. Hence we have just one unknown coefficient (say, A_l) for each l,[91] and may rewrite Eq. (3.218) in an even simpler form:

$$\mathcal{R}_l(r) = A_l\left[j_l(kr) + iy_l(kr)\right] \equiv A_l h_l^{(1)}(kr), \quad \text{at } r \gg R, \tag{3.219}$$

and use Eqs. (3.213) and (3.216) to write the following expression for the scattered wave at large distances:

$$\psi_s \approx \frac{a_i}{kr}e^{ikr}\sum_{l=0}^{\infty}(-i)^{l+1}A_l P_l(\cos\theta), \quad \text{at } r \gg R, \frac{1}{k}, \frac{l}{k}. \tag{3.220}$$

Comparing this expression with the general Eq. (3.81), we see that for a spherically-symmetric, localized scatterer,

$$f = \frac{1}{k}\sum_{l=0}^{\infty}(-i)^{l+1}A_l P_l(\cos\theta), \tag{3.221}$$

so that the differential cross-section (3.84) is

$$\frac{d\sigma}{d\Omega} = \frac{1}{k^2}\left|\sum_{l=0}^{\infty}(-i)^{l+1}A_l P_l(\cos\theta)\right|^2 \equiv \frac{1}{k^2}\sum_{l,l'=0}^{\infty}i^{l'-l}A_l A_{l'}^* P_l(\cos\theta)P_{l'}(\cos\theta). \tag{3.222}$$

---

[91] Moreover, using the conservation of the orbital momentum, to be discussed in section 5.6, it is possible to show that this *complex* coefficient may be further reduced to just one *real* parameter, usually recast as the *partial phase shift* δ_l between the lth spherical harmonics of the incident and scattered waves. However, I will not use this notion, because practical calculations are more physically transparent (and not more complex) without it.

The last expression is more convenient for the calculation of the total cross-section (3.59):

$$\sigma = \oint_{4\pi} \frac{d\sigma}{d\Omega} d\Omega = 2\pi \int_{-1}^{+1} \frac{d\sigma}{d\Omega} d(\cos\theta) = \frac{2\pi}{k^2} \sum_{l,l'=0}^{\infty} i^{l'-l} A_l A_{l'}^* \int_{-1}^{+1} P_l(\xi)P_{l'}(\xi)d\xi, \quad (3.223)$$

where $\xi \equiv \cos\theta$, because this result may be much simplified by using Eq. (3.167):

$$\sigma = \sum_{l=0}^{\infty} \sigma_l, \quad \text{with } \sigma_l = \frac{4\pi}{k^2} \frac{|A_l|^2}{2l+1}; \quad (3.224)$$

physically, this reduction of the double sum to a single one means that due to the orthogonality of the spherical functions, the total scattering probability outflows due to each partial wave just add up.

Hence the solution of the scattering problem is reduced to the calculation of the partial wave amplitudes A_l—and for the total cross-section, merely of their magnitudes. This task is facilitated by using the following *Rayleigh formula* for the expansion of the incident plane wave's exponent into a series over the Legendre polynomials[92],

$$e^{ikz} \equiv e^{ikr\cos\theta} = \sum_{l=0}^{\infty} i^l(2l+1)j_l(kr)P_l(\cos\theta). \quad (3.225)$$

As the simplest example, let us calculate the scattering by a completely opaque and 'hard' (meaning sharp-boundary) sphere, which may be described by the following potential:

$$U(r) = \begin{cases} +\infty, & \text{for } r < R, \\ 0, & \text{for } R < r. \end{cases} \quad (3.226)$$

In this case, the total wavefunction has to vanish at $r \leqslant R$, and hence for the external problem ($r \geqslant R$) the sphere enforces the boundary condition $\psi \equiv \psi_0 + \psi_s = 0$ for all values of θ, at $r = R$. With Eqs. (3.213), (3.220) and (3.225), this condition becomes

$$a_i \sum_{l=0}^{\infty} \left[\mathcal{R}_l(R) + i^l(2l+1)j_l(kR) \right] P_l(\cos\theta) = 0. \quad (3.227)$$

Due to the orthogonality of the Legendre polynomials, this condition may be satisfied for all angles θ only if all the coefficients before all $P_l(\cos\theta)$ vanish, i.e. if

$$\mathcal{R}_l(R) = -i^l(2l+1)j_l(kR). \quad (3.228)$$

---

[92] It may be proved using the Rodrigues formula (3.165) and integration by parts—the task left for the reader's exercise.

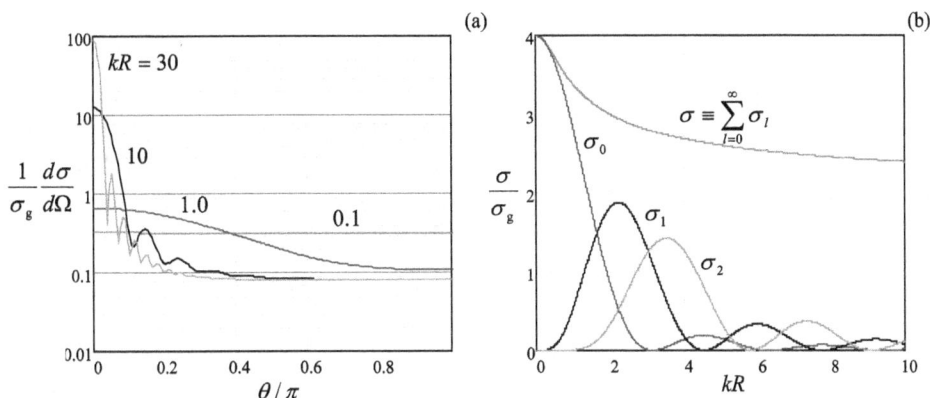

Figure 3.25. Particle scattering by an opaque, hard sphere: (a) the differential cross-section, normalized by the geometric cross-section $\sigma_g \equiv \pi R^2$ of the sphere, as a function of the scattering angle θ, and (b) the (similarly normalized) total cross-section and its lowest spherical components, as functions of the dimensionless product $kR \propto E^{1/2}$.

On the other hand, for $r > R$, $U(r) = 0$, so that Eq. (3.183) is valid, and its outward-wave solution (3.219) has to be valid even at $r \to R$, giving

$$\mathcal{R}_l(R) = A_l\big[j_l(kR) + iy_l(kR)\big].\tag{3.229}$$

Requiring the two last expressions to give the same result, we get

$$A_l = -i^l(2l+1)\frac{j_l(kR)}{j_l(kR) + iy_l(kR)},\tag{3.230}$$

so that Eqs. (3.222) and (3.224) yield:

$$\frac{d\sigma}{d\Omega} = \frac{1}{k^2}\left| \sum_{l=0}^{\infty}(2l+1)\frac{j_l(kR)}{j_l(kR) + iy_l(kR)}P_l(\cos\theta)\right|^2,$$

$$\sigma_l = \frac{4\pi(2l+1)}{k^2}\frac{j_l^2(kR)}{j_l^2(kR) + y_l^2(kR)}.\tag{3.231}$$

As figure 3.25a shows, the first of these results describes an angular structure of the scattered de Broglie wave, qualitatively similar to that given by the Born approximation—cf Eq. (3.98) and figure 3.10. Namely, at low particle's energies ($kR \ll 1$), the scattering is essentially isotropic[93], while in the opposite, high-energy limit $kR \gg 1$, it is mostly confined to small angles $\theta \sim \pi/kR \ll 1$, and exhibits numerous local destructive-interference minima at angles $\theta_n \sim \pi n/kR$. However, in our current (exact!) theory these minima are always finite, because the theory describes the effective bending of the de Broglie waves along the back side of the sphere, which smears the interference pattern.

---

[93] In this limit, the scattering is dominated by the lowest spherical-harmonic component with $l = 0$—see, e.g. figure 3.25b.

Such bending is also responsible for a rather counter-intuitive fact, described by the second of Eqs. (3.231), and clearly visible in figure 3.25b: even at $kR \to \infty$, the total cross-section σ of scattering tends to $2\sigma_g \equiv 2\pi R^2$, rather than to σ_g as in the purely-classical scattering theory. (The fact that at $kR \ll 1$, the cross-section is also larger than σ_g, approaching $4\sigma_g$ at $kR \to 0$, is much less surprising, because in this limit the de Broglie wavelength $\lambda = 2\pi/k$ is much longer than the sphere's radius R, so that the wave's propagation is affected by the whole sphere.)

The above analysis may be readily generalized to the case of a uniform, *step-like*, but finite potential (3.97)—a problem left for the reader's exercise. For a finite and *smooth* scattering potential $U(r)$, plugging Eq. (3.225) into Eq. (3.213) and the latter one into Eq. (3.66), requiring the coefficients before each angular function $P_l(\cos\theta)$ to be balanced, and canceling the common coefficient a_0, we get the following inhomogeneous generalization of Eq. (3.181) for the radial functions defined by Eq. (3.213):

$$[E - U(r)]\mathcal{R}_l + \frac{\hbar^2}{2mr^2}\left[\frac{d}{dr}\left(r^2\frac{d}{dr}\right) - l(l+1)\right]\mathcal{R}_l(r) = U(r)\,i^l(2l+1)j_l(kr). \quad (3.232)$$

This differential equation has to be solved, in the whole scatterer volume (i.e. for all $r \sim R$) with the boundary conditions for the functions $\mathcal{R}_l(r)$ to be finite at $r \to 0$, and to tend to the asymptotic form (3.219) at $r \gg R$. The last requirement enables the evaluation of the coefficients A_l that are needed for spelling out Eqs. (3.222) and (3.224). Unfortunately, due to the lack of time, I have to refer the reader interested in such cases to special literature[94].

3.9 Problems

Problem 3.1. A particle of energy E is incident (see figure below, within the plane of drawing) on a sharp potential step:

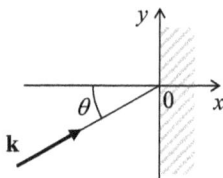

$$U(\mathbf{r}) = \begin{cases} 0, & \text{for } x < 0, \\ U_0, & \text{for } 0 < x. \end{cases}$$

Calculate the particle's reflection probability $\mathcal{R}$ as a function of the incidence angle θ; sketch and discuss this function for various magnitudes and signs of U_0.

Problem 3.2.* Analyze how the Landau levels (3.50) are modified by an additional uniform electric field $\mathbf{E}$, directed along the plane of the particle's motion. Contemplate the physical meaning of your result, and its implications for the

---

[94] See, e.g. [16].

quantum Hall effect in a gate-defined Hall bar. (The area $l \times w$ area of such a bar (see figure 3.6) is defined by metallic gate electrodes parallel to the 2D electron gas plane—see figure below. The negative voltage V_g, applied to the gates, squeezes the 2D gas from the area under the gates into the complementary, Hall-bar part of the plane.)

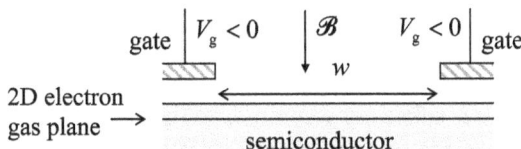

Problem 3.3. Analyze how the Landau levels (3.50) are modified if a 2D particle is confined in an additional 1D potential well $U(x) = m\omega_0^2 x^2/2$.

Problem 3.4. Find the eigenfunctions of a spinless, charged 3D particle moving in 'crossed' (mutually perpendicular), uniform electric and magnetic fields, with $\mathscr{E} \ll c\mathscr{B}$. For each eigenfunction, calculate the expectation value of the particle's velocity in the direction perpendicular to both fields, and compare the result with the solution of the corresponding classical problem.

Hint: Generalize Landau's solution for 2D particles, discussed in section 3.2.

Problem 3.5. Use the Born approximation to calculate the angular dependence and the full cross-section of scattering of an incident plane wave, propagating along the x-axis, by the following pair of point inhomogeneities:

$$U(\mathbf{r}) = \mathscr{W}\left[\delta\left(\mathbf{r} - \mathbf{n}_z \frac{a}{2}\right) + \delta\left(\mathbf{r} + \mathbf{n}_z \frac{a}{2}\right)\right].$$

Analyze the results in detail. Derive the condition of the Born approximation's validity for such delta-functional scatterers.

Problem 3.6. Complete the analysis of the Born scattering by a uniform spherical potential (3.97), started in section 3.3, by calculation of its total cross-section. Analyze the result in the limits $kR \ll 1$ and $kR \gg 1$.

Problem 3.7. Use the Born approximation to calculate the differential cross-section of particle scattering by a very thin spherical shell, whose potential may be approximated as

$$U(r) = \mathscr{W}\,\delta(r - R).$$

Analyze the results in the limits $kR \ll 1$ and $kR \gg 1$, and compare them with those for a uniform sphere considered in section 3.3.

Problem 3.8. Use the Born approximation to calculate the differential and full cross-sections of electron scattering by a screened Coulomb field of a point charge Ze, with the electrostatic potential

$$\phi(\mathbf{r}) = \frac{Ze}{4\pi\varepsilon_0 r} e^{-\lambda r},$$

neglecting spin interaction effects, and analyze the result's dependence on the screening parameter λ. Compare the results with those given by the classical ('Rutherford') formula[95] for the unscreened Coulomb potential ($\lambda \to 0$), and formulate the condition of Born approximation's validity in this limit.

Problem 3.9. A quantum particle with electric charge Q is scattered by a localized distributed charge with a spherically-symmetric density $\rho(r)$, and zero total charge. Use the Born approximation to calculate the differential cross-section of the *forward scattering* (with the scattering angle $\theta = 0$), and evaluate it for the scattering of electrons by a hydrogen atom in its ground state.

Problem 3.10. Reformulate the Born approximation for the 1D case. Use the result to find the scattering and transfer matrices of a 'rectangular' (flat-top) scatterer

$$U(x) = \begin{cases} U_0, & \text{for } |x| < d/2, \\ 0, & \text{otherwise.} \end{cases}$$

Compare the results with the those of the exact calculations carried out earlier in chapter 2, and analyze how their relation changes in the eikonal approximation.

Problem 3.11. In the tight-binding approximation, calculate the lowest eigenenergies and the corresponding eigenstates of a particle placed into a system of three similar, weakly coupled potential wells located in the vertices of an equilateral triangle.

Problem 3.12. The figure below shows a fragment of a periodic 2D lattice, with the red and blue points showing the location of different local potentials—say, different atoms.

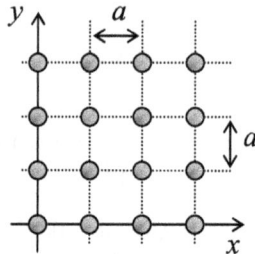

(i) Find the reciprocal lattice and the 1st Brillouin zone.
(ii) Calculate the wave number k of the monochromatic de Broglie wave incident along axis x, at which the lattice creates the first-order diffraction peak within the $[x, y]$ plane, and the direction toward this peak.
(iii) Semi-qualitatively, describe the evolution of the intensity of the peak when the local potentials, represented by the different points, become similar.

---

[95] See, e.g. *Part CM* section 3.5, in particular Eq. (3.73).

Problem 3.13. For the 2D hexagonal lattice (figure 3.12b):

(i) find the reciprocal lattice $\mathbf{Q}$ and the 1st Brillouin zone;
(ii) use the tight-binding approximation to calculate the dispersion relation $E(\mathbf{q})$ for a 2D particle moving in a potential with such periodicity, with an energy close to the eigenenergy of the axially-symmetric states quasi-localized at the potential minima;
(iii) analyze and sketch (or plot) the resulting dispersion relation $E(\mathbf{q})$ inside the 1st Brillouin zone.

Problem 3.14. Complete the tight-binding approximation calculation of the band structure of the honeycomb lattice, started in the end of section 3.4. Analyze the results; in particular prove that the Dirac points $\mathbf{q}_D$ are located in the corners of the 1st Brillouin zone, and express the velocity v_n, participating in Eq. (3.122), in terms of the coupling energy δ_n. Show that the final results do not change if the quasi-localized wavefunctions are not axially-symmetric, but are proportional to $\exp\{im\varphi\}$—as they are, with $m = 1$, for the $2p_z$ electrons of carbon atoms in graphene, which are responsible for its transport properties.

Problem 3.15. Examine basic properties of the so-called *Wannier functions* defined as

$$\phi_{\mathbf{R}}(\mathbf{r}) \equiv \text{const} \times \int_{\text{BZ}} \psi_{\mathbf{q}}(\mathbf{r})e^{-i\mathbf{q}\cdot\mathbf{R}}d^3q,$$

where $\psi_{\mathbf{q}}(\mathbf{r})$ is the Bloch wavefunction (3.108), $\mathbf{R}$ is any vector of the Bravais lattice, and the integration over the quasi-momentum $\mathbf{q}$ is extended over any (e.g. the first) Brillouin zone.

Problem 3.16. Evaluate the long-range electrostatic interaction (the so-called *London dispersion force*) between two similar, electrically-neutral atoms or molecules, modeling each of them as an isotropic 3D harmonic oscillator with the electric dipole moment $\mathbf{d} = q\mathbf{s}$, where $\mathbf{s}$ is the oscillator's displacement from its equilibrium position.

Hint: Represent the total Hamiltonian of the system as a sum of Hamiltonians of independent 1D harmonic oscillators, and calculate their total ground-state energy as a function of distance between the dipoles[96].

Problem 3.17. Derive expressions for the eigenfunctions and the corresponding eigenenergies of a 2D particle of mass m, free to move inside a thin round disk of radius R. What is the degeneracy of each energy level? Calculate five lowest energy levels with an accuracy better than 1%.

---

[96] This explanation of the interaction between electrically-neutral atoms was put forward in 1930 by F London, on the background of a prior (1928) work by C Wang. Note that in some texts this interaction is (rather inappropriately) referred to as the 'van der Waals force', though it is only one, long-range component of the van der Waals model—see, e.g. *Part SM* section 4.1.

Problem 3.18. Calculate the ground-state energy of a 2D particle of mass m, localized in a very shallow flat-bottom potential well

$$U(\rho) = \begin{cases} -U_0, & \text{for } \rho < R, \\ 0, & \text{for } \rho > R, \end{cases} \quad \text{with } 0 < U_0 \ll \frac{\hbar^2}{mR^2}.$$

Problem 3.19. Estimate the energy E of the localized ground state of a particle of mass m, in an axially-symmetric 2D potential well of a finite radius R, with an arbitrary but very small potential $U(\rho)$. (Quantify this condition.)

Problem 3.20. Spell out the explicit form of the spherical harmonics $Y_4^0(\theta, \varphi)$ and $Y_4^4(\theta, \varphi)$.

Problem 3.21. Calculate $\langle x \rangle$ and $\langle x^2 \rangle$ in the ground state of the planar and spherical rotators of radius R. What can you say about the averages $\langle p_x \rangle$ and $\langle p_x^2 \rangle$?

Problem 3.22. A spherical rotator, with $r \equiv (x^2 + y^2 + z^2)^{1/2} = R = \text{const}$, of mass m is in a state with the following wavefunction: $\psi = \text{const} \times (1/3 + \sin^2\theta)$. Calculate its energy.

Problem 3.23. According to the discussion in the beginning of section 3.5, eigenfunctions of a 3D harmonic oscillator may be calculated as products of three 1D 'Cartesian oscillators'—see, in particular Eq. (3.125), with $d = 3$. However, according to the discussion in section 3.6, the wavefunctions of the type (3.200), proportional to the spherical harmonics Y_l^m, are also eigenstates of this spherically-symmetric system. Represent the wavefunctions (3.200) of:

(i) the ground state of the oscillator, and
(ii) each of its lowest excited states,

as linear combinations of products of 1D oscillator's wavefunctions. Also, calculate the degeneracy of the nth energy level of the oscillator.

Problem 3.24. Calculate the smallest depth U_0 of a spherical, flat-bottom potential well

$$U(\mathbf{r}) = \begin{cases} -U_0, & \text{for } r < R, \\ 0, & \text{for } R < r, \end{cases}$$

at which it has a bound (localized) eigenstate. Does such a state exist for a very narrow and deep well $U(\mathbf{r}) = -\mathcal{W}\,\delta(\mathbf{r})$, with a positive and finite $\mathcal{W}$?

Problem 3.25. A 3D particle of mass m is placed into a spherically-symmetric potential well with $-\infty < U(r) \leqslant U(\infty) = 0$. Relate its ground-state energy to that of a 1D particle of the same mass, moving in the following potential well:

$$U'(x) = \begin{cases} U(x), & \text{for } x \geqslant 0, \\ +\infty, & \text{for } x \leqslant 0. \end{cases}$$

In the light of the found relation, discuss the origin of the difference between the solutions of the previous problem and problem 2.17.

Problem 3.26. Calculate the smallest value of the parameter U_0, for that the following spherically-symmetric potential well,

$$U(r) = -U_0 e^{-r/R}, \quad \text{with } U_0, \ R > 0,$$

has a bound (localized) eigenstate.

Hint: You may like to introduce the following new variables: $f \equiv r\mathcal{R}$ and $\xi \equiv Ce^{-r/2R}$, with an appropriate choice of the constant C.

Problem 3.27. A particle moving in a certain central potential $U(r)$, with $U(r) \to 0$ at $r \to \infty$, has a stationary state with the following wavefunction:

$$\psi = Cr^\alpha e^{-\beta r} \cos \theta,$$

where C, α, and $\beta > 0$ are constants. Calculate:

  (i) the probabilities of all possible values of the quantum numbers m and l,
 (ii) the confining potential, and
(iii) the state's energy.

Problem 3.28. Use the variational method to estimate the ground-state energy of a particle of mass m, moving in the following spherically-symmetric potential:

$$U(\mathbf{r}) = a\, r^4.$$

Problem 3.29. Use the variational method, with the trial wavefunction $\psi_{\text{trial}} = \text{const}/(r + a)^b$, where both $a > 0$ and $b > 1$ are fitting parameters, to estimate the ground-state energy of the hydrogen-like atom/ion with the nuclear charge $+Ze$. Compare the solution with the exact result.

Problem 3.30. Calculate the energy spectrum of a particle moving in a monotonic, but otherwise arbitrary attractive spherically-symmetric potential $U(r) < 0$, in the approximation of very large orbital quantum numbers l. Formulate the quantitative condition(s) of validity of your theory. Check that for the Coulomb potential $U(r) = -C/r$, your result agrees with Eq. (3.201).

Hint: Try to solve Eq. (3.181) approximately, introducing the same new function, $f(r) \equiv r\mathcal{R}(r)$, as was already used in section 3.1 and in the solutions of a few earlier problems.

Problem 3.31. An electron had been in the ground state of a hydrogen-like atom/ion with nuclear charge Ze, when the charge suddenly changed to $(Z + 1)e$.[97] Calculate the probabilities for the electron of the changed system to be:

 (i) in the ground state, and
(ii) in the lowest excited state.

---

[97] Such a fast change happens, for example, at the beta-decay, when one of a nucleus's neutrons suddenly turns into a proton, emitting a high-energy electron and a neutrino, which leave the system very fast (instantly on the atomic time scale), and do not affect directly the atom transition's dynamics.

Problem 3.32. Due to a very short pulse of an external force, the nucleus of a hydrogen-like atom, initially at rest in its ground state, starts moving with velocity **v**. Calculate the probability W_g that the atom remains in its ground state. Evaluate the energy to be given, by the pulse, to a hydrogen atom in order to reduce W_g to 50%.

Problem 3.33. Calculate $\langle x^2 \rangle$ and $\langle p_x^2 \rangle$ in the ground state of a hydrogen-like atom/ion. Compare the results with the Heisenberg's uncertainty relation. What do these results tell us about the electron's velocity in the system?

Problem 3.34. Use the Hellmann–Feynman theorem (see problem 1.5) to prove:

 (i) the first of Eqs. (3.211), and
 (ii) the fact that for a spinless particle in an arbitrary spherically-symmetric attractive potential $U(r)$, the ground state is always an *s*-state (with the orbital quantum number $l = 0$).

Problem 3.35. For the ground state of a hydrogen atom, calculate the expectation values of $\mathscr{E}$ and $\mathscr{E}^2$, where $\mathscr{E}$ is the electric field created by the atom, at distances $r \gg r_0$ from its nucleus. Interpret the resulting relation between $\langle \mathscr{E} \rangle^2$ and $\langle \mathscr{E}^2 \rangle$, at the same observation point.

Problem 3.36. Calculate the condition at which a particle of mass m, moving in the field of a very thin spherically-symmetric shell, with

$$U(\mathbf{r}) = w\delta(r - R),$$

and $w < 0$, has at least one localized ('bound') stationary state.

Problem 3.37. Calculate the lifetime of the lowest metastable state of a particle in the same spherical-shell potential as in the previous problem, but now with $w > 0$, for sufficiently large w. (Quantify this condition.)

Problem 3.38. A particle of mass m and energy E is incident on a very thin spherical shell of radius R, whose localized states were the subject of two previous problems.

 (i) Derive the general expressions for the differential and total cross-sections of scattering.
 (ii) Spell out the contribution σ_0 to the full cross-section σ, due to the spherically-symmetric component of the scattered de Broglie wave.
 (iii) Analyze the result for σ_0 in the limits of very small and very large magnitudes of w, for both signs of this parameter. In particular, in the limit $w \to +\infty$, relate the result to the metastable state's lifetime τ calculated in the previous problem.

Problem 3.39. Calculate the spherically-symmetric contribution σ_0 to the total cross-section of particle scattering by a uniform sphere of radius R, described by the potential

$$U(r) = \begin{cases} U_0, & \text{for } r < R, \\ 0, & \text{otherwise.} \end{cases}$$

Analyze the result, and give an interpretation of it most remarkable features.

Problem 3.40. Use the finite difference method with the step $h = a/2$ to calculate as many eigenenergies as possible, for a particle confined to the interior of:

(i) a square with side a, and
(ii) a cube with side a,

with hard walls. For the square, repeat the calculations, using a finer step: $h = a/3$. Compare the results for different values of h with each other and with the exact formulas.

Hint: It is advisable to first solve (or review the solution of) the similar 1D problem in chapter 1, or start from reading about the finite difference method[98]. Also: try to exploit the problem's symmetry.

References

[1] Cronin A, Schmiedmayer J and Pritchard D 2009 *Rev. Mod. Phys.* **81** 1051
[2] Tonomura A *et al* 1986 *Phys. Rev. Lett.* **565** 792
[3] Novoselov K *et al* 2007 *Science* **315** 1379
[4] Götz M *et al* 2018 *Appl. Phys. Lett.* **112** 072102
[5] Yoshioka D 1998 *The Quantum Hall Effect* (Springer)
[6] Yennie D 1987 *Rev. Mod. Phys.* **59** 781
[7] Shapere A and Wilczek F 1992 *Geometric Phases in Physics* (World Scientific)
[8] Bohm A *et al* 2003 *The Geometric Phase in Quantum Systems* (Springer)
[9] Ashcroft N and Mermin N 1976 *Solid State Physics* (Saunders College)
[10] Stephens P and Goldman A 1991 *Sci. Amer.* **264** 24
[11] Neto A *et al* 2009 *Rev. Mod. Phys.* **81** 109
[12] Lu X *et al* 2017 *Appl. Phys. Rev.* **4** 021306
[13] Aleiner I and Larkin A 1997 *Phys. Rev.* E **55** R1243
[14] Gutzwiller M 1991 *Chaos in Classical and Quantum Mechanics* (Springer)
[15] Lévy L *et al* 1990 *Phys. Rev. Lett.* **64** 2074
[16] Taylor J 2006 *Scattering Theory* (Dover)

---

[98] See, e.g. *Part CM* section 8.5 or *Part EM* section 2.11.

IOP Publishing

Quantum Mechanics
Lecture notes
Konstantin K Likharev

Chapter 4

Bra–ket formalism

The objective of this chapter is to describe the Dirac's 'bra–ket' formalism of quantum mechanics, which not only overcomes some inconveniences of wave mechanics, but also allows a natural description of such intrinsic properties of particles as their spin. In the course of discussion of the formalism I will give only a few simple examples of its application, leaving more involved cases for the following chapters.

4.1 Motivation

As the reader could see from the previous chapter of these notes, wave mechanics gives many results of primary importance. Moreover, it is mostly sufficient for many applications, for example, for solid state electronics and device physics. However, in the course of our survey we have filed several grievances about this approach. Let me briefly summarize these complaints:

(i) Attempts to analyze *temporal* evolution of quantum systems within this approach beyond the trivial time behavior of the stationary states, described by Eq. (1.62), run into technical difficulties. For example, we could derive Eq. (2.151) describing the metastable state's decay, or Eq. (2.181) describing the quantum oscillations in coupled wells, only for the simplest potential profiles, though it is intuitively clear that such simple results should be common for all problems of this kind. Solving such problems for more complex potential profiles would entangle the time evolution analysis with the calculation of the spatial distribution of the evolving wavefunctions—which (as we could see in sections 2.9 and 3.6) may be rather complex even for simple time-independent potentials. Some separation of the spatial and temporal dependences is possible using perturbation approaches (to be discussed in chapter 6), but even those would lead, in the wavefunction language, to very cumbersome formulas.

doi:10.1088/2053-2563/aaf3a3ch4

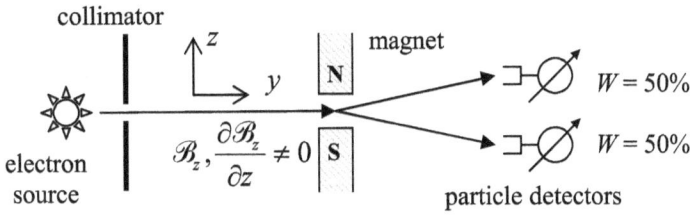

Figure 4.1. The simplest Stern–Gerlach experiment.

(ii) The last statement is also correct for other issues that are conceptually addressable within the wave mechanics, e.g. the Feynman path integral approach, coupling to environment, etc. Pursuing them in wave mechanics would lead to formulas so bulky that I had postponed their discussion until we have got a more compact formalism on hand.

(iii) In the discussion of several key problems (for example the harmonic oscillator and spherically-symmetric potentials) we have run into rather complicated eigenfunctions coexisting with very simple energy spectra—that infer some simple background physics. It is very important to get this physics revealed.

(iv) In the wave mechanics postulates formulated in section 1.2, the quantum mechanical operators of the coordinate and momentum are treated rather unequally—see Eqs. (1.26b). However, some key expressions, e.g. for the fundamental eigenfunction of a free particle,

$$\exp\left\{i\frac{\mathbf{p}\cdot\mathbf{r}}{\hbar}\right\},\tag{4.1}$$

or the harmonic oscillator's Hamiltonian,

$$\hat{H} = \frac{1}{2m}\hat{p}^2 + \frac{m\omega_0^2}{2}\hat{r}^2,\tag{4.2}$$

just beg for a similar treatment of coordinates and momenta.

However, the strongest motivation for a more general formalism comes from wave mechanics' conceptual inability to describe elementary particles' *spins* and other internal quantum degrees of freedom, such as quark flavors or lepton numbers. In this context, let us review the basic facts on spin (which is a very representative and experimentally the most accessible of all internal quantum numbers), to understand what a more general formalism has to explain—as a minimum.

Figure 4.1 shows the conceptual scheme of the simplest spin-revealing experiment, first carried out by O Stern and W Gerlach in 1922.[1] A collimated beam of

---

[1] To the best of my knowledge, the concept of spin as a measure of the internal rotation of a particle was first suggested by R Kronig, then a 20 year-old student, in January 1925, a few months before two other students, G Uhlenbeck and S Goudsmit—to whom the idea is usually attributed. The concept was then accepted and developed quantitatively by W Pauli.

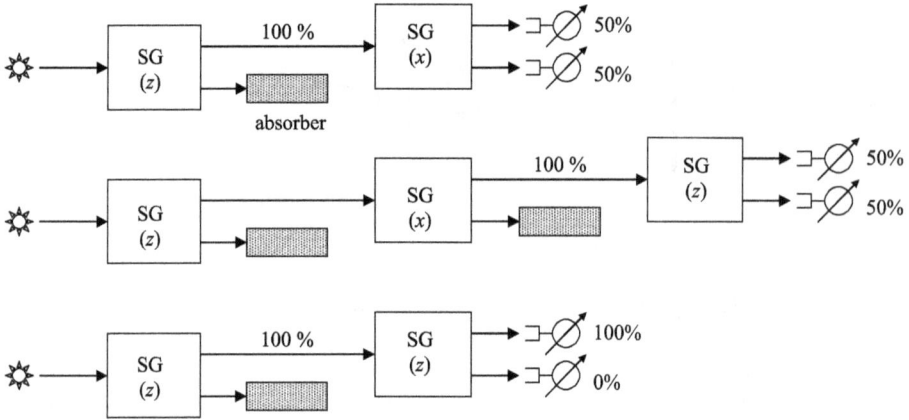

Figure 4.2. Three multi-stage Stern–Gerlach experiments. The boxes SG (…) denote magnets similar to one shown in figure 4.1, with the field oriented in the indicated direction.

electrons from a natural source such as a heated cathode, is passed through a gap between the poles of a strong magnet, whose magnetic field $\mathscr{B}$, (in figure 4.1, directed along the z-axis) is non-uniform, so that both $\mathscr{B}_z$ and $d\mathscr{B}_z/dz$ are not equal to zero. The experiment shows that the beam splits into two parts of equal intensity.

This result may be semi-quantitatively explained on classical, though somewhat phenomenological grounds by assuming that each electron has an intrinsic, permanent magnetic dipole moment **m**. Indeed, classical electrodynamics[2] tells us that the potential energy U of a magnetic dipole in an external magnetic field $\mathscr{B}$ is equal to $(-\mathbf{m} \cdot \mathscr{B})$, so that the force acting on the particle,

$$\mathbf{F} = -\nabla U = -\nabla(-\mathbf{m} \cdot \mathscr{B}), \tag{4.3}$$

has a nonvanishing vertical component

$$F_z = -\frac{\partial}{\partial z}(-m_z \cdot \mathscr{B}_z) \equiv m_z \frac{\partial \mathscr{B}_z}{\partial z}. \tag{4.4}$$

Hence if we further assume that the electron has an intrinsic magnetic moment, with two equally probable discrete values of $m_z = \pm\mu$ (though such discreteness does not follow from any classical model of the particle), this may explain the Stern–Gerlach effect qualitatively. The quantitative explanation of the beam splitting angle requires the magnitude of μ to be equal (or close) to the so-called *Bohr magneton*[3]

$$\mu_{\mathrm{B}} \equiv \frac{\hbar e}{2m_{\mathrm{e}}} \approx 0.9274 \times 10^{-23} \frac{\mathrm{J}}{\mathrm{T}}. \tag{4.5}$$

However, as we will see below, this value cannot be explained by any internal motion of the electron, say its rotation about the axis z. More importantly, this semi-classical phenomenology cannot explain, even qualitatively, the results of the set of

---

[2] See, e.g. *Part EM* section 5.4, in particular Eq. (5.100).
[3] A good mnemonic rule is that it is close to 1 K T^{-1}. In the Gaussian units, $\mu_{\mathrm{B}} \equiv \hbar e/2m_{\mathrm{e}}c \approx 0.9274 \times 10^{-20}$ erg G^{-1}.

multi-stage Stern–Gerlach experiments, shown in figure 4.2. In the first of the experiments, the electron beam is first passed through a magnetic field (and its gradient) oriented along the z-axis, just as in figure 4.1. Then one of the two resulting beams is absorbed (or otherwise removed from the setup), while the other one is passed through a similar but x-oriented field. The experiment shows that this beam is split again into two components of equal intensity. A classical explanation of this experiment would require an even more unnatural assumption that the initial electrons had random but discrete components of the magnetic moment simultaneously in *two* directions, z and x.

However, even this assumption cannot explain the results of the three-stage Stern–Gerlach experiment shown on the middle panel of figure 4.2. Here, the previous two-state setup is complemented with one more absorber and one more magnet, now with the z-orientation again. Completely counter-intuitively, it again gives two beams of equal intensity, as if we have not yet filtered out the electrons with m_z corresponding to the lower beam, at the first, z-stage. The only way to save the classical explanation here is to say that maybe, electrons *somehow* interact with the magnetic field, so that the x-polarized (non-absorbed) beam becomes spontaneously depolarized again somewhere between magnetic stages. But any hope for such an explanation is ruined by the control experiment shown in the bottom panel of figure 4.2, whose results indicate that no such depolarization happens.

We will see below that all these (and many more) results find a natural explanation in the *matrix mechanics* pioneered by W Heisenberg, M Born and P Jordan in 1925. However, the matrix formalism is inconvenient for the solution of most problems discussed in chapters 1–3, and for a short time it was eclipsed by the Schrödinger's wave mechanics, which had been put forward just a few months later. However, very soon P A M Dirac introduced a more general *bra–ket formalism* of quantum mechanics, which provides a generalization of both approaches and proves their equivalence. Let me describe it, begging for the reader's patience, because (in a contrast with my usual style), I will not be able to give particular examples for a while—until all the basic notions of the formalism have been introduced.

4.2 States, state vectors, and linear operators

The basic notion of the general formulation of quantum mechanics is the *quantum state* of a system[4]. To get some gut feeling of this notion, if a quantum state α of a particle may be adequately described by wave mechanics, this description is given by the corresponding wavefunction $\Psi_\alpha(\mathbf{r}, t)$. Note, however, the state as such is *not* a mathematical object (such as a function)[5], and can participate in mathematical formulas only as a 'pointer'—e.g. the index of the function Ψ_α. On the other hand, the wavefunction is *not* a state, but a mathematical object (a complex function of

---

[4] An attentive reader could notice my smuggling the term 'system' instead of 'particle', which was used in the previous chapters. Indeed, the bra–ket formalism allows the description of quantum systems much more complex than a single spinless particle that is a typical (though not the only possible) subject of wave mechanics.

[5] As was expressed nicely by A Peres, one of pioneers of the quantum information theory, 'quantum phenomena do not occur in the Hilbert space, they occur in a laboratory'.

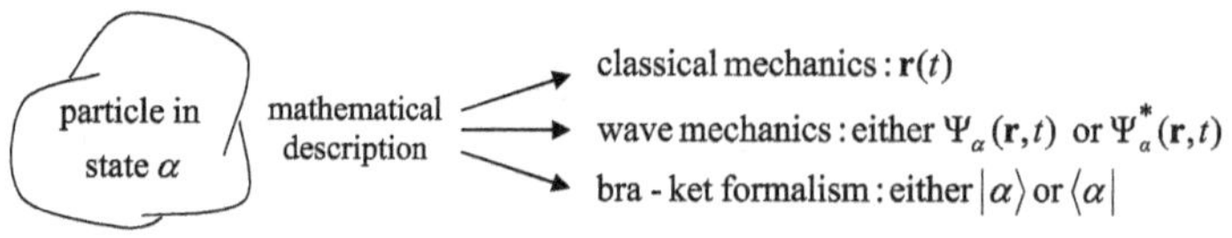

Figure 4.3. A particle's state and its descriptions.

space and time) giving a quantitative description of the state—just as the classical radius-vector **r** as a function of time is a mathematical object describing the motion of a classical particle—see figure 4.3. Similarly, in the Dirac formalism a certain quantum state α is described by either of two mathematical objects, called the *state vectors*: the *ket-vector* $|\alpha\rangle$ and *bra-vector* $\langle\alpha|$,[6] whose relation is close to that between the wavefunction Ψ_α and its complex conjugate Ψ_α^*.

One should be cautions with the term 'vector' here. The usual 'geometric' vectors, such as **r**, are defined in the usual geometric (say, Euclidean) space. In contrast, the bra- and ket-vectors are defined in a more abstract *Hilbert space* of a given system— the full set of its possible bra- and ket-vectors[7]. So, despite certain similarities with the geometric vectors, the bra- and ket-vectors are different mathematical objects, so that we need to define the rules of their handling. The primary rules are essentially postulates and are justified only by the correct description of all experimental observations. While there is a general consensus among physicists what the corollaries are, there are many possible ways to carve from them the basic postulate sets. Just as in section 1.2, I will not try too hard to beat the number of the postulates to the smallest possible minimum, trying instead to keep their physical meaning transparent.

(i) *Ket-vectors*. Let us start with *ket-vectors*—sometimes called just *kets* for short. Their most important property is the *linear superposition*. Namely, if several ket-vectors $|\alpha_j\rangle$ describe possible states of a quantum system, numbered by the index j, then any linear combination (*superposition*)

$$|\alpha\rangle = \sum_j c_j |\alpha_j\rangle, \tag{4.6}$$

where c_j are any (possibly complex) c-numbers, also describes a possible state of the same system. (One may say that vector $|\alpha\rangle$ belongs to the same Hilbert space as all $|\alpha_j\rangle$.) Actually, since ket-vectors are new mathematical objects, the exact meaning of the right-hand side of Eq. (4.6) becomes clear only after we have postulated the following rules of summation of these vectors,

$$|\alpha_j\rangle + |\alpha_{j'}\rangle = |\alpha_{j'}\rangle + |\alpha_j\rangle, \tag{4.7}$$

---

[6] The terms *bra* and *ket* were suggested to reflect the fact that the pair $\langle\beta|$ and $|\alpha\rangle$ may be considered as the parts   Q2
of combinations like $\langle\beta|\alpha\rangle$ (see below), which recall an expression in the usual angle *brackets*.
[7] I have to confess that this is a bit superficial definition, and I will rectify it soon.

and their multiplication by an arbitrary c-number:

$$c|\alpha_j\rangle = |\alpha_j\rangle c. \tag{4.8}$$

Note that in the set of wave mechanics postulates, the statements parallel to (4.7) and (4.8) were unnecessary, because the wavefunctions are the usual (albeit complex) functions of space and time, and we know from the usual algebra that such relations are valid.

As evident from Eq. (4.6), the complex coefficient c_j may be interpreted as the 'weight' of the state α_j in the linear superposition α. One important particular case is $c_j = 0$, showing that the state α_j does not participate in the superposition α. The corresponding term of the sum (4.6), i.e. the product

$$0|\alpha_j\rangle, \tag{4.9}$$

has a special name: the *null-state* vector. (It is important to avoid confusion between the null-state corresponding to vector (4.9), and the ground state of the system, which is frequently denoted by the ket-vector $|0\rangle$. In some sense, the null-state does not exist at all, while the ground state does—and frequently is the most important quantum state of the system.)

(ii) *Bra-vectors and inner ('scalar') products.* Bra-vectors $\langle\alpha|$, which obey the rules similar to Eqs. (4.7) and (4.8), are not new, independent objects: a ket-vector $|\alpha\rangle$ and the corresponding bra-vector $\langle\alpha|$ describe the same state. In other words, there is a *unique dual correspondence* between $|\alpha\rangle$ and $\langle\alpha|$,[8] very similar (though not identical) to that between a wavefunction Ψ and its complex conjugate Ψ^*. The correspondence between these vectors is described by the following rule: if a ket-vector of a linear superposition is described by Eq. (4.6), then the corresponding bra-vector is

$$\langle\alpha| = \sum_j c_j^*\langle\alpha_j| = \sum_j \langle\alpha_j|c_j^*. \tag{4.10}$$

The mathematical convenience of using two types of vectors, rather than just one, becomes clear from the notion of their *inner product* (due to its second, shorthand form, also called the *short bracket*):

$$(\langle\beta|)(|\alpha\rangle) \equiv \langle\beta|\alpha\rangle. \tag{4.11}$$

This is a (generally, complex[9]) scalar c-number (frequently called just the *scalar*), whose main property is the linearity with respect to any of its component vectors. For example, if a linear superposition α is described by the ket-vector (4.6), then

---

[8] Mathematicians like to say that the ket- and bra-vectors of the same quantum system are defined in two *isomorphic Hilbert spaces*.

[9] This is one of the differences of bra- and ket-vectors from the 'usual' (geometrical) vectors whose inner (scalar) products are always *real* scalars.

$$\langle \beta | \alpha \rangle = \sum_j c_j \langle \beta | \alpha_j \rangle, \tag{4.12}$$

while if Eq. (4.10) is true, then

$$\langle \alpha | \beta \rangle = \sum_j c_j^* \langle \alpha_j | \beta \rangle. \tag{4.13}$$

In plain English, c-numbers may be moved either into, or out of the inner products. The second key property of the inner product is

$$\langle \alpha | \beta \rangle = \langle \beta | \alpha \rangle^*. \tag{4.14}$$

It is compatible with Eq. (4.10); indeed, the complex conjugation of both parts of Eq. (4.12) gives:

$$\langle \beta | \alpha \rangle^* = \sum_j c_j^* \langle \beta | \alpha_j \rangle^* = \sum_j c_j^* \langle \alpha_j | \beta \rangle = \langle \alpha | \beta \rangle. \tag{4.15}$$

Finally, one more rule: the inner product of the bra- and ket-vectors describing the same state (called the *norm squared*) is real and non-negative,

$$\|\alpha\|^2 \equiv \langle \alpha | \alpha \rangle \geqslant 0. \tag{4.16}$$

In order to give the reader some feeling about the meaning of this rule: we will see below that if some state α may be described by the wavefunction $\Psi_\alpha(\mathbf{r}, t)$, then

$$\langle \alpha | \alpha \rangle = \int \Psi_\alpha^* \Psi_\alpha d^3 r \geqslant 0. \tag{4.17}$$

Hence the role of the bracket is very similar to the complex conjugation of the wavefunction, and Eq. (4.10) emphasizes this similarity. (Note that, by convention, there is no conjugation sign in the bra-part of the inner product; its role is played by the angular bracket inversion.)

(iii) *Operators.* One more key notion of the Dirac formalism is quantum-mechanical *linear operators*. Just as for the operators discussed in wave mechanics, the function of an operator is 'generation' of one state from another: if $|\alpha\rangle$ is a possible ket of the system, and $\hat{A}$ is a legitimate operator[10], then the following combination,

$$\hat{A} | \alpha \rangle, \tag{4.18}$$

is also a ket-vector describing a possible state of the system, i.e. a ket-vector in the same Hilbert space as the initial vector $|\alpha\rangle$. Another formulation of the same rule is

---

[10] Here the term 'legitimate' means 'having a clear sense in the bra–ket formalism'. Some examples of 'illegitimate' expressions are: $|\alpha\rangle \hat{A}$, $\hat{A}\langle\alpha|$, $|\alpha\rangle|\beta\rangle$, and $\langle\alpha|\langle\beta|$. Note, however, that the last two expressions may be legitimate if α and β are states of different systems, i.e. if their state vectors belong to different Hilbert spaces. We will run into such *direct products* of the bra- and ket-vectors (sometimes denoted, respectively, as $|\alpha\rangle \otimes |\beta\rangle$ and $\langle\alpha| \otimes \langle\beta|$) in chapters 6–8 and 10.

the following clarification of the notion of the Hilbert space: for the given set of the linear operators of a system[11], its Hilbert state includes all vectors that may be obtained from each other using the operations of the type (4.18).

As follows from the adjective 'linear', the main rules governing the operators is their linearity with respect to both any superposition of vectors:

$$\hat{A}\left(\sum_j c_j |\alpha_j\rangle\right) = \sum_j c_j \hat{A} |\alpha_j\rangle, \tag{4.19}$$

and any superposition of operators:

$$\left(\sum_j c_j \hat{A}_j\right) |\alpha\rangle = \sum_j c_j \hat{A}_j |\alpha\rangle. \tag{4.20}$$

These rules are evidently similar to Eqs. (1.53) and (1.54) of wave mechanics.

The above rules imply that an operator 'acts' on the ket-vector on its right; however, a combination of the type $\langle\alpha|\,\hat{A}$ is also legitimate and represents a new bra-vector. It is important that, generally, this vector does *not* represent the same state as ket-vector (4.18); instead, the bra-vector isomorphic to the ket-vector (4.18) is

$$\langle\alpha|\hat{A}^\dagger. \tag{4.21}$$

This statement serves as the definition of the *Hermitian conjugate* (also called the 'Hermitian adjoint') $\hat{A}^\dagger$ of the initial operator $\hat{A}$. For an important class of operators, called the *Hermitian operators*, the conjugation is inconsequential, i.e. for them

$$\hat{A}^\dagger = \hat{A}. \tag{4.22}$$

(This equality, as well as any other operator equation below, means that these operators act similarly on any bra- or ket-vector of the given Hilbert space[12].)

To proceed further, we need one more additional postulate, sometimes called the *associative axiom of multiplication*: just as an ordinary product of scalars, *any* legitimate bra–ket expression, not including an explicit summation, does not change from an insertion or removal of parentheses—meaning as usual that the operation inside the parentheses has to be performed first. The first two examples of this

---

[11] Such an operator set usually (if not always) implies a certain approximate model of the system. For example, if the coupling of the spin and orbital degrees of freedom of a particle may be ignored in a particular problem (as it may be for a non-relativistic particle in the absence of an external magnetic field), we may describe the spin dynamics of the particle using spin operators only. In this case, the set of possible spin vectors of the particle forms a Hilbert space separate from that of the orbital-state vectors of that particle.

[12] If we consider c-numbers as a particular type of operators (which is legitimate for any Hilbert space), then according to Eqs. (4.11) and (4.21), for them the Hermitian conjugation is equivalent to the simple complex conjugation, so that only real c-numbers may be considered as a particular type of the Hermitian operators (4.22).

postulate are given by Eqs. (4.19) and (4.20), but the associative axiom is more general and means, for example, that

$$\langle\beta|(\hat{A}|\alpha\rangle) = ((\langle\beta|\hat{A})|\alpha\rangle \equiv \langle\beta|\hat{A}|\alpha\rangle, \tag{4.23}$$

This last equality serves as the definition of the last form, called the *long bracket* (evidently, also a scalar), with an operator sandwiched between a bra-vector and a ket-vector. This definition, when combined with the definition of the Hermitian conjugate and Eq. (4.14), yields an important corollary:

$$\langle\beta|\hat{A}|\alpha\rangle = \langle\beta|(\hat{A}|\alpha\rangle) = ((\langle\alpha|\hat{A}^\dagger)|\beta\rangle)^* = \langle\alpha|\hat{A}^\dagger|\beta\rangle^*, \tag{4.24}$$

which is most frequently rewritten as

$$\langle\alpha|\hat{A}|\beta\rangle^* = \langle\beta|\hat{A}^\dagger|\alpha\rangle. \tag{4.25}$$

The associative axiom also enables one to explore the following definition of one more, *outer* product of bra- and ket-vectors:

$$|\beta\rangle\langle\alpha|. \tag{4.26}$$

In contrast to the inner product (4.12), which is a *scalar*, this mathematical construct is an *operator*. Indeed, the associative axiom allows us to remove the parentheses in the following expression:

$$(|\beta\rangle\langle\alpha|)\,|\gamma\rangle = |\beta\rangle\langle\alpha|\gamma\rangle. \tag{4.27}$$

But the last bracket is just a scalar; hence the mathematical object (4.26) acting on a ket-vector (in this case, $|\gamma\rangle$) gives a new ket-vector, which is the essence of operator's action. Very similarly,

$$\langle\delta|(|\beta\rangle\langle\alpha|) = \langle\delta|\beta\rangle\langle\alpha| \tag{4.28}$$

—again a typical operator's action on a bra-vector. So, Eq. (4.26) defines an operator.

Now let us perform the following calculation. We may use the parentheses' insertion into the bra–ket equality following from Eq. (4.14),

$$\langle\gamma|\alpha\rangle\langle\beta|\delta\rangle = ((\langle\delta|\beta\rangle\langle\alpha|\gamma\rangle)^*, \tag{4.29}$$

to transform it to the following form:

$$\langle\gamma|(|\alpha\rangle\langle\beta|)|\delta\rangle = ((\langle\delta|(|\beta\rangle\langle\alpha|)|\gamma\rangle)^*. \tag{4.30}$$

Since this equation should be valid for any vectors $\langle\gamma|$ and $|\beta\rangle$, its comparison with Eq. (4.25) gives the following operator equality

$$(|\alpha\rangle\langle\beta|)^\dagger = |\beta\rangle\langle\alpha|. \tag{4.31}$$

This is the conjugate rule for outer products; it recalls the rule (4.14) for inner products, but involves the Hermitian (rather than the usual complex) conjugation.

The associative axiom is also valid for the operator 'multiplication':

$$(\hat{A}\hat{B})|\alpha\rangle = \hat{A}(\hat{B}|\alpha\rangle), \quad \langle\beta|(\hat{A}\hat{B}) = (\langle\beta|\hat{A}) \hat{B}, \tag{4.32}$$

showing that the action of an operator product on a state vector is nothing more than the sequential action of the operands. However, we have to be rather careful with the operator products; generally they do not commute: $\hat{A}\hat{B} \neq \hat{B}\hat{A}$. This is why the *commutator*—the operator defined as

$$[\hat{A}, \hat{B}] \equiv \hat{A}\hat{B} - \hat{B}\hat{A}, \tag{4.33}$$

is a non-trivial and very useful notion. Another similar notion is the *anticommutator*[13]:

$$\{\hat{A}, \hat{B}\} \equiv \hat{A}\hat{B} + \hat{B}\hat{A}. \tag{4.34}$$

Finally, the bra–ket formalism broadly uses two special operators. The *null operator* $\hat{0}$, defined by the following relations:

$$\hat{0}|\alpha\rangle \equiv 0|\alpha\rangle, \quad \langle\alpha|\hat{0} \equiv \langle\alpha|0, \tag{4.35}$$

for an arbitrary state α; we may say that the null operator 'kills' *any* state, turning it into the null-state.

Another useful notion is the *identity operator*, which is defined by its following action (or rather 'inaction') on an arbitrary state vector:

$$\hat{I}|\alpha\rangle \equiv |\alpha\rangle, \quad \langle\alpha|\hat{I} \equiv \langle\alpha|. \tag{4.36}$$

4.3 State basis and matrix representation

While some operations in quantum mechanics may be carried out in the general bra–ket formalism outlined above, many calculations are done for specific quantum systems that feature at least one *full and orthonormal* set $\{u\}$ of states u_j, frequently called a *basis*. These terms mean that any state vector of the system (i.e. of its Hilbert space) may be represented as a unique sum of the type (4.6) or (4.10) over its basis vectors:

$$|\alpha\rangle = \sum_j \alpha_j|u_j\rangle, \quad \langle\alpha| = \sum_j \alpha_j^*\langle u_j|, \tag{4.37}$$

(so that, in particular, if α is one of the basis states, say $u_{j'}$, then $\alpha_j = \delta_{jj'}$), and that

$$\langle u_j|u_{j'}\rangle = \delta_{jj'}. \tag{4.38}$$

For the systems that may be described by wave mechanics, examples of the full orthonormal bases are represented by any orthonormal set of eigenfunctions calculated in the previous three chapters—as the simplest example, see Eq. (1.87).

---

[13] Another popular notation for the anticommutator is $[\hat{A}, \hat{B}]_+$; it will not be used in these notes.

Due to the uniqueness of the expansion (4.37), the full set of coefficients α_j gives a complete description of the state α (in a fixed basis $\{u\}$)—just as the usual Cartesian components A_x, A_y, and A_z give a complete description of a usual geometric 3D vector $\mathbf{A}$ (in a fixed reference frame). Still, let me emphasize some differences between such a representation of the quantum-mechanical bra- and ket-vectors and the usual geometric vectors:

(i) a basis may have a large or even infinite number of states u_j, and
(ii) the expansion coefficients α_j may be complex.

With these reservations in mind, the analogy with geometric vectors may be pushed further. Let us inner-multiply both parts of the first of Eqs. (4.37) by a bra-vector $\langle u_{j'}|$ and then transform the relation using the linearity rules discussed in the previous section, and Eq. (4.38):

$$\left\langle u_{j'}|\alpha\right\rangle = \left\langle u_{j'}\left|\sum_j \alpha_j\right|u_j\right\rangle = \sum_j \alpha_j\left\langle u_{j'}|u_j\right\rangle = \alpha_{j'}, \tag{4.39}$$

Together with Eq. (4.14), this means that any of the expansion coefficients in Eqs. (4.37) may be represented as an inner product:

$$\alpha_j = \left\langle u_j|\alpha\right\rangle, \qquad \alpha_j^* = \left\langle \alpha|u_j\right\rangle; \tag{4.40}$$

these relations are analogs of equalities $A_j = \mathbf{n}_j \cdot \mathbf{A}$ of the usual vector algebra. Using these important relations (which we will use on numerous occasions), the expansions (4.37) may be rewritten as

$$|\alpha\rangle = \sum_j |u_j\rangle\langle u_j|\alpha\rangle \equiv \sum_j \hat{\Lambda}_j|\alpha\rangle, \quad \langle\alpha| = \sum_j \langle\alpha|u_j\rangle\langle u_j| \equiv \sum_j \langle\alpha|\hat{\Lambda}_j, \tag{4.41}$$

This relation shows that the outer product defined as

$$\hat{\Lambda}_j \equiv |u_j\rangle\langle u_j|, \tag{4.42}$$

is a legitimate linear operator. Such an operator, acting on any state vector of the type (4.37), singles out just one of its components, for example,

$$\hat{\Lambda}_j|\alpha\rangle = |u_j\rangle\langle u_j|\alpha\rangle = \alpha_j|u_j\rangle, \tag{4.43}$$

i.e. 'kills' all components of the linear superposition but one. In the geometric analogy, this operator 'projects' the state vector on the jth 'direction', hence its name—the *projection operator*. Probably the most important property of the projection operators, called the *closure* (or *completeness*) *relation*, immediately follows from Eq. (4.41): their sum over the full basis is equivalent to the identity operator:

$$\sum_j |u_j\rangle\langle u_j| = \hat{I}. \tag{4.44}$$

This means in particular that we may insert the left-hand side of Eq. (4.44) into any bra–ket relation, at any place—the trick that we will use again and again.

Now let us see how the expansions (4.37) transform the key notions introduced in the last section, starting from the short bracket (4.11), i.e. the inner product of two state vectors:

$$\langle \beta | \alpha \rangle = \sum_{j,j'} \langle u_j | \beta_j^* \alpha_{j'} | u_{j'} \rangle = \sum_{j,j'} \beta_j^* \alpha_{j'} \delta_{jj'} = \sum_j \beta_j^* \alpha_j. \tag{4.45}$$

Besides the complex conjugation, this expression is similar to the scalar product of the usual vectors. Now, let us explore the long bracket (4.23):

$$\langle \beta | \hat{A} | \alpha \rangle = \sum_{j,j'} \beta_j^* \langle u_j | \hat{A} | u_{j'} \rangle \alpha_{j'} \equiv \sum_{j,j'} \beta_j^* A_{jj'} \alpha_{j'}. \tag{4.46}$$

Here, the last step uses the very important notion of *matrix elements* of the operator, defined as

$$A_{jj'} \equiv \langle u_j | \hat{A} | u_{j'} \rangle. \tag{4.47}$$

As evident from Eq. (4.46), the full set of the matrix elements completely characterizes the operator, just as the full set of the expansion coefficients (4.40) fully characterizes a quantum state. The term 'matrix' means, first of all, that it is convenient to represent the full set of $A_{jj'}$ as a square table (*matrix*), with the linear dimension equal to the number of basis states u_j of the system under consideration. By the way, this number (which may be infinite) is called the *dimensionality* of its Hilbert space.

As two simplest examples, all matrix elements of the null-operator, defined by Eqs. (4.35), are evidently equal to zero (in any basis), and hence it may be represented as a matrix of zeros (called the *null matrix*):

$$0 \equiv \begin{pmatrix} 0 & 0 & \dots \\ 0 & 0 & \dots \\ \dots & \dots & \dots \end{pmatrix}, \tag{4.48}$$

while for the identity operator $\hat{I}$, defined by Eqs. (4.36), we readily get

$$I_{jj'} = \langle u_j | \hat{I} | u_{j'} \rangle = \langle u_j | u_{j'} \rangle = \delta_{jj'}, \tag{4.49}$$

i.e. its matrix (naturally called the *identity matrix*) is diagonal—also in any basis:

$$I \equiv \begin{pmatrix} 1 & 0 & \dots \\ 0 & 1 & \dots \\ \dots & \dots & \dots \end{pmatrix}. \tag{4.50}$$

The convenience of the matrix language extends well beyond the representation of particular operators. For example, let us use the definition (4.47) to calculate matrix elements of a product of two operators:

$$(AB)_{jj''} = \langle u_j | \hat{A} \hat{B} | u_{j''} \rangle. \tag{4.51}$$

Here we may use Eq. (4.44) for the first (but not the last!) time, inserting the identity operator between the two operators, and then expressing it via a sum of projection operators:

$$(AB)_{jj''} = \langle u_j | \hat{A}\hat{B} | u_{j''} \rangle = \langle u_j | \hat{A}\hat{I}\hat{B} | u_{j''} \rangle$$
$$= \sum_{j'} \langle u_j | \hat{A} | u_{j'} \rangle \langle u_{j'} | \hat{B} | u_{j''} \rangle = \sum_{j'} A_{jj'} B_{j'j''}. \tag{4.52}$$

This result corresponds to the standard 'row by column' rule of calculation of an arbitrary element of the matrix product

$$AB = \begin{pmatrix} A_{11} & A_{12} & \cdots \\ A_{21} & A_{22} & \cdots \\ \cdots & \cdots & \cdots \end{pmatrix} \begin{pmatrix} B_{11} & B_{12} & \cdots \\ B_{21} & B_{22} & \cdots \\ \cdots & \cdots & \cdots \end{pmatrix}. \tag{4.53}$$

Hence the product of operators may be represented (in a fixed basis!) by that of their matrices (in the same basis). This is so convenient that the same language is often used to represent not only the long brackets,

$$\langle \beta | \hat{A} | \alpha \rangle = \sum_{j'} \beta_j^* A_{jj'} \alpha_{j'} = (\beta_1^*, \beta_2^*, \cdots) \begin{pmatrix} A_{11} & A_{12} & \cdots \\ A_{21} & A_{22} & \cdots \\ \cdots & \cdots & \cdots \end{pmatrix} \begin{pmatrix} \alpha_1 \\ \alpha_2 \\ \cdots \end{pmatrix}, \tag{4.54}$$

but even the simpler short brackets:

$$\langle \beta | \alpha \rangle = \sum_{j} \beta_j^* \alpha_j = (\beta_1^*, \beta_2^*, \cdots) \begin{pmatrix} \alpha_1 \\ \alpha_2 \\ \cdots \end{pmatrix}, \tag{4.55}$$

although these equalities require the use of non-square matrices: rows of (complex-conjugate!) expansion coefficients for the representation of bra-vectors, and columns of these coefficients for the representation of ket-vectors. With that, the mapping of states and operators on matrices becomes completely general.

Now let us have a look at the outer product operator (4.26). Its matrix elements are just

$$(|\alpha\rangle\langle\beta|)_{jj'} = \langle u_j | \alpha \rangle \langle \beta | u_{j'} \rangle = \alpha_j \beta_{j'}^*. \tag{4.56}$$

These are the elements of a very special square matrix, whose filling requires the knowledge of just $2N$ scalars (where N is the basis set size), rather than N^2 scalars as for an arbitrary operator. However, a simple generalization of such an outer product may represent an arbitrary operator. Indeed, let us insert two identity operators (4.44), with different summation indices, on both sides of any operator:

$$\hat{A} = \hat{I}\hat{A}\hat{I} = \left(\sum_{j} |u_j\rangle\langle u_j| \right) \hat{A} \left(\sum_{j'} |u_{j'}\rangle\langle u_{j'}| \right), \tag{4.57}$$

and use the associative axiom to rewrite this expression as

$$\hat{A} = \sum_{j,j'} |u_j\rangle \left(\langle u_j | \hat{A} | u_{j'} \rangle \right) \langle u_{j'} |. \tag{4.58}$$

But the expression in the middle long bracket is just the matrix element (4.47), so that we may write

$$\hat{A} = \sum_{j,j'} |u_j\rangle A_{jj'} \langle u_{j'} |. \tag{4.59}$$

The reader has to agree that this formula, which is a natural generalization of Eq. (4.44), is extremely elegant.

The matrix representation is so convenient that it makes sense to move it by one level lower—from state vector products to the 'bare' state vectors resulting from operator's action upon a given state. For example, let us use Eq. (4.59) to represent the ket-vector (4.18) as

$$|\alpha'\rangle \equiv \hat{A}|\alpha\rangle = \left(\sum_{j,j'} |u_j\rangle A_{jj'} \langle u_{j'} | \right) |\alpha\rangle = \sum_{j,j'} |u_j\rangle A_{jj'} \langle u_{j'} | \alpha \rangle. \tag{4.60}$$

According to Eq. (4.40), the last short bracket is just $\alpha_{j'}$, so that

$$|\alpha'\rangle = \sum_{j,j'} |u_j\rangle A_{jj'} \alpha_{j'} = \sum_j \left(\sum_{j'} A_{jj'} \alpha_{j'} \right) |u_j\rangle \tag{4.61}$$

But the expression in middle parentheses is just the coefficient α'_j of the expansion (4.37) of the resulting ket-vector (4.60) in the same basis, so that

$$\alpha'_j = \sum_{j'} A_{jj'} \alpha_{j'}. \tag{4.62}$$

This result corresponds to the usual rule of multiplication of a matrix by a column, so that we may represent any ket-vector by its column matrix, with the operator's action looking like

$$\begin{pmatrix} \alpha'_1 \\ \alpha'_2 \\ \dots \end{pmatrix} = \begin{pmatrix} A_{11} & A_{12} & \dots \\ A_{21} & A_{22} & \dots \\ \dots & \dots & \dots \end{pmatrix} \begin{pmatrix} \alpha_1 \\ \alpha_2 \\ \dots \end{pmatrix}. \tag{4.63}$$

Absolutely similarly, the operator action on the bra-vector (4.21), represented by its row-matrix, is

$$\left(\alpha'^*_1, \alpha'^*_2, \dots \right) = \left(\alpha^*_1, \alpha^*_2, \dots \right) \begin{pmatrix} (A^\dagger)_{11} & (A^\dagger)_{12} & \dots \\ (A^\dagger)_{21} & (A^\dagger)_{22} & \dots \\ \dots & \dots & \dots \end{pmatrix}. \tag{4.64}$$

By the way, Eq. (4.64) naturally raises the following question: what are the elements of the matrix on its right-hand side, or more exactly, what is the relation between the matrix elements of an operator and its Hermitian conjugate? The simplest way to get an answer is to use Eq. (4.25) with two arbitrary states (say, u_j and $u_{j'}$) of the same basis in the role of α and β. Together with the orthonormality relation (4.38), this immediately gives[14]

$$(\hat{A}^{\dagger})_{jj'} = (A_{j'j})^{*}. \tag{4.65}$$

Thus, the matrix of the Hermitian-conjugate operator is the *complex conjugated and transposed* matrix of the initial operator. This result exposes very clearly the difference between the Hermitian and the complex conjugation. It also shows that for the Hermitian operators, defined by Eq. (4.22),

$$A_{jj'} = A_{j'j}^{*}, \tag{4.66}$$

i.e. any pair of their matrix elements, symmetric about the main diagonal, should be complex conjugate of each other. As a corollary, their main-diagonal elements have to be real:

$$A_{jj} = A_{jj}^{*}, \quad \text{i.e. Im } A_{jj} = 0. \tag{4.67}$$

(The matrix (4.50) evidently satisfies Eq. (4.66), so that the identity operator is Hermitian.)

In order to fully appreciate the special role played by Hermitian operators in the quantum theory, let us introduce the key notions of *eigenstates* a_j (described by their *eigenvectors* $\langle a_j|$ and $|a_j\rangle$) and *eigenvalues* (*c*-numbers) A_j of an operator $\hat{A}$, defined by the equation they have to satisfy[15]:

$$\hat{A}|a_j\rangle = A_j|a_j\rangle. \tag{4.68}$$

Let us prove that eigenvalues of any Hermitian operator are real[16],

$$A_j = A_j^{*}, \quad \text{for } j = 1, 2, \ldots, N, \tag{4.69}$$

while the eigenstates corresponding to different eigenvalues are orthogonal:

$$\langle a_j|a_{j'}\rangle = 0, \quad \text{if } A_j \neq A_{j'}. \tag{4.70}$$

---

[14] For the sake of formula compactness, below I will use the shorthand notation in that the operands of this equality are just $A^{\dagger}_{jj'}$ and $A^{*}_{j'j}$. I believe that it leaves little chance for confusion, because the Hermitian conjugation sign † may pertain only to an operator (or its matrix), while the complex conjugation sign *, pertains to a scalar—say a matrix element.

[15] This equation should look familiar to the reader—see the stationary Schrödinger equation (1.60), which was the focus of our studies in the first three chapters. We will see soon that that equation is just a particular (coordinate) representation of Eq. (4.68) for the Hamiltonian as the operator of energy.

[16] The reciprocal statement is also true: if all eigenvalues of an operator are real, it is Hermitian (in any basis). This statement may be readily proved by applying Eq. (4.93) below to the case when $A_{kk'} = A_k\delta_{kk'}$, with $A_k^{*} = A_k$.

The proof of both statements is surprisingly simple. Let us inner-multiply both sides of Eq. (4.68) by the bra-vector $\langle a_{j'}|$. On the right-hand side of the result, the eigenvalue A_j, as a c-number, may be taken out of the bracket, giving

$$\langle a_{j'}|\hat{A}|a_j\rangle = A_j\langle a_{j'}|a_j\rangle. \tag{4.71}$$

This equality has to hold for any pair of eigenstates, so that we may swap the indices in Eq. (4.71), and complex-conjugate the result:

$$\langle a_j|\hat{A}|a_{j'}\rangle^* = A_{j'}^*\langle a_j|a_{j'}\rangle^*. \tag{4.72}$$

Now using Eqs. (4.14) and (4.25), together with the Hermitian operator's definition (4.22), we may transform Eq. (4.72) into the following form:

$$\langle a_{j'}|\hat{A}|a_j\rangle = A_{j'}^*\langle a_{j'}|a_j\rangle. \tag{4.73}$$

Subtracting this equation from Eq. (4.71), we get

$$0 = \left(A_j - A_{j'}^*\right)\langle a_{j'}|a_j\rangle. \tag{4.74}$$

There are two possibilities to satisfy this relation. If the indices j and j' are equal (denote the same eigenstate), then the bracket is the state's norm squared, and cannot be equal to zero. In this case the left parentheses (with $j = j'$) have to be zero, proving Eq. (4.69). On the other hand, if j and j' correspond to different eigenvalues of A, the parentheses cannot equal zero (we have just proved that all A_j are real!), and hence the state vectors indexed by j and j' should be orthogonal, e.g. Eq. (4.70) is valid.

As will be discussed below, these properties make Hermitian operators suitable, in particular, for the description of physical observables.

4.4 Change of basis, and matrix diagonalization

From the discussion of the last section, it may seem that the matrix language is fully similar to, and in many instances more convenient than the general bra–ket formalism. In particular, Eqs. (4.52), (4.54) and (4.55) show that any part of any bra–ket expression may be directly mapped on the similar matrix expression, with the only slight inconvenience of using not only columns, but also rows (with their elements complex-conjugated), for state vector presentation. In this context, why do we need the bra–ket language at all? The answer is that the elements of the matrices depend on the particular choice of the basis set, very much as the Cartesian components of a usual vector depend on the particular choice of reference frame orientation (figure 4.4), and very frequently, at problem solution, it is convenient to use two or more different basis sets for the same system. (Just a bit of patience—numerous examples will follow soon.)

With this motivation, let us study what happens if we switch from one basis, $\{u\}$, to another one, $\{v\}$—both full and orthonormal. First of all, let us prove that for each such pair of bases, there exists such an operator $\hat{U}$ that, first,

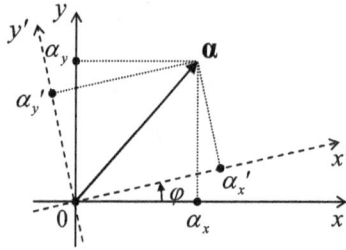

Figure 4.4. The transformation of components of a 2D vector at a reference frame's rotation.

$$|v_j\rangle = \hat{U}|u_j\rangle, \tag{4.75}$$

and, second,

$$\hat{U}\hat{U}^\dagger = \hat{U}^\dagger\hat{U} = \hat{I}. \tag{4.76}$$

(Due to the last property[17], $\hat{U}$ is called a *unitary operator*, and Eq. (4.75), a *unitary transformation*.)

A very simple proof of both statements may be achieved by construction. Indeed, let us take

$$\hat{U} \equiv \sum_{j'}|v_{j'}\rangle\langle u_{j'}|, \tag{4.77}$$

—an evident generalization of Eq. (4.44). Then

$$\hat{U}|u_j\rangle = \sum_{j'}|v_{j'}\rangle\langle u_{j'}|u_j\rangle = \sum_{j'}|v_{j'}\rangle\delta_{j'j} = |v_j\rangle, \tag{4.78}$$

so that Eq. (4.75) has been proved. Now, applying Eq. (4.31) to each term of the sum (4.77), we get

$$\hat{U}^\dagger \equiv \sum_{j'}|u_{j'}\rangle\langle v_{j'}|, \tag{4.79}$$

so that

$$\hat{U}\hat{U}^\dagger = \sum_{j,j'}|v_j\rangle\langle u_j|u_{j'}\rangle\langle v_{j'}| = \sum_{j,j'}|v_j\rangle\delta_{jj'}\langle v_{j'}| = \sum_j|v_j\rangle\langle v_j|. \tag{4.80}$$

But according to the closure relation (4.44), the last expression is just the identity operator, q.e.d.[18] (The proof of the second equality in Eq. (4.76) is absolutely similar.)

As a by-product of our proof, we have also got another important expression—Eq. (4.79). It implies, in particular, that while, according to Eq. (4.77), the operator $\hat{U}$

---

[17] An alternative way to express Eq. (4.76) is to write $\hat{U}^\dagger = \hat{U}^{-1}$, but I will try to avoid this language.

[18] *Quod erat demonstrandum* (Lat.)—what needed to be proved.

performs the transform from the 'old' basis $\{u\}$ to the 'new' basis $\{v\}$, its Hermitian adjoint $\hat{U}^\dagger$ performs the reciprocal unitary transform:

$$\hat{U}^\dagger \left|v_j\right\rangle = \sum_{j'}\left|u_{j'}\right\rangle \delta_{j'j} = \left|u_j\right\rangle. \tag{4.81}$$

Now let us see what the matrix elements of the unitary transform operators look like. Generally, as was discussed above, the operator's elements may depend on the basis we calculate them in, so we should be careful—initially. For example, let us calculate the desired elements in the basis $\{u\}$:

$$U_{jj'}|_{\text{in } u} \equiv \left\langle u_j\middle|\hat{U}\middle|u_{j'}\right\rangle = \left\langle u_j\middle|\left(\sum_k \left|v_k\right\rangle\left\langle u_k\right|\right)\middle|u_{j'}\right\rangle = \left\langle u_j\middle|v_{j'}\right\rangle. \tag{4.82}$$

Now performing a similar calculation in the basis $\{v\}$, we get

$$U_{jj'}|_{\text{in } v} \equiv \left\langle v_j\middle|\hat{U}\middle|v_{j'}\right\rangle = \left\langle v_j\middle|\left(\sum_k \left|v_k\right\rangle\left\langle u_k\right|\right)\middle|v_{j'}\right\rangle = \left\langle u_j\middle|v_{j'}\right\rangle. \tag{4.83}$$

Surprisingly, the result is the same! This is of course true for the Hermitian conjugate (4.79) as well:

$$U_{jj'}^\dagger|_{\text{in } u} = U_{jj'}^\dagger|_{\text{in } v} = \left\langle v_j\middle|u_{j'}\right\rangle. \tag{4.84}$$

These expressions may be used, first of all, to rewrite Eq. (4.75) in a more direct form. Applying the first of Eqs. (4.41) to a state $v_{j'}$ of the 'new' basis, we get

$$\left|v_{j'}\right\rangle = \sum_j \left|u_j\right\rangle\left\langle u_j\middle|v_{j'}\right\rangle = \sum_j U_{jj'}\left|u_j\right\rangle. \tag{4.85}$$

Similarly, the reciprocal transform is

$$\left|u_{j'}\right\rangle = \sum_j \left|v_j\right\rangle\left\langle v_j\middle|u_{j'}\right\rangle = \sum_j U_{jj'}^\dagger\left|v_j\right\rangle. \tag{4.86}$$

These equalities are very convenient for applications; we will use them later in this section.

Next, we may use Eqs. (4.83) and (4.84) to express the effect of the unitary transform on the expansion coefficients α_j of the vectors of an *arbitrary* state α, defined by Eq. (4.37). In the 'old' basis $\{u\}$, they are given by Eq. (4.40). Similarly, in the 'new' basis $\{v\}$,

$$\alpha_j|_{\text{in } v} = \left\langle v_j\middle|\alpha\right\rangle. \tag{4.87}$$

Again inserting the identity operator in the form of the closure (4.44), with the internal index j', and then using Eq. (4.84), we get

$$\alpha_j|_{\text{in }v} = \langle v_j| \left(\sum_{j'} |u_{j'}\rangle\langle u_{j'}| \right) |\alpha\rangle = \sum_{j'} \langle v_j|u_{j'}\rangle\langle u_{j'}|\alpha\rangle$$
$$= \sum_{j'} U^\dagger_{jj'}\langle u_{j'}|\alpha\rangle = \sum_{j'} U^\dagger_{jj'}\alpha_{j'}|_{\text{in }u}. \tag{4.88}$$

The reciprocal transform is (of course) performed by matrix elements of the operator $\hat{U}$:

$$\alpha_j|_{\text{in }u} = \sum_{j'} U_{jj'}\alpha_{j'}|_{\text{in }v}. \tag{4.89}$$

Both structurally and philosophically, these expressions are similar to the transformation of components of a usual vector at coordinate frame rotation. For example, for a 2D vector whose actual position in space is fixed (figure 4.4):

$$\begin{pmatrix} \alpha_x' \\ \alpha_y' \end{pmatrix} = \begin{pmatrix} \cos\varphi & \sin\varphi \\ -\sin\varphi & \cos\varphi \end{pmatrix}\begin{pmatrix} \alpha_x \\ \alpha_y \end{pmatrix}. \tag{4.90}$$

(In this analogy, the equality of the determinant of the rotation matrix in Eq. (4.90) to 1 corresponds to the unitary property (4.76) of the unitary transform operators.) Please pay attention here: while the transform (4.75) from the 'old' basis $\{u\}$ to the 'new' basis $\{v\}$ is performed by the unitary operator, the change (4.88) of a state vectors components at this transformation requires its Hermitian conjugate. Actually, this is also natural from the point of view of the geometric analog of the unitary transform (figure 4.4): if the 'new' reference frame $\{x', y'\}$ is obtained by a *counterclockwise* rotation of the 'old' frame $\{x, y\}$ by some angle φ, for the observer rotating with the frame, the vector $\boldsymbol{\alpha}$ rotates *clockwise*[19].

Due to the analogy between expressions (4.88) and (4.89) on one hand, and our old friend Eq. (4.62) on the other hand, it is tempting to skip indices in these new results by writing

$$|\alpha\rangle_{\text{in }v} = \hat{U}^\dagger |\alpha\rangle_{\text{in }u}, \qquad |\alpha\rangle_{\text{in }u} = \hat{U} |\alpha\rangle_{\text{in }v}. \quad \text{SYMBOLIC ONLY!} \tag{4.91}$$

Since the matrix elements of $\hat{U}$ and $\hat{U}^\dagger$ do not depend on the basis, such language is not too bad and mnemonically useful. However, since in the bra–ket formalism (or at least its version presented in this course), the state vectors are basis-independent, Eq. (4.91) has to be treated as a symbolic one, and should not be confused with the

---

[19] In the formal geometry, such vector $\boldsymbol{\alpha}$ is called *contravariant*, while the reference frame vectors, *covariant*. (A brief discussion of these terms may be found in *Part EM* section 9.4.)

strict Eqs. (4.88) and (4.89), and with the *strict* (basis-independent) vector and operator equalities discussed in section 4.2.

Now let us use the same trick of identity operator insertion, repeated twice, to find the transformation rule for matrix elements of an arbitrary operator:

$$A_{jj'}|_{\text{in } v} \equiv \langle v_j | \hat{A} | v_{j'} \rangle = \langle v_j | \left(\sum_k |u_k\rangle\langle u_k| \right) \hat{A} \left(\sum_{k'} |u_{k'}\rangle\langle u_{k'}| \right) | v_{j'} \rangle$$
$$= \sum_{k,k'} U_{jk}^\dagger A_{kk'}|_{\text{in } u} U_{k'j'} ; \tag{4.92}$$

absolutely similarly, we can get

$$A_{jj'}|_{\text{in } u} \equiv \sum_{k,k'} U_{jk} A_{kk'}|_{\text{in } v} U_{k'j'}^\dagger . \tag{4.93}$$

In the spirit of Eq. (4.91), we may represent these results symbolically as well, in a compact form:

$$\hat{A}|_{\text{in } v} = \hat{U}^\dagger \hat{A}|_{\text{in } u} \hat{U}, \qquad \hat{A}|_{\text{in } u} = \hat{U} \hat{A}|_{\text{in } v} \hat{U}^\dagger. \quad \text{SYMBOLIC ONLY!} \tag{4.94}$$

As a sanity check, let us apply this general equality to the identity operator:

$$\hat{I}|_{\text{in } v} = (\hat{U}^\dagger \hat{I} \hat{U})_{\text{in } u} = (\hat{U}^\dagger \hat{U})_{\text{in } u} = \hat{I}|_{\text{in } u} \tag{4.95}$$

—as it should be. One more (strict rather than symbolic) invariant of the basis change is the *trace* of any operator, defined as the sum of the diagonal terms of its matrix in a certain basis:

$$\text{Tr } \hat{A} \equiv \text{Tr} A \equiv \sum_j A_{jj}. \tag{4.96}$$

The (easy) proof of this fact, using the relations we have already discussed, is left for the reader's exercise.

So far, I have implied that both state bases {u} and {v} are known, and the natural question is where does this information comes from in quantum mechanics of actual physical systems. To get a partial answer to this question, let us return to Eq. (4.68), which defines the eigenstates and the eigenvalues of an operator. Let us assume that the eigenstates a_j of a certain operator $\hat{A}$ form a full and orthonormal set, and calculate the matrix elements of the operator in the basis of these states. For that, it is sufficient to inner-multiply both sides of Eq. (4.68), written for some index j', by the bra-vector of an arbitrary state a_j of the same set:

$$\langle a_j | \hat{A} | a_{j'} \rangle = \langle a_j | A_{j'} | a_{j'} \rangle. \tag{4.97}$$

The left-hand side of this equality is the matrix element $A_{jj'}$ we are looking for, while its right-hand side is just $A_{j'}\delta_{jj'}$. As a result, we see that the matrix is diagonal, with the diagonal consisting of the operator's eigenvalues:

$$A_{jj'} = A_j \delta_{jj'}. \tag{4.98}$$

In particular, in the eigenstate basis (but not necessarily in an arbitrary basis!), A_{jj} means the same as A_j. Thus the important problem of finding the eigenvalues and eigenstates of an operator is equivalent to the *diagonalization* of its matrix[20], i.e. finding the basis in which the corresponding operator acquires the diagonal form (4.98); then the diagonal elements are the eigenvalues, and the basis itself is the desirable set of eigenstates.

To see how this is done in practice, let us inner-multiply Eq. (4.68) by a bra-vector of a *different* basis (say, $\{u\}$), in that we have happened to know the matrix elements $A_{jj'}$. The multiplication gives

$$\langle u_k | \hat{A} | a_j \rangle = \langle u_k | A_j | a_j \rangle. \tag{4.99}$$

On the left-hand side we can (as usual) insert the identity operator between the operator $\hat{A}$ and the ket-vector, and then use the closure relation (4.44) in the same basis $\{u\}$, while on the right-hand side, we can move the eigenvalue A_j out of the bracket, and then insert a summation over the same index as in the closure, compensating it with the proper Kronecker delta symbol:

$$\langle u_k | \hat{A} \sum_{k'} |u_{k'}\rangle \langle u_{k'} | a_j \rangle = A_j \sum_{k'} \langle u_{k'} | a_j \rangle \, \delta_{kk'}. \tag{4.100}$$

Moving out the sign of summation over k', and using the definition (4.47) of the matrix elements, we get

$$\sum_{k'} (A_{kk'} - A_j \delta_{kk'}) \langle u_{k'} | a_j \rangle = 0. \tag{4.101}$$

But the set of such equalities, for all N possible values of the index k, is just a system of linear, homogeneous equations for unknown c-numbers $\langle u_{k'} | a_j \rangle$. According to Eqs. (4.82)–(4.84), these numbers are nothing else than the matrix elements $U_{k'j}$ of a unitary matrix providing the required transformation from the initial basis $\{u\}$ to the basis $\{a\}$ that diagonalizes the matrix A. The system may be represented in the matrix form:

$$\begin{pmatrix} A_{11} - A_j & A_{12} & \dots \\ A_{21} & A_{22} - A_j & \dots \\ \dots & \dots & \dots \end{pmatrix} \begin{pmatrix} U_{1j} \\ U_{2j} \\ \dots \end{pmatrix} = 0, \tag{4.102}$$

and the usual condition of its consistency,

---

[20] Note that the expression 'matrix diagonalization' is a common but dangerous jargon. (Formally, a matrix is just a matrix, an ordered set of c-numbers, and cannot be 'diagonalized'.) It is OK to use this jargon if you remember clearly what it actually means—see the definition above.

$$\begin{vmatrix} A_{11} - A_j & A_{12} & \cdots \\ A_{21} & A_{22} - A_j & \cdots \\ \cdots & \cdots & \cdots \end{vmatrix} = 0, \tag{4.103}$$

plays the role of the characteristic equation of the system. This equation has N roots A_j—the eigenvalues of the operator $\hat{A}$; plugging each of them back into system (4.102), we can use it to find N matrix elements U_{kj} ($k = 1, 2, ...N$) corresponding to this particular eigenvalue. However, since equations (4.103) are homogeneous, they allow finding U_{kj} only to a constant multiplier. In order to ensure their normalization, i.e. the unitary character of the matrix U, we may use the condition that all eigenvectors are normalized (just as the basis vectors are):

$$\langle a_j | a_j \rangle \equiv \sum_k \langle a_j | u_k \rangle \langle u_k | a_j \rangle \equiv \sum_k |U_{kj}|^2 = 1, \tag{4.104}$$

for each j. This normalization completes the diagonalization[21].

Now (at last!) I can give the reader some examples. As a simple but very important case, let us diagonalize the operators described (in a certain 2-function basis $\{u\}$, i.e. in a two-dimensional Hilbert space) by the so-called *Pauli matrices*

$$\sigma_x \equiv \begin{pmatrix} 0 & 1 \\ 1 & 0 \end{pmatrix}, \qquad \sigma_y \equiv \begin{pmatrix} 0 & -i \\ i & 0 \end{pmatrix}, \qquad \sigma_z \equiv \begin{pmatrix} 1 & 0 \\ 0 & -1 \end{pmatrix}. \tag{4.105}$$

Though introduced by a physicist, with a specific purpose to describe the electron's spin, these matrices have a general mathematical significance, because together with the 2×2 identity matrix, they provide a full, linearly-independent system—meaning that an arbitrary 2×2 matrix may be represented as

$$\begin{pmatrix} A_{11} & A_{12} \\ A_{21} & A_{22} \end{pmatrix} = bI + c_x \sigma_x + c_y \sigma_y + c_z \sigma_z, \tag{4.106}$$

with a unique set of four c-number coefficients b, c_x, c_y, and c_z.

Since the matrix σ_z is already diagonal, with the evident eigenvalues ± 1, let us start with diagonalizing the matrix σ_x. For it, the characteristic equation (4.103) is evidently

$$\begin{vmatrix} -A_j & 1 \\ 1 & -A_j \end{vmatrix} = 0, \tag{4.107}$$

and has two roots, $A_{1,2} = \pm 1$. (Again, the numbering is arbitrary!) So the eigenvalues of the matrix σ_x are the same as of the matrix σ_z. (The reader may readily check that the eigenvalues of the matrix σ_y are also the same.) However, the eigenvectors of the operators corresponding to all these matrices are different. To find them for σ_x, let us

---

[21] A possible slight complication here is degenerate cases when the characteristic equation gives equal eigenvalues for certain groups of different eigenvectors. In this case the requirement of the mutual orthogonality of these states should be additionally enforced.

plug its first eigenvalue, $A_1 = +1$, back into equations (4.101), spelled out for this particular case:

$$-\langle u_1|a_1\rangle + \langle u_2|a_1\rangle = 0,$$
$$\langle u_1|a_1\rangle - \langle u_2|a_1\rangle = 0. \tag{4.108}$$

The equations are compatible (of course, because the used eigenvalue $A_1 = +1$ satisfies the characteristic equation), and any of them gives

$$\langle u_1|a_1\rangle = \langle u_2|a_1\rangle, \quad \text{i.e.} \quad U_{11} = U_{21}. \tag{4.109}$$

With that, the normalization condition (4.104) yields

$$|U_{11}|^2 = |U_{21}|^2 = \frac{1}{2}. \tag{4.110}$$

Although the normalization is insensitive to the simultaneous multiplication of U_{11} and U_{21} by the same phase factor $\exp\{i\varphi\}$ with any real φ, it is convenient to keep the coefficients real, for example taking $\varphi = 0$, i.e. to get

$$U_{11} = U_{21} = \frac{1}{\sqrt{2}}. \tag{4.111}$$

Performing an absolutely similar calculation for the second characteristic value, $A_2 = -1$, we get $U_{12} = -U_{22}$, and we may choose the common phase to get

$$U_{12} = -U_{22} = \frac{1}{\sqrt{2}}, \tag{4.112}$$

so that the whole unitary matrix for diagonalization of the operator corresponding to σ_x is[22]

$$\mathsf{U}_x = \mathsf{U}_x^\dagger = \frac{1}{\sqrt{2}}\begin{pmatrix} 1 & 1 \\ 1 & -1 \end{pmatrix}. \tag{4.113}$$

For what follows, it will be convenient to have this result expressed in the ket-relation form—see Eqs. (4.85) and (4.86):

$$|a_1\rangle = U_{11}|u_1\rangle + U_{21}|u_2\rangle = \frac{1}{\sqrt{2}}(|u_1\rangle + |u_2\rangle),$$
$$|a_2\rangle = U_{12}|u_1\rangle + U_{22}|u_2\rangle = \frac{1}{\sqrt{2}}(|u_1\rangle - |u_2\rangle), \tag{4.114a}$$

$$|u_1\rangle = U_{11}^\dagger|a_1\rangle + U_{21}^\dagger|a_2\rangle = \frac{1}{\sqrt{2}}(|a_1\rangle + |a_2\rangle),$$
$$|u_2\rangle = U_{12}^\dagger|a_1\rangle + U_{22}^\dagger|a_2\rangle = \frac{1}{\sqrt{2}}(|a_1\rangle - |a_2\rangle). \tag{4.114b}$$

---

[22] Note that though this particular unitary matrix is Hermitian, this is not true for an arbitrary choice of phases φ.

Let us put our general discussion on hold, in order to show that these results are already sufficient to understand the Stern–Gerlach experiments described in section 4.1—with two additional postulates. The first of them is that a particle's interaction with the external magnetic field may be described by the following vector operator of the dipole magnetic moment[23]:

$$\hat{\mathbf{m}} = \gamma \hat{\mathbf{S}}, \tag{4.115}$$

where the coefficient γ, specific for every particle type, is called the *gyromagnetic ratio*[24], and $\hat{\mathbf{S}}$ is the *vector operator*[25] *of spin*. For the so-called *spin-½ particles* (including the electron), this operator may be expressed very simply, as

$$\hat{\mathbf{S}} = \frac{\hbar}{2}\hat{\boldsymbol{\sigma}}, \tag{4.116}$$

via the *Pauli vector operator* $\hat{\boldsymbol{\sigma}}$. In turn, in the so-called *z-basis*, the latter operator is represented by the following 3D vector of the Pauli matrices (4.105):

$$\boldsymbol{\sigma} = \mathbf{n}_x \sigma_x + \mathbf{n}_y \sigma_y + \mathbf{n}_z \sigma_z \equiv \begin{pmatrix} \mathbf{n}_z & \mathbf{n}_x - i\mathbf{n}_y \\ \mathbf{n}_x + i\mathbf{n}_y & -\mathbf{n}_z \end{pmatrix}, \tag{4.117}$$

and $\mathbf{n}_{x,y,z}$ are the usual Cartesian unit vectors in the geometric 3D space. (In the quantum-mechanics sense, they are just *c*-numbers, or rather '*c*-vectors'.) The *z*-basis, in which Eq. (4.177) is valid, is *defined* as an orthonormal basis of certain two states, frequently denoted ↑ an ↓, in which the *z*-component of the vector-operator $\hat{\boldsymbol{\sigma}}$ is diagonal, with eigenvalues, respectively, $+1$ and -1, and hence the vector-operator (4.116) of spin is also diagonal, with the eigenvalues $+\hbar/2$ and $-\hbar/2$. Note that we do not 'understand' what exactly the states ↑ and ↓ are[26], but loosely associate them with a certain internal rotation of the electron about the *z*-axis, with either positive or negative angular momentum component S_z. However, any attempt to use such classical interpretation for quantitative predictions runs into fundamental difficulties —see section 4.6 below.

The second necessary postulate describes the general relation between the bra–ket formalism and experiment. Namely, in quantum mechanics, each real observable A is represented by a Hermitian operator $\hat{A} = \hat{A}^\dagger$, and a result of its measurement[27] in

---

[23] This was the key point in the electron spin's description, developed by W Pauli in 1925–7.

[24] For the electron, with its negative charge $q = -e$, the gyromagnetic ratio is negative: $\gamma_e = -g_e e/2m_e$, where $g_e \approx 2$ is the dimensionless g-*factor*. Due to quantum-electrodynamic (relativistic) effects, this g-factor is slightly higher than 2: $g_e = 2(1 + \alpha/2\pi + ...) \approx 2.002\ 319\ 304...$, where $\alpha \equiv e^2/4\pi\varepsilon_0\hbar c \equiv (E_H/m_ec^2)^{1/2} \approx 1/137$ is the so-called *fine structure constant*. (The origin of its name will be clear from the discussion in section 6.3.)

[25] Such vector operators are transformed as usual vectors at all geometric operations—see below.

[26] If you think about it, the word 'understand' typically means that we can explain a new, more complex notion in terms of those discussed earlier and considered 'known'. In our example, we cannot express the spin states by some wavefunction $\psi(\mathbf{r})$, or any other mathematical notion discussed earlier. The bra–ket formalism has been invented exactly to enable mathematical analyses of such 'new' quantum states we do not initially 'understand', though gradually get accustomed to, and eventually, as we know more and more about their properties, start to treat them as 'known'.

[27] Here again, just like in section 1.2, the statement implies the abstract notion of 'ideal experiments', deferring the discussion of real (physical) measurements until chapter 10.

a quantum state α, described by a linear superposition of the eigenstates a_j of the operator,

$$|\alpha\rangle = \sum_j \alpha_j |a_j\rangle, \qquad \text{with } \alpha_j = \langle a_j | \alpha \rangle, \tag{4.118}$$

may be only one of the corresponding eigenvalues A_j.[28] Specifically, if the ket (4.118) and all eigenkets $|a_j\rangle$ are normalized to 1,

$$\langle \alpha | \alpha \rangle = 1, \qquad \langle a_j | a_j \rangle = 1, \tag{4.119}$$

then the probability of a certain measurement outcome A_j is[29]

$$W_j = |\alpha_j|^2 \equiv \alpha_j^* \alpha_j \equiv \langle \alpha | a_j \rangle \langle a_j | \alpha \rangle, \tag{4.120}$$

This relation is evidently a generalization of Eq. (1.22) in wave mechanics. As a sanity check, let us assume that the set of the eigenstates a_j is full, and calculate the sum of all the probabilities:

$$\sum_j W_j = \sum_j \langle \alpha | a_j \rangle \langle a_j | \alpha \rangle = \langle \alpha | \hat{I} | \alpha \rangle = 1. \tag{4.121}$$

Now returning to the Stern–Gerlach experiment, conceptually the description of the first (z-oriented) experiment shown in figure 4.1 is the hardest for us, because the statistical ensemble describing the unpolarized electron beam at its input is *mixed* ('incoherent'), and cannot be described by a *pure* ('coherent') superposition of the type (4.6) that have been the subject of our studies so far. (We will discuss the mixed ensembles in chapter 7.) However, it is intuitively clear that its results are compatible with the description of the two output beams as sets of electrons in the pure states ↑ and ↓, respectively. The absorber following that first stage (figure 4.2) just takes all spin-down electrons out of the picture, producing an output beam of polarized electrons in the definite ↑ state. For such a beam, the probabilities (4.120) are $W_\uparrow = 1$ and $W_\downarrow = 0$. This is certainly compatible with the result of the 'control' experiment shown on the bottom panel of figure 4.2: the repeated SG (z) stage does not split such a beam, keeping the probabilities the same.

Now let us discuss the double Stern–Gerlach experiment shown on the top panel of figure 4.2. For that, let us represent the z-polarized beam in another basis of two states (I will denote them as → and ←) in which, by definition, the matrix of operator $\hat{S}_x$ is diagonal. But this is exactly the set we called $a_{1,2}$ in the σ_x matrix diagonalization problem solved above. On the other hand, the states ↑ and ↓ are exactly what we called $u_{1,2}$ in that problem, because in this basis, we know the

---

[28] As a reminder, at the end of section 4.3 we have already proved that such eigenstates corresponding to different A_j are orthogonal. If any of these values is degenerate, i.e. corresponds to several different eigenstates, they should be also selected orthogonal, in order for Eq. (4.118) to be valid.

[29] This key relation, in particular, explains the most common term for the (generally, complex) coefficients α_j, the *probability amplitudes*.

explicit matrix **σ** (and hence **S**)—see Eqs. (4.116) and (4.117). Hence, in the application to the electron spin problem, we may rewrite Eqs. (4.114) as

$$|\rightarrow\rangle = \frac{1}{\sqrt{2}}(|\uparrow\rangle + |\downarrow\rangle), \qquad |\leftarrow\rangle = \frac{1}{\sqrt{2}}(|\uparrow\rangle - |\downarrow\rangle), \qquad (4.122)$$

$$|\uparrow\rangle = \frac{1}{\sqrt{2}}(|\rightarrow\rangle + |\leftarrow\rangle), \qquad |\downarrow\rangle = \frac{1}{\sqrt{2}}(|\rightarrow\rangle - |\leftarrow\rangle), \qquad (4.123)$$

Currently, for us the first of Eqs. (4.123) is most important, because it shows that the quantum state of electrons entering the SG (x) stage may be represented as a coherent superposition of electrons with $S_x = +\hbar/2$ and $S_x = -\hbar/2$. Notice that the beams have equal probability amplitude moduli, so that according to Eq. (4.122), the split beams $\rightarrow$ and $\leftarrow$ have equal intensities, in accordance with experimental results. (The minus sign before the second ket-vector is of no consequence here, but it may have an impact on outcome of other experiments—for example if the coherent $\rightarrow$ and $\leftarrow$ beams are brought together again.)

Now, let us discuss the most mysterious (from the classical point of view) multi-stage SG experiment shown on the middle panel of figure 4.2. After the second absorber has taken out all electrons in, say, the $\leftarrow$ state, the remaining electrons, all in the state $\rightarrow$, are passed to the final, SG (z), stage. But according to the first of Eqs. (4.122), this state may be represented as a (coherent) linear superposition of the $\uparrow$ and $\downarrow$ states, with equal amplitudes. The final stage separates these two states into separate beams, with equal probabilities $W_\uparrow = W_\downarrow = \frac{1}{2}$ to find an electron in each of them, thus explaining the experimental results.

To conclude our discussion of the multistage Stern–Gerlach experiment, let me note that though it cannot be explained in terms of the wave mechanics (which operates with *scalar* de Broglie waves), it has an analogy in classical theories of *vector* fields, such as classical electrodynamics. Indeed, let a plane electromagnetic wave propagate perpendicular to the plane of drawing in figure 4.5, and pass through the linear polarizer 1. Similarly to the output of the initial SG (z) stages (including the absorbers) shown in figure 4.2, the output is a wave linearly polarized in one direction—the vertical direction in figure 4.5. Now its electric field vector has no horizontal component—as may be revealed by the wave's full absorption in a perpendicular polarizer 3. However, let us pass the wave through polarizer 2 first. In this case, the output wave does acquire a horizontal component, as can be, again,

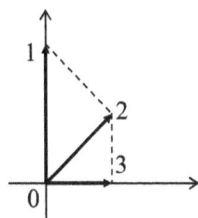

Figure 4.5. A light polarization sequence similar to the three-stage Stern–Gerlach experiment shown on the middle panel of figure 4.2.

revealed by passing it through polarizer 3. If the angles between the polarization directions 1 and 2, and between 2 and 3, are both equal to $\pi/4$, each polarizer reduces the wave amplitude by a factor of $\sqrt{2}$, and hence the intensity by a factor of 2, exactly like in the multistage SG experiment, with polarizer 2 playing the role of the SG (x) stage. The 'only' difference is that the necessary angle is $\pi/4$, rather than by $\pi/2$ for the Stern–Gerlach experiment. In quantum electrodynamics (see chapter 9 below), which confirms the classical predictions for this experiment, this difference is explained by that between the *integer* spin of the electromagnetic field quanta, photons, and the *half-integer* spin of electrons.

4.5 Observables: expectation values and uncertainties

After this particular (and hopefully very inspiring) example, let us discuss the general relation between the Dirac formalism and experiment in more detail. The expectation value of an observable over *any* statistical ensemble (not necessarily coherent) may be always calculated using the general rule (1.37). For the particular case of a coherent superposition (4.118), we can combine that rule with Eq. (4.120) and the second of Eqs. (4.118), and then use Eqs. (4.59) and (4.98) to write

$$
\langle A \rangle = \sum_j A_j W_j = \sum_j \alpha_j^* A_j \alpha_j = \sum_j \langle \alpha | a_j \rangle A_j \langle a_j | \alpha \rangle
$$
$$
= \sum_{j,j'} \langle \alpha | a_j \rangle \langle a_j | \hat{A} | a_{j'} \rangle \langle a_{j'} | \alpha \rangle. \tag{4.124}
$$

Now using the closure relation (4.44) twice, with indices j and j', we arrive at a very simple and important formula[30]

$$
\langle A \rangle = \langle \alpha | \hat{A} | \alpha \rangle. \tag{4.125}
$$

This is a clear analog of the wave-mechanics formula (1.23)—and as we will see soon, may be used to derive it. A huge advantage of Eq. (4.125) is that it does not explicitly involve the eigenvector set of the corresponding operator, and allows the calculation to be performed in any convenient basis[31].

For example, let us consider an arbitrary state α of spin-½,[32] and calculate the expectation values of its components. The calculations are easiest in the z-basis, because we know the matrix elements of the spin operator components in that basis. Representing the ket- and bra-vectors of the given state as linear superpositions of the corresponding vectors of the basis states $\uparrow$ and $\downarrow$,

---

[30] This equality reveals the full beauty of the Dirac's notation. Indeed, initially the quantum-mechanical brackets just *recalled* the angular brackets used for the statistical averaging. Now we see that in this particular (but most important) case, the angular brackets of these two types may be indeed *equal* to each other!

[31] Note also that Eq. (4.120) may be rewritten in the form similar to Eq. (4.125): $W_j = \langle \alpha | \hat{\Lambda}_j | \alpha \rangle$, where $\hat{\Lambda}_j \equiv | a_j \rangle \langle a_j |$ is the operator (4.42) of the state's projection upon the jth eigenstate a_j.

[32] For clarity, the noun 'spin-½' is used, here and below, to denote the spin degree of freedom of a spin-½ particle, independent of its orbital motion.

$$|\alpha\rangle = \alpha_\uparrow |\uparrow\rangle + \alpha_\downarrow |\downarrow\rangle, \qquad \langle\alpha| = \langle\uparrow|\alpha_\uparrow^* + \langle\downarrow|\alpha_\downarrow^*. \tag{4.126}$$

and plugging these expressions to Eq. (4.125), written for the observable S_z, we get

$$\begin{aligned}\langle S_z\rangle &= \left((\langle\uparrow|\alpha_\uparrow^* + \langle\downarrow|\,\alpha_\downarrow^*)\hat{S}_z(\alpha_\uparrow |\uparrow\rangle + \alpha_\downarrow |\downarrow\rangle)\right)\\ &= \alpha_\uparrow\alpha_\uparrow^*\langle\uparrow|\,\hat{S}_z\,|\uparrow\rangle + \alpha_\downarrow\alpha_\downarrow^*\langle\downarrow|\,\hat{S}_z\,|\downarrow\rangle \\ &\quad + \alpha_\uparrow\alpha_\downarrow^*\langle\downarrow|\,\hat{S}_z\,|\uparrow\rangle + \alpha_\downarrow\alpha_\uparrow^*\langle\uparrow|\,\hat{S}_z\,|\downarrow\rangle.\end{aligned} \tag{4.127}$$

Now there are two equivalent ways (both simple) to calculate the long brackets in this expression. The first one is to represent each of them in the matrix form in the z-basis, in which the bra- and ket-vectors of states $\uparrow$ and $\downarrow$ are the matrix-rows (1, 0) and (0, 1), or the similar matrix-columns. Another (perhaps more elegant) way is to use the general Eq. (4.59), in the z-basis, to write

$$\hat{S}_x = \frac{\hbar}{2}(|\uparrow\rangle\langle\downarrow| + |\downarrow\rangle\langle\uparrow|), \quad \hat{S}_y = -i\frac{\hbar}{2}(|\uparrow\rangle\langle\downarrow| - |\downarrow\rangle\langle\uparrow|),$$
$$\hat{S}_z = \frac{\hbar}{2}(|\uparrow\rangle\langle\uparrow| - |\downarrow\rangle\langle\downarrow|). \tag{4.128}$$

For our particular calculation, we may plug the last of these expressions into Eq. (4.127), and to use the orthonormality conditions (4.38):

$$\langle\uparrow|\uparrow\rangle = \langle\downarrow|\downarrow\rangle = 1, \quad \langle\uparrow|\downarrow\rangle = \langle\downarrow|\uparrow\rangle = 0. \tag{4.129}$$

Both approaches give (of course) the same result:

$$\langle S_z\rangle = \frac{\hbar}{2}(\alpha_\uparrow\alpha_\uparrow^* - \alpha_\downarrow\alpha_\downarrow^*). \tag{4.130}$$

This particular result might be also obtained using Eq. (4.120) for the probabilities $W_\uparrow = \alpha_\uparrow\alpha_\uparrow^*$ and $W_\downarrow = \alpha_\downarrow\alpha_\downarrow^*$, namely:

$$\langle S_z\rangle = W_\uparrow\left(+\frac{\hbar}{2}\right) + W_\downarrow\left(-\frac{\hbar}{2}\right) = \alpha_\uparrow\alpha_\uparrow^*\left(+\frac{\hbar}{2}\right) + \alpha_\downarrow\alpha_\downarrow^*\left(-\frac{\hbar}{2}\right). \tag{4.131}$$

The formal way (4.127), based on using Eq. (4.125), has, however, an advantage of being applicable, without any change, to finding the observables whose operators are not diagonal in the z-basis, as well. In particular, the absolutely similar calculations give

$$\begin{aligned}\langle S_x\rangle &= \alpha_\uparrow\alpha_\uparrow^*\langle\uparrow|\hat{S}_x|\uparrow\rangle + \alpha_\downarrow\alpha_\downarrow^*\langle\downarrow|\hat{S}_x|\downarrow\rangle + \alpha_\uparrow\alpha_\downarrow^*\langle\downarrow|\hat{S}_x|\uparrow\rangle \\ &\quad + \alpha_\downarrow\alpha_\uparrow^*\langle\uparrow|\hat{S}_x|\downarrow\rangle = \frac{\hbar}{2}(\alpha_\uparrow\alpha_\downarrow^* + \alpha_\downarrow\alpha_\uparrow^*),\end{aligned} \tag{4.132}$$

$$\begin{aligned}\langle S_y\rangle &= \alpha_\uparrow\alpha_\uparrow^*\langle\uparrow|\hat{S}_y|\uparrow\rangle + \alpha_\downarrow\alpha_\downarrow^*\langle\downarrow|\hat{S}_y|\downarrow\rangle + \alpha_\uparrow\alpha_\downarrow^*\langle\downarrow|\hat{S}_y|\uparrow\rangle \\ &\quad + \alpha_\downarrow\alpha_\uparrow^*\langle\uparrow|\hat{S}_y|\downarrow\rangle = i\frac{\hbar}{2}(\alpha_\uparrow\alpha_\downarrow^* - \alpha_\downarrow\alpha_\uparrow^*),\end{aligned} \tag{4.133}$$

Similarly, we can express, via the same coefficients $\alpha_\uparrow$ and $\alpha_\downarrow$, the rms fluctuations of all spin components. For example, let us have a good look at the particular spin-up state $\uparrow$. According to Eq. (4.126), in this state $\alpha_\uparrow = 1$ and $\alpha_\downarrow = 0$, so that Eqs. (4.130)–(4.133) yield:

$$\langle S_z \rangle = \frac{\hbar}{2}, \qquad \langle S_x \rangle = \langle S_y \rangle = 0. \tag{4.134}$$

Now let us use the same Eq. (4.125) to calculate the spin component uncertainties. According to Eqs. (4.105), (4.116) and (4.117), the operator of each spin component squared is equal to $(\hbar/2)^2 \hat{I}$, so that the general Eq. (1.33) yields

$$(\delta S_z)^2 = \langle S_z^2 \rangle - \langle S_z \rangle^2 = \langle \uparrow | \hat{S}_z^2 | \uparrow \rangle - \left(\frac{\hbar}{2}\right)^2$$
$$= \left(\frac{\hbar}{2}\right)^2 \langle \uparrow | \hat{I} | \uparrow \rangle - \left(\frac{\hbar}{2}\right)^2 = 0, \tag{4.135a}$$

$$(\delta S_x)^2 = \langle S_x^2 \rangle - \langle S_x \rangle^2 = \langle \uparrow | \hat{S}_x^2 | \uparrow \rangle - 0$$
$$= \left(\frac{\hbar}{2}\right)^2 \langle \uparrow | \hat{I} | \uparrow \rangle = \left(\frac{\hbar}{2}\right)^2, \tag{4.135b}$$

$$(\delta S_y)^2 = \langle S_y^2 \rangle - \langle S_y \rangle^2 = \langle \uparrow | \hat{S}_y^2 | \uparrow \rangle - 0$$
$$= \left(\frac{\hbar}{2}\right)^2 \langle \uparrow | \hat{I} | \uparrow \rangle = \left(\frac{\hbar}{2}\right)^2. \tag{4.135c}$$

While Eqs. (4.134) and (4.135a) are compatible with the classical notion of the spin being definitely in the $\uparrow$ state, this correspondence should not be overstretched to the interpretation of this state as a certain (z) orientation of electron's magnetic moment **m**, because such a classical picture cannot explain Eqs. (4.135b) and (4.135c). The best (but still imprecise!) classical image I can offer is the magnetic moment **m** oriented, on the average, in the z-direction, but still having x- and y-components strongly 'wobbling' (fluctuating) about their zero average values.

It is straightforward to verify that in the x-polarized and y-polarized states the situation is similar, with the corresponding change of axis indices. Thus, in neither state may all three components of the spin have exact values. Let me show that this is not just an occasional fact, but reflects the perhaps most profound property of quantum mechanics, the *uncertainty relations*. For that, let us consider two observables, A and B, that may be measured in the same quantum system. There are two possibilities here. If the (Hermitian!) operators corresponding to these observables commute,

$$[\hat{A}, \hat{B}] = 0, \tag{4.136}$$

then all the matrix elements of the commutator in any orthogonal basis (in particular, in the basis of eigenstates a_j of the operator $\hat{A}$) also equal zero. From here, we get

$$\left\langle a_j \left| [\hat{A}, \hat{B}] \right| a_{j'} \right\rangle \equiv \left\langle a_j \left| \hat{A}\hat{B} \right| a_{j'} \right\rangle - \left\langle a_j \left| \hat{B}\hat{A} \right| a_{j'} \right\rangle = 0. \tag{4.137}$$

In the first bracket of the middle expression, let us act by the Hermitian operator $\hat{A}$ on the bra-vector, while in the second one, on the ket-vector. According to Eq. (4.68), such an action turns the operators into the corresponding eigenvalues, so that we get

$$A_j \left\langle a_j \left| \hat{B} \right| a_{j'} \right\rangle - A_{j'} \left\langle a_j \left| \hat{B} \right| a_{j'} \right\rangle \equiv (A_j - A_{j'}) \left\langle a_j \left| \hat{B} \right| a_{j'} \right\rangle = 0. \tag{4.138}$$

This means that if eigenstates of the operator $\hat{A}$ are non-degenerate (i.e. $A_j \neq A_{j'}$ if $j \neq j'$), the matrix of the operator $\hat{B}$ has to be diagonal in the basis $\{a\}$, i.e. the eigenstate sets of operators $\hat{A}$ and $\hat{B}$ coincide. Such pairs of observables (and their operators) that share their eigenstates, are called *compatible*. For example, in the wave mechanics of a particle, its momentum (1.26) and the kinetic energy (1.27) are compatible, sharing the eigenfunctions (1.29). Now we see that this is not occasional, because each Cartesian component of the kinetic energy is proportional to the square of the corresponding component of the momentum, and any operator commutes with an arbitrary integer power of itself:

$$[\hat{A}, \hat{A}^n] \equiv \left[\hat{A}, \underbrace{\hat{A}\hat{A}...\hat{A}}_{n}\right] = \hat{A}\underbrace{\hat{A}\hat{A}...\hat{A}}_{n} - \underbrace{\hat{A}\hat{A}...\hat{A}}_{n}\hat{A} = 0. \tag{4.139}$$

Now, what if the operators $\hat{A}$ and $\hat{B}$ do *not* commute? Then the following *general uncertainty relation* is valid[33]:

$$\delta A \, \delta B \geqslant \frac{1}{2} \left| \langle [\hat{A}, \hat{B}] \rangle \right|, \tag{4.140}$$

where all the expectation values are for the same, but otherwise arbitrary state of the system. The proof of Eq. (4.140) may be divided into two steps, the first one proving the so-called *Schwartz inequality* for any two possible states, say α and β:[34]

$$\langle \alpha | \alpha \rangle \langle \beta | \beta \rangle \geqslant | \langle \alpha | \beta \rangle |^2. \tag{4.141}$$

The proof may be started by using the postulate (4.16)—that the norm of any legitimate state of the system cannot be negative. Let us apply this postulate to the state with the following ket-vector:

$$|\delta\rangle \equiv |\alpha\rangle - \frac{\langle \beta | \alpha \rangle}{\langle \beta | \beta \rangle} |\beta\rangle, \tag{4.142}$$

---

[33] Note that each side of Eq. (4.140) is state-specific; the uncertainty relation statement is that this inequality should be valid for *any* possible quantum state of the system; in this sense, this is an operator relation.

[34] This inequality is the quantum-mechanical analog of the usual vector algebra's result $\alpha^2 \beta^2 \geqslant |\boldsymbol{\alpha} \cdot \boldsymbol{\beta}|^2$.

where α and β are possible, non-null states of the system, so that the denominator in Eq. (4.142) is not equal to zero. For this case, Eq. (4.16) gives

$$\left(\langle\alpha| - \frac{\langle\alpha|\beta\rangle}{\langle\beta|\beta\rangle}\langle\beta| \right)\left(|\alpha\rangle - \frac{\langle\beta|\alpha\rangle}{\langle\beta|\beta\rangle}|\beta\rangle \right) \geq 0. \tag{4.143}$$

Opening the parentheses, we get

$$\langle\alpha|\alpha\rangle - \frac{\langle\alpha|\beta\rangle}{\langle\beta|\beta\rangle}\langle\beta|\alpha\rangle - \frac{\langle\beta|\alpha\rangle}{\langle\beta|\beta\rangle}\langle\alpha|\beta\rangle + \frac{\langle\alpha|\beta\rangle\langle\beta|\alpha\rangle}{\langle\beta|\beta\rangle^2}\langle\beta|\beta\rangle \geq 0. \tag{4.144}$$

After the cancellation of one inner product $\langle\beta|\beta\rangle$ in the numerator and the denominator of the last term, it cancels with the 2nd (or the 3rd) term. What remains is the Schwartz inequality (4.141).

Now let us apply this inequality to the states

$$|\alpha\rangle \equiv \hat{\tilde{A}}|\gamma\rangle \quad \text{and} \quad |\beta\rangle \equiv \hat{\tilde{B}}|\gamma\rangle, \tag{4.145}$$

where, in both relations, γ is the same (but otherwise arbitrary) possible state of the system, and the deviation operators are defined similarly to the observable deviations (see section 1.2):

$$\hat{\tilde{A}} \equiv \hat{A} - \langle A\rangle, \quad \hat{\tilde{B}} \equiv \hat{B} - \langle B\rangle. \tag{4.146}$$

With this substitution, and taking into account again that the observable operators $\hat{A}$ and $\hat{B}$ are Hermitian, Eq. (4.141) yields

$$\langle\gamma|\hat{\tilde{A}}^2|\gamma\rangle\langle\gamma|\hat{\tilde{B}}^2|\gamma\rangle \geq \left|\langle\gamma|\hat{\tilde{A}}\hat{\tilde{B}}|\gamma\rangle\right|^2. \tag{4.147}$$

Since the state γ is arbitrary, we may use Eq. (4.125) to rewrite this relation as an operator inequality:

$$\delta A\delta B \geq \left|\left\langle\hat{\tilde{A}}\hat{\tilde{B}}\right\rangle\right|. \tag{4.148}$$

Actually, this is already an uncertainty relation, even 'better' (stronger) than its standard form (4.140); moreover, it is more convenient in some cases. In order to proceed to Eq. (4.140), we need a couple more steps. First, let us notice that the operator product in Eq. (4.148) may be recast as

$$\hat{\tilde{A}}\hat{\tilde{B}} = \frac{1}{2}\left\{\hat{\tilde{A}},\hat{\tilde{B}}\right\} - \frac{i}{2}\hat{C}, \quad \text{where} \quad \hat{C} \equiv i\left[\hat{\tilde{A}},\hat{\tilde{B}}\right]. \tag{4.149}$$

Any anticommutator of Hermitian operators, including that in Eq. (4.149), is a Hermitian operator, and its eigenvalues are purely real, so that its expectation value (in any state) is also purely real. On the other hand, the commutator part of Eq. (4.149) is just

$$\hat{C} \equiv i \left[\hat{A}, \, \hat{B} \right] \equiv i \, (\hat{A} - \langle A \rangle)(\hat{B} - \langle B \rangle) - i(\hat{B} - \langle B \rangle)(\hat{A} - \langle A \rangle)$$
$$= i \, (\hat{A}\hat{B} - \hat{B}\hat{A}) \equiv i \, [\hat{A}, \, \hat{B}]. \tag{4.150}$$

Second, according to Eqs. (4.52) and (4.65), the Hermitian conjugate of any product of Hermitian operators $\hat{A}$ and $\hat{B}$ is just the product of these operators swapped. Using the fact, we may write

$$\hat{C}^{\dagger} = (i[\hat{A}, \, \hat{B}])^{\dagger} = -i \, (\hat{A}\hat{B})^{\dagger} + i \, (\hat{B}\hat{A})^{\dagger}$$
$$= -i\hat{B}\hat{A} + i\hat{A}\hat{B} = i[\hat{A}, \, \hat{B}] = \hat{C}, \tag{4.151}$$

so that the operator $\hat{C}$ is also Hermitian, i.e. its eigenvalues are also real, and thus its average is purely real as well. As a result, the square of the expectation value of the operator product (4.149) may be represented as

$$\left\langle \hat{A}\hat{B} \right\rangle^2 = \left\langle \frac{1}{2}\{\hat{A}, \, \hat{B}\} \right\rangle^2 + \left\langle \frac{1}{2}\hat{C} \right\rangle^2. \tag{4.152}$$

Since the first term on the right-hand side of this equality cannot be negative, we may write

$$\left\langle \hat{A}\hat{B} \right\rangle^2 \geqslant \left\langle \frac{1}{2}\hat{C} \right\rangle^2 = \left\langle \frac{i}{2}[\hat{A}, \, \hat{B}] \right\rangle^2, \tag{4.153}$$

and hence continue Eq. (4.148) as

$$\delta A \delta B \geqslant \left| \left\langle \hat{A}\hat{B} \right\rangle \right| \geqslant \frac{1}{2} \, |\langle[\hat{A}, \, \hat{B}]\rangle|, \tag{4.154}$$

thus proving Eq. (4.140).

For the particular case of operators $\hat{x}$ and $\hat{p}_x$ (or a similar pair of operators for another Cartesian coordinate), we may readily combine Eq. (4.140) with Eq. (2.14b) and to prove the original Heisenberg's uncertainty relation (2.13). For the spin-½ operators defined by Eqs. (4.116) and (4.117), it is very simple (and highly recommended to the reader) to show that

$$\left[\hat{\sigma}_j, \, \hat{\sigma}_{j'} \right] = 2i\varepsilon_{jj'j''}\hat{\sigma}_{j''}, \quad \text{i.e.} \quad \left[\hat{S}_j, \, \hat{S}_{j'} \right] = i\varepsilon_{jj'j''}\hbar\hat{S}_{j''}, \tag{4.155}$$

where $\varepsilon_{jj'j''}$ is the *Levi-Civita permutation symbol*—see, e.g. Eq. (A.83). As a result, the uncertainty relations (4.140) for all spin-½ systems are, for example

$$\delta S_x \delta S_y \geqslant \frac{\hbar}{2}|\langle S_z \rangle|, \text{ etc.} \tag{4.156}$$

In particular, as we already know, in the ↑ state the right-hand side of this relation equals $(\hbar/2)^2 > 0$, and neither of the uncertainties δS_x, δS_y can equal zero. As a reminder, our direct calculation earlier in this section has shown that each of these uncertainties is equal to $\hbar/2$, i.e. their product equals the *lowest* value allowed by the uncertainty relation (4.156)—just as the Gaussian wave packets (2.16) provide the lowest possible value of the product $\delta x \delta p_x$, allowed by the Heisenberg relation (2.13).

4.6 Quantum dynamics: three pictures

So far in this chapter, I shied away from the discussion of the system's dynamics, implying that the bra- and ket-vectors of the system are their 'snapshots' at a certain instant t. Now we are sufficiently prepared to examine their time dependence. One of the most beautiful features of quantum mechanics is that the time evolution may be described using either of three alternative 'pictures', giving exactly the same final results for the expectation values of all observables.

From the standpoint of our wave mechanics experience, the *Schrödinger picture* is the most natural. In this picture, the operators corresponding to time-independent observables (e.g. to the Hamiltonian function H of an isolated system) are also constant in time, while the bra- and ket-vectors of the quantum state of the system evolve in time as

$$\langle \alpha(t)| = \langle \alpha(t_0)|\hat{u}^\dagger(t, t_0), \qquad |\alpha(t)\rangle = \hat{u}(t, t_0)|\alpha(t_0)\rangle. \qquad (4.157a)$$

Here $\hat{u}(t, t_0)$ is the *time-evolution operator*, which obeys the following differential equation:

$$i\hbar \frac{\partial}{\partial t}\hat{u} = \hat{H}\hat{u}, \qquad (4.157b)$$

where $\hat{H}$ is the Hamiltonian operator of the system (which is always Hermitian: $\hat{H}^\dagger = \hat{H}$). This equation remains valid even if the Hamiltonian depends on time explicitly. Differentiating the second of Eqs. (4.157a) over t, and then using Eq. (4.157b) twice, we can merge these relations into a single equation, without an explicit use of the time-evolution operator:

$$i\hbar \frac{\partial}{\partial t}|\alpha(t)\rangle = \hat{H}|\alpha(t)\rangle, \qquad (4.158)$$

which is frequently more convenient. (However, for some purposes the notion of the time-evolution operator, together with Eq. (4.157b), are useful—as we will see in a minute.) While Eq. (4.158) is a very natural replacement of the wave-mechanical equation (1.25), and is also frequently called the *Schrödinger equation*[35], it still should be considered as a new, more general postulate, which finds its final justification (as it is usual in physics) in the agreement between its corollaries with experiment—more exactly, in the absence of a single credible contradiction with experiment.

---

[35] Moreover, we will be able to *derive* Eq. (1.25) from Eq. (4.158)—see below.

Starting the discussion of Eq. (4.158), let us first consider the case of a time-independent Hamiltonian, whose eigenstates a_n and eigenvalues E_n obey Eq. (4.68) for this operator[36]:

$$\hat{H} |a_n\rangle = E_n|a_n\rangle, \tag{4.159}$$

and hence are also time-independent. (Similarly to the wavefunctions ψ_n defined by Eq. (1.60), a_n are called the *stationary states* of the system.) Let us use Eqs. (4.158) and (4.159) to calculate the law of time evolution of the expansion coefficients α_n (i.e. the probability amplitudes) defined by Eq. (4.118), in the stationary state basis:

$$\dot{\alpha}_n(t) = \frac{d}{dt}\langle a_n|\alpha(t)\rangle = \left\langle a_n\left|\frac{d}{dt}\right|\alpha(t)\right\rangle$$

$$= \langle a_n|\frac{1}{i\hbar}\hat{H}|\alpha(t)\rangle = \frac{E_n}{i\hbar}\langle a_n|\alpha(t)\rangle = -\frac{i}{\hbar}E_n\alpha_n. \tag{4.160}$$

This is the same simple equation as Eq. (1.61), and its integration yields a similar result—cf Eq. (1.62), just with the initial time t_0 rather than 0:

$$\alpha_n(t) = \alpha_n(t_0)\exp\left\{-\frac{i}{\hbar}E_n(t - t_0)\right\}. \tag{4.161}$$

In order to illustrate how this result works, let us consider the spin-½ dynamics in a time-independent, uniform external magnetic field $\mathscr{B}$. To construct the system's Hamiltonian, we may apply the correspondence principle to the classical expression for the energy of a magnetic moment $\mathbf{m}$ in the external magnetic field $\mathscr{B}$,[37]

$$U = -\mathbf{m} \cdot \mathscr{B}. \tag{4.162}$$

In quantum mechanics, the operator corresponding to the moment $\mathbf{m}$ is given by Eqs. (4.115) and (4.116) (suggested by W Pauli), so that the spin-field interaction is described by the so-called *Pauli Hamiltonian*, which may be, due to Eqs. (4.115)–(4.117), represented in several equivalent forms:

$$\hat{H} = -\hat{\mathbf{m}} \cdot \mathscr{B} \equiv -\gamma\hat{\mathbf{S}} \cdot \mathscr{B}. \tag{4.163a}$$

If the z-axis is aligned with field's direction, this expression is reduced to

$$\hat{H} = -\gamma\mathscr{B}\hat{S}_z \equiv -\gamma\mathscr{B}\frac{\hbar}{2}\hat{\sigma}_z. \tag{4.163b}$$

According to Eq. (4.117), in the z-basis of the spin states ↑ and ↓, the matrix of the operator (4.163b) is

---

[36] Here I have intentionally switched the state index from j to n, which was used for numbering stationary states in chapter 1, to emphasize the special role played by the stationary states a_n in the quantum dynamics.
[37] See, e.g. *Part EM* Eq. (5.100). As a reminder, we have already used this expression for the derivation of Eq. (4.3).

$$H = -\frac{\gamma\hbar\mathscr{B}}{2}\sigma_z \equiv \frac{\hbar\Omega}{2}\sigma_z, \quad \text{where} \quad \Omega \equiv -\gamma\mathscr{B}. \tag{4.164}$$

The constant Ω so defined coincides with the classical frequency of the precession, about axis z, of an axially-symmetric rigid body (the so-called *symmetric top*), with an angular momentum $\mathbf{S}$ and the magnetic moment $\mathbf{m} = \gamma\mathbf{S}$, induced by the external torque $\tau = \mathbf{m} \times \mathscr{B}$.[38] (For an electron, with its negative gyromagnetic ratio $\gamma_e = -g_e e/2m_e$, neglecting the tiny difference of the g_e-factor from 2, we get

$$\Omega = \frac{e}{m_e}\mathscr{B}, \tag{4.165}$$

so that according to Eq. (3.48), the frequency Ω coincides with the electron's cyclotron frequency ω_c.)

In order to apply the general Eq. (4.161), at this stage we would need to find the eigenstates a_n and eigenenergies E_n of our Hamiltonian. However, with our (smart :-) choice of the z-axis, the Hamiltonian matrix is already diagonal:

$$H = \frac{\hbar\Omega}{2}\sigma_z \equiv \frac{\hbar\Omega}{2}\begin{pmatrix} 1 & 0 \\ 0 & -1 \end{pmatrix}, \tag{4.166}$$

meaning that the states $\uparrow$ and $\downarrow$ are the eigenstates of this system, with the eigenenergies, respectively,

$$E_\uparrow = +\frac{\hbar\Omega}{2} \quad \text{and } E_\downarrow = -\frac{\hbar\Omega}{2}. \tag{4.167}$$

Note that their difference,

$$\Delta E \equiv |E_\uparrow - E_\downarrow| = \hbar\,|\Omega| = \hbar\,|\gamma\mathscr{B}|, \tag{4.168}$$

corresponds to the classical energy $2|m\mathscr{B}|$ of flipping a magnetic dipole with the moment's magnitude $m = \gamma\hbar/2$, oriented along the direction of the field $\mathscr{B}$. Note also that if the product $\gamma\mathscr{B}$ is positive, then Ω is negative, so that $E_\uparrow$ is negative, while $E_\downarrow$ is positive. This is in the agreement with the classical picture of a magnetic dipole $\mathbf{m}$ having a negative potential energy when it is aligned with the external magnetic field $\mathscr{B}$—see Eq. (4.162).

For the time evolution of the probability amplitudes of these states, Eq. (4.161) immediately yields the following expressions:

$$\alpha_\uparrow(t) = \alpha_\uparrow(t_0)\exp\left\{-\frac{i}{2}\Omega\,(t - t_0)\right\}, \quad \alpha_\downarrow(t) = \alpha_\downarrow(t_0)\exp\left\{\frac{i}{2}\Omega\,(t - t_0)\right\}, \tag{4.169}$$

allowing a ready calculation of the time evolution of the expectation values of any observable. In particular, we can calculate the expectation value of S_z as a function of time by applying Eq. (4.130) to the (arbitrary) time moment t:

---

[38] See, e.g. *Part CM* section 4.5, in particular Eq. (4.72), and *Part EM* section 5.5, in particular Eq. (5.114) and its discussion.

$$\langle S_z \rangle(t) = \frac{\hbar}{2} \Big[\alpha_\uparrow(t) \alpha_\uparrow^*(t) - \alpha_\downarrow(t) \alpha_\downarrow^*(t) \Big]$$

$$= \frac{\hbar}{2} \Big[\alpha_\uparrow(0) \alpha_\uparrow^*(0) - \alpha_\downarrow(0) \alpha_\downarrow^*(0) \Big] = \langle S_z \rangle(0). \tag{4.170}$$

Thus the expectation value of the spin component parallel to the applied magnetic field remains constant in time, regardless of the initial state of the system. However, this is not true for the components perpendicular to the field. For example, Eq. (4.132), applied to the moment t, gives

$$\langle S_x \rangle(t) = \frac{\hbar}{2} \Big[\alpha_\uparrow(t) \, \alpha_\downarrow^*(t) + \alpha_\downarrow(t) \, \alpha_\uparrow^*(t) \Big]$$

$$= \frac{\hbar}{2} \Big[\alpha_\uparrow(0) \, \alpha_\downarrow^*(0) \, e^{-i\Omega(t-t_0)} + \alpha_\downarrow(0) \, \alpha_\uparrow^*(0) \, e^{i\Omega(t-t_0)} \Big]. \tag{4.171}$$

Clearly, this expression describes sinusoidal oscillations with frequency (4.164). The amplitude and the phase of these oscillations depend on initial conditions. Indeed, solving Eqs. (4.132) and (4.133) for the probability amplitude products, we get the following relations:

$$\hbar \alpha_\downarrow(t) \alpha_\uparrow^*(t) = \langle S_x \rangle(t) + i \langle S_y \rangle(t), \qquad \hbar \alpha_\uparrow(t) \alpha_\downarrow^*(t) = \langle S_x \rangle(t) - i \langle S_y \rangle(t), \tag{4.172}$$

valid for any time t. Plugging their values for $t = t_0 = 0$ into Eq. (4.171), we get

$$\langle S_x \rangle(t) = \frac{1}{2} \Big[\langle S_x \rangle(0) + i \langle S_y \rangle(0) \Big] \, e^{+i\Omega(t-t_0)} + \frac{1}{2} \Big[\langle S_x \rangle(0) - i \langle S_y \rangle(0) \Big] \, e^{-i\Omega(t-t_0)}$$

$$\equiv \langle S_x \rangle(0) \cos \Omega t - \langle S_y \rangle(0) \sin \Omega t. \tag{4.173}$$

An absolutely similar calculation using Eq. (4.133) gives

$$\langle S_y \rangle(t) = \langle S_y \rangle(0) \cos \Omega t + \langle S_x \rangle(0) \sin \Omega t. \tag{4.174}$$

These formulas show, for example, if at moment $t = 0$ the spin's state was $\uparrow$, i.e. $\langle S_x \rangle(0) = \langle S_y \rangle(0) = 0$, then the oscillation amplitudes of the both 'lateral' components of spin vanish. On the other hand, if the spin was initially in the state $\rightarrow$, i.e. had the definite, maximum possible value of S_x, equal to $\hbar/2$ (in classics, we would say 'the spin $\hbar/2$ was oriented in direction x'), then both expectation values $\langle S_x \rangle$ and $\langle S_y \rangle$ oscillate in time[39] with this amplitude, with the phase shift $\pi/2$ between them.

---

[39] This is one more (hopefully, redundant) illustration of the difference between averaging over the statistical ensemble and over time: in Eqs. (4.170), (4.173) and (4.174), and quite a few relations below, only the former averaging has been performed, so the results are still functions of time.

So, the quantum-mechanical results for the expectation values of the Cartesian components of spin-½ are indistinguishable from the classical results for the precession, with the frequency $\Omega = -\gamma\mathcal{B}$,[40] of a symmetric top with the angular momentum of magnitude $S = \hbar/2$, about the field's direction (our axis z), under the effect of an external torque $\boldsymbol{\tau} = \mathbf{m} \times \mathcal{B}$ exerted by the field $\mathcal{B}$ on the magnetic moment $\mathbf{m} = \gamma\mathbf{S}$. Note, however, that the classical language does not describe the large quantum-mechanical uncertainties of the components, specified by Eqs. (4.156), which are absent in the classical picture—at least when it starts from a definite orientation of the angular momentum vector. Also, as we have seen in section 3.5, the component L_z of an angular momentum at the orbital motion of particles is always a multiple of $\hbar$—see, e.g. Eq. (3.139). As a result, the angular momentum of a spin-½ particle, with $S_z = \pm\hbar/2$, cannot be explained by any addition of orbital angular moments of its hypothetical components, i.e. by any internal rotation of the particle about its axis.

Now let us return to the discussion of the general Schrödinger equation (4.157b) and prove the following fascinating fact: it is possible to write the general solution of this *operator* equation. In the easiest case when the Hamiltonian is time-independent, this solution is an exact analog of Eq. (4.161),

$$\hat{u}(t, t_0) = \hat{u}(t_0, t_0)\exp\left\{-\frac{i}{\hbar}\hat{H}(t - t_0)\right\} = \exp\left\{-\frac{i}{\hbar}\hat{H}(t - t_0)\right\}. \qquad (4.175)$$

To start its proof we should, first of all, understand what does a function (in this case, the exponent) of an operator mean. In the operator (and matrix) algebra, such functions are *defined* by their Taylor expansions; in particular, Eq. (4.175) means that

$$\hat{u}(t, t_0) = \hat{I} + \sum_{k=1}^{\infty}\frac{1}{k!}\left[-\frac{i}{\hbar}\hat{H}(t - t_0)\right]^k$$

$$\equiv \hat{I} + \frac{1}{1!}\left(-\frac{i}{\hbar}\right)\hat{H}(t - t_0) + \frac{1}{2!}\left(-\frac{i}{\hbar}\right)^2\hat{H}^2(t - t_0)^2 \qquad (4.176)$$

$$+ \frac{1}{3!}\left(-\frac{i}{\hbar}\right)^3\hat{H}^3(t - t_0)^3 + \dots ,$$

where $\hat{H}^2 \equiv \hat{H}\hat{H}$, $\hat{H}^3 \equiv \hat{H}\hat{H}\hat{H}$, etc. Working with such series of operator products is not as hard as one could imagine, due to their regular structure. For example, let us differentiate both parts of Eq. (4.176) over t:

---

[40] Note that according to this (classical!) relation, the gyromagnetic ratio γ may be interpreted just as the angular frequency of the spin precession per unit magnetic field—hence the name. In particular, for electrons, $|\gamma_e| \approx 1.761 \times 10^{11}$ s^{-1} T^{-1}; for protons, the ratio is much smaller, $\gamma_p \approx 2.675 \times 10^8$ s^{-1} T^{-1}, mostly because of their larger mass m_p, at a g-factor of the same order as for the electron: $g_p \equiv (2m_p/e)\gamma_p \approx 5.586$. For larger spin-½ particles, e.g. atomic nuclei with such spin, the values of γ are smaller still—e.g. $\gamma \approx 8.681 \times 10^6$ s^{-1} T^{-1} for the ^{57}Fe nucleus.

$$\frac{\partial}{\partial t}\hat{u}(t, t_0) = \hat{0} + \frac{1}{1!}\left(-\frac{i}{\hbar}\right)\hat{H} + \frac{1}{2!}\left(-\frac{i}{\hbar}\right)^2\hat{H}^2 2(t - t_0)$$

$$+ \frac{1}{3!}\left(-\frac{i}{\hbar}\right)^2\hat{H}^3 3(t - t_0)^2 + \ldots$$

$$\equiv \left(-\frac{i}{\hbar}\right)\hat{H}\left[\hat{I} + \frac{1}{1!}\left(-\frac{i}{\hbar}\right)\hat{H}(t - t_0) + \frac{1}{2!}\left(-\frac{i}{\hbar}\right)^2\hat{H}^2(t - t_0)^2\right] + \ldots$$

$$\equiv -\frac{i}{\hbar}\hat{H}\hat{u}\,(t, t_0),$$

(4.177)

so that the differential equation (4.158) is indeed satisfied. On the other hand, Eq. (4.175) also satisfies the initial condition

$$\hat{u}(t_0, t_0) = \hat{u}^\dagger(t_0, t_0) = \hat{I},$$

(4.178)

which immediately follows from the definition (4.157a) of the evolution operator. Thus, Eq. (4.175) indeed gives the (unique) solution for the time evolution operator—in the Schrödinger picture.

Now let us allow the operator $\hat{H}$ to be a function of time, but with the condition that its 'values' (in fact, operators) at different instants commute with each other:

$$[\hat{H}(t'), \hat{H}(t'')] = 0, \quad \text{for any } t', t''.$$

(4.179)

(An important non-trivial example of such a Hamiltonian is the time-dependent part of the Hamiltonian of a particle, due to the effect of a classical, time-dependent, but position-independent force $\mathbf{F}(t)$,

$$\hat{H}_F = -\mathbf{F}(t) \cdot \hat{\mathbf{r}}.$$

(4.180)

Indeed, the radius-vector operator $\hat{\mathbf{r}}$ does not depend explicitly on time and hence commutes with itself, as well as with the c-numbers $\mathbf{F}(t')$ and $\mathbf{F}(t'')$.) In this case it is sufficient to replace, in all above formulas, the product $\hat{H}(t - t_0)$ with the corresponding integral over time; in particular, Eq. (4.175) is generalized as

$$\hat{u}(t, t_0) = \exp\left\{-\frac{i}{\hbar}\int_{t_0}^t \hat{H}(t')dt'\right\}.$$

(4.181)

This replacement means that the first form of Eq. (4.176) should be replaced with

$$\hat{u}(t, t_0) = \hat{I} + \sum_{k=1}^\infty \frac{1}{k!}\left(-\frac{i}{\hbar}\right)^k\left(\int_{t_0}^t \hat{H}(t')dt'\right)^k$$

$$\equiv \hat{I} + \sum_{k=1}^\infty \left(-\frac{i}{\hbar}\right)^k \int_{t_0}^t dt_1 \int_{t_0}^t dt_2 \ldots \int_{t_0}^t dt_k \hat{H}(t_1)\hat{H}(t_2)\ldots\hat{H}(t_k).$$

(4.182)

The proof that Eq. (4.182) satisfies Eq. (4.158) is absolutely similar to the one carried out above.

We may now use Eq. (4.181) to show that the time-evolution operator is unitary at any moment, even for the time-dependent Hamiltonian, if it satisfies Eq. (4.179). Indeed, Eq. (4.181) yields

$$\hat{u}(t, t_0)\hat{u}^\dagger(t, t_0) = \exp\left\{-\frac{i}{\hbar}\int_{t_0}^t \hat{H}(t')dt'\right\}\exp\left\{+\frac{i}{\hbar}\int_{t_0}^t \hat{H}(t'')dt''\right\}. \quad (4.183)$$

Since each of the exponents may be represented with the Taylor series (4.182), and, thanks to Eq. (4.179), different components of these sums may be swapped at will, the expression (4.183) may be manipulated exactly as the product of c-number exponents, in particular rewritten as

$$\hat{u}(t, t_0)\hat{u}^\dagger(t, t_0) = \exp\left\{-\frac{i}{\hbar}\left[\int_{t_0}^t \hat{H}(t')dt' - \int_{t_0}^t \hat{H}(t'')dt''\right]\right\}$$
$$= \exp\{\hat{0}\} = \hat{I}. \quad (4.184)$$

This property ensures, in particular, that the system state's normalization does not depend on time:

$$\langle\alpha(t)|\alpha(t)\rangle = \langle\alpha(t_0)|\hat{u}^\dagger(t, t_0)\hat{u}(t, t_0)|\alpha(t_0)\rangle = \langle\alpha(t_0)|\alpha(t_0)\rangle. \quad (4.185)$$

The most difficult cases for the explicit solution of Eq. (4.158) are those when Eq. (4.179) is violated[41]. It may be proven that in these cases the integral limits in the last form of Eq. (4.182) should be truncated, giving the so-called *Dyson series*

$$\hat{u}(t, t_0) = \hat{I} + \sum_{k=1}^\infty \left(-\frac{i}{\hbar}\right)^k \int_{t_0}^t dt_1 \int_{t_0}^{t_1} dt_2 ... \int_{t_0}^{t_{k-1}} dt_k \hat{H}(t_1)\hat{H}(t_2)...\hat{H}(t_k). \quad (4.186)$$

Since we would not have time/space to use this relation in our course, I will skip its proof[42].

Let me now return to the general discussion of quantum dynamics to outline its alternative, *Heisenberg picture*. For that, let us recall that according to Eq. (4.125), in quantum mechanics the expectation value of any observable A is a long bracket. Other quantities (for example, the rates of quantum transitions between pairs of different states, say α and β) may also be measured in experiment; the general form of such observable quantities is the following long bracket:

$$\langle\alpha|\hat{A}|\beta\rangle. \quad (4.187)$$

As was discussed above, in the Schrödinger picture the bra- and ket-vectors of the states evolve in time, while the operators of observables remain time-independent (if the corresponding variables do not explicitly depend on time), so that Eq. (4.187), applied to a moment t, may be represented as

$$\langle\alpha(t)|\hat{A}_S|\beta(t)\rangle, \quad (4.188)$$

---

[41] We will run into such situations in chapter 7, but will not need to apply Eq. (4.186).
[42] It may be found, for example, in chapter 5 of J Sakurai's textbook—see References.

where the index 'S' is added to emphasize the Schrödinger picture. Let us apply the evolution law (4.157a) to the bra- and ket-vectors in this expression:

$$\langle \alpha | \hat{A} | \beta \rangle = \langle \alpha(t_0) | \, \hat{u}^\dagger(t, t_0) \hat{A}_S \hat{u} \, (t, t_0) | \beta(t_0) \rangle. \tag{4.189}$$

This equality means that if we form a long bracket with bra- and ket-vectors of the initial-time states, together with the following time-dependent *Heisenberg operator*[43]

$$\hat{A}_H(t) \equiv \hat{u}^\dagger(t, t_0) \hat{A}_S \hat{u}(t, t_0) = \hat{u}^\dagger(t, t_0) \hat{A}_H(t_0) \hat{u}(t, t_0), \tag{4.190}$$

all experimentally measurable results will remain the same as in the Schrödinger picture:

$$\langle \alpha | \hat{A} | \beta \rangle = \langle \alpha(t_0) | \, \hat{A}_H(t, t_0) | \beta(t_0) \rangle. \tag{4.191}$$

For full clarity, let us see how the Heisenberg picture works for the same simple (but very important!) problem of the spin-½ precession in a z-oriented magnetic field, described (in the z-basis) by the Hamiltonian matrix (4.164). In that basis, Eq. (4.157b) for the time-evolution operator becomes

$$i\hbar \frac{\partial}{\partial t} \begin{pmatrix} u_{11} & u_{12} \\ u_{21} & u_{22} \end{pmatrix} = \frac{\hbar\Omega}{2} \begin{pmatrix} 1 & 0 \\ 0 & -1 \end{pmatrix} \begin{pmatrix} u_{11} & u_{12} \\ u_{21} & u_{22} \end{pmatrix} \equiv \frac{\hbar\Omega}{2} \begin{pmatrix} u_{11} & u_{12} \\ -u_{21} & -u_{22} \end{pmatrix}. \tag{4.192}$$

We see that in this simple case the equations for different matrix elements of the evolution operator matrix are decoupled, and readily solvable, using the universal initial conditions (4.178):[44]

$$u(t, 0) = \begin{pmatrix} e^{-i\Omega t/2} & 0 \\ 0 & e^{i\Omega t/2} \end{pmatrix} \equiv I \cos \frac{\Omega t}{2} - i\sigma_z \sin \frac{\Omega t}{2}. \tag{4.193}$$

Now we can use Eq. (4.190) to calculate the Heisenberg-picture operators of spin components—still in the z-basis. Dropping the index 'H' for brevity (the Heisenberg-picture operators are clearly marked by their dependence on time anyway), we get

$$S_x(t) = u^\dagger(t, 0) S_x(0) u(t, 0) = \frac{\hbar}{2} u^\dagger(t, 0) \sigma_x u(t, 0)$$

$$= \frac{\hbar}{2} \begin{pmatrix} e^{i\Omega t/2} & 0 \\ 0 & e^{-i\Omega t/2} \end{pmatrix} \begin{pmatrix} 0 & 1 \\ 1 & 0 \end{pmatrix} \begin{pmatrix} e^{-i\Omega t/2} & 0 \\ 0 & e^{i\Omega t/2} \end{pmatrix} \tag{4.194}$$

$$= \frac{\hbar}{2} \begin{pmatrix} 0 & e^{i\Omega t} \\ e^{-i\Omega t} & 0 \end{pmatrix} = \frac{\hbar}{2} \left[\sigma_x \cos \Omega t - \sigma_y \sin \Omega t \right]$$

$$\equiv S_x(0) \cos \Omega t - S_y(0) \sin \Omega t.$$

---

[43] Note that this *strict* relation is similar in structure to first of the *symbolic* Eqs. (4.94), with the bases $\{v\}$ and $\{u\}$ loosely associated, respectively, with the time moments t and t_0.

[44] We could of course use this solution, together with Eq. (4.157), to obtain all the above results for this system within the Schrödinger picture. In our simple case, the use of Eqs. (4.161) for this purpose was more straightforward, but in some cases (e.g. for time-dependent Hamiltonians) an explicit calculation of the time-evolution matrix may be the only practicable way to proceed.

Absolutely similar calculations of the other spin components yield

$$S_y(t) = \frac{\hbar}{2}\begin{pmatrix} 0 & -ie^{i\Omega t} \\ ie^{-i\Omega t} & 0 \end{pmatrix} = \frac{\hbar}{2}[\sigma_y \cos \Omega t + \sigma_x \sin \Omega t] \tag{4.195}$$

$$\equiv S_y(0)\cos \Omega t + S_x(0)\sin \Omega t,$$

$$S_z(t) = \frac{\hbar}{2}\begin{pmatrix} 1 & 0 \\ 0 & -1 \end{pmatrix} = \frac{\hbar}{2}\sigma_z = S_z(0) \ . \tag{4.196}$$

A practical advantage of these formulas is that they describe system's evolution for *arbitrary* initial conditions, thus making the analysis of the initial state effects very simple. Indeed, since in the Heisenberg picture the expectation values of observables are calculated using Eq. (4.191) (with $\beta = \alpha$), with time-independent bra- and ket vectors, such averaging of Eqs. (4.194)–(4.196) immediately returns us to Eqs. (4.170), (4.173), and (4.174), which had been obtained above in the Schrödinger picture. Moreover, these equations for the Heisenberg operators formally coincide with the classical equations of the torque-induced precession for *c*-number variables. (Below we will see that the same exact mapping is valid for the Heisenberg picture of the orbital motion.)

In order to see that the last fact is by no means a coincidence, let us combine Eqs. (4.157b) and (4.190) to form an explicit differential equation of the Heisenberg operator's evolution. For that, let us differentiate Eq. (4.190) over time:

$$\frac{d}{dt}\hat{A}_H = \frac{\partial \hat{u}^\dagger}{\partial t}\hat{A}_S\hat{u} + \hat{u}^\dagger\frac{\partial \hat{A}_S}{\partial t}\hat{u} + \hat{u}^\dagger\hat{A}_S\frac{\partial \hat{u}}{\partial t}. \tag{4.197}$$

Plugging in the derivatives of the time evolution operator from Eq. (4.157b) and its Hermitian conjugate, and multiplying both parts of the equation by $i\hbar$, we get

$$i\hbar\frac{d}{dt}\hat{A}_H = -\hat{u}^\dagger\hat{H}\hat{A}_S\hat{u} + i\hbar\hat{u}^\dagger\frac{\partial \hat{A}_S}{\partial t}\hat{u} + \hat{u}^\dagger\hat{A}_S\hat{H}\hat{u} \ . \tag{4.198a}$$

If for the Schrödinger-picture's Hamiltonian the condition similar to Eq. (4.179) is satisfied, then, according to Eqs. (4.177) or (4.182), the Hamiltonian commutes with the time evolution operator and its Hermitian conjugate, and may be swapped with any of them[45]. Hence, we may rewrite Eq. (4.198a) as

$$i\hbar\frac{d}{dt}\hat{A}_H = -\hat{H}\hat{u}^\dagger\hat{A}_S\hat{u} + i\hbar\hat{u}^\dagger\frac{\partial \hat{A}_S}{\partial t}\hat{u} + \hat{u}^\dagger\hat{A}_S\hat{u}\hat{H}$$

$$= i\hbar\hat{u}^\dagger\frac{\partial \hat{A}_S}{\partial t}\hat{u} + [\hat{u}^\dagger\hat{A}_S\hat{u}, \hat{H}]. \tag{4.198b}$$

---

[45] Due to the same reason, $\hat{H}_H \equiv \hat{u}^\dagger\hat{H}_S\hat{u} = \hat{u}^\dagger\hat{u}\hat{H}_S = \hat{H}_S$; this is why the index of the Hamiltonian operator may be dropped in Eqs. (4.198) and (4.199).

Now using the definition (4.190) again, for both terms on the right-hand side, we may write

$$i\hbar\frac{d}{dt}\hat{A}_\mathrm{H} = i\hbar\left(\frac{\partial\hat{A}}{\partial t}\right)_\mathrm{H} + [\hat{A}_\mathrm{H}, \hat{H}]. \tag{4.199}$$

This is the so-called *Heisenberg equation of motion*[46].

Let us see how this equation looks for the same problem of the spin-½ precession in a z-oriented, time-independent magnetic field, described in the z-basis by the Hamiltonian matrix (4.164), which does not depend on time. In this basis, Eq. (4.199) for the vector-operator of spin reads[47]

$$i\hbar\begin{pmatrix}\dot{S}_{11} & \dot{S}_{12} \\ \dot{S}_{21} & \dot{S}_{22}\end{pmatrix} = \frac{\hbar\Omega}{2}\left[\begin{pmatrix}S_{11} & S_{12} \\ S_{21} & S_{22}\end{pmatrix}, \begin{pmatrix}1 & 0 \\ 0 & -1\end{pmatrix}\right] = \hbar\Omega\begin{pmatrix}0 & -S_{12} \\ S_{21} & 0\end{pmatrix}. \tag{4.200}$$

Once again, the equations for different matrix elements are decoupled, and their solution is elementary:

$$\begin{aligned}S_{11}(t) &= S_{11}(0) = \text{const}, \quad S_{22}(t) = S_{22}(0) = \text{const}, \\ S_{12}(t) &= S_{12}(0)e^{i\Omega t}, \quad S_{21}(t) = S_{21}(0)e^{-i\Omega t}.\end{aligned} \tag{4.201}$$

According to Eq. (4.190), the initial 'values' of the Heisenberg-picture matrix elements are just the Schrödinger-picture ones, so that using Eq. (4.117) we may rewrite this solution in either of two forms:

$$\begin{aligned}S(t) &= \frac{\hbar}{2}\left[\mathbf{n}_x\begin{pmatrix}0 & e^{i\Omega t} \\ e^{-i\Omega t} & 0\end{pmatrix} + \mathbf{n}_y\begin{pmatrix}0 & -ie^{i\Omega t} \\ ie^{-i\Omega t} & 0\end{pmatrix} + \mathbf{n}_z\begin{pmatrix}1 & 0 \\ 0 & -1\end{pmatrix}\right] \\ &\equiv \frac{\hbar}{2}\begin{pmatrix}\mathbf{n}_z & \mathbf{n}_-e^{i\Omega t} \\ \mathbf{n}_+e^{-i\Omega t} & -\mathbf{n}_z\end{pmatrix}, \quad \text{with } \mathbf{n}_\pm \equiv \mathbf{n}_x \pm i\mathbf{n}_y.\end{aligned} \tag{4.202}$$

The simplicity of the last expression is spectacular. (Remember, it covers *any* initial conditions, and *all* three spatial components of spin!) On the other hand, for some purposes the previous form may be more convenient; in particular, its Cartesian components give our earlier results (4.194)–(4.196).

One of the advantages of the Heisenberg picture is that it provides a clearer link between classical and quantum mechanics. Indeed, analytical classical mechanics may be used to derive the following equation of time evolution of an arbitrary

---

[46] Reportedly, this equation was derived by P A M Dirac, who was so generous that he himself gave the name of his junior colleague to this key result, because 'Heisenberg was saying something like this'.

[47] Using the commutation relations (4.155), this equation may be readily generalized to the case of an arbitrary magnetic field $\mathscr{B}(t)$ and an arbitrary state basis—the exercise highly recommended to the reader.

function $A(q_j, p_j, t)$ of the generalized coordinates q_j and momenta p_j of the system, and time t:[48]

$$\frac{dA}{dt} = \frac{\partial A}{\partial t} - \{A, H\}, \qquad (4.203)$$

where H is the classical Hamiltonian function of the system, and $\{..,..\}$ is the so-called *Poisson bracket* defined, for two arbitrary functions $A(q_j, p_j, t)$ and $B(q_j, p_j, t)$, as

$$\{A, B\} \equiv \sum_j \left(\frac{\partial A}{\partial p_j} \frac{\partial B}{\partial q_j} - \frac{\partial A}{\partial q_j} \frac{\partial B}{\partial p_j} \right). \qquad (4.204)$$

Comparing Eq. (4.203) with Eq. (4.199), we see that the correspondence between classical and quantum mechanics (in the Heisenberg picture) is provided by the following symbolic relation[49]

$$\{A, B\} \leftrightarrow \frac{1}{-i\hbar}[\hat{A}, \hat{B}]. \qquad (4.205)$$

This relation may be used, in particular, for finding appropriate operators for system's observables, if their form is not immediately evident from the correspondence principle. We will develop this argumentation further when we revisit the wave mechanics in section 4.7, and also in chapter 9.

Finally, let us discuss one more alternative picture of quantum dynamics. It is also attributed to P A M Dirac, and is called either the 'Dirac picture', or (more frequently) the *interaction picture*. The last name stems from the fact that this picture is very useful for the *perturbative* (approximate) approaches to systems whose Hamiltonians may be partitioned into two parts,

$$\hat{H} = \hat{H}_0 + \hat{H}_{\text{int}}, \qquad (4.206)$$

where $\hat{H}_0$ is the sum of relatively simple Hamiltonians of the component subsystems, while the second term in Eq. (4.206) represents their weak interaction. (Note, however, that all relations in the balance of this section are exact and not directly based on the interaction weakness.) In this case, it is natural to consider, together with the genuine unitary operator $\hat{u}(t, t_0)$ of the time evolution of the system, which obeys Eq. (4.157b), a similarly defined unitary operator of evolution of the 'unperturbed system' described by the Hamiltonian $\hat{H}_0$ alone:

---

[48] See, e.g. *Part CM* Eq. (10.17). Also, please excuse my use, for the Poisson bracket, of the same (traditional) symbol $\{...,...\}$ as for the anticommutator. We will not run into the Poisson brackets again in the course, minimizing any chance of confusion.

[49] Since we have run into the commutator of Heisenberg-picture operators, let me note emphasize again that the 'values' of the same operator at different moments of time may not commute. Perhaps the simplest example is the operator $\hat{x}$ of coordinate of a free 1D particle, with the time-independent Hamiltonian $\hat{H} = \hat{p}^2/2m$. Indeed, for this case Eq. (4.199) yields $i\hbar\dot{\hat{x}} = [\hat{x}, \hat{H}] = i\hbar\hat{p}/m$ and $i\hbar\dot{\hat{p}} = [\hat{p}, \hat{H}] = 0$, with the following simple solutions (similar to those for the classical motion of the corresponding observables): $\hat{p}(t) = \text{const} = \hat{p}(0)$, $\hat{x}(t) = \hat{x}(0) + \hat{p}(0)\,t/m$, so that $[\hat{x}(0), \hat{x}(t)] = [\hat{x}(0), \hat{p}(0)]\,t/m \equiv [\hat{x}_S, \hat{p}_S]\,t/m = i\hbar t/m \neq 0$, if $t \neq 0$.

$$i\hbar\frac{\partial}{\partial t}\hat{u}_0 = \hat{H}_0\hat{u}_0, \tag{4.207}$$

and also the following interaction evolution operator,

$$\hat{u}_I \equiv \hat{u}_0^\dagger\hat{u}. \tag{4.208}$$

The reason for the last definition becomes more clear if we insert the reciprocal relation,

$$\hat{u} \equiv \hat{u}_0\,\hat{u}_0^\dagger\hat{u} = \hat{u}_0\hat{u}_I, \tag{4.209}$$

and its Hermitian conjugate,

$$\hat{u}^\dagger = (\hat{u}_0\hat{u}_I)^\dagger = \hat{u}_I^\dagger\hat{u}_0^\dagger, \tag{4.210}$$

into the basic Eq. (4.189)—which is valid in any picture:

$$\langle\alpha|\hat{A}|\beta\rangle \equiv \langle\alpha(t_0)|\,\hat{u}^\dagger(t, t_0)\hat{A}_S\hat{u}\,(t, t_0)|\beta(t_0)\rangle$$
$$= \langle\alpha(t_0)|\,\hat{u}_I^\dagger(t, t_0)\hat{u}_0^\dagger(t, t_0)\hat{A}_S\hat{u}_0(t, t_0)\hat{u}_I(t, t_0)\,|\beta(t_0)\rangle. \tag{4.211}$$

This relation shows that all calculations of the observable expectation values and transition rates (i.e. all the results of quantum mechanics that may be experimentally verified) are expressed by the following formula, with the standard long bracket structure (4.187),

$$\langle\alpha|\hat{A}|\beta\rangle = \langle\alpha_I(t)|\,\hat{A}_I(t)|\beta_I(t)\rangle, \tag{4.212}$$

if we assume that both the state vectors and the operators depend on time, with the vectors evolving only due to the *interaction operator* $\hat{u}_I$,

$$\langle\alpha_I(t)| \equiv \langle\alpha(t_0)|\hat{u}_I^\dagger(t, t_0), \qquad |\beta_I(t)\rangle \equiv \hat{u}_I(t, t_0)\,|\beta(t_0)\rangle, \tag{4.213}$$

and the operators' evolution being governed by the *unperturbed operator* $\hat{u}_0$:

$$\hat{A}_I(t) \equiv \hat{u}_0^\dagger(t, t_0)\hat{A}_S\hat{u}_0(t, t_0). \tag{4.214}$$

These relations describe the interaction picture of quantum dynamics. Let me defer an example of its use until the perturbative analysis of open quantum systems in section 7.6, and end this section by a proof that the interaction evolution operator (4.208) satisfies the following equation,

$$i\hbar\frac{\partial}{\partial t}\hat{u}_I = \hat{H}_I\hat{u}_I, \tag{4.215}$$

where $\hat{H}_I$ is the interaction Hamiltonian transformed in accordance with the same rule (4.214):

$$\hat{H}_I(t) \equiv \hat{u}_0^\dagger(t, t_0)\hat{H}_{\text{int}}\hat{u}_0(t, t_0). \tag{4.216}$$

The proof is very straightforward: first using the definition (4.208), and then Eqs. (4.157b) and the Hermitian conjugate of Eq. (4.207), we may write

$$i\hbar\frac{\partial}{\partial t}\hat{u}_I \equiv i\hbar\frac{\partial}{\partial t}\left(\hat{u}_0^\dagger\hat{u}\right) \equiv i\hbar\frac{\partial\hat{u}_0^\dagger}{\partial t}\hat{u} + \hat{u}_0^\dagger i\hbar\frac{\partial\hat{u}}{\partial t} = -\hat{H}_0\hat{u}_0^\dagger\hat{u} + \hat{u}_0^\dagger\hat{H}\hat{u}$$
$$= -\hat{H}_0\hat{u}_0^\dagger\hat{u} + \hat{u}_0^\dagger(\hat{H}_0 + \hat{H}_{int})\hat{u} \equiv -\hat{H}_0\hat{u}_0^\dagger\hat{u} + \hat{u}_0^\dagger\hat{H}_0\hat{u} + \hat{u}_0^\dagger\hat{H}_{int}\hat{u} \qquad (4.217)$$
$$\equiv \left(-\hat{H}_0\hat{u}_0^\dagger + \hat{u}_0^\dagger\hat{H}_0\right)\hat{u} + \hat{u}_0^\dagger\hat{H}_{int}\hat{u}\,.$$

Since $\hat{u}_0^\dagger$ may be represented as an integral of an exponent of $\hat{H}_0$ over time (similar to Eq. (4.181) relating $\hat{u}$ and $\hat{H}$), these operators commute, so that the parentheses in the last form of Eq. (4.217) vanish. Now plugging $\hat{u}$ from the last form of Eq. (4.209), we get the equation,

$$i\hbar\frac{\partial}{\partial t}\hat{u}_I = \hat{u}_0^\dagger\hat{H}_{int}\hat{u}_0 u_I \equiv \left(\hat{u}_0^\dagger\hat{H}_{int}\hat{u}_0\right)\hat{u}_I, \qquad (4.218)$$

which is clearly equivalent to the combination of Eqs. (4.215) and (4.216).

As Eq. (4.215) shows, if the energy scale of the interaction H_{int} is much smaller than that of the background Hamiltonian H_0, the interaction evolution operators $\hat{u}_I$ and $\hat{u}_I^\dagger$, and hence the state vectors (4.213) evolve relatively slowly, without fast background oscillations. This is very convenient for the perturbative approaches to complex interacting systems, in particular to the 'open' quantum systems that weakly interact with their environment—see section 7.6.

4.7 Coordinate and momentum representations

Now let me show that in application to the orbital motion of a particle, the bra–ket formalism naturally reduces to the notions and postulates of wave mechanics, which were discussed in chapter 1. For that, we first have to modify some of the above formulas for the case of a basis with a continuous spectrum of eigenvalues—such as the particle's coordinate. In that case, it is more appropriate to remove discrete indices, such as the indices j, j', etc. broadly used above, replacing them with the corresponding eigenvalue[50]. For example, the key Eq. (4.68), defining the eigenkets and eigenvalues of an operator, may be conveniently rewritten in the form

$$\hat{A}|a_A\rangle = A|a_A\rangle. \qquad (4.219)$$

More substantially, all sums over such continuous eigenstate sets should be replaced with integrals. For example, for a full and orthonormal set of the eigenstate $|a_A\rangle$, the closure relation (4.44) should be replaced with

$$\int dA|a_A\rangle\langle a_A| = \hat{I}, \qquad (4.220)$$

---

[50] Actually, such notation was already used earlier—see, e.g. Eqs. (1.88), (2.20), etc.

where the integral should be taken over the whole interval of possible eigenvalues of the observable A.[51] Applying this relation to the ket-vector of an arbitrary state α, we get the following replacement of Eq. (4.37):

$$|\alpha\rangle \equiv \hat{I}|\alpha\rangle = \int dA |a_A\rangle\langle a_A|\alpha\rangle = \int dA \langle a_A|\alpha\rangle|a_A\rangle. \qquad (4.221)$$

For the particular case when $|\alpha\rangle = |a_{A'}\rangle$, this relation requires

$$\langle a_A|a_{A'}\rangle = \delta(A - A'); \qquad (4.222)$$

this formula replaces the orthonormality condition (4.38).

According to Eq. (4.221), in the continuous case the bracket $\langle a_A|\alpha\rangle$ continues to play the role of the probability amplitude, i.e. the complex of c-number whose modulus squared determines the state a_A's probability—see the last form of Eq. (4.120). However, for a continuous observable, the probability to find the system exactly in a particular state is infinitesimal; instead we should speak about the *probability density* $w(A) \propto |\langle a_A|\alpha\rangle|^2$ of finding the observable within a *small interval* $dA \ll A$ about a certain value A. The coefficient in that relation may be found by making the similar change from the summation to integration (without any additional coefficients) in the normalization condition (4.121):

$$\int dA \langle\alpha|a_A\rangle\langle a_A|\alpha\rangle = 1. \qquad (4.223)$$

Since the total probability of the system to be in *some* state should also equal to $\int w(A)dA$, this means that

$$w(A) = \langle\alpha|a_A\rangle\langle a_A|\alpha\rangle = |\langle\alpha|a_A\rangle|^2. \qquad (4.224)$$

Now let us see how we can calculate the expectation values of continuous observables, i.e. their ensemble averages. If we speak about the same observable A whose eigenstates are used as the continuous basis (or any compatible observable), everything is simple. Indeed, inserting Eq. (4.224) into the general statistical relation

$$\langle A \rangle = \int w(A)AdA, \qquad (4.225)$$

which is just the obvious continuous version of Eq. (1.37), we get

$$\langle A \rangle = \int \langle\alpha|a_A\rangle A \langle a_A|\alpha\rangle dA. \qquad (4.226)$$

Inserting a delta-function to represent this expression as a formally double integral,

$$\langle A \rangle = \int dA \int dA' \langle\alpha|a_A\rangle A \delta(A - A')\langle a_{A'}|\alpha\rangle, \qquad (4.227)$$

---

[51] The generalization to the case when the eigenvalue spectrum consists of both a continuum range (or ranges) plus some discrete values is straightforward, but leads to cumbersome formulas.

and using the continuous-spectrum version of Eq. (4.98),

$$\langle a_A | \hat{A} | a_{A'} \rangle = A \delta(A - A'), \tag{4.228}$$

we may write

$$\langle A \rangle = \int dA \int dA' \langle \alpha | a_A \rangle \langle a_A | \hat{A} | a_{A'} \rangle \langle a_{A'} | \alpha \rangle \equiv \langle \alpha | \hat{A} | \alpha \rangle, \tag{4.229}$$

so that Eq. (4.125) remains valid in the continuous-spectrum case without any changes.

The situation is a bit more complicated for the expectation values of an operator that does not commute with the basis-generating operator, because its matrix in that basis may not be diagonal. We will consider (and overcome) this technical difficulty very soon, but otherwise we are ready for the discussion of the relation between the bra–ket formalism and the wave mechanics. (For the notation simplicity I will discuss its 1D version; the generalization to the 2D and 3D cases is straightforward.)

Let us postulate the (intuitively almost evident) existence of a quantum state basis, whose ket-vectors will be called $|x\rangle$, corresponding to a certain, exactly defined value x of the particle's coordinate. Writing the following trivial identity:

$$x|x\rangle = x \,|x\rangle, \tag{4.230}$$

and comparing this relation with Eq. (4.219), we see that they do not contradict each other if we assume that x on the left-hand side of this relation is considered as the (Hermitian) operator $\hat{x}$ of particle's coordinate, whose action on a ket- (or bra-) vector is just its multiplication by the c-number x. (This looks like a proof, but is actually a separate, independent postulate, no matter how plausible.) This means that we may consider the set of vectors $|x\rangle$ as the eigenstates of the operator $\hat{x}$. Let me hope that the reader will excuse me if I do not pursue here a strict proof that this set is full and orthogonal[52], so that we may apply to them Eq. (4.222):

$$\langle x|x'\rangle = \delta(x - x'). \tag{4.231}$$

Using this basis is called the *coordinate representation*—the term which was already used in the end of section 1.1, but without explanation.

In the basis of the x-states, the inner product $\langle a_A | \alpha(t) \rangle$ becomes $\langle x | \alpha(t) \rangle$, and Eq. (4.223) takes the form

$$w(x, t) = \langle \alpha(t) | x \rangle \langle x | \alpha(t) \rangle \equiv \langle x | \alpha(t) \rangle^* \langle x | \alpha(t) \rangle. \tag{4.232}$$

Comparing this formula with the basic postulate (1.22) of wave mechanics, we see that they coincide if the wavefunction of a time-dependent state α is identified with that bracket[53]:

---

[52] Actually such a proof is rather involved mathematically, but physically this fact should be evident.

[53] I do not quite like expressions like $\langle x | \Psi \rangle$ used in some papers and even textbooks. Of course, one is free to replace α with any other letter (Ψ including) to denote a quantum state, but then it is better not to use the same letter to denote the wavefunction, i.e. an inner product of two state vectors, to avoid confusion.

$$\Psi_\alpha(x,\, t) \equiv \langle x | \alpha(t) \rangle. \tag{4.233}$$

This key formula provides the desired connection between the bra–ket formalism and the wave mechanics, and should not be too surprising for the (thoughtful :-) reader. Indeed, Eq. (4.45) shows that any inner product of the state vectors describing two states is a measure of their coincidence—just as the scalar product of two geometric vectors is; the orthonormality condition (4.38) is a particular manifestation of this fact. In this language, the particular value (4.233) of a wavefunction Ψ_α at point x and moment t characterizes 'how much of a particular coordinate x' does the state α contain at this particular instance. (Of course this informal language is too crude to describe the fact that $\Psi_\alpha(x,\, t)$ is generally a complex function, which has not only a modulus, but also a phase.)

Now let us rewrite the most important formulas of the bra–ket formalism in the wave mechanics notation. Inner-multiplying both parts of Eq. (4.219) by the ket-vector $\langle x |$, and then inserting into the left-hand side of the relation the identity operator in the form (4.222), we get

$$\int dx' \langle x | \hat{A} | x' \rangle \langle x' | a_A \rangle = A \langle x | a_A \rangle, \tag{4.234}$$

i.e. using the wavefunction's definition (4.233),

$$\int dx' \langle x | \hat{A} | x' \rangle \, \Psi_A(x') = A \Psi_A(x), \tag{4.235}$$

where, for the notation brevity, the time dependence of the wavefunction is just implied (with the capital Ψ serving as a reminder), and will be restored when needed.

For a general operator, we would have to stop here, because if it does not commute with the coordinate operator, its matrix in the x-basis is not diagonal, and the integral on the left-hand side of Eq. (4.235) cannot be worked out explicitly. However, virtually all quantum-mechanical operators discussed in this course[54] are (*space-*) *local*: they depend on only one spatial coordinate, say x. For such operators, the left-hand side of Eq. (4.235), for an arbitrary wavefunction, may be further transformed as

$$\int \langle x | \hat{A} | x' \rangle \Psi(x') dx' = \int \langle x | x' \rangle \hat{A} \Psi(x') dx'$$
$$\equiv \hat{A} \int \delta(x - x') \Psi(x') dx' = \hat{A} \Psi(x). \tag{4.236}$$

The first step in this transformation may appear as elementary as the last two, with the ket-vector $| x' \rangle$ swapped with the operator depending only on x; however, due to the delta-functional character of the bracket (4.231), this step is in fact an additional postulate, so that the second equality in Eq. (4.236) essentially *defines* the coordinate representation of the local operator, whose explicit form still needs to be determined.

---

[54] The only substantial exception is the statistical operator $\hat{w}(x,\, x')$, to be discussed in chapter 7.

Let us consider, for example, the 1D version of the Hamiltonian (1.41),

$$\hat{H} = \frac{\hat{p}_x^2}{2m} + U(\hat{x}), \qquad (4.237)$$

which was the basis of all our discussions in chapter 2. Its potential-energy part U (which may be time-dependent as well) commutes with the operator $\hat{x}$, i.e. its matrix in the x-basis has to be diagonal. For this operator, the transformation (4.236) is indeed trivial, and its coordinate representation is given merely by the c-number function $U(x)$.

The situation of the momentum operator $\hat{p}_x$ (and hence the kinetic energy $\hat{p}_x^2/2m$), not commuting with $\hat{x}$, is less evident. Let me show that its coordinate representation is given by the 1D version of Eq. (1.26), if we *postulate* that the commutation relation (2.14),

$$[\hat{x}, \hat{p}] = i\hbar\hat{I}, \qquad \text{i.e.} \quad \hat{x}\hat{p}_x - \hat{p}_x\hat{x} = i\hbar\hat{I}, \qquad (4.238)$$

is valid in *any* representation[55]. For that, let us consider the following matrix element, $\langle x | \hat{x}\hat{p}_x - \hat{p}_x\hat{x} | x' \rangle$. On one hand, we may use Eq. (4.238), and then Eq. (4.231), to write

$$\langle x|\hat{x}\hat{p}_x - \hat{p}_x\hat{x}|x'\rangle = \langle x| i\hbar\hat{I}|x'\rangle = i\hbar\langle x|x'\rangle = i\hbar\delta(x - x'). \qquad (4.239)$$

On the other hand, since $\hat{x}|x'\rangle = x'|x'\rangle$ and $\langle x|\hat{x} = \langle x|x$, we may represent the same matrix element as

$$\langle x|\hat{x}\hat{p}_x - \hat{p}_x\hat{x}|x'\rangle = \langle x|x\hat{p}_x - \hat{p}_x x'|x'\rangle = (x - x')\langle x| \hat{p}_x|x'\rangle. \qquad (4.240)$$

Comparing Eqs. (4.239) and (4.240), we get

$$\langle x|\hat{p}_x|x'\rangle = i\hbar\frac{\delta(x - x')}{x - x'}. \qquad (4.241)$$

As follows from the definition of the delta-function[56], all expressions involving it acquire final sense only at their integration, in our current case, at that described by Eq. (4.236). Plugging Eq. (4.241) into the left-hand side of that relation, we get

$$\int \langle x|\hat{p}_x|x'\rangle\Psi(x')dx' = i\hbar \int \frac{\delta(x - x')}{x - x'}\Psi(x')dx'. \qquad (4.242)$$

Since this integral is evidently contributed only by an infinitesimal vicinity of the point $x' = x$, we may calculate it by expanding the continuous wavefunction $\Psi(x')$ into the Taylor series in small $(x' - x)$, and keeping only two leading terms of the series, so that Eq. (4.242) is reduced to

---

[55] Another popular approach to the wave mechanics axiomatics is to *derive* Eq. (4.238) by *postulating* the form, $\hat{\mathcal{T}}_X = \exp\{-i\hat{p}_x X/\hbar\}$, of the operator that shifts any wavefunction by distance X along the axis x. In my approach, this expression with be *derived* when we need it (in section 5.5), while Eq. (4.238) is *postulated*.
[56] If necessary, please revisit section A.14.

$$\int \langle x|\hat{p}_x|x'\rangle \Psi(x', t)dx' = i\hbar \left[\Psi(x) \int \frac{\delta(x - x')}{x - x'} dx' \right.$$
$$\left. - \int \delta(x - x') \frac{\partial \Psi(x')}{\partial x'} \Big|_{x'=x} dx' \right].$$
(4.243)

Since the delta-function may be always understood as an even function of its argument, in our case of $(x - x')$, the first term on the right-hand side is proportional to an integral of an odd function in symmetric limits, and is equal to zero, and we get[57]

$$\int \langle x|\hat{p}_x|x'\rangle \Psi(x', t)dx' = -i\hbar \frac{\partial \Psi}{\partial x}.$$
(4.244)

Comparing this expression with the right-hand side of (4.236), we see that in the coordinate representation we indeed get the 1D version of Eq. (1.26), which was used so much in chapter 2,[58]

$$\hat{p}_x = -i\hbar \frac{\partial}{\partial x}.$$
(4.245)

It is straightforward to show (and is virtually evident) that the coordinate representation of any operator function $f(\hat{p}_x)$ is

$$f\left(-i\hbar \frac{\partial}{\partial x}\right).$$
(4.246)

In particular, this pertains to the kinetic energy operator in Eq. (4.237), so the coordinate representation of this Hamiltonian also takes the very familiar form:

$$\hat{H} = \frac{1}{2m}\left(-i\hbar \frac{\partial}{\partial x}\right)^2 + U(x, t) \equiv -\frac{\hbar^2}{2m}\frac{\partial^2}{\partial x^2} + U(x, t).$$
(4.247)

Now returning to the discussion of the general Eq. (4.235), and comparing its last form with that of Eq. (4.236), we see that for a local operator in the coordinate representation, the eigenproblem (4.219) takes the form

$$\hat{A}\Psi_A(x) = A\Psi_A(x),$$
(4.248)

even if the operator $\hat{A}$ does not commute with the operator $\hat{x}$. The most important case of this coordinate-representation form of the eigenproblem (4.68) is the familiar Eq. (1.60) for the eigenvalues E_n of the energy of a system with a time-independent Hamiltonian.

---

[57] One more useful expression of this type, which may be proved similarly, is $(\partial/\partial x)\delta(x - x') = \delta(x - x')\partial/\partial x'$.
[58] This means, in particular, that in the sense of Eq. (4.236), the operator of differentiation is local, despite the fact that its action on a function f may be interpreted as the limit of the fraction $\Delta f/\Delta x$, involving *two* points. (In some axiomatic systems, local operators are *defined* as arbitrary polynomials of functions and their derivatives.)

The operator locality also simplifies the expression for its expectation value. Indeed, plugging the completeness relation in the form (4.231) into the general Eq. (4.125) twice (written in the first case for x and in the second case for x'), we get

$$
\begin{aligned}
\langle A \rangle &= \int dx \int dx' \langle \alpha(t)|x\rangle\langle x\,|\hat{A}|x'\rangle\langle x'|\alpha(t)\rangle \\
&= \int dx \int dx'\, \Psi_\alpha^*(x,\,t)\langle x\,|\hat{A}|x'\rangle\Psi_\alpha(x',\,t).
\end{aligned}
\tag{4.249}
$$

Now, Eq. (4.236) reduces this result to just

$$
\begin{aligned}
\langle A \rangle &= \int dx \int dx' \Psi_\alpha^*(x,\,t)\hat{A}\Psi_\alpha(x,\,t)\delta(x-x') \\
&\equiv \int \Psi_\alpha^*(x,\,t)\hat{A}\Psi_\alpha(x,\,t)dx.
\end{aligned}
\tag{4.250}
$$

i.e. to Eq. (1.23), which had to be postulated in chapter 1.

Now let us discuss the time evolution of the wavefunction, in the Schrödinger picture. For that, we may use Eq. (4.233) to calculate the (partial) time derivative of the wavefunction of some state α:

$$
i\hbar\frac{\partial\Psi_\alpha}{\partial t} = i\hbar\frac{\partial}{\partial t}\langle x|\alpha(t)\rangle.
\tag{4.251}
$$

Since the coordinate operator $\hat{x}$ does not depend on time explicitly, its eigenstates x are stationary, and we can swap the time derivative and the time-independent bra-vector $\langle x|$. Now using the Schrödinger-picture Eq. (4.158), and then inserting the identity operator in the continuous form (4.220) of the closure relation, written for the coordinate eigenstates,

$$
\int dx'\,|x'\rangle\langle x'| = \hat{I},
\tag{4.252}
$$

we may continue to develop the right-hand side of Eq. (4.251) as

$$
\begin{aligned}
\langle x|i\hbar\frac{\partial}{\partial t}\,|\alpha(t)\rangle &= \langle x|\hat{H}|\alpha(t)\rangle = \int dx'\langle x|\hat{H}|x\rangle\langle x'|\alpha(t)\rangle \\
&= \int dx'\langle x|\hat{H}|x'\rangle\Psi_\alpha(x'),
\end{aligned}
\tag{4.253}
$$

If the Hamiltonian operator is local, we may apply Eq. (4.236) to the last expression, to get the familiar form (1.28) of the Schrödinger equation:

$$
i\hbar\frac{\partial\Psi_\alpha}{\partial t} = \hat{H}\Psi_\alpha,
\tag{4.254}
$$

in which the coordinate representation of the operator $\hat{H}$ is implied.

So, for the local operators that obey Eq. (4.236), we have been able to derive all the basic notions and postulates of the wave mechanics from the bra–ket formalism. Moreover, the formalism has allowed us to get the very useful equation (4.248) for an arbitrary local operator, which will be repeatedly used below. (In the first three chapters of this course, we have only used its particular case (1.60) for the Hamiltonian operator.)

Now let me deliver on my promise to develop a more balanced view at the monochromatic de Broglie waves (4.1), which would be more respectful to the evident $\mathbf{r} \leftrightarrow \mathbf{p}$ symmetry of the coordinate and momentum. Let us do this for the 1D case when the wave may be represented as

$$\psi_p(x) = a_p \exp\left\{ i\frac{px}{\hbar} \right\}, \qquad \text{for all } -\infty < x < +\infty. \qquad (4.255)$$

(For the sake of brevity, from this point to the end of the section, I am dropping the index x in the notation of the momentum—just as was done in chapter 2.)

Let us have a good look at this function. Since it satisfies Eq. (4.248) for the 1D momentum operator (4.245),

$$\hat{p}\psi_p = p\psi_p, \qquad (4.256)$$

ψ_p is an eigenfunction of that operator. But this means that we can also write Eq. (4.219) for the corresponding ket-vector:

$$\hat{p} |p\rangle = p |p\rangle, \qquad (4.257)$$

and according to Eq. (4.233), the wavefunction (4.255) may be represented as

$$\psi_p(x) = \langle x|p \rangle. \qquad (4.258)$$

This expression is quite remarkable in its $x \leftrightarrow p$ symmetry—which may be pursued further on. Before doing that, however, we have to discuss the normalization of such wavefunctions. Indeed, in this case, the probability density $w(x)$ (4.18) is constant, so that its integral

$$\int_{-\infty}^{+\infty} w(x)dx = \int_{-\infty}^{+\infty} \psi_p(x)\psi_p^*(x)dx \qquad (4.259)$$

diverges if $a_p \neq 0$. Earlier in the course, we discussed two ways to avoid this divergence. One is to use a very large but finite integration volume—see Eq. (1.31). Another way is to work with wave packets of the type (2.20), possibly of a very large length and hence a very narrow spread of the momentum values. Then the integral (4.54) may be required to equal 1 without any conceptual problem.

However, both these methods, convenient for the solution of many particular problems, violate the $x \leftrightarrow p$ symmetry, and hence are inconvenient for our current conceptual discussion. Instead, let us continue to identify the eigenvectors $\langle p|$ and $|p\rangle$ of the momentum with the bra- and ket-vectors $\langle a_A|$ and $|a_A\rangle$ of the general theory, developed in the beginning of this section. Then the normalization condition (4.222) becomes

$$\langle p|p'\rangle = \delta(p - p'). \qquad (4.260)$$

Inserting the identity operator in the form (4.252) (with the integration variable x' replaced by x) into the left-hand side of this equation, and using Eq. (4.258), we can translate this normalization rule to the wavefunction language:

$$\int dx \langle p|x\rangle\langle x|p'\rangle \equiv \int dx \psi_p^*(x)\psi_{p'}(x) = \delta(p - p').$$ (4.261)

For the wavefunction (4.255), this requirement turns into the following condition:

$$a_p^* a_{p'} \int_{-\infty}^{+\infty} \exp\left\{i\frac{(p - p')x}{\hbar}\right\}dx \equiv |a_p|^2\, 2\pi\hbar\delta(p - p') = \delta(p - p'),$$ (4.262)

so that, finally, $a_p = e^{i\phi}/(2\pi\hbar)^{1/2}$, where ϕ is an arbitrary (real) phase, and Eq. (4.255) becomes[59]

$$\psi_p(x) = \langle x|p\rangle = \frac{1}{(2\pi\hbar)^{1/2}}\exp\left\{i\left(\frac{px}{\hbar} + \phi\right)\right\},$$ (4.263)

Now let us represent an arbitrary wavefunction $\psi(x)$ as a wave packet of the type (2.20), based on the wavefunctions (4.263), taking $\phi = 0$ for the notation brevity, because the phase factor may be incorporated into the (generally, complex) envelope function $\varphi(p)$:

$$\psi(x) = \frac{1}{(2\pi\hbar)^{1/2}}\int \varphi(p)\exp\left\{i\frac{px}{\hbar}\right\}dp.$$ (4.264)

From the mathematical point of view, this is just a 1D Fourier spatial transform, and its reciprocal is

$$\varphi(p) \equiv \frac{1}{(2\pi\hbar)^{1/2}}\int \psi(x)\exp\left\{-i\frac{px}{\hbar}\right\}dx.$$ (4.265)

These expressions are completely symmetric, and represent the same wave packet; this is why the functions $\psi(x)$ and $\varphi(p)$ are frequently called the *reciprocal representations* of a quantum state of the particle—respectively, its *coordinate* (x-) and *momentum* (p-) *representations*. Using Eq. (4.258), and Eq. (4.263) with $\phi = 0$, they may be recast into simpler forms,

$$\psi(x) = \int \varphi(p)\langle x|p\rangle dp, \qquad \varphi(p) = \int \psi(x)\langle p|x\rangle dx,$$ (4.266)

in which the scalar products satisfy the basic postulate (4.14) of the bra–ket formalism:

$$\langle p|x\rangle = \frac{1}{(2\pi\hbar)^{1/2}}\exp\left\{-i\frac{px}{\hbar}\right\} = \langle x|p\rangle^*.$$ (4.267)

---

[59] Repeating this calculation for each Cartesian component of a plane monochromatic wave of arbitrary dimensionality d, we get $\psi_\mathbf{p} = (2\pi\hbar)^{-d/2}\exp\{i(\mathbf{p}\cdot\mathbf{r}/\hbar + \varphi)\}$.

Next, we already know that in the x-representation, i.e. in the usual wave mechanics, the coordinate operator $\hat{x}$ is reduced to the multiplication by x, and the momentum operator is proportional to the partial derivative over the coordinate:

$$\hat{x}|_{\text{in } x} = x, \qquad \hat{p}|_{\text{in } x} = -i\hbar\frac{\partial}{\partial x}. \qquad (4.268)$$

It is natural to guess that in the p-representation, the expressions for operators would be reciprocal:

$$\hat{x}|_{\text{in } p} = i\hbar\frac{\partial}{\partial p}, \qquad \hat{p}|_{\text{in } p} = p, \qquad (4.269)$$

with the difference of one sign, which is due to the opposite signs of the Fourier exponents in Eqs. (4.264) and (4.265). The proof of Eqs. (4.269) is straightforward; for example, acting by the momentum operator on the arbitrary wavefunction (4.264), we get

$$\begin{aligned}
\hat{p}\psi(x) &= -i\hbar\frac{\partial}{\partial x}\psi(x) \\
&= \frac{1}{(2\pi\hbar)^{1/2}}\int \varphi(p)\left(-i\hbar\frac{\partial}{\partial x}\exp\left\{i\frac{px}{\hbar}\right\}\right)dp \qquad (4.270) \\
&= \frac{1}{(2\pi\hbar)^{1/2}}\int p\varphi(p)\exp\left\{i\frac{px}{\hbar}\right\}dp,
\end{aligned}$$

and similarly for the operator $\hat{x}$ acting on the function $\varphi(p)$. Comparing the final expression with the initial Eq. (4.264), we see that the action of the operators (4.268) on the wavefunction ψ (i.e. the state's x-representation) gives the same results as the action of the operators (4.269) on the function φ (i.e. its p-representation).

It is illuminating to have one more, different look at this coordinate-momentum duality. For that, notice that according to Eqs. (4.82)–(4.84), we may consider the bracket $\langle x|p \rangle$ as an element of the (infinite-size) matrix U_{xp} of the unitary transform from the x-basis to the p-basis. Let us use this fact to derive the general operator transform rule that would be a continuous version of Eq. (4.92). Say, we want to calculate the general matrix element of some operator, known in the x-representation, in the p-representation:

$$\langle p|\hat{A}|p' \rangle. \qquad (4.271)$$

Inserting two identity operators (4.252) into this bracket, and then using Eq. (4.258) and its complex conjugate, and also Eq. (4.236) (again, valid only for space-local operators!), we get

$$\langle p|\hat{A}|p'\rangle = \int dx \int dx' \langle p|x\rangle\langle x|\hat{A}|x'\rangle\langle x'|p'\rangle$$

$$= \int dx \int dx' \psi_p^*(x)\langle x|\hat{A}|x'\rangle\psi_{p'}(x')$$

$$= \frac{1}{2\pi\hbar}\int dx \int dx' \exp\left\{-i\frac{px}{\hbar}\right\}\delta(x-x')\hat{A}\exp\left\{i\frac{p'x'}{\hbar}\right\} \quad (4.272)$$

$$= \frac{1}{2\pi\hbar}\int dx \exp\left\{-i\frac{px}{\hbar}\right\}\hat{A}\exp\left\{i\frac{p'x}{\hbar}\right\}.$$

As a sanity check, for the momentum operator itself, this relation yields:

$$\langle p|\hat{p}|p'\rangle = \frac{1}{2\pi\hbar}\int dx \exp\left\{-i\frac{px}{\hbar}\right\}\left(-i\hbar\frac{\partial}{\partial x}\right)\exp\left\{i\frac{p'x}{\hbar}\right\}$$

$$= \frac{p'}{2\pi\hbar}\int_{-\infty}^{+\infty}\exp\left\{i\frac{(p'-p)x}{\hbar}\right\}dx = p'\delta(p'-p). \quad (4.273)$$

Due to Eq. (4.257), this result is equivalent to the second of Eqs. (4.269).

From a thoughtful reader, I anticipate the following natural question: why is the momentum representation used much less frequently than the coordinate representation—i.e. the wave mechanics? The answer is purely practical: besides the special case of the 1D harmonic oscillator (to be revisited in section 5.4), in most systems the orbital-motion Hamiltonian (4.237) is not $x \leftrightarrow p$ symmetric, with the potential energy $U(\mathbf{r})$ typically being a more complex function than the kinetic energy $p^2/2m$. Because of that, it is easier to analyze such systems treating this potential energy operator just as a c-number multiplier, as it is in the coordinate representation—as was done in chapters 1–3.

The most significant exception is the motion in a periodic potential in the presence of an external force $\mathbf{F}$. As was discussed in sections 2.7 and 3.4, in such periodic systems the eigenenergies $E_n(\mathbf{q})$, playing the role of the effective kinetic energy of the particle, may be rather involved functions of its quasi-momentum $\hbar\mathbf{q}$, while its effective potential energy $U_{ef} = -\mathbf{F}(t)\cdot\mathbf{r}$, due to the additional force $\mathbf{F}(t)$, is a simple function of coordinates. This is why detailed analyses of the quantum effects briefly discussed in section 2.8 (the Bloch oscillations, etc) and also such statistical phenomena as drift, diffusion, etc[60], in solid state theory are typically based on the momentum (or rather the quasi-momentum) representation.

4.8 Problems

Problem 4.1. Prove that if $\hat{A}$ and $\hat{B}$ are linear operators, and C is a c-number, then:

(i) $(\hat{A}^\dagger)^\dagger = \hat{A}$;
(ii) $(C\hat{A})^\dagger = C^*\hat{A}^\dagger$;

---

[60] In this series, a brief discussion of these effects may be found in *Part SM* chapter 6.

(iii) $(\hat{A}\hat{B})^\dagger = \hat{B}^\dagger\hat{A}^\dagger$;

(iv) the operators $\hat{A}\hat{A}^\dagger$ and $\hat{A}^\dagger\hat{A}$ are Hermitian.

Problem 4.2. Prove that for any linear operators $\hat{A}$, $\hat{B}$, $\hat{C}$, and $\hat{D}$,

$$[\hat{A}\hat{B}, \hat{C}\hat{D}] = \hat{A}\{\hat{B}, \hat{C}\}\hat{D} - \hat{A}\hat{C}\{\hat{B}, \hat{D}\} + \{\hat{A}, \hat{C}\}\hat{D}\hat{B} - \hat{C}\{\hat{A}, \hat{D}\}\hat{B}.$$

Problem 4.3. Calculate all possible binary products $\sigma_j\sigma_{j'}$ (for $j, j' = x, y, z$) of the Pauli matrices, defined by Eqs. (4.105), and their commutators and anticommutators (defined similarly to those of the corresponding operators). Summarize the results, using the Kronecker delta and Levi-Civita permutation symbols[61].

Problem 4.4. Calculate the following expressions

(i) $(\mathbf{c} \cdot \boldsymbol{\sigma})^n$, and then
(ii) $(b\mathbf{I} + \mathbf{c} \cdot \boldsymbol{\sigma})^n$,

for the scalar product $\mathbf{c} \cdot \boldsymbol{\sigma}$ of the Pauli matrix vector $\boldsymbol{\sigma} \equiv \mathbf{n}_x\sigma_x + \mathbf{n}_y\sigma_y + \mathbf{n}_z\sigma_z$ by an arbitrary c-number vector $\mathbf{c}$, where $n \geqslant 0$ is an integer, and b is an arbitrary scalar c-number.

Hint: For task (ii), you may like to use the binomial theorem[62], and then transform the result in a way enabling you to use the same theorem backwards.

Problem 4.5. Use the solution of the previous problem to derive Eqs. (2.191) for the transparency $\mathscr{T}$ of a system of N similar, equidistant, delta-functional potential barriers.

Problem 4.6. Use the solution of problem 4.4(i) to spell out the following matrix: $\exp\{i\theta\mathbf{n} \cdot \boldsymbol{\sigma}\}$, where $\boldsymbol{\sigma}$ is the vector of Pauli matrices, $\mathbf{n}$ is a c-number vector of unit length, and θ is a c-number scalar.

Problem 4.7. Use the solution of problem 4.4(ii) to calculate $\exp\{A\}$, where A is an arbitrary 2×2 matrix.

Problem 4.8. Express elements of the matrix $B = \exp\{A\}$ explicitly via those of the 2×2 matrix A. Spell out your result for the following matrices:

$$A = \begin{pmatrix} a & a \\ a & a \end{pmatrix}, \qquad A' = \begin{pmatrix} i\varphi & i\varphi \\ i\varphi & i\varphi \end{pmatrix},$$

with real a and φ.

Problem 4.9. Prove that for arbitrary square matrices A and B,

$$\text{Tr}\,(AB) = \text{Tr}\,(BA).$$

---

[61] See, e.g. Eqs. (A.82) and (A.83).
[62] See, e.g. Eq. (A.12).

Is each diagonal element $(AB)_{jj}$ necessarily equal to $(BA)_{jj}$?

Problem 4.10. Calculate the trace of the following 2×2 matrix:

$$A \equiv (\mathbf{a} \cdot \boldsymbol{\sigma})(\mathbf{b} \cdot \boldsymbol{\sigma})(\mathbf{c} \cdot \boldsymbol{\sigma}),$$

where $\boldsymbol{\sigma}$ is the Pauli matrix vector, while $\mathbf{a}$, $\mathbf{b}$, and $\mathbf{c}$ are arbitrary c-number vectors.

Problem 4.11. Prove that the matrix trace of an arbitrary operator does not change at its arbitrary unitary transformation.

Problem 4.12. Prove that for any two full and orthonormal bases $\{u\}$, $\{v\}$ of the same Hilbert space,

$$\text{Tr}\left(|u_j\rangle\langle v_{j'}|\right) = \langle v_{j'}|u_j\rangle.$$

Problem 4.13. Is the 1D scattering matrix S, defined by Eq. (2.124), unitary? What about the 1D transfer matrix T defined by Eq. (2.125)?

Problem 4.14. Calculate the trace of the following matrix:

$$\exp\{i\mathbf{a} \cdot \boldsymbol{\sigma}\} \exp\{i\mathbf{b} \cdot \boldsymbol{\sigma}\},$$

where $\boldsymbol{\sigma}$ is the Pauli matrix vector, while $\mathbf{a}$ and $\mathbf{b}$ are c-number geometric vectors.

Problem 4.15. Prove the following vector-operator identity:

$$(\boldsymbol{\sigma} \cdot \hat{\mathbf{r}})(\boldsymbol{\sigma} \cdot \hat{\mathbf{p}}) = I\,\hat{\mathbf{r}} \cdot \hat{\mathbf{p}} + i\boldsymbol{\sigma} \cdot (\hat{\mathbf{r}} \times \hat{\mathbf{p}}),$$

where I is the 2×2 identity matrix.

Hint: Take into account that the vector operators $\hat{\mathbf{r}}$ and $\hat{\mathbf{p}}$ are defined in the orbital-motion Hilbert space, independent of that of the Pauli vector-matrix $\boldsymbol{\sigma}$, and hence commute with it—even though they do not commute with each other.

Problem 4.16. Let A_j be eigenvalues of some operator $\hat{A}$. Express the following two sums,

$$\Sigma_1 \equiv \sum_j A_j, \qquad \Sigma_2 \equiv \sum_j A_j^2,$$

via the matrix elements $A_{jj'}$ of this operator in an arbitrary basis.

Problem 4.17. Calculate $\langle \sigma_z \rangle$ of a spin-½ in a quantum state with the following ket-vector:

$$|\alpha\rangle = \text{const} \times (|\uparrow\rangle + |\downarrow\rangle + |\rightarrow\rangle + |\leftarrow\rangle),$$

where $(\uparrow, \downarrow)$ and $(\rightarrow, \leftarrow)$ are the eigenstates of the Pauli matrices σ_z and σ_x, respectively.

Hint: Double-check whether the solution you are giving is general.

Problem 4.18. A spin-½ is fully polarized in the positive z-direction. Calculate the probabilities of the alternative outcomes of a perfect Stern–Gerlach experiment with the magnetic field oriented in an arbitrary different direction, performed on a particle in this spin state.

Problem 4.19. In a certain basis, the Hamiltonian of a two-level system is described by the matrix

$$H = \begin{pmatrix} E_1 & 0 \\ 0 & E_2 \end{pmatrix}, \qquad \text{with } E_1 \neq E_2,$$

while the operator of some observable A of this system, by the matrix

$$A = \begin{pmatrix} 1 & 1 \\ 1 & 1 \end{pmatrix}.$$

For the system's state with the energy definitely equal to E_1, find the possible results of measurements of the observable A and the probabilities of the corresponding measurement outcomes.

Problem 4.20. Certain states $u_{1,2,3}$ form an orthonormal basis of a system with the following Hamiltonian

$$\hat{H} = -\delta(|u_1\rangle\langle u_2| + |u_2\rangle\langle u_3| + |u_3\rangle\langle u_1|) + \text{h.c.},$$

where δ is a real constant, and h.c. means the Hermitian conjugate of the previous expression. Calculate its stationary states and energy levels. Can you relate this system with any other(s) discussed earlier in the course?

Problem 4.21. Guided by Eq. (2.203) of the lecture notes, and by the solutions of problems 3.11 and 4.20, suggest a Hamiltonian describing particle's dynamics in an infinite 1D chain of similar potential wells in the tight-binding approximation, in the bra–ket formalism. Verify that its eigenstates and eigenvalues correspond to those discussed in section 2.7.

Problem 4.22. Calculate the eigenvectors and eigenvalues of the following matrices:

$$A = \begin{pmatrix} 0 & 1 & 0 \\ 1 & 0 & 1 \\ 0 & 1 & 0 \end{pmatrix}, \qquad B = \begin{pmatrix} 0 & 0 & 0 & 1 \\ 0 & 0 & 1 & 0 \\ 0 & 1 & 0 & 0 \\ 1 & 0 & 0 & 0 \end{pmatrix}.$$

Problem 4.23. A certain state γ is an eigenstate of each of two operators $\hat{A}$ and $\hat{B}$. What can be said about the corresponding eigenvalues a and b, if the operators anticommute?

Problem 4.24. Derive a differential equation for the time evolution of the expectation value of an observable, using both the Schrödinger picture and the Heisenberg picture of quantum dynamics.

Problem 4.25. At $t = 0$, a spin-½ system whose interaction with an external field is described by the Hamiltonian

$$\hat{H} = \mathbf{c} \cdot \hat{\boldsymbol{\sigma}} \equiv c_x\hat{\sigma}_x + c_y\hat{\sigma}_y + c_z\hat{\sigma}_z,$$

(where $c_{x,y,z}$ are real and time-independent c-numbers, and $\hat{\sigma}_{x,\,y,\,z}$ are the Pauli operators), was in the state ↑, one of the two eigenstates of the operator $\hat{\sigma}_z$. In the Schrödinger picture, calculate the time evolution of:

  (i) the ket-vector $|\alpha\rangle$ of the system (in any time-independent basis you like),
  (ii) the probabilities to find the system in the states ↑ and ↓, and
  (iii) the expectation values of all three Cartesian components ($\hat{S}_x$, etc) of the spin vector operator $\hat{\mathbf{S}} = (\hbar/2)\hat{\boldsymbol{\sigma}}$.

Analyze and interpret the results for the particular case $c_y = c_z = 0$.

Hint: Think about the best basis to use for the solution.

Problem 4.26. For the same system as in the previous problem, use the Heisenberg picture to calculate the time evolution of:

  (i) all three Cartesian components of the spin operator $\hat{\mathbf{S}}_H(t)$, and
  (ii) the expectation values of the spin components.

Compare the latter results with those of the previous problem.

Problem 4.27. For the same system as in two last problems, calculate the matrix elements of the operator $\hat{\sigma}_z$ in the basis of the stationary states of the system.

Problem 4.28. In the Schrödinger picture of quantum dynamics, certain three operators satisfy the following commutation relation:

$$[\hat{A}, \hat{B}] = \hat{C}.$$

What is their relation in the Heisenberg picture, at a certain time instant t?

Problem 4.29. Prove the Bloch theorem, given by either Eq. (3.107) or Eq. (3.108).

Hint: Consider the *translation operator* $\hat{\mathcal{T}}_\mathbf{R}$, defined by the following result of its action on an arbitrary function $f(\mathbf{r})$:

$$\hat{\mathcal{T}}_\mathbf{R} f(\mathbf{r}) = f(\mathbf{r} + \mathbf{R}),$$

for the case when $\mathbf{R}$ is an arbitrary vector of the Bravais lattice (3.106). In particular, analyze the commutation properties of this operator, and apply them to an eigenfunction $\psi(\mathbf{r})$ of the stationary Schrödinger equation for a particle moving in the 3D periodic potential described by Eq. (3.105).

Problem 4.30. A constant force F is applied to an (otherwise free) 1D particle of mass m. Calculate the eigenfunctions of the problem in:

(i) the coordinate representation, and
(ii) the momentum representation.

Discuss the relation between the results.

Problem 4.31. Use the momentum representation to re-solve the problem discussed in the beginning of section 2.6, i.e. calculate the eigenenergy of a 1D particle of mass m, localized in a very short potential well, of the 'area' $\mathcal{w}$.

Problem 4.32. In the momentum representation, an operator of the 1D orbital motion equals p^{-1}. Find its coordinate representation.

Problem 4.33.* For a particle moving in a 3D periodic potential, develop the bra–ket formalism for the **q**-representation, in which a complex amplitude similar to a_q in Eq. (2.234) (but generalized to 3D and all energy bands) plays the role of the wavefunction. In particular, calculate the operators **r** and **v** in this representation, and use the result to prove Eq. (2.237) for the 1D case in the low-field limit.

Problem 4.34. A uniform, time-independent magnetic field $\mathcal{B}$ is induced in one semi-space, while the other semi-space is field-free, with a sharp, plane boundary between these two regions. A monochromatic beam of electrically-neutral spin-½ particles with a nonvanishing gyromagnetic ratio γ,[63] in a certain spin state, and with a kinetic energy E, is incident on this boundary, from the field-free side, under angle θ—see figure below. Calculate the coefficient of particle reflection from the boundary.

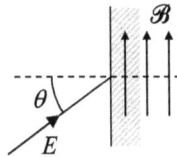

---

[63] The fact that γ may be different from zero even for electrically-neutral particles, such as neutrons, is explained by the Standard Model of the elementary particles, in which a neutron 'consists' (in a broad sense of the word) of three electrically-charged quarks.

IOP Publishing

Quantum Mechanics
Lecture notes
Konstantin K Likharev

Chapter 5

Some exactly solvable problems

The objective of this chapter is to describe several relatively simple but very important applications of the bra–ket formalism, notably including a few core wave-mechanics problems we have already started to discuss in chapters 2 and 3.

5.1 Two-level systems

The discussion of the bra–ket formalism in the previous chapter was peppered with numerous illustrations of its main concepts on the example of 'spin-½'—the system with the smallest non-trivial, two-dimensional Hilbert space, in which the bra- and ket-vectors of an arbitrary quantum state α may be represented as a linear superposition of just two basis vectors, for example

$$|\alpha\rangle = \alpha_\uparrow |\uparrow\rangle + \alpha_\downarrow |\downarrow\rangle, \tag{5.1}$$

where the states $\uparrow$ and $\downarrow$ are defined as the eigenstates of the Pauli matrix σ_z—see Eq. (4.105). This particular system is described by the Pauli Hamiltonian (4.163), and the states $\uparrow$ and $\downarrow$ are the Hamiltonian's stationary 'spin-up' and 'spin-down' states, with the corresponding two energy levels (4.167) split by the applied magnetic field. However, as was mentioned above, and will be proved in the next chapter, an approximate but adequate quantum description of some other important systems may also be given in such a two-dimensional Hilbert space. For example, as was mentioned in section 2.6, weak coupling of two space-localized orbital states is sufficient for an approximate description of quantum oscillations of a spin-free particle between two potential wells. A similar coupling of two traveling waves explains the energy band splitting in the weak-potential approximation of the band theory—section 2.7.

As will be shown in chapter 6, in systems with time-independent Hamiltonians, such a situation almost unavoidably appears each time when two energy levels are much closer to each other than to other levels. Moreover, as will be shown in section

doi:10.1088/2053-2563/aaf3a3ch5

6.5, a similar truncated description is adequate in cases when two levels E_n and $E_{n'}$ of an unperturbed system are not close to each other, but the corresponding states become coupled by an applied ac field of a frequency ω very close to the difference $(E_n - E_{n'})/\hbar$. Such *two-level systems* (alternatively called 'spin-½-like' systems) are nowadays the focus of additional attention in the view of prospects of their use for quantum information processing and encryption[1]. For this reason let me spend a bit more time reviewing the main properties of a two-level system.

First, the most general form of the Hamiltonian (and of any operator) of the two-level system, in an arbitrary basis, is given by a 2 × 2 matrix

$$\mathrm{H} = \begin{pmatrix} H_{11} & H_{12} \\ H_{21} & H_{22} \end{pmatrix}. \tag{5.2}$$

According to the discussion in sections 4.3–4.5, since the Hamiltonian operator has to be Hermitian, the diagonal elements of the matrix H have to be real, and its off-diagonal elements be complex conjugates of each other: $H_{21} = H_{12}^{*}$. As a result, we may represent H via a combination of the identity matrix and the Pauli matrices—just as in Eq. (4.106):

$$\mathrm{H} = \begin{pmatrix} b + c_z & c_x - ic_y \\ c_x + ic_y & b - c_z \end{pmatrix} \equiv b\mathrm{I} + c_x\sigma_x + c_y\sigma_y + c_z\sigma_z \equiv b\mathrm{I} + \mathbf{c} \cdot \boldsymbol{\sigma}, \tag{5.3}$$

where the scalar b and the Cartesian components of the vector $\mathbf{c}$ are real c-number coefficients:

$$b = \frac{H_{11} + H_{22}}{2}, \qquad c_x = \frac{H_{12} + H_{21}}{2} \equiv \mathrm{Re}\, H_{21},$$

$$c_y = \frac{H_{21} - H_{12}}{2i} \equiv \mathrm{Im} H_{21}, \qquad c_z = \frac{H_{11} - H_{22}}{2}. \tag{5.4}$$

If the Hamiltonian does not depend on time, the corresponding characteristic equation (4.103) for the system's energy levels $E_\pm$,

$$\begin{vmatrix} b + c_z - E & c_x - ic_y \\ c_x + ic_y & b - c_z - E \end{vmatrix} = 0, \tag{5.5}$$

is a simple quadratic equation, with the solutions

$$E_\pm = b \pm c \equiv b \pm \left(c_x^2 + c_y^2 + c_z^2\right)^{1/2}$$

$$\equiv \frac{H_{11} + H_{22}}{2} \pm \left[\left(\frac{H_{11} - H_{22}}{2}\right)^2 + |H_{21}|^2\right]^{1/2}. \tag{5.6}$$

---

[1] In the last context, to be discussed in section 8.5, the two-level systems are usually called *qubits*.

The parameter $b = (H_{11} + H_{22})/2$ evidently gives the average energy $E^{(0)}$ of the system, which does not contribute to the level splitting

$$\Delta E \equiv E_+ - E_- = 2c \equiv 2\left(c_x^2 + c_y^2 + c_z^2\right)^{1/2} \equiv [(H_{11} - H_{22})^2 + 4\,|\,H_{21}\,|^2]^{1/2}. \quad (5.7)$$

So, the splitting is a hyperbolic function of the coefficient $c_z = (H_{11} - H_{22})/2$. A plot of this function is the famous level-anticrossing diagram (figure 5.1), which has already been discussed in section 2.7 in a particular context of the weak-potential limit of the 1D band theory.

The physics of the diagram becomes especially clear if the two states of the basis used to write the matrix (5.2) may be interpreted as the stationary states of two potentially independent subsystems, with the energies, respectively, H_{11} and H_{22}. (For example, in the case of two weakly coupled potential wells, discussed in section 2.6, these are the ground-state energies of two distant wells.) Then the off-diagonal elements H_{12} and $H_{21} = H_{12}^*$ describe the subsystem coupling, and the level anticrossing diagram shows how the energies of the coupled system depend (at fixed coupling) on the difference of the subsystems. As was already discussed in section 2.7, the most striking feature of the diagram is that any non-zero coupling changes the topology of the energies of the coupled states (shown in figure 5.1 with dashed lines), creating a gap between its two branches, with the minimum width $\Delta E_{min} = 2|H_{21}| \equiv 2|H_{12}|$, which is reached at $c_z = 0$, i.e. when the subsystems become similar, with $H_{11} = H_{22}$.

As follows from the discussions of particular two-level systems in sections 2.6 and 4.6, their time dynamics has one general feature—quantum oscillations. Namely, if we put any two-level system into an initial state different from one of its eigenstates ±, and then let it evolve on its own, the probability of its finding the system in any of the 'partial' states exhibits oscillations with the frequency

$$\Omega = \frac{\Delta E}{\hbar} \equiv \frac{E_+ - E_-}{\hbar}, \quad (5.8)$$

lowest at the exact subsystem symmetry ($H_{11} = H_{22}$), when it is proportional to the coupling strength: $\Omega_{min} = 2|H_{21}|/\hbar$. In the case discussed in section 2.6, these are the oscillations of a particle between the two coupled potential wells (or rather the

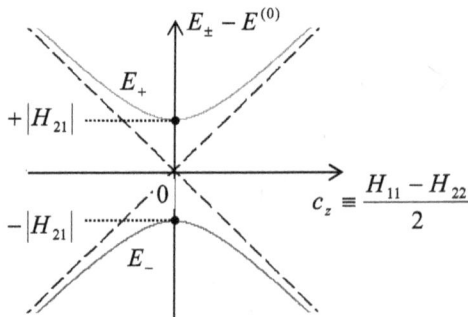

Figure 5.1. The level-anticrossing diagram for an arbitrary two-level system.

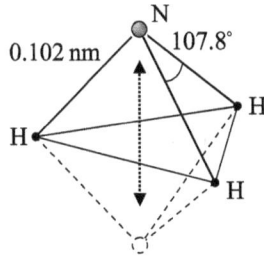

Figure 5.2. An ammonia molecule and its inversion.

probabilities to find it in either well)– see, e.g. Eqs. (2.181). On the other hand, for a spin-½ in an external magnetic field, these oscillations take the form of the spin precession in the plane normal to the field, with the periodic oscillations of its Cartesian components (or rather their expectation values)—see, e.g. Eqs. (4.173) and (4.174).

Some other examples of the quantum oscillations in two-level systems may be rather unexpected; for example, the ammonium molecule NH_3 (figure 5.2) has two symmetric states that differ by the inversion of the nitrogen atom relative to the plane of the three hydrogen atoms, and are coupled due to quantum-mechanical tunneling of the nitrogen atom through the plane of the hydrogen atoms[2]. Since for this particular molecule, in the absence of external fields, the level splitting ΔE corresponds to an experimentally convenient frequency $\Omega/2\pi \approx 24$ GHz, it played an important historic role in the initial development of the atomic frequency standards and microwave quantum generators (*masers*) in the early 1950s,[3] which paved the way toward the development of the laser technology.

Let us now return to the convenient spin-½ notation, ↑ and ↓, of the basis states to discuss a very convenient geometric representation of an arbitrary state α of (any!) two-level system. As Eq. (5.1) shows, it is completely described by two complex coefficients (*c*-numbers)—say, $\alpha_\uparrow$ and $\alpha_\downarrow$. If the vectors of the basis states ↑ and ↓ are normalized, then these coefficients must obey the following restriction:

$$W_\Sigma = \langle \alpha | \alpha \rangle = \left(\langle \uparrow | \, \alpha_\uparrow^* + \langle \downarrow | \, \alpha_\downarrow^* \right) \left(\alpha_\uparrow \, | \uparrow \rangle + \alpha_\downarrow \, | \downarrow \rangle \right)$$
$$= \alpha_\uparrow^* \alpha_\uparrow + \alpha_\downarrow^* \alpha_\downarrow = |\alpha_\uparrow|^2 + |\alpha_\downarrow|^2 = 1. \tag{5.9}$$

This requirement is automatically satisfied if we take the moduli of $\alpha_\uparrow$ and $\alpha_\downarrow$ equal to the sine and cosine of the same (real) angle. Thus we may write, for example,

$$\alpha_\uparrow = \cos\frac{\theta}{2} e^{i\gamma}, \qquad \alpha_\downarrow = \sin\frac{\theta}{2} e^{i(\gamma+\varphi)}. \tag{5.10}$$

---

[2] Since the hydrogen atoms are much lighter, it is fairer to speak about the tunneling of their triangle around the (nearly immobile) nitrogen atom.

[3] In particular, these molecules were used in the demonstration of the first maser by C Townes' group in 1954.

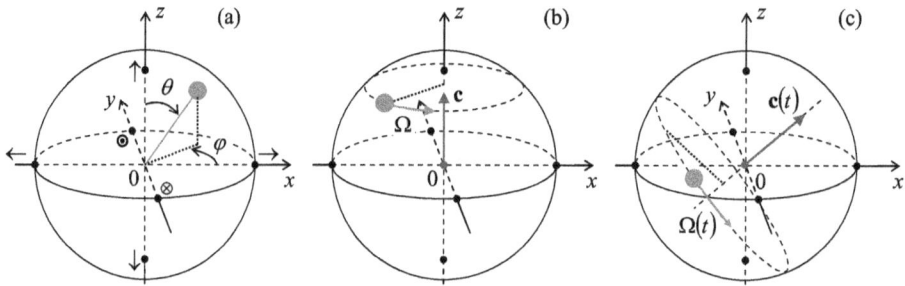

Figure 5.3. The Bloch sphere: (a) notation and the basic state representations, and (b, c) the two-level system's evolution: (b) in a constant 'field' **c** directed along the axis z, and (c) in an arbitrary field.

Moreover, according to the general Eq. (4.125), if we deal with just one system[4], the common phase factor $\exp\{i\gamma\}$ drops out of the calculation of any expectation value, so that we may take $\gamma = 0$, and Eq. (5.10) is reduced to

$$\alpha_\uparrow = \cos\frac{\theta}{2}, \quad \alpha_\downarrow = \sin\frac{\theta}{2}e^{i\varphi}. \tag{5.11}$$

The reason why the argument of these sine and cosine functions is usually taken in the form $\theta/2$, becomes clear from figure 5.3a: Eq. (5.11) conveniently maps each state α of a two-level system on a certain *representation point* of a unit-radius *Bloch sphere*[5], with the polar angle θ and the azimuthal angle φ.

In particular, the basis state $\uparrow$, described by Eq. (5.1) with $\alpha_\uparrow = 1$ and $\alpha_\downarrow = 0$, corresponds to the North Pole of the sphere ($\theta = 0$), while the opposite state $\downarrow$, with $\alpha_\uparrow = 0$ and $\alpha_\downarrow = 1$, to its South Pole ($\theta = \pi$). Similarly, the eigenstates $\rightarrow$ and $\leftarrow$ of the matrix σ_x, described by Eqs. (4.122), i.e. having $\alpha_\uparrow = 1/\sqrt{2}$ and $\alpha_\downarrow = \pm 1/\sqrt{2}$, correspond to the equator ($\theta = \pi/2$) points with, respectively, $\varphi = 0$ and $\varphi = \pi$. Two more special points (denoted in figure 5.3a as $\odot$ and $\otimes$) are also located on the sphere's equator, at $\theta = \pi/2$ and $\varphi = \pm\pi/2$; it is easy to check that they correspond to the eigenstates of the matrix σ_y (in the same z-basis). In order to understand why such mutually perpendicular locations of these three special point pairs on the Bloch sphere are not occasional, let us plug Eqs. (5.11) into Eqs. (4.131)–(4.133) for the expectation values of the spin-½ components. In terms of the Pauli vector $\boldsymbol{\sigma} \equiv \mathbf{S}/(\hbar/2)$,

$$\langle\sigma_x\rangle = \sin\theta\cos\varphi, \quad \langle\sigma_y\rangle = \sin\theta\sin\varphi, \quad \langle\sigma_z\rangle = \cos\theta, \tag{5.12}$$

showing that the radius-vector of any representation point is just the expectation value of $\boldsymbol{\sigma}$.

---

[4] If you need a reminder why this condition is crucial, please revisit the discussion in the end of section 1.6. Note also that the mutual phase shifts between different qubits are important, in particular, for quantum information processing (see section 8.5 below), so that most discussions of these applications have to start from Eq. (5.10) rather than Eq. (5.11).

[5] Named after the same F Bloch who pioneered the energy band theory, which was discussed in chapters 2–3.

Now let us use Eq. (5.3) to see how the representation point moves in various cases, ignoring the term $b\mathrm{I}$—which, again, describes the offset of the total energy of the system relative to some reference level, and does not affect its dynamics. First of all, according to Eq. (4.158), in the case $\mathbf{c} = 0$ (when the Hamiltonian operator vanishes, and hence the state vectors do not depend on time) the point does not move at all, and its position is determined by initial conditions, i.e. by the system state's preparation. If $\mathbf{c} \neq 0$, we may re-use some results of section 4.6, obtained for the Pauli Hamiltonian (4.163a), which coincides with Eq. (5.3) with[6]

$$\mathbf{c} = -\gamma \mathscr{B} \frac{\hbar}{2}. \tag{5.13}$$

In particular, if the field $\mathscr{B}$, and hence the vector $\mathbf{c}$ are directed along the z-axis and are time-independent, Eqs. (4.170), (4.173) and (4.174) show that the representation point $\langle \boldsymbol{\sigma} \rangle$ on the Bloch sphere rotates within a plane normal to this axis (see figure 5.3b) with the angular velocity

$$\frac{d\varphi}{dt} \equiv \Omega = -\gamma \mathscr{B}_z \equiv \frac{2c_z}{\hbar}. \tag{5.14}$$

Almost evidently, since the selection of the coordinate axes is arbitrary, this picture should remain valid for any orientation of the vector $\mathbf{c}$, with the representation point rotating, on the Bloch sphere, around its direction, with the angular speed $|\Omega| = 2c\hbar$—see figure 5.3c. This fact may be proved using any picture of the quantum dynamics, discussed in section 4.6. Actually, the reader may already have done that by solving problems 4.25 and 4.26, just to see that even for the particular, simple initial state of the system ($\uparrow$), the results for the Cartesian components of the vector $\langle \boldsymbol{\sigma} \rangle$ are rather bulky. However, this description may be readily simplified, even for an arbitrary time dependence of the 'field' vector $\mathbf{c}(t)$ in Eq. (5.3), using the (geometric) vector language.

Indeed, let us rewrite Eq. (5.3) (again, with $b = 0$) in the operator form,

$$\hat{H} = \mathbf{c}(t) \cdot \hat{\boldsymbol{\sigma}}, \tag{5.15}$$

valid in an arbitrary basis. According to Eq. (4.199), the corresponding Heisenberg equation of motion for the jth Cartesian components of the vector-operator $\hat{\boldsymbol{\sigma}}$ (which does not depend on time explicitly, so that $\partial \hat{\boldsymbol{\sigma}} / \partial t = 0$) is

$$i\hbar \dot{\hat{\sigma}}_j = \left[\hat{\sigma}_j, \hat{H} \right] \equiv \left[\hat{\sigma}_j, \mathbf{c}(t) \cdot \hat{\boldsymbol{\sigma}} \right] \equiv \left[\hat{\sigma}_j, \sum_{j'=1}^{3} c_{j'}(t) \hat{\sigma}_{j'} \right] \equiv \sum_{j'=1}^{3} c_{j'}(t) \left[\hat{\sigma}_j, \hat{\sigma}_{j'} \right]. \tag{5.16}$$

---

[6] This correspondence justifies using the use of term 'field' for the vector $\mathbf{c}$.

Now using the commutation relations (4.155), which are valid in any basis, and in any picture of time evolution[7], we get

$$i\hbar\dot{\hat{\sigma}}_j = 2i\sum_{j'=1}^{3}c_{j'}(t)\hat{\sigma}_{j''}\varepsilon_{jj'j''}. \tag{5.17}$$

But it is straightforward to use the definition of the Levi-Civita symbol[8] to verify that the usual vector product of two 3D vectors may be represented in a similar Cartesian-component form:

$$(\mathbf{a} \times \mathbf{b})_j = \begin{vmatrix} \mathbf{n}_1 & \mathbf{n}_2 & \mathbf{n}_3 \\ a_1 & a_2 & a_3 \\ b_1 & b_2 & b_3 \end{vmatrix}_j = \sum_{j'=1}^{3}aH_{j'}b_{j''}\varepsilon_{jj'j''}. \tag{5.18}$$

As a result, Eq. (5.17) may be rewritten in a vector form—or rather several equivalent forms:

$$i\hbar\dot{\hat{\sigma}}_j = 2i[\mathbf{c}(t) \times \hat{\boldsymbol{\sigma}}]_j, \quad \text{i.e. } i\hbar\dot{\hat{\boldsymbol{\sigma}}} = 2i\mathbf{c}(t) \times \hat{\boldsymbol{\sigma}}, \quad \text{or} \quad \dot{\hat{\boldsymbol{\sigma}}} = \frac{2}{\hbar}\mathbf{c}(t) \times \hat{\boldsymbol{\sigma}}, \tag{5.19}$$

$$\text{or} \quad \dot{\hat{\boldsymbol{\sigma}}} = \boldsymbol{\Omega}(t) \times \hat{\boldsymbol{\sigma}},$$

where the vector $\boldsymbol{\Omega}$ is defined as

$$\hbar\boldsymbol{\Omega}(t) \equiv 2\mathbf{c}(t) \tag{5.20}$$

—an evident generalization of Eq. (5.14).[9] As we have seen in section 4.6, any linear relation between two Heisenberg operators is also valid for the expectation values of the corresponding observables, so that the last form of Eq. (5.19) yields:

$$\left\langle \dot{\boldsymbol{\sigma}} \right\rangle = \boldsymbol{\Omega}(t) \times \langle\boldsymbol{\sigma}\rangle. \tag{5.21}$$

But this is the well-known kinematic formula[10] for the rotation of a constant-length vector $\langle\boldsymbol{\sigma}\rangle$ around the instantaneous direction of the vector $\boldsymbol{\Omega}(t)$, with the instantaneous angular velocity $\boldsymbol{\Omega}(t)$. So, the time evolution of the representation point on the Bloch sphere is quite simple, especially in the case of a time-independent

---

[7] Indeed, if three operators in the Schrödinger picture are related as $[\hat{A}_S, \hat{B}_S] = \hat{C}_S$, then according to Eq. (4.190), in the Heisenberg picture

$$[\hat{A}_H, \hat{B}_H] = [\hat{u}^\dagger\hat{A}_H\hat{u}, \hat{u}^\dagger\hat{B}_H\hat{u}] \equiv \hat{u}^\dagger\hat{A}_H\hat{u}\hat{u}^\dagger\hat{B}_H\hat{u} - \hat{u}^\dagger\hat{B}_H\hat{u}\hat{u}^\dagger\hat{A}_H\hat{u}$$
$$\equiv \hat{u}^\dagger[\hat{A}_S, \hat{B}_S]\hat{u} \equiv \hat{u}^\dagger\hat{C}_S\hat{u} = \hat{C}_H.$$

[8] See, e.g. Eq. (A.57).

[9] It is also easy to check that in the particular case $\boldsymbol{\Omega} = \Omega\mathbf{n}_z$, Eqs. (5.19) are reduced to Eqs. (4.200) for the spin-½ vector $\mathbf{S} = (\hbar/2)\boldsymbol{\sigma}$.

[10] See, e.g. Part CM section 4.1, in particular Eq. (4.8).

[11] The bulkiness of the solutions of problems 4.25 and 4.26 (which were offered just as useful exercises in quantum dynamic formalisms) reflects just the awkward expression of the resulting circular motion of the vector $\langle\boldsymbol{\sigma}\rangle$ (see figure 5.3c) in its Cartesian components.

c, and hence $\mathbf{\Omega}$—see figure 5.3c.[11] Note that it is sufficient to turn off the field to stop the precession instantly. (Since Eq. (5.21) is the first-order differential equation, the representation point has no effective inertia[12].) Hence, changing the direction and the magnitude of the effective external field, it is possible to drive the representation point of a two-level system from any initial position to any final position on the Bloch sphere, i.e. make the system take any of its possible quantum states.

In the particular case of a spin-$\frac{1}{2}$ in a magnetic field $\mathscr{B}(t)$, it is more customary to use Eqs. (5.13) and (5.20) to rewrite Eq. (5.21) as the following equation for the expectation value of the spin vector $\mathbf{S} = (\hbar/2)\boldsymbol{\sigma}$:

$$\left\langle \dot{\mathbf{S}} \right\rangle \equiv \gamma\langle\mathbf{S}\rangle \times \mathscr{B}(t). \tag{5.22}$$

As we know from the discussion in chapter 4, such a classical description of the spin's evolution does not give a full picture of the quantum reality; in particular, it does not describe the possible large uncertainties of its components—see, e.g. Eqs. (4.135). The situation, however, is different for a collection of $N \gg 1$ similar, non-interacting spins, initially prepared to be in the same state—for example by polarizing all spins with a strong external field $\mathscr{B}_0$, at relatively low temperature T, with $k_B T \ll \gamma\mathscr{B}_0\hbar$. (A practically important example of such a collection is a set of nuclear spins in a macroscopic condensed matter sample, where the spin interaction with each other and the environment is typically very small.) For such an collection, Eq. (5.22) is still valid, while the relative uncertainty of the resulting sample's magnetization $\mathbf{M} = n\langle\mathbf{m}\rangle = n\gamma\langle\mathbf{S}\rangle$ (where $n \equiv N/V$ is the spin density) is proportional to $1/N^{1/2} \ll 1$. Thus, the evolution of magnetization may be described, with good precision, by the essentially classical equation (valid for any spin):

$$\dot{\mathbf{M}} = \gamma\mathbf{M} \times \mathscr{B}(t). \tag{5.23}$$

This equation, or the equivalent set of three *Bloch equations*[13] for its Cartesian components, with the right-hand side augmented with small terms describing the effects of dephasing and relaxation (to be discussed in chapter 7), is used, in particular, for the description of *magnetic resonance*, taking place when the frequency (4.164) of the spin's precession in a strong constant magnetic field approaches the frequency of an additional ac field—see the next chapter.

5.2 The Ehrenfest theorem

In section 4.7, we have derived all the basic relations of wave mechanics from the bra–ket formalism, which will also enable us to get some important additional results in that area. One of them is a pair of very interesting relations, together called the *Ehrenfest theorem*. In order to derive them, for the simplest case of the 1D orbital motion, let us calculate the following commutator:

---

[12] This is also true for the classical angular momentum **L** at its torque-induced precession—see, e.g. *Part CM* section 4.5.

[13] They were introduced by the same F Bloch in 1946.

$$\left[\hat{x}, \hat{p}_x^2\right] \equiv \hat{x}\hat{p}_x\hat{p}_x - \hat{p}_x\hat{p}_x\hat{x}. \tag{5.24}$$

Let us apply the commutation relation (4.238),

$$\hat{x}\hat{p}_x = \hat{p}_x\hat{x} + i\hbar\hat{I}, \tag{5.25}$$

to the first term of the right-hand side of Eq. (5.24) twice, in order to sequentially move the coordinate operator to the right:

$$\begin{aligned}
\hat{x}\hat{p}_x\hat{p}_x &= \left(\hat{p}_x\hat{x} + i\hbar\hat{I}\right)\hat{p}_x \equiv \hat{p}_x\hat{x}\hat{p}_x + i\hbar\hat{p}_x \\
&= \hat{p}_x\left(\hat{p}_x\hat{x} + i\hbar\hat{I}\right) + i\hbar\hat{p}_x \equiv \hat{p}_x\hat{p}_x\hat{x} + 2i\hbar\hat{p}_x.
\end{aligned} \tag{5.26}$$

The first term of this result cancels with the second term of Eq. (5.24), so that the commutator becomes rather simple:

$$\left[\hat{x}, \hat{p}_x^2\right] = 2i\hbar\hat{p}_x. \tag{5.27}$$

Let us use this equality to calculate the Heisenberg-picture equation of motion of the operator $\hat{x}$, by applying the general Heisenberg equation (4.199) to the 1D orbital motion described by the Hamiltonian (4.237), but possibly with a time-dependent potential energy U:

$$\frac{d\hat{x}}{dt} = \frac{1}{i\hbar}[\hat{x}, \hat{H}] = \frac{1}{i\hbar}\left[\hat{x}, \frac{\hat{p}_x^2}{2m} + U(\hat{x}, t)\right]. \tag{5.28}$$

The potential energy operator commutes with the coordinate operator. Thus, the right-hand side of Eq. (5.28) is proportional to the commutator (5.27), and we get

$$\frac{d\hat{x}}{dt} = \frac{\hat{p}_x}{m}. \tag{5.29}$$

In this *operator* equality, we readily recognize the full analog of the classical relation between the particle's momentum and its velocity.

Now let us see what a similar procedure gives for the momentum's derivative:

$$\frac{d\hat{p}_x}{dt} = \frac{1}{i\hbar}\left[\hat{p}_x, \hat{H}\right] = \frac{1}{i\hbar}\left[\hat{p}_x, \frac{\hat{p}_x^2}{2m} + U(\hat{x}, t)\right]. \tag{5.30}$$

The kinetic energy operator commutes with the momentum operator, and hence drops from the right-hand side of this equation. In order to calculate the remaining commutator of the momentum and potential energy, let us use the fact that any smooth (infinitely differentiable) function may be represented by its Taylor expansion:

$$U(\hat{x}, t) = \sum_{k=0}^{\infty} \frac{1}{k!}\frac{\partial^k U}{\partial \hat{x}^k}\hat{x}^k, \tag{5.31}$$

where the derivatives of U may be understood as c-numbers (evaluated at $x = 0$, and the given time t), so that we may write

$$[\hat{p}_x, U(\hat{x}, t)] = \sum_{k=0}^{\infty} \frac{1}{k!} \frac{\partial^k U}{\partial \hat{x}^k} [\hat{p}_x, \hat{x}^k] = \sum_{k=0}^{\infty} \frac{1}{k!} \frac{\partial^k U}{\partial \hat{x}^k} \left(\hat{p}_x \underbrace{\hat{x}\hat{x}..\hat{x}}_{k \text{ times}} - \underbrace{\hat{x}\hat{x}...\hat{x}}_{k \text{ times}} \hat{p}_x \right). \quad (5.32)$$

Applying Eq. (5.25) k times to the last term in the parentheses, exactly as we did it in Eq. (5.26), we get

$$[\hat{p}_x, U(\hat{x}, t)] = -\sum_{k=1}^{\infty} \frac{1}{k!} \frac{\partial^k U}{\partial \hat{x}^k} ik\hbar\hat{x}^{k-1} = -i\hbar \sum_{k=1}^{\infty} \frac{1}{(k-1)!} \frac{\partial^k U}{\partial \hat{x}^k} \hat{x}^{k-1}. \quad (5.33)$$

But the last sum is just the Taylor expansion of the derivative $\partial U/\partial x$. Indeed,

$$\frac{\partial U}{\partial \hat{x}} = \sum_{k'=0}^{\infty} \frac{1}{k'!} \frac{\partial^{k'}}{\partial \hat{x}^{k'}} \left(\frac{\partial U}{\partial \hat{x}} \right) \hat{x}^{k'} = \sum_{k'=0}^{\infty} \frac{1}{k'!} \frac{\partial^{k'+1}U}{\partial \hat{x}^{k'+1}} \hat{x}^{k'} = \sum_{k=1}^{\infty} \frac{1}{(k-1)!} \frac{\partial^k U}{\partial \hat{x}^k} \hat{x}^{k-1}, \quad (5.34)$$

where at the last step the summation index was changed from k' to $k - 1$. As a result, Eq. (5.30) yields:

$$\frac{d\hat{p}_x}{dt} = -\frac{\partial}{\partial \hat{x}} U(\hat{x}, t). \quad (5.35)$$

This equation again coincides with the classical equation of motion! Moreover, averaging Eqs. (5.29) and (5.39) over the initial state (as Eq. (4.191) prescribes), we get similar results for the expectation values[14]:

$$\frac{d\langle x \rangle}{dt} = \frac{\langle p_x \rangle}{m}, \qquad \frac{d\langle p_x \rangle}{dt} = -\left\langle \frac{\partial U}{\partial x} \right\rangle. \quad (5.36)$$

However, it is important to remember that the equivalence between these quantum-mechanical equations and similar equations of classical mechanics is superficial, and the degree of the similarity between the two mechanics very much depends on the problem. As one extreme, let us consider the case when a particle's state, at any moment between t_0 and t, may be accurately represented by one, relatively p_x-narrow wave packet. Then we may interpret Eqs. (5.36) as the equations of the essentially classical motion of the wave packet's center, in accordance with the correspondence principle. However, even in this case it is important to remember about the purely quantum mechanical effects of nonvanishing wave packet width and its spread in time, which were discussed in section 2.2.

As the opposite extreme case, let us revisit the 'leaky' potential well discussed in section 2.5—see figure 2.15. Since both the potential $U(x)$ and the state of the system are symmetric relative to point $x = 0$ at all times, the right-hand sides of both Eqs. (5.36) identically equal zero. Of course, the result (that the average values of

---

[14] The set of equations (5.36) constitute the *Ehrenfest theorem*, named after its author, P Ehrenfest.

both the coordinate and the momentum stay equal to zero at all times) is correct, but this fact does not tell us too much about the rich dynamics of the system (the finite lifetime of the metastable state, the formation of two wave packets, their waveform and propagation speed, see figure 2.17), and about the important insight the solution gives for the quantum measurement theory and the system's irreversibility. Another similar example is the energy band theory (section 2.7), with its purely quantum effect of the allowed energy bands and forbidden gaps, of which Eq. (5.36) gives no clue.

To summarize, the Ehrenfest theorem is important as an illustration of the correspondence principle, but its predictive power should not be exaggerated.

5.3 The Feynman path integral

As has been already mentioned, even within the realm of wave mechanics, the bra–ket language allows one to simplify some calculations that would be very bulky using the notation used in chapters 1–3. Probably the best example is the famous alternative, *path-integral* formulation of quantum mechanics, developed in 1948 by R Feynman[15]. I will review this important concept—admittedly cutting one math corner for the sake of brevity[16]. (This shortcut will be clearly marked below.)

Let us inner-multiply both parts of Eq. (4.157a), which is essentially the definition of the time-evolution operator, by the bra-vector of state x,

$$\langle x|\alpha(t)\rangle = \langle x|\hat{u}(t, t_0)|\alpha(t_0)\rangle, \tag{5.37}$$

insert the identity operator before the ket-vector on the right-hand side, and then use the closure condition in the form of Eq. (4.252), with x' replaced by x_0:

$$\langle x|\alpha(t)\rangle = \int dx_0 \langle x|\hat{u}(t, t_0)|x_0\rangle\langle x_0|\alpha(t_0)\rangle. \tag{5.38}$$

According to Eq. (4.233), this equality may be represented as

$$\Psi_\alpha(x, t) = \int dx_0 \langle x|\hat{u}(t, t_0)|x_0\rangle\Psi_\alpha(x_0, t_0). \tag{5.39}$$

Comparing this expression with Eq. (2.44), we see that the bracket in this relation is nothing other than the 1D propagator, which was discussed in section 2.2, i.e.

$$G(x, t; x_0, t_0) = \langle x|\hat{u}(t, t_0)|x_0\rangle. \tag{5.40}$$

Let me hope that the reader sees that this equality corresponds to the physical sense of the propagator.

Now let us break the time segment $[t_0, t]$ into N (for the time being, not necessarily equal) parts, by inserting $(N - 1)$ intermediate points (figure 5.4)

---

[15] According to Feynman's memories, his work was motivated by a 'mysterious' remark by P A M Dirac in his pioneering 1930 textbook on quantum mechanics.

[16] A more thorough discussion of the path-integral approach may be found in the famous text R Feynman and A Hibbs, *Quantum Mechanics and Path Integrals* first published in 1965. (For its latest edition by Dover in 2010, the book was emended by D Styler.) For a more recent monograph, which reviews more applications, see [1].

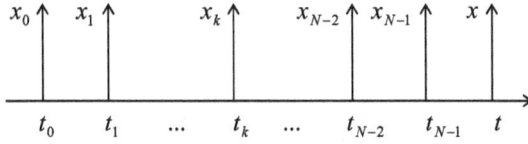

Figure 5.4. Time partition and coordinate notation at the initial stage of the Feynman path integral's derivation.

$$t_0 < t_1 < \ldots < t_k < \ldots < t_{N-1} < t, \tag{5.41}$$

and use the definition (4.157) of the time evolution operator to write

$$\hat{u}(t, t_0) = \hat{u}(t, t_{N-1})\hat{u}(t_{N-1}, t_{N-2}) \ldots \hat{u}(t_2, t_1)\hat{u}(t_1, t_0). \tag{5.42}$$

After plugging Eq. (5.42) into Eq. (5.40), let us insert the identity operator, again in the closure form (4.252), but written for x_k rather than x', between each of the two partial evolution operators including the time argument t_k. The result is

$$G(x, t; x_0, t_0) = \int dx_{N-1} \int dx_{N-2} \ldots \int dx_1 \langle x|\hat{u}(t, t_{N-1})|x_{N-1}\rangle$$
$$\times \langle x_{N-1}|\hat{u}(t_{N-1}, t_{N-2})|x_{N-2}\rangle \ldots \langle x_1|\hat{u}(t_1, t_0)|x_0\rangle. \tag{5.43}$$

The physical sense of each integration variable x_k is the wavefunction's argument at time t_k—see figure 5.4.

The key Feynman's breakthrough was the realization that if all intervals are taken similar and sufficiently small, $t_k - t_{k-1} = d\tau \to 0$, all the partial brackets participating in Eq. (5.43) may be expressed via the free-particle's propagator, given by Eq. (2.49), even if the particle is not free, but moves in a stationary potential profile $U(x)$. To show that, let us use either Eq. (4.175) or Eq. (4.181), which, for a small time interval $d\tau$, give the same result:

$$\hat{u}(\tau + d\tau, \tau) = \exp\left\{-\frac{i}{\hbar}\hat{H}d\tau\right\} = \exp\left\{-\frac{i}{\hbar}\left(\frac{\hat{p}^2}{2m}d\tau + U(\hat{x})d\tau\right)\right\}. \tag{5.44}$$

Generally, an exponent of a sum of two operators may be treated as that of c-number arguments, and in particular factored into a product of two exponents, only if the operators commute. (In this case we can use all the standard algebra for the exponents of c-number arguments.) In our case, this is not so, because the operator $\hat{p}^2/2m$ does not commute with $\hat{x}$, and hence with $U(\hat{x})$. However, it may be shown[17] that for an infinitesimal time interval $d\tau$, the nonvanishing commutator

$$\left[\frac{\hat{p}^2}{2m}d\tau, U(\hat{x})d\tau\right] \neq 0, \tag{5.45}$$

---

[17] This is exactly the mathematical corner I am going to cut, because a strict proof of this (intuitively evident) statement would take more space and time than I can afford.

proportional to $(d\tau)^2$, may be ignored in the first, linear approximation in $d\tau$. As a result, we may factorize the right-hand side in Eq. (5.44) by writing

$$\hat{u}(\tau + d\tau, \tau)_{d\tau \to 0} \to \exp\left\{-\frac{i}{\hbar}\frac{\hat{p}^2}{2m}d\tau\right\}\exp\left\{-\frac{i}{\hbar}U(\hat{x})d\tau\right\}. \qquad (5.46)$$

(This approximation is very much similar in spirit to the trapezoidal-rule approximation in the usual 1D integration[18], which is also asymptotically impeachable.)

Since the second exponential function on the right-hand side of Eq. (5.46) commutes with the coordinate operator, we may move it out of each partial bracket participating in Eq. (5.43), with $U(x)$ turning into a c-number function:

$$\langle x_{\tau+d\tau}|\hat{u}(\tau + d\tau, \tau)|x_\tau\rangle = \langle x_{\tau+d\tau}|\exp\left\{-\frac{i}{\hbar}\frac{\hat{p}^2}{2m}d\tau\right\}|x_\tau\rangle$$
$$\times \exp\left\{-\frac{i}{\hbar}U(x)d\tau\right\}. \qquad (5.47)$$

But the remaining bracket is just the propagator of a free particle, and we can use Eq. (2.49) for it:

$$\langle x_{\tau+d\tau}|\exp\left\{-\frac{i}{\hbar}\frac{\hat{p}^2}{2m}d\tau\right\}|x_\tau\rangle = \left(\frac{m}{2\pi i\hbar d\tau}\right)^{1/2}\exp\left\{i\frac{m(dx)^2}{2\hbar d\tau}\right\}. \qquad (5.48)$$

As the result, the full propagator (5.43) takes the form

$$G(x, t; x_0, t_0) = \lim_{\substack{d\tau \to 0 \\ N \to \infty}}\int dx_{N-1}\int dx_{N-2}..\int dx_1 \left(\frac{m}{2\pi i\hbar d\tau}\right)^{N/2}$$
$$\times \exp\left\{\sum_{k=1}^{N}\left[i\frac{m(dx)^2}{2\hbar d\tau} - i\frac{U(x)}{\hbar}d\tau\right]\right\}. \qquad (5.49)$$

At $N \to \infty$ and hence $d\tau \equiv (t - t_0)/N \to 0$, the sum under the exponent in this expression may be approximated with the corresponding integral:

$$\sum_{k=1}^{N}\frac{i}{\hbar}\left[\frac{m}{2}\left(\frac{dx}{d\tau}\right)^2 - U(x)\right]_{\tau=t_k}d\tau \to \frac{i}{\hbar}\int_{t_0}^{t}\left[\frac{m}{2}\left(\frac{dx}{d\tau}\right)^2 - U(x)\right]d\tau, \qquad (5.50)$$

and the expression in the square brackets is just the particle's Lagrangian function $\mathscr{L}$.[19] The integral of this function over time is the classical action $\mathscr{S}$ calculated along a particular 'path' $x(\tau)$.[20] As a result, defining the (1D) *path integral* as

$$\int(...)D[x(\tau)] \equiv \lim_{\substack{d\tau \to 0 \\ N \to \infty}}\left(\frac{m}{2\pi i\hbar d\tau}\right)^{N/2}\int dx_{N-1}\int dx_{N-2}..\int dx_1 (...), \qquad (5.51a)$$

---

[18] See, e.g. Eq. (A.26).
[19] See, e.g. *Part CM* section 2.1.
[20] See, e.g. *Part CM* section 10.3.

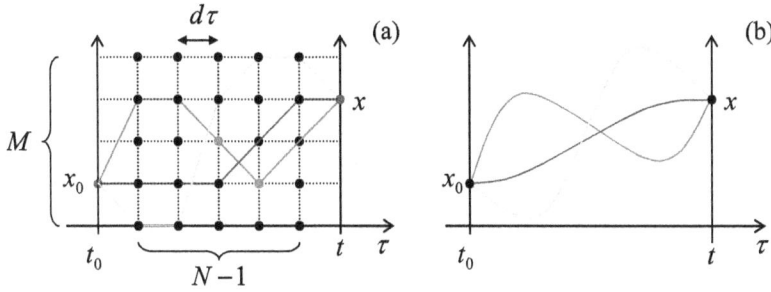

Figure 5.5. Several 1D classical paths: (a) in the discrete approximation and (b) in the continuous limit.

we can bring our result to a superficially simple form

$$G(x, t; x_0, t_0) = \int \exp\left\{\frac{i}{\hbar} \mathcal{S}[x(\tau)]\right\} D[x(\tau)]. \tag{5.51b}$$

The name 'path integral' for the mathematical construct (5.51a) may be readily explained if we keep the number N of time intervals large but finite, and also approximate each of the enclosed integrals by a sum over $M \gg 1$ discrete points along the coordinate axis—see figure 5.5a. Then the path integral is a product of $(N - 1)$ sums corresponding to different values of time τ, each of them with M terms, each of those representing the function under the integral at a particular spatial point. Multiplying those $(N - 1)$ sums, we get a sum of $(N - 1)M$ terms, each evaluating the function at a specific spatial-temporal point $[x, \tau]$. These terms may be now grouped to represent all possible different continuous classical paths $x[\tau]$ from the initial point $[x_0, t_0]$ to the finite point $[x, t]$. It is evident that the last interpretation remains true even in the continuous limit $N, M \to \infty$—see figure 5.5b.

Why does such path representation of the sum have sense? This is because in the classical limit the particle follows just a certain path, corresponding to the minimum of the action $\mathcal{S}$. Hence, for all close trajectories, the difference $(\mathcal{S} - \mathcal{S}_{cl})$ is proportional to the square of the deviation from the classical trajectory. Hence, for a quasiclassical motion, with $\mathcal{S}_{cl} \gg \hbar$, there is a substantial bunch of close trajectories, with $(\mathcal{S} - \mathcal{S}_{cl}) \ll \hbar$, that give similar contributions to the path integral. On the other hand, strongly non-classical trajectories, with $(\mathcal{S} - \mathcal{S}_{cl}) \gg \hbar$, give phases $\mathcal{S}/\hbar$ rapidly oscillating from one trajectory to the next one, and their contributions to the path integral are averaged out[21]. As a result, for the quasi-classical motion, the propagator's exponent may be evaluated on the classical path only:

---

[21] This fact may be proved expanding the difference $(\mathcal{S} - \mathcal{S}_{cl})$ in the Taylor series in path variations (leaving only the leading quadratic terms) and working out the resulting Gaussian integrals. It is interesting that the integration, together with the pre-exponential coefficient in Eq. (5.51a), gives exactly the pre-exponential factor that we have already found in section 2.4, refining the **WKB** approximation.

$$G_{cl} \propto \exp\left\{\frac{i}{\hbar}\mathcal{S}_{cl}\right\} = \exp\left\{\frac{i}{\hbar}\int_{t_0}^{t}\left[\frac{m}{2}\left(\frac{dx}{d\tau}\right)^2 - U(x)\right]d\tau\right\}. \qquad (5.52)$$

The sum of the kinetic and potential energies is the full energy E of the particle, that remains constant for motion in a stationary potential $U(x)$, so that we may rewrite the expression under this integral as[22]

$$\left[\frac{m}{2}\left(\frac{dx}{d\tau}\right)^2 - U(x)\right]d\tau = \left[m\left(\frac{dx}{d\tau}\right)^2 - E\right]d\tau = m\frac{dx}{d\tau}dx - Ed\tau. \qquad (5.53)$$

With this replacement, Eq. (5.52) yields

$$\begin{aligned} G_{cl} &\propto \exp\left\{\frac{i}{\hbar}\int_{x_0}^{x} m\frac{dx}{d\tau}dx\right\}\exp\left\{-\frac{i}{\hbar}E(t-t_0)\right\} \\ &= \exp\left\{\frac{i}{\hbar}\int_{x_0}^{x} p(x)dx\right\}\exp\left\{-\frac{i}{\hbar}E(t-t_0)\right\}, \end{aligned} \qquad (5.54)$$

where p is the classical momentum of the particle. But (at least, leaving the pre-exponential factor alone) this is the WKB approximation result that was derived and studied in detail in chapter 2!

One may question the value of such a complicated calculation that yields the results that could be readily obtained from the Schrödinger's wave mechanics. The Feynman's approach is indeed not used too often, but it has its merits. First, it has an important philosophical (and hence heuristic) value. Indeed, Eq. (5.51) may be interpreted by saying that the essence of quantum mechanics is the exploration, by the system, of all possible paths $x(\tau)$, each of them classical-like, in the sense that the particle's coordinate x and velocity $dx/d\tau$ are exactly defined simultaneously at each point. The resulting contributions to the path integral are added up coherently to form the final propagator G, and via it, the final probability $W \propto |G|^2$ of the particle's propagation from $[x_0, t_0]$ to $[x, t]$. As the scale of action (i.e. the energy-by-time product) of the motion decreases and becomes comparable to $\hbar$, more and more paths produce a substantial contribution to this sum, and hence to W, ensuring a larger and larger difference between the quantum and classical properties of the system.

Second, the path integral provides a justification for some simple explanations of quantum phenomena. A typical example is the quantum interference effects discussed in section 3.1—see, e.g. figure 3.1 and the corresponding text. At that discussion, we used the Huygens principle to argue that at the two-slit interference, the WKB approximation might be restricted to contributions from two paths that pass through different slits, but otherwise consist of straight-line segments. To have another look at that assumption, let us generalize the path integral to multi-

---

[22] The same trick is often used in analytical classical mechanics—say, for proving the Hamilton principle, and for the derivation of the Hamilton–Jacobi equations (see, e.g. *Part CM* sections 10.3–10.4).

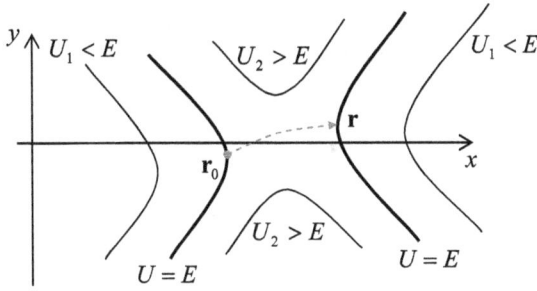

Figure 5.6. A saddle-type 2D potential profile and the instanton trajectory of a particle of energy E (schematically).

dimensional geometries. Fortunately, the simple structure of Eq. (5.51b) makes such generalization virtually evident:

$$G(\mathbf{r}, t; \mathbf{r}_0, t_0) = \int \exp\left\{\frac{i}{\hbar} \mathcal{S}[\mathbf{r}(\tau)]\right\} D[\mathbf{r}(\tau)],$$

$$\mathcal{S} \equiv \int_{t_0}^{t} \mathcal{L}\left(\mathbf{r}, \frac{d\mathbf{r}}{d\tau}\right) d\tau = \int_{t_0}^{t} \left[\frac{m}{2}\left(\frac{d\mathbf{r}}{d\tau}\right)^2 - U(\mathbf{r})\right] d\tau.$$

(5.55)

where definition (5.51a) of the path integral should be also modified correspondingly. (I will not go into these technical details.) For the Young-type experiment (figure 3.1), where a classical particle could reach the detector only after passing through one of the slits, the classical paths are the straight-line segments shown in figure 3.1, and if they are much longer than the de Broglie wavelength, the propagator may be well approximated by the sum of two integrals of $\mathcal{L}d\tau = i\mathbf{p}(\mathbf{r}) \cdot d\mathbf{r}/\hbar$—as was done in section 3.1.

Last but not least, the path integral allows simple solutions of some problems that would be hard to obtain by other methods. As the simplest example, let us consider the problem of tunneling in multi-dimensional space, sketched in figure 5.6 for the 2D case—just for the graphics' simplicity. Here, the potential profile $U(x, y)$ has the 'saddle' shape. (Another helpful image is a mountain path between two summits, in figure 5.6 located on the top and at the bottom of the shown region.) A particle of energy E may move classically in the left and right regions with $U(x, y) < E$, but if E is not sufficiently high, it can pass from one of these regions to another one only via the quantum-mechanical tunneling under the pass. Let us calculate the transparency of this potential barrier in the WKB approximation, ignoring the possible pre-exponential factor[23].

---

[23] Actually, one can argue that the pre-exponential factor should be close to 1, just like in Eq. (2.117), especially if the potential is smooth, in the sense of Eq. (2.107), in all spatial directions. (Let me remind the reader that for most practical problems of quantum tunneling, the pre-exponential factor is of minor importance.)

According to the evident multi-dimensional generalization Eq. (5.54), for the classically forbidden region, where $E < U(x, y)$, and hence $\mathbf{p}(\mathbf{r})/\hbar = i\boldsymbol{\kappa}(\mathbf{r})$, the contributions to the propagator (5.55) are proportional to

$$e^{-I} \exp\left\{-\frac{i}{\hbar}E(t - t_0)\right\}, \qquad \text{where } I \equiv \int_{\mathbf{r}_0}^{\mathbf{r}} \boldsymbol{\kappa}(\mathbf{r}) \cdot d\mathbf{r}, \qquad (5.56)$$

where the magnitude of the vector $\boldsymbol{\kappa}$ at each point may be calculated just in the 1D case—cf Eq. (2.97):

$$\frac{\hbar^2 \kappa^2(\mathbf{r})}{2m} = U(\mathbf{r}) - E. \qquad (5.57)$$

Hence the path integral in this region is actually much simpler than in the classically-allowed region, because the spatial exponents are purely real and there is no complex interference between them. Because of the minus sign before I in the exponent (5.56), the *largest* contribution to G evidently comes from the trajectory (or rather a narrow bundle of trajectories) for which the integral I has the *smallest* value, so that the barrier transparency may be calculated as

$$\mathscr{T} \approx |G|^2 \approx e^{-2I} \equiv \exp\left\{-2\int_{\mathbf{r}_0}^{\mathbf{r}} \boldsymbol{\kappa}(\mathbf{r}') \cdot d\mathbf{r}'\right\}, \qquad (5.58)$$

where $\mathbf{r}$ and $\mathbf{r}_0$ are certain points on the opposite classical turning-point surfaces: $U(\mathbf{r}) = U(\mathbf{r}_0) = E$—see figure 5.6.

Thus the barrier transparency problem is reduced to finding the trajectory (including the points $\mathbf{r}$ and $\mathbf{r}_0$) that connects the two surfaces and minimizes the functional I. This is of course a well-known problem of the calculus of variations[24], but it is interesting that the path integral provides a simple alternative way of solving it. Let us consider an auxiliary problem of a particle's motion in the potential profile $U_{\mathrm{inv}}(\mathbf{r})$ that is *inverted* relative to the particle's energy E, i.e. is defined by the following equality:

$$U_{\mathrm{inv}}(\mathbf{r}) - E \equiv E - U(\mathbf{r}). \qquad (5.59)$$

As was discussed above, at fixed energy E, the path integral for the WKB motion in the classically allowed region of potential $U_{\mathrm{inv}}(x, y)$ (that coincides with the classically forbidden region of the original problem) is dominated by the classical trajectory corresponding to the minimum of

$$\mathscr{S}_{\mathrm{inv}} = \int_{\mathbf{r}_0}^{\mathbf{r}} \mathbf{p}_{\mathrm{inv}}(\mathbf{r}') \cdot d\mathbf{r}' = \hbar \int_{\mathbf{r}_0}^{\mathbf{r}} \mathbf{k}_{\mathrm{inv}}(\mathbf{r}') \cdot d\mathbf{r}, \qquad (5.60)$$

where $\mathbf{k}_{\mathrm{inv}}$ should be determined from the WKB relation

$$\frac{\hbar^2 k_{\mathrm{inv}}^2(\mathbf{r})}{2m} \equiv E - U_{\mathrm{inv}}(\mathbf{r}). \qquad (5.61)$$

---

[24] For a concise introduction to the field see, e.g. [2] or [3].

But comparing Eqs. (5.57), (5.59), and (5.61), we see that $\mathbf{k}_{inv} = \boldsymbol{\kappa}$ at each point! This means that the tunneling path (in the WKB limit) corresponds to the classical (so-called *instanton*[25]) trajectory of the same particle moving in the inverted potential $U_{inv}(\mathbf{r})$. If the initial point $\mathbf{r}_0$ is fixed, this trajectory may be readily found by the means of classical mechanics. (Note that the initial kinetic energy, and hence the initial velocity of the instanton launched from point $\mathbf{r}_0$ should be zero, because by the classical turning point definition, $U_{inv}(\mathbf{r}_0) = U(\mathbf{r}_0) = E$.) Thus the problem is further reduced to a simpler task of maximizing the transparency (5.58) by choosing the optimal position of $\mathbf{r}_0$ on the equipotential surface $U(\mathbf{r}_0) = E$—see figure 5.6. Moreover, for many symmetric potentials, the position of this point may be readily guessed even without calculations—as it is in the problem given for the reader's exercise.

Note that besides the calculation of the potential barrier's transparency, the instanton trajectory has one more important implication: the so-called *traversal time* τ_t of the classical motion along it, from the point $\mathbf{r}_0$ to the point $\mathbf{r}$, in the inverted potential defined by Eq. (5.59), plays the role of the most important (though not the only) time scale of the particle's tunneling under the barrier[26].

5.4 Revisiting harmonic oscillator

Now let us return to the 1D harmonic oscillator, now understood as any system, regardless of its physical nature, described by the Hamiltonian (4.237) with the potential energy (2.111):

$$\hat{H} = \frac{\hat{p}^2}{2m} + \frac{m\omega_0^2\hat{x}^2}{2}. \tag{5.62}$$

In section 2.9 we have used a 'brute-force' (wave-mechanics) approach to analyze the eigenfunctions $\psi_n(x)$ and eigenvalues E_n of this Hamiltonian, and found that, unfortunately, this approach required a relatively complex mathematics, which does not enable an easy calculation of its key characteristics. Fortunately, the bra–ket formalism helps to make such calculations.

First, introducing normalized (dimensionless) operators of coordinates and momentum[27]:

$$\hat{\xi} \equiv \frac{\hat{x}}{x_0}, \qquad \hat{\zeta} \equiv \frac{\hat{p}}{m\omega_0 x_0}, \tag{5.63}$$

where $x_0 \equiv (\hbar/m\omega_0)^{1/2}$ is the natural coordinate scale, discussed in detail in section 2.9, we can represent the Hamiltonian (5.62) in a very simple and $x \leftrightarrow p$ symmetric form:

---

[25] In the quantum field theory, the instanton concept may be formulated somewhat differently, and has more complex applications—see, e.g. [4].

[26] For more on this interesting issue see, e.g. [5] and references therein.

[27] This normalization is not really necessary, it just makes the following calculations less bulky—and thus more aesthetically appealing.

$$\hat{H} = \frac{\hbar\omega_0}{2}\left(\hat{\xi}^2 + \hat{\zeta}^2\right). \tag{5.64}$$

This symmetry, as well as the discussion of the very similar coordinate and momentum representations, hints that much may be gained by treating the operators $\hat{\xi}$ and $\hat{\zeta}$ on an equal footing. Inspired by this clue, let us introduce a new operator

$$\hat{a} \equiv \frac{\hat{\xi} + i\hat{\zeta}}{\sqrt{2}} \equiv \left(\frac{m\omega_0}{2\hbar}\right)^{1/2}\left(\hat{x} + i\frac{\hat{p}}{m\omega_0}\right). \tag{5.65a}$$

Since both operators $\hat{\xi}$ and $\hat{\zeta}$ correspond to real observables, i.e. have real eigenvalues and hence are Hermitian (self-adjoint), the Hermitian conjugate of the operator $\hat{a}$ is simply its complex conjugate:

$$\hat{a}^\dagger \equiv \frac{\hat{\xi} - i\hat{\zeta}}{\sqrt{2}} \equiv \left(\frac{m\omega_0}{2\hbar}\right)^{1/2}\left(\hat{x} - i\frac{\hat{p}}{m\omega_0}\right). \tag{5.65b}$$

Because of the reason that will be clear very soon, $\hat{a}^\dagger$ and $\hat{a}$ (in this order!) are called the *creation and annihilation operators*.

Now solving the simple system of two equations (5.65) for $\hat{\xi}$ and $\hat{\zeta}$, we get the following reciprocal relations:

$$\hat{\xi} = \frac{\hat{a} + \hat{a}^\dagger}{\sqrt{2}}, \qquad \hat{\zeta} = \frac{\hat{a} - \hat{a}^\dagger}{\sqrt{2}\,i},$$
$$\text{i.e.} \quad \hat{x} = \left(\frac{\hbar}{m\omega_0}\right)^{1/2}\frac{\hat{a} + \hat{a}^\dagger}{\sqrt{2}}, \qquad \hat{p} = (\hbar m\omega_0)^{1/2}\frac{\hat{a} - \hat{a}^\dagger}{\sqrt{2}\,i}. \tag{5.66}$$

Our Hamiltonian (5.64) includes squares of these operators. Calculating them, we have to be careful to avoid swapping the new operators, because they do not commute. Indeed, for the normalized operators (5.63), Eq. (2.14) gives

$$[\hat{\xi}, \hat{\zeta}] \equiv \frac{1}{x_0^2 m\omega_0}[\hat{x}, \hat{p}] = i\hat{I}, \tag{5.67}$$

so that Eqs. (5.65) yield

$$[\hat{a}, \hat{a}^\dagger] = \frac{1}{2}[(\hat{\xi} + i\hat{\zeta}), (\hat{\xi} - i\hat{\zeta})] = -\frac{i}{2}([\hat{\xi}, \hat{\zeta}] - [\hat{\zeta}, \hat{\xi}]) = \hat{I}. \tag{5.68}$$

With such due caution, Eq. (5.66) gives

$$\hat{\xi}^2 = \frac{1}{2}(\hat{a}^2 + \hat{a}^{\dagger 2} + \hat{a}\hat{a}^\dagger + \hat{a}^\dagger\hat{a}), \qquad \hat{\zeta}^2 = -\frac{1}{2}(\hat{a}^2 + \hat{a}^{\dagger 2} - \hat{a}\hat{a}^\dagger - \hat{a}^\dagger\hat{a}). \tag{5.69}$$

Plugging these expressions back into Eq. (5.64), we get

$$\hat{H} = \frac{\hbar\omega_0}{2}(\hat{a}\hat{a}^\dagger + \hat{a}^\dagger\hat{a}). \tag{5.70}$$

This expression is elegant enough, but may be recast into an even more convenient form. For that, let us rewrite the commutation relation (5.68) as

$$\hat{a}\hat{a}^\dagger = \hat{a}^\dagger\hat{a} + \hat{I},\tag{5.71}$$

and plug it into Eq. (5.70). The result is

$$\hat{H} = \frac{\hbar\omega_0}{2}(2\hat{a}^\dagger\hat{a} + \hat{I}) = \hbar\omega_0\left(\hat{N} + \frac{1}{2}\hat{I}\right),\tag{5.72}$$

where, in the last form, one more (evidently, Hermitian) operator,

$$\hat{N} \equiv \hat{a}^\dagger\hat{a},\tag{5.73}$$

has been introduced. Since, according to Eq. (5.72), the operators $\hat{H}$ and $\hat{N}$ differ only by the addition of the identity operator and the multiplication by a c-number, these operators commute. Hence, according to the general arguments of section 4.5, they share the set of stationary ('Fock') eigenstates n, and we can write the standard eigenproblem (4.68) for the new operator as

$$\hat{N}|n\rangle = N_n|n\rangle,\tag{5.74}$$

where N_n are some eigenvalues that, according to Eq. (5.72), determine also the energy spectrum of the oscillator:

$$E_n = \hbar\omega_0\left(N_n + \frac{1}{2}\right).\tag{5.75}$$

So far, we know only that all eigenvalues N_n are real; in order to calculate them, let us carry out the following calculation—splendid in its simplicity and efficiency. Consider the result of action of the operator $\hat{N}$ on the ket-vector $\hat{a}^\dagger|n\rangle$. Using the definition (5.73) and then the associative rule of the bra–ket formalism, we may write

$$\hat{N}(\hat{a}^\dagger|n\rangle) \equiv (\hat{a}^\dagger\hat{a})(\hat{a}^\dagger|n\rangle) = \hat{a}^\dagger(\hat{a}\hat{a}^\dagger)|n\rangle.\tag{5.76}$$

Now using the commutation relation (5.71), and then Eq. (5.74), we may continue as

$$\hat{a}^\dagger(\hat{a}\hat{a}^\dagger)|n\rangle = \hat{a}^\dagger(\hat{a}^\dagger\hat{a} + \hat{I})|n\rangle = \hat{a}^\dagger(\hat{N} + \hat{I})|n\rangle$$
$$= \hat{a}^\dagger(N_n + 1)|n\rangle = (N_n + 1)(\hat{a}^\dagger|n\rangle).\tag{5.77}$$

Let us summarize the result of this calculation:

$$\hat{N}(\hat{a}^\dagger|n\rangle) = (N_n + 1)(\hat{a}^\dagger|n\rangle).\tag{5.78}$$

Performing an absolutely similar calculation with the operator $\hat{a}$, we get a similar formula:

$$\hat{N}(\hat{a}|n\rangle) = (N_n - 1)(\hat{a}|n\rangle).\tag{5.79}$$

It is time to stop and translate these results into plain English: if $|n\rangle$ is an eigenket of the operator $\hat{N}$ with eigenvalue N_n, then $\hat{a}^\dagger|n\rangle$ and $\hat{a}|n\rangle$ are also eigenkets of that

eigenket   ...   eigenvalue of $\hat{N}$

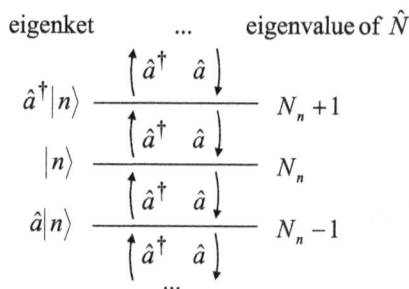

$\hat{a}^\dagger|n\rangle$ ———————— $N_n + 1$

$|n\rangle$ ———————— N_n

$\hat{a}|n\rangle$ ———————— $N_n - 1$

Figure 5.7. The 'ladder diagram' hierarchy of eigenstates of a 1D harmonic oscillator. Arrows show the actions of the creation and annihilation operators on the eigenstates.

operator, with the eigenvalues $(N_n + 1)$, and $(N_n - 1)$, respectively. This statement may be vividly represented with the so-called *ladder diagram* shown in figure 5.7. The operator $\hat{a}^\dagger$ moves the system one step up this ladder, while the operator $\hat{a}$ brings it one step down. In other words, the former operator creates a new *excitation* of the system[28], while the latter operator kills ('annihilates') this excitation. This is exactly why $\hat{a}^\dagger$ is called the creation operator, and $\hat{a}$, the annihilation operator. On the other hand, according to Eq. (5.74) inner-multiplied by the bra-vector $\langle n|$, the operator $\hat{N}$ does not change the state of the system, but 'counts' its position on the ladder:

$$\langle n \mid \hat{N} \mid n\rangle = \langle n \mid N_n \mid n\rangle = N_n. \tag{5.80}$$

This is why $\hat{N}$ is called the *number operator*, in our current context meaning the number of the elementary excitations of the oscillator.

This calculation still needs completion. Indeed, we still do not know whether the ladder shown in figure 5.7 shows *all* eigenstates of the oscillator, and what exactly the numbers N_n are. Fascinatingly enough, both questions may be answered by exploring just a single paradox. Let us start with some state n (a step of the ladder), and keep going down it, applying the operator $\hat{a}$ again and again. Each time, the eigenvalue N_n is decreased by one, so that eventually it should become negative. However, this cannot happen, because any real eigenstate, including the states represented by kets $|d\rangle \equiv \hat{a}|n\rangle$ and $|n\rangle$, should have a positive norm—see Eq. (4.16). Comparing the norms,

$$\|n\|^2 = \langle n|n\rangle, \qquad \|d\|^2 = \langle n|\hat{a}^\dagger \hat{a}|n\rangle = \langle n|\hat{N}|n\rangle = N_n\langle n|n\rangle, \tag{5.81}$$

we see that the both of them cannot be positive simultaneously if N_n is negative.

To resolve this paradox let us notice that the action of the creation and annihilation operators on the stationary states n may consist in not only their promotion to another step of the ladder diagram, but also by their multiplication by some *c*-numbers:

---

[28] For the electromagnetic field oscillators, such excitations are called *photons*; for the mechanical wave oscillators, *phonons*, etc.

$$\hat{a}\,|n\rangle = A_n\,|\,n-1\rangle, \qquad \hat{a}^\dagger\,|n\rangle = A'_n|\,n+1\rangle. \tag{5.82}$$

(The linear relations (5.78)–(5.79) clearly allow that.) Let us calculate the coefficients A_n assuming, for convenience, that all eigenstates, including the states n and $(n-1)$, are normalized:

$$\langle n|n\rangle = 1, \qquad \langle n-1|n-1\rangle = \langle n\,|\,\frac{\hat{a}^\dagger\,\hat{a}}{A_n^*\,A_n}\,|n\rangle$$

$$= \frac{1}{A_n^*A_n}\langle n\,|\,\hat{N}\,|n\rangle = \frac{N_n}{A_n^*A_n}\langle n|n\rangle = 1. \tag{5.83}$$

From here, we get $|A_n| = (N_n)^{1/2}$, i.e.

$$\hat{a}\,|n\rangle = N_n^{1/2}e^{i\varphi_n}\,|\,n-1\rangle, \tag{5.84}$$

where φ_n is an arbitrary real phase. Now let us consider what happens if all numbers N_n are integers. (Because of the definition of N_n, given by Eq. (5.74), it is convenient to call these integers n, i.e. to use the same letter as for the corresponding eigenstate.) Then when we have come down to state with $n = 0$, an attempt to make one more step down gives

$$\hat{a}\,|\,0\rangle = 0\,|\,-1\rangle. \tag{5.85}$$

But in accordance with Eq. (4.9), the state on the right-hand side of this equation is the 'null-state', i.e. does not exist[29]. This gives the (only known :-) resolution of the state ladder paradox: the ladder has the lowest step with $N_n = n = 0$.

As a by-product of our discussion, we have obtained a very important relation $N_n = n$, which means, in particular, that the state ladder shown in figure 5.7 includes *all* eigenstates of the oscillator. Plugging this relation into Eq. (5.75), we see that the *full* spectrum of eigenenergies of the harmonic oscillator is described by the simple formula

$$E_n = \hbar\omega_0\!\left(n + \frac{1}{2}\right), \qquad n = 0,\,1,\,2..., \tag{5.86}$$

which was already discussed in section 2.9. It is rather remarkable that the bra–ket formalism has allowed us to derive it without calculation of the corresponding (rather cumbersome) wavefunctions $\psi_n(x)$—see Eqs. (2.284).

Moreover, the formalism may be also used to calculate virtually any matrix element pertaining to the oscillator, without using $\psi_n(x)$. However, in order to do that, we should first calculate the coefficient A'_n participating in the second of Eqs. (5.82). This may be done absolutely similarly to the above calculation of A_n; alternatively, since we already know that $|A_n| = (N_n)^{1/2} = n^{1/2}$, we may notice that

---

[29] Please note again the radical difference between the null-state on the right-hand side of Eq. (5.85) and the state described by ket-vector $|0\rangle$ on the left-hand side of that relation. The latter state *does* exist and, moreover, represents the most important, ground state of the system, with $n = 0$—see Eqs. (2.274) and (2.275).

according to Eqs. (5.73) and (5.82), the eigenproblem (5.74), which in our new notation for N_n becomes

$$\hat{N} |n\rangle = n |n\rangle, \tag{5.87}$$

may be rewritten as

$$n |n\rangle = \hat{a}^\dagger \hat{a} |n\rangle = \hat{a}^\dagger A_n |n - 1\rangle = A_n A'_{n-1} |n\rangle. \tag{5.88}$$

Comparing the first and the last form of this equality, we see that $|A'_{n-1}| = n/|A_n| = n^{1/2}$, so that $A'_n = (n + 1)^{1/2} \exp(i\varphi_n')$. Taking all phases φ_n and φ_n' equal to zero for simplicity, we may spell out Eqs. (5.82) as[30]

$$\hat{a}^\dagger |n\rangle = (n + 1)^{1/2} | n + 1\rangle, \qquad \hat{a} |n\rangle = n^{1/2} | n - 1\rangle. \tag{5.89}$$

Now we can use these formulas to calculate, for example, the matrix elements of the operator $\hat{x}$ in the Fock state basis:

$$
\begin{aligned}
\langle n'| \hat{x} |n\rangle &\equiv x_0 \langle n'| \hat{\xi} |n\rangle = \frac{x_0}{\sqrt{2}} \langle n'| (\hat{a} + \hat{a}^\dagger) |n\rangle \\
&= \frac{x_0}{\sqrt{2}} (\langle n'| \hat{a} |n\rangle + \langle n'| \hat{a}^\dagger |n\rangle) \\
&= \frac{x_0}{\sqrt{2}} [n^{1/2}\langle n'|n - 1\rangle + (n + 1)^{1/2}\langle n'|n - 1\rangle].
\end{aligned}
\tag{5.90}
$$

Taking into account the Fock state orthonormality:

$$\langle n'|n\rangle = \delta_{n'n}, \tag{5.91}$$

this result becomes

$$
\begin{aligned}
\langle n'| \hat{x} |n\rangle &= \frac{x_0}{\sqrt{2}} [n^{1/2}\delta_{n',n-1} + (n + 1)^{1/2}\delta_{n',n+1}] \\
&\equiv \left(\frac{\hbar}{2m\omega_0}\right)^{1/2} [n^{1/2}\delta_{n',n-1} + (n + 1)^{1/2}\delta_{n',n+1}].
\end{aligned}
\tag{5.92}
$$

Acting absolutely similarly, for the momentum's matrix elements we get a similar expression:

$$\langle n'| \hat{p} |n\rangle = i\left(\frac{\hbar m\omega_0}{2}\right)^{1/2} [-n^{1/2}\delta_{n',n-1} + (n + 1)^{1/2}\delta_{n',n+1}]. \tag{5.93}$$

Hence the matrices of both operators in the Fock-state basis have only two diagonals, adjacent to the main diagonal; all other elements (including the main-diagonal ones) are zeros.

---

[30] A useful mnemonic rule for these key relations is that the c-number coefficient in any of them is equal to the square root of the *largest* number of the two states it relates.

The matrix elements of higher powers of these operators, as well as their products, may be handled similarly, though the higher the power, the bulkier the result. For example[31],

$$
\begin{aligned}
\langle n'| \, \hat{x}^2 \, |n\rangle = \langle n'| \, \hat{x}\hat{x} \, |n\rangle &= \sum_{n''=0}^{\infty} \langle n'| \, \hat{x} \, | n''\rangle \langle n''| \, \hat{x} \, |n\rangle \\
&= \frac{x_0^2}{2} \sum_{n''=0}^{\infty} \left[(n'')^{1/2}\delta_{n',n''-1} + (n''+1)^{1/2}\delta_{n',n''+1} \right] \\
&\quad \times \left[n^{1/2}\delta_{n'',n-1} + (n+1)^{1/2}\delta_{n'',n+1} \right] \\
&= \frac{x_0^2}{2} \{ [n(n-1)]^{1/2}\delta_{n',n-2} \\
&\quad + [(n+1)(n+2)]^{1/2}\delta_{n',n+2} + (2n+1)\delta_{n',n} \}.
\end{aligned}
\tag{5.94}
$$

For applications, the most important of these matrix elements are those on its main diagonal:

$$
\langle x^2 \rangle \equiv \langle n \, | \, \hat{x}^2 \, |n\rangle = \frac{x_0^2}{2}(2n+1).
\tag{5.95}
$$

This expression shows, in particular, that the expectation value of the oscillator's potential energy in the nth Fock state is

$$
\langle U \rangle \equiv \frac{m\omega_0^2}{2}\langle x^2 \rangle = \frac{\hbar\omega_0}{2}\left(n+\frac{1}{2}\right).
\tag{5.96}
$$

This is exactly ½ of the total energy (5.86) of the oscillator. As a sanity check, an absolutely similar calculation for the momentum squared, and hence for the kinetic energy $p^2/2m$, yields

$$
\langle p^2 \rangle = \langle n \, | \, \hat{p}^2 \, |n\rangle = (m\omega_0)^2\frac{x_0^2}{2}(2n+1) \equiv \hbar m\omega_0\left(n+\frac{1}{2}\right), \quad \text{so that}
$$
$$
\left\langle \frac{p^2}{2m} \right\rangle = \frac{\hbar\omega_0}{2}\left(n+\frac{1}{2}\right),
\tag{5.97}
$$

i.e. both partial energies are equal to $E_n/2$, just as in a classical oscillator[32].

Note that according to Eqs. (5.92) and (5.93), the expectation values of both x and p in any Fock state are equal zero:

$$
\langle x \rangle \equiv \langle n|\hat{x}|n\rangle = 0, \qquad \langle p \rangle \equiv \langle n|\hat{p}|n\rangle = 0,
\tag{5.98}
$$

---

[31] The first line of Eq. (5.94), evidently valid for any time-independent system, is the simplest of the so-called *sum rules*, which will be repeatedly discussed below.

[32] Still note that operators of the partial (potential and kinetic) energies do *not* commute either with each other or with the full-energy (Hamiltonian) operator, so that the Fock states n are *not* their eigenstates. This fact maps on the well-known oscillations of these partial energies (with the frequency $2\omega_0$) in a classical oscillator, at the full energy staying constant.

This is why, according to the general Eqs. (1.33) and (1.34), the results (5.95) and (5.97) also give the variances of the coordinate and the momentum, i.e. the squares of their uncertainties, $(\delta x)^2$ and $(\delta p)^2$—see the general Eq. (1.34). In particular, for the ground state ($n = 0$), these uncertainties are

$$\delta x = \frac{x_0}{\sqrt{2}} \equiv \left(\frac{\hbar}{2m\omega_0}\right)^{1/2}, \quad \delta p = \frac{m\omega_0 x_0}{\sqrt{2}} \equiv \left(\frac{\hbar m\omega_0}{2}\right)^{1/2}. \tag{5.99}$$

In the theory of measurements, these expressions are called the *standard quantum limit*.

5.5 Glauber states and squeezed states

There is evidently a huge difference between a quantum stationary (Fock) state of the oscillator and its classical state. Indeed, let us write the well-known classical equations of motion[33] of the oscillator (using capital letters to distinguish the classical variables from the arguments of quantum wavefunctions):

$$\dot{X} = \frac{P}{m}, \quad \dot{P} = -\frac{\partial U}{\partial x} = -m\omega_0^2 X. \tag{5.100}$$

On the so-called *phase plane*, with the Cartesian coordinates x and p, these equations describe a clockwise rotation of the representation point $\{X(t), P(t)\}$ along an elliptic trajectory starting from the initial point $\{X(0), P(0)\}$. (The normalization of the momentum by $m\omega_0$, similar to the one performed by the second of Eqs. (5.63), makes the trajectory pleasingly circular, with a constant radius equal to the oscillation's amplitude A, reflecting the constant full energy

$$E = \frac{m\omega_0^2}{2}A^2, \quad A^2 = [X(t)]^2 + \left[\frac{P(t)}{m\omega_0}\right]^2 = \text{const} = [X(0)]^2 + \left[\frac{P(0)}{m\omega_0}\right]^2, \tag{5.101}$$

determined by the initial conditions—see figure 5.8.)

For the forthcoming comparison with quantum states, it is convenient to describe this classical motion by the following dimensionless complex variable

$$\alpha(t) \equiv \frac{1}{\sqrt{2}x_0}\left[X(t) + i\frac{P(t)}{m\omega_0}\right], \tag{5.102}$$

which is essentially the standard complex-number representation of system's position on the 2D phase plane, with $|\alpha| \equiv A/\sqrt{2}x_0$. With this definition, Eqs. (5.100) are conveniently merged into one equation,

$$\dot{\alpha} = -i\omega_0\alpha, \tag{5.103}$$

---

[33] If Eqs. (5.100) are not evident, please consult a classical mechanics course—e.g. *Part CM* sections 3.2 and/or 10.1.

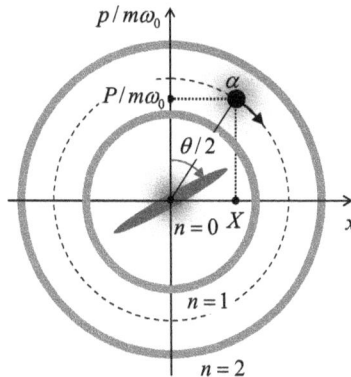

Figure 5.8. Representations of various states of a harmonic oscillator on the phase plane. The bold black point represents a classical state with a complex amplitude α, with the dashed line showing its trajectory. The (very imperfect) classical images of the Fock states with $n = 0$, 1, and 2 are shown in blue. The blurred red spot is the (equally schematic) image of a Glauber state α. Finally, the magenta elliptical spot is a classical image of a squeezed ground state. Arrows show the direction of the states' evolution in time.

with an evident, very simple solution

$$\alpha(t) = \alpha(0)\exp\{-i\omega_0 t\}, \tag{5.104}$$

where the constant $\alpha(0)$ may be complex, and is just the (normalized) classical complex amplitude of oscillations[34]. This equation describes sinusoidal oscillations of both $X(t) \propto \mathrm{Re}[\alpha(t)]$ and $P \propto \mathrm{Im}[\alpha(t)]$, with a phase shift of $\pi/2$ between them.

On the other hand, according to the basic Eq. (4.161), the time dependence of a Fock state, as of a stationary state of the oscillator, is limited to the phase factor $\exp\{-iE_n t/\hbar\}$ not in observables, but rather in the probability amplitude, and as a result, gives time-independent expectation values of x, p, or of any function thereof. (Moreover, as Eqs. (5.98) show, $\langle x \rangle = \langle p \rangle = 0$.) Taking into account Eqs. (5.96) and (5.97), the closest (though very imperfect) geometric image[35] for such a state on the phase plane is a circle of the radius $A_n = x_0(2n + 1)^{1/2}$, along which the wavefunction is uniformly spread as a wave—see the blue rings in figure 5.8. For the ground state ($n = 0$), with the wavefunction (2.275), a better image may be a blurred round spot, of a radius $\sim x_0$, at the origin. (It is easy to criticize such blurring, intended to represent the nonvanishing spreads (5.99), because it fails to reflect the fact that the total energy of the oscillator in the state, $E_0 = \hbar\omega_0/2$ is defined exactly, without any uncertainty.)

---

[34] See, e.g. *Part CM* chapter 5, especially Eq. (5.4).

[35] I have to confess that such geometric mapping of a quantum state on the phase plane [x, p] is not exactly defined; you may think about colored areas in figure 5.8 as the regions of the observable pairs $\{x, p\}$ most probably obtained in measurements. A quantitative definition of such a mapping will be given in section 7.3 using the Wigner function, though, as we will see, even such imaging has certain internal contradictions. Still such cartoons as figure 5.8 have a substantial cognitive/heuristic value, if their limitations are kept in mind.

So, the difference between a classical state of the oscillator and its Fock state n is very profound. However, the Fock states are not the only possible quantum states of the oscillator: according to the basic Eq. (4.6), any state described by the ket-vector

$$|\alpha\rangle = \sum_{n=0}^{\infty} \alpha_n |n\rangle \tag{5.105}$$

with an arbitrary set of (complex) c-numbers α_n, is also its legitimate state, subject only to the normalization condition $\langle\alpha|\alpha\rangle = 1$, giving

$$\sum_{n=0}^{\infty} |\alpha_n|^2 = 1. \tag{5.106}$$

It is natural to ask: could we select the coefficients α_n in such a special way that the state properties would be closer to the classical one; in particular the expectation values $\langle x\rangle$ and $\langle p\rangle$ of the coordinate and momentum would evolve in time as the classical values $X(t)$ and $P(t)$, while the uncertainties of these observables would be, just as in the ground state, given by Eqs. (5.99), and hence have the smallest possible uncertainty product, $\delta x \delta p = \hbar/2$. Let me show that such a *Glauber state*[36], which is schematically represented in figure 5.8 by a blurred red spot around the classical point $\{X(t), P(t)\}$, is indeed possible.

Conceptually the simplest way to find the corresponding coefficients α_n would be to calculate $\langle x\rangle$, $\langle p\rangle$, δx and δp for an arbitrary set of α_n, and then try to optimize these coefficients to reach our goal. However, this problem may be solved much more easily using the wave mechanics. Indeed, let us consider the following wavefunction

$$\Psi_a(x, t) = \left(\frac{m\omega_0}{\pi\hbar}\right)^{1/4} \exp\left\{-\frac{m\omega_0}{2\hbar}[x - X(t)]^2 + i\frac{P(t)x}{\hbar}\right\}, \tag{5.107}$$

Its comparison with Eqs. (2.275) shows that this is just the ground-state wavefunction, but with the center shifted from the origin into the classical point $\{X(t), P(t)\}$. A straightforward (though a bit bulky) differentiation over x and t shows that it satisfies the oscillator's Schrödinger equation, provided that the functions $X(t)$ and $P(t)$ obey the classical equations (5.100). Moreover, a similar calculation shows that the wavefunction (5.107) also satisfies the Schrödinger equation of an oscillator under the effect of a pulse of classical force $F(t)$, provided that the oscillator initially was in its ground state, and that the classical evolution law $\{X(t), P(t)\}$ in Eq. (5.107) takes this force into account[37]. Since for many incarnations of the harmonic

---

[36] Named after R J Glauber who studied these states in detail in 1965, though they had been discussed in brief by E Schrödinger as early as in 1926. Another popular name, 'coherent', for the Glauber states is very misleading, because all the quantum states we have studied so far (including the Fock states) may be represented as coherent (pure) superpositions of the basis states.
[37] For its description, it is sufficient to solve Eqs. (5.100), with $F(t)$ added to the right-hand side of the second of these equations.

oscillator, the ground state may be readily formed (for example, by providing a weak coupling of the oscillator to a low-temperature environment), the Glauber state is usually easier to form experimentally than any Fock state with $n > 0$. This is why the Glauber states are so important, and deserve more discussion.

In such discussion, there is a substantial place for the bra–ket formalism. For example, in order to calculate the corresponding coefficients in the expansion (5.105) by wave-mechanical means,

$$a_n = \langle n|\alpha \rangle = \int dx \langle n|x\rangle \langle x|\alpha\rangle = \int \psi_n^*(x)\ \Psi_\alpha(x, 0)dx, \qquad (5.108)$$

we would need to use not only the simple Eq. (5.107), but also the Fock state wavefunctions $\psi_n(x)$, which are not very appealing—see Eq. (2.284) again. Instead, this calculation may be readily done in the bra–ket formalism, giving us one important byproduct result as well.

Let us start from expressing the double shift of the ground state (by X and P), that has led us to Eq. (5.107), in the operator language. Forgetting about the P for a minute, let us find the *translation operator* $\hat{\mathcal{T}}_X$ that would produce the desirable shift of an arbitrary quantum state by the c-number distance X along the coordinate argument x. In the coordinate representation, this means

$$\hat{\mathcal{T}}_X \psi(x) \equiv \psi(x - X). \qquad (5.109)$$

Representing the wavefunction as the standard wave packet (4.264), we see that

$$\begin{aligned}\hat{\mathcal{T}}_X\psi(x) &= \frac{1}{(2\pi\hbar)^{1/2}}\int \varphi(p)\exp\left\{i\frac{p(x-X)}{\hbar}\right\}dp \\ &\equiv \frac{1}{(2\pi\hbar)^{1/2}}\int \varphi(p)\exp\left\{i\frac{px}{\hbar}\right\}\exp\left\{-i\frac{pX}{\hbar}\right\}dp.\end{aligned} \qquad (5.110)$$

Hence, the shift may be achieved by the multiplication of each Fourier component of the packet, with the momentum p, by $\exp\{-ipX/\hbar\}$. This gives us a hint that the general form of the translation operator, valid in *any* representation, should be

$$\hat{\mathcal{T}}_X = \exp\left\{-i\frac{\hat{p}X}{\hbar}\right\}. \qquad (5.111)$$

The proof of this formula is provided merely by the fact that, as we know from chapter 4, any operator is uniquely determined by the set of its matrix elements in any full and orthogonal basis, in particular the basis of momentum states p. According to Eq. (5.110), the analog of Eq. (4.235) for the p-representation, applied to the translation operator (which is evidently local), is

$$\int dp\langle p|\hat{\mathcal{T}}_X|p'\rangle\varphi(p') = \exp\left\{-i\frac{pX}{\hbar}\right\}\varphi(p), \qquad (5.112)$$

so that the operator (5.111) does exactly the job we need it to.

The operator that provides the shift of momentum by a c-number P is absolutely similar—with the opposite sign under the exponent, due to the opposite sign of the exponent in the reciprocal Fourier transform, so that the simultaneous shift by both X and P may be achieved by the following translation operator:

$$\hat{\mathcal{T}}_\alpha = \exp\left\{i\frac{P\hat{x} - \hat{p}X}{\hbar}\right\}. \tag{5.113}$$

As we already know, for a harmonic oscillator the creation–annihilation operators are more natural, so that we may use Eqs. (5.65) to recast Eq. (5.113) as

$$\hat{\mathcal{T}}_\alpha = \exp\{\alpha\hat{a}^\dagger - \alpha^*\hat{a}\}, \qquad \text{so that } \hat{\mathcal{T}}_\alpha^\dagger = \exp\{\alpha^*\hat{a} - \alpha\hat{a}^\dagger\}, \tag{5.114}$$

where α (which, generally, may be a function of time) is the c-number defined by Eq. (5.102). Now, according to Eq. (5.107), we may form the Glauber state's ket-vector just as

$$|\alpha\rangle = \hat{\mathcal{T}}_\alpha |0\rangle. \tag{5.115}$$

This formula, valid in any representation, is very elegant, but using it for practical calculations (say, of the expectation values of observables) is not too easy because of the exponent-of-operators form of the translation operator. Fortunately, it turns out that a much simpler representation for the Glauber state is possible. To show this, let us start with the following general (and very useful) property of exponential functions of an operator argument: if

$$\left[\hat{A}, \hat{B}\right] = \mu\hat{I}, \tag{5.116}$$

(where $\hat{A}$ and $\hat{B}$ are arbitrary linear operators, and μ is a c-number), then[38]

$$\exp\{+\hat{A}\}\hat{B}\exp\{-\hat{A}\} = \hat{B} + \mu\hat{I}. \tag{5.117}$$

Let us apply Eqs. (5.116) and (5.117) to two cases, both with

$$\hat{A} = \alpha^*\hat{a} - \alpha\hat{a}^\dagger, \quad \text{so that } \exp\{+\hat{A}\} = \hat{\mathcal{T}}_\alpha^\dagger, \quad \exp\{-\hat{A}\} = \hat{\mathcal{T}}_\alpha. \tag{5.118}$$

First, let us take $\hat{B} = \hat{I}$; then Eq. (5.116) is valid with $\mu = 0$, and Eq. (5.107) yields

$$\hat{\mathcal{T}}_\alpha^\dagger \hat{\mathcal{T}}_\alpha = \hat{I}, \tag{5.119}$$

This equality means that the translation operator is unitary—not a big surprise, because if we shift a classical point on the phase plane by a complex number $(+\alpha)$ and then by $(-\alpha)$, we certainly must come back to the initial position. Eq. (5.119) means merely that this fact is true for any quantum state as well.

---

[38] A proof of Eq. (5.117) may be readily achieved by expanding the operator $\hat{f}(\lambda) \equiv \exp\{+\lambda\hat{A}\}\hat{B}\exp\{-\lambda\hat{A}\}$ in the Taylor series with respect to the c-number parameter λ, and then evaluating the result at $\lambda = 1$.

Second, let us take $\hat{B} = \hat{a}$; in order to find the corresponding parameter μ, let us calculate the commutator on the left-hand side of Eq. (5.116) for this case. Using, at the due stage of the calculation, Eq. (5.68), we get

$$[\hat{A}, \hat{B}] = [\alpha^*\hat{a} - \alpha\hat{a}^\dagger, \hat{a}] = -\alpha[\hat{a}^\dagger, \hat{a}] = \alpha\hat{I}, \tag{5.120}$$

so that in this case $\mu = \alpha$, and Eq. (5.117) yields

$$\hat{\mathcal{T}}_\alpha^\dagger \hat{a} \hat{\mathcal{T}}_\alpha = \hat{a} + \alpha\hat{I}. \tag{5.121}$$

We have approached the summit of this beautiful calculation. Let us consider the operator

$$\hat{\mathcal{T}}_\alpha \hat{\mathcal{T}}_\alpha^\dagger \hat{a} \hat{\mathcal{T}}_\alpha. \tag{5.122}$$

Using Eq. (5.119), we may reduce this product to $\hat{a}\hat{\mathcal{T}}_\alpha$, while the application of Eq. (5.121) to the same expression (5.122) yields $\hat{\mathcal{T}}_\alpha\hat{a} + \alpha\hat{\mathcal{T}}_\alpha$. Hence, we get the following operator equality:

$$\hat{a}\hat{\mathcal{T}}_\alpha = \hat{\mathcal{T}}_\alpha\hat{a} + \alpha\hat{\mathcal{T}}_\alpha, \tag{5.123}$$

which may be applied to any state. Now acting by both sides of this equality on the ground state $|0\rangle$, and using the fact that $\hat{a}|0\rangle$ is the null-state, while $\hat{\mathcal{T}}_\alpha|0\rangle \equiv |\alpha\rangle$, we finally get a very simple and elegant result[39]:

$$\hat{a}|\alpha\rangle = \alpha|\alpha\rangle. \tag{5.124}$$

Thus any Glauber state α is one of eigenstates of the annihilation operator, namely the one with the eigenvalue equal to the c-number parameter α of the state, i.e. to the complex representation (5.102) of the classical point which is the center of the Glauber state's wavefunction[40]. This fact makes the calculations of all Glauber state properties much simpler. As an example, let us calculate $\langle x \rangle$ in the Glauber state with some c-number α:

$$\langle x \rangle = \langle\alpha|\hat{x}|\alpha\rangle = \frac{x_0}{\sqrt{2}}\langle\alpha|(\hat{a} + \hat{a}^\dagger)|\alpha\rangle$$
$$= \frac{x_0}{\sqrt{2}}\left(\langle\alpha|\hat{a}|\alpha\rangle + \langle\alpha|\hat{a}^\dagger|\alpha\rangle\right) \tag{5.125}$$

In the first term in the parentheses, we can apply Eq. (5.124) directly, while in the second term, we can use the bra-counterpart of that relation, $\langle\alpha|\hat{a}^\dagger = \langle\alpha|\alpha^*$. Now

---

[39] This result is also rather counter-intuitive. Indeed, according to Eq. (5.89), the annihilation operator $\hat{a}$, acting upon a Fock state n, 'beats it down' to the lower-energy state $(n - 1)$. However, according to Eq. (5.124), the action of the same operator on a Glauber state α does *not* lead to the state change and hence to any energy change! The resolution of this paradox is given by the representation of the Glauber state as a series of Fock states—see Eq. (5.134) below. The operator $\hat{a}$ indeed transfers each Fock component of this series to a lower-energy state, but it also re-weighs each term, so that the complete energy of the Glauber state remains constant.

[40] Note that the spectrum of eigenvalues α of Eq. (5.124), viewed as an eigenproblem, is continuous—it may be any complex number.

assuming that the Glauber state is normalized, $\langle \alpha | \alpha \rangle = 1$, and using Eq. (5.102), we get

$$\langle x \rangle = \frac{x_0}{\sqrt{2}}(\langle \alpha | \alpha | \alpha \rangle + \langle \alpha | \alpha^* | \alpha \rangle) = \frac{x_0}{\sqrt{2}}(\alpha + \alpha^*) = X, \qquad (5.126)$$

Acting absolutely similarly, we may readily extend this sanity check to verify that $\langle p \rangle = P$, and that δx and δp do indeed obey Eq. (5.99).

As one more sanity check, let us use Eq. (5.124) to re-calculate the Glauber state's wavefunction. Inner-multiplying both sides of that relation by the bra-vector $\langle x |$, and using the definition (5.65a) of the annihilation operator, we get

$$\frac{1}{\sqrt{2}\,x_0}\langle x | \left(\hat{x} + i\frac{\hat{p}}{m\omega_0} \right) | \alpha \rangle = \alpha \langle x | \alpha \rangle. \qquad (5.127)$$

Since $\langle x |$ is the bra-vector of the eigenstate of the Hermitian operator $\hat{x}$, they may be swapped, with the operator giving its eigenvalue x; acting on that bra-vector by the (local!) operator of momentum, we have to use it in the coordinate representation—see Eq. (4.245). As a result, we get

$$\frac{1}{\sqrt{2}\,x_0}\left(x\langle x | \alpha \rangle + \frac{\hbar}{m\omega_0}\frac{\partial}{\partial x}\langle x | \alpha \rangle \right) = \alpha \langle x | \alpha \rangle. \qquad (5.128)$$

But $\langle x | \alpha \rangle$ is nothing else than the Glauber state's wavefunction Ψ_α, so that Eq. (5.128) gives for it a first-order differential equation

$$\frac{1}{\sqrt{2}\,x_0}\left(x\Psi_\alpha + \frac{\hbar}{m\omega_0}\frac{\partial}{\partial x}\Psi_\alpha \right) = \alpha \Psi_\alpha. \qquad (5.129)$$

Chasing Ψ_α and x to the opposite sides of the equation, and using the definition (5.102) of the parameter α, we bring this equation to the form

$$\frac{\partial \Psi_\alpha}{\Psi_\alpha} = \frac{m\omega_0}{\hbar}\left[-x + \left(X + i\frac{P}{m\omega_0} \right) \right]\partial x. \qquad (5.130)$$

Integrating both parts, we return to Eq. (5.107) that had been derived earlier by wave-mechanics means.

Now that we can use Eq. (5.124) for finding the coefficients α_n in the expansion (5.105) of the Glauber state α in the series over the Fock states n. Plugging Eq. (5.105) into both sides of Eq. (5.124), using the second of Eqs. (5.89) on the left-hand side, and requiring the coefficients at each ket-vector $|n\rangle$ in both parts of the resulting relation to be equal, we get the following recurrence relation for the coefficients α_n:

$$\alpha_{n+1} = \frac{\alpha}{(n+1)^{1/2}}\alpha_n. \qquad (5.131)$$

Assuming some value of α_0, and applying the relation sequentially for $n = 1, 2,$ etc, we get

$$\alpha_n = \frac{\alpha^n}{(n\,!)^{1/2}}\alpha_0. \tag{5.132}$$

Now we can find α_0 from the normalization requirement (5.106), getting

$$|\alpha_0|^2 \sum_{n=0}^{\infty} \frac{|\alpha|^{2n}}{n!} = 1. \tag{5.133}$$

In this sum, we may readily recognize the Taylor expansion of the function $\exp\{|\alpha|^2\}$, so that the final result (besides an arbitrary common phase multiplier) is

$$|\alpha\rangle = \exp\left\{-\frac{|\alpha|^2}{2}\right\}\sum_{n=0}^{\infty} \frac{\alpha^n}{(n!)^{1/2}}\,|n\rangle. \tag{5.134}$$

It means in particular that if the oscillator is in the Glauber state α, the probabilities $W_n \equiv \alpha_n\alpha_n^*$ of finding the system on the nth energy level (5.86) obey the well-known Poisson distribution (figure 5.9):

$$W_n = \frac{\langle n\rangle^n}{n!}e^{-\langle n\rangle}, \tag{5.135}$$

where $\langle n\rangle$ is the statistical average of n (which is not necessarily integer!):

$$\langle n\rangle = \sum_{n=0}^{\infty} n\,W_n; \tag{5.136}$$

in our particular case

$$\langle n\rangle = |\alpha|^2. \tag{5.137}$$

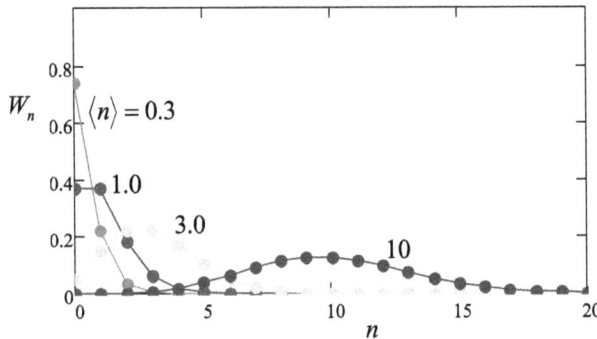

Figure 5.9. The Poisson distribution (5.135) for several values of $\langle n\rangle$. Note that W_n are defined only for integer values of n; the lines are only guides for the eye.

For applications, perhaps the most important mathematical property of this distribution is

$$\langle \tilde{n}^2 \rangle \equiv \langle (n - \langle n \rangle)^2 \rangle = \langle n \rangle, \qquad \text{so that } \delta n \equiv \langle \tilde{n}^2 \rangle^{1/2} = \langle n \rangle^{1/2}. \qquad (5.138)$$

Another important property is that at $\langle n \rangle \gg 1$, the Poisson distribution approaches the Gaussian ('normal') one, with a small relative rms fluctuation: $\delta n / \langle n \rangle \ll 1$—a trend visible in figure 5.9.

Now let us discuss the Glauber state's evolution in time. In the wave mechanics language, it is completely described by the dynamics (5.100) of the c-number shifts $X(t)$ and $P(t)$ participating in the wavefunction (5.107). Note again that, in contrast to the spread of the wave packet of a free particle, discussed in section 2.2, in the harmonic oscillator the Gaussian packet of the special width (5.99) does not spread at all!

An alternative and equivalent way of dynamics description is to use the Heisenberg equation of motion. As Eqs. (5.29) and (5.35) tell us, such equations for the Heisenberg operators of coordinate and momentum have to be similar to the classical equations (5.100):

$$\dot{\hat{x}}_{\mathrm{H}} = \frac{\hat{p}_{\mathrm{H}}}{m}, \qquad \dot{\hat{p}}_{\mathrm{H}} = -m\omega_0^2 \hat{x}_{\mathrm{H}}. \qquad (5.139)$$

Now using Eqs. (5.66), for the Heisenberg-picture creation and annihilation operators we get the equations

$$\dot{\hat{a}}_{\mathrm{H}} = -i\omega_0 \hat{a}_{\mathrm{H}}, \qquad \dot{\hat{a}}_{\mathrm{H}}^\dagger = +i\omega_0 \hat{a}_{\mathrm{H}}^\dagger, \qquad (5.140)$$

which are completely similar to the classical Eq. (5.103) for the c-number parameter α and its complex conjugate, and hence have the solutions identical to Eq. (5.104):

$$\hat{a}_{\mathrm{H}}(t) = \hat{a}_{\mathrm{H}}(0)e^{-i\omega_0 t}, \qquad \hat{a}_{\mathrm{H}}^\dagger(t) = \hat{a}_{\mathrm{H}}^\dagger(0)e^{i\omega_0 t}. \qquad (5.141)$$

As was discussed in section 4.6, such equations are very convenient, because they enable simple calculation of time evolution of observables for any initial state of the oscillator (Fock, Glauber, or any other) using Eq. (4.191). In particular, Eq. (5.141) shows that regardless of the initial state, the oscillator always returns to it *exactly* with the period $2\pi/\omega_0$.[41] Applied to the Glauber state with $\alpha = 0$, i.e. the ground state of the oscillator, such a calculation confirms that the Gaussian wave packet of the special width (5.99) does not spread in time at all—even temporarily.

Now let me briefly mention the states whose initial wave packet is still Gaussian, but has a different width, say $\delta x < x_0/\sqrt{2}$. As we already know from section 2.2, the momentum spread δp will be correspondingly larger, still with the smallest possible

---

[41] Actually, this fact is also evident from the Schrödinger picture of the oscillator's time evolution: due to the exactly equal distances $\hbar\omega_0$ between the eigenenergies (5.86), the time functions $a_n(t)$ in the fundamental expansion (1.69) of its wavefunction oscillate with frequencies $n\omega_0$, and hence they all share the same time period $2\pi/\omega_0$.

uncertainty product: $\delta x \delta p = \hbar/2$. Such *squeezed ground state* ζ, with zero expectation values of x and p, may be generated from the Fock/Glauber ground state:

$$| \zeta \rangle = \hat{S}_\zeta | 0 \rangle, \qquad (5.142a)$$

using the so-called *squeezing operator*,

$$\hat{S}_\zeta \equiv \exp \left\{ \frac{1}{2} (\zeta^* \hat{a} \hat{a} - \zeta \hat{a}^\dagger \hat{a}^\dagger) \right\}, \qquad (5.142b)$$

which depends on a complex c-number parameter $\zeta = re^{i\theta}$, where r and θ are real. The parameter's modulus r determines the squeezing degree; if ζ is real (i.e. $\theta = 0$), then

$$\delta x = \frac{x_0}{\sqrt{2}} e^{-r}, \qquad \delta p = \frac{m\omega_0 x_0}{\sqrt{2}} e^r, \qquad \text{so that } \delta x \delta p = \frac{m\omega_0 x_0^2}{2} \equiv \frac{\hbar}{2}. \qquad (5.143)$$

On the phase plane (figure 5.8), this state, with $r > 0$, may be represented by an oval spot squeezed along the axis x (hence the state's name) and stretched by the same factor e^r along the axis p; the same formulas but with $r < 0$ describe the opposite squeezing. On the other hand, the phase θ of the squeezing parameter ζ determines the angle $\theta/2$ of the oval's turn about the phase plane origin—see the magenta ellipse in figure 5.8; if $\theta \neq 0$, Eqs. (5.143) are valid for the variables $\{x', p'\}$ obtained from $\{x, p\}$ via clockwise rotation by that angle. For any of such origin-centered squeezed states, the time evolution is reduced to an increase of the angle with the rate ω_0, i.e. to the clockwise rotation of the ellipse, without its deformation, with the angular velocity ω_0—see the magenta arrows in figure 5.8. As a result, the uncertainties δx and δp oscillate in time with the double frequency $2\omega_0$. Such squeezed ground states have and may be formed, for example, by a parametric excitation of the oscillator[42], with a parameter modulation depth close to, but still below the threshold of the excitation of degenerate parametric oscillations.

By action of an additional external force, the center of a squeezed state may be displaced from the origin to an arbitrary point $\{X, P\}$. Such a displaced squeezed state may be described by the action of the translation operator (5.113) upon the ground squeezed state, i.e. by the action of the operator product $\hat{\mathcal{T}}_\alpha \hat{S}_\zeta$ on the usual (Fock/Glauber, i.e. non-squeezed) ground state. Calculations similar to those that led us from Eq. (5.114) to Eq. (5.124), show that such a displaced squeezed state is an eigenstate of the following mixed operator:

$$\hat{b} \equiv \hat{a} \cosh r + \hat{a}^\dagger e^{i\theta} \sinh r, \qquad (5.144)$$

with the same parameters r and θ, with the eigenvalue

$$\beta = \alpha \cosh r + \alpha^* e^{i\theta} \sinh r, \qquad (5.145)$$

---

[42] For a discussion and classical theory of this effect, see, e.g. *Part CM* section 5.5.

thus generalizing Eq. (5.124), which corresponds to $r = 0$. For the particular case $\alpha = 0$, Eq. (5.145) yields $\beta = 0$, i.e. the action of the operator (5.144) on the squeezed ground state ζ yields the null-state. Just as Eq. (5.124) in the case of the Glauber states, Eqs. (5.144) and (5.145) makes the calculation of the basic properties of the squeezed states (for example, the proof of Eqs. (5.143) for the case $\alpha = \theta = 0$) very straightforward.

Unfortunately, I do not have time here for a further discussion of the squeezed states—which have important implications for sensitive quantum measurements[43].

5.6 Revisiting spherically-symmetric systems

One more blank spot to fill has been left by our study, in section 3.6, of the wave mechanics of particle motion in spherically-symmetric 3D potentials. Indeed, while the azimuthal components of the eigenfunctions (the spherical harmonics) of such systems are very simple,

$$\psi_m = \frac{1}{(2\pi)^{1/2}} e^{im\varphi}, \quad \text{with } m = 0, \pm 1, \pm 2, \dots, \tag{5.146}$$

their polar components include the associated Legendre functions $P_l^m(\cos\theta)$, which may be expressed via elementary functions only indirectly—see Eqs. (3.165) and (3.168). This makes all the calculations less than transparent and, in particular, does not allow a clear insight into the origin of the very simple energy spectrum of such systems—see, e.g. Eq. (3.163). The bra–ket formalism, applied to the angular momentum operator, not only enables such insight and produces a very convenient tool for many calculations involving spherically-symmetric potentials, but also opens a clear way toward the unification of the orbital momentum with the particle's spin—the task to be addressed in the next section.

Let us start from using the correspondence principle to spell out the quantum-mechanical vector operator of the orbital angular momentum $\mathbf{L} \equiv \mathbf{r} \times \mathbf{p}$ of a point particle:

$$\hat{\mathbf{L}} \equiv \hat{\mathbf{r}} \times \hat{\mathbf{p}} = \begin{vmatrix} \mathbf{n}_x & \mathbf{n}_y & \mathbf{n}_z \\ \hat{r}_1 & \hat{r}_2 & \hat{r}_3 \\ \hat{p}_1 & \hat{p}_2 & \hat{p}_3 \end{vmatrix}, \quad \text{i.e. } \hat{L}_j \equiv \sum_{j'=1}^{3} \hat{r}_{j'}\hat{p}_{j''}\varepsilon_{jj'j''}, \tag{5.147}$$

where each of the indices j and j' and j'' may take any of the values 1, 2, and 3, and $\varepsilon_{jj'j''}$ is the Levi-Civita permutation symbol, which we have already used in section 4.5, and also in section 5.1, in the similar expression (5.18). From this definition, we

---

[43] See a brief discussion of this issue in section 10.1 below. For more on the squeezed states see, e.g. chapter 7 in the monograph by Gerry and Knight [6]. Also, note the spectacular measurements of the Glauber and squeezed states of electromagnetic (optical) oscillators by Breitenbach et al [7], a large (ten-fold) squeezing achieved in such oscillators by Vahlbruch et al [8], and first results on the ground state squeezing in micromechanical oscillators, with resonance frequencies $\omega_0/2\pi$ as low as a few MHz, using their parametric coupling to microwave electromagnetic oscillators—see, e.g. [9] and/or [10].

can readily calculate the commutation relations for all Cartesian components of operators $\hat{\mathbf{L}}$, $\hat{\mathbf{r}}$, and $\hat{\mathbf{p}}$; for example,

$$\left[\hat{L}_j, \hat{r}_{j'}\right] = \left[\sum_{k=1}^{3} \hat{r}_k \hat{p}_{j''} \varepsilon_{jkj''}, \hat{r}_{j'}\right] \equiv -\sum_{k=1}^{3} \hat{r}_k \left[\hat{r}_{j'}, \hat{p}_{j''}\right] \varepsilon_{jkj''}$$

$$= -i\hbar \sum_{k=1}^{3} \hat{r}_k \delta_{j'j''} \varepsilon_{jkj''} \equiv i\hbar \sum_{k=1}^{3} \hat{r}_k \varepsilon_{jj'k} \equiv i\hbar \hat{r}_{j''} \varepsilon_{jj'j''}.$$

(5.148)

The summary of all these calculations may be represented in a similar compact form:

$$\left[\hat{L}_j, \hat{r}_{j'}\right] = i\hbar \hat{r}_{j''} \varepsilon_{jj'j''}, \quad \left[\hat{L}_j, \hat{p}_{j'}\right] = i\hbar \hat{p}_{j''} \varepsilon_{jj'j''}, \quad \left[\hat{L}_j, \hat{L}_{j'}\right] = i\hbar \hat{L}_{j''} \varepsilon_{jj'j''}; \quad (5.149)$$

they show, in particular, that the commutator of $\hat{L}_j$ with a different (j'th) Cartesian component of this vector-operator is proportional to the complementary component (with the number $j'' \neq j, j'$) of that operator.

Also introducing in the natural way a (scalar!) operator of the observable $L^2 \equiv |\mathbf{L}|^2$,

$$\hat{L}^2 \equiv \hat{L}_x^2 + \hat{L}_y^2 + \hat{L}_z^2 \equiv \sum_{j=1}^{3} L_j^2, \quad (5.150)$$

it is straightforward to check that this operator commutes with each of the Cartesian components:

$$\left[\hat{L}^2, \hat{L}_j\right] = 0. \quad (5.151)$$

This result, at first sight, may seem to contradict the last of Eqs. (5.149). Indeed, haven't we learned in section 4.5 that commuting operators (e.g. $\hat{L}^2$ and any of $\hat{L}_j$) share their eigenstate sets? If yes, doesn't this set have to be common for all four operators? The resolution in this paradox may be found in the condition that was mentioned just after Eq. (4.138), but (sorry!) was not sufficiently emphasized there. According to that relation, if an operator has *degenerate* eigenstates (i.e. if $A_j = A_{j'}$ even for $j \neq j'$), they should not be necessarily all shared by another compatible operator.

This is exactly the situation with the orbital angular momentum operators, which may be schematically shown on a *Venn diagram* (figure 5.10):[44] the eigenstates of the operator $\hat{L}^2$ are highly degenerate[45], and their set broader than those of the

---

[44] This is just a particular example of the Venn diagrams (introduced in the 1880s by J Venn) that show possible relations (such as intersections, unions, complements, etc) between various sets of objects, and are a very useful tool in the general set theory.

[45] Note that this particular result is consistent with the classical picture of the angular momentum vector: even when its length is fixed, the vector may be oriented in various directions, corresponding to different values of its Cartesian components. However, in the classical picture, all these components may be fixed simultaneously, while in the quantum picture this is not true.

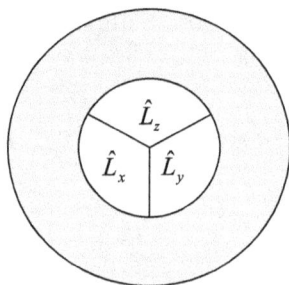

Figure 5.10. The Venn diagram showing the partitioning of the set of eigenstates of the operator $\hat{L}^2$. Each inner sector corresponds to the states shared with one of Cartesian component operators $\hat{L}_j$, while the outer (shaded) ring represents the eigenstates of $\hat{L}^2$ that are not shared with either of $\hat{L}_j$—for example, all linear combinations of the eigenstates of different component operators.

component operators $\hat{L}_j$ (that, as will be shown below, are non-degenerate until we consider a particle's spin).

Let us focus on just one of these three joint sets of eigenstates—by tradition, of the operators $\hat{L}^2$ and $\hat{L}_z$. (This tradition stems from the canonical form of the spherical coordinates, in which the polar angle is measured from the z-axis. Indeed, in the coordinate representation we may write

$$\hat{L}_z \equiv \hat{x}p_y - \hat{y}p_x = x\left(-i\hbar\frac{\partial}{\partial y}\right) - y\left(-i\hbar\frac{\partial}{\partial x}\right) = -i\hbar\frac{\partial}{\partial\varphi}. \tag{5.152}$$

Writing the standard eigenproblem for the operator in this representation, $\hat{L}_z\psi_m = L_z\psi_m$, we see that it is satisfied by the eigenfunctions (5.146), with eigenvalues $L_z = \hbar m$—which was already conjectured in section 3.5.) More specifically, let us consider a set of eigenstates $\{l, m\}$ corresponding to a certain degenerate eigenvalue of the operator $\hat{L}^2$, but all possible eigenvalues of operator $\hat{L}_z$, i.e. all possible quantum numbers m. (At this point, l is just some parameter that determines the eigenvalue of $\hat{L}^2$; it will be defined more explicitly in a minute.) In order to analyze this set, it is instrumental to introduce the so-called *ladder* (also called, respectively, 'raising' and 'lowering') *operators*[46]

$$\hat{L}_\pm \equiv \hat{L}_x \pm i\hat{L}_y. \tag{5.153}$$

It is simple to use this definition and the last of Eqs. (5.149) to calculate the following commutators:

$$[\hat{L}_+, \hat{L}_-] = 2\hbar\hat{L}_z, \quad \text{and} \quad [\hat{L}_z, \hat{L}_\pm] = \pm\hbar\hat{L}_\pm, \tag{5.154}$$

and also to use Eqs. (5.149) and (5.150) to prove two other important relations:

$$\hat{L}^2 = \hat{L}_z^2 + \hat{L}_+\hat{L}_- - \hbar\hat{L}_z, \qquad \hat{L}^2 = \hat{L}_z^2 + \hat{L}_-\hat{L}_+ + \hbar\hat{L}_z. \tag{5.155}$$

---

[46] Note a substantial similarity between this definition and Eqs. (5.65) for the creation/annihilation operators—defined in a different (harmonic oscillator's) Hilbert space.

Now let us rewrite the last of Eqs. (5.154) as

$$\hat{L}_z\hat{L}_\pm = \hat{L}_\pm\hat{L}_z \pm \hbar\hat{L}_\pm, \tag{5.156}$$

and act by both of its sides upon the ket-vector $|l, m\rangle$ of an arbitrary common eigenstate specified above:

$$\hat{L}_z\hat{L}_\pm|l, m\rangle = \hat{L}_\pm\hat{L}_z|l, m\rangle \pm \hbar\hat{L}_\pm|l, m\rangle. \tag{5.157}$$

Since the eigenvalues of the operator $\hat{L}_z$ are equal to $\hbar m$, in the first term of the right-hand side of Eq. (5.157) we may write

$$\hat{L}_z|l, m\rangle = \hbar m|l, m\rangle. \tag{5.158}$$

With that, Eq. (5.157) may be recast as

$$\hat{L}_z(\hat{L}_\pm|l, m\rangle) = \hbar(m \pm 1)(\hat{L}_\pm|l, m\rangle). \tag{5.159}$$

In a spectacular similarity with Eqs. (5.78) and (5.79) for the harmonic oscillator, Eq. (5.159) means that the states $\hat{L}_\pm|l, m\rangle$ are also eigenstates of the operator $\hat{L}_z$, corresponding to the eigenvalues $(m \pm 1)$. Thus the ladder operators act exactly as the creation and annihilation operators of a harmonic oscillator, moving the system up or down a ladder of eigenstates—see figure 5.11.

The most significant difference is that now the state ladder must end in both directions, because an infinite increase of $|m|$, with whatever sign of m, would cause the expectation values of the operator

$$\hat{L}_x^2 + \hat{L}_y^2 \equiv \hat{L}^2 - \hat{L}_z^2, \tag{5.160}$$

which corresponds to a non-negative observable, becoming negative. Hence there have to be two states on both ends of the ladder, with ket-vectors $|l, m_{max}\rangle$ and $|l, m_{min}\rangle$, such that

$$\hat{L}_+|l, m_{max}\rangle = 0, \qquad \hat{L}_-|l, m_{min}\rangle = 0. \tag{5.161}$$

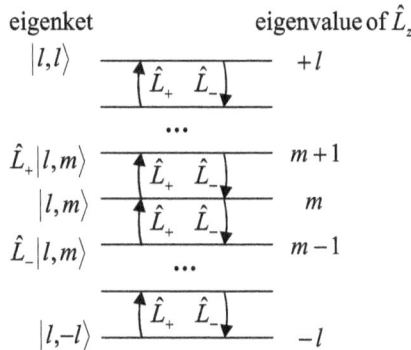

Figure 5.11. The ladder diagram hierarchy of the common eigenstates of the operators $\hat{L}^2$ and $\hat{L}_z$.

Due to the symmetry of the whole problem with respect to the replacement $m \to -m$, we should have $m_{min} = -m_{max}$. This m_{max} is exactly the quantum number traditionally called l, so that

$$-l \leqslant m \leqslant +l. \tag{5.162}$$

Evidently, this relation of quantum numbers m and l is semi-quantitatively compatible with the classical image of the angular momentum vector **L**, of the same length L, pointing in various directions, thus affecting the value of its component L_z. In this classical picture, however, L^2 would be equal to the square of $(L_z)_{max}$, i.e. to $(\hbar l)^2$; however, this is not so. Indeed, applying both parts of the second of the operator equalities (5.155) to the top state's vector $|l, m_{max}\rangle \equiv |l, l\rangle$, we get

$$\begin{aligned}\hat{L}^2 |l, l\rangle &= \hbar \hat{L}_z |l, l\rangle + \hat{L}_z^2 |l, l\rangle + \hat{L}_- \hat{L}_+ |l, l\rangle \\ &= \hbar^2 l |l, l\rangle + \hbar^2 l^2 |l, l\rangle + 0 = \hbar^2 l(l + 1) |l, l\rangle.\end{aligned} \tag{5.163}$$

Since by our initial assumption, all eigenvectors $|l, m\rangle$ correspond to the same eigenvalue of operator $\hat{L}^2$, this result means that all these eigenvalues are equal to $\hbar^2 l(l + 1)$. Just as in case of the spin-½ vector operators discussed in section 4.5, the deviation of this result from $\hbar^2 l^2$ may be interpreted as the result of unavoidable uncertainties ('fluctuations') of the x- and y-components of the angular momentum, which give non-zero positive contributions to $\langle L_x^2 \rangle$ and $\langle L_y^2 \rangle$, and hence to $\langle L^2 \rangle$ even if the angular momentum vector is aligned with the z-axis in the best possible way.

(For various applications of the ladder operators (5.153), one more relation is convenient:

$$\hat{L}_{\pm}|l, m\rangle = \hbar[l(l + 1) - m(m \pm 1)]|l, m \pm 1\rangle. \tag{5.164}$$

It may be readily proved from the above relations in the same way as the parallel Eqs. (5.89) for the harmonic-oscillator operators (5.65) were proved in section 5.4; due to this similarity, the proof is left for the reader's exercise[47].)

Now let us compare our results with those of section 3.6. Using the expression of Cartesian coordinates via the spherical ones exactly as this was done in Eq. (5.152), we get the following expressions for the ladder operators (5.153) in the coordinate representation:

$$\hat{L}_{\pm} = \hbar e^{\pm i\varphi} \left(\pm \frac{\partial}{\partial \theta} + i \cotan\theta \frac{\partial}{\partial \varphi} \right). \tag{5.165}$$

---

[47] The reader is also challenged to use the commutation relations discussed above to prove one more important property of the common eigenstates of $\hat{L}_z$ and $\hat{L}^2$:

$$\langle l, m|\hat{r}_j|l', m'\rangle = 0, \quad \text{unless } l' = l \pm 1 \text{ } and \text{ } m' = \text{ either } m \pm 1 \text{ or } m.$$

This property gives the *selection rule* for the orbital electric-dipole quantum transitions, to be discussed later in the course, especially in section 9.3. (The final selection rules at these transitions may be affected by the particle's spin—see the next section.)

Now plugging this relation, together with Eq. (5.152), into any of Eqs. (5.155), we get

$$\hat{L}^2 = -\hbar^2 \left[\frac{1}{\sin\theta} \frac{\partial}{\partial\theta} \left(\sin\theta \frac{\partial}{\partial\theta} \right) + \frac{1}{\sin^2\theta} \frac{\partial^2}{\partial\varphi^2} \right]. \tag{5.166}$$

But this is exactly the operator (besides its division by the constant parameter $2mR^2$) that stands on the left-hand side of Eq. (3.156). Hence that equation, which was explored by the 'brute-force' (wave-mechanical) approach in section 3.6, may be understood as the eigenproblem for the operator $\hat{L}^2$ in the coordinate representation, with the eigenfunctions $Y_l^m(\theta,\varphi)$ corresponding to the eigenkets $|l, m\rangle$, and the eigenvalues $L^2 = 2mR^2E$. As a reminder, the main result of that, rather involved analysis was expressed by Eq. (3.163), which now may be rewritten as

$$L_l^2 \equiv 2mR^2E_l = \hbar^2 l(l + 1), \tag{5.167}$$

in full agreement with Eq. (5.163), which was obtained by much more efficient means based on the bra–ket formalism. In particular, it is fascinating to see how easy it is to operate with the eigenvectors $|l, m\rangle$, while the coordinate representations of these ket-vectors, the spherical harmonics $Y_l^m(\theta,\varphi)$, may be only expressed by rather complicated functions—please have one more look at Eq. (3.171) and figure 3.20.

Note that all relations considered in this section are not conditioned by any particular Hamiltonian of the system under analysis, though they (as well as those discussed in the next section) are especially important for particles moving in spherically-symmetric potentials.

5.7 Spin and its addition to orbital angular momentum

Surprisingly, the theory described in the last section is useful for much more than the orbital motion analysis. In particular, it helps to generalize the spin-½ results discussed in chapter 4 to other values of spin s—the parameter still to be defined. For that, let us notice that the commutation relations (4.155) for spin-½, which were derived from the Pauli matrix properties, may be rewritten in exactly the same form as Eqs. (5.149) and (5.151) for the orbital momentum:

$$\left[\hat{S}_j, \hat{S}_{j'} \right] = i\hbar \hat{S}_{j''} \varepsilon_{jj'j''}, \quad \left[\hat{S}^2, \hat{S}_j \right] = 0. \tag{5.168}$$

It had been postulated (and then confirmed by numerous experiments) that these relations hold for quantum particles with any spin. Now note that all the calculations of the last section have been based *almost* exclusively on such relations—the only exception will be discussed imminently. Hence, we may repeat them for the spin operators, and get the relations similar to Eqs. (5.158) and (5.163):

$$\hat{S}_z|s, m_s\rangle = \hbar m_s|s, m_s\rangle, \quad \hat{S}^2|s, m_s\rangle = \hbar^2 s(s + 1)|s, m_s\rangle,$$
$$0 \leqslant s, \quad -s \leqslant m_s \leqslant +s, \tag{5.169}$$

where m_s is a quantum number parallel to the orbital magnetic number m, and the non-negative constant s is defined as the maximum value of $|m_s|$. This parameter s is exactly what is called the *particle's spin*.

Now let us return to the only part of our orbital moment calculations that has *not* been derived from the commutation relations. This was the fact, based on the solution (5.146) of the orbital motion problems, that the quantum number m (the analog of m_s) may be only integer. For the spin, we do not have such a solution, so that the spectrum of numbers m_s (and hence its limits $\pm s$) should be found from the more loose requirement that the eigenstate ladder, extending from $-s$ to $+s$, has an integer number of steps. Hence, $2s$ has to be integer, i.e. the spin s of a quantum particle may be either *integer* (as it is, for example, for photons, gluons, and massive bosons $W^\pm$ and Z^0), or *half-integer* (e.g. for all quarks and leptons, notably including electrons)[48]. For $s = \frac{1}{2}$, this picture yields all properties of the spin-½, which were derived in chapter 4 from Eqs. (4.115)–(4.117). In particular, the operators $\hat{S}^2$ and $\hat{S}_z$ have two common eigenstates (↑ and ↓), with $S_z = \hbar m_s = \pm\hbar/2$, both with $S^2 = s(s+1)\hbar^2 = (3/4)\hbar^2$.

Note that this analogy with the angular momentum sheds new light on the symmetry properties of spin-½. Indeed, the fact that m in Eq. (5.146) is integer was derived in section 3.5 from the requirement that making a full circle around axis z, we should find a similar value of wavefunction ψ_m, which differs from the initial one by an inconsequential factor $\exp\{2\pi i m\} = +1$. With the replacement $m \to m_s = \pm\frac{1}{2}$, such operation would multiply the wavefunction by $\exp\{\pm\pi i\} = -1$, i.e. reverse its sign. Of course, spin cannot be described by a usual wavefunction, but this *odd parity* of electrons, and all other spin-½ particles, is clearly revealed in properties of multiparticle systems (see chapter 8 below), and as a result, in their statistics (see, e.g. *Part SM* chapter 2).

Now we are sufficiently equipped to analyze the situations in which a particle has both the orbital momentum and the spin—as an electron in an atom. In classical mechanics, such an object, with the spin **S** interpreted as the angular moment of its internal rotation, would be characterized by the *total angular momentum* vector **J** = **L** + **S**. Following the correspondence principle, we may make an assumption that quantum-mechanical properties of this observable may be described by the similarly defined vector operator:

$$\hat{\mathbf{J}} \equiv \hat{\mathbf{L}} + \hat{\mathbf{S}}, \tag{5.170}$$

with the Cartesian components

$$\hat{J}_z \equiv \hat{L}_z + \hat{S}_z, \text{ etc.}, \tag{5.171}$$

and the magnitude squared equal to

$$\hat{J}^2 \equiv \hat{J}_x^2 + \hat{J}_y^2 + \hat{J}_z^2. \tag{5.172}$$

---

[48] As a reminder, in the Standard Model of particle physics, such hadrons as mesons and baryons (notably including protons and neutrons) are essentially composite particles. However, at non-relativistic energies, protons and neutrons may be considered fundamental particles with $s = \frac{1}{2}$.

Let us examine the properties of this vector operator. Since its two components (5.170) describe different degrees of freedom of the particle, i.e. belong to different Hilbert spaces, they have to be completely commuting:

$$\left[\hat{L}_j, \hat{S}_{j'}\right] = 0, \qquad \left[\hat{L}^2, \hat{S}^2\right] = 0. \qquad (5.173)$$

The above equalities are sufficient to derive the commutation relations for the operator $\hat{\mathbf{J}}$, and unsurprisingly, they turn out to be absolutely similar to those of its components:

$$\left[\hat{J}_j, \hat{J}_{j'}\right] = i\hbar \hat{J}_{j''}\varepsilon_{jj'j''}, \qquad \left[\hat{J}^2, \hat{J}_j\right] = 0. \qquad (5.174)$$

Now repeating all the arguments of the last section, we may derive the following expressions for the common eigenstates of operators $\hat{J}^2$ and $\hat{J}_z$:

$$J_z\left|j, m_j\right\rangle = \hbar m_j\left|j, m_j\right\rangle, \qquad \hat{J}^2\left|j, m_j\right\rangle = \hbar^2 j(j+1)\left|j, m_j\right\rangle,$$
$$0 \leqslant j, \quad -j \leqslant m_j \leqslant +j, \qquad (5.175)$$

where j and m_j are new quantum numbers[49]. Repeating the arguments just made for s and m_s, we may conclude that j and m_j may be either integer or half-integer.

Before we proceed, one remark on notation: it is very convenient to use the same letter m for numbering eigenstates of all momentum components participating in Eq. (5.171), with corresponding indices (j, l, and s), in particular, to replace what we called m with m_l. With this replacement, the main results of the last section may be summarized in the form similar to Eqs. (5.168), (5.169), (5.174), and (5.175):

$$\left[\hat{L}_j, \hat{L}_{j'}\right] = i\hbar \hat{L}_{j''}\varepsilon_{jj'j''}, \qquad \left[\hat{L}^2, \hat{L}_j\right] = 0, \qquad (5.176)$$

$$\hat{L}_z\left|l, m_l\right\rangle = \hbar m_l\left|l, m_l\right\rangle, \qquad \hat{L}^2\left|l, m_l\right\rangle = \hbar^2 l(l+1)\left|l, m_l\right\rangle,$$
$$0 \leqslant l, \quad -l \leqslant m_l \leqslant +l. \qquad (5.177)$$

In order to understand which eigenstates participating in Eqs. (5.169), (5.175), and (5.177) are compatible with each other, it is straightforward to use Eq. (5.172), together with Eqs. (5.168), (5.173), (5.174), and (5.176) to get the following relations:

$$\left[\hat{J}^2, \hat{L}^2\right] = 0, \qquad \left[\hat{J}^2, \hat{S}^2\right] = 0, \qquad (5.178)$$

$$\left[\hat{J}^2, \hat{L}_z\right] \neq 0, \qquad \left[\hat{J}^2, \hat{S}_z\right] \neq 0. \qquad (5.179)$$

This result is represented schematically in the Venn diagram shown in figure 5.12, in which the crossed arrows indicate the only *non*-commuting pairs of operators.

---

[49] Let me hope that the difference between the quantum number j, and the indices j, j', j'' numbering the Cartesian components in the relations like Eqs. (5.168) or (5.174), is absolutely clear from the context.

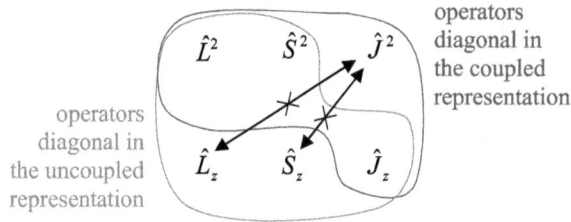

Figure 5.12. The Venn diagram of angular momentum operators, and their mutually-commuting groups.

This means that there are eigenstates shared by two groups of operators encircled with colored lines in figure 5.12. The first group (encircled red), consists of all these operators but $\hat{J}^2$. This means that there are eigenstates shared by the five remaining operators, and these states correspond to certain values of the corresponding quantum numbers: l, m_l, s, m_s, and m_j. Actually, only four of these numbers are independent, because due to Eq. (5.171) for these compatible operators, for each eigenstate of this group, their 'magnetic' quantum numbers m have to satisfy the following relation:

$$m_j = m_l + m_s. \tag{5.180}$$

Hence the common eigenstates of the operators of this group are fully defined by just four quantum numbers, for example, l, m_l, s, and m_s. For some calculations, especially those for the systems whose Hamiltonians include only the operators of this group, it is convenient[50] to the use this set of eigenstates as the basis; frequently this is called the *uncoupled representation*.

However, in some situations we cannot ignore interactions between the orbital and spin degrees of freedom (in the common jargon, the *spin–orbit coupling*), which leads in particular to splitting (called the *fine structure*) of the atomic energy levels even in the absence of external magnetic field. I will discuss these effects in detail in the next chapter, and now will only note that they may be described by a term proportional to the product $\hat{\mathbf{L}} \cdot \hat{\mathbf{S}}$, in the system's Hamiltonian. If this term is substantial, the uncoupled representation becomes inconvenient. Indeed, writing

$$\hat{J}^2 = (\hat{\mathbf{L}} + \hat{\mathbf{S}})^2 = \hat{L}^2 + \hat{S}^2 + 2\hat{\mathbf{L}} \cdot \hat{\mathbf{S}}, \quad \text{so that} \quad 2\hat{\mathbf{L}} \cdot \hat{\mathbf{S}} = \hat{J}^2 - \hat{L}^2 - \hat{S}^2, \tag{5.181}$$

and looking at figure 5.12 again, we see that the operator $\hat{\mathbf{L}} \cdot \hat{\mathbf{S}}$, describing the spin–orbit coupling, does not commute with the operators $\hat{L}_z$ and $\hat{S}_z$. This means that stationary states of the system with this term in the Hamiltonian do not belong to the uncoupled representation's basis. On the other hand, Eq. (5.181) shows that the operator $\hat{\mathbf{L}} \cdot \hat{\mathbf{S}}$ does commute with all four operators of another group, encircled blue in figure 5.12. According to Eqs. (5.178), (5.179), and (5.181), all operators of that group also commute with each other, so that they have common eigenstates,

---

[50] This is especially true for motion in spherically-symmetric potentials, whose stationary states correspond to definite l and m_l; however, the relations discussed in this section are important for some other problems as well.

described by quantum numbers, l, s, j, and m_j. This group is the basis for the so-called *coupled representation* of particle's states.

Excluding, for notation briefness, the quantum numbers l and s, common for the both groups, it is convenient to denote the common ket-vectors of each group as, respectively,

$$|\, m_l, m_s\rangle, \quad \text{for the uncolpled representation's basis,}$$
$$|\, j, m_j\rangle, \quad \text{for the coupled representation's basis.} \tag{5.182}$$

As we will see in the next chapter, for solution of some important problems (e.g. the fine structure of atomic spectra, and the Zeeman effect), we will need the relation between the kets $|\, j, m_j\rangle$ and the kets $|m_l, m_s\rangle$. This relation may be represented as the usual linear superposition,

$$\left|\, j, m_j\right\rangle = \sum_{m_l, m_s} |\, m_l, m_s\rangle\left\langle m_l, m_s\middle|\, j, m_j\right\rangle, \tag{5.183}$$

whose brackets (*c*-numbers), essentially the elements of the unitary matrix of the transformation between two eigenstate bases (5.182), are called the *Clebsch–Gordan coefficients*.

The best (though imperfect) classical interpretation of Eq. (5.183) I can offer is as follows. If the lengths of the vectors $\mathbf{L}$ and $\mathbf{S}$ (in quantum mechanics associated with the numbers l and s, respectively), and also their scalar product $\mathbf{L} \cdot \mathbf{S}$, are all fixed, then so is the length of the vector $\mathbf{J} = \mathbf{L} + \mathbf{S}$—whose length in quantum mechanics is described by the number j. Hence, the classical image of a specific eigenket $|\, j, m_j\rangle$, in which l, s, j, and m_j are all fixed, is a state in which L^2, S^2, J^2, and J_z are fixed. However, this fixation still allows for an arbitrary rotation of the pair of vectors $\mathbf{L}$ and $\mathbf{S}$ (with a fixed angle between them, and hence fixed $\mathbf{L} \cdot \mathbf{S}$ and J^2) about the direction of vector $\mathbf{J}$—see figure 5.13.

Hence the components L_z and S_z in these conditions are *not* fixed, and in classical mechanics may take a continuum of values, two of which (with the largest and smallest possible values of S_z) are shown in figure 5.13. In quantum mechanics, these components are quantized, with their states represented by eigenkets $|m_l, m_s\rangle$, so that a linear combination of such kets is necessary to represent a ket $|\, j, m_j\rangle$. This is exactly what Eq. (5.183) does.

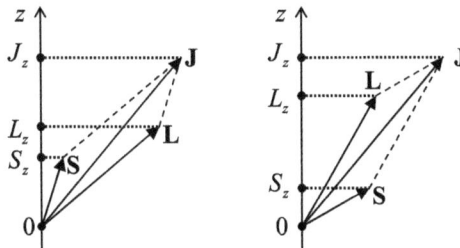

Figure 5.13. A classical image of two quantum states with the same l, s, j, and m_j, but different m_l and m_s.

Some of properties of the Clebsch–Gordan coefficients $\langle m_l, m_s | j, m_j \rangle$ may be readily established. For example, the coefficients do not vanish only if the involved magnetic quantum numbers satisfy Eq. (5.180). (In our current case, this relation is not an elementary corollary of Eq. (5.171), because in the Clebsch–Gordan coefficients, with the quantum numbers m_l, m_s in one state vector, and m_j in the other state vector, characterize the relation between different groups of the basis states, so we need to prove this fact.) All matrix elements of the null-operator

$$\hat{J}_z - (\hat{L}_z + \hat{S}_z) = \hat{0} \qquad (5.184)$$

should equal zero in any basis; in particular

$$\langle j, m_j | \hat{J}_z - (\hat{L}_z + \hat{S}_z) | m_l, m_s \rangle = 0. \qquad (5.185)$$

Acting by the operator $\hat{J}_z$ upon the bra-vector, and by the sum $(\hat{L}_z + \hat{S}_z)$ upon the ket-vector, we get

$$\left[m_j - (m_l + m_s) \right] \langle j, m_j | m_l, m_s \rangle = 0, \qquad (5.186)$$

thus proving that

$$\langle m_l, m_s | j, m_s \rangle \equiv \langle m_l, m_s | j, m_s \rangle^* = 0, \quad \text{if } m_j \neq m_l + m_s. \qquad (5.187)$$

For the most important case of spin-½ particles (with $s = $ ½, and hence $m_s = \pm$½), whose uncoupled representation basis includes $2 \times (2l + 1)$ states, the restriction (5.187) enables the representation of all nonvanishing Clebsch–Gordan coefficients on the simple 'rectangular' diagram shown in figure 5.14. Indeed, each coupled-representation eigenket $|j, m_j\rangle$, with $m_j = m_l + m_s = m_l \pm$ ½, may be related by non-zero Clebsch–Gordan coefficients to at most two uncoupled-representation eigenstates $|m_l, m_s\rangle$. Since m_l may only take integer values from $-l$ to $+l$, m_j may only take semi-integer values on the interval $[-l - $½$, l + $½$]$. Hence, by the definition of j as $(m_j)_{max}$, its maximum value has to be $l + $ ½, and for $m_j = l + $ ½, this is the only possible value with this j. This means that the uncoupled state with $m_l = l$ and $m_s = $ ½ should be identical to the coupled-representation state with $j = l + $ ½ and $m_j = l + $ ½:

$$\left| j = l + \tfrac{1}{2}, m_j = l + \tfrac{1}{2} \right\rangle = \left| m_l = m_j - \tfrac{1}{2}, m_s = +\tfrac{1}{2} \right\rangle. \qquad (5.188)$$

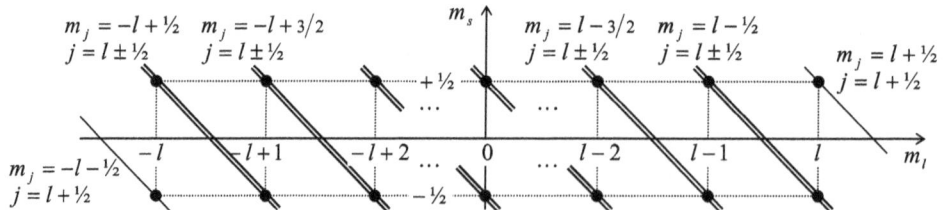

Figure 5.14. A graphical representation of possible basis states of a spin-½ particle with a fixed l. Each dot corresponds to an uncoupled-representation ket-vector $|m_l, m_s\rangle$, while each sloped line corresponds to one coupled-representation ket-vector $|j, m_j\rangle$, related by Eq. (5.183) to the kets $|m_l, m_s\rangle$ whose dots it connects.

In figure 5.14, these identical states are represented with the top-rightmost point (the uncoupled representation) and the sloped line passing through it (the coupled representation).

However, already the next value of this quantum number, $m_j = l - \frac{1}{2}$, is compatible with two values of j, so that each $|m_l, m_s\rangle$ ket has to be related to two $|j, m_j\rangle$ kets by two Clebsch–Gordan coefficients. Since j changes in unit steps, these values of j have to be $l \pm \frac{1}{2}$. This choice,

$$j = l \pm \frac{1}{2}, \qquad\qquad (5.189)$$

evidently satisfies all lower values of m_j as well—see figure 5.14.[51] (Again, only one value, $j = l + \frac{1}{2}$, is necessary to represent the state with the lowest $m_j = -l - \frac{1}{2}$—see the bottom leftmost point of that diagram.) Note that the total number of the coupled-representation states is $1 + 2 \times 2l + 1 \equiv 2(2l + 1)$, i.e. the same as those in the uncoupled representation. So, for spin-$\frac{1}{2}$ systems, each sum (5.183), for fixed j and m_j (plus the fixed common parameter l, plus the common $s = \frac{1}{2}$), has at most two terms, i.e. involves at most two Clebsch–Gordan coefficients.

These coefficients may be calculated in a few steps, all but the last one rather simple even for an arbitrary spin s. First, the similarity of the vector operators $\hat{\mathbf{J}}$ and $\hat{\mathbf{S}}$ to the operator $\hat{\mathbf{L}}$, expressed by Eqs. (5.169), (5.175), and (5.177), may be used to argue that the matrix elements of the operators $\hat{S}_\pm$ and $\hat{J}_\pm$, defined similarly to $\hat{L}_\pm$, have the matrix elements similar to those given by Eq. (5.164). Next, acting by the operator $\hat{J}_\pm = \hat{L}_\pm + \hat{S}_\pm$ upon both parts of Eq. (5.183), and then inner-multiplying the result by the bra vector $\langle m_l, m_s|$ and using the above matrix elements, we may get recurrence relations for the Clebsch–Gordan coefficients with adjacent values of m_l, m_s, and m_j. Finally, these relations may be recurrently applied to the adjacent states in both representations, starting from any of the two states common for them—for example, from state with the ket-vectors (5.188), corresponding to the top right point in figure 5.14. Let me leave these straightforward but somewhat tedious calculations for the reader's exercise, and just cite the final result of this procedure for $s = \frac{1}{2}$:[52]

$$\left\langle m_l = m_j - \tfrac{1}{2}, m_s = +\tfrac{1}{2} \middle| j = l \pm \tfrac{1}{2}, m_j \right\rangle = \pm \left(\frac{l \pm m_j + \tfrac{1}{2}}{2l + 1} \right)^{1/2},$$

$$\left\langle m_l = m_j + \tfrac{1}{2}, m_s = -\tfrac{1}{2} \middle| j = l \pm \tfrac{1}{2}, m_j \right\rangle = + \left(\frac{l \mp m_j + \tfrac{1}{2}}{2l + 1} \right)^{1/2}. \qquad (5.190)$$

---

[51] Eq. (5.189) allows a semi-qualitative classical interpretation in terms of the vector diagrams shown in figure 5.13: since, according to Eq. (5.169), $\hbar s$ gives the scale of the length of the vector $\mathbf{S}$, if it is small ($s = \frac{1}{2}$), the length of vector $\mathbf{J}$ (similarly scaled by $\hbar j$) cannot deviate much from the length of the vector $\mathbf{L}$ (scaled by $\hbar l$) for any spatial orientation of these vectors, so that j cannot differ from l too much. Note also that for a fixed m_j, the alternating sign in Eq. (5.189) is independent of the sign of m_s—see also Eqs. (5.190).
[52] For arbitrary spin s, the calculations and even the final expressions for the Clebsch–Gordan coefficients are rather bulky. They may be found, typically in a table form, mostly in special monographs—see, e.g. [11].

In this course, this relation will be used in section 6.4 for an analysis of the anomalous Zeeman effect. Moreover, most of the angular momentum addition theory described above is also valid for the addition of angular momenta of multiparticle system components, so we will revisit it in chapter 8.

To conclude this section, I have to note that the Clebsch–Gordan coefficients (for arbitrary s) participate also in the so-called *Wigner–Eckart theorem* that expresses the matrix elements of *spherical tensor operators*, in the coupled-representation basis $|j, m_j\rangle$, via a reduced set of matrix elements. This theorem may be useful, for example, for calculation of the rate of quantum transitions to/from high-n states in spherically-symmetric potentials. Unfortunately, a discussion of this theorem and its applications would require a higher mathematical background than I can expect from my readers, and more time/space than I can afford[53].

5.8 Problems

Problem 5.1. Use the discussion in section 5.1 to find an alternative solution of problem 4.18.

Problem 5.2. A spin-½ is placed into an external magnetic field, with a time-independent orientation, its magnitude $\mathscr{B}(t)$ being an arbitrary function of time. Find explicit expressions for the Heisenberg operators and the expectation values of all three Cartesian components of the spin, as functions of time, in a coordinate system of your choice.

Problem 5.3. A two-level system is in a quantum state α described by the ket-vector $|\alpha\rangle = \alpha_\uparrow|\uparrow\rangle + \alpha_\downarrow|\downarrow\rangle$, with given (generally, complex) c-number coefficients $\alpha_{\uparrow\downarrow}$. Prove that we can always select a three-component vector $\mathbf{c} = \{c_x, c_y, c_z\}$ of real c-numbers, such that α is an eigenstate of the operator $\mathbf{c} \cdot \hat{\boldsymbol{\sigma}}$, where $\hat{\boldsymbol{\sigma}}$ is the Pauli vector-operator. Find all possible values of $\mathbf{c}$ satisfying this condition, and the second eigenstate (orthogonal to α) of the operator $\mathbf{c} \cdot \hat{\boldsymbol{\sigma}}$. Give a Bloch-sphere interpretation of your result.

Problem 5.4.* Analyze the statistics of the spacing $S \equiv E_+ - E_-$ between the energy levels of a two-level system, assuming that all elements $H_{jj'}$ of its Hamiltonian matrix (5.2) are independent random numbers, with equal and constant probability densities within the energy interval of interest. Compare the result with that for a purely diagonal matrix, with the similar probability distribution of the random diagonal elements.

Problem 5.5. For a periodic motion of a single particle in a confining potential $U(\mathbf{r})$, the *virial theorem* of non-relativistic classical mechanics[54] is reduced to the following equality:

---

[53] For the interested reader I can recommend either section 17.7 in [12] or section 3.10 in [13].
[54] See, e.g. *Part CM* problem 1.12.

$$\overline{T} = \frac{1}{2}\overline{\mathbf{r} \cdot \nabla U},$$

where T is particle's kinetic energy, and the top bar means averaging over the time period of motion. Prove the quantum-mechanical version of the theorem for an arbitrary stationary quantum state, in the absence of spin effects:

$$\langle T \rangle = \frac{1}{2}\langle \mathbf{r} \cdot \nabla U \rangle,$$

where the angular brackets mean the expectation values of the observables.

Hint: Mimicking the proof of the classical virial theorem, consider the time evolution of the following operator:

$$\hat{G} \equiv \hat{\mathbf{r}} \cdot \hat{\mathbf{p}}.$$

Problem 5.6. Calculate, in the WKB approximation, the transparency $\mathscr{T}$ of tunneling of a 2D particle with energy $E < U_0$ through a saddle-shaped potential 'pass'

$$U(x, y) = U_0\left(1 + \frac{xy}{a^2}\right),$$

where $U_0 > 0$ and a are real constants.

Problem 5.7. Calculate the so-called *Gamow factor*[55] for the alpha decay of atomic nuclei, i.e. the exponential factor in the transparency of the potential barrier resulting from the following simple model of the alpha-particle's potential energy as a function of its distance from the nuclear center:

$$U(r) = \begin{cases} U_0 < 0, & \text{for } r < R, \\ \dfrac{ZZ'e^2}{4\pi\varepsilon_0 r}, & \text{for } R < r, \end{cases}$$

(where $Ze = 2e > 0$ is the charge of the particle, $Z'e > 0$ is that of the nucleus after the decay, and R is the nucleus' radius), in the WKB approximation.

Problem 5.8. Use the WKB approximation to calculate the average time of ionization of a hydrogen atom, initially in its ground state, made metastable by the application of an additional weak, uniform, constant electric field $\mathscr{E}$. Formulate the conditions of validity of your result.

Problem 5.9. For a 1D harmonic oscillator with mass m and frequency ω_0, calculate:

---

[55] Named after G Gamow, who made this calculation as early as in 1928.

(i) all matrix elements $\langle n|\hat{x}^3|n'\rangle$, and

(ii) the diagonal matrix elements $\langle n|\hat{x}^4|n\rangle$,

where n and n' are arbitrary Fock states.

Problem 5.10. Calculate the sum (over all $n > 0$) of the so-called *oscillator strengths*,

$$f_n \equiv \frac{2m}{\hbar^2}(E_n - E_0)\,|\langle n|\hat{x}|0\rangle|^2,$$

(i) for a 1D harmonic oscillator, and

(ii) for a 1D particle confined in an arbitrary stationary potential.

Problem 5.11. Prove the so-called *Bethe sum rule*,

$$\sum_{n'}(E_{n'} - E_n)|\langle n|e^{ik\hat{x}}|n'\rangle|^2 = \frac{\hbar^2 k^2}{2m},$$

valid for a 1D particle moving in an arbitrary time-independent potential $U(x)$, and discuss its relation with the *Thomas–Reiche–Kuhn sum rule* whose derivation was the subject of the previous problem.

Hint: Calculate the expectation value, in a stationary state n, of the following double commutator,

$$\hat{D} \equiv [[\hat{H}, e^{ik\hat{x}}], e^{-ik\hat{x}}],$$

in two ways—first, just spelling out both commutators, and, second, using the commutation relations between operators $\hat{p}$ and $e^{ik\hat{x}}$, and compare the results.

Problem 5.12. Given Eq. (5.116), prove Eq. (5.117), using the hint given in the accompanying footnote.

Problem 5.13. Use Eqs. (5.116) and (5.117) to simplify the following operators:

(i) $\exp\{+ia\hat{x}\}\,\hat{p}_x\,\exp\{-ia\hat{x}\}$, and

(ii) $\exp\{+ia\hat{p}_x\}\,\hat{x}\,\exp\{-ia\hat{p}_x\}$,

where a is a c-number.

Problem 5.14. For a 1D harmonic oscillator, calculate:

(i) the expectation value of energy, and

(ii) the time evolution of the expectation values of the coordinate and momentum, provided that in the initial moment ($t = 0$) it was in the state described by the following ket-vector:

$$|\alpha\rangle = \frac{1}{\sqrt{2}}(|\,31\rangle + |\,32\rangle),$$

where $|n\rangle$ are the ket-vectors of the stationary (Fock) states of the oscillator.

*Problem 5.15.** Re-derive the London dispersion force's potential of interaction of two isotropic 3D harmonic oscillators (already calculated in problem 3.16), using the language of mutually-induced polarization.

Problem 5.16. An external force pulse $F(t)$, of a finite time duration τ, has been exerted on a 1D harmonic oscillator, initially in its ground state. Use the Heisenberg-picture equations of motion to calculate the expectation value of the oscillator's energy at the end of the pulse.

Problem 5.17. Use Eqs. (5.144) and (5.145) to calculate the uncertainties δx and δp of a squeezed ground state, and in particular prove Eqs. (5.143) for the case $\theta = 0$.

Problem 5.18. Calculate the energy of a harmonic oscillator in the squeezed ground state ζ.

*Problem 5.19.** Prove that the squeezed ground state, described by Eqs. (5.142), (5.144) and (5.145), may be sustained by a sinusoidal modulation of a harmonic oscillator's parameter, and calculate the squeezing factor r as a function of the parameter modulation depth, assuming that the depth is small, and the oscillator's damping is negligible.

Problem 5.20. Use Eqs. (5.148) to prove that the operators $\hat{L}_j$ and $\hat{L}^2$ commute with the Hamiltonian of a spinless particle placed in any central potential field.

Problem 5.21. Use Eqs. (5.149), (5.150) and (5.153) to prove Eqs. (5.155).

Problem 5.22. Derive Eq. (5.164), using any of the prior formulas.

Problem 5.23. In the basis of common eigenstates of the operators $\hat{L}_z$ and $\hat{L}^2$, described by kets $|l, m\rangle$:

  (i) calculate the matrix elements $\langle l, m_1|\hat{L}_x|l, m_2\rangle$ and $\langle l, m_1|\hat{L}_x^{\,2}|l, m_2\rangle$;
  (ii) spell out your results for diagonal matrix elements (with $m_1 = m_2$) and their y-axis counterparts; and
  (iii) calculate the diagonal matrix elements $\langle l, m\,|\hat{L}_x\hat{L}_y|\,l, m\rangle$ and $\langle l, m\,|\hat{L}_y\hat{L}_x|\,l, m\rangle$.

Problem 5.24. For the state described by the common eigenket $|l, m\rangle$ of the operators $\hat{L}_z$ and $\hat{L}^2$ in a reference frame $\{x, y, z\}$, calculate the expectation values $\langle L_{z'}\rangle$ and $\langle L_{z'}^{\,2}\rangle$ in the reference frame whose axis z' forms angle θ with the axis z.

Problem 5.25. Write down the matrices of the following angular momentum operators: $\hat{L}_x$, $\hat{L}_y$, $\hat{L}_z$, and $\hat{L}_\pm$, in the z-basis of the $\{l, m\}$ states with $l = 1$.

Problem 5.26. Calculate the angular factor of the orbital wavefunction of a particle with a definite value of L^2, equal to $6\hbar^2$, and the largest possible value of L_x. What is this value?

Problem 5.27. For the state with the wavefunction $\psi = Cxye^{-\lambda r}$, with a real, positive λ, calculate:

(i) the expectation values of the observables L_x, L_y, L_z and L^2, and
(ii) the normalization constant C.

Problem 5.28. The angular state of a spinless particle is described by the following ket-vector:

$$|\alpha\rangle = \frac{1}{\sqrt{2}}(|\, l = 3, m = 0\rangle + |\, l = 3, m = 1\rangle).$$

Calculate the expectation values of the x- and y-components of its angular momentum. Is the result sensitive to a possible phase shift between two component eigenkets?

Problem 5.29. A particle is in the state α with the orbital wavefunction proportional to the spherical harmonic $Y_1^1(\theta, \varphi)$. Find the angular dependence of the wavefunctions corresponding to the following ket-vectors:

(i) $\hat{L}_x|\alpha\rangle$,
(ii) $\hat{L}_y|\alpha\rangle$,
(iii) $\hat{L}_z|\alpha\rangle$,
(iv) $\hat{L}_+\hat{L}_-|\alpha\rangle$, and
(v) $\hat{L}^2|\alpha\rangle$.

Problem 5.30. A charged, spinless 2D particle of mass m is trapped in a soft in-plane potential well $U(x, y) = m\omega_0^2(x^2 + y^2)/2$. Calculate its energy spectrum in the presence of a uniform magnetic field $\mathcal{B}$, normal to the plane.

Problem 5.31. Solve the previous problem for a spinless 3D particle, placed (in addition to a uniform magnetic field $\mathcal{B}$) into a spherically-symmetric potential well $U(\mathbf{r}) = m\omega_0^2 r^2/2$.

Problem 5.32. Calculate the spectrum of rotational energies of an axially-symmetric, rigid body.

Problem 5.33. Simplify the following double commutator: $[\hat{r}_j, [\hat{L}^2, \hat{r}_{j'}]]$.

Problem 5.34. Prove the following commutation relation:

$$\left[\hat{L}^2, \left[\hat{L}^2, \hat{r}_j\right]\right] = 2\hbar^2\left(\hat{r}_j\hat{L}^2 + \hat{L}^2\hat{r}_j\right).$$

Problem 5.35. Use the commutation relation proved in the previous problem, and Eq. (5.148), to prove the orbital electric-dipole selection rules mentioned in section 5.6 of the lecture notes.

Problem 5.36. Express the commutators listed in Eq. (5.179), $[\hat{J}^2, \hat{L}_z]$ and $[\hat{J}^2, \hat{S}_z]$, via $\hat{L}_j$ and $\hat{S}_j$.

Problem 5.37. Find the operator $\hat{\mathcal{T}}_\phi$ describing a quantum state's rotation by angle ϕ about a certain axis, using the similarity of this operation with the shift of a Cartesian coordinate, discussed in section 5.5. Then use this operator to calculate the probabilities of measurements of spin-½ components of a beam of particles with z-polarized spin, by a Stern–Gerlach instrument turned by angle θ within the $[z, x]$ plane, where y is the axis of particle propagation—see figure 4.1.[56]

Problem 5.38. The rotation ('angle translation') operator $\hat{\mathcal{T}}_\phi$ analyzed in the previous problem, and the coordinate translation operator $\hat{\mathcal{T}}_X$ discussed in section 5.5, have a similar structure:

$$\hat{\mathcal{T}}_\lambda = \exp\{-i\hat{C}\lambda/\hbar\},$$

where λ is a real c-number, characterizing the shift, and $\hat{C}$ is a Hermitian operator, which does not explicitly depend on time.

(i) Prove that such operators $\hat{\mathcal{T}}_\lambda$ are unitary.

(ii) Prove that if the shift by λ, induced by the operator $\hat{\mathcal{T}}_\lambda$, leaves the Hamiltonian of some system unchanged for any λ, then $\langle C \rangle$ is a constant of motion for any initial state of the system.

(iii) Discuss what does the last conclusion mean for the particular operators $\hat{\mathcal{T}}_X$ and $\hat{\mathcal{T}}_\phi$.

Problem 5.39. A particle with spin s is in a state with definite quantum numbers l and j. Prove that the observable $\mathbf{L} \cdot \mathbf{S}$ also has a definite value, and calculate it.

Problem 5.40. For a spin-½ in a state with definite quantum numbers l, m_l, and m_s, calculate the expectation value of the observable J^2, and the probabilities of all its possible values. Interpret your results in the terms of the Clebsh–Gordan coefficients (5.190).

Problem 5.41. Derive the general recurrence relations for the Clebsh–Gordan coefficients.

Hint: Using the similarity of the commutation relations discussed in section 5.7, write the relations similar to Eqs. (5.164), for other components of the angular momentum, and apply them to Eq. (5.170).

---

[56] Note that the last task is just a particular case of problem 4.18 (see also problem 5.1).

Problem 5.42. Use the recurrence relations derived in the previous problem to prove Eqs. (5.190) for the spin-½ Clebsh–Gordan coefficients.

Problem 5.43. A spin-½ particle is in a state with definite values of L^2, J^2, and J_z. Find all possible values of the observables S^2, S_z, and L_z, the probability of each listed value, and the expectation value for each of these observables.

Problem 5.44. Re-solve the Landau-level problem, discussed in section 3.2, for a spin-½ particle. Discuss the result for the particular case of an electron, with the g-factor equal to 2.

Problem 5.45. In the Heisenberg picture of quantum dynamics, find an explicit expression for the operator of acceleration,

$$\hat{\mathbf{a}} \equiv d\hat{\mathbf{v}}/dt,$$

of a spin-½ particle with electric charge q, moving in an arbitrary external electro-magnetic field. Compare the result with the corresponding classical expression.

Hint: For the orbital motion's description, you may use Eq. (3.26).

Problem 5.46. A byproduct of the solution of problem 5.41 is the following relation for the spin operators (valid for any spin s):

$$\langle m_s \pm 1 \mid \hat{S}_\pm \mid m_s \rangle = \hbar[(s \pm m_s + 1)(s \mp m_s)]^{1/2}.$$

Use this result to spell out the matrices S_x, S_y, S_z, and S^2 of a particle with $s = 1$, in the z-basis—defined as the basis in which the matrix S_z is diagonal.

Problem 5.47.* For a particle with an arbitrary spin s, find the ranges of the quantum numbers m_j and j that are necessary to describe, in the coupled-representation basis:

 (i) all states with a definite quantum number l, and
 (ii) a state with definite values of not only l, but also m_l and m_s.

Give an interpretation of your results in terms of the classical geometric vector diagram (figure 5.13).

Problem 5.48. A particle of mass m, with electric charge q and spin s, free to move along a plane ring of a radius R, is placed into a constant, uniform magnetic field $\mathscr{B}$, directed normally to the ring's plane. Calculate the energy spectrum of the system. Explore and interpret the particular form the result takes when the particle is an electron with the g-factor $g_e = 2$.

References

[1] Schulman L 1981 *Techniques and Applications of Path Integration* (Wiley)
[2] Gelfand I and Fomin S 2000 *Calculus of Variations* (Dover)

[3] Elsgolc L 2007 *Calculus of Variations* (Dover)

[4] Rajaraman R 1987 *Solitons and Instantons* (North Holland)

[5] Buttiker M and Landauer R 1982 *Phys. Rev. Lett.* **49** 1739

[6] Gerry C and Knight P 2005 *Introductory Quantum Optics* (Cambridge University Press)

[7] Breitenbach G *et al* 1997 *Nature* **387** 471

[8] Vahlbruch H *et al* 2008 *Phys. Rev. Lett.* **100** 033602

[9] Wollman E *et al* 2015 *Science* **349** 952

[10] Pirkkalainen J-M *et al* 2015 *Phys. Rev. Lett.* **115** 243601

[11] Edmonds A 1957 *Angular Momentum in Quantum Mechanics* (Princeton University Press)

[12] Merzbacher E 1998 *Quantum Mechanics* 3rd ed (Wiley)

[13] Sakurai J 1994 *Modern Quantum Mechanics* (Addison-Wesley)

IOP Publishing

Quantum Mechanics
Lecture notes
Konstantin K Likharev

Chapter 6

Perturbative approaches

This chapter discusses various perturbative approaches to problems of quantum mechanics, and their simplest but important applications including the fine structure of atomic levels, and the effects of external dc and ac electric and magnetic fields of these levels. It continues with a discussion of the perturbation theory of transitions to continuous spectrum and the Golden Rule of quantum mechanics, which will naturally bring us to the issue of open quantum systems—to be discussed in the next chapter.

6.1 Eigenproblems

Unfortunately, only a few problems of quantum mechanics may be solved exactly in the analytical form. Actually, in the previous chapters we have solved a substantial part of such problems for a single particle, and for multiparticle problems the exactly solvable cases are even more rare. However, most practical problems of physics feature a certain small parameter, and this smallness may be exploited by various approximate analytical methods. Earlier in the course, we have explored one of them, the WKB approximation, which is adequate for a particle moving through a soft potential profile. In this chapter we will discuss other techniques that are more suitable for other cases. The historic name for these techniques is *the perturbation theory*, but it is fairer to speak about several *perturbative approaches*, because they are substantially different for different cases.

The simplest version of the perturbation theory addresses the eigenproblem for systems described by time-independent Hamiltonians of the type

$$\hat{H} = \hat{H}^{(0)} + \hat{H}^{(1)}, \qquad (6.1)$$

where the operator $\hat{H}^{(1)}$, describing the system's 'perturbation', is relatively small—in the sense that its addition to the unperturbed operator $\hat{H}^{(0)}$ results in a relatively small change of the eigenenergies E_n of the system, and the corresponding

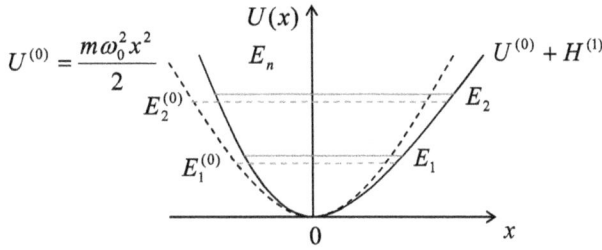

Figure 6.1. The simplest application of the perturbation theory: a weakly anharmonic 1D oscillator. (Dashed lines characterize the unperturbed, harmonic oscillator.)

eigenstates. A typical problem of this type is the 1D *weakly anharmonic oscillator* (figure 6.1), described by the Hamiltonian (6.1) with

$$\hat{H}^{(0)} = \frac{\hat{p}^2}{2m} + \frac{m\omega_0^2}{2}\hat{x}^2, \qquad \hat{H}^{(1)} = \alpha\hat{x}^3 + \beta\hat{x}^4 + \ldots \tag{6.2}$$

with small coefficients α, β, ….

I will use this system as our first particular example, but let me start from describing the perturbative approach to the general time-independent Hamiltonian (6.1). In the bra–ket formalism, the eigenproblem for the perturbed system is

$$(\hat{H}^{(0)} + \hat{H}^{(1)})|n\rangle = E_n|n\rangle. \tag{6.3}$$

Let the eigenstates and eigenvalues of the unperturbed Hamiltonian, which satisfy the equation

$$\hat{H}^{(0)}|n^{(0)}\rangle = E_n^{(0)}|n^{(0)}\rangle, \tag{6.4}$$

be considered as known. In this case, the solution of problem (6.3) means finding, first, its perturbed eigenvalues E_n and, second, the coefficients $\langle n'^{(0)}|n\rangle$ of the expansion of the perturbed state's vectors $|n\rangle$ in series over the unperturbed ones, $|n'^{(0)}\rangle$:

$$|n\rangle = \sum_{n'}|n'^{(0)}\rangle\langle n'^{(0)}|n\rangle. \tag{6.5}$$

Let us plug Eq. (6.5), with the summation index n' replaced with n'' (just to have a more compact notation in our forthcoming result), into the both parts of Eq. (6.3):

$$\sum_{n''}\langle n''^{(0)}|n\rangle\,\hat{H}^{(0)}|n''^{(0)}\rangle + \sum_{n''}\langle n''^{(0)}|n\rangle\,\hat{H}^{(1)}|n''^{(0)}\rangle = \sum_{n''}\langle n''^{(0)}|n\rangle E_n|n''^{(0)}\rangle, \tag{6.6}$$

and then inner-multiply all terms by an arbitrary unperturbed bra-vector $\langle n'^{(0)}|$ of the system. Assuming that the unperturbed eigenstates are orthonormal, $\langle n'^{(0)}|n''^{(0)}\rangle = \delta_{n'n''}$, and using Eq. (6.4) in the first term of the left-hand side, we get the following system of linear equations

$$\sum_{n''} \langle n''^{(0)}|n\rangle H^{(1)}_{n'n''} = \langle n'^{(0)}|n\rangle \left(E_n - E^{(0)}_{n'}\right), \tag{6.7}$$

where the matrix elements of the perturbation are calculated, by definition, in the *unperturbed* brackets:

$$H^{(1)}_{n'n''} \equiv \langle n'^{(0)}| \, \hat{H}^{(1)} \, |n''^{(0)}\rangle. \tag{6.8}$$

The linear equation system (6.7) is still exact[1], and is frequently used for numerical calculations. (Since the matrix coefficients (6.8) typically decrease when n' and/or n'' become sufficiently large, the sum on the left-hand side of Eq. (6.7) may usually be truncated, still giving an acceptable accuracy of the solution.) For getting analytical results we need to make approximations. In the simple perturbation theory we are discussing now, this is achieved by the expansion of both the eigenenergies and the expansion coefficients into the Taylor series in a certain small parameter μ of the problem:

$$E_n = E^{(0)}_n + E^{(1)}_n + E^{(2)}_n..., \tag{6.9}$$

$$\langle n'^{(0)}|n\rangle = \langle n'^{(0)}|n\rangle^{(0)} + \langle n'^{(0)}|n\rangle^{(1)} + \langle n''^{(0)}|n\rangle^{(2)}..., \tag{6.10}$$

where

$$E^{(k)}_n \propto \langle n'^{(0)}|n\rangle^{(k)} \propto \mu^k. \tag{6.11}$$

In order to explore the 1st-order approximation, which ignores all terms $O(\mu^2)$ and higher, let us plug only the two first terms of the expansions (6.9) and (6.10) into the basic equation (6.7):

$$\sum_{n''} H^{(1)}_{n'n''}(\delta_{n''n} + \langle n''^{(0)}|n\rangle^{(1)}) = (\delta_{n'n} + \langle n'^{(0)}|n\rangle^{(1)})\left(E^{(0)}_n + E^{(1)}_n - E^{(0)}_{n'}\right). \tag{6.12}$$

Now let us open the parentheses, and disregard all the remaining terms $O(\mu^2)$. The result is

$$H^{(1)}_{n'n} = \delta_{n'n}E^{(1)}_n + \langle n'^{(0)}|n\rangle^{(1)}\left(E^{(0)}_n - E^{(0)}_{n'}\right), \tag{6.13}$$

This relation is valid for any set of indices n and n'; let us start from the case $n = n'$ and immediately get a very simple (and practically, the most important!) result:

$$E^{(1)}_n = H^{(1)}_{nn} \equiv \langle n^{(0)}| \, \hat{H}^{(1)} \, |n^{(0)}\rangle. \tag{6.14}$$

---

[1] Please note the similarity of Eq. (6.7) with Eq. (2.215) of the 1D band theory. Indeed, the latter equation is not much more than a particular form of Eq. (6.7) for the 1D wave mechanics, and a specific (periodic) potential $U(x)$ considered as the perturbation. Moreover, the whole approximate treatment of the weak-potential limit in section 2.7 is essentially a particular case of the perturbation theory we are discussing now (in the 1st approximation).

For example, let us see what this result gives for two first perturbation terms in the weakly anharmonic oscillator (6.2):

$$E_n^{(1)} = \alpha \langle n^{(0)} | \hat{x}^3 | n^{(0)} \rangle + \beta \langle n^{(0)} | \hat{x}^4 | n^{(0)} \rangle. \tag{6.15}$$

As the reader knows (or should know :-) from the solution of problem 5.9, the first bracket equals zero, while the second one yields[2]

$$E_n^{(1)} = \frac{3}{4}\beta x_0^4 (2n^2 + 2n + 1). \tag{6.16}$$

Naturally, there should be some contribution to the energies from the (typically, larger) term proportional to α, so we need to explore the 2nd approximation of the perturbation theory. However, before doing that, let us complete our discussion of its 1st order.

For $n' \neq n$, Eq. (6.13) may be used to calculate the eigenstates rather than the eigenvalues:

$$\langle n'^{(0)} | n \rangle^{(1)} = \frac{H_{n'n}^{(1)}}{E_n^{(0)} - E_{n'}^{(0)}}, \qquad \text{for } n' \neq n. \tag{6.17}$$

This means that the eigenket's expansion (6.5), in the 1st order, may be represented as

$$|n^{(1)}\rangle = C |n^{(0)}\rangle + \sum_{n' \neq n} \frac{H_{n'n}^{(1)}}{E_n^{(0)} - E_{n'}^{(0)}} |n'^{(0)}\rangle. \tag{6.18}$$

The coefficient C cannot be found from Eq. (6.17); however, requiring the final state n to be normalized, we see that other terms may provide only corrections $O(\mu^2)$, so that in the 1st order we should take $C = 1$. The most important feature of Eq. (6.18) is its denominator: the closer the unperturbed eigenenergies of two states, the larger is their mutual 'interaction' due to the perturbation.

This feature also affects the 1st approximation's validity condition, which may be quantified using Eq. (6.17): the magnitudes of the brackets it describes have to be much less than the unperturbed bracket $\langle n|n \rangle^{(0)} = 1$, so that all elements of the perturbation matrix have to be much less than the difference between the corresponding unperturbed energies. For the anharmonic oscillator's energy corrections (6.16), this requirement is reduced to $E_n^{(1)} \ll \hbar\omega_0$.

Now we are ready to go after the 2nd-order approximation to Eq. (6.7). Let us focus on the case $n' = n$, because as we already know, only this term will give us a correction to the eigenenergies. Moreover, since the left-hand side of Eq. (6.7) already has a small factor $H^{(1)}_{n'n''} \propto \mu$, the bracket coefficients in that part may be taken from the 1st-order result (6.17). As a result, we get

[2] A useful exercise for the reader: analyze the relation between Eq. (6.16) and the result of the classical theory of such weakly anharmonic ('nonlinear') oscillator—see, e.g. *Part CM* section 5.2, in particular, Eq. (5.49).

$$E_n^{(2)} = \sum_{n''} \langle n''^{(0)} | n \rangle^{(1)} H_{nn''}^{(1)} = \sum_{n'' \neq n} \frac{H_{n''n}^{(1)} H_{nn''}^{(1)}}{E_n^{(0)} - E_{n''}^{(0)}}. \tag{6.19}$$

Since $\hat{H}^{(1)}$ represents an observable (energy), and hence has to be Hermitian, we may rewrite this expression as

$$E_n^{(2)} = \sum_{n' \neq n} \frac{|H_{n'n}^{(1)}|^2}{E_n^{(0)} - E_{n'}^{(0)}} \equiv \sum_{n' \neq n} \frac{\left| \langle n'^{(0)} | \hat{H}^{(1)} | n^{(0)} \rangle \right|^2}{E_n^{(0)} - E_{n'}^{(0)}}. \tag{6.20}$$

This is the much-celebrated 2nd-order perturbation result, which frequently (in sufficiently symmetric problems) is the first nonvanishing correction to the state energy—for example, from the cubic term (proportional to α) in our weakly anharmonic oscillator problem (6.2). In order to calculate the corresponding correction, we may use another result of the solution of problem 5.9:

$$\langle n' | \hat{x}^3 | n \rangle = \left(\frac{x_0}{\sqrt{2}} \right)^3 \times \{[n(n-1)(n-2)]^{1/2} \delta_{n', n-3} + 3n^{3/2} \delta_{n', n-1} \tag{6.21}$$
$$+ 3(n+1)^{3/2} \delta_{n', n+1} + [(n+1)(n+2)(n+3)]^{1/2} \delta_{n', n+3}\}.$$

So, according to Eq. (6.20), we need to calculate

$$E_n^{(2)} = \alpha^2 \left(\frac{x_0}{\sqrt{2}} \right)^6$$
$$\times \sum_{n' \neq n} \frac{\left\{ \begin{array}{l} [n(n-1)(n-2)]^{1/2} \delta_{n', n-3} + 3n^{3/2} \delta_{n', n-1} \\ + 3(n+1)^{3/2} \delta_{n', n+1} + [(n+1)(n+2)(n+3)]^{1/2} \delta_{n', n+3} \end{array} \right\}^2}{\hbar \omega_0 (n - n')}. \tag{6.22}$$

The summation is actually not as cumbersome as it may look, because all mixed products are proportional to the products of different Kronecker deltas and hence vanish, so that we need to sum up only the squares of each term, getting:

$$E_n^{(2)} = -\frac{15}{4} \frac{\alpha^2 x_0^6}{\hbar \omega_0} \left(n^2 + n + \frac{11}{30} \right). \tag{6.23}$$

This formula shows that all energy level corrections are negative, regardless of the sign of α.[3] In contrast, the 1st order correction $E_n^{(1)}$, given by Eq. (6.16), does depend on the sign of β, so that the net correction, $E_n^{(1)} + E_n^{(2)}$, may be of any sign.

The results (6.18) and (6.20) are clearly inapplicable to the degenerate case where, in the absence of perturbation, several states correspond to the same energy level,

---

[3] Note this is correct for the *ground-state* energy correction $E_g^{(2)}$ of *any* system, because for this state, the denominators of all terms of the sum (6.20) are negative, while their numerators are always positive.

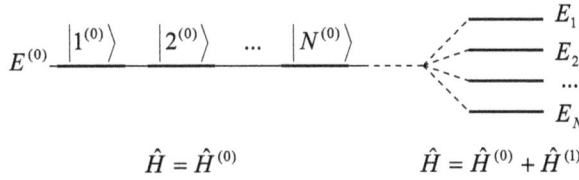

Figure 6.2. Lifting the energy level degeneracy by a perturbation (schematically).

because of the divergence of their denominators[4]. This divergence hints that the largest effect of the perturbation in that case is the *degeneracy lifting*, e.g. splitting of the initially degenerate energy level $E^{(0)}$ (figure 6.2), and that for the analysis of this case we can, to the first approximation, ignore the effect of all other energy levels. (A careful analysis shows that this is indeed the case until the level splitting becomes comparable with the distance to other energy levels.)

Limiting the summation in Eq. (6.7) to the group of N degenerate states with equal $E_{n'}^{(0)} \equiv E^{(0)}$, we reduce it to

$$\sum_{n''=1}^{N} \langle n''^{(0)}|n\rangle H_{n'n''}^{(1)} = \langle n'^{(0)}|n\rangle(E_n - E^{(0)}). \tag{6.24}$$

where now n' and n'' number the N states of the degenerate group[5]. For $n = n'$, Eq. (6.24) may be rewritten as

$$\sum_{n''=1}^{N} \left(H_{n'n''}^{(1)} - E_{n''}^{(1)}\delta_{n'n''}\right)\langle n''^{(0)}|n'\rangle = 0, \quad \text{where } E_n^{(1)} \equiv E_n - E^{(0)}. \tag{6.25}$$

For each $n' = 1, 2, \ldots N$, this is a system of N linear, homogenous equations (with N terms each) for N unknown coefficients $\langle n''^{(0)}|n'\rangle$. In this problem, we readily recognize the problem of diagonalization of the perturbation matrix $\mathsf{H}^{(1)}$—cf section 4.4 and in particular Eq. (4.101). As in the general case, the condition of self-consistency of the system is:

$$\begin{vmatrix} H_{11}^{(1)} - E_n^{(1)} & H_{12}^{(1)} & \cdots \\ H_{21}^{(1)} & H_{22}^{(1)} - E_n^{(1)} & \cdots \\ \cdots & \cdots & \cdots \end{vmatrix} = 0, \tag{6.26}$$

where now the index n numbers the N roots of this equation, in an arbitrary order. According to the definition (6.25) of $E_n^{(1)}$, the resulting N energy levels E_n may be

---

[4] This is exactly the reason why such perturbation theory runs into serious problems for systems with continuous spectrum, and other techniques (such as the WKB approximation) are often necessary.
[5] Note that here the choice of the basis is to some extent arbitrary, because due to the linearity of equations of quantum mechanics, any linear combination of the states $n''^{(0)}$ is also an eigenstate of the unperturbed Hamiltonian. However, for using Eq. (6.25), these combinations have to be orthonormal, as was supposed at the derivation of Eq. (6.7).

found as $E^{(0)} + E_n^{(1)}$. If the perturbation matrix is diagonal in the chosen basis $n^{(0)}$, the result is extremely simple,

$$E_n - E^{(0)} \equiv E_n^{(1)} = H_{nn}^{(1)},\tag{6.27}$$

and formally coincides with Eq. (6.14) for the non-degenerate case, but now it may give a different result for each of N previously degenerate states n.

Let us see what this theory gives for several important examples. First of all, let us consider a system with two degenerate states with an energy far enough from all other levels. Then, in the basis of these two degenerate states, the most general perturbation matrix is

$$\mathrm{H}^{(1)} = \begin{pmatrix} H_{11}^{(1)} & H_{12}^{(1)} \\ H_{21}^{(1)} & H_{22}^{(1)} \end{pmatrix}\tag{6.28}$$

Besides the upper index, this matrix coincides with the general matrix (5.2) of a two-level system. Hence, we come to the very important conclusion: for a weak perturbation, all properties of double-degenerate system are identical to those of the genuine two-level systems, which were the subject of numerous discussions in chapter 4 and again in section 5.1. In particular, its eigenenergies are given by Eq. (5.6), and may be described by the level-anticrossing diagram shown in figure 5.1.

6.2 The Stark effect

As a more involved example of the level degeneracy lifting by a perturbation, let us discuss the *linear Stark effect*[6]—the atomic level splitting by an external electric field. Let us study this effect, in the linear approximation, for a hydrogen-like atom. Taking the direction of the external electric field $\mathscr{E}$ (which is practically always uniform on the atomic scale) for the z-axis, the perturbation may be represented by the following Hamiltonian:

$$\hat{H}^{(1)} = -F\hat{z} = -q\mathscr{E}\hat{z} = -q\mathscr{E}r\cos\theta.\tag{6.29}$$

(In the last form, the operator sign is dropped, because we will work in the coordinate representation.)

As you (should :-) remember, energy levels of a hydrogen-like atom depend only on the principal quantum number n—see Eq. (3.201); hence all the states, besides the ground state $n = 1$ ('1s' in the spectroscopic nomenclature) in which $l = m = 0$, have some degeneracy, which grows rapidly with n. I will carry out the calculations only for the lowest degenerate level with $n = 2$. Since, according to Eq. (3.203), $0 \leqslant l \leqslant n - 1$, at this level the orbital quantum number l may be equal either 0 (one 2s state,

---

[6] The effect was discovered experimentally in 1913 by J Stark and (independently) by A Lo Surdo, and is sometimes (and more fairly) called the 'Stark–Lo Surdo effect'. Sometimes this name is used with the qualifier 'dc' to distinguish it from the *ac Stark* effect—the energy level shift under the effect of an ac field—see section 6.5.

with $m = 0$) or 1 (three $2p$ states, with $m = 0, \pm1$). Due to this four-fold degeneracy, $\mathrm{H}^{(1)}$ is a 4×4 matrix with 16 elements:

$$
\mathrm{H}^{(1)} =
\overbrace{\phantom{\begin{matrix}m=0\end{matrix}}}^{l=0}\;\overbrace{\phantom{\begin{matrix}m=0\;m=+1\;m=-1\end{matrix}}}^{l=1}
$$

$$
\begin{matrix} m = 0 & m = 0 & m = +1 & m = -1 \end{matrix}
$$

$$
\mathrm{H}^{(1)} =
\begin{pmatrix}
H_{11} & H_{12} & H_{13} & H_{14} \\
H_{21} & H_{22} & H_{23} & H_{24} \\
H_{31} & H_{32} & H_{33} & H_{34} \\
H_{41} & H_{42} & H_{43} & H_{44}
\end{pmatrix}
\begin{matrix}
m = 0, & l = 0, \\
m = 0, & \\
m = +1, & \left.\begin{matrix}\\\\\end{matrix}\right\} l = 1. \\
m = -1, &
\end{matrix}
\tag{6.30}
$$

However, please do not be scared. First, due to the Hermitian nature of the operator, only 10 of the matrix elements (four diagonal ones and six off-diagonal elements) may be substantially different. Moreover, due to a high symmetry of the problem, there are a lot of zeros even among these elements. Indeed, let us have a look at the angular components Y_l^m of the corresponding wavefunctions, with $l = 0$ and $l = 1$, described by Eqs. (3.174) and (3.175). For the states with $m = \pm1$, the azimuthal parts of wavefunctions are proportional to $\exp\{\pm i\varphi\}$; hence the off-diagonal elements H_{34} and H_{43} of the matrix (6.30), relating these functions, are proportional to

$$
\oint d\Omega\, Y_1^{\pm*}\hat{H}^{(1)} Y_1^{\mp} \propto \int_0^{2\pi} d\varphi\, (e^{\pm i\varphi})^*(e^{\mp i\varphi}) = 0.
\tag{6.31}
$$

The azimuthal-angle symmetry also kills the off-diagonal elements H_{13}, H_{14}, H_{23}, H_{24} (and hence their complex conjugates H_{31}, H_{41}, H_{32}, and H_{42}), because they relate the states with $m = 0$ and $m \neq 0$, and are proportional to

$$
\oint d\Omega\, Y_1^{0*}\hat{H}^{(1)} Y_1^{\pm1} \propto \int_0^{2\pi} d\varphi\, e^{\pm i\varphi} = 0.
\tag{6.32}
$$

For the diagonal elements H_{33} and H_{44}, corresponding to $m = \pm1$, the azimuthal-angle integral does not vanish, but since the corresponding spherical functions depend on the polar angle as $\sin\theta$, the matrix elements are proportional to

$$
\oint d\Omega\, Y_1^{\pm*}\hat{H}^{(1)} Y_1^{\pm} \propto \int_0^{\pi} \sin\theta d\theta\, \sin\theta \cos\theta \sin\theta =
$$
$$
\int_{-1}^{+1} \cos\theta(1 - \cos^2\theta) \times d(\cos\theta),
\tag{6.33}
$$

and are equal to zero—as any limit-symmetric integral of an odd function. Finally, for the states $2s$ and $2p$ with $m = 0$, the diagonal elements H_{11} and H_{22} are also killed by the polar-angle integration:

$$
\oint d\Omega\, Y_0^{0*}\hat{H}^{(1)} Y_0^{0} \propto \int_0^{\pi} \sin\theta d\theta\, \cos\theta = \int_{-1}^{1} \cos\theta\, d(\cos\theta) = 0,
\tag{6.34}
$$

$$
\oint d\Omega\, Y_0^{1*}\hat{H}^{(1)} Y_0^{1} \propto \int_0^{\pi} \sin\theta d\theta\, \cos^3\theta = \int_{-1}^{+1} \cos^3\theta\, d(\cos\theta) = 0.
\tag{6.35}
$$

Hence, the only nonvanishing elements of the matrix (6.30) are two off-diagonal elements H_{12} and H_{21}, which relate two states with the same $m = 0$, but different $l = 0, 1$, because they are proportional to

$$\oint d\Omega\, Y_0^{0*} \cos\theta\, Y_1^0 = \frac{\sqrt{3}}{4\pi} \int_0^{2\pi} d\varphi \int_0^{\pi} \sin\theta d\theta \cos^2\theta = \frac{1}{\sqrt{3}} \neq 0. \qquad (6.36)$$

What remains is to use Eqs. (3.209) for the radial parts of these functions to complete the calculation of those two matrix elements:

$$H_{12} = H_{21} = -\frac{q\mathscr{E}}{\sqrt{3}} \int_0^{\infty} r^2 dr\, \mathcal{R}_{2,\,0}(r)\, r\, \mathcal{R}_{2,\,1}(r). \qquad (6.37)$$

Due to the additive structure of the function $\mathcal{R}_{2,\,0}(r)$, the integral falls into a sum of two table integrals, both of the type (A.34d), finally giving

$$H_{12} = H_{21} = 3q\mathscr{E}r_0, \qquad (6.38)$$

where r_0 is the scale given by Eq. (3.192); for the hydrogen atom it is just the Bohr radius r_B (1.10).

Thus, the perturbation matrix (6.30) is reduced to

$$\mathsf{H}^{(1)} = \begin{pmatrix} 0 & 3q\mathscr{E}r_0 & 0 & 0 \\ 3q\mathscr{E}r_0 & 0 & 0 & 0 \\ 0 & 0 & 0 & 0 \\ 0 & 0 & 0 & 0 \end{pmatrix}, \qquad (6.39)$$

so that the condition (6.26) of the self-consistency of the system (6.25),

$$\begin{vmatrix} -E_n^{(1)} & 3q\mathscr{E}r_0 & 0 & 0 \\ 3q\mathscr{E}r_0 & -E_n^{(1)} & 0 & 0 \\ 0 & 0 & -E_n^{(1)} & 0 \\ 0 & 0 & 0 & -E_n^{(1)} \end{vmatrix} = 0, \qquad (6.40)$$

gives a very simple characteristic equation

$$\left(E_n^{(1)}\right)^2 \left[\left(E_n^{(1)}\right)^2 - (3q\mathscr{E}r_0)^2\right] = 0. \qquad (6.41)$$

with the roots

$$E_{1,\,2}^{(1)} = 0, \qquad E_{3,\,4}^{(1)} = \pm 3q\mathscr{E}r_0. \qquad (6.42)$$

so that the degeneracy is only partly lifted—see figure 6.3.

Generally, in order to understand the nature of states corresponding to these levels, we should return to Eq. (6.25) with each calculated value of $E_n^{(1)}$, and find the corresponding expansion coefficients $\langle n''^{(0)}|n'\rangle$, which describe the perturbed states. However, in our simple case the outcome of this procedure is clear in advance. Indeed, since the states with $l = 1$ and $m = \pm 1$ are not affected by the perturbation at

$$|+\rangle = \frac{1}{\sqrt{2}}\left(|2s\rangle + |2p\rangle\right)$$

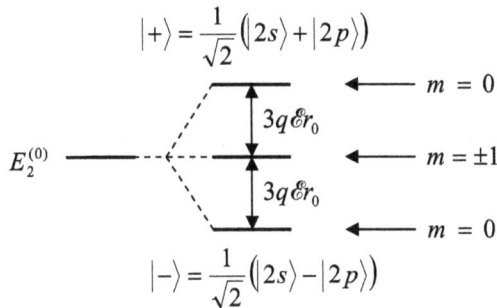

$$E_2^{(0)} \qquad\qquad 3q\mathscr{E}r_0 \qquad\qquad m = 0$$

$$3q\mathscr{E}r_0 \qquad\qquad m = \pm 1$$

$$m = 0$$

$$|-\rangle = \frac{1}{\sqrt{2}}\left(|2s\rangle - |2p\rangle\right)$$

Figure 6.3. The linear Stark effect for the level $n = 2$ of a hydrogen-like atom.

all (in the linear approximation in the electric field), their degeneracy is not lifted, and energy unaffected—see the middle line in figure 6.3. On the other hand, the partial perturbation matrix connecting the states $2s$ and $2p$, i.e. the top left 2×2 part of the full matrix (6.39), is proportional to the Pauli matrix σ_x, and we already know the result of its diagonalization—see Eqs. (4.113) and (4.114). This means that the upper and lower split levels correspond to very simple linear combinations of the previously degenerate states with $m = 0$,

$$|\pm\rangle = \frac{1}{\sqrt{2}}(|2s\rangle \pm |2p\rangle). \tag{6.43}$$

Finally, let us estimate the magnitude of the linear Stark effect for a hydrogen atom. For a very high electric field of $\mathscr{E} = 3 \times 10^6$ V m^{-1},[7] $|q| = e \approx 1.6 \times 10^{-19}$ C, and $r_0 = r_B \approx 0.5 \times 10^{-10}$ m, we get a level splitting of $3q\mathscr{E}r_0 \approx 0.8 \times 10^{-22}$ J $\approx$ 0.5 meV. This number is much lower than the unperturbed energy of the level, $E_2 = -E_H/(2 \times 2^2) \approx -3.4$ eV, so that the perturbative result is quite applicable. On the other hand, the calculated splitting is much larger than the resolution limit imposed by the natural linewidth ($\sim 10^{-7} E_2$, see chapter 9), so that the effect is quite observable even in substantially lower electric fields.

6.3 Fine structure of atomic levels

Now let us use the same perturbation theory analyze, also for the simplest case of a hydrogen-like atom, the so-called *fine structure* of atomic levels—their degeneracy lifting even in the absence of external fields. In the limit when the effective speed v of the electron motion is much smaller than the speed of light c (as it is in the hydrogen atom), the fine structure may be analyzed as a sum of two small relativistic effects. To analyze the first of them, let us expand the well-known classical relativistic expression[8] for the kinetic energy $T = E - mc^2$ of a free particle with the rest mass m,[9]

---

[7] This value approximately corresponds to the threshold of electric breakdown in air at ambient conditions, due to the impact ionization on typical metallic electrode surfaces. (Reducing the air pressure only enhances the ionization and lowers the breakdown threshold.) As a result, experiments with higher dc fields are rather difficult.

[8] See, e.g. *Part EM* Eq. (9.78)—or any undergraduate text on the relativity theory.

[9] This fancy font is used, as in sections 3.5–3.8, to distinguish the mass m from the magnetic quantum number m.

$$T = (m^2c^4 + p^2c^2)^{1/2} - mc^2 \equiv mc^2\left[\left(1 + \frac{p^2}{m^2c^2}\right)^{1/2} - 1\right], \tag{6.44}$$

into the Taylor series with respect to the small ratio $(p/mc)^2 \approx (v/c)^2$:

$$T = mc^2\left[1 + \frac{1}{2}\left(\frac{p}{mc}\right)^2 - \frac{1}{8}\left(\frac{p}{mc}\right)^4 + \ldots -1\right] \equiv \frac{p^2}{2m} - \frac{p^4}{8m^3c^2} + \ldots, \tag{6.45}$$

and drop all the terms besides the first (non-relativistic) one and the next spelled-out term, which represents the first relativistic correction to T.

Following the correspondence principle, the quantum-mechanical problem in this approximation may be described by the perturbative Hamiltonian (6.1), where the unperturbed (non-relativistic) Hamiltonian of the problem, whose eigenstates and eigenenergies were discussed in section 3.5, is

$$\hat{H}^{(0)} = \frac{\hat{p}^2}{2m} + \hat{U}(r), \qquad \hat{U}(r) = -\frac{C}{r}, \tag{6.46}$$

while the kinetic-relativistic perturbation is

$$\hat{H}^{(1)} = -\frac{\hat{p}^4}{8m^3c^2} \equiv -\frac{1}{2mc^2}\left(\frac{\hat{p}^2}{2m}\right)^2. \tag{6.47}$$

Using Eq. (6.46), we may rewrite the last formula as

$$\hat{H}^{(1)} = -\frac{1}{2mc^2}[\hat{H}^{(0)} - \hat{U}(r)]^2, \tag{6.48}$$

so that its matrix elements participating in the characteristic equation (6.25) for a given degenerate energy level (3.201), i.e. a given principal quantum number n, are

$$\langle nlm \mid \hat{H}^{(1)} \mid nl'm' \rangle = -\frac{1}{2mc^2}\langle nlm \mid [\hat{H}^{(0)} - \hat{U}(r)][\hat{H}^{(0)} - \hat{U}(r)] \mid nl'm' \rangle, \tag{6.49}$$

where the bra- and ket-vectors describe the unperturbed eigenstates, whose eigen-functions (in the coordinate representation) are given by Eq. (3.200): $\psi_{n,l,m} = \mathcal{R}_{n,l}(r)Y_l^m(\theta, \varphi)$.

It is straightforward (and hence left for the reader) to prove that all off-diagonal elements of the set (6.49) are equal to 0. Thus we may use Eq. (6.27) for each set of the quantum numbers $\{n, l, m\}$:

$$E_{n,l,m}^{(1)} \equiv E_{n,l,m} - E_n^{(0)} = \langle nlm \mid \hat{H}^{(1)} \mid nlm \rangle$$

$$= -\frac{1}{2mc^2} \left\langle (\hat{H}^{(0)} - \hat{U}(r))^2 \right\rangle_{n,l,m}$$

$$= -\frac{1}{2mc^2} \left(E_n^2 - 2E_n \langle \hat{U} \rangle_{n,l} + \langle \hat{U}^2 \rangle_{n,l} \right) \qquad (6.50)$$

$$= -\frac{1}{2mc^2} \left(\frac{E_0^2}{4n^4} - \frac{E_0}{n^2} C \left\langle \frac{1}{r} \right\rangle_{n,l} + C^2 \left\langle \frac{1}{r^2} \right\rangle_{n,l} \right),$$

where the index m has been dropped, because the radial wavefunctions $\mathcal{R}_{n,l}(r)$, which affect these expectation values, do not depend on that quantum number. Now using Eqs. (3.191), (3.201) and the first two of Eqs. (3.211), we finally get

$$E_{n,l}^{(1)} = -\frac{mC^2}{2\hbar^2 c^2 n^4} \left(\frac{n}{l + \frac{1}{2}} - \frac{3}{4} \right) \equiv -\frac{2E_n^2}{mc^2} \left(\frac{n}{l + \frac{1}{2}} - \frac{3}{4} \right). \qquad (6.51)$$

Let us discuss this result. First of all, its last form confirms that that the correction (6.51) is indeed much smaller than the unperturbed energy E_n (and hence the perturbation theory is valid) if the latter is much smaller than the relativistic rest energy mc^2 of the particle—as it is for the hydrogen atom. Next, since in the Bohr problem's solution $n \geqslant l + 1$, the first fraction in the parentheses of Eq. (6.53) is always larger than 1, and hence than 3/4, so that the kinetic relativistic correction to energy is negative for all n and l. (Actually, this could be predicted already from Eq. (6.47), which shows that the perturbation's Hamiltonian is a negatively defined form.) Finally, at a fixed principal number n, the negative correction's magnitude decreases with the growth of l. This fact may be classically interpreted using Eq. (3.210): the larger l is (at fixed n), the smaller is the particle's average distance from the center, and hence the smaller is its effective velocity, i.e. the smaller is the magnitude of the quantum-mechanical average of the negative relativistic correction (6.47) to the kinetic energy.

The result (6.51) is valid for the Coulomb interaction $U(r) = -C/r$ of any physical nature. However, if we speak specifically about hydrogen-like atoms/ions, there is also another relativistic correction to energy, due to the so-called *spin–orbit interaction* (alternatively called the 'spin–orbit coupling'). Its physics may be understood from the following semi-qualitative, classical reasoning: from the 'the point of view' of an electron rotating about the nucleus at constant distance r with velocity v, it is the nucleus, of the electric charge Ze, that rotates about the electron with the velocity $(-v)$ and hence the time period $\mathcal{T} = 2\pi r/v$. From the point of view of magnetostatics, such circular motion of the electric charge $Q = Ze$, is equivalent to a circular dc electric current $I = Q/\mathcal{T} = (Ze)(v/2\pi r)$ which creates, at the electron's location, i.e. in the center of the current loop, the magnetic field with the following magnitude[10]:

---

[10] See, e.g. *Part EM* section 5.1, in particular, Eq. (5.24). Note that such effective magnetic field is induced by any motion of electrons, in particular that in solids, leading to a variety of spin–orbit level-splitting effects there—see, e.g. a concise review by R Winkler *et al*, in [1], p 211.

$$\mathscr{B}_a = \frac{\mu_0}{2r}I = \frac{\mu_0}{2r}\frac{Zev}{2\pi r} \equiv \frac{\mu_0 Zev}{4\pi r^2}. \tag{6.52}$$

The field's direction $\mathbf{n}$ is perpendicular to the apparent plane of the nucleus's rotation (i.e. that of the real rotation of the electron), and hence its vector may be readily expressed via the similarly directed vector $\mathbf{L} = m_e v r \mathbf{n}$ of the electron's angular (orbital) momentum:

$$\mathscr{B}_a = \frac{\mu_0 Zev}{4\pi r^2}\mathbf{n} \equiv \frac{\mu_0 Ze}{4\pi r^3 m_e}m_e v r \mathbf{n} \equiv \frac{\mu_0 Ze}{4\pi r^3 m_e}\mathbf{L} \equiv \frac{Ze}{4\pi\varepsilon_0 r^3 m_e c^2}\mathbf{L}, \tag{6.53}$$

where the last step used the basic relation between the SI unit constants: $\mu_0 \equiv 1/c^2\varepsilon_0$.

A more careful (but still classical) analysis of the problem[11] brings both good and bad news. The bad news is that the result (6.53) is wrong by the so-called *Thomas factor* of 2 even for the circular motion, because the electron moves with acceleration, and the reference frame bound to it cannot be considered inertial (as was implied in the above reasoning), so that the effective magnetic field felt by the electron is actually

$$\mathscr{B} = \frac{Ze}{8\pi\varepsilon_0 r^3 m_e c^2}\mathbf{L}. \tag{6.54}$$

The good news is that, so corrected, the result is valid not only for circular but for an arbitrary orbital motion in the Coulomb field $U(r)$. Hence from the discussion in sections 4.1 and 4.4 we may expect that the quantum-mechanical description of the interaction between this effective magnetic field and the electron's spin moment (4.115) is given by the following perturbation Hamiltonian[12]

$$\hat{H}^{(1)} = -\hat{\mathbf{m}} \cdot \hat{\mathscr{B}} = -\gamma_e\hat{\mathbf{S}} \cdot \left(\frac{Ze}{8\pi\varepsilon_0 r^3 m_e c^2}\hat{\mathbf{L}}\right) \equiv \frac{1}{2m_e^2 c^2}\frac{Ze^2}{4\pi\varepsilon_0}\frac{1}{r^3}\hat{\mathbf{S}} \cdot \hat{\mathbf{L}}, \tag{6.55}$$

where at spelling out the electron's gyromagnetic ratio $\gamma_e \equiv -g_e e/2m_e$, the small correction to the value $g_e = 2$ of the electron's g-factor (see section 4.4) has been ignored, because Eq. (6.55) is already a small correction. This expression is confirmed by the fully-relativistic Dirac theory, to be discussed in section 9.7 below: it yields, for an arbitrary central potential $U(r)$, the following spin–orbit coupling Hamiltonian:

$$\hat{H}^{(1)} = \frac{1}{2m_e^2 c^2}\frac{1}{r}\frac{dU(r)}{dr}\hat{\mathbf{S}} \cdot \hat{\mathbf{L}}. \tag{6.56}$$

For the Coulomb potential $U(r) = -Ze^2/4\pi\varepsilon_0 r$, this formula is reduced to Eq. (6.55).

---

[11] It was carried out first by L Thomas in 1926; for a simple review see, e.g. [2].
[12] In the Gaussian units, Eq. (6.55) is valid without the factor $4\pi\varepsilon_0$ in the denominator; while Eq. (6.56), 'as is'.

As we already know from the discussion in section 5.7, the angular factor of this Hamiltonian commutes with all the operators of the coupled-representation group (inside the blue line in figure 5.12): $\hat{L}^2$, $\hat{S}^2$, $\hat{J}^2$, and $\hat{J}_z$, and hence is diagonal in the coupled-representation basis with definite quantum numbers l, j, and m_j (and of course $s = \frac{1}{2}$). Hence, using Eq. (5.181) to rewrite Eq. (6.56) as

$$\hat{H}^{(1)} = \frac{1}{2m_e^2 c^2} \frac{Ze^2}{4\pi\varepsilon_0} \frac{1}{r^3} \frac{1}{2} (\hat{J}^2 - \hat{L}^2 - \hat{S}^2), \tag{6.57}$$

we may again use Eq. (6.27) for each set $\{s, l, j, m_j\}$, with the common n:

$$E_{n,j,l}^{(1)} = \frac{1}{2m_e^2 c^2} \frac{Ze^2}{4\pi\varepsilon_0} \left\langle \frac{1}{r^3} \right\rangle_{n,l} \frac{1}{2} \langle \hat{J}^2 - \hat{L}^2 - \hat{S}^2 \rangle_{j,s}, \tag{6.58}$$

where the indices irrelevant for each particular factor have been dropped. Now using the last of Eqs. (3.211), and similar expressions (5.169), (5.175), and (5.177), we get an explicit expression for the spin–orbit corrections[13]

$$\begin{aligned}
E_{n,j,l}^{(1)} &= \frac{1}{2m_e^2 c^2} \frac{Ze^2}{4\pi\varepsilon_0} \frac{\hbar^2}{2r_0^3} \frac{j(j+1) - l(l+1) - \frac{3}{4}}{n^3 l(l+1/2)(l+1)} \\
&\equiv \frac{E_n^2}{m_e c^2} n \frac{j(j+1) - l(l+1) - \frac{3}{4}}{l(l+1/2)(l+1)},
\end{aligned} \tag{6.59}$$

with l and j related by Eq. (5.189).

The last form of its result shows clearly that this correction has the same scale as the kinetic correction (6.51).[14] In the 1st order of the perturbation theory, they may be just added (with $m = m_e$), giving a surprisingly simple formula for the net fine structure of the nth energy level:

$$E_{\text{fine}}^{(1)} = \frac{E_n^2}{2m_e c^2} \left(3 - \frac{4n}{j + \frac{1}{2}} \right). \tag{6.60}$$

This simplicity, as well as the independence of the result of the orbital quantum number l, will become less surprising when (in section 9.7) we see that this formula follows in one shot from the Dirac theory, in which the Bohr atom's energy spectrum in numbered only with n and j, but not l.

Let us recall that for an electron ($s = \frac{1}{2}$), according to Eq. (5.189) with $0 \leqslant l \leqslant n - 1$, the quantum number j may take n positive half-integer values, from $\frac{1}{2}$ to $n - \frac{1}{2}$. Hence, Eq. (6.60) shows that the fine structure of the nth Bohr's energy level has n sub-levels—see figure 6.6.

---

[13] The factor l in the denominator does not give a divergence at $l = 0$, because in this case $j = s = \frac{1}{2}$, so that $j(j+1) = 3/4$, and the numerator turns into 0 as well. A careful analysis of this case (which may be found, e.g. in [3]), as well as the exact analysis of the hydrogen atom using the Dirac theory (see section 9.7), show that Eq. (6.60), which does not include l, is valid even in this case.

[14] This is natural, because the magnetic interaction of charged particles is essentially a relativistic effect, of the same order ($\sim v^2/c^2$) as the kinetic correction (6.47)—see, e.g. Part EM section 5.1, in particular Eq. (5.3).

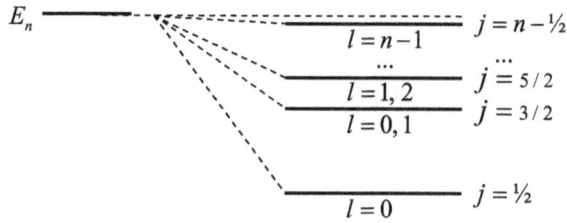

Figure 6.4. The fine structure of a hydrogen-like atom's level.

Please note that according to Eq. (5.175), each of these sub-levels is still $(2j + 1)$-times degenerate in the quantum number m_j. This degeneracy is very natural, because in the absence of external field the system is still isotropic. Moreover, on each fine-structure level, besides the lowest ($j = \frac{1}{2}$) and the highest ($j = n - \frac{1}{2}$) ones, each of the m_j-states is doubly-degenerate in the orbital quantum number $l = j \mp \frac{1}{2}$—see the labels of l in figure 6.4. (According to Eq. (5.190), each of these states, with fixed j and m_j, may be represented as a linear combination of two states with adjacent values of l, and hence different electron spin orientations, $m_s = \pm\frac{1}{2}$, weighed with the Clebsch–Gordan coefficients.)

These details aside, one may crudely say that the relativistic corrections combined make the total eigenenergy grow with l, contributing to the effect already mentioned with our analysis of the periodic table of elements in section 3.7. The relative scale of this increase may be scaled by the largest deviation from the unperturbed energy E_n, reached for the state with $j = \frac{1}{2}$ (and hence $l = 0$):

$$\frac{|E_{\max}^{(1)}|}{E_n} = \frac{E_n}{m_e c^2}\left(2n - \frac{3}{2}\right) \equiv \left(\frac{Ze^2}{4\pi\varepsilon_0\hbar c}\right)^2\left(\frac{1}{n} - \frac{3}{4n^2}\right) \equiv Z^2\alpha^2\left(\frac{1}{n} - \frac{3}{4n^2}\right). \quad (6.61)$$

where α is the *fine structure* ('Sommerfeld's') *constant*,

$$\alpha \equiv \frac{e^2}{4\pi\varepsilon_0\hbar c} \approx \frac{1}{137}, \quad (6.62)$$

already mentioned in section 4.4, which characterizes the strength (or rather the weakness :-) of the electromagnetic effects in quantum mechanics—which in particular makes the perturbative quantum electrodynamics possible[15]. These expressions show that the fine-structure splitting is a very small effect ($\sim\alpha^2 \sim 10^{-6}$) for the hydrogen atom, but it rapidly grows (as Z^2) with the nuclear charge (i.e. the atomic number) Z, and becomes rather substantial for the heaviest stable atoms with $Z \sim 100$.

---

[15] The alternative expression $\alpha = E_H/m_e c^2$, where E_H is the Hartree energy (1.13), i.e. the scale of all energies E_n, is also very revealing.

6.4 The Zeeman effect

Now, we are ready to review the *Zeeman effect*—the atomic level splitting by an external magnetic field[16]. Using Eq. (3.26), with $q = -e$, for the description of the electron's orbital motion in the field, and the Pauli Hamiltonian (4.163), with $\gamma = -e/m_e$, for the electron spin's interaction with the field, we see that even for a hydrogen-like (i.e. single-electron) atom/ion, neglecting the relativistic effects, the full Hamiltonian is rather involved:

$$\hat{H} = \frac{1}{2m_e}(\hat{\mathbf{p}} + e\hat{\mathbf{A}})^2 - \frac{Ze^2}{4\pi\varepsilon_0 r} + \frac{e}{m_e}\mathscr{B} \cdot \hat{\mathbf{S}}. \qquad (6.63)$$

There are several simplifications we may make. First, let us assume that the external field is spatial-uniform on the atomic scale (which is a very good approximation for most cases), so that we can take the vector-potential in an axially-symmetric gauge—cf Eq. (3.132):

$$\mathbf{A} = \frac{1}{2}\mathscr{B} \times \mathbf{r}. \qquad (6.64)$$

Second, let us neglect the terms proportional to $\mathscr{B}^2$, which are small in practical magnetic fields of the order of a few tesla[17]. The remaining term in the effective kinetic energy, describing the interaction with the magnetic field, is linear in the momentum operator, so that we may repeat the standard classical calculation[18] to reduce it to the product of $\mathscr{B}$ by the orbital magnetic moment's component $m_z = -eL_z/2m_e$—besides that both m_z and L_z should be understood as operators now. As a result, the Hamiltonian (6.63) reduces to Eq. (6.1), $\hat{H}^{(0)} + \hat{H}^{(1)}$, where $\hat{H}^{(0)}$ is that of the atom at $\mathscr{B} = 0$, and

$$\hat{H}^{(1)} = \frac{e\mathscr{B}}{2m_e}(\hat{L}_z + 2\hat{S}_z). \qquad (6.65)$$

This expression immediately reveals the major complication with the Zeeman effect's analysis. Namely, in comparison with the equal orbital and spin contributions to the total angular momentum (5.171) of the electron, the spin produces a twice larger contribution to the magnetic moment, so that the right-hand side of Eq. (6.65) is *not* proportional to the total angular moment **J**. As a result, the effect's description is simple only in two limits.

If the magnetic field is so *high* that its effects are much stronger than the relativistic (fine-structure) effects discussed in the previous section, we may treat the

---

[16] It was discovered experimentally in 1896 by P Zeeman who, amazingly, was fired from the University of Leiden for an unauthorized use of lab equipment for this work—just to receive a Nobel Prize for it in a few years.

[17] Despite its smallness, the quadratic term is necessary for description of the negative contribution of the orbital motion to the magnetic susceptibility χ_m (the so-called *orbital diamagnetism*, see *Part EM* section 5.5), whose analysis, using Eq. (6.63), is left for the reader's exercise.

[18] See, e.g. *Part EM* section 5.4, in particular Eqs. (5.95) and (5.100).

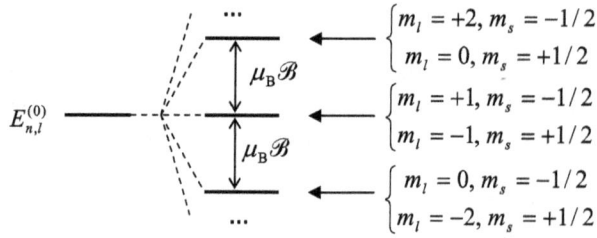

Figure 6.5. The Paschen–Back effect.

two terms in Eq. (6.65) as independent perturbations of different (orbital and spin) degrees of freedom. Since each of the perturbation matrices is diagonal in its own z-basis, we can again use Eq. (6.27) to write

$$
\begin{aligned}
E - E^{(0)} &= \frac{e\mathscr{B}}{2m_e}\left(\langle n, l, m_l \mid \hat{L}_z \mid n, l, m_l\rangle + 2\langle m_s \mid \hat{S}_z \mid m_s\rangle\right)\\
&= \frac{e\mathscr{B}}{2m_e}(\hbar m_l + 2\hbar m_s)\\
&\equiv \mu_B \mathscr{B}(m_l \pm 1).
\end{aligned}
\tag{6.66}
$$

This result describes the splitting of each $2\times(2l + 1)$-degenerate energy level, with certain n and l, into $(2l + 3)$ levels (figure 6.5), with the adjacent level splitting of $\mu_B\mathscr{B}$, of the order of $\sim 10^{-23}$ J $\sim 10^{-4}$ eV per tesla. Note that all the levels, these besides the top and bottom one, remain doubly degenerate. This limit of the Zeeman effect is sometimes called the *Paschen–Back effect*—whose simplicity was recognized only in the 1920s, due to the need in very high magnetic fields for its observation.

In the opposite limit of relatively *low* magnetic fields, the Zeeman effect takes place on the background of the much larger fine-structure splitting. As was discussed in section 6.3, at $\mathscr{B} = 0$ each split sub-level has a $2(2l + 1)$-fold degeneracy corresponding to $(2l + 1)$ different values of the half-integer quantum number m_j, ranging from $-j$ to $+j$, and 2 values of the integer $l = j \mp \frac{1}{2}$—see figure 6.4. The magnetic field lifts this degeneracy[19]. Indeed, in the coupled representation discussed in section 5.7, the perturbation (6.65) is described by the matrix with elements

$$
\begin{aligned}
H^{(1)} &= \frac{e\mathscr{B}}{2m_e}\langle j, m_j \mid \hat{L}_z + 2\hat{S}_z \mid j', m_{j'}\rangle \equiv \frac{e\mathscr{B}}{2m_e}\langle j, m_j \mid \hat{J}_z + \hat{S}_z \mid j', m_{j'}\rangle\\
&= \frac{e\mathscr{B}}{2m_e}\left(\hbar m_j \delta_{m_j m_{j'}} + \langle j, m_j \mid \hat{S}_z \mid j', m_{j'}\rangle\right).
\end{aligned}
\tag{6.67}
$$

To spell out the second term, let us use the general expansion (5.183) for the particular case $s = \frac{1}{2}$, when (as was discussed in the end of section 5.7) it has at most two nonvanishing terms, with the Clebsch–Gordan coefficients (5.190):

---

[19] In almost-hydrogen-like, but more complex atoms (such as those of alkali metals), the degeneracy in l is lifted by the electron–electron Coulomb interaction even in the absence of the external magnetic field.

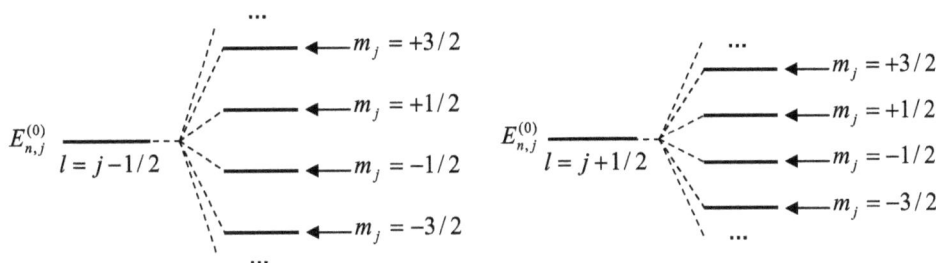

Figure 6.6. The anomalous Zeeman effect in a hydrogen-like atom.

$$\left| j = l \pm \tfrac{1}{2},\, m_j \right\rangle$$

$$= \pm \left(\frac{l \pm m_j + \tfrac{1}{2}}{2l + 1} \right)^{1/2} \left| m_l = m_j - \tfrac{1}{2},\, m_s = +\tfrac{1}{2} \right\rangle \tag{6.68}$$

$$+ \left(\frac{l \mp m_j + \tfrac{1}{2}}{2l + 1} \right)^{1/2} \left| m_l = m_j + \tfrac{1}{2},\, m_s = -\tfrac{1}{2} \right\rangle.$$

Taking into account that the operator $\hat{S}_z$ gives non-zero brackets only for $m_s = m_{s'}$, the 2×2 matrix of elements $\langle m_l = m_j \pm \tfrac{1}{2}, m_s = \mp\tfrac{1}{2} | \hat{S}_z | m_l = m_j \pm \tfrac{1}{2}, m_s = \mp\tfrac{1}{2} \rangle$ is diagonal, so we may use Eq. (6.27) to get

$$E - E^{(0)} = \frac{e\mathcal{B}}{2m_e} \left[\hbar m_j + \frac{\hbar}{2} \frac{(l \pm m_j + 1/2)}{2l + 1} - \frac{\hbar}{2} \frac{(l \mp m_j + 1/2)}{2l + 1} \right]$$

$$\equiv \frac{e\mathcal{B}}{2m_e} \hbar m_j \left(1 \pm \frac{1}{2l + 1} \right) \equiv \mu_B \mathcal{B} m_j \left(1 \pm \frac{1}{2l + 1} \right), \tag{6.69}$$

$$\text{for} - j \leqslant m_j \leqslant +j,$$

where the two signs correspond to the two possible values of $l = j \mp \tfrac{1}{2}$—see figure 6.6.

We see that the magnetic field splits each sub-level of the fine structure, with a given l, into $2j + 1$ equidistant levels, with the distance between the levels depending on l. In the late 1890s, when the Zeeman effect was first observed, there was no notion of spin at all, so that this puzzling result was called the *anomalous Zeeman effect*. (In this terminology, the *normal Zeeman effect* is the one with no spin splitting, i.e. without the second terms in the parentheses of Eqs. (6.66), (6.67), and (6.69); it may be observed experimentally in atoms with zero net spin.)

The strict quantum-mechanical analysis of the anomalous Zeeman effect for arbitrary s (which is important for applications to multi-electron atoms) is conceptually not complex, but requires explicit expressions for the corresponding Clebsch–Gordan coefficients, which are rather bulky. Let me just cite the unexpectedly simple result of this analysis:

$$\Delta E = \mu_B \mathcal{B} m_j g, \tag{6.70a}$$

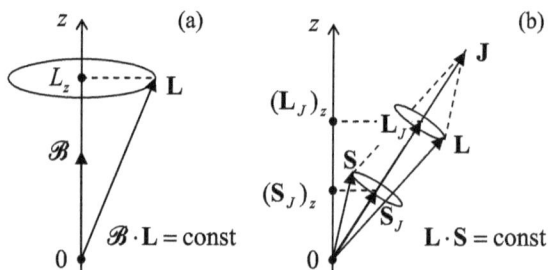

Figure 6.7. Classical images of (a) the orbital angular momentum's quantization in a magnetic field, and (b) the fine-structure level splitting.

where g is the so-called *Lande factor*[20]:

$$g = 1 + \frac{j(j + 1) + s(s + 1) - l(l + 1)}{2j(j + 1)}. \tag{6.70b}$$

For $s = \frac{1}{2}$ (and hence $j = l \pm \frac{1}{2}$), this factor is reduced to the parentheses in the last form of Eq. (6.69).

It is remarkable that Eqs. (6.70) may be readily derived using very plausible classical arguments, similar to those used in section 5.7—see figure 5.13 and its discussion. As was discussed in section 5.6, in the absence of spin, the quantization of the observable L_z is an extension of the classical picture of the torque-induced precession of the vector **L** about the magnetic field's direction, so that the interaction energy, proportional to $\mathscr{B}L_z = \mathscr{B} \cdot \mathbf{L}$, remains constant—see figure 6.7a. At the spin–orbit interaction without an external magnetic field, the Hamiltonian function of the system includes the product **S·L**, so that it has to be constant, together with J^2, L^2, and S^2. Hence, this system's classical image is a rapid precession of vectors **S** and **L** about the direction of the vector $\mathbf{J} = \mathbf{L} + \mathbf{S}$, so that the spin–orbit interaction energy, proportional to the product **L·S**, remains constant (figure 6.7b). On this backdrop, the anomalous Zeeman effect in a relatively weak magnetic field $\mathscr{B} = Bn_z$ corresponds to a slow precession of the vector **J** ('dragging' the rapidly rotating vectors **L** and **S** with it) about the axis z.

This picture allows us to conjecture that what is important for the slow precession rate are only the vectors **L** and **S** averaged over the period of the much faster precession about vector **J**—in other words, only their components $\mathbf{L}_J$ and $\mathbf{S}_J$ along the vector **J**. Classically, these components may be calculated as

$$\mathbf{L}_J = \frac{\mathbf{L} \cdot \mathbf{J}}{J^2}\mathbf{J}, \qquad \text{and} \qquad \mathbf{S}_J = \frac{\mathbf{S} \cdot \mathbf{J}}{J^2}\mathbf{J}. \tag{6.71}$$

The scalar products participating in these expressions may be readily expressed via the squared lengths of the vectors, using the following evident formulas:

---

[20] This formula is frequently used with capital letters J, S, and L, which denote the quantum numbers of the atom as a whole.

$$S^2 = (\mathbf{J} - \mathbf{L})^2 \equiv J^2 + L^2 - 2\mathbf{L}\cdot\mathbf{J}, \quad L^2 = (\mathbf{J} - \mathbf{S})^2 \equiv J^2 + S^2 - 2\mathbf{J}\cdot\mathbf{S}. \quad (6.72)$$

As a result, we get the following time average:

$$\overline{L_z + 2S_z} = (\mathbf{L}_J + 2\mathbf{S}_J)_z = \left(\frac{\mathbf{L}\cdot\mathbf{J}}{J^2}\mathbf{J} + 2\frac{\mathbf{S}\cdot\mathbf{J}}{J^2}\mathbf{J}\right)_z = \frac{J_z}{J^2}(\mathbf{L}\cdot\mathbf{J} + 2\mathbf{S}\cdot\mathbf{J})$$

$$= J_z\frac{(J^2 + L^2 - S^2) + 2(J^2 + S^2 - L^2)}{2J^2} \quad (6.73)$$

$$\equiv J_z\left(1 + \frac{J^2 + S^2 - L^2}{2J^2}\right).$$

The last move is to smuggle in some quantum mechanics by using, instead of the vector lengths squared, and the z-component of J_z, their eigenvalues given by Eqs. (5.169), (5.175), and (5.177). As a result, we immediately arrive at the exact result given by Eqs. (6.70). This coincidence encourages thinking about quantum mechanics of angular momenta in the classical terms of torque-induced precession, which turns out to be very fruitful in more complex problems of atomic and molecular physics.

The high-field limit and low-field limits of the Zeeman effect, described respectively by Eqs. (6.66) and (6.69), are separated by a medium field range, in which the Zeeman splitting is of the order of the fine-structure splitting analyzed in section 6.3. There is no time in this course for a quantitative analysis of this crossover[21].

6.5 Time-dependent perturbations

Now let us proceed to the case when the perturbation $\hat{H}^{(1)}$ in Eq. (6.1) is a function of time, while $\hat{H}^{(0)}$ is time-independent. The adequate perturbative approach to this problem, and its results, depend critically on the relation between the characteristic frequency (or the characteristic reciprocal time) ω of the perturbation and the distance between the initial system's energy levels:

$$\hbar\omega \leftrightarrow |E_n - E_{n'}|. \quad (6.74)$$

In the case when all essential frequencies of a perturbation are very small in the sense of Eq. (6.74), we are dealing with the so-called *adiabatic* change of parameters, that may be treated essentially as a time-independent perturbation (see the previous sections of this chapter). The most interesting observation here is that the adiabatic perturbation does not allow any significant transfer of the system's probability from one eigenstate to another. For example, in the WKB limit of the orbital motion, the Bohr–Sommerfeld quantization rule (2.110), and its multi-dimensional generalizations, guarantee that the integral

---

[21] For a more complete discussion of the Stark, Zeeman, and fine-structure effects in atoms, I can recommend, for example, either the monograph by G Woolgate, cited above, or the one by I Sobelman [4].

$$\oint_C \mathbf{p} \cdot d\mathbf{r}, \tag{6.75}$$

taken along the particle's classical trajectory, is an *adiabatic invariant*, i.e. does not change at a slow change of system's parameters. (It is curious that classical mechanics also guarantees the invariance of the integral (6.75), but its proof there[22] is much harder than the quantum-mechanical derivation of this fact, carried out in section 2.4.) This is why even if the perturbation becomes large with time (while changing sufficiently slowly), we can expect the eigenstate and eigenvalue classification to persist.

Let me proceed to the case when both sides of Eq. (6.74) are comparable, using for this discussion the Schrödinger picture of quantum dynamics, given by Eq. (4.158). Combining it with Eq. (6.1), we get the Schrödinger equation in the form

$$i\hbar \frac{\partial}{\partial t} |\alpha(t)\rangle = \left(\hat{H}^{(0)} + H^{(1)}(t) \right) |\alpha(t)\rangle \tag{6.76}$$

Very much in the spirit of our treatment of the time-independent case in section 6.1, let us represent the time-dependent ket-vector of the system with its expansion,

$$|\alpha(t)\rangle = \sum_n |n\rangle \langle n|\alpha(t)\rangle, \tag{6.77}$$

over the full and orthonormal set of the unperturbed, stationary ket-vectors defined by equation

$$\hat{H}^{(0)} |n\rangle = E_n |n\rangle. \tag{6.78}$$

(Note that these kets $|n\rangle$ are exactly what was called $|n^{(0)}\rangle$ in section 6.1; we just may afford a less bulky notation in this section.) Plugging the expansion (6.77), with n replaced with n', into both sides of Eq. (6.76), and then inner-multiplying both its sides by the bra-vector $\langle n|$ of another unperturbed (and hence time-independent) state of the system, we get a set of linear, ordinary differential equations for the expansion coefficients:

$$i\hbar \frac{d}{dt} \langle n|\alpha(t)\rangle = E_n \langle n|\alpha(t)\rangle + \sum_{n'} H^{(1)}_{nn'}(t) \langle n'|\alpha(t)\rangle, \tag{6.79}$$

where the matrix elements of the perturbation in the unperturbed state basis, defined similarly to Eq. (6.8), are now functions of time:

$$H^{(1)}_{nn'}(t) \equiv \langle n|\hat{H}^{(1)}(t)|n'\rangle. \tag{6.80}$$

The set of differential equations (6.79), which are still exact, may be useful for numerical calculations[23]. However, Eq. (6.79) has a certain technical inconvenience,

---

[22] See, e.g. *Part CM* section 10.2.

[23] Even if the problem under analysis may be described by the wave-mechanics Schrödinger equation (1.25), a direct numerical integration of that *partial* differential equation is typically less convenient than that of the *ordinary* differential equations (6.79).

which becomes clear if we consider its (evident) solution in the absence of perturbation[24]:

$$\langle n|\alpha(t)\rangle = \langle n|\alpha(0)\rangle \exp\left\{-i\frac{E_n}{\hbar}t\right\}. \tag{6.81}$$

We see that the solution oscillates very fast, and its numerical modeling may represent a challenge for even the fastest computers. These spurious oscillations (whose frequency, in particular, depends of the energy reference level) may be partly tamed by looking for the general solution of Eqs. (6.79) in a form inspired by Eq. (6.81):

$$\langle n|\alpha(t)\rangle \equiv a_n(t)\exp\left\{-i\frac{E_n}{\hbar}t\right\}. \tag{6.82}$$

Here $a_n(t)$ are new functions of time (essentially, the stationary states' probability amplitudes), which may be used, in particular, to calculate the time-dependent *level occupancies*, i.e. the probabilities W_n to find the perturbed system on the corresponding energy levels of the unperturbed system:

$$W_n(t) = |\langle n|\alpha(t)\rangle|^2 = |a_n(t)|^2. \tag{6.83}$$

Plugging Eq. (6.82) into Eq. (6.79), for these functions we readily get a slightly modified system of equations:

$$i\hbar\dot{a}_n = \sum_{n'}a_{n'}H^{(1)}_{nn'}(t)e^{i\omega_{nn'}t}, \tag{6.84}$$

where the factors $\omega_{nn'}$, defined by the relation

$$\hbar\omega_{nn'} \equiv E_n - E_{n'} \tag{6.85}$$

have the physical sense of frequencies of *potential* quantum transitions between the nth and n'th energy levels of the unperturbed system. (The conditions when such transitions indeed take place will be clear soon.) The advantages of Eq. (6.84) over Eq. (6.79), for both analytical and numerical calculations, is their independence of the energy reference, and lower frequencies of oscillations of the right-hand side terms, especially when the energy levels of interest are close to each other[25].

In order to continue our analytical treatment, let us restrict ourselves to a particular but very important case of a sinusoidal perturbation *turned on* at some moment—which may be taken for $t = 0$:

$$\hat{H}^{(1)}(t) = \begin{cases} 0, & \text{for } t < 0, \\ \hat{A}e^{-i\omega t} + \hat{A}^\dagger e^{+i\omega t}, & \text{for } t \geqslant 0, \end{cases} \tag{6.86}$$

---

[24] This is of course just a more general form of Eq. (1.62) of the wave mechanics of time-independent systems.
[25] Note that the relation of Eq. (6.84) to the initial Eq. (6.79) is very close to the relation of the interaction picture of quantum dynamics, discussed in the end of section 4.6, to its Schrödinger picture, with the perturbation Hamiltonian playing the role of the interaction one—compare Eqs. (6.1) and Eq. (4.206). Indeed, Eq. (6.84) could be readily obtained from the interaction picture, and I did not do this just to avoid using this heavy bra–ket artillery for our (relatively) simple problem, and hence to keep its physics more transparent.

where the *perturbation amplitude operators* $\hat{A}$ and $\hat{A}^\dagger$,[26] and hence their matrix elements,

$$\langle n|\hat{A}|n'\rangle \equiv A_{nn'}, \qquad \langle n|\hat{A}^\dagger|n'\rangle = A^*_{n'n},\qquad (6.87)$$

are time-independent. In this case, Eq. (6.84) yields

$$i\hbar\dot{a}_n = \sum_{n'} a_{n'}\left[A_{nn'}e^{i(\omega_{nn'}-\omega)t} + A^*_{n'n}e^{i(\omega_{nn'}+\omega)t}\right], \qquad \text{for } t > 0. \qquad (6.88)$$

This is, generally, still a nontrivial system of coupled differential equations; however, it allows simple and explicit solutions in two very important limits. First, let us assume that our system initially was definitely in one eigenstate n' (usually, though not necessarily, in the ground state), and that the occupancies W_n of all other levels stay very low all the time. (We will find the condition when the second assumption is valid *a posteriori*—from the solution.) With the corresponding assumption

$$a_{n'} = 1; \qquad |a_n| \ll 1, \quad \text{for } n \neq n', \qquad (6.89)$$

Eq. (6.88) may be readily integrated, giving

$$a_n = -\frac{A_{nn'}}{\hbar(\omega_{nn'} - \omega)}[e^{i(\omega_{nn'}-\omega)t} - 1] - \frac{A^*_{n'n}}{\hbar(\omega_{nn'} + \omega)}[e^{i(\omega_{nn'}+\omega)t} - 1], \text{ for } n \neq n'. \;(6.90)$$

This expression describes what is colloquially called the *ac excitation of* (other) *energy levels*. Qualitatively, it shows that the probability W_n (6.83) of finding the system in each state ('on each energy level') of the system does not tend to any constant value but rather oscillates in time—the so-called *Rabi oscillations*. It also shows that that the ac-field-induced transfer of the system from one state to the other one has a clearly resonant character: the maximum occupancy W_n of a level grows infinitely when the corresponding *detuning*[27]

$$\Delta_{nn'} \equiv \omega - \omega_{nn'}, \qquad (6.91)$$

tends to zero. This conclusion is clearly unrealistic, and is due to our initial assumption (6.89); according to Eq. (6.90), it is satisfied only if[28]

$$|A_{nn'}| \ll \hbar|\omega \pm \omega_{nn'}|, \qquad (6.92)$$

and hence does not allow a more deep analysis of the resonance excitation.

---

[26] The notation of the amplitude operators in Eq. (6.86) is justified by the fact that the perturbation Hamiltonian has to be self-adjoint (Hermitian), and hence each term on the right-hand side of that relation has to be a Hermitian conjugate of its counterpart, which is evidently true only if the amplitude operators are also the Hermitian conjugates of each other. Note, however, that each of these amplitude operators is generally *not* Hermitian.

[27] The notion of detuning is also very useful in the classical theory of oscillations (see, e.g. *Part CM* chapter 5), where the role of $\omega_{nn'}$ is played by the own frequency ω_0 of the oscillator.

[28] Strictly speaking, one more condition is that the number of 'resonance' levels is also not too high—see section 6.6.

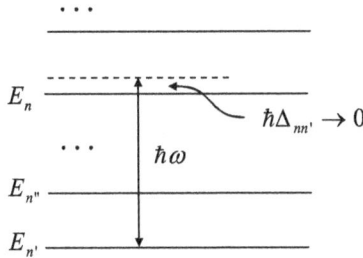

Figure 6.8. The resonant excitation of an energy level.

In order to overcome this limitation, we may perform the following trick—very similar to the one we used for the transfer to the degenerate case in section 6.1. Let us assume that for a certain level n,

$$|\Delta_{nn'}| \ll \omega, |\omega \pm \omega_{n''n}|, |\omega \pm \omega_{n''n'}|, \qquad \text{for all } n'' \neq n, n' \qquad (6.93)$$

—the condition illustrated in figure 6.8. Then, according to Eq. (6.90), we may ignore the occupancy of all but two levels, n and n', and also the second, non-resonant term with frequency $\omega_{nn'} + \omega \approx 2\omega \gg |\Delta_{nn'}|$ in Eqs. (6.88),[29] now written for two probability amplitudes, a_n and $a_{n'}$.

The result is the following system of two linear equations:

$$i\hbar \dot{a}_n = a_{n'} A e^{-i\Delta t}, \qquad i\hbar \dot{a}_{n'} = a_n A^* e^{i\Delta t}, \qquad (6.94)$$

which uses the shorthand notation $A \equiv A_{nn'}$ and $\Delta \equiv \Delta_{nn'}$. (I will use it for a while—until other energy levels become involved, in the beginning of the next section.) This system may be readily reduced to a form without an explicit time independence—for example, by introducing the following new probability amplitudes, which the same moduli:

$$b_n \equiv a_n e^{i\Delta t/2}, \qquad b_{n'} \equiv a_{n'} e^{-i\Delta t/2}, \qquad (6.95)$$

so that

$$a_n = b_n e^{-i\Delta t/2}, \qquad a_{n'} = b_{n'} e^{i\Delta t/2}. \qquad (6.96)$$

Plugging these relations into Eq. (6.94), we get two usual linear first-order differential equations:

$$i\hbar \dot{b}_n = -\frac{\hbar\Delta}{2} b_n + A b_{n'}, \qquad i\hbar \dot{b}_{n'} = A^* b_n + \frac{\hbar\Delta}{2} b_{n'}. \qquad (6.97)$$

As the reader knows very well by now, the general solution of such a system is a linear combination of two exponential functions, $\exp\{\lambda_\pm t\}$, with the exponents $\lambda_\pm$ that may be found by plugging any of these functions into Eq. (6.97), and requiring

---

[29] The second assumption, i.e. the omission of non-resonant terms in the equations for the amplitudes is called the *Rotating Wave Approximation* (RWA); the same idea in the classical theory of oscillations is the basis of what is usually called the *van der Pol method*, and its result, the *reduced equations*—see, e.g. *Part CM* sections 5.3–5.5.

the consistency of two resulting linear algebraic equations. In our case, the consistency condition (i.e. the characteristic equation of the system) is

$$\begin{vmatrix} -\hbar\Delta/2 - i\hbar\lambda & A \\ A^* & \hbar\Delta/2 - i\hbar\lambda \end{vmatrix} = 0, \tag{6.98}$$

with roots $\lambda_\pm = \pm i\Omega$, where

$$\Omega \equiv \left(\frac{\Delta^2}{4} + \frac{|A|^2}{\hbar^2} \right)^{1/2}, \qquad \text{i.e. } 2\Omega = \left(\Delta^2 + 4\frac{|A|^2}{\hbar^2} \right)^{1/2}. \tag{6.99}$$

The coefficients at the exponents are determined by initial conditions. If, as was assumed before, the system was completely on the level n' initially, at $t = 0$, i.e. $a_{n'}(0) = 1$, $a_n(0) = 0$, so that $b_{n'}(0) = 1$, $b_n(0) = 0$ as well, then Eqs. (6.97) yield, in particular:

$$b_n(t) = -i\frac{A}{\hbar\Omega} \sin \Omega t, \tag{6.100}$$

so that the nth level occupancy is

$$W_n = |b_n|^2 = \frac{|A|^2}{\hbar^2\Omega^2} \sin^2 \Omega t \equiv \frac{|A|^2}{|A|^2 + (\hbar\Delta/2)^2} \sin^2 \Omega t. \tag{6.101}$$

This is the famous *Rabi formula*[30]. If the detuning is large in comparison with $|A|/\hbar$, though still small in the sense of Eq. (6.93), the frequency 2Ω of the Rabi oscillations is completely determined by the detuning, and their amplitude is small:

$$W_n(t) = 4\frac{|A|^2}{\hbar^2\Delta^2} \sin^2 \frac{\Delta t}{2} \ll 1, \qquad \text{for } |A|^2 \ll (\hbar\Delta)^2, \tag{6.102}$$

—the result which could be obtained directly from Eq. (6.90), just neglecting the second term on its right-hand side. However, now we may also analyze the results of an increase of the perturbation amplitude $|A|$: it leads not only to an increase of the amplitude of the probability oscillations, but also of their frequency—see figure 6.9. Ultimately, at $|A| \gg \hbar|\Delta|$ (for example, at the exact resonance, $\Delta = 0$, i.e. $\omega_{nn'} = \omega$, so that $E_n = E_{n'} + \hbar\omega$), Eqs. (6.101) and (6.102) give $\Omega = |A|/\hbar$ and $(W_n)_{max} = 1$, i.e. describe a periodic, full 'repumping' of the system from one level to another and back, with a frequency proportional to the perturbation amplitude[31].

This effect is a close analog of the quantum oscillations in two-level systems with time-independent Hamiltonians, which were discussed in sections 2.6 and 5.1. Indeed, let us revisit, for a moment, their discussion started in the end of section 6.1, now paying more attention to the time evolution of the system under the perturbation. As was argued in that section, the most general perturbation

---

[30] It was derived in 1952 by I Rabi, in the context of his group's pioneering experiments with the ac (practically, microwave) excitation of quantum states, using molecular beams in vacuum.

[31] As Eqs. (6.82), (6.96), and (6.99) show, the lowest frequency in the system is $\omega_1 = \omega_{n'} - \Delta/2 + \Omega$, so that at $A \to 0$, $\hbar\omega_1 \approx \hbar\omega_{n'} + 2|A|^2/\hbar\Delta$. This effective shift of the lowest energy level (which may be measured by another 'probe' field of a different frequency) is a particular case of the *ac Stark effect*, which was already mentioned in section 6.2.

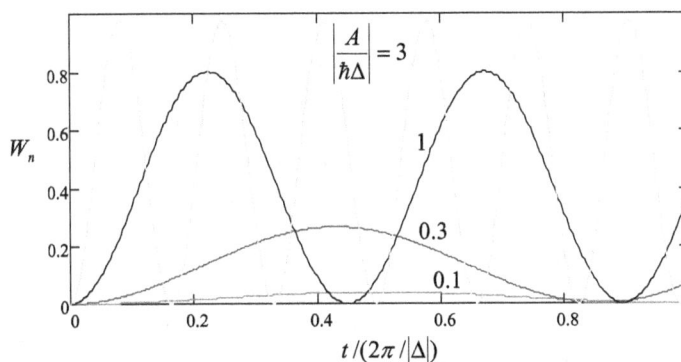

Figure 6.9. The Rabi oscillations.

Hamiltonian lifting the two-fold degeneracy of an energy level, in an arbitrary basis, has the matrix (6.28). Let us describe the system's dynamics using, again, the Schrödinger picture, representing the ket-vector of an arbitrary state of the system in the form (5.1), where ↑ and ↓ are the time-independent states of the basis in which Eq. (6.28) is written (now without any obligation to associate these states with the z-basis of a spin-½). Then, the Schrödinger equation (4.158) yields

$$
i\hbar \begin{pmatrix} \dot{\alpha}_\uparrow \\ \dot{\alpha}_\downarrow \end{pmatrix} = \mathrm{H}^{(1)} \begin{pmatrix} \alpha_\uparrow \\ \alpha_\downarrow \end{pmatrix} \equiv \begin{pmatrix} H_{11}^{(1)} & H_{12}^{(1)} \\ H_{21}^{(1)} & H_{22}^{(1)} \end{pmatrix} \begin{pmatrix} \alpha_\uparrow \\ \alpha_\downarrow \end{pmatrix} \equiv \begin{pmatrix} H_{11}^{(1)}\alpha_\uparrow + H_{12}^{(1)}\alpha_\downarrow \\ H_{21}^{(1)}\alpha_\uparrow + H_{22}^{(1)}\alpha_\downarrow \end{pmatrix}. \tag{6.103}
$$

As we know (for example, from the discussion in section 5.1), the average of the diagonal elements of the matrix gives just a common shift of the system's energy; for the purpose of the dynamics analysis, it may be absorbed into the energy reference level. Also, the Hamiltonian operator has to be Hermitian, so that the off-diagonal elements of its matrix have to be complex-conjugate. With this, Eqs. (6.103) are reduced to the form,

$$
i\hbar\dot{\alpha}_\uparrow = -\frac{\xi}{2}\alpha_\uparrow + H_{12}\alpha_\downarrow, \quad i\hbar\dot{\alpha}_\downarrow = H_{12}^*\dot{\alpha}_\uparrow + \frac{\xi}{2}\alpha_\downarrow, \quad \text{with } \hbar\xi \equiv H_{22}^{(1)} - H_{22}^{(1)}, \tag{6.104}
$$

which is absolutely similar to Eqs. (6.97). In particular, these equations describe the quantum oscillations of the probabilities $W_\uparrow = |\alpha_\uparrow|^2$ and $W_\downarrow = |\alpha_\downarrow|^2$ with the frequency[32]

$$
2\Omega = \left(\xi^2 + 4\frac{|H_{12}^{(1)}|^2}{\hbar^2} \right)^{1/2}. \tag{6.105}
$$

The similarity of Eqs. (6.97) and (6.104), and hence of Eqs. (6.99) and (6.105), shows that the Rabi oscillations and the 'usual' quantum oscillations have essentially

---

[32] By the way, Eq. (6.105) gives a natural generalization of the relations obtained for the frequency of such oscillations in section 2.6, where the coupled potential wells were assumed to be exactly similar, so that $\xi = 0$. Moreover, Eqs. (6.104) give a long-promised proof of Eqs. (2.201), and hence a better justification of Eqs. (2.203).

the same physical nature, besides that in the former case the external ac signal quantum $\hbar\omega$ bridges the separated energy levels, effectively reducing their difference $(E_n - E_{n'})$ to a much smaller difference $-\Delta \equiv (E_n - E_{n'})-\hbar\omega$. Also, since the Hamiltonian (6.28) is similar to that given by Eq. (5.2), the dynamics of such a system with two ac-coupled energy levels, within the limits (6.93) of the perturbation theory, is completely similar to that of a time-independent two-level system. In particular, its state may be similarly represented by a point on the Bloch sphere shown in figure 5.3, with its dynamics described, in the Heisenberg picture, by Eq. (5.19). This fact is very convenient for the experimental implementation of quantum information systems (to be discussed in more detail in section 8.5), because it enables one to manipulate qubits using a broad variety of physical systems, with well separated energy levels, using controlled external ac (usually microwave or optical) sources.

Note, however, that according to Eq. (6.90), if the system has energy levels other than n and n', they also become occupied to some extent. Since the sum of all occupancies equals 1, this means that $(W_n)_{\max}$ may approach 1 only if the excitation amplitude is very small, and hence the state manipulation time scale $\mathcal{T} = 2\pi/\Omega = 2\pi\hbar/|A|$ is very long. The ultimate limit in this sense is provided by the harmonic oscillator where all energy levels are equidistant, and the probability repumping between all of them occurs with the same rate. In particular, in this system the implementation of the full Rabi oscillations is impossible even at the exact resonance[33].

However, I would not like these quantitative details to obscure from the reader the most important qualitative (OK, maybe semi-quantitative :-) conclusion of this section's analysis: a resonant increase of the interlevel transition intensity at $\omega \to \omega_{nn'}$. As will be shown later in the course, this increase is accompanied by a sharp increase of the external field's *absorption*, which may be readily measured. This effect has numerous practical applications including systems based on the electron paramagnetic resonance (EPR) and nuclear magnetic resonance (NMR) spectroscopies, which are broadly used in material science, chemistry, and medicine. Unfortunately, I will not have time to discuss the related technical issues and methods (in particular, interesting ac pulsing techniques, including the so-called *Ramsey interferometry*) in detail, and have to refer the reader to special literature[34].

6.6 Quantum-mechanical golden rule

One of the results of the past section, Eq. (6.102), may be used to derive one of the most important and nontrivial results of quantum mechanics. For that, let us consider the case when the perturbation causes quantum transitions from a *discrete* energy level $E_{n'}$ into a group of eigenstates with a dense (essentially continuous) spectrum E_n—see figure 6.10a.

---

[33] From section 5.5, we already know what happens to the ground state of an oscillator at its external sinusoidal (or any other) excitation: it turns into a Glauber state, i.e. a superposition of *all* Fock states—see Eq. (5.134).
[34] For introductions see, e.g. [5, 6].

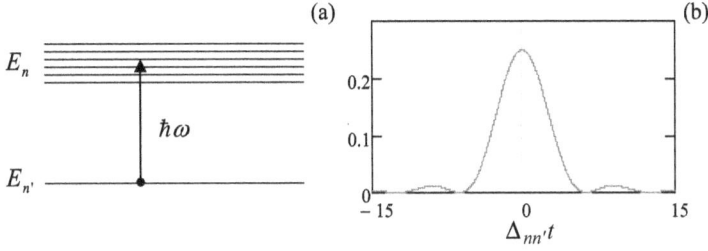

Figure 6.10. Deriving the Golden Rule: (a) the energy level scheme, and (b) the function under the integral (6.108).

If, for all states n of the group, the following conditions are satisfied

$$|A_{nn'}|^2 \ll (\hbar\Delta_{nn'})^2 \ll (\hbar\omega_{nn'})^2, \tag{6.106}$$

then Eq. (6.102) coincides with the result that would follow from Eq. (6.90). This means that we may apply Eq. (6.102), with indices n and n' duly restored, to any level n of our tight group. As a result, the total probability of having our system transferred from the initial level n' to that group is

$$W_\Sigma(t) = \sum_n W_n(t) = \frac{4}{\hbar^2} \sum_n \frac{|A_{nn'}|^2}{\Delta_{nn'}^2} \sin^2 \frac{\Delta_{nn'}t}{2}. \tag{6.107}$$

Now comes the main, absolutely beautiful trick: let us assume that the summation over n is limited to a tight group of very similar states whose matrix elements $A_{nn'}$ are virtually similar (we will check the validity of this assumption later on), so that we can take it out of the sum (6.107) and then replace the sum with the corresponding integral:

$$\begin{aligned} W_\Sigma(t) &= \frac{4\,|A_{nn'}|^2}{\hbar^2} \int \frac{1}{\Delta_{nn'}^2} \sin^2 \frac{\Delta_{nn'}t}{2}\,dn \\ &\equiv \frac{4\,|A_{nn'}|^2\,\rho_n t}{\hbar} \int \frac{1}{(\Delta_{nn'}t)^2} \sin^2 \frac{\Delta_{nn'}t}{2}\,d(-\Delta_{nn'}t), \end{aligned} \tag{6.108}$$

where ρ_n is the density of the states n on the energy axis:

$$\rho_n \equiv \frac{dn}{dE_n}. \tag{6.109}$$

This density, as well as the matrix element $A_{nn'}$, have to be evaluated at $\Delta_{nn'} = 0$, i.e. at energy $E_n = E_{n'} + \hbar\omega$, and are assumed to be constant within the final state group. At fixed $E_{n'}$, the function under integral (6.108) is even and decreases fast at $|\Delta_{nn'}t| \gg 1$—see figure 6.10b. Hence we may introduce a dimensionless integration variable $\xi \equiv \Delta_{nn'}t$, and extend the integration over it formally from $-\infty$ to $+\infty$. Then the integral in Eq. (6.108) is reduced to a table one[35], and yields

---

[35] See, e.g. Eq. (A.39).

$$W_\Sigma(t) = \frac{4\,|A_{nn'}|^2\,\rho_n t}{\hbar}\int_{-\infty}^{+\infty}\frac{1}{\xi^2}\sin^2\frac{\xi}{2}d\xi = \frac{4\,|A_{nn'}|^2\,\rho_n t}{\hbar}\frac{\pi}{2} \equiv \Gamma t, \tag{6.110}$$

where the constant

$$\Gamma = \frac{2\pi}{\hbar}\,|A_{nn'}|^2\,\rho_n. \tag{6.111}$$

is the called the *transition rate*[36].

This is one of the most famous and useful results of quantum mechanics, its *Golden Rule* (sometimes, rather unfairly, called the 'Fermi Golden Rule'[37]), which deserves much discussion. First of all, let us reproduce the reasoning already used in section 2.5 to show that the meaning of the rate Γ is much deeper than Eq. (6.110) seems to imply. Indeed, due to the conservation of the total probability, $W_{n'} + W_\Sigma = 1$, we can rewrite that equation as

$$\dot{W}_{n'}\,|_{t=0} = -\Gamma. \tag{6.112}$$

Evidently, this result cannot be true for all times, otherwise the probability $W_{n'}$ would become negative. The reason for this apparent contradiction is that Eq. (6.110) was obtained in the assumption that initially the system was completely on the level n': $W_{n'}(0) = 1$. Now, if in the initial moment the value of $W_{n'}$ is different, the result (6.110) has to be multiplied by that number, due to the linear relation (6.88) between da_n/dt and $a_{n'}$. Hence, instead of Eq. (6.112) we get a differential equation similar to Eq. (2.159),

$$\dot{W}_{n'}\,|_{t\geqslant 0} = -\Gamma W_{n'}, \tag{6.113}$$

which, for a time-independent Γ, has the evident solution,

$$W_{n'}(t) = W_{n'}(0)e^{-\Gamma t}, \tag{6.114}$$

describing the exponential decay of the initial state's occupancy, with the time constant $\tau = 1/\Gamma$.

I am inviting the reader to review this fascinating result again: by the summation of *periodic oscillations* (6.102) over many levels n, we have got an *exponential decay* (6.114) of the probability. This becomes possible because the effective range ΔE_n of the state energies E_n giving substantial contributions into the integral (6.108), shrinks with time: $\Delta E_n \sim \hbar/t$.[38] By the way, since most of the decay takes place

---

[36] In some texts, the density of states in Eq. (6.111) is replaced with a formal expression $\Sigma_n \delta(E_n - E_{n'} - \hbar\omega)$. Indeed, applied to a finite energy interval ΔE_n with $\Delta n \gg 1$ levels, it gives the same result: $\Delta n \equiv (dn/dE_n)\Delta E_n \equiv \rho_n\Delta E_n$. Such replacement may be technically useful in some cases, but is incorrect for $\Delta n \sim 1$, and hence should be used with utmost care, so that for most applications the more explicit form (6.111) is preferable.

[37] Actually, this result was developed mostly by the same P A M Dirac in 1927; E Fermi's role was not much more than advertising it, under the name of 'Golden Rule No. 2', in his lecture notes on nuclear physics, which were published much later, in 1950. (To be fair to Fermi, he has never tried to pose as the Golden Rule's author.)

[38] Here we have run again, in a more general context, into the 'energy-time uncertainty relation' which was already discussed in the end of section 2.5.

within the time interval of the order of $\tau \equiv 1/\Gamma$, the range of the participating final energies may be estimated as

$$\Delta E_n \sim \frac{\hbar}{\tau} \equiv \hbar\Gamma. \qquad (6.115)$$

This estimate is very instrumental for the formulation of conditions of the Golden Rule's validity. First, we have assumed that the matrix elements of the perturbation and the density of states are independent of the energy within the interval (6.115). This gives the following requirement

$$\Delta E_n \sim \hbar\Gamma \ll E_n - E_{n'} \sim \hbar\omega, \qquad (6.116)$$

Second, for the transfer from the sum (6.107) to the integral (6.108), we need the number of states within that energy interval, $\Delta N_n = \rho_n \Delta E_n$, to be much larger than 1. Merging Eq. (6.116) with Eq. (6.92) for all the energy levels $n'' \neq n$, n' not participating in the resonant transition, we may summarize all conditions of the Golden Rule validity as

$$\rho_n^{-1} \ll \hbar\Gamma \ll \hbar \, |\omega \pm \omega_{n'n''}|. \qquad (6.117)$$

(The reader may ask whether I have neglected the condition expressed by the first of Eqs. (6.106). However, for $\Delta_{nn'} \sim \Delta E_n/\hbar \sim \Gamma$, this condition is just $|A_{nn'}|^2 \ll (\hbar\Gamma)^2$, so that plugging it into Eq. (6.111),

$$\Gamma \ll \frac{2\pi}{\hbar}(\hbar\Gamma)^2 \rho_n, \qquad (6.118)$$

and canceling one Γ and one $\hbar$, we see that it coincides with the left relation in Eq. (6.117) above.)

Let us have a look at whether these conditions may be satisfied in practice, at least in some cases. For example, let us consider the optical ionization of an atom, with the released electron confined in a volume of the order of 1 $cm^3 \equiv 10^{-6} \ m^3$. According to Eq. (1.90), with E of the order of the atomic ionization energy $E_n - E_{n'} = \hbar\omega \sim 1$ eV, the density of electron states in that volume is of the order of 10^{17} 1/eV, while the right-hand side of Eq. (6.117) is of the order of $E_n \sim 1$ eV. Thus the conditions (6.117) provide an approximately 17 orders-of magnitude range for acceptable values of $\hbar\Gamma$. This illustration should give the reader a taste of why the Golden Rule is applicable to so many situations.

Finally, the physical picture of the initial state's decay (which will also be the key for our discussion of quantum-mechanical 'open' systems in the next chapter) is also very important. According to Eq. (6.114), the external excitation transfers the system into the continuous spectrum of levels n, and it never comes back to the initial level n'. However, it was derived from the quantum mechanics of Hamiltonian systems, whose equations are invariant with respect to time reversal. This paradox is a result of the generalization (6.113) of the exact result (6.112), which breaks the time reversal symmetry. This is a trick of course, but a trick absolutely adequate for the physics under study. Indeed, some gut feeling of the physical sense of this irreversibility may be obtained from the following observation. As Eq. (1.86) illustrates, the distance between the adjacent orbital energy levels tends to zero only if the system's size goes to infinity. This means that the assumption of the continuous energy spectrum of the finial states n essentially requires these states to be broadly extended in space—being free, or essentially free de Broglie waves. Thus the Golden Rule corresponds to the (physically justified) assumption that in an infinitely large system, the traveling de Broglie waves excited by a local source and propagating outward from it, would never come back, and even if they did, the unpredictable phase shifts introduced by the uncontrollable perturbations on their way would never allow them to sum up in the coherent way necessary to bring the system back into the initial state n'. (This is essentially the same situation which was discussed, for a particular 1D wave-mechanical system, in section 2.5.[39])

To get a better feeling of the Golden Rule, let us apply it to the following simple problem—which is a toy model of the photoelectric effect, briefly discussed in section 1.1(ii). A 1D particle is initially trapped in the ground state of a narrow potential well, described by Eq. (2.158):

$$U(x) = -\mathcal{W}\delta(x), \qquad \text{with } \mathcal{W} > 0. \tag{6.119}$$

Let us calculate the rate Γ of the particle's 'ionization' (i.e. its excitation into a group of extended, delocalized states) by a weak classical sinusoidal force of amplitude F_0 and frequency ω, suddenly turned on at some instant.

As a reminder, the initial, localized state (in our current notation, n') of such a particle was already found in section 2.6:

$$\psi_{n'}(x) = \kappa^{1/2}\exp\{-\kappa|x|\}, \quad \text{with} \quad \kappa \equiv \frac{m\mathcal{W}}{\hbar^2}, \quad E_{n'} \equiv -\frac{\hbar^2\kappa^2}{2m} = -\frac{m\mathcal{W}^2}{2\hbar^2}. \tag{6.120}$$

The final, extended states n, with a continuous spectrum, for this problem exist only at energies $E_n > 0$, so that the excitation rate is different from zero only for frequencies

$$\omega > \omega_{\min} \equiv \frac{|E_{n'}|}{\hbar} = \frac{m\mathcal{W}^2}{2\hbar^3}. \tag{6.121}$$

The weak sinusoidal force may be described by the following perturbation Hamiltonian,

$$\hat{H}^{(1)} = -F(t)\hat{x} = -F_0\hat{x}\cos\omega t = -\frac{F_0}{2}\hat{x}\left(e^{i\omega t} + e^{-i\omega t}\right), \qquad \text{for } t > 0, \tag{6.122}$$

so that according to Eq. (6.86), which serves as the amplitude operator definition, in this case

---

[39] This situation is also very much similar to the entropy increase in macroscopic systems, which is postulated in thermodynamics, and justified in statistical physics, even though it is based on time-reversible laws of mechanics—see, e.g. *Part SM* sections 1.2 and 2.2.

$$\hat{A} = \hat{A}^\dagger = -\frac{F_0}{2}\hat{x}. \tag{6.123}$$

Now the matrix elements $A_{nn'}$ that participate in Eq. (6.111) may be calculated in the coordinate representation:

$$A_{nn'} = \int_{-\infty}^{+\infty} \psi_n^*(x)\hat{A}(x)\,\psi_{n'}(x)dx = -\frac{F_0}{2}\int_{-\infty}^{+\infty} \psi_n^*(x)x\,\psi_{n'}(x)dx. \tag{6.124}$$

Since, according to Eq. (6.120), the initial $\psi_{n'}$ is a *symmetric* function of x, nonvanishing contributions to this integral are given only by *asymmetric* functions $\psi_n(x)$, proportional to $\sin k_n x$, with the wavenumber k_n related to the final energy by the well-familiar equality (1.89):

$$\frac{\hbar^2 k_n^2}{2m} = E_n. \tag{6.125}$$

As we know from section 2.6 (see in particular Eq. (2.167) and its discussion), such asymmetric functions, with $\psi_n(0) = 0$, are not affected by the zero-centered delta-functional potential (6.119), so that their density ρ_n is the same as that in a completely free space, and we can use Eq. (1.100). (Actually, since that relation was derived for traveling waves, it is more prudent to repeat the calculation that has led to that result, confining the waves on an artificial segment $[-l/2, +l/2]$—so long,

$$k_n l,\ \kappa l \gg 1, \tag{6.126}$$

that it does not affect the initial localized state and the excitation process. Then the confinement requirement $\psi_n(\pm l/2) = 0$ immediately yields the condition $k_n l/2 = n\pi$, so that Eq. (1.100) is indeed valid, but only for positive values of k_n, because $\sin k_n x$ with $k_n \to -k_n$ does not describe an independent standing-wave eigenstate.) Hence the final state density is

$$\rho_n \equiv \frac{dn}{dE_n} = \frac{dn}{dk_n}\bigg/\frac{dE_n}{dk_n} = \frac{l}{2\pi}\bigg/\frac{\hbar^2 k_n}{m} \equiv \frac{lm}{2\pi\hbar^2 k_n}. \tag{6.127}$$

It may look troubling that the density of states depends on the artificial segment's length l, but the same l also participates in the final wavefunctions' normalization factor[40],

$$\psi_n = \left(\frac{2}{l}\right)^{1/2}\sin k_n x, \tag{6.128}$$

and hence in the matrix element (6.124):

---

[40] The normalization to infinite volume, using Eq. (4.263), is also possible, but physically less transparent.

$$A_{nn'} = -\frac{F_0}{2}\left(\frac{2\kappa}{l}\right)^{1/2} \int_{-l}^{+l} \sin k_n x \; e^{-\kappa|x|}x dx.$$

$$= -\frac{F_0}{2i}\left(\frac{2\kappa}{l}\right)^{1/2}\left(\int_0^l e^{(ik_n-\kappa)x}x dx - \int_0^l e^{-(ik_n+\kappa)x}x dx\right).$$

(6.129)

These two integrals may be readily worked out by parts. Taking into account that, according to the condition (6.126), their upper limits may be extended to ∞, the result is

$$A_{nn'} = -\left(\frac{2\kappa}{l}\right)^{1/2}F_0\frac{2k_n\kappa}{(k_n^2+\kappa^2)^2}.$$

(6.130)

Note that the matrix element is a smooth function of k_n (and hence of E_n), so that the main assumption of the Golden Rule, the virtual constancy of $A_{nn'}$ in the interval $\Delta E_n \sim \hbar\Gamma \ll E_n$, is satisfied. So, Eq. (6.111) is reduced to the expression

$$\Gamma = \frac{2\pi}{\hbar}\left[\left(\frac{2\kappa}{l}\right)^{1/2}F_0\frac{2k_n\kappa}{(k_n^2+\kappa^2)^2}\right]^2 \frac{lm}{2\pi\hbar^2 k_n} \equiv \frac{8F_0^2 m k_n\kappa^3}{\hbar^3(k_n^2+\kappa^2)^4},$$

(6.131)

which is independent of the artificially introduced l—thus justifying its use.

Note that due to the above definitions of k_n and κ, the expression in parentheses in the denominator of the last expression does not depend on the potential well's 'area' $\mathcal{W}$, and is a function of only the excitation frequency ω (and the particle's mass):

$$\frac{\hbar^2(k_n^2+\kappa^2)}{2m} = E_n - E_{n'} = \hbar\omega.$$

(6.132)

As a result, Eq. (6.131) may be recast simply as

$$\hbar\Gamma = \frac{F_0^2\mathcal{W}^3 k_n}{2(\hbar\omega)^4}.$$

(6.133)

What is still hidden here is that k_n, defined by Eq. (6.125) with $E_n = E_{n'} + \hbar\omega$, is a function of the external force's frequency, changing as $\omega^{1/2}$ at $\omega \gg \omega_{\min}$ (so that Γ drops as $\omega^{-7/2}$ at $\omega \to \infty$), and as $(\omega - \omega_{\min})^{1/2}$ when ω approaches the 'red boundary' (6.121) of the ionization effect, so that $\Gamma \propto (\omega - \omega_{\min})^{1/2} \to 0$ in that limit as well. So, our toy model does describe this main feature of the photoelectric effect, whose explanation by A Einstein was essentially the starting point of quantum mechanics—see section 1.1.

The conceptually very similar, but a bit more involved analysis of this effect in a more realistic 3D case, namely the hydrogen atom's ionization by an optical wave, is left for the reader's exercise.

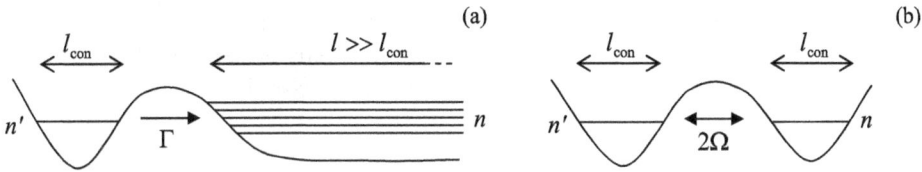

Figure 6.11. Tunneling coupling of a discrete-energy state n' to: (a) a state continuum, and (b) another discrete-energy state n.

6.7 Golden rule for step-like perturbations

Now let us reuse some of our results for a perturbation being turned on at $t = 0$, but after that time-independent:

$$\hat{H}^{(1)}(t) = \begin{cases} 0, & t < 0, \\ \hat{H}^{(1)} = \text{const}, & t \geqslant 0. \end{cases} \tag{6.134}$$

A superficial comparison of this equality and the former Eq. (6.86) seems to indicate that we may use all our previous results, taking $\omega = 0$ and replacing $\hat{A} + \hat{A}^\dagger$ with $\hat{H}^{(1)}$. However, that conclusion (which would give us a wrong factor of 2 in the result) does not take into account the fact that in analyzing both the two-level approximation in section 6.5, and the Golden Rule in section 6.6, we have dropped the second (non-resonant) term in Eq. (6.90). This is why it is more prudent to use the general Eq. (6.84),

$$i\hbar \dot{a}_n = \sum_{n'} a_{n'} H^{(1)}{}_{nn'} e^{i\omega_{nn'}t}, \tag{6.135}$$

in which the matrix element of the perturbation is now time-independent at $t > 0$. We see that it is formally equivalent to Eq. (6.88) with only the first (resonant) term kept, if we make the following replacements:

$$\hat{A} \to \hat{H}^{(1)}, \quad \Delta_{nn'} \equiv \omega - \omega_{nn'} \to -\omega_{nn'}. \tag{6.136}$$

Let us use this equivalency to consider the results of coupling between a discrete-energy state n', into which the particle is initially placed, and a dense group of states with a quasi-continuum spectrum, in the same energy range. Figure 6.11a shows an example of such a system: a particle is initially (say, at $t = 0$) placed into a potential well separated by a penetrable potential barrier from a formally infinite region with a continuous energy spectrum. Let me hope that the physical discussion in the last section makes the outcome of such an experiment evident: the particle will gradually and irreversibly tunnel out of the well, so that the probability $W_{n'}(t)$ of its still residing in the well will decay in accordance with Eq. (6.114). The rate of this decay may be found by making the replacements (6.136) in Eq. (6.111):

$$\Gamma = \frac{2\pi}{\hbar} \left| H^{(1)}_{nn'} \right|^2 \rho_n, \tag{6.137}$$

where the states n and n' now have virtually the same energy[41].

---

[41] The condition of validity of Eq. (6.137) is again given by Eq. (6.117), but with $\omega = 0$ in the upper limit.

It is very informative to compare this result, semi-quantitatively, with Eq. (6.105) for a symmetric ($E_n = E_{n'}$) system of two potential barrier separated by a similar potential barrier—see figure 6.11b. For the symmetric case, i.e. $\xi = 0$, Eq. (6.105) is reduced to simply

$$\Omega = \frac{1}{\hbar} \, |H_{nn'}^{(1)}|_{\mathrm{con}} \qquad (6.138)$$

Here I have used index 'con' (from 'confinement') to emphasize that this matrix element is somewhat different from the one participating in Eq. (6.137), even if the potential barriers are similar. Indeed, in the latter case, the matrix element,

$$H_{nn'}^{(1)} = \langle n|\hat{H}^{(1)}|n'\rangle = \int \psi_{n'}^* \hat{H}^{(1)} \psi_n dx, \qquad (6.139)$$

has to be calculated for two wavefunctions ψ_n and $\psi_{n'}$ confined to spatial intervals of the same scale l_{con}, while in Eq. (6.137), the wavefunctions ψ_n are extended to a much larger distance $l \gg l_{\mathrm{con}}$—see figure 6.11. As Eq. (6.129) tells us, in the 1D model this means an additional small factor of the order of $(l_{\mathrm{con}}/l)^{1/2}$. Now using Eq. (6.128) as a crude but suitable model for the final-state wavefunctions, we arrive at the following estimate, independent of the artificially introduced length l:

$$\hbar\Gamma \sim 2\pi \, |H_{nn'}^{(1)}|_{\mathrm{con}}^2 \, \frac{l_{\mathrm{con}}}{l} \rho_n \sim 2\pi \, |H_{nn'}^{(1)}|_{\mathrm{con}}^2 \, \frac{l_{\mathrm{con}}}{l} \frac{lm}{2\pi\hbar^2 k_n} \sim \frac{|H_{nn'}^{(1)}|_{\mathrm{con}}^2}{\Delta E_{n'}} \equiv \frac{(\hbar\Omega)^2}{\Delta E_{n'}}, \qquad (6.140)$$

where $\Delta E_{n'} \sim \hbar^2/m l_{\mathrm{con}}^2$ is the scale of the differences between the eigenenergies of the particle in an unperturbed potential well. Since the condition of validity of Eq. (6.138) is $\hbar\Omega \ll \Delta E_{n'}$, we see that

$$\hbar\Gamma \sim \frac{\hbar\Omega}{\Delta E_n} \hbar\Omega \ll \hbar\Omega. \qquad (6.141)$$

This (sufficiently general[42]) perturbative result confirms the conclusion of a more particular analysis carried out in the end of section 2.6: the rate of the (irreversible) quantum tunneling into a state continuum is always *much lower* than the frequency of (reversible) quantum oscillations between discrete states separated with the same potential barrier—at least for the case when both are much lower than $\Delta E_{n'}/\hbar$, so that the perturbation theory is valid. A very handwaving interpretation of this result is that the confined particle oscillates between the confined state in the well and some state behind the barrier many times before finally 'deciding' to perform an irreversible transition into the unconfined continuum. This picture is consistent with the experimentally observable effects of dispersive electromagnetic environment on the electron tunneling[43].

---

[42] It is straightforward to check that the estimate (6.141) is valid for similar problems of any spatial dimensionality, not just the 1D case we have analyzed.
[43] See, e.g. [7].

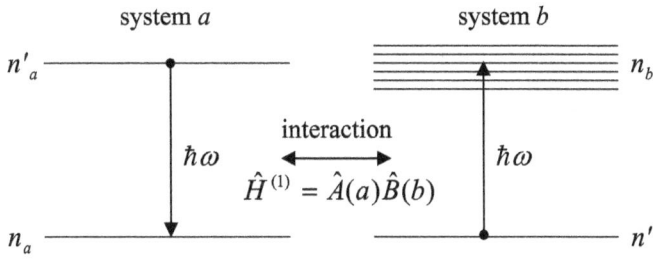

Figure 6.12. Energy relaxation in system a due to its weak coupling to system b (which serves as the environment of a).

Let me conclude this section (and this chapter) with the application of Eq. (6.137) to a very important case, which will provide a smooth transition to the next chapter's topic. Consider a *composite* system consisting of two parts ('components'), a and b, with the energy spectra sketched in figure 6.12. Let the systems be completely independent initially. The independence means that in the absence of their coupling (say, at $t < 0$), the total Hamiltonian of the system may be represented as a sum of two operators:

$$\hat{H}^{(0)} = \hat{H}_a(a) + \hat{H}_b(b), \tag{6.142}$$

where the arguments a and b symbolize the non-overlapping sets of the degrees of freedom of the two systems. Such operators, belonging to their individual, different Hilbert spaces, naturally commute. Similarly, the eigenkets of the system may be naturally factored as

$$|n\rangle = |n_a\rangle \otimes |n_b\rangle \tag{6.143}$$

The *direct product* sign $\otimes$ is used here (and below) to denote the formation of a joint ket-vector from the kets of the independent systems, belonging to different Hilbert spaces. Evidently, the order of operands in such a product may be changed at will. As a result, its eigenenergies separate into a sum, just as the Hamiltonian (6.142) does:

$$\hat{H}^{(0)} |n\rangle = (\hat{H}_a + \hat{H}_b) |n_a\rangle \otimes |n_b\rangle$$
$$\equiv (\hat{H}_a |n_a\rangle) \otimes |n_b\rangle + (\hat{H}_b |n_b\rangle) \otimes |n_a\rangle = (E_{na} + E_{nb})|n\rangle . \tag{6.144}$$

In such a composite systems, the relatively weak interaction of its components may be usually represented as a *bilinear* product of two Hermitian operators, each depending only on the degrees of freedom of only one component system:

$$\hat{H}^{(1)} = \hat{A}(a)\hat{B}(b) \tag{6.145}$$

A very common example of such an interaction is the electric-dipole interaction between an atomic-scale system (with a linear size of the order of the Bohr radius

$r_B \sim 10^{-10}$ m) and the electromagnetic field at optical frequencies $\omega \sim 10^{16}$ s^{-1}, with the wavelength $\lambda = 2\pi c/\omega \sim 10^{-6}$ m $\gg r_B$: [44]

$$\hat{H}^{(1)} = -\hat{\mathbf{d}} \cdot \hat{\mathscr{E}}, \qquad \text{with} \quad \hat{\mathbf{d}} \equiv \sum_k q_k \hat{\mathbf{r}}_k, \tag{6.146}$$

where the dipole electric moment $\mathbf{d}$ depends only on the positions $\mathbf{r}_k$ of the charged particles (numbered with index k) of the atomic system, while that of electric field $\mathscr{E}$ is a function of only the electromagnetic field's degrees of freedom—to be discussed in chapter 9 below.

Returning to the general situation shown in figure 6.12, if the component system a was initially in an excited state n'_a, the interaction (6.145) may bring it into another discrete state n_a of a lower energy—for example, the ground state. In the process of this transition, the released energy, in the form of an energy quantum

$$\hbar\omega \equiv E_{n'a} - E_{na}, \tag{6.147}$$

is picked up by the system b:

$$E_{nb} = E_{n'b} + \hbar\omega \equiv E_{n'b} + (E_{n'a} - E_{na}), \tag{6.148}$$

so that the total energy $E = E_a + E_b$ of the system does not change. (If the states n_a and n'_b are the ground states of the two component systems, as they are in most applications of this analysis, and we take the ground state energy $E_g = E_{na} + E_{n'b}$ of the composite system for the reference, then Eq. (6.148) gives merely $E_{nb} = E_{n'a}$.) If the final state n_b of the system b is inside a state group with a quasi-continuous energy spectrum (figure 6.12), the process has the exponential character (6.114)[45] and may be interpreted as the effect of *energy relaxation* of the system a, with the released energy quantum $\hbar\omega$ absorbed by the system b. Note that since the quasi-continuous spectrum essentially requires a system of a large spatial size, such a model is very convenient for description of the *environment* b of the quantum system a. (In physics, the 'environment' typically means all the Universe—less the system under consideration.)

If the relaxation rate Γ is sufficiently low, it may be described by the Golden Rule (6.137). Since the perturbation (6.145) does not depend on time explicitly, and the total energy E does not change, this relation, with the account of Eqs. (6.143) and (6.145), takes the form

$$\Gamma = \frac{2\pi}{\hbar} |A_{nn'}|^2 |B_{nn'}|^2 \rho_n, \qquad \text{where } A_{nn'} \equiv \langle n_a | \hat{A} | n'_a \rangle, \tag{6.149}$$

$$\text{and } B_{nn'} = \langle n_b | \hat{B} | n'_b \rangle,$$

where ρ_n is the density of the final states of the system b at the relevant energy (6.148). In particular, Eq. (6.149), with the dipole Hamiltonian (6.146), enables a

---

[44] See, e.g. *Part EM* section 3.1, in particular Eq. (3.16), in which letter $\mathbf{p}$ is used for the electric dipole moment.
[45] Such a process is *spontaneous*: it does not require any external agent, and starts as soon as either the interaction (6.145) has been turned on, or (if it is always on) as soon as the system a is placed into the excited state n'_a.

very straightforward calculation of the natural linewidth of atomic electric-dipole transitions. However, such calculation has to be postponed until chapter 9, in which we will discuss the electromagnetic field quantization—i.e., the exact nature of the states n_b and n'_b for this problem, and hence will be able to calculate $B_{nn'}$ and ρ_n. Instead, I will now proceed to a general discussion of the effects of interaction of quantum systems with their environment, toward which the situation shown in figure 6.12 provides a clear conceptual path.

6.8 Problems

Problem 6.1. Use Eq. (6.14) to prove the following general form of the Hellmann–Feynman theorem (whose proof in the wave-mechanics domain was the task of problem 1.5):

$$\frac{\partial E_n}{\partial \lambda} = \langle n | \frac{\partial \hat{H}}{\partial \lambda} | n \rangle,$$

where λ is an arbitrary c-number parameter.

Problem 6.2. Establish a relation between Eq. (6.16) and the result of the classical theory of weakly anharmonic ('nonlinear') oscillations for negligible damping.

Hint: Use the N Bohr's reasoning, discussed in problem 1.1.

Problem 6.3. A weak, time-independent additional force F is exerted on a 1D particle that was placed into a hard-wall potential well

$$U(x) = \begin{cases} 0, & \text{for } 0 < x < a, \\ +\infty, & \text{otherwise.} \end{cases}$$

Calculate, sketch, and discuss the 1st-order perturbation of its ground-state wavefunction.

Problem 6.4. A time-independent force $\mathbf{F} = \mu(\mathbf{n}_x y + \mathbf{n}_y x)$, where μ is a small constant, is applied to a 3D harmonic oscillator of mass m and frequency ω_0. Calculate, in the first order of the perturbation theory, the effect of the force upon the ground state energy of the oscillator, and its lowest excited energy level. How small should the constant μ be for your results to be quantitatively correct?

Problem 6.5. A 1D particle of mass m is localized at a narrow potential well that may be approximated with a delta-function:

$$U(x) = -\mathcal{W}\delta(x), \qquad \text{with } \mathcal{W} > 0.$$

Calculate the change of its ground state's energy by an additional weak, time-independent force F, in the first nonvanishing approximation of the perturbation theory. Discuss the limits of validity of this result, taking into account that at $F \neq 0$, the localized state of the particle is metastable.

Problem 6.6. Use the perturbation theory to calculate the eigenvalues of the observable L^2 (where **L** is the orbital angular momentum) in the limit $|m| \approx l \gg 1$, by purely wave-mechanical means.

Hint: Try the following substitution: $\Theta(\theta) = f(\theta)/\sin^{1/2}\theta$.

Problem 6.7. In the first nonvanishing order of the perturbation theory, calculate the shift of the ground-state energy of an electrically charged spherical rotator (i.e. a particle of mass m, free to move over a spherical surface of radius R) due to a weak, uniform, time-independent electric field $\mathscr{E}$.

Problem 6.8. Use the perturbation theory to evaluate the effect of a time-independent, uniform electric field $\mathscr{E}$ on the ground state energy E_g of a hydrogen atom. In particular:

  (i) calculate the 2nd-order shift of E_g, neglecting the extended unperturbed states with $E > 0$, and bring the result to the simplest analytical form you can,
  (ii) find the lower and the upper bounds on the shift, and
  (iii) discuss the simplest manifestations of this *quadratic Stark effect*.

Problem 6.9. A particle of mass m, with electric charge q, is in its ground s-state with a given energy $E_g < 0$, being localized by a very short-range, spherically-symmetric potential well. Calculate its static electric polarizability α.

Problem 6.10. In some atoms, the charge-screening effect of other electrons on the motion of each of them may be reasonably well approximated by the replacement of the Coulomb potential (3.190), $U = -C/r$, with the so-called *Hulthén potential*

$$U = -\frac{C/a}{\exp\{r/a\} - 1} \rightarrow -C \times \begin{cases} 1/r, & \text{for } r \ll a, \\ \exp\{-r/a\}/a, & \text{for } a \ll r, \end{cases}$$

where a is the effective screening radius. Assuming that $a \gg r_0$, use the perturbation theory to calculate the energy spectrum in this model, in the lowest order needed to lift the l-degeneracy of the levels.

Problem 6.11. In the first nonvanishing order of the perturbation theory, calculate the correction to energies of the ground state and all lowest excited states of a hydrogen-like atom/ion, due to electron's penetration into its nucleus, modeling it as a spinless, uniformly charged sphere of a radius $R \ll r_B/Z$.

Problem 6.12. Prove that the kinetic-relativistic correction operator (6.48) indeed has only diagonal matrix elements in the basis of unperturbed Bohr atom states (3.200).

Problem 6.13. Calculate the lowest-order relativistic correction to the ground-state energy of a 1D harmonic oscillator.

Problem 6.14. Use the perturbation theory to calculate the contribution to the magnetic susceptibility χ_m of a dilute gas, that is due to the orbital motion of a single electron inside each gas's particle. Spell out your result for a spherically-isotropic ground state of the electron, and give an estimate of the magnitude of this *orbital susceptibility*.

Problem 6.15. How to calculate the energy level degeneracy lifting, by a time-independent perturbation, in the 2nd order of the perturbation in $\hat{H}^{(1)}$, assuming that it is not lifted in the 1st order? Carry out such calculation for a plane rotator of mass m and radius R, carrying electric charge q, and placed into a weak, uniform, constant electric field $\mathscr{E}$.

Problem 6.16.* The Hamiltonian of a quantum system is slowly changed in time.

(i) Develop a theory of quantum transitions in the system, and spell out its result in the 1st order in the speed of the change.
(ii) Use the 1st-order result to calculate the probability that a finite-time pulse of a slowly changing force $F(t)$ drives a 1D harmonic oscillator, initially in its ground state, into an excited state.
(iii) Compare the last result with the exact one.

Problem 6.17. Use the single-particle approximation to calculate the complex electric permittivity $\varepsilon(\omega)$ of a dilute gas of similar atoms, due to their induced electric polarization by a weak external ac field, for a field frequency ω very close to one of quantum transition frequencies $\omega_{nn'}$. Based on the result, calculate and estimate the absorption cross-section of each atom.

Hint: In the single-particle approximation, an atom's properties are determined by Z similar, non-interacting electrons, each moving in a similar static attracting potential, generally different from the Coulomb one, because it is contributed not only by the nucleus, but also by other electrons.

Problem 6.18. Use the solution of the previous problem to generalize the expression for the London dispersion force between two atoms (whose calculation in the harmonic oscillator model was the subject of problems 3.16 and 5.15) to the single-particle model with an arbitrary energy spectrum.

Problem 6.19. Use the solution of the previous problem to calculate the potential energy of interaction of two hydrogen atoms, both in their ground state, separated by distance $r \gg r_B$.

Problem 6.20. In a certain quantum system, distances between three lowest energy levels are slightly different—see figure below ($|\xi| \ll \omega_{1,2}$). Assuming that the involved matrix elements of the perturbation Hamiltonian are known, and are all proportional to the external ac field's amplitude, find the time necessary to populate the first excited level almost completely (with a given precision $\varepsilon \ll 1$), using the Rabi oscillation effect, if at $t = 0$ the system is completely in its ground state.

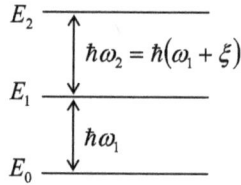

$$E_2 \quad \rule{2cm}{0.4pt}$$
$$\hbar\omega_2 = \hbar(\omega_1 + \xi)$$
$$E_1 \quad \rule{2cm}{0.4pt}$$
$$\hbar\omega_1$$
$$E_0 \quad \rule{2cm}{0.4pt}$$

Problem 6.21.* Analyze the possibility of a slow transfer of a system from one of its energy levels to another one (in the figure below, from level 1 to level 3), using the scheme shown in that figure, in which the monochromatic external excitation amplitudes A_+ and A_- may be slowly changed at will.

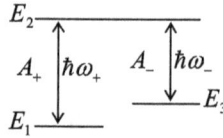

$$E_2 \quad \rule{2cm}{0.4pt} \qquad \rule{1.5cm}{0.4pt}$$
$$A_+ \; \big| \hbar\omega_+ \qquad A_- \; \big| \hbar\omega_-$$
$$\qquad\qquad\qquad\qquad E_3$$
$$E_1 \quad \rule{1cm}{0.4pt}$$

Problem 6.22. A weak external force pulse $F(t)$, of a finite time duration, is applied to a 1D harmonic oscillator that initially was in its ground state.

(i) Calculate, in the lowest nonvanishing order of the perturbation theory, the probability that the pulse drives the oscillator into its lowest excited state.
(ii) Compare the result with the exact solution of the problem.
(iii) Spell out the perturbative result for a Gaussian-shaped waveform,

$$F(t) = F_0 \exp\{-t^2/\tau^2\},$$

and analyze its dependence on the scale τ of the pulse duration.

Problem 6.23. A spatially-uniform, but time-dependent external electric field $\mathscr{E}(t)$ is applied, starting from $t = 0$, to a charged plane rotator, initially in its ground state.

(i) Calculate, in the lowest nonvanishing order in the field's strength, the probability that by time $t > 0$, the rotator is in its nth excited state.
(ii) Spell out and analyze your results for a constant-magnitude field rotating, with a constant angular velocity ω, within the rotator's plane.
(iii) Do the same for a monochromatic field of frequency ω, with a fixed polarization.

Problem 6.24. A spin-½ with a gyromagnetic ratio γ is placed into a magnetic field including a time-independent component $\mathscr{B}_0$, and a perpendicular field of a constant magnitude $\mathscr{B}_r$, rotated with a constant angular velocity ω. Can this *magnetic resonance* problem be reduced to one already discussed in chapter 6?

Problem 6.25. Develop the general theory of quantum excitations of the higher levels of a discrete-spectrum system, initially in the ground state, by a weak time-dependent perturbation, up to the 2nd order. Spell out and discuss the result for the case of a monochromatic excitation, with a nearly perfect tuning of its frequency ω to *the half* of a certain quantum transition frequency $\omega_{n0} \equiv (E_n - E_0)/\hbar$.

Problem 6.26. A heavy, relativistic particle, with electric charge $q = Ze$, passes by a hydrogen atom, initially in its ground state, with an impact parameter (the shortest distance) b within the range $r_B \ll b \ll r_B/\alpha$, where $\alpha \approx 1/137$ is the fine structure constant. Calculate the probability of atom's transition to its lowest excited states.

Problem 6.27.* A particle of mass m is initially in the localized ground state, with the known energy $E_g < 0$, of a very small, spherically-symmetric potential well. Calculate the rate of its delocalization by an applied force $\mathbf{F}(t)$ with time-independent amplitude F_0, frequency ω, and direction $\mathbf{n}_F$.

Problem 6.28.* Calculate the rate of ionization of a hydrogen atom, initially in its ground state, by a classical, linearly polarized electromagnetic wave with an electric field's amplitude $\mathscr{E}_0$, and a frequency ω within the range

$$\hbar/m_e r_B^2 \ll \omega \ll c/r_B,$$

where r_B is the Bohr radius. Recast your result in terms of the cross-section of this electromagnetic wave absorption process. Discuss briefly what changes of the theory would be necessary if either of the above conditions had been violated.

Problem 6.29.* Use the quantum-mechanical Golden Rule to derive the general expression for the electric current I through a weak tunnel junction between two conductors, biased with dc voltage V, treating the conductors as Fermi gases of electrons, with negligible direct interaction. Simplify the result in the low-voltage limit.

Hint: The electric current flowing through a weak tunnel junction is so low that it does not substantially perturb the electron states inside each conductor.

Problem 6.30.* Generalize the result of the previous problem to the case when a weak tunnel junction is biased with voltage $V(t) = V_0 + A\cos\omega t$, with $\hbar\omega$ generally comparable with eV_0 and eA.

Problem 6.31.* Use the quantum-mechanical Golden Rule to derive the Landau–Zener formula (2.257).

References

[1] Kramer D (ed) 2001 *Advances in Solid State Physics* vol 41 (Berlin: Springer) p 211
[2] Harr R and Curtis L 1987 *Am. J. Phys.* **55** 1044
[3] Woolgate G 1983 *Elementary Atomic Structure* 2nd ed (Oxford University Press)
[4] Sobelman I 2006 *Theory of Atomic Spectra* (Alpha Science)
[5] Wertz J and Bolton J 2007 *Electron Spin Resonance* 2nd ed (Wiley)
[6] Keeler J 2010 *Understanding NMR Spectroscopy* 2nd ed (Wiley)
[7] Delsing P *et al* 1989 *Phys. Rev. Lett.* **63** 1180

IOP Publishing

Quantum Mechanics
Lecture notes
Konstantin K Likharev

Chapter 7

Open quantum systems

This chapter discusses the effects of weak interaction of a quantum system with its environment. Some part of this material is on the fine line between the quantum mechanics and the (quantum) statistical physics. Here I will only cover those aspects of the latter field[1] that are of key importance for the major goals of this course, including the discussion of quantum measurements in chapter 10.

7.1 Open systems, and the density matrix

All the way until the very end of the previous chapter, we have discussed quantum systems isolated from their environment. Indeed, from the very beginning we have assumed that we are dealing with the statistical ensembles of systems as similar to each other as is only allowed by the laws of quantum mechanics. Each member of such an ensemble, called *pure* or *coherent*, may be described by the same quantum state α—in the wave mechanics case, by the same wavefunction Ψ_α. Even our discussion in the end of the last chapter, in which one component system (in figure 6.13, system b) may be used as a model of the environment of its counterpart (system a), was still based on the assumption of a pure initial state (6.143) of the composite system. If the interaction of two component systems is described by a certain Hamiltonian (the one given by Eq. (6.145), for example), for its state α at an arbitrary instant we may write

$$|\alpha\rangle = \sum_n \alpha_n |n\rangle = \sum_n \alpha_n |n_a\rangle \otimes |n_b\rangle, \tag{7.1}$$

with a *unique* correspondence between the eigenstates n_a and n_b.

---

[1] For a broader discussion of statistical mechanics and physical kinetics, including those of quantum systems, the reader is referred to the *Part SM* of this series.

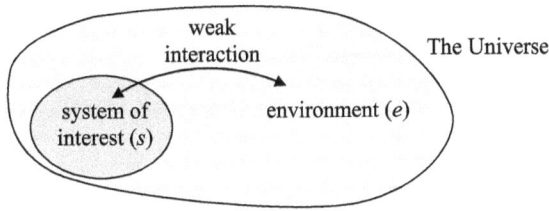

Figure 7.1. A quantum system and its environment (VERY schematically).

However, in many important cases our knowledge of a quantum system's state is incomplete[2]. These cases fall into two different categories. The first case is when a relatively simple quantum system s of our interest (say, an electron or an atom) is in a weak[3] but substantial contact with its environment e—here understood in the most general sense, say, as the whole Universe less system s—see figure 7.1. Then there is virtually no chance of making two or more experiments with exactly the same composite system, because it would imply a repeated preparation of the whole environment (including the experimenter :-) in a certain quantum state—a rather challenging task, to put it mildly. It makes much more sense to consider a statistical ensemble of another kind—a *mixed ensemble*, with *random* quantum states of the environment, though possibly with its macroscopic parameters (e.g. temperature, pressure, etc) known with a high precision. Such ensembles will be the focus of the analysis in this chapter.

Much of this analysis will pertain also to another category of cases, namely when the system of interest is isolated from its environment, at present, with an acceptable precision, but our knowledge of its state is still incomplete by some other reason. The most common of such reasons is that the system and its environment *had been* in contact at some previous time. So, this second category of cases may be considered as a particular case of the first one, and may be described by the results of its analysis, with certain simplifications—which will be spelled out in appropriate places of my narrative.

In classical physics, the analysis of mixed statistical ensembles is based on the notion of the probability W (or the probability density w) of each detailed ('microscopic') state of the system of interest[4]; let us see how this ensemble may be described in quantum mechanics. In the case when the coupling between the system of our interest and its environment is sufficiently weak, so that they may be clearly separated, we can, as in the perturbation theory, use the bra- and ket-vectors of their unperturbed states, defined in completely different Hilbert spaces. Then the

---

[2] Indeed, I am unaware of a single case when a system would be *exactly* coherent, though in many cases, such as the ones discussed in the previous chapters, deviations from the coherence may be ignored *with acceptable accuracy*.

[3] If the interaction between a system and its environment is very strong, their very partition is impossible.

[4] See, e.g. *Part SM* section 2.1.

most general quantum state of the whole Universe, still assumed to be pure[5], may be described as the following linear superposition:

$$|\alpha\rangle = \sum_{j,k} \alpha_{jk} |s_j\rangle \otimes |e_k\rangle. \tag{7.2}$$

The 'only' difference between the description of such an *entangled* state, and the superposition of separable states described by Eq. (7.1), is that the coefficients α_{jk} on the right-hand side of Eq. (7.2) are numbered with two indices: the index j listing the quantum states of the system s, and the index k numbering the (enormously large) set of quantum states of the environment. So, in a mixed ensemble[6], a certain quantum state s_j of the system of interest may coexist with different states of its environment. Of course, the enormity of the Hilbert space of the environment, i.e. the number of the k-factors in the superposition (7.2), strips us of any practical opportunity to make direct calculations using that sum. For example, according to the basic Eq. (4.125), in order to find the expectation value of an arbitrary observable A in the state (7.2), we would need to calculate the long bracket

$$\langle A \rangle = \langle \alpha | \hat{A} | \alpha \rangle \equiv \sum_{j,j';k,k'} \alpha_{jk}^* \alpha_{j'k'} \langle e_k | \otimes \langle s_j | \hat{A} | s_{j'} \rangle \otimes | e_{k'} \rangle. \tag{7.3}$$

Even if we assume that each of the sets $\{s\}$ and $\{e\}$ is full and orthonormal, Eq. (7.3) still includes a double sum over the enormous basis state set of the environment!

However, let us consider a limited, but the most important subset of operators— those of *intrinsic* observables, which depend only on the degrees of freedom of the system of our interest (s). These operators do not act upon environment's degrees of freedom, and hence in Eq. (7.3) we may move the environment bra-vectors $\langle e_k |$ over all the way to the ket-vectors $|e_{k'}\rangle$. Assuming, again, that the set of environmental eigenstates is full and orthonormal, Eq. (7.3) is now reduced to

$$\langle A \rangle = \sum_{j,j';k,k'} \alpha_{jk}^* \alpha_{j'k'} \langle s_j | \hat{A} | s_{j'} \rangle \langle e_k | e_{k'} \rangle = \sum_{jj'} A_{jj'} \sum_k \alpha_{jk}^* \alpha_{j'k}. \tag{7.4}$$

This is already a big relief, because we have 'only' a single sum over k, but the main trick[7] is still ahead. After the summation over k, the second sum in the last form of Eq. (7.4) is some function w of the indices j and j', so that, according to Eq. (4.96), this relation may be represented as

$$\langle A \rangle = \sum_{jj'} A_{jj'} w_{j'j} = \text{Tr}(Aw) , \tag{7.5}$$

where the matrix w, with the elements

---

[5] Whether this assumption is true is an interesting issue, still being debated (more by philosophers than by physicists), but it is widely believed that its solution is not critical for the validity of the results of this approach.
[6] Besides its 'pathological' particular case when it may be reduced to the pure ensemble (7.1), i.e. when the overwhelming majority of the coefficients α_{jk} vanish.
[7] It was suggested in 1927 by J von Neumann.

$$w_{j'j} \equiv \sum_k \alpha^*_{jk}\alpha_{j'k}, \quad \text{i.e. } w_{jj'} \equiv \sum_k \alpha_{jk}\alpha^*_{j'k}, \tag{7.6}$$

is called the *density matrix* of the system. Most importantly, Eq. (7.5) shows that the knowledge of this matrix allows the calculation of the expectation value of *any* intrinsic observable A (and, according to the general Eqs. (1.33) and (1.34), its rms fluctuation as well, if necessary), even for the very general mixed statistical ensemble (7.2). For this reason let us have a very good look at the density matrix.

First of all, we know from the general discussion in chapter 4, fully applicable to the pure state (7.2), the expansion coefficients in superpositions of this type may be always expressed as brackets; in our current case, we may write

$$\alpha_{jk} = \left(\langle e_k| \otimes \langle s_j|\right)|\alpha\rangle. \tag{7.7}$$

Plugging this expression into Eq. (7.6), we get

$$w_{jj'} \equiv \sum_k \alpha_{jk}\alpha^*_{j'k} = \langle s_j| \otimes \left(\sum_k \langle e_k|\alpha\rangle\langle\alpha|e_k\rangle\right) \otimes |s_{j'}\rangle = \langle s_j|\hat{w}|s_{j'}\rangle. \tag{7.8}$$

We see that from the point of our system (i.e. in its Hilbert space whose basis states may be numbered by the index j only), the density matrix is indeed just the matrix of some construct[8],

$$\hat{w} \equiv \sum_k \langle e_k|\alpha\rangle\langle\alpha|e_k\rangle, \tag{7.9}$$

which is called the *density* (or 'statistical') *operator*. As follows from the definition (7.9), in contrast to the density matrix this operator does not depend on the choice of a particular basis s_j—just as all linear operators considered earlier in this course. In contrast to them, the density operator *does* depend on the composite system's state α, including the state of the system s as well. However, in the j-space it is mathematically still just an operator whose matrix elements obey all relations of the bra–ket formalism.

In particular, due to its definition (7.6), the density operator is Hermitian:

$$w^*_{jj'} = \sum_k \alpha^*_{jk}\alpha_{j'k} = \sum_k \alpha_{j'k}\alpha^*_{jk} = w_{j'j}, \tag{7.10}$$

so that according to the general analysis of section 4.3, in the Hilbert space of the system s, there should be a certain basis $\{w\}$ in which the matrix of this operator is diagonal:

$$w_{jj'}\big|_{\text{in } w} = w_j\delta_{jj'}. \tag{7.11}$$

---

[8] Of course the 'brackets' in this expression are not c-numbers, because state α is defined in a larger Hilbert space (of the environment plus the system of interest) than the basis states e_k (of the environment only).

Since any operator, in any basis, may be represented in the form (4.59), in the basis $\{w\}$ we may write

$$\hat{w} = \sum_j |w_j\rangle \, w_j \, \langle w_j|. \tag{7.12}$$

This expression recalls, but is not equivalent to Eq. (4.44) for the identity operator, that has been used so many times in this course, and in the basis w_j has the form

$$\hat{I} = \sum_j |w_j\rangle\langle w_j|. \tag{7.13}$$

In order to comprehend the meaning of the coefficients w_j participating in Eq. (7.12), let us use Eq. (7.5) to calculate the expectation value of any observable A whose eigenstates coincide with those of the special basis set $\{w\}$:

$$\langle A \rangle = \mathrm{Tr}\,(\mathrm{Aw}) = \sum_{jj'} A_{jj'} w_j \delta_{jj'} = \sum_j A_j w_j, \tag{7.14}$$

where A_j is just the expectation value of observable A in the state w_j. Hence, in order to comply with the general Eq. (1.37), the real c-number w_j must have the physical sense of the probability W_j of finding the system in the state j. As a result, we may rewrite Eq. (7.12) in the form

$$\hat{w} = \sum_j |w_j\rangle \, W_j \, \langle w_j|. \tag{7.15}$$

In one ultimate case when only one of the probabilities (say, $W_{j''}$) is different from zero,

$$W_j = \delta_{jj''}, \tag{7.16}$$

the system is in a coherent (pure) state $w_{j''}$. Indeed, it is fully described by one ket-vector $|w_{j''}\rangle$, and we can use the general rule (4.86) to represent it in another (arbitrary) basis $\{s\}$ as a coherent superposition

$$|w_{j''}\rangle = \sum_{j'} (U^\dagger)_{jj'} |s_{j'}\rangle = \sum_{j'} U^*_{jj'} |s_{j'}\rangle, \tag{7.17}$$

where U is the unitary matrix of transform from the basis $\{w\}$ to the basis $\{s\}$. According to Eqs. (7.11) and (7.16), in such a pure state the density matrix is diagonal in the $\{w\}$ basis,

$$w_{jj'}|_{\text{in } w} = \delta_{j,j''}\delta_{j',j''}, \tag{7.18a}$$

but not in an arbitrary basis. Indeed, using the general rule (4.92), we get

$$w_{jj'}|_{\text{in } s} = \sum_{l,l'} U^\dagger_{jl} w_{ll'}|_{\text{in } w} \, U_{l'j'} = U^\dagger_{jj''} U_{j''j'} = U^*_{j''j} U_{j''j'}. \tag{7.18b}$$

To make this result more transparent, let us denote the matrix elements $U_{j''j} = \langle w_{j''}|s_j \rangle$ (which, for a fixed j'', depend on just one index j) by α_j; then

$$w_{jj'}\Big|_{\text{in } s} = \alpha_j^* \alpha_{j'}, \tag{7.19}$$

so that N^2 elements of the whole $N \times N$ matrix is determined by just *one* string of N c-numbers α_j. For example, for a two-level system ($N = 2$),

$$\mathsf{w}|_{\text{in } s} = \begin{pmatrix} \alpha_1 \alpha_1^* & \alpha_2 \alpha_1^* \\ \alpha_1 \alpha_2^* & \alpha_2 \alpha_2^* \end{pmatrix}. \tag{7.20}$$

We see that the off-diagonal terms are, colloquially, 'as large as the diagonal ones', in the following sense:

$$w_{12} w_{21} = w_{11} w_{22}. \tag{7.21}$$

Since the diagonal terms have the sense of the probabilities $W_{1,2}$ to find the system in the corresponding state, we may represent Eq. (7.20) in the form

$$\mathsf{w}|_{\text{pure state}} = \begin{pmatrix} W_1 & (W_1 W_2)^{1/2} e^{i\varphi} \\ (W_1 W_2)^{1/2} e^{-i\varphi} & W_2 \end{pmatrix}. \tag{7.22}$$

The physical sense of the (real) constant φ is the phase shift between the coefficients in the linear superposition (7.17), which represents the pure state $w_{j''}$ in the basis $\{s_{1,2}\}$.

Now let us consider a different statistical ensemble of two-level systems, that includes the member states identical in all aspects (including similar probabilities $W_{1,2}$ in the same basis $s_{1,2}$), besides that the phase shifts φ are random, with the phase probability uniformly distributed over the trigonometric circle. Then the ensemble averaging is equivalent to the averaging over φ from 0 to 2π,[9] which kills the off-diagonal terms of the density matrix (7.22), so that the matrix becomes diagonal:

$$\mathsf{w}|_{\text{classical mixture}} = \begin{pmatrix} W_1 & 0 \\ 0 & W_2 \end{pmatrix}. \tag{7.23}$$

The mixed statistical ensemble with the density matrix diagonal in the stationary state basis is called the *classical mixture*, and represents the limit opposite to the pure (coherent) state.

After this example, the reader should not be much shocked by the main claim[10] of statistical mechanics that any large ensemble of similar systems in *thermodynamic* (or 'thermal') *equilibrium* is exactly such a classical mixture. Moreover, for systems in the

---

[9] For a system with a time-independent Hamiltonian, such averaging is especially plausible in the basis of the stationary states n of the system, in which the phase φ is just the difference of integration constants in Eq. (4.158), and its randomness may be naturally produced by minor fluctuations of the energy difference $E_1 - E_2$. In section 7.3 below, we will study the dynamics of this *dephasing* process.

[10] This fact follows from the basic postulate of statistical physics, called the *microcanonical distribution*—see, e.g. *Part SM* section 2.2.

thermal equilibrium with a much larger environment of a fixed temperature T (such an environment is usually called a *heat bath* or a *thermostat*) the statistical physics gives[11] a very simple expression, called the *Gibbs distribution*, for the probabilities W_n:

$$W_n = \frac{1}{Z} \exp\left\{-\frac{E_n}{k_B T}\right\}, \qquad \text{with } Z \equiv \sum_n \exp\left\{-\frac{E_n}{k_B T}\right\}. \qquad (7.24)$$

where E_n is the eigenenergy of the corresponding stationary state, and the normalization coefficient Z is called the *statistical sum*.

A detailed analysis of classical and quantum ensembles in thermodynamic equilibrium is a major focus of statistical physics courses (such as *Part SM* of this series) rather than this course of quantum mechanics. However, I would still like to attract reader's attention to the key fact that, in contrast with the similar-looking Boltzmann distribution for single particles[12], the Gibbs distribution is general and is *not* limited to classical statistics. In particular, for a quantum gas of indistinguishable particles, it is absolutely compatible with the quantum statistics (such as the Bose–Einstein or Fermi–Dirac distributions) of the component particles. For example, if we use Eq. (7.24) to calculate the average energy of a 1D harmonic oscillator of frequency ω_0 in thermal equilibrium, we easily get[13]

$$W_n = \exp\left\{-n\frac{\hbar\omega_0}{k_B T}\right\}\left(1 - \exp\left\{-\frac{\hbar\omega_0}{k_B T}\right\}\right),$$

$$Z = \exp\left\{-\frac{\hbar\omega_0}{2k_B T}\right\}\bigg/\left(1 - \exp\left\{-\frac{\hbar\omega_0}{k_B T}\right\}\right). \qquad (7.25)$$

$$\langle E \rangle \equiv \sum_{n=0}^{\infty} W_n E_n = \frac{\hbar\omega_0}{2} \coth\frac{\hbar\omega_0}{2k_B T} \equiv \frac{\hbar\omega_0}{2} + \frac{\hbar\omega_0}{\exp\{\hbar\omega_0/k_B T\} - 1}. \qquad (7.26a)$$

An alternative form of the last result,

$$\langle E \rangle = \frac{\hbar\omega_0}{2} + \hbar\omega_0\langle n \rangle, \quad \text{with}$$

$$\langle n \rangle = \frac{1}{\exp\{\hbar\omega_0/k_B T\} - 1} \rightarrow \begin{cases} 0, & \text{for } k_B T \ll \hbar\omega_0, \\ k_B T/\hbar\omega_0, & \text{for } \hbar\omega_0 \ll k_B T, \end{cases} \qquad (7.26b)$$

may be interpreted as the addition, to the ground-state energy $\hbar\omega_0/2$, of the average number $\langle n \rangle$ of thermally-induced excitations, with the energy $\hbar\omega_0$ each. In the harmonic oscillator, whose energy levels are equidistant, such a language is completely appropriate, because the transfer of the system from any level to the

---

[11] See. e.g. *Part SM* section 2.4. The Boltzmann constant k_B is only needed if temperature is measured in non-energy units, say in kelvins.

[12] See, e.g. *Part SM* section 2.8.

[13] See, e.g. *Part SM* section 2.5—but mind a different energy reference level, $E_0 = \hbar\omega_0/2$, used for example in *Part SM* Eqs. (2.68) and (2.69), affecting the expression for Z. Actually, the calculation, using Eqs. (7.24) and (5.86), is so straightforward that it is highly recommended to the reader as a simple exercise.

one just above it adds the same amount of energy, $\hbar\omega_0$. Note that the above expression for $\langle n \rangle$ is actually the Bose–Einstein distribution (for the particular case of zero chemical potential); we see that it does not contradict the Gibbs distribution (7.24) of the total energy of the system, but rather immediately follows from it[14].

7.2 Coordinate representation, and the Wigner function

For many applications of the density operator, its coordinate representation is convenient. (I will only discuss it for 1D case; the generalization to multi-dimension case is straightforward.) Following Eq. (4.47), it is natural to define the following function of two arguments (traditionally, also called the *density matrix*):

$$w(x, x') \equiv \langle x|\hat{w}| x' \rangle. \qquad (7.27)$$

Inserting, into the right-hand side of this definition, two closure conditions (4.44) for an arbitrary (but full and orthonormal) basis $\{s\}$, and then using Eq. (4.233), we get[15]

$$w(x, x') = \sum_{j,j'}\langle x|s_j\rangle \langle s_j|\hat{w}|s_{j'}\rangle \langle s_{j'}|x'\rangle = \sum_{j,j'}\psi_j(x)w_{jj'}|_{\text{lin }s}\psi_{j'}^*(x'). \qquad (7.28)$$

In the special basis $\{w\}$, in which the density matrix is diagonal, this expression is reduced to

$$w(x, x') = \sum_{j}\psi_j(x)W_j \psi_j^*(x'). \qquad (7.29)$$

Let us discuss the properties of this function. At coinciding arguments, $x' = x$, this is just the probability density[16]:

$$w(x, x) = \sum_{j}\psi_j(x)W_j \psi_j^*(x) = \sum_{j}w_j(x)W_j = w(x). \qquad (7.30)$$

However, the density matrix gives more information about the system than just the probability density. As the simplest example, let us consider a pure quantum state, with $W_j = \delta_{j,j'}$, so that $\psi(x) = \psi_{j'}(x)$, and

$$w(x, x') = \psi_{j'}(x)\psi_{j'}^*(x') \equiv \psi(x)\psi^*(x'). \qquad (7.31)$$

We see that the density matrix carries the information not only about the modulus, but also the phase of the wavefunction. (Of course one may argue rather

---

[14] Because of the fundamental importance of Eq. (7.26) for virtually all fields of physics, let me draw the reader's attention to its main properties. At low temperatures, $k_BT \ll \hbar\omega_0$, there are virtually no excitations, $\langle n \rangle \to 0$, and the average energy of the oscillator is dominated by that of its ground state. In the opposite limit of high temperatures, $\langle n \rangle \to k_BT/\hbar\omega_0 \gg 1$, and $\langle E \rangle$ approaches the classical value k_BT (complying with the classical *equipartition theorem* that assigns energy $k_BT/2$ to each quadratic contribution to the system's Hamiltonian function—in the 1D oscillator case, one to the potential-energy term and one to the kinetic-energy term).

[15] For now, I will focus on a fixed time instant (say, $t = 0$), and hence write $\psi(x)$ instead of $\Psi(x, t)$.

[16] This fact is the origin of the density matrix' name.

convincingly that in this ultimate limit the density-matrix description is redundant, because all this information is contained in the wavefunction itself.)

How may the density matrix be interpreted? In the simple case (7.31), we can write

$$|w(x, x')|^2 \equiv w(x, x')w^*(x, x') = \psi(x)\psi^*(x)\psi(x')\psi^*(x') = w(x)w(x'), \qquad (7.32)$$

so that the modulus squared of the density matrix is just as the joint probability density to find the system at point x and point x'. For example, a simple wave packet with a spatial extent δx, $w(x, x')$ has an appreciable magnitude only if both points are not farther than δx from the packet center, and hence from each other. The interpretation becomes more complex if we deal with an incoherent mixture of several wavefunctions, for example the classical mixture describing the thermodynamic equilibrium. In this case, we can use Eq. (7.24) to rewrite Eq. (7.29) as follows:

$$w(x, x') = \sum_n \psi_n(x) W_n \psi_n^*(x') = \frac{1}{Z} \sum_n \psi_n(x) \exp\left\{ -\frac{E_n}{k_B T} \right\} \psi_n^*(x'). \qquad (7.33)$$

As the simplest example, let us see what is the density matrix of a *free* (1D) particle in the thermal equilibrium. As we know very well by now, in this case, the set of energies $E_p = p^2/2m$ of stationary states (monochromatic waves) forms a continuum, so that we need to replace the sum (7.33) with an integral, using for example the 'delta-normalized' traveling-wave eigenfunctions (4.264):

$$w(x, x') = \frac{1}{2\pi\hbar Z} \int_{-\infty}^{+\infty} \exp\left\{ -\frac{ipx}{\hbar} \right\} \exp\left\{ -\frac{p^2}{2mk_B T} \right\} \exp\left\{ \frac{ipx'}{\hbar} \right\} dp. \qquad (7.34)$$

This is a usual Gaussian integral, and may be worked out, as we have done repeatedly in chapter 2 and beyond, by complementing the exponent to the full square of the momentum p plus a constant. The statistical sum Z may be also readily calculated[17],

$$Z = (2\pi m k_B T)^{1/2}, \qquad (7.35)$$

However, for what follows it is more useful to write the result for the product wZ (the so-called *un-normalized density matrix*):

$$w(x, x')Z = \left(\frac{mk_B T}{2\pi\hbar^2} \right)^{1/2} \exp\left\{ -\frac{mk_B T(x - x')^2}{2\hbar^2} \right\}. \qquad (7.36)$$

---

[17] Due to the delta-normalization of the eigenfunction, the density matrix (7.34) for the free particle (and any system with a continuous eigenvalue spectrum) is normalized as

$$\int_{-\infty}^{+\infty} w(x, x')Z dx' = \int_{-\infty}^{+\infty} w(x, x')Z dx = 1.$$

This is a very interesting result: the density matrix depends only on the difference of its arguments, dropping to zero fast as the distance between the points x and x' exceeds the following characteristic scale (called the *correlation length*)

$$x_c \equiv \langle (x - x')^2 \rangle^{1/2} = \frac{\hbar}{(mk_BT)^{1/2}}. \tag{7.37}$$

This length may be interpreted in the following way. It is straightforward to use Eq. (7.24) to verify that the average energy $\langle E \rangle = \langle p^2/2m \rangle$ of a free particle in the thermal equilibrium, i.e. in the classical mixture (7.33), equals $k_BT/2$. Hence the average momentum's magnitude may be estimated as

$$p_c \equiv \langle p^2 \rangle^{1/2} = (2m\langle E \rangle)^{1/2} = (mk_BT)^{1/2}, \tag{7.38}$$

so that x_c is of the order of the minimal length allowed by the Heisenberg-like 'uncertainty relation':

$$x_c = \hbar/p_c. \tag{7.39}$$

Note that with the growth of temperature, the correlation length (7.37) goes to zero, and the density matrix (7.36) tends to a delta-function:

$$w(x, x')Z|_{T\to\infty} \to \delta(x - x'). \tag{7.40}$$

Since in this limit the average kinetic energy of the particle is not smaller than its potential energy in *any* fixed potential profile, Eq. (7.40) is the general property of the density matrix (7.33).

Let us discuss the following curious feature of Eq. (7.36): if we replace k_BT with $\hbar/i$ $(t - t_0)$, and x' with x_0, the un-normalized density matrix wZ for a free particle turns into the particle's propagator—cf Eq. (2.49). This is not just an occasional coincidence. Indeed, in chapter 2 we saw that the propagator of a system with an arbitrary stationary Hamiltonian may be expressed via the stationary eigenfunctions as

$$G(x, t; x_0, t_0) = \sum_n \psi_n(x)\exp\left\{-i\frac{E_n}{\hbar}(t - t_0)\right\}\psi_n^*(x_0). \tag{7.41}$$

Comparing this expression with Eq. (7.33), we see that the replacements

$$\frac{i(t - t_0)}{\hbar} \to \frac{1}{k_BT}, \qquad x_0 \to x', \tag{7.42}$$

turn the pure-state propagator G into the un-normalized density matrix wZ of the same system in thermodynamic equilibrium. This important fact, rooted in the formal similarity of the Gibbs distribution (7.24) with the Schrödinger equation's solution (1.69), enables a theoretical technique of the so-called *thermodynamic Green's functions*, which is especially productive in condensed matter physics[18].

For our current purposes, we can employ Eq. (7.42) to re-use some of the wave mechanics results, in particular the following formula for the harmonic oscillator's propagator

$$G(x, t; x_0, t_0) = \left(\frac{m\omega_0}{2\pi i\hbar \, \sin[\omega_0(t - t_0)]}\right)^{1/2}$$
$$\times \exp\left\{-\frac{m\omega_0[(x^2 + x_0^2)\cos[\omega_0(t - t_0)] - 2xx_0]}{2i\hbar \, \sin[\omega_0(t - t_0)]}\right\}. \tag{7.43}$$

which may be readily proved to satisfy the Schrödinger equation for the Hamiltonian (5.62), with the appropriate initial condition: $G(x, t_0; x_0, t_0) = \delta(x - x_0)$. Making the substitution (7.42), we immediately get

$$w(x, x')Z = \left(\frac{m\omega_0}{2\pi\hbar \, \sinh[\hbar\omega_0/k_BT]}\right)^{1/2}$$
$$\times \exp\left\{-\frac{m\omega_0[(x^2 + x'^2)\cosh[\hbar\omega_0/k_BT] - 2xx']}{2\hbar \, \sinh[\hbar\omega_0/k_BT]}\right\}. \tag{7.44}$$

As a sanity check, at very low temperatures, $k_BT \ll \hbar\omega_0$, both hyperbolic functions participating in this expression are very large and nearly equal, and it yields

$$w(x, x')Z|_{T\to0} \to \left[\left(\frac{m\omega_0}{\pi\hbar}\right)^{1/4} \exp\left\{-\frac{m\omega_0 x^2}{\hbar}\right\}\right] \times \exp\left\{-\frac{\hbar\omega_0}{2k_BT}\right\}$$
$$\times \left[\left(\frac{m\omega_0}{\pi\hbar}\right)^{1/4} \exp\left\{-\frac{m\omega_0 x'^2}{\hbar}\right\}\right]. \tag{7.45}$$

In each of the expressions in square brackets we can readily recognize the ground state's wavefunction (2.275) of the oscillator, while the middle exponent is just the statistical sum (7.24) in the low-temperature limit, when it is dominated by the ground-level contribution:

$$Z|_{T\to0} \to \exp\left\{-\frac{\hbar\omega_0}{2k_BT}\right\}. \tag{7.46}$$

As a result, Z in both parts of Eq. (7.45) may be cancelled, and the density matrix in this limit is described by Eq. (7.31), with the ground state as the only state of the system. This is natural when temperature is too low for the thermal excitation of any other state.

---

[18] I will have no time to discuss this technique, and have to refer the interested reader to special literature. Probably, the most famous text of that field is A Abrikosov, L Gor'kov, and I Dzyaloshinski, *Methods of Quantum Field Theory in Statistical Physics*, Prentice-Hall, 1963. (Later reprintings are available from Dover.)

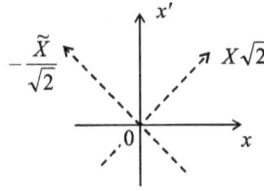

Figure 7.2. The coordinates X and $\tilde{X}$ employed in the Weyl–Wigner transform (7.50). They differ from the coordinates obtained by the rotation of the reference frame by the angle $\pi/2$ only by coefficients $\sqrt{2}$, describing a scale stretch.

Returning to arbitrary temperatures, Eq. (7.44) in coinciding arguments gives the following expression for the probability density[19]:

$$w(x, x)Z = w(x)Z = \left(\frac{m\omega_0}{2\pi\hbar \sinh[\hbar\omega_0/k_BT]} \right)^{1/2} \exp\left\{ -\frac{m\omega_0 x^2}{\hbar} \tanh\frac{\hbar\omega_0}{2k_BT} \right\}. \quad (7.47)$$

This is just a Gaussian function of x, with the following variance:

$$\langle x^2 \rangle = \frac{\hbar}{2m\omega_0} \coth\frac{\hbar\omega_0}{2k_BT}. \quad (7.48)$$

In order to compare this result with our earlier ones, it is useful to recast it as

$$\langle U \rangle = \frac{m\omega_0^2}{2}\langle x^2 \rangle = \frac{\hbar\omega_0}{4} \coth\frac{\hbar\omega_0}{2k_BT}. \quad (7.49)$$

Comparing this expression with Eq. (7.26), we see that the average value of potential energy is exactly one half of the total energy—the other half being the average kinetic energy. This is what we could expect, because according to Eqs. (5.96) and (5.97), such a relation holds for each Fock state and hence should also hold for their classical mixture.

Unfortunately, besides the trivial case (7.30) of coinciding arguments, it is hard to give a straightforward interpretation of the density function in terms of the system's measurements. This is a fundamental difficulty, which has been well explored in terms of the *Wigner function* (sometimes called the 'Wigner–Ville distribution')[20] defined as

$$W(X, P) \equiv \frac{1}{2\pi\hbar} \int w\left(X + \frac{\tilde{X}}{2}, X - \frac{\tilde{X}}{2} \right) \exp\left\{ -\frac{iP\tilde{X}}{\hbar} \right\} d\tilde{X}. \quad (7.50)$$

---

[19] I have to confess that this notation is imperfect, because from the point of view of rigorous mathematics, $w(x, x')$ and $w(x)$ are different functions, and so are $w(p, p')$ and $w(p)$ used below. In the perfect world, I would use different letters for them all, but I desperately want to stay with 'w' for all the probability densities, and there are not so many good different fonts for this letter. Let me hope that the difference between these functions is clear from their arguments, and from the context.

[20] It was introduced in 1932 by E Wigner on the basis of a general (*Weyl–Wigner*) transform suggested by H Weyl in 1927, and then re-derived in 1948 by J Ville on a different mathematical basis.

From the mathematical standpoint, this is just the Fourier transform of the density matrix in one of two new coordinates, defined by the relations illustrated in figure 7.2:

$$X \equiv \frac{x + x'}{2}, \quad \tilde{X} \equiv x - x', \quad \text{so that} \quad x \equiv X + \frac{\tilde{X}}{2}, \quad x' \equiv X - \frac{\tilde{X}}{2}. \quad (7.51)$$

Physically, the new argument X may be interpreted as the average position of the particle during the time interval $(t - t')$, while $\tilde{X}$, as the distance passed by it during that time interval, so that P characterizes the momentum of the particle during that motion. As a result, the Wigner function is a mathematical construct intended to characterize the system's distribution simultaneously in the coordinate and the momentum space—for 1D systems, on the phase plane $[X, P]$, which we had discussed earlier—see figure 5.8. Let us see how fruitful this intention is.

First of all, we may write the Fourier transform reciprocal to Eq. (7.50):

$$w\left(X + \frac{\tilde{X}}{2}, X - \frac{\tilde{X}}{2}\right) = \int W(X, P) \exp\left\{\frac{iP\tilde{X}}{\hbar}\right\} dP. \quad (7.52)$$

For the particular case $\tilde{X} = 0$, this relation yields

$$w(X) \equiv w(X, X) = \int W(X, P) dP. \quad (7.53)$$

Hence the integral of the Wigner function over the momentum P gives the probability density to find the system at point X—just as it does for a classical distribution function $w_{cl}(X, P)$.[21]

Next, the Wigner function has the similar property for integration over X. To prove this fact, we may first introduce the momentum representation of the density matrix, in the full analogy with its coordinate representation (7.27):

$$w(p, p') \equiv \langle p | \hat{w} | p' \rangle. \quad (7.54)$$

Inserting, as usual, two identity operators, in the form given by Eq. (4.252), into the right hand part of this equality, we can get the following relation between the momentum and coordinate representations:

$$\begin{aligned}
w(p, p') &= \int\int dx dx' \langle p | x \rangle \langle x | \hat{w} | x' \rangle \langle x' | p' \rangle \\
&= \frac{1}{2\pi\hbar} \int\int dx dx' \exp\left\{-\frac{ipx}{\hbar}\right\} w(x, x') \exp\left\{\frac{ip'x'}{\hbar}\right\}.
\end{aligned} \quad (7.55)$$

This is of course nothing other than the unitary transform of an operator from the x-basis to p-basis, similar to the first form of Eq. (4.272).[22] For coinciding arguments, $p = p'$, Eq. (7.55) is reduced to

---

[21] Such a function, used to express the probability dW to find the system in a small area of the phase plane as $dW = w_{cl}(X, P) dX dP$, is the basic notion of (1D) classical statistics—see, e.g. *Part SM* section 2.1.

[22] Note that the last line of Eq. (4.272) is invalid for the density operator $\hat{w}$, because it is *not* local!

$$w(p) \equiv w(p, p) = \frac{1}{2\pi\hbar} \int\int dx dx' w(x, x') \exp\left\{-\frac{ip(x - x')}{\hbar}\right\}. \quad (7.56)$$

Using Eq. (7.29) and then Eq. (4.265), this function may be represented as

$$w(p) = \frac{1}{2\pi\hbar} \sum_j W_j \int\int dx dx' \; \psi_j(x) \psi_j^*(x) \exp\left\{-\frac{ip(x - x')}{\hbar}\right\}$$
$$= \sum_j W_j \varphi_j(p) \varphi_j^*(p), \quad (7.57)$$

and hence interpreted as the probability density of the particle's momentum at point p. Now, in the variables (7.51), Eq. (7.56) has the form

$$w(p) = \frac{1}{2\pi\hbar} \int\int w\left(X + \frac{\tilde{X}}{2}, X - \frac{\tilde{X}}{2}\right) \exp\left\{-\frac{ip\tilde{X}}{\hbar}\right\} d\tilde{X} dX. \quad (7.58)$$

Comparing this equality with the definition (7.50) of the Wigner function, we see that

$$w(P) = \int W(X, P) dX. \quad (7.59)$$

Thus, according to Eqs. (7.53) and (7.59), the integrals of the Wigner function over either the coordinate or momentum give the probability densities to find them at certain values of these variables. This is of course the main requirement to any quantum-mechanical candidate for the best analog of the classical probability density, $w_{cl}(X, P)$.

Let us see how the Wigner function looks for the simplest systems in the thermodynamic equilibrium. For a free 1D particle, we can use Eq. (7.34), ignoring for simplicity the normalization issues:

$$W(X, P) \propto \int_{-\infty}^{+\infty} \exp\left\{-\frac{mk_BT\tilde{X}^2}{2\hbar^2}\right\} \exp\left\{-\frac{iP\tilde{X}}{\hbar}\right\} d\tilde{X}. \quad (7.60)$$

The usual Gaussian integration yields:

$$W(X, P) = \text{const} \times \exp\left\{-\frac{P^2}{2mk_BT}\right\}. \quad (7.61)$$

We see that the function is independent of X (as it should be for this translational-invariant system), and coincides with the Gibbs distribution (7.24). We could get the same result directly from the classical statistics. This is natural, because as we know from section 2.2, the free motion is essentially not quantized—at least in terms of its energy and momentum.

Now let us consider a substantially quantum system, the harmonic oscillator. Plugging Eq. (7.44) into Eq. (7.50), for that system in thermal equilibrium it is easy

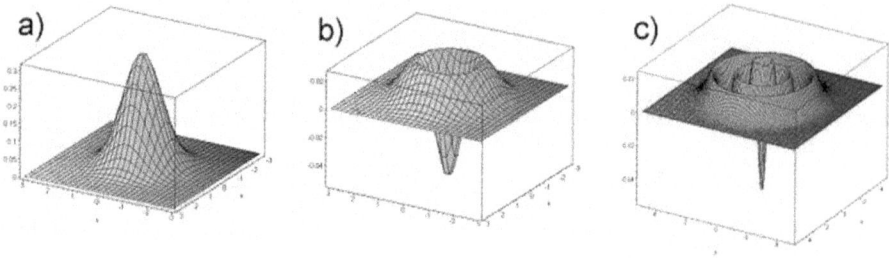

Figure 7.3. The Wigner functions $W(X, P)$ of a harmonic oscillator, in a few of its stationary (Fock) states n: (a) $n = 0$, (b) $n = 1$; (c) $n = 5$. Graphics by J S Lundeen; adapted from http://en.wikipedia.org/wiki/Wigner_function as public-domain material.

(and hence is left for reader's exercise) to show that the Wigner function is also Gaussian, now in both its arguments:

$$W(X, P) = \text{const} \times \exp\left\{-C\left[\frac{m\omega_0^2 X^2}{2} + \frac{P^2}{2m}\right]\right\},\tag{7.62}$$

though the coefficient C is now different from $1/k_\text{B}T$, and tends to that limit only at high temperatures, $k_\text{B}T \gg \hbar\omega_0$. Moreover, for the Glauber state the Wigner function also gives a very plausible result—a Gaussian distribution similar to Eq. (7.62), but shifted to the central point of the state—see section 5.5.[23] Unfortunately, for some other possible states of the harmonic oscillator, e.g. any pure Fock state with $n > 0$, the Wigner function takes negative values in some regions of the $[X, P]$ plane—see figure 7.3.[24] (Such plots were the basis of my, admittedly very imperfect, classical images of the Fock states in figure 5.8.)

The same is true for most other quantum systems and their states. Indeed, this fact could be predicted just by looking at the definition (7.50) applied to a pure quantum state, in which the density function may be factored—see Eq. (7.31):

$$W(X, P) = \frac{1}{2\pi\hbar}\int \psi\left(X + \frac{\tilde{X}}{2}\right)\psi^*\left(X - \frac{\tilde{X}}{2}\right)\exp\left\{-\frac{iP\tilde{X}}{\hbar}\right\}d\tilde{X}.\tag{7.63}$$

Changing the argument P (say, at fixed X), we are essentially changing the spatial 'frequency' (wave number) of the wavefunction product's Fourier component we are calculating, and we know that Fourier images typically change sign as the frequency is changed. Hence the wavefunctions should have some high-symmetry properties to avoid this effect. Indeed, the Gaussian functions (describing, for example, the Glauber states, and as the particular case, the ground state of the harmonic oscillator) have such a symmetry, but many other functions do not.

---

[23] Please note that in notations of that section, the capital letters X and P mean not the arguments of the Wigner function, but the Cartesian coordinates of the central point (5.102), i.e. the classical complex amplitude of the oscillations.

[24] Spectacular experimental measurements of this function (for $n = 0$ and $n = 1$) were carried out recently by E Bimbard *et al* [1].

Hence if the Wigner function was taken seriously as the quantum-mechanical analog of the classical probability density $w_{cl}(X, P)$, we would need to interpret the negative probability of finding the particle in certain elementary intervals $dXdP$—which is hard to do. However, the function is still used for a semi-quantitative interpretation of mixed states of quantum systems.

7.3 Open system dynamics: dephasing

So far we have discussed the density operator as something *given* at a particular time instant. Now let us discuss how is it *formed*, i.e. its evolution in time, starting from the simplest case when the probabilities W_j participating in Eq. (7.15) are *time-independent*—by this or that reason, to be discussed below. In this case, in the Schrödinger picture, we may rewrite Eq. (7.15) as

$$\hat{w}(t) = \sum_j |w_j(t)\rangle \, W_j \, \langle w_j(t)|. \tag{7.64}$$

Taking a time derivative of both parts of this equation, multiplying them by $i\hbar$, and applying Eq. (4.158) to the basis states w_j, with the account of the fact that the Hamiltonian operator is Hermitian, we get

$$i\hbar\hat{\dot{w}} = i\hbar \sum_j [|\dot{w}_j(t)\rangle \, W_j \, \langle w_j(t)| + |w_j(t)\rangle \, W_j \, \langle \dot{w}_j(t)|]$$

$$= \sum_j [\hat{H}|w_j(t)\rangle \, W_j \, \langle w_j(t)| - |w_j(t)\rangle \, W_j \, \langle w_j(t)|\hat{H}] \tag{7.65}$$

$$\equiv \hat{H} \sum_j |w_j(t)\rangle \, W_j \, \langle w_j(t)| - \sum_j |w_j(t)\rangle \, W_j \, \langle w_j(t)|\hat{H}.$$

Now using Eq. (7.64) again (twice), we get the so-called *von Neumann equation*[25]

$$i\hbar\hat{\dot{w}} = [\hat{H}, \hat{w}]. \tag{7.66}$$

Note that this equation is similar in structure to Eq. (4.199) describing the time evolution of time-independent operators in the Heisenberg picture operators:

$$i\hbar\hat{\dot{A}} = [\hat{A}, \hat{H}], \tag{7.67}$$

besides the opposite order of the operators order in the commutator. (This change, of course, changes only the sign of the right-hand side.) This should not be too surprising, because Eq. (7.66) belongs to the Schrödinger picture, while Eq. (7.67) belongs to the Heisenberg picture of the quantum dynamics.

The most important case when the von Neumann equation is (approximately) valid is when the 'own' Hamiltonian $\hat{H}_s$ of the system s of our interest is time-independent, and its interaction with the environment is so small that its effect on the

---

[25] In some texts, it is called the 'Liouville equation', due to the philosophical proximity to the classical Liouville theorem for the distribution function $w_{cl}(X, P)$—see, e.g. *Part SM* section 6.1, in particular Eq. (6.5).

system's evolution during the considered time interval is negligible, but it had lasted so long that it gradually put the system into a non-pure state—for example, but not necessarily, into the classical mixture (7.24).[26] (This is an example of the second case discussed in section 7.1, when we need the mixed-ensemble description of the system, even if its current contact with the environment is negligible.) If the interaction with the environment is stronger, and hence is not negligible at the considered time interval, Eq. (7.66) is generally *not* valid[27], because the probabilities W_j may change in time. However, this equation may still be used for a discussion of one major effect of the environment, namely *dephasing* (also called 'decoherence'), within a simple model.

Let us start with the following general model a system interacting with its environment, which will be used throughout this chapter:

$$\hat{H} = \hat{H}_s + \hat{H}_e\{\lambda\} + \hat{H}_{\text{int}}, \qquad (7.68)$$

where $\{\lambda\}$ denotes the (huge) set of the degrees of freedom of the environment[28]. Evidently, this model is useful only if we may somehow tame the enormous size of the Hilbert space of these degrees of freedom, and so work out the calculations all the way to a practicably simple result. This turns out to be possible mostly if the elementary act of interaction of the system and its environment is in some sense small. Below, I will describe several cases when this is true; the classical example is the *Brownian particle* interacting with the molecules of the surrounding gas or fluid[29]. (In this example, a single hit by a molecule changes the particle's momentum by a minor faction.) On the other hand, the model (7.68) is not very productive for a particle interacting with the environment consisting of *similar* particles, when a single collision may change its momentum dramatically. In such cases, the methods discussed in the next chapter are more relevant.

Now let us analyze a very simple model of a open two-level quantum system, with its 'own' Hamiltonian having the form

---

[26] In the last case, the statistical operator is diagonal in the stationary state basis and hence commutes with the Hamiltonian. Hence the right-hand side of Eq. (7.66) vanishes, and it shows that in this basis, the density matrix in completely time-independent.

[27] Very unfortunately, this fact is not explained in some textbooks, which cite the von Neumann equation without proper qualifications.

[28] Note that by writing Eq. (7.68), we are treating the whole system, including the environment, as a Hamiltonian one. This can always be done if the accounted part of the environment is large enough, so that the processes in the system s of our interest do not depend on the type of boundary between this part and the 'external' (even larger) environment; in particular we may assume the total system closed, i.e. Hamiltonian.

[29] The theory of the Brownian motion, the effect first observed experimentally by biologist R Brown in the early 1800s, was pioneered by A Einstein in 1905 and developed in detail by M Smoluchowski in 1906–7, and A Fokker in 1913. Due to this historic background, in some older texts the whole approach described in this chapter is called the 'quantum theory of Brownian motion'. Let me, however, emphasize that due to the later progress of experimental techniques, quantum-mechanical behaviors, including the environmental effects on them, have been observed in a growing number of various quasi-macroscopic systems—for example, the micromechanical oscillators mentioned in section 2.9, each with many more than 10^{10} atoms. Moreover, Eq. (7.68) is adequate for most systems explored as possible qubits of prospective quantum computing systems —see section 8.5 below.

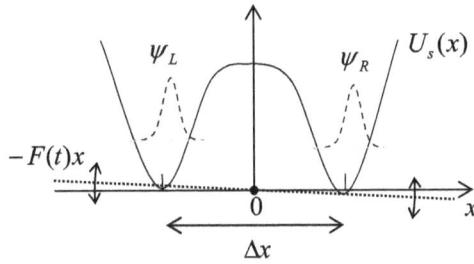

Figure 7.4. Dephasing in a double-well system.

$$\hat{H}_s = c_z \hat{\sigma}_z, \qquad (7.69)$$

similar to the Pauli Hamiltonian (4.163),[30] and a factorable, bilinear interaction—cf Eq. (6.145) and its discussion:

$$\hat{H}_{\text{int}} = \hat{f}\{\lambda\}\hat{\sigma}_z, \qquad (7.70)$$

where $\hat{f}$ is a Hermitian operator depending only on the set $\{\lambda\}$ of environmental degrees of freedom ('coordinates'), describing the Hilbert space different from that of the two-level system. As a result, the operators $\hat{f}\{\lambda\}$ and $\hat{H}_e\{\lambda\}$ commute with $\hat{\sigma}_z$—and with any other intrinsic operator of the two-level system. Of course, any realistic $\hat{H}_e\{\lambda\}$ is extremely complex, so that how much we will be able to achieve without specifying it, may be a pleasant surprise for the reader.

Before we proceed to the analysis, let me recognize two examples of two-level systems that may be described by this model. The first example is a spin-½ in an external magnetic field of a fixed direction (taken for the axis z), which includes both an average component $\bar{\mathscr{B}}$ and a random (fluctuating) component $\tilde{\mathscr{B}}_z(t)$ induced by the environment. As follows from Eq. (4.163b), it may be described by the Hamiltonian (7.68)–(7.70) with

$$c_z = -\frac{\hbar\gamma}{2}\bar{\mathscr{B}}_z, \quad \hat{f} = -\frac{\hbar\gamma}{2}\hat{\tilde{\mathscr{B}}}_z \quad (t). \qquad (7.71)$$

Another important example is a particle in a symmetric double-well potential U_s (figure 7.4), with a barrier between them sufficiently high to be practically impenetrable, and an additional force $F(t)$, including the fluctuations exerted by the environment, so that the total potential energy is $U(x, t) = U_s(x) - F(t)x$. If the force is sufficiently weak, we can neglect its effects on the shape of potential wells and hence on the localized wavefunctions $\psi_{L,R}$, so that the force effect is reduced to the variation of the difference $E_L - E_R = F(t)\Delta x$ between the eigenenergies. As a result, the system may described by Eqs. (7.68)–(7.70) with

$$c_z = -\bar{F}\Delta x/2; \qquad \hat{f} = -\hat{\tilde{F}}(t)\Delta x/2. \qquad (7.72)$$

---

[30] As we know from sections 4.6 and 5.1, such Hamiltonian is sufficient to lift the energy level degeneracy.

Let us start from writing the equation of motion for the Heisenberg operator $\hat{\sigma}_z$:

$$i\hbar\dot{\hat{\sigma}}_z = [\hat{\sigma}_z, \hat{H}] = (c_z + \hat{f})[\hat{\sigma}_z, \hat{\sigma}_z] = 0, \tag{7.73}$$

showing that in our simple model (7.68)–(7.70), the operator $\hat{\sigma}_z$ does not evolve in time. What does this mean for the observables? For an arbitrary density matrix of any two-level system,

$$w = \begin{pmatrix} w_{11} & w_{12} \\ w_{21} & w_{22} \end{pmatrix}, \tag{7.74}$$

we can readily calculate the trace of the operator $\hat{\sigma}_z$. Indeed, since the operator traces are basis-independent, we can do this in any basis, in particular in the usual z-basis:

$$\mathrm{Tr}(\hat{\sigma}_z \hat{w}) = \mathrm{Tr}(\sigma_z w) = \mathrm{Tr}\left[\begin{pmatrix} 1 & 0 \\ 0 & -1 \end{pmatrix}\begin{pmatrix} w_{11} & w_{12} \\ w_{21} & w_{22} \end{pmatrix}\right] = w_{11} - w_{22} = W_1 - W_2. \tag{7.75}$$

Since, according to Eq. (7.5), $\hat{\sigma}_z$ may be considered the operator for the observable $W_1 - W_2$, in the case (7.73) the difference $W_1 - W_2$ does not depend on time, and since the sum of the probabilities is also fixed, $W_1 + W_2 = 1$, both of them are constant. The physics of this result is especially clear for the model shown in figure 7.4: since the potential barrier separating the potential wells is so high that tunneling through it is negligible, the interaction with environment cannot move the system from one well into another one. It may look like nothing interesting may happen in such simple situation, but in a minute we will see that this is not true.

Due to the time independence of W_1 and W_2 in this particular system, we may use the von Neumann equation (7.66) to describe its density matrix evolution—now in the Schrödinger picture. In the usual z-basis:

$$i\hbar\dot{w} \equiv i\hbar\begin{pmatrix} \dot{w}_{11} & \dot{w}_{12} \\ \dot{w}_{21} & \dot{w}_{22} \end{pmatrix} = [H, w] \equiv (c_z + \hat{f})\,[\sigma_z, w]$$

$$\equiv (c_z + \hat{f})\left[\begin{pmatrix} 1 & 0 \\ 0 & -1 \end{pmatrix}, \begin{pmatrix} w_{11} & w_{12} \\ w_{21} & w_{22} \end{pmatrix}\right] = (c_z + \hat{f})\begin{pmatrix} 0 & 2w_{12} \\ -2w_{21} & 0 \end{pmatrix}. \tag{7.76}$$

This result means that though the diagonal elements, i.e. the probabilities of the states, do not evolve in time (as we already know), the off-diagonal coefficients do change; for example,

$$i\hbar\dot{w}_{12} = 2(c_z + \hat{f})w_{12}, \tag{7.77}$$

with a similar but complex-conjugate equation for w_{21}. The solution of the linear differential equation (7.77) is straightforward, and yields

$$w_{12}(t) = w_{12}(0)\exp\left\{-i\frac{2c_z}{\hbar}t\right\}\exp\left\{-i\frac{2}{\hbar}\int_0^t \hat{f}(t')dt'\right\}. \tag{7.78}$$

The first exponent is a deterministic c-number factor, while in the second one $\hat{f}(t) \equiv \hat{f}\{\lambda(t)\}$ is still an operator in the Hilbert space of the environment, and, from the point of view of the two-level system of our interest, a random function of time. The time-average part of this force may be included into c_z, so in what follows, we will assume that it equals zero.

Let us start from the limit when the environment behaves classically[31]. In this case, the operator in Eq. (7.78) may be considered as a *classical* random function of time $f(t)$, provided that we average its effects over a statistical ensemble of many functions $f(t)$ describing many (macroscopically similar) experiments. For a small time interval $t = dt \to 0$, we can use the Taylor expansion of the exponent, truncating it after the quadratic term:

$$
\begin{aligned}
\left\langle \exp\left\{-i\frac{2}{\hbar}\int_0^{dt} f(t')dt'\right\}\right\rangle &\approx 1 + \left\langle -i\frac{2}{\hbar}\int_0^{dt} f(t')dt'\right\rangle \\
&+ \left\langle \frac{1}{2}\left(-i\frac{2}{\hbar}\int_0^{dt} f(t')dt'\right)\left(-i\frac{2}{\hbar}\int_0^{dt} f(t'')dt''\right)\right\rangle \\
&\equiv 1 - i\frac{2}{\hbar}\int_0^{dt}\langle f(t')\rangle dt' \\
&- \frac{2}{\hbar^2}\int_0^{dt} dt'\int_0^{dt} dt''\langle f(t')f(t'')\rangle \\
&\equiv 1 - \frac{2}{\hbar^2}\int_0^{dt} dt'\int_0^{dt} dt'' K_f(t'-t'').
\end{aligned}
\tag{7.79}
$$

Here we have used the fact that the statistical average of $f(t)$ is equal to zero, while the second average, called the *correlation function*, in a statistically- (i.e. macroscopically-) stationary state of any environment may only depend on the time difference $\tau \equiv t' - t''$:

$$
\langle f(t')f(t'')\rangle = K_f(t'-t'') \equiv K_f(\tau).
\tag{7.80}
$$

If this difference is much larger than some time scale τ_c, called the *correlation time* of the environment, the values $f(t')$ and $f(t'')$ are completely independent (*uncorrelated*), as illustrated in figure 7.5a, so that the correlation function has to tend to zero. On the other hand, at $\tau = 0$, i.e. $t' = t''$, the correlation function is just the variance of f:

$$
K_f(0) = \langle f^2\rangle,
\tag{7.81}
$$

and has to be positive. As a result, the function looks (semi-quantitatively) as shown in figure 7.5b.

Hence, if we are only interested in time differences τ much longer than τ_c, which is typically very short, we may approximate $K_f(\tau)$ with a delta-function of the time difference. Let us take it in the following convenient form

---

[31] This assumption is not in contradiction with the need for the quantum treatment of the two-level system, because a typical environment is large, and hence has a very dense energy spectrum, with the distances adjacent levels that may be readily bridged by thermal excitations of small energies, often making its essentially classical.

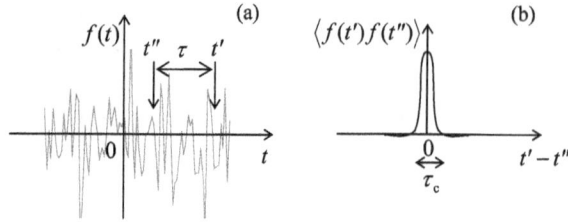

Figure 7.5. (a) A typical random process and (b) its correlation function—schematically.

$$K_f(\tau) \approx \hbar^2 D_\varphi \delta(\tau), \tag{7.82}$$

where D_φ is a positive constant called the *phase diffusion coefficient*. The origin of this term stems from the very similar effect of classical diffusion of the Brownian particles in a highly viscous medium. Indeed, the particle's velocity in such a medium is approximately proportional to the external force. Hence, if the random hits of a particle by the molecules may be described by a force that obeys a law similar to Eq. (7.82), the velocity (along any Cartesian coordinate) is also *delta-correlated*:

$$\langle v(t) \rangle = 0, \quad \langle v(t')v(t'') \rangle = 2D\delta(t' - t''). \tag{7.83}$$

Now we can integrate the kinematic relation $\dot{x} = v$, to calculate particle's displacement from its initial position, and its variance:

$$x(t) - x(0) = \int_0^t v(t')dt', \tag{7.84}$$

$$\langle (x(t) - x(0))^2 \rangle = \left\langle \int_0^t v(t')dt' \int_0^t v(t'')dt'' \right\rangle = \int_0^t dt' \int_0^t dt'' \langle v(t')v(t'') \rangle$$
$$= \int_0^t dt' \int_0^t dt'' 2D\delta(t' - t'') = 2Dt. \tag{7.85}$$

This is the famous law of diffusion, showing that the rms deviation of the particle from the initial point grows with time as $(2Dt)^{1/2}$, where the constant D is called the *diffusion coefficient*.

Returning to the diffusion of the quantum-mechanical phase, with Eq. (7.82) the last double integral in Eq. (7.79) yields $\hbar^2 D_\varphi dt$, so that the statistical average of Eq. (7.78) is

$$\langle w_{12}(dt) \rangle = w_{12}(0)\exp\left\{ -i\frac{2c_z}{\hbar}dt \right\}(1 - 2D_\varphi dt). \tag{7.86}$$

Applying this formula to sequential time intervals,

$$\langle w_{12}(2dt) \rangle = \langle w_{12}(dt) \rangle \exp\left\{ -i\frac{2c_z}{\hbar}dt \right\}(1 - 2D_\varphi dt)$$
$$= w_{12}(0)\exp\left\{ -i\frac{2c_z}{\hbar}2dt \right\}(1 - 2D_\varphi dt)^2, \tag{7.87}$$

etc, for a finite time $t = Ndt$, in the limit $N \to \infty$ and $dt \to 0$ (at fixed t) we get

$$\langle w_{12}(t) \rangle = w_{12}(0)\exp\left\{-i\frac{2c_z}{\hbar}t\right\} \times \lim_{N\to\infty}\left(1 - 2D_\varphi t\frac{1}{N}\right)^N. \tag{7.88}$$

By the definition of the natural logarithm base e,[32] this limit is just $\exp\{-2D_\varphi t\}$, so that, finally:

$$\langle w_{12}(t) \rangle = w_{12}(0)\exp\left\{-i\frac{2a}{\hbar}t\right\}\exp\{-2D_\varphi t\}$$
$$\equiv w_{12}(0)\exp\left\{-i\frac{2a}{\hbar}t\right\}\exp\left\{-\frac{t}{T_2}\right\}. \tag{7.89}$$

So, due to coupling to environment, the off-diagonal elements of the density matrix decay with some *dephasing time* $T_2 = 1/2D_\varphi$, providing a natural evolution from the density matrix (7.22) of a pure state to the diagonal matrix (7.23), with the same probabilities $W_{1,2}$, describing a fully dephased (incoherent) classical mixture[33].

This simple model offers a very clear look at the nature of the decoherence: the random 'force' $f(t)$, exerted by the environment, 'shakes' the energy difference between two eigenstates of the system and hence the instantaneous velocity $2(c_z + f)/\hbar$ of their mutual phase shift $\varphi(t)$—cf Eq. (7.22). Due to the randomness of the force, $\varphi(t)$ performs a random walk around the trigonometric circle, so that the average of its trigonometric functions $\exp\{\pm i\varphi\}$ over time gradually tends to zero, killing the off-diagonal elements of the density matrix. Our analysis, however, has left open two important issues:

(i) Is this approach valid for *quantum* description of a typical environment?
(ii) If yes, what is physically the D_φ that was formally defined by Eq. (7.82)?

7.4 Fluctuation–dissipation theorem

Similar questions may be asked about a more general situation, when the Hamiltonian $\hat{H}_s$ of the system of interest (s), in the composite Hamiltonian (7.68), is not specified at all, but the interaction between that system and its environment still has the bilinear form similar to Eqs. (7.70) and (6.130):

$$\hat{H}_{int} = -\hat{F}\{\lambda\}\hat{x}, \tag{7.90}$$

where x is some observable of the subsystem s (say, its generalized coordinate or generalized momentum). It may look incredible that in this very general situation one may make a very simple and powerful statement about the statistical properties

---

[32] See, e.g. Eq. (A.2a) with $n = -N/2D_\varphi t$.
[33] Note that this result is valid only if the approximation (7.82) may be applied at time interval dt which, in turn, should be much smaller than the T_2 in Eq. (7.88), i.e. if the dephasing time is much longer than the environment's correlation time τ_c. This requirement may be always satisfied by making the coupling to environment sufficiently weak. In addition in typical environments, τ_c is very short. For example, in the original Brownian motion experiments with a-few-μm ink particles in water, it is of the order of the average interval between sequential molecular impacts, of the order of 10^{-21} s.

of the generalized external force F, under only two (interrelated) conditions—which are satisfied in a huge number of cases of interest:

(i) the coupling of system s of interest to environment e is weak—in the sense that the perturbation theory (see chapter 6) is applicable, and

(ii) the environment may be considered as staying in thermodynamic equilibrium, with a certain temperature T, regardless of the process in the system of interest[34].

This famous statement is called the *fluctuation–dissipation theorem* (FDT)[35]. Due to the importance of this fundamental result, let me derive it[36]. Since by writing Eq. (7.68) we treat the whole system $(s + e)$ as a Hamiltonian one, we may use the Heisenberg equation (4.199) to write

$$i\hbar \dot{\hat{F}} = [\hat{F}, \hat{H}] = [\hat{F}, \hat{H}_e], \qquad (7.91)$$

because, as was discussed in the last section, the operator $\hat{F}\{\lambda\}$ commutes with the operators $\hat{H}_s$ and $\hat{x}$. Generally, very little may be done with this equation, because the time evolution of the environment's Hamiltonian depends, in turn, on that of the force. This is where the perturbation theory becomes indispensable. Let us decompose the external force's operator into the following sum:

$$\hat{F}\{\lambda\} = \langle \hat{F} \rangle + \tilde{\hat{F}}(t), \quad \text{with } \langle \tilde{\hat{F}}(t) \rangle = 0, \qquad (7.92)$$

where (here and on, until further notice) the sign $\langle ... \rangle$ means the statistical averaging over the environment alone, i.e. over an ensemble in similar quantum states to system s, but random states of its environment[37]. From the point of view of the system s, the first term of the sum (still an operator!) describes the average response of the environment to the system dynamics (possibly, including such irreversible effects as friction), and has to be calculated with account of their interaction—as we will do later in this section. On the other hand, the second term in Eq. (7.92)

---

[34] The most frequent example of violation of these conditions is the environment's overheating by the energy flow from the subsystem. I leave it to the reader to estimate the overheating of a standard physical laboratory room by a typical dissipative quantum process—the emission of an optical photon by an atom. (*Hint*: it is extremely small.)

[35] The FDT was first derived by H Callen and T Welton in 1951, on the background of an earlier derivation of its classical limit by H Nyquist in 1928, and the already mentioned pioneering 1905 work by A Einstein.

[36] The FDT may be proved in several ways that are shorter than the one given below—see, e.g. either the proof in *Part SM* sections 5.5 and 5.6 (based on H Nyquist's arguments), or the original paper by H Callen and T Welton [2]—wonderful in its clarity. The longer approach I describe here, besides giving the important *Green–Kubo formula* (7.109) as a byproduct, is a very useful exercise in the operator manipulation and the perturbation theory in its integral form—different from the differential form used in chapter 6. If the reader is not interested in this exercise, (s)he may skip the derivation and jump straight to the result, expressed by Eq. (7.134), which uses the notions defined by Eqs. (7.114) and (7.123).

[37] For usual ('ergodic') environments, without intrinsic long-term memories, this statistical averaging over an ensemble of environments is equivalent to averaging over relatively short times—much longer than the correlation time τ_c of the environment, but still much shorter than the characteristic time of evolution of the system under analysis, such as the dephasing time T_2 and the energy relaxation time T_1—both still to be calculated.

represents *fluctuations* of the environment, which exist even in the absence of system *s*. Hence, in the first nonvanishing approximation in the interaction strength, the fluctuation part may be calculated ignoring the interaction, i.e. treating the environment as being in the thermodynamic equilibrium:

$$i\hbar \dot{\tilde{F}} = [\tilde{F}, \hat{H}_{e|eq}].$$

(7.93)

Since in this approximation the environment's Hamiltonian does not have an explicit dependence on time, the solution of this equation may be written combining Eqs. (4.190) and (4.175):

$$\tilde{F}(t) = \exp\left\{\frac{i}{\hbar}\hat{H}_{e|eq}t\right\}\hat{F}(0)\exp\left\{-\frac{i}{\hbar}\hat{H}_{e|eq}t\right\}.$$

(7.94)

Let us use this relation to calculate the correlation function of the fluctuations *F(t)*, defined similarly to Eq. (7.80), but taking care of the order of the time arguments (very soon we will see why):

$$\langle \tilde{F}(t)\tilde{F}(t')\rangle = \left\langle \exp\left\{\frac{i}{\hbar}\hat{H}_e t\right\}\hat{F}(0)\exp\left\{-\frac{i}{\hbar}\hat{H}_e t\right\}\right.$$
$$\left. \times \exp\left\{\frac{i}{\hbar}\hat{H}_e t'\right\}\hat{F}(0)\exp\left\{-\frac{i}{\hbar}\hat{H}_e t'\right\}\right\rangle.$$

(7.95)

(Here, for the notation brevity, the thermal equilibrium of the environment is just implied.) We are at will to calculate this expectation value in any basis, and the best choice is evident: in the environment's stationary-state basis, the density operator of the environment, its Hamiltonian, and hence the exponents in Eq. (7.95) are all represented by diagonal matrices. Using Eq. (7.5), the correlation function becomes

$$\langle \tilde{F}(t)\tilde{F}(t')\rangle = \text{Tr}\left[\hat{w}\exp\left\{\frac{i}{\hbar}\hat{H}_e t\right\}\hat{F}(0)\exp\left\{-\frac{i}{\hbar}\hat{H}_e t\right\}\right.$$
$$\left. \times \exp\left\{\frac{i}{\hbar}\hat{H}_e t'\right\}\hat{F}(0)\exp\left\{-\frac{i}{\hbar}\hat{H}_e t'\right\}\right]$$
$$\equiv \sum_n\left[\hat{w}\exp\left\{\frac{i}{\hbar}\hat{H}_e t\right\}\hat{F}(0)\exp\left\{-\frac{i}{\hbar}\hat{H}_e t\right\}\right.$$
$$\left. \times \exp\left\{\frac{i}{\hbar}\hat{H}_e t'\right\}\hat{F}(0)\exp\left\{-\frac{i}{\hbar}\hat{H}_e t'\right\}\right]_{nn}$$

(7.96)

$$= \sum_{n,n'} W_n\exp\left\{\frac{i}{\hbar}E_n t\right\}\hat{F}_{nn'}\exp\left\{-\frac{i}{\hbar}E_{n'}t\right\}$$
$$\times \exp\left\{\frac{i}{\hbar}E_{n'}t'\right\}\hat{F}_{n'n}\exp\left\{-\frac{i}{\hbar}E_n t'\right\}$$
$$\equiv \sum_{n,n'} W_n|F_{nn'}|^2\exp\left\{\frac{i}{\hbar}(E_n - E_{n'})(t-t')\right\}.$$

Here W_n are the Gibbs distribution probabilities given by Eq. (7.24), with the environment's temperature T, and $F_{nn'} \equiv F_{nn'}(0)$ are the Schrödinger-picture matrix elements of the interaction force operator.

We see that though the correlator (7.96) is a function of the difference $\tau \equiv t - t'$ only (as it should be for fluctuations in a macroscopically stationary system), it may depend on the order of its arguments. For this reason let us mark this particular correlation function with the upper index '+',

$$K_F^+(\tau) \equiv \langle \tilde{F}(t)\tilde{F}(t') \rangle = \sum_{n,n'} W_n |F_{nn'}|^2 \exp\left\{ \frac{i\tilde{E}\tau}{\hbar} \right\}, \quad \text{where } \tilde{E} \equiv E_n - E_{n'}, \quad (7.97)$$

and its counterpart, with the swapped times t and t', with the upper index '-':

$$K_F^-(\tau) \equiv K_F^+(-\tau) = \langle \tilde{F}(t')\tilde{F}(t) \rangle = \sum_{n,n'} W_n |F_{nn'}|^2 \exp\left\{ -\frac{i\tilde{E}\tau}{\hbar} \right\}. \quad (7.98)$$

So, in contrast with classical processes, in quantum mechanics the correlation function of fluctuations $\tilde{F}$ is not necessarily time-symmetric:

$$K_F^+(\tau) - K_F^-(\tau) \equiv K_F^+(\tau) - K_F^+(-\tau) = \langle \tilde{F}(t)\tilde{F}(t') - \tilde{F}(t')\tilde{F}(t) \rangle$$
$$= 2i \sum_{n,n'} W_n |F_{nn'}|^2 \sin \frac{\tilde{E}\tau}{\hbar} \neq 0, \quad (7.99)$$

so that $\hat{F}(t)$ gives one more example of a Heisenberg-picture operator whose 'values', taken in different moments of time, generally do not commute—see footnote 49 in chapter 4. (A good sanity check here is that at $\tau = 0$, i.e. at $t = t'$, the difference (7.99) between K_F^+ and K_F^- vanishes.)

Now let us return to the force's decomposition (7.92), and calculate its first (average) component. In order to do that, let us write the formal solution of Eq. (7.91) as follows:

$$\hat{F}(t) = \frac{1}{i\hbar} \int_{-\infty}^{t} [\hat{F}(t'), \hat{H}_e(t')] \, dt'. \quad (7.100)$$

On the right-hand side of this relation, we still cannot treat the Hamiltonian of the environment as an unperturbed (equilibrium) one, even if the effect of our system (s) on the environment is very weak, because this would have zero statistical average of the force $F(t)$. Hence, we should make one more step of our perturbative treatment, taking into account the effect of the force on the environment. To do this, let us use Eqs. (7.68) and (7.90) to write the (so far, exact) Heisenberg equation of motion for the environment's Hamiltonian,

$$i\hbar\dot{\hat{H}}_e = [\hat{H}_e, \hat{H}] = -\hat{x}[\hat{H}_e, \hat{F}], \quad (7.101)$$

and its formal solution, similar to Eq. (7.100), but for the time t' rather than t:

$$\hat{H}_e(t') = -\frac{1}{i\hbar} \int_{-\infty}^{t'} \hat{x}(t'') [\hat{H}_e(t''), \hat{F}(t'')] \, dt''. \quad (7.102)$$

Plugging this equality into the right-hand side of Eq. (7.100), and averaging the result (again, over the environment only!), we get

$$\langle \hat{F}(t) \rangle = \frac{1}{\hbar^2} \int_{-\infty}^{t} dt' \int_{-\infty}^{t'} dt'' \; \hat{x}(t'') \, \langle [\hat{F}(t'), [\hat{H}_e(t''), \hat{F}(t'')]\,] \rangle. \tag{7.103}$$

This is still an exact result, but now it is ready for an approximate treatment, implemented by averaging its right-hand side over the unperturbed (thermal-equilibrium) state of the environment. This may be done absolutely similarly to that in Eq. (7.96), at the last step using Eq. (7.94):

$$\langle [\hat{F}(t'), [\hat{H}_e(t''), \hat{F}(t'')]\,] \rangle = \mathrm{Tr}\{w[F(t'), [H_e F(t'')]]\}$$

$$\equiv \mathrm{Tr}\{w[F(t')H_e F(t'') - F(t')F(t'')H_e - H_e F(t'')F(t') + F(t'')H_e F(t')]\}$$

$$= \sum_{n,n'} W_n[F_{nn'}(t')E_n F_{n'n}(t'') - F_{nn'}(t')F_{n'n}(t'')E_n$$

$$- E_n F_{nn'}(t'')F_{n'n}(t') + F_{nn'}(t'')E_n F_{n'n}(t'')] \tag{7.104}$$

$$\equiv -\sum_{n,n'} W_n \tilde{E} \; |F_{nn'}|^2 \left[\exp\left\{ \frac{i\tilde{E}(t'-t'')}{\hbar} \right\} + \text{c.c.} \right].$$

Now, if we try to integrate each term of this sum, as Eq. (7.103) seems to require, we will see that the lower-limit substitution (at t', $t'' \to -\infty$) is uncertain, because the exponents oscillate without decay. This mathematical difficulty may be overcome by the following physical reasoning. As illustrated by the example considered in the previous section, coupling to a disordered environment makes the 'memory horizon' of the system of our interest (s) finite: its current state does not depend on its history beyond a certain time scale[38]. As a result, the functions under the integrals of Eq. (7.103), i.e. the sum (7.104), should self-average at a certain finite time. A simplistic technique for expressing this fact mathematically is just dropping the lower-limit substitution; this would give the correct result for Eq. (7.103). However, a better (mathematically more acceptable) trick is to first multiply the function under each integral by, respectively, $\exp\{\varepsilon(t - t')\}$ and $\exp\{\varepsilon(t' - t'')\}$, where ε is a very small positive constant, then carry out the integration, and after that take the limit $\varepsilon \to 0$. The physical justification of this procedure may be provided by saying that system's behavior should not be affected if its interaction with the environment was not kept constant but rather turned on gradually—say, exponentially with an infinitesimal rate ε. With this modification, Eq. (7.103) becomes

$$\langle \hat{F}(t) \rangle = -\frac{1}{\hbar^2} \sum_{n,n'} W_n \tilde{E} \; |F_{nn'}|^2 \; \lim_{\varepsilon \to 0} \int_{-\infty}^{t} dt' \int_{-\infty}^{t'} dt'' \; \hat{x}(t'')$$

$$\times \left[\exp\left\{ \frac{i\tilde{E}(t'-t'')}{\hbar} + \varepsilon(t''-t) \right\} + \text{c.c.} \right]. \tag{7.105}$$

---

[38] Actually, this is true for virtually any real physical system—in contrast to idealized models such as a dissipation-free oscillator that swings for ever and ever with the same amplitude and phase, thus 'remembering' the initial conditions.

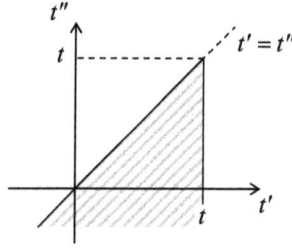

Figure 7.6. The 2D integration area in Eqs. (7.105) and (7.106).

This double integration is over the area shaded in figure 7.6, so that the order of integration may be changed to the opposite one as

$$\int_{-\infty}^{t} dt' \int_{-\infty}^{t'} dt'' \ldots = \int_{-\infty}^{t} dt'' \int_{t''}^{t} dt' \ldots = \int_{-\infty}^{t} dt'' \int_{t''-t}^{0} d(t'-t)\ldots$$
$$= \int_{-\infty}^{t} dt'' \int_{0}^{\tau} d\tau' \ldots,$$

(7.106)

where $\tau' \equiv t - t'$, and $\tau \equiv t - t''$.

As a result, Eq. (7.105) may be rewritten as a single integral,

$$\langle \hat{F}(t) \rangle = \int_{-\infty}^{t} G(t - t'')\, \hat{x}(t'')dt'' \equiv \int_{0}^{\infty} G(\tau)\, \hat{x}(t - \tau)d\tau,$$

(7.107)

whose kernel,

$$G(\tau > 0) \equiv -\frac{1}{\hbar^2} \sum_{n,n'} W_n \tilde{E} \, |F_{nn'}|^2 \lim_{\varepsilon \to 0} \int_{0}^{\tau} \left[\exp\left\{ \frac{i\tilde{E}(\tau - \tau')}{\hbar} - \varepsilon\tau \right\} + \text{c.c.} \right] d\tau'$$
$$= \lim_{\varepsilon \to 0} \frac{2}{\hbar} \sum_{n,n'} W_n \, |F_{nn'}|^2 \sin \frac{\tilde{E}\tau}{\hbar} e^{-\varepsilon\tau}$$

(7.108)

$$\equiv \frac{2}{\hbar} \sum_{n,n'} W_n \, |F_{nn'}|^2 \sin \frac{\tilde{E}\tau}{\hbar},$$

does not depend on the particular law of evolution of the subsystem (s) under study, i.e. provides a general characterization of its coupling to the environment.

In Eq. (7.107) we may readily recognize the most general form of the linear response of a system (in our case, the environment), taking into account the causality principle, where $G(\tau)$ is the *response function* (also called the 'temporal Green's function') of the environment. Now comparing Eq. (7.108) with Eq. (7.99), we get a wonderfully simple universal relation,

$$\langle [\hat{F}(\tau), \hat{F}(0)] \rangle = i\hbar G(\tau).$$

(7.109)

that emphasizes once again the quantum nature of the correlation function's time asymmetry. (This relation, called the *Green–Kubo* (or just 'Kubo') *formula* after the works by M Green (1954) and R Kubo (1957), does not come up in the easier derivations of the FDT, mentioned at the beginning of this section.)

However, the relation between $G(\tau)$ and the force's *anti*-commutator,

$$\langle\{\hat{F}(t + \tau), \hat{F}(t)\}\rangle \equiv \langle\hat{F}(t + \tau)\hat{F}(t) + \hat{F}(t)\hat{F}(t + \tau)\rangle$$

$$\equiv K_F^+(\tau) + K_F^-(\tau), \tag{7.110}$$

is much more important, because of the following reason. Eqs. (7.97) and (7.98) show that the so-called *symmetrized correlation function*,

$$K_F(\tau) \equiv \frac{K_F^+(\tau) + K_F^-(\tau)}{2} = \frac{1}{2}\langle\{\hat{F}(\tau), \hat{F}(0)\}\rangle$$

$$= \lim_{\varepsilon\to 0}\sum_{n,n'} W_n |F_{nn'}|^2 \cos\frac{\tilde{E}\tau}{\hbar}e^{-2\varepsilon|\tau|} \tag{7.111}$$

$$\equiv \sum_{n,n'} W_n |F_{nn'}|^2 \cos\frac{\tilde{E}\tau}{\hbar},$$

which is an even function of the time difference τ, looks very similar to the response function (7.108), 'only' with another trigonometric function under the sum, and a constant front factor[39]. This similarity may be used to obtain an *exact* algebraic relation between the Fourier images of these two functions of τ. Indeed, the function (7.111) may be represented as the Fourier transform[40]

$$K_F(\tau) = \int_{-\infty}^{+\infty} S_F(\omega)e^{-i\omega\tau}d\omega = 2\int_0^{+\infty} S_F(\omega)\cos\omega\tau\, d\omega, \tag{7.112}$$

with the reciprocal transform

$$S_F(\omega) = \frac{1}{2\pi}\int_{-\infty}^{+\infty} K_F(\tau)e^{i\omega\tau}d\tau = \frac{1}{\pi}\int_0^{+\infty} K_F(\tau)\cos\omega\tau\, d\tau, \tag{7.113}$$

of the *symmetrized spectral density* of variable F, defined as

$$S_F(\omega)\delta(\omega - \omega') \equiv \frac{1}{2}\langle\hat{F}_\omega\hat{F}_{-\omega'} + \hat{F}_{-\omega'}\hat{F}_\omega\rangle \equiv \frac{1}{2}\langle\{\hat{F}_\omega, \hat{F}_{-\omega'}\}\rangle, \tag{7.114}$$

where the function $\hat{F}_\omega$ (also an operator rather than a *c*-number!) is defined as

$$\hat{F}_\omega \equiv \frac{1}{2\pi}\int_{-\infty}^{+\infty} \hat{F}(t)e^{i\omega t}dt, \qquad \text{so that } \hat{F}(t) = \int_{-\infty}^{+\infty} \hat{F}_\omega e^{-i\omega t}d\omega. \tag{7.115}$$

---

[39] For the heroic reader who has suffered through the calculations up to this point: our conceptual work is done! What remains is just some simple math to bring the relation between Eqs. (7.108) and (7.111) to an explicit form.

[40] Due to their practical importance, and certain mathematical issues of their justification for random functions, Eqs. (7.112) and (7.113) have their own grand name, the *Wiener–Khinchin theorem*, though the math rigor aside, they are just a straightforward corollary of the standard Fourier integral transform (7.115).

The physical meaning of the function $S_F(\omega)$ becomes clear if we write Eq. (7.112) for the particular case $\tau = 0$:

$$K_F(0) \equiv \langle \hat{F}^2 \rangle = \int_{-\infty}^{+\infty} S_F(\omega) d\omega = 2 \int_0^{+\infty} S_F(\omega) d\omega. \quad (7.116)$$

This formula infers that if we pass the function $F(t)$ through a linear filter cutting from its frequency spectrum a narrow band $d\omega$ of real (positive) frequencies, then the variance $\langle F_f^2 \rangle$ of the filtered signal $F_f(t)$ would be equal to $2S_F(\omega)d\omega$—hence the name 'spectral density'[41].

Let us use Eqs. (7.111) and (7.113) to calculate the spectral density of fluctuations $\tilde{F}(t)$ in our model, using the same ε-trick as at the deviation of Eq. (7.108), to quench the upper-limit substitution:

$$S_F(\omega) = \sum_{n,n'} W_n |F_{nn'}|^2 \frac{1}{2\pi} \lim_{\varepsilon \to 0} \int_{-\infty}^{+\infty} \cos \frac{\tilde{E}\tau}{\hbar} e^{-\varepsilon|\tau|} e^{i\omega\tau} d\tau$$

$$\equiv \frac{1}{2\pi} \sum_{n,n'} W_n |F_{nn'}|^2 \lim_{\varepsilon \to 0} \int_0^{+\infty} \left[\exp\left\{ \frac{i\tilde{E}\tau}{\hbar} \right\} + \text{c.c.} \right] e^{-\varepsilon\tau} e^{i\omega\tau} d\tau \quad (7.117)$$

$$= \frac{1}{2\pi} \sum_{n,n'} W_n |F_{nn'}|^2 \lim_{\varepsilon \to 0} \left[\frac{1}{i(\tilde{E}/\hbar + \omega) - \varepsilon} + \frac{1}{i(-\tilde{E}/\hbar + \omega) - \varepsilon} \right].$$

Now it is a convenient time to recall that each of the two summations here is over the eigenenergies of the environment, whose spectrum is virtually continuous because of its large size, so that we may transform each sum into an integral—just as was done in section 6.6:

$$\sum_n \dots \to \int \dots dn = \int \dots \rho(E_n) dE_n, \quad (7.118)$$

where $\rho(E) \equiv dn/dE$ is the density of environment's states at a given energy. This transformation yields

$$S_F(\omega) = \frac{1}{2\pi} \lim_{\varepsilon \to 0} \int dE_n W(E_n)\rho(E_n) \int dE_{n'}\rho(E_{n'})|F_{nn'}|^2$$
$$\times \left[\frac{1}{i(\tilde{E}/\hbar - \omega) - \varepsilon} + \frac{1}{i(-\tilde{E}/\hbar - \omega) - \varepsilon} \right]. \quad (7.119)$$

Since the expression inside the square bracket depends only on a specific linear combination of two energies, namely $\tilde{E} \equiv E_n - E_{n'}$, it is convenient to introduce also another, linearly-independent combination of the energies, for example, the average energy $\bar{E} = (E_n + E_{n'})/2$, so that the state energies may be represented as

---

[41] An alternative popular measure of the spectral density of a process $F(t)$ is $\mathcal{S}_F(\nu) \equiv \langle F_f^2 \rangle/d\nu = 4\pi S_F(\omega)$, where $\nu = \omega/2\pi$ is the 'cyclic' frequency (measured in Hz).

$$E_n = \bar{E} + \frac{\tilde{E}}{2}, \quad E_{n'} = \bar{E} - \frac{\tilde{E}}{2}. \tag{7.120}$$

With this notation, Eq. (7.119) becomes

$$S_F(\omega) = -\frac{\hbar}{2\pi} \lim_{\varepsilon \to 0} \int d\tilde{E} \ |F_{nn'}|^2$$

$$\times \left[\int W\left(\bar{E} + \frac{\tilde{E}}{2}\right) \rho\left(\bar{E} + \frac{\tilde{E}}{2}\right) \rho\left(\bar{E} - \frac{\tilde{E}}{2}\right) \frac{1}{i\,(\tilde{E} - \hbar\omega) - \hbar\varepsilon} \right. \tag{7.121}$$

$$\left. + \int W\left(\bar{E} + \frac{\tilde{E}}{2}\right) \rho\left(\bar{E} + \frac{\tilde{E}}{2}\right) \rho\left(\bar{E} - \frac{\tilde{E}}{2}\right) \frac{1}{i\,(-\tilde{E} - \hbar\omega) - \hbar\varepsilon} \right].$$

Due to the smallness of the parameter $\hbar\varepsilon$ (which should be much less than all genuine energies of the problem, including $k_B T$, $\hbar\omega$, E_n, and $E_{n'}$), each of the internal integrals is dominated by an infinitesimal vicinity of one point, $\tilde{E}_\pm = \pm\hbar\omega$, in which the state densities, the matrix elements, and the Gibbs probabilities do not change considerably, and may be taken out of the integral, which may be then worked out explicitly[42]:

$$S_F(\omega) = -\frac{\hbar}{2\pi} \lim_{\varepsilon \to 0} \int d\bar{E} \rho_+ \rho_- \left[W_+ |F_+|^2 \int_{-\infty}^{+\infty} \frac{d\tilde{E}}{i\,(\tilde{E} - \hbar\omega) - \hbar\varepsilon} \right.$$

$$\left. + W_- |F_-|^2 \int_{-\infty}^{+\infty} \frac{d\tilde{E}}{i\,(-\tilde{E} - \hbar\omega) - \hbar\varepsilon} \right]$$

$$= -\frac{\hbar}{2\pi} \lim_{\varepsilon \to 0} \int d\bar{E} \rho_+ \rho_- \left[W_+ |F_+|^2 \int_{-\infty}^{+\infty} \frac{-i\,(\tilde{E} - \hbar\omega) - \hbar\varepsilon}{(\tilde{E} - \hbar\omega)^2 + (\hbar\varepsilon)^2} d\tilde{E} \right. \tag{7.122}$$

$$\left. + W_- |F_-|^2 \int_{-\infty}^{+\infty} \frac{i\,(\tilde{E} + \hbar\omega) - \hbar\varepsilon}{(\tilde{E} + \hbar\omega)^2 + (\hbar\varepsilon)^2} d\tilde{E} \right]$$

$$= \frac{\hbar}{2} \int \rho_+ \rho_- [W_+ |F_+|^2 + W_- |F_-|^2] \, d\bar{E},$$

where the indices $\pm$ mark the functions' values at the special points $\tilde{E}_\pm = \pm\hbar\omega$, i.e. $E_n = E_{n'} \pm \hbar\omega$. The physics of these points becomes simple if we interpret the state n, the argument of the equilibrium Gibbs distribution function W_n, as the initial state of the environment, and n' as its final state. Then the top-sign point corresponds to

---

[42] Using, e.g. Eq. (A.32a). (The imaginary parts of the integrals vanish, because the integration in infinite limits may be always re-centered to the finite points $\pm\hbar\omega$.) A math-enlightened reader may have noticed that the integrals might be taken without the introduction of small ε, using the Cauchy theorem—see Eq. (A.91).

$E_{n'} = E_n - \hbar\omega$, i.e. to the result of an emission of one energy quantum $\hbar\omega$ of the 'observation' frequency ω by the environment into the system s of our interest, while the bottom-sign point $E_{n'} = E_n + \hbar\omega$, corresponds to the absorption of such a quantum by the environment. As Eq. (7.122) shows, both processes give similar, positive contributions into the force fluctuations.

The situation is different for the Fourier image of the response function $G(\tau)$,[43]

$$\chi(\omega) \equiv \int_0^{+\infty} G(\tau)\, e^{i\omega\tau} d\tau, \tag{7.123}$$

that is usually called either the *generalized susceptibility* or the *response function*—in our case, of the environment. Its physical meaning is that according to Eq. (7.107), the complex function $\chi(\omega) = \chi'(\omega) + i\chi''(\omega)$ relates the Fourier amplitudes of the generalized coordinate and the generalized force[44]:

$$\langle \hat{F}_\omega \rangle = \chi(\omega)\hat{x}_\omega. \tag{7.124}$$

The physics of its imaginary part $\chi''(\omega)$ is especially clear. Indeed, if x_ω represents a sinusoidal classical process, say

$$x(t) = x_0 \cos\omega t \equiv \frac{x_0}{2}e^{-i\omega t} + \frac{x_0}{2}e^{+i\omega t}, \quad \text{i.e. } x_\omega = x_{-\omega} = \frac{x_0}{2}, \tag{7.125}$$

then, in accordance with the correspondence principle, Eq. (7.124) should hold for the c-number complex amplitudes F_ω and x_ω, enabling us to calculate the time dependence of the force as

$$\begin{aligned}
F(t) &= F_\omega e^{-i\omega t} + F_{-\omega}e^{+i\omega t} = \chi(\omega)x_\omega e^{-i\omega t} + \chi(-\omega)x_{-\omega}e^{+i\omega t} \\
&= \frac{x_0}{2}[\chi(\omega)e^{-i\omega t} + \chi^*(\omega)e^{+i\omega t}] \\
&= \frac{x_0}{2}[(\chi' + i\chi'')e^{-i\omega t} + (\chi' - i\chi'')e^{+i\omega t}] \\
&\equiv x_0[\chi'(\omega)\cos\omega t + \chi''(\omega)\sin\omega t].
\end{aligned} \tag{7.126}$$

We see that $\chi''(\omega)$ weighs the force's part (frequently called *quadrature*) that is $\pi/2$-shifted from the coordinate, i.e. is in phase with the velocity, and hence characterizes

---

[43] The integration in Eq. (7.123) may be extended to the whole time axis, $-\infty < \tau < +\infty$, if we complement the definition (7.107) of the function $G(\tau)$ for $\tau > 0$ with its definition as $G(\tau) = 0$ for $\tau < 0$, in correspondence with the causality principle.

[44] In order to prove this relation, it is sufficient to plug expression $\hat{x}_s = \hat{x}_\omega e^{-i\omega t}$, or any sum of such exponents, into Eqs. (7.107) and then use the definition (7.123). This (simple) exercise is highly recommended to the reader.

the time-average power flow from the system into its environment, i.e. the *energy dissipation* rate[45]:

$$\bar{\mathscr{P}} = \overline{- F(t)\dot{x}(t)} = \overline{-x_0[\chi'(\omega)\cos\omega t + \chi''(\omega)\sin\omega t](-\omega x_0 \sin\omega t)}$$

$$= \frac{x_0^2}{2}\omega\chi''(\omega).$$

(7.127)

Let us calculate this function from Eqs. (7.108) and (7.123), just as we have done for the spectral density of fluctuations:

$$\chi''(\omega) = \mathrm{Im}\int_0^{+\infty} G(\tau)\,e^{i\omega\tau}d\tau$$

$$= \frac{2}{\hbar}\sum_{n,n'} W_n\,|F_{nn'}|^2 \lim_{\varepsilon\to 0} \mathrm{Im}\int_0^{+\infty} \frac{1}{2i}\left(\exp\left\{i\frac{\tilde{E}\tau}{\hbar}\right\} - \mathrm{c.c.}\right) e^{i\omega\tau}e^{-\varepsilon\tau}d\tau$$

$$= \sum_{n,n'} W_n\,|F_{nn'}|^2 \lim_{\varepsilon\to 0}\mathrm{Im}\left(\frac{1}{-\tilde{E}-\hbar\omega - i\hbar\varepsilon} - \frac{1}{\tilde{E}-\hbar\omega - i\hbar\varepsilon}\right)$$

$$\equiv \sum_{n,n'} W_n\,|F_{nn'}|^2 \lim_{\varepsilon\to 0}\left(\frac{\hbar\varepsilon}{(\tilde{E}+\hbar\omega)^2 + (\hbar\varepsilon)^2} - \frac{\hbar\varepsilon}{(\tilde{E}-\hbar\omega)^2 + (\hbar\varepsilon)^2}\right).$$

(7.128)

Making the transfer (7.118) from the double sum to the double integral, and then the integration variable transfer (7.120), we get

$$\chi''(\omega) = \lim_{\varepsilon\to 0}\int d\bar{E}\,|F_{nn'}|^2$$

$$\times\left[\int_{-\infty}^{+\infty} W\left(\bar{E} + \frac{\tilde{E}}{2}\right)\rho\left(\bar{E} + \frac{\tilde{E}}{2}\right)\rho\left(\bar{E} - \frac{\tilde{E}}{2}\right)\frac{\hbar\varepsilon}{(\tilde{E}+\hbar\omega)^2 + (\hbar\varepsilon)^2}d\tilde{E}\right.$$

$$\left. - \int_{-\infty}^{+\infty} W\left(\bar{E} + \frac{\tilde{E}}{2}\right)\rho\left(\bar{E} + \frac{\tilde{E}}{2}\right)\rho\left(\bar{E} - \frac{\tilde{E}}{2}\right)\frac{\hbar\varepsilon}{(\tilde{E}-\hbar\omega)^2 + (\hbar\varepsilon)^2}d\tilde{E}\right].$$

(7.129)

Now using the same argument about the smallness of parameter ε as above, we may take the spectral densities, matrix elements of force, and the Gibbs probabilities out of the integrals, and work out the integrals, getting a result very similar to Eq. (7.122):

$$\chi''(\omega) = \pi\int\rho_+\rho_-[W_-\,|F_-|^2 - W_+\,|F_+|^2]\,d\bar{E}\ .$$

(7.130)

---

[45] The sign minus in Eq. (7.127) is due to the fact that according to Eq. (7.90), F is the force exerted *on* our system (*s*) by the environment, so that the force exerted *by* our system on the environment is $-F$. With this sign clarification, the expression $\mathscr{P} = -F\dot{x} = -Fv$ for the instant power flow is evident if x is the usual Cartesian coordinate of a 1D particle. However, according to analytical mechanics (see, e.g. *Part CM* chapters 2 and 10), it is also valid for any [generalized coordinate–generalized force] pair which forms the interaction Hamiltonian (7.90).

In order to relate these two results, it is sufficient to notice that according to Eq. (7.24), the Gibbs probabilities $W_\pm$ are related by a coefficient depending on only the temperature T and observation frequency ω:

$$
\begin{aligned}
W_\pm &\equiv W\left(\bar{E} + \frac{\tilde{E}_\pm}{2}\right) \\
&\equiv W\left(\bar{E} \pm \frac{\hbar\omega}{2}\right) = \frac{1}{Z}\exp\left\{-\frac{\bar{E} \pm \hbar\omega/2}{k_BT}\right\} = W(\bar{E})\exp\left\{\mp\frac{\hbar\omega}{2k_BT}\right\},
\end{aligned}
\tag{7.131}
$$

so that both the spectral density (7.122) and the dissipative part (7.130) of the generalized susceptibility may be expressed via the same integral over environment energies:

$$
S_F(\omega) = \hbar\cosh\left(\frac{\hbar\omega}{2k_BT}\right)\int \rho_+\rho_- W(\bar{E})\,[\,|F_+|^2 + |F_-|^2\,]\,d\bar{E},
\tag{7.132}
$$

$$
\chi''(\omega) = 2\pi\sinh\left(\frac{\hbar\omega}{2k_BT}\right)\int \rho_+\rho_- W(\bar{E})\,[\,|F_+|^2 + |F_-|^2\,]\,d\bar{E},
\tag{7.133}
$$

and hence are universally related as

$$
S_F(\omega) = \frac{\hbar}{2\pi}\chi''(\omega)\coth\frac{\hbar\omega}{2k_BT}.
\tag{7.134}
$$

This is, finally, the famous Callen–Welton's fluctuation–dissipation theorem (FDT). It reveals a fundamental, intimate relation between these two effects of the environment ('no dissipation without fluctuation')—hence the name. A curious feature of the FDT is that Eq. (7.134) includes exactly the same function of temperature as the average energy (7.26) of a quantum oscillator of frequency ω, though, as the reader could witness, the notion of the oscillator was by no means used in its derivation. As we will see in the next section, this fact leads to rather interesting consequences and even conceptual opportunities.

In the classical limit, $\hbar\omega \ll k_BT$, the FDT is reduced to

$$
S_F(\omega) = \frac{\hbar}{2\pi}\chi''(\omega)\frac{2k_BT}{\hbar\omega} = \frac{k_BT}{\pi}\frac{\operatorname{Im}\chi(\omega)}{\omega}.
\tag{7.135}
$$

In most systems of interest, the last fraction tends to a finite (positive) constant in a substantial range of relatively low frequencies. Indeed, expanding the right-hand side of Eq. (7.123) into the Taylor series in small ω, we get

$$
\chi(\omega) = \chi(0) + i\omega\eta + \dots, \quad \text{with}
$$

$$
\chi(0) = \int_0^\infty G(\tau)\,d\tau, \quad \text{and} \quad \eta \equiv \int_0^\infty G(\tau)\,\tau d\tau.
\tag{7.136}
$$

Since the temporal Green's function G is real by definition, the Taylor expansion of $\chi''(\omega) \equiv \operatorname{Im}\chi(\omega)$ starts with the linear term $\omega\eta$, where η is a certain real coefficient, and unless $\eta = 0$, is dominated by this term at small ω. The physical sense of the constant η becomes clear if we consider an environment that provides friction described by a simple, well-known kinematic friction law

$$\langle \hat{F} \rangle = -\eta \, \hat{x}, \qquad \text{with } \eta \geqslant 0, \tag{7.137}$$

where η is called the *drag coefficient*. For the Fourier images of coordinate and force this gives the relation $F_\omega = i\omega\eta x_\omega$, so that according to Eq. (7.124),

$$\chi(\omega) = i\omega\eta, \quad \text{i.e.} \quad \frac{\chi''(\omega)}{\omega} \equiv \frac{\text{Im } \chi(\omega)}{\omega} = \eta \geqslant 0. \tag{7.138}$$

Within this approximation, and in the classical limit, the FDT (7.134) is reduced to the well-known *Nyquist formula*[46]:

$$S_F(\omega) = \frac{k_{\mathrm{B}}T}{\pi}\eta, \qquad \text{i.e.} \quad \langle F_{\mathrm{f}}^2 \rangle = 4k_{\mathrm{B}}T\eta d\upsilon. \tag{7.139}$$

According to Eq. (7.112), if such a constant spectral density[47] persisted at all frequencies, it would correspond to a delta-correlated process $F(t)$, with

$$K_F(\tau) = 2\pi \, S_F(0)\delta(\tau) = 2k_{\mathrm{B}}T\eta\delta(\tau), \tag{7.140}$$

cf. Eqs. (7.82) and (7.83). Since in the classical limit the right-hand side of Eq. (7.109) is negligible, and the correlation function may be considered an even function of time, the symmetrized function under the integral in Eq. (7.113) may be rewritten just as $\langle F(\tau)F(0) \rangle$. In the limit of relatively low observation frequencies (in the sense that ω is much smaller than not only the quantum frontier $k_{\mathrm{B}}T/\hbar$, but also the frequency scale of the function $\chi''(\omega)/\omega$), Eq. (7.138) may be used to recast Eq. (7.135) in the form[48]

$$\eta \equiv \lim_{\omega \to 0}\frac{\chi''(\omega)}{\omega} = \frac{1}{k_{\mathrm{B}}T}\int_0^\infty \langle F(\tau)F(0) \rangle d\tau. \tag{7.141}$$

---

[46] Actually, the 1928 work by H Nyquist was about the electronic noise in resistors, just discovered experimentally by his Bell Labs colleague J Johnson. For an Ohmic resistor, as the dissipative 'environment' of the electric circuit it is connected with, Eq. (7.137) is just the Ohm's law, and may be recast as either $\langle V \rangle = -R(dQ/dt) = RI$, or $\langle I \rangle = -G(d\Phi/dt) = GV$. Thus for the voltage V across an open circuit, η corresponds to its resistance R, and for current I in a short circuit, to its conductance $G = 1/R$. In this case, the fluctuations described by Eq. (7.139) are referred to as the *Johnson–Nyquist noise*. (Because of this important application, any model leading to Eq. (7.138) is commonly referred to as the *Ohmic dissipation*, even if the physical nature of the variables x and F is quite different.)

[47] A random process whose spectral density may be reasonably approximated by a constant is frequently called the *white noise*, because it is a random mixture of all possible sinusoidal components with equal weights, recalling natural white light's composition.

[48] Note that in some fields (especially in physical kinetics and chemical physics), this particular limit of the Nyquist formula is called the Green–Kubo (or just 'Kubo') formula. However, in the view of the FDT development history, discussed above, it is much more reasonable to associate these names with Eq. (7.109)—as is done in most fields of physics.

To conclude this section, let me return for a minute to the questions formulated in our earlier discussion of dephasing in the two-level model. In that problem, the dephasing time scale is $T_2 = 1/2D_\varphi$. Hence the classical approach to the dephasing, used in section 7.3, is adequate if $\hbar D_\varphi \ll k_B T$. Next, we may identify the operators $\hat{f}$ and $\hat{\sigma}_z$ participating in Eq. (7.70) with, respectively, the operators $-\hat{F}$ and $\hat{x}$ of the general Eq. (7.90). Then the comparison of Eqs. (7.82), (7.89) and (7.140) yields

$$\frac{1}{T_2} \equiv 2D_\varphi = \frac{4k_B T}{\hbar^2}\eta, \tag{7.142}$$

so that, for the model described by Eq. (7.137) with a temperature-independent drag coefficient η, the rate of dephasing by a classical environment is proportional to its temperature.

7.5 The Heisenberg–Langevin approach

The fluctuation–dissipation theorem opens a very simple and efficient, though limited way for the analysis of the system of interest (s in figure 7.1). It is to write its Heisenberg equations (4.199) of motion of the relevant operators, which would now include the environmental force operator, and explore these equations using the Fourier transform and the Wiener–Khinchin theorem (7.112)–(7.113). This approach to classical equations of motion is commonly associated with the name of Langevin[49], so that its extension to dynamics of Heisenberg-picture operators is frequently referred to as the *Heisenberg–Langevin* (or 'quantum Langevin', or 'Langevin–Lax'[50]) *approach* to open system analysis.

Perhaps the best way to describe this method is to demonstrate how it works for the very important case of a 1D harmonic oscillator, so that the generalized coordinate x of section 7.4 is just the oscillator's coordinate. For the sake of simplicity, let us assume that the environment provides the simple Ohmic dissipation described by Eq. (7.137)—which is a good approximation in many cases. As we already know from chapter 5, the Heisenberg equations of motion for operators of coordinate and momentum of the oscillator, in the presence of an external force $F(t)$, are

$$\dot{\hat{x}} = \frac{\hat{p}}{m}, \qquad \dot{\hat{p}} = -m\omega_0^2 \hat{x} + \hat{F}, \tag{7.143}$$

so that using Eqs. (7.92) and (7.137), we get

$$\dot{\hat{x}} = \frac{\hat{p}}{m}, \qquad \dot{\hat{p}} = -m\omega_0^2 \hat{x} - \eta\dot{\hat{x}} + \hat{F}(t). \tag{7.144}$$

---

[49] A 1908 work by P Langevin was the first systematic development of the Einstein's ideas (1905) on the Brownian motion, using the random force language, as an alternative to M Smoluchowski's approach using the probability density language—see section 7.6 below.

[50] Indeed, perhaps the largest credit for the extension of the Langevin approach to quantum systems belongs to M Lax, whose work in the early 1960s was motivated mostly by quantum electronics applications—see, e.g. his monograph M Lax, *Fluctuation and Coherent Phenomena in Classical and Quantum Physics*, Gordon and Breach, 1968, and references therein.

Combining Eqs. (7.144), we may write their system as a single differential equation

$$m\ddot{\hat{x}} + \eta\dot{\hat{x}} + m\omega_0^2\hat{x} = \hat{F}(t), \qquad (7.145)$$

which is absolutely similar to the well-known classical equation of motion of a damped oscillator under the effect of an external force. In the view of Eqs. (5.29) and (5.35), whose corollary the Ehrenfest theorem (5.36) is, this may look not surprising, but please note again that the approach discussed in the previous section justifies a qualitative description of the drag force in quantum mechanics—necessarily in parallel with the accompanying fluctuation force.

For the Fourier images of the operators, defined similarly to Eq. (7.115), Eq. (7.145) gives the following relation,

$$\hat{x}_\omega = \frac{\hat{F}_\omega}{m(\omega_0^2 - \omega^2) - i\eta\omega}, \qquad (7.146)$$

which should be also well known to the reader from the classical theory of forced oscillations[51]. However, since these Fourier components are still Heisenberg-picture operators, and their 'values' for different ω generally do not commute, we have to tread carefully. The best way to proceed is to write a copy of Eq. (7.146) for the frequency $(-\omega')$, and then combine these equations to form a symmetrical combination similar that used in Eq. (7.114). The result is

$$\frac{1}{2}\langle\hat{x}_\omega\hat{x}_{-\omega'} + \hat{x}_{-\omega'}\hat{x}_\omega\rangle = \frac{1}{\left|m\left(\omega_0^2 - \omega^2\right) - i\eta\omega\right|^2}\frac{1}{2}\langle\hat{F}_\omega\hat{F}_{-\omega'} + \hat{F}_{-\omega'}\hat{F}_\omega\rangle. \qquad (7.147)$$

Since the spectral density definition similar to Eq. (7.114) is valid for any observable, in particular for x, Eq. (7.147) allows us to relate the symmetrized spectral densities of coordinate and force:

$$S_x(\omega) = \frac{S_F(\omega)}{\left|m\left(\omega_0^2 - \omega^2\right) - i\eta\omega\right|^2} = \frac{S_F(\omega)}{m^2\left(\omega_0^2 - \omega^2\right)^2 + (\eta\omega)^2}. \qquad (7.148)$$

Now using an analog of Eq. (7.116) for x, we can calculate the coordinate's variance:

$$\langle x^2\rangle = K_x(0) = \int_{-\infty}^{+\infty} S_x(\omega)d\omega = 2\int_0^{+\infty} \frac{S_F(\omega)d\omega}{m^2\left(\omega_0^2 - \omega^2\right)^2 + (\eta\omega)^2}, \qquad (7.149)$$

where now, in contrast to the notation used in section 7.4, the sign $\langle...\rangle$ means the averaging over the usual statistical ensemble of many systems of interest—in our current case, of many harmonic oscillators.

If the coupling to the environment is so weak that drag coefficient η is small (in the sense that the oscillator's dimensionless Q-factor is large, $Q \equiv m\omega_0/\eta \gg 1$), this integral

---

[51] If necessary, see *Part CM* section 5.1.

is dominated by the resonance peak in a narrow vicinity, $|\omega - \omega_0| \equiv |\xi| \ll \omega_0$, of its resonance frequency, and we can take the relatively smooth function $S_F(\omega)$ out of the integral, thus reducing it to a table form[52]:

$$
\begin{aligned}
\langle x^2 \rangle &\approx 2S_F(\omega_0) \int_0^{+\infty} \frac{d\omega}{m^2 \left(\omega_0^2 - \omega^2 \right)^2 + (\eta\omega)^2} \\
&\approx 2S_F(\omega_0) \int_{-\infty}^{+\infty} \frac{d\xi}{(2m\omega_0\xi)^2 + (\eta\omega_0)^2} \\
&= 2S_F(\omega_0) \frac{1}{(\eta\omega_0)^2} \int_{-\infty}^{+\infty} \frac{d\xi}{(2m\xi/\eta)^2 + 1} \\
&= 2S_F(\omega_0) \frac{1}{(\eta\omega_0)^2} \frac{\pi\eta}{2m} = \frac{\pi S_F(\omega_0)}{\eta m \omega_0^2}.
\end{aligned}
\tag{7.150}
$$

With the account of the FDT (7.134) and of Eq. (7.138), this gives[53]

$$
\langle x^2 \rangle = \frac{\pi}{\eta m \omega_0^2} \frac{\hbar}{2\pi} \eta\omega_0 \coth\frac{\hbar\omega_0}{2k_BT} = \frac{\hbar}{2m\omega_0} \coth\frac{\hbar\omega_0}{2k_BT}.
\tag{7.151}
$$

But this is exactly Eq. (7.48), which was derived in section 7.2 from the Gibbs distribution, without any explicit account of the environment—though keeping it in mind by using the notion of the thermally-equilibrium ensemble[54]. (Notice that the drag coefficient η, which characterizes the oscillator-to-environment interaction strength, has cancelled!) Does this mean that in section 7.4 we toiled in vain?

By no means. First of all, the result (7.150), augmented by the FDT (7.134), has an important conceptual value. For example, let us consider the low-temperature limit $k_BT \ll \hbar\omega_0$, when Eq. (7.151) is reduced to

$$
\langle x^2 \rangle = \frac{\hbar}{2m\omega_0} \equiv \frac{x_0^2}{2}.
\tag{7.152}
$$

Let us ask a naïve question: What exactly is the origin of this coordinate's uncertainty? From the point of view of the usual quantum mechanics of absolutely closed (Hamiltonian) systems, there is no doubt: this nonvanishing variance of the coordinate is the result of the final spatial extension of the ground-state wave-function, reflecting the Heisenberg's uncertainty relation (which in turn results from the fact that the operators of coordinate and momentum do not commute)—see either Eq. (2.275), or Eq. (5.95) with $n = 0$. However, from the point of view of the Heisenberg–Langevin equation (7.145), the variance (7.152) is an unalienable part of

---

[52] See, e.g. Eq. (A.32a).

[53] Note that this calculation remains correct even if the dissipation's dispersion deviates from the Ohmic model (7.138), provided if the drag coefficient η is replaced with its effective value $\text{Im}\chi(\omega_0)/\omega_0$, because the effects of the environment are only felt, by the oscillator, at its oscillation frequency.

[54] By the way, the simplest way to calculate $S_F(\omega)$, i.e. to derive the FDT, is to *require* that Eqs. (7.48) and (7.150) give the same result for an oscillator with any eigenfrequency ω. This is exactly the approach used by H Nyquist (for the classical case)—see also *Part SM* section 5.5.

the oscillator's response to the fluctuation force $\tilde{F}(t)$ *exerted by the environment* at frequencies $\omega \approx \omega_0$. Though it is impossible to refute the former, absolutely legitimate point of view, in many applications it is much easier to subscribe to the latter standpoint, and treat the coordinate's uncertainty as the result of the so-called *quantum noise* of the environment—which tends to the fixed value (7.152) when the coupling to the environment tends to zero. This notion has received numerous confirmations in experiments that did not include *any* oscillators with eigenfrequencies ω_0 close to the noise measurement frequency ω.[55]

The second advantage of the Heisenberg–Langevin approach is that it is possible to use Eq. (7.148) to calculate the (experimentally measurable!) distribution $S_x(\omega)$, i.e. decompose the fluctuations into their spectral components. This procedure is not restricted to the limit of small values of η (i.e. to large Q-factors); for any damping we may just plug the FDT (7.134) into Eq. (7.148). As an example, let us have a look at the so-called *quantum diffusion*. A free 1D particle, moving in a medium with the Ohmic damping (7.137), may be considered as the particular case of a 1D harmonic oscillator (7.145) with $\omega_0 = 0$, so that combining Eqs. (7.134) and (7.149), we get

$$
\begin{aligned}
\langle x^2 \rangle &= 2 \int_0^{+\infty} \frac{S_F(\omega)d\omega}{(m\omega^2)^2 + (\eta\omega)^2} \\
&= 2\eta \int_0^{+\infty} \frac{1}{(m\omega^2)^2 + (\eta\omega)^2} \frac{\hbar\omega}{2\pi} \coth\frac{\hbar\omega}{2k_{\mathrm{B}}T} d\omega.
\end{aligned}
\tag{7.153}
$$

This integral has two divergences. The first one, of the type $\int d\omega/\omega^2$ at the lower limit, is just a classical effect: according to Eq. (7.85), the particle's displacement variance grows with time, so it cannot have a finite time-independent value that Eq. (7.153) tries to calculate. However, we still can use that result to single out the quantum effects on diffusion—say, by comparing it with a similar but purely classical case. These effects are prominent at high frequencies, especially if the quantum noise overcomes the thermal noise before the dynamic cut-off, i.e. if

$$
\frac{k_{\mathrm{B}}T}{\hbar} \ll \frac{\eta}{m}.
\tag{7.154}
$$

In this case there is a broad range of frequencies where the quantum noise gives a substantial contribution to the integral:

$$
\langle x^2 \rangle_Q \approx 2\eta \int_{k_{\mathrm{B}}T/\hbar}^{\eta/m} \frac{1}{(\eta\omega)^2} \frac{\hbar\omega}{2\pi} d\omega \equiv \frac{\hbar}{\pi\eta} \int_{k_{\mathrm{B}}T/\hbar}^{\eta/m} \frac{d\omega}{\omega} = \frac{\hbar}{\pi\eta} \ln\frac{\hbar\eta}{mk_{\mathrm{B}}T} \sim \frac{\hbar}{\eta}.
\tag{7.155}
$$

Formally, this contribution diverges at either $m \to 0$ or $T \to 0$, but this logarithmic (i.e. extremely weak) divergence is readily quenched by almost any change of the environment model at very high frequencies, where the 'Ohmic' approximation given by Eq. (7.136) becomes unrealistic.

---

[55] See, for example, [3].

The Heisenberg–Langevin approach is very powerful, because its straightforward generalizations enable analyses of fluctuations in virtually arbitrary linear systems, i.e. the systems described by linear differential (or integro-differential) equations of motion, including those with many degrees of freedom, and distributed systems (*continua*), and such systems prevail in many fields of physics. However, this approach also has its limitations. The main one of them is that if the equations of motion of the Heisenberg operators are *not* linear, there is no linear relation, such as Eq. (7.146), between the Fourier images of the generalized forces and the generalized coordinates, and as the result there is no simple relation, such as Eq. (7.148), between their spectral densities. In other words, if the Heisenberg equations of motion are nonlinear, there is no regular simple way to use them to calculate statistical properties of the observables.

For example, let us return to the dephasing problem described by Eqs. (7.68)–(7.70), and assume that the deterministic and the fluctuating parts of the effective force $-f$, exerted by the environment, are characterized by relations similar, respectively, to Eqs. (7.124) and (7.134). Now writing the Heisenberg equations of motion for the two remaining spin operators, and using the commutation relations between them, we get

$$
\begin{aligned}
\dot{\hat{\sigma}}_x &= \frac{1}{i\hbar}[\hat{\sigma}_x, \hat{H}] = \frac{1}{i\hbar}[\hat{\sigma}_x, (c_z + \hat{f})\hat{\sigma}_z] = -\frac{2}{\hbar}\hat{\sigma}_y(c_z + \hat{f}) \\
&= -\frac{2}{\hbar}\hat{\sigma}_y(c_z + \eta\hat{\sigma}_z + \hat{f}),
\end{aligned}
\tag{7.156}
$$

and a similar equation for $\hat{\sigma}_y$. Such nonlinear equations cannot be used to calculate the statistical properties of the Pauli operators exactly—at least analytically.

For some calculations, this problem may be circumvented by *linearization*: if we are only interested in small fluctuations of the observables, their nonlinear Heisenberg equations of motion, such as Eq. (7.156), may be linearized with respect to small deviations of the operators about their (generally, time-dependent) deterministic 'values', and the resulting linear equations for the operator variations solved either as has been demonstrated above, or (if the deterministic 'values' evolve in time) using their Fourier expansions. Sometimes such an approach gives relatively simple and important results[56], but for many other problems, this approach is insufficient.

7.6 Density matrix approach

The main alternative approach to the dynamics of open quantum systems, which is essentially a generalization of the one discussed in section 7.2, is to extract the final results from the dynamics of the density operator of our subsystem s of interest. Let us discuss this approach in detail[57].

We already know that the density matrix allows the calculation of the expectation value of any observable of system s—see Eq. (7.5). However, our initial recipe (7.6)

---

[56] For example, the formula used for processing of the experimental results by R Koch *et al* (mentioned above), had been derived in this way. (This derivation is suggested to the reader as an exercise.)

[57] As in section 7.4, the reader not interested in the derivation of the basic Eq. (7.181) of the density matrix evolution may immediately jump to the discussion of this equation and its applications.

for the density matrix element calculation, which requires the knowledge of the exact state (7.2) of the whole Universe, is not too practicable, while the von Neumann equation (7.66) for the density matrix evolution is limited to cases in which probabilities W_j of the system states are fixed—thus excluding such important effects as the energy relaxation. However, such effects may be analyzed using a different assumption—that the system of interest interacts only with a local environment that is very close to its thermally-equilibrium state, described, in the stationary-state basis, by a diagonal density matrix with the elements (7.24).

This calculation is facilitated by the following general observation. Let us number the basis states of the full local system (the system of our interest plus its local environment) by l, and use Eq. (7.5) to write

$$\langle A \rangle = \text{Tr}(\hat{A}\hat{w}_l) \equiv \sum_{l,l'} A_{ll'} w_{l'l} = \sum_{l,l'} \langle l|\hat{A}|l'\rangle \langle l'|\hat{w}_l|l\rangle, \tag{7.157}$$

where $\hat{w}_l$ is the density operator of this full local system. At a weak interaction between the system s and the local environment e, their states reside in different Hilbert spaces, so that we can write

$$|l\rangle = |s_j\rangle \otimes |e_k\rangle. \tag{7.158}$$

and if the observable A depends only on the coordinates of the system s of our interest, we may reduce Eq. (7.157) to Eq. (7.5):

$$\langle A \rangle = \sum_{j,j';k,k'} \langle e_k| \otimes \langle s_j|\hat{A}|s_{j'}\rangle \otimes |e_{k'}\rangle\langle e_{k'}| \otimes \langle s_{j'}|\hat{w}_l|s_j\rangle \otimes |e_k\rangle$$

$$= \sum_{j,j'} A_{jj'} \langle s_{j'}| \left(\sum_k \otimes \langle e_k|\hat{w}_l|e_k\rangle \otimes \right) |s_j\rangle = \text{Tr}_j(\hat{A}\hat{w}), \tag{7.159}$$

where

$$\hat{w} \equiv \sum_k \langle e_k|\hat{w}_l|e_k\rangle = \text{Tr}_k \hat{w}_l, \tag{7.160}$$

showing how exactly the density operator $\hat{w}$ of the system s may be calculated from $\hat{w}_l$.

Now comes the key physical assumption of this approach: since we may select the local environment e to be much larger than the system s of our interest, we may consider the composite system $(s + e)$ as Hamiltonian, with time-independent probabilities of its stationary states, so that for the description of time evolution of *its* full density operator $\hat{w}_l$ (again, in contrast to that, $\hat{w}$, of the system of our interest) we *may* use the von Neumann equation (7.66). Partitioning its right-hand side in accordance with Eq. (7.68), we get:

$$i\hbar \dot{\hat{w}}_l = [\hat{H}_s, \hat{w}_l] + [\hat{H}_e, \hat{w}_l] + [\hat{H}_{\text{int}}, \hat{w}_l]. \tag{7.161}$$

The next step is to use the perturbation theory to solve this equation in the lowest order in $\hat{H}_{\text{int}}$, that would yield, for the evolution of w, a nonvanishing contribution due to the interaction. For that, Eq. (7.161) is not very convenient, because its right-hand side contains two other terms, of a much larger scale than the interaction Hamiltonian. To mitigate this technical difficulty, the interaction picture that was discussed in the end of section 4.6, is very natural. (It is not necessary though, and I will use this picture mostly as an exercise of its application—unfortunately, the only example I can afford in this course.)

As a reminder, in that picture (whose entities will be marked with index I, with the unmarked operators assumed to be in the Schrödinger picture), both the operators *and* the state vectors (and hence the density operator) depend on time. However, the time evolution of the operator of any observable A is described by an equation similar to Eq. (7.67), but with the *unperturbed* part of the Hamiltonian only—see Eq. (4.214). In the model (7.68), this means

$$i\hbar \dot{\hat{A}}_I = [\hat{A}_I, \hat{H}_0].\tag{7.162}$$

where the unperturbed Hamiltonian consists of two parts defined in different Hilbert spaces:

$$\hat{H}_0 \equiv \hat{H}_s + \hat{H}_e.\tag{7.163}$$

On the other hand, the state vector's dynamics is governed by the interaction evolution operator $\hat{u}_I$ that obeys Eqs. (4.215). Since this equation, using the interaction-picture Hamiltonian (4.216),

$$\hat{H}_I \equiv \hat{u}_0^{\dagger} \hat{H}_{\text{int}} \hat{u}_0,\tag{7.164}$$

is absolutely similar to the ordinary Schrödinger equation using the full Hamiltonian, we may repeat all arguments given in the beginning of section 7.3 to prove that the dynamics of the density operator in the interaction picture of a Hamiltonian system is governed by the following analog of the von Neumann equation (7.66):

$$i\hbar \dot{\hat{w}}_I = [\hat{H}_I, \hat{w}_I],\tag{7.165}$$

where the index I is dropped for the notation simplicity. Since this equation is similar in structure (with the opposite sign) to the Heisenberg equation (7.67), we may use the solution Eq. (4.190) of the latter equation to write its analog:

$$\hat{w}_I(t) = \hat{u}_I(t, 0)\hat{w}_I(0)\hat{u}_I^{\dagger}(t. 0).\tag{7.166}$$

It is also straightforward to verify that in this picture, the expectation value of any observable A may be found from an expression similar to the basic Eq. (7.5):

$$\langle A \rangle = \text{Tr}(\hat{A}_I \hat{w}_I),\tag{7.167}$$

showing again that the interaction and Schrödinger pictures give the same final results.

In the most frequent case of the factorable interaction (7.90),[58] Eq. (7.162) is simplified for both operators participating in that product—for each one in its own way. In particular, for $\hat{A} = \hat{x}$, it yields

$$i\hbar\dot{\hat{x}}_I = [\hat{x}_I, \hat{H}_0] \equiv [\hat{x}_I, \hat{H}_s] + [\hat{x}_I, \hat{H}_e]. \tag{7.168}$$

Since the coordinate operator is defined in the Hilbert space of the system s, it commutes with the Hamiltonian of the environment, so that we finally get

$$i\hbar\dot{\hat{x}}_I = [\hat{x}_I, \hat{H}_s]. \tag{7.169}$$

On the other hand, if $\hat{A} = \hat{F}$, this operator is defined in the Hilbert space of the environment, and commutes with the Hamiltonian of the unperturbed system s. As a result, we get

$$i\hbar\dot{\hat{F}}_I = [\hat{F}_I, \hat{H}_e]. \tag{7.170}$$

This means that with our time-independent unperturbed Hamiltonians, $\hat{H}_s$ and $\hat{H}_e$, the time evolution of the interaction-picture operators is rather simple. In particular, the analogy between Eqs. (7.170) and (7.93) allows us to immediately write the following analog of Eq. (7.94):

$$\hat{F}_I(t) = \exp\left\{\frac{i}{\hbar}\hat{H}_e t\right\}\hat{F}(0)\exp\left\{-\frac{i}{\hbar}\hat{H}_e t\right\}, \tag{7.171}$$

so that in the stationary-state basis n of the environment,

$$
\begin{aligned}
(\hat{F}_I)_{nn'}(t) &= \exp\left\{\frac{i}{\hbar}E_n t\right\}F_{nn'}(0)\exp\left\{-\frac{i}{\hbar}E_{n'}t\right\} \\
&\equiv F_{nn'}(0)\exp\left\{-i\frac{E_n - E_{n'}}{\hbar}t\right\},
\end{aligned}
\tag{7.172}
$$

and similarly (but in the basis of the eigenstates of the system s) for operator $\hat{x}$. As a result, the right-hand side of Eq. (7.164) may be also factored:

$$
\begin{aligned}
\hat{H}_I(t) &\equiv \hat{u}_0^\dagger(t, 0)\hat{H}_{\text{int}}\hat{u}_0(t, 0) \\
&= \exp\left\{\frac{i}{\hbar}(\hat{H}_s + \hat{H}_e)\,t\right\}(-\hat{x}\hat{F})\exp\left\{-\frac{i}{\hbar}(\hat{H}_s + \hat{H}_e)\,t\right\} \\
&= -\left(\exp\left\{\frac{i}{\hbar}\hat{H}_s t\right\}\hat{x}\exp\left\{-\frac{i}{\hbar}\hat{H}_s t\right\}\right) \\
&\quad \times\left(\exp\left\{\frac{i}{\hbar}\hat{H}_e t\right\}\hat{F}(0)\exp\left\{-\frac{i}{\hbar}\hat{H}_e t\right\}\right) \\
&\equiv -\hat{x}_I(t)\hat{F}_I(t).
\end{aligned}
\tag{7.173}
$$

---

[58] A similar analysis of a more general case, when the interaction with environment has to be represented as a sum of several products of the type (7.90), may be found, for example, in the monograph by K Blum [4].

So, the transfer to the interaction picture has taken some time, but now it enables a smooth ride[59]. Indeed, just as in section 7.4, we may rewrite Eq. (7.165) in the integral form:

$$\hat{w}_I(t) = \frac{1}{i\hbar} \int_{-\infty}^{t} [\hat{H}_I(t'), \hat{w}_I(t')] \, dt';$$

(7.174)

plugging this result, for time t', into the right-hand side of Eq. (7.174) again, we get

$$
\begin{aligned}
\hat{w}_I(t) &= -\frac{1}{\hbar^2} \int_{-\infty}^{t} [\hat{H}_I(t), [\hat{H}_I(t'), \hat{w}_I(t')]] \, dt' \\
&= -\frac{1}{\hbar^2} \int_{-\infty}^{t} [\hat{x}(t)\hat{F}(t), [\hat{x}(t')\hat{F}(t'), \hat{w}_I(t')]] \, dt',
\end{aligned}
$$

(7.175)

where, for the notation's brevity, from this point on I will strip the operators $\hat{x}$ and $\hat{F}$ of their index I. (Their time dependence indicates the interaction picture clearly enough.)

So far, this equation is exact (and cannot be solved analytically), but this is a good time to notice that even if we take the density operator on its right-hand side equal to its unperturbed, factorable 'value' (corresponding to no interaction between the system s and its thermally-equilibrium environment e)[60],

$$\hat{w}_I(t') \rightarrow \hat{w}(t') \, \hat{w}_e, \qquad \text{with } \langle e_n | \hat{w}_e | e_{n'} \rangle = W_n \delta_{nn'},$$

(7.176)

where e_n are the stationary states of the environment, and W_n are the Gibbs probabilities (7.24), Eq. (7.175) still describes a nontrivial time evolution of the density operator. This is exactly the first nonvanishing approximation (in the weak interaction) we are looking for. Now using Eq. (7.160), we find the equation of evolution of the density operator of the system of our interest:

$$\hat{w}(t) = -\frac{1}{\hbar^2} \int_{-\infty}^{t} \text{Tr}_n [\hat{x}(t)\hat{F}(t), [\hat{x}(t')\hat{F}(t'), \hat{w}(t')\hat{w}_e]] \, dt',$$

(7.177)

where the trace is over the stationary states of the environment. In order to spell out the right-hand side of Eq. (7.177), note again that the coordinate and force operators commute with each other (but not with themselves at different time moments!) and hence may be swapped at will, so that we may write

---

[59] If we used either the Schrödinger or the Heisenberg picture instead, the forthcoming Eq. (7.175) would pick up a rather annoying multitude of fast-oscillating exponents, of different time arguments, on its right-hand side.

[60] For the notation simplicity, the fact that here (and in all following formulas) the density operator $\hat{w}$ of the system s of our interest is taken in the interaction picture, is just implied.

$$\mathrm{Tr}_n[\ldots, [\ldots, \ldots]] = \hat{x}(t)\hat{x}(t')\hat{w}(t')\mathrm{Tr}_n[\hat{F}(t)\hat{F}(t')\hat{w}_e]$$
$$- \hat{x}(t)\hat{w}(t')\hat{x}(t')\mathrm{Tr}_n[\hat{F}(t)\hat{w}_e\hat{F}(t')]$$
$$- \hat{x}(t')\hat{w}(t')\hat{x}(t)\mathrm{Tr}_n[\hat{F}(t')\hat{w}_e\hat{F}(t)]$$
$$+ \hat{w}(t')\hat{x}(t')\hat{x}(t)\mathrm{Tr}_n[\hat{w}_e\hat{F}(t')\hat{F}(t)]$$
$$= \hat{x}(t)\hat{x}(t')\hat{w}(t')\sum_{n,n'}F_{nn'}(t)F_{n'n}(t')W_n$$

$$- \hat{x}(t)\hat{w}(t')\hat{x}(t')\sum_{n,n'}F_{nn'}(t)W_{n'}F_{n'n}(t')$$

$$- \hat{x}(t')\hat{w}(t')\hat{x}(t)\sum_{n,n'}F_{nn'}(t')W_{n'}F_{n'n}(t)$$

$$+ \hat{w}(t')\hat{x}(t')\hat{x}(t)\sum_{n,n'}W_nF_{nn'}(t')F_{n'n}(t). \tag{7.178}$$

Since the summation on both indices n and n' in this expression is over the same energy level set (of all eigenstates of the environment), we may swap these indices in any of the sums. Doing this only in the terms including the factors $W_{n'}$, we turn them into W_n, so that this factor becomes common:

$$\mathrm{Tr}_n[\ldots, [\ldots, \ldots]] = \sum_{n,n'}W_n[\hat{x}(t)\hat{x}(t')\hat{w}(t')F_{nn'}(t)F_{n'n}(t')$$
$$- \hat{x}(t)\hat{w}(t')\hat{x}(t')F_{n'n}(t)F_{nn'}(t')$$
$$- \hat{x}(t')\hat{w}\hat{x}(t)F_{n'n}(t')F_{nn'}(t) + \hat{w}\hat{x}(t')\hat{x}(t)F_{nn'}(t')F_{n'n}(t)]. \tag{7.179}$$

Now using Eq. (7.172), we get

$$\mathrm{Tr}_n[\ldots, [\ldots, \ldots]] = \sum_{n,n'}W_n\,|F_{nn'}|^2$$

$$\times \left[\begin{matrix} \hat{x}(t)\hat{x}(t')\hat{w}(t')\exp\left\{\dfrac{i\tilde{E}(t-t')}{\hbar}\right\} - \hat{x}(t)\hat{w}(t')\hat{x}(t')\exp\left\{-\dfrac{i\tilde{E}(t-t')}{\hbar}\right\} \\ -\hat{x}(t')\hat{w}(t')\hat{x}(t)\exp\left\{\dfrac{i\tilde{E}(t-t')}{\hbar}\right\} + \hat{w}(t')\hat{x}(t')\hat{x}(t)\exp\left\{-\dfrac{i\tilde{E}(t-t')}{\hbar}\right\} \end{matrix} \right] \tag{7.180}$$

$$\equiv \sum_{n,n'}W_n\,|F_{nn'}|^2\,\cos\frac{\tilde{E}(t-t')}{\hbar}\,[\hat{x}(t), [\hat{x}(t'), \hat{w}(t')]]$$

$$+ i\sum_{n,n'}W_n\,|F_{nn'}|^2\,\sin\frac{\tilde{E}(t-t')}{\hbar}[\hat{x}(t), \{\hat{x}(t'), \hat{w}(t')\}].$$

Comparing the two double sums participating in this expression with Eqs. (7.108) and (7.111), we see that they are nothing other than, respectively, the symmetrized correlation function and the temporal Green's function (multiplied by $\hbar/2$) of the time-difference argument $\tau = t - t' \geqslant 0$. As a result, Eq. (7.177) takes a compact form:

$$
\begin{aligned}
\dot{\hat{w}}(t) = &-\frac{1}{\hbar^2} \int_{-\infty}^{t} K_F(t - t') \left[\hat{x}(t), [\hat{x}(t'), \hat{w}(t')]\right] dt' \\
&-\frac{i}{2\hbar} \int_{-\infty}^{t} G(t - t') \left[\hat{x}(t), \{\hat{x}(t'), \hat{w}(t')\}\right] dt'.
\end{aligned}
\tag{7.181}
$$

Let me hope that readers (especially the ones who have braved through this derivation) enjoy this beautiful result as much as I do. It gives an equation for the time evolution of the density operator of the system of our interest (s), with the effects of its environment represented only by two real, c-number functions of τ—one (K_F) describing the fluctuation force exerted by the environment, and another one (G) representing its ensemble-averaged response to the system's evolution. And most spectacularly, these are exactly the same functions as participate in the alternative, Heisenberg–Langevin approach to the problem, and hence related to each other by the fluctuation–dissipation theorem (7.134).

After a short celebration, let us acknowledge that Eq. (7.181) is still an integro-differential equation, and needs to be solved together with Eq. (7.169) for the system coordinate's evolution. Such equations do not allow explicit analytical solutions, with the exception of very simple (and not very interesting) cases. For most applications, further simplifications should be made. One of them is based on the fact (which was already discussed in section 7.3) that both environmental functions participating in Eq. (7.181) tend to zero when their argument τ becomes larger than the environment's correlation time τ_c, independent of the system-to-environment coupling strength. If the coupling is sufficiently weak, the time scales $T_{nn'}$ of the evolution of the density matrix elements, following from Eq. (7.181), are much longer than this correlation time, and also the characteristic time scale of the coordinate operator's evolution. In this limit, all arguments t' of the density operator, giving substantial contributions to the right-hand side of Eq. (7.181), are so close to t that it does not matter whether its argument is t' or just t. This simplification, $w(t') \approx w(t)$, is known as the *Markov approximation*[61].

However, this approximation alone is still insufficient for finding the general solution of Eq. (7.181). Substantial further progress is possible in two important cases. The most important of them is when the intrinsic Hamiltonian $\hat{H}_s$ of the system s of our of interest is time-independent, and has a *discrete* eigenenergy spectrum E_n,[62] with well-separated levels:

---

[61] Named after A Markov (1856–1922; in older literature, 'Markoff'), a mathematician famous for his general theory of the so-called *Markov process*, whose future development is completely determined by its present state, rather than its pre-history.

[62] Here, rather reluctantly, I will use this standard notation, E_n, for the eigenenergies of our system of interest (s), in the hope that the reader would not confuse these discrete energy levels with the quasi-continuous energy levels of its environment (e), participating in particular in Eqs. (7.108) and (7.111). As a reminder, by this stage of our calculations the environment levels have disappeared from our formulas, leaving behind their functionals $K_F(\tau)$ and $G(\tau)$.

$$|E_n - E_{n'}| \gg \frac{\hbar}{T_{nn'}}. \tag{7.182}$$

Let us see what this condition yields for Eq. (7.181), rewritten for the matrix elements in the stationary state basis:

$$
\begin{aligned}
\dot{w}_{nn'} = &-\frac{1}{\hbar^2} \int_{-\infty}^{t} K_F(t - t')[\hat{x}(t), [\hat{x}(t'), \hat{w}]]_{nn'} \, dt' \\
&-\frac{i}{2\hbar} \int_{-\infty}^{t} G(t - t')[\hat{x}(t), \{\hat{x}(t'), \hat{w}\}]_{nn'} \, dt'.
\end{aligned}
\tag{7.183}
$$

After spelling out the commutators, the right-hand side of this expression includes four operator products, which differ 'only' by the operator order. Let us first have a look at one of these products,

$$[\hat{x}(t)\hat{x}(t')\hat{w}]_{nn'} \equiv \sum_{m,m'} x_{nm}(t)x_{mm'}(t')w_{m'n'}, \tag{7.184}$$

where the indices m and m' run over the same set of eigenenergies of the system s of our interest as the indices n and n'. According to Eq. (7.169) with a time-independent H_s, the matrix elements $x_{nn'}$ (in the stationary state basis) oscillate in time as $\exp\{i\omega_{nn'}t\}$, so that

$$[\hat{x}(t)\hat{x}(t')\hat{w}]_{nn'} = \sum_{m,m'} x_{nm}x_{mm'} \exp\{i(\omega_{nm}t + \omega_{mm'}t')\}w_{m'n'}, \tag{7.185}$$

where on the right-hand side, the coordinate matrix elements are in the Schrödinger picture, and the usual notation (6.85) is used for the quantum transition frequencies:

$$\hbar\omega_{nn'} \equiv E_n - E_{n'}. \tag{7.186}$$

According to the condition (7.182), frequencies $\omega_{nn'}$ with $n \neq n'$ are much higher than the speed of evolution of the density matrix elements (in the interaction picture!)—on both the left-hand and right-hand sides of Eq. (7.183). Hence, on the right-hand side of Eq. (7.183) we may keep only the terms that do not oscillate with these frequencies $\omega_{nn'}$, because rapidly-oscillating terms would give negligible contributions to the density matrix dynamics[63]. For that, in the double sum (7.185) we should save only the terms proportional to the difference $(t - t')$, because they will give (after the integration over t') a slowly changing contribution to the right-hand side[64]. These terms should have $\omega_{nm} + \omega_{mm'} = 0$, i.e. $(E_n - E_m) + (E_m - E_{m'}) \equiv E_n - E_{m'} = 0$. For a non-degenerate energy spectrum, this requirement means $m' = n$; as a result, the double sum is reduced to a single one:

---

[63] This is essentially the same rotating-wave approximation (RWA) as was used in section 6.5.

[64] As was already discussed in section 7.4, the lower-limit substitution ($t' = -\infty$) in the integrals participating in Eq. (7.183) gives zero, due to the finite-time 'memory' of the system, expressed by the decay of the correlation and response functions at large values of the time delay $\tau = t - t'$.

$$[\hat{x}(t)\hat{x}(t')\hat{w}]_{nn'} \approx w_{nn'} \sum_m x_{nm}x_{mn} \exp\{i\omega_{nm}(t - t')\}$$

$$\equiv w_{nn'} \sum_m |x_{nm}|^2 \exp\{i\omega_{nm}(t - t')\}. \tag{7.187}$$

Another product, $[\hat{w}\hat{x}(t')\hat{x}(t)]_{nn'}$, which appears on the right-hand side of Eq. (7.183), may be simplified absolutely similarly, giving

$$[\hat{w}\hat{x}(t')\hat{x}(t)]_{nn'} \approx \sum_m |x_{n'm}|^2 \exp\{i\omega_{n'm}(t' - t)\} w_{nn'}. \tag{7.188}$$

These expressions hold whether n and n' are equal or not. The situation is different for two other products on the right-hand side of Eq. (7.183), with w sandwiched between x and x'. For example,

$$[\hat{x}(t)\hat{w}\hat{x}(t')]_{nn'} = \sum_{m,m'} x_{nm}(t)w_{mm'}x_{m'n'}(t')$$

$$= \sum_{m,m'} x_{nm}w_{mm'}x_{m'n'} \exp\{i(\omega_{nm}t + \omega_{m'n'}t')\}. \tag{7.189}$$

For this term, the same requirement of having a fast oscillating function of $(t - t')$ only, yields a different condition: $\omega_{nm} + \omega_{m'n'} = 0$, i.e.

$$(E_n - E_m) + (E_{m'} - E_{n'}) = 0. \tag{7.190}$$

Here the double sum's reduction is possible only if we make an additional assumption that all interlevel energy distances are unique, i.e. our system of interest has no equidistant levels (such as in the harmonic oscillator). For the diagonal elements ($n = n'$), the RWA requirement is reduced to $m = m'$, giving sums over all diagonal elements of the density matrix:

$$[\hat{x}(t)\hat{w}\hat{x}(t')]_{nn} = \sum_m |x_{nm}|^2 \exp\{i\omega_{nm}(t - t')\} w_{mm}. \tag{7.191}$$

(Another similar term, $[\hat{x}(t')\hat{w}\hat{x}(t)]_{nn}$, is just a complex conjugate of Eq. (7.191).) However, for off-diagonal matrix elements ($n \neq n'$), the situation is different: Eq. (7.190) may be satisfied only if $m = n$ and also $m' = n'$, so that the double sum is reduced to just one, non-oscillating term:

$$[\hat{x}(t)\hat{w}\hat{x}(t')]_{nn'} = x_{nn}w_{nn'}x_{n'n'}, \quad \text{for } n \neq n'. \tag{7.192}$$

The second similar term, $[\hat{x}(t')\hat{w}\hat{x}(t)]_{nn}$, is exactly the same, so that in one of the integrals of Eq. (7.183), these terms add up, while in the second one, they cancel.

This is why the final equations of evolution look differently for diagonal and off-diagonal elements of the density matrix. For the former case ($n = n'$), Eq. (7.183) is

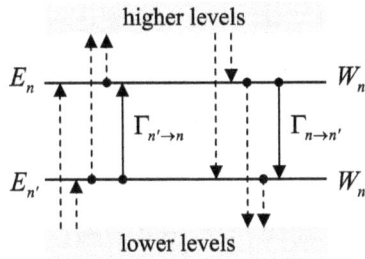

Figure 7.7. Probability dynamics in a discrete-spectrum system: solid arrows: the exchange between two energy levels, n and n', described by one term in the master equation (7.194); dashed arrows—other transitions to/from these levels.

reduced to the so-called *master equation*[65] relating diagonal elements w_{nn} of the density matrix, i.e. the energy level occupancies W_n:[66]

$$\dot{W}_n = \sum_{m \neq n} |x_{nm}|^2 \int_0^\infty \left[-\frac{1}{\hbar^2} K_F(\tau)(W_n - W_m)(\exp\{i\omega_{nm}\tau\} + \exp\{-i\omega_{nm}\tau\}) \right.$$
$$\left. -\frac{i}{2\hbar} G(\tau)(-W_n - W_m)(\exp\{i\omega_{nm}\tau\} - \exp\{-i\omega_{nm}\tau\})d\tau, \right] \tag{7.193}$$

where $\tau \equiv t - t'$. Changing the summation index notation from m to n', we may rewrite the master equation in its canonical form

$$\dot{W}_n = \sum_{n' \neq n} (\Gamma_{n' \to n} W_{n'} - \Gamma_{n \to n'} W_n), \tag{7.194}$$

where the coefficients

$$\Gamma_{n' \to n} \equiv |x_{nn'}|^2 \int_0^\infty \left[\frac{2}{\hbar^2} K_F(\tau) \cos \omega_{nn'}\tau - \frac{1}{\hbar} G(\tau) \sin \omega_{nn'}\tau \right] dt', \tag{7.195}$$

are called the *interlevel transition rates*[67]. Eq. (7.194) has a very clear physical meaning of the level occupancy dynamics (i.e. the balance of the probability flows ΓW) due to the quantum transitions between the energy levels (see figure 7.7), in our current case caused by the interaction between the system of our interest and its environment.

The Fourier transforms (7.113) and (7.123) enable us to express the two integrals in Eq. (7.195) via, respectively, the symmetrized spectral density $S_F(\omega)$ of environment force fluctuations and the imaginary part $\chi''(\omega)$ of the generalized

---

[65] The master equations, first introduced to quantum mechanics in 1928 by W Pauli, are sometimes called the 'Pauli master equations', or 'kinetic equations', or 'rate equations'.

[66] As Eq. (7.193) shows, the term with $m = n$ would vanish, and thus may be legitimately excluded from the sum.

[67] As Eq. (7.193) shows, the result for $\Gamma_{n \to n'}$ is described by Eq. (7.195) as well, provided that indices n and n' are swapped in all components of its right-hand side, including the swap $\omega_{nn'} \to \omega_{n'n} = -\omega_{nn'}$.

susceptibility, both at frequency $\omega = \omega_{nn'}$. After that we may use the fluctuation–dissipation theorem (7.134) to exclude the former function, getting finally[68]

$$
\begin{aligned}
\Gamma_{n' \to n} &= \frac{1}{\hbar} \, |x_{nn'}|^2 \chi''(\omega_{nn'}) \left(\coth\frac{\hbar\omega_{nn'}}{2k_B T} - 1 \right) \\
&\equiv \frac{2}{\hbar} \, |x_{nn'}|^2 \frac{\chi''(\omega_{nn'})}{\exp\{(E_n - E_{n'})/k_B T\} - 1}.
\end{aligned}
\tag{7.196}
$$

Note that since the imaginary part of the generalized susceptibility is an odd function of frequency, Eq. (7.196) is in compliance with the Gibbs distribution for arbitrary temperature. Indeed, according to this equation, the ratio of the 'up' and 'down' rates for each pair of levels equals

$$
\begin{aligned}
\frac{\Gamma_{n' \to n}}{\Gamma_{n \to n'}} &= \frac{\chi''(\omega_{nn'})}{\exp\{(E_n - E_{n'})/k_B T\} - 1} \Big/ \frac{\chi''(\omega_{n'n})}{\exp\{(E_{n'} - E_n)/k_B T\} - 1} \\
&= \exp\left\{ -\frac{E_n - E_{n'}}{k_B T} \right\}.
\end{aligned}
\tag{7.197}
$$

On the other hand, according to the Gibbs distribution (7.24), in the thermal equilibrium the level populations should be in the same proportion. Hence, Eq. (7.196) is in compliance with the so-called *detailed balance equation*,

$$
W_n \Gamma_{n \to n'} = W_{n'} \Gamma_{n' \to n},
\tag{7.198}
$$

valid in the equilibrium for each pair $\{n, n'\}$, so that all right-hand sides of all Eqs. (7.194), and hence the time derivatives of all W_n vanish—as they should. Thus, the stationary solution of the master equations indeed describes the thermal equilibrium.

The system of master equations (7.194), frequently complemented by additional terms on their right-hand sides, which describe interlevel transitions due to other factors (e.g. by an external ac force with a frequency close to one of $\omega_{nn'}$), is the key starting point for practical analyses of many quantum systems, notably including optical quantum amplifiers and generators (lasers). It is important to remember that they are strictly valid only in the rotating-wave approximation, i.e. if Eq. (7.182) is well satisfied for all n and n' of substance.

For a particular but very important case of a two-level system (with, say, $E_1 > E_2$), the rate $\Gamma_{1 \to 2}$ may be interpreted (especially in the low-temperature limit $k_B T \ll \hbar\omega_{12} = E_1 - E_2$, when $\Gamma_{1 \to 2} \gg \Gamma_{2 \to 1}$) as the reciprocal characteristic time $1/T_1 \equiv \Gamma_{1 \to 2}$ of the *energy relaxation* process that brings the *diagonal* elements of the density matrix

---

[68] It is straightforward (and highly recommended to the reader) to show that at low temperatures ($k_B T \ll |E_{n'} - E_n|$), Eq. (7.196) gives the same result as the Golden Rate formula (6.111), with $A = x$. (The low temperature limit is necessary to ensure that the initial occupancy of the excited level n is negligible, as was assumed at the derivation of Eq. (6.111).)

to their thermally-equilibrium values (7.24). For the Ohmic dissipation described by Eqs. (7.137) and (7.138), Eq. (7.196) yields

$$\frac{1}{T_1} \equiv \Gamma_{1\to2} = \frac{2}{\hbar^2}|x_{12}|^2\eta \times \begin{cases} \hbar\omega_{12}, & \text{for } k_BT \ll \hbar\omega_{12}, \\ k_BT, & \text{for } \hbar\omega_{12} \ll k_BT. \end{cases} \tag{7.199}$$

This relaxation time T_1 should not be confused with the characteristic time T_2 of the *off-diagonal* element decay, i.e. dephasing, which was already discussed in section 7.3. In this context, let us see what do Eqs. (7.183) say about the dephasing rates. Taking into account our intermediate results (7.187)–(7.192), and merging the non-oscillating components (with $m = n$ and $m = n'$) of the sums Eqs. (7.187) and (7.188) with the terms (7.192), which also do not oscillate in time, we get the following equation[69]:

$$\dot{w}_{nn'} = -\left\{\int_0^\infty \left[\frac{1}{\hbar^2}K_F(\tau)\right.\right.$$

$$\times \left(\sum_{m\neq n}|x_{nm}|^2\exp\{i\omega_{nm}\tau\} + \sum_{m\neq n'}|x_{n'm}|^2\exp\{-i\omega_{n'm}\tau\} + (x_{nn}-x_{n'n'})^2\right) \tag{7.200}$$

$$\left.\left.+ \frac{i}{2\hbar}G(\tau)\left(\sum_{m\neq n}|x_{nm}|^2\exp\{i\omega_{nm}\tau\} - \sum_{m\neq n'}|x_{n'm}|^2\exp\{-i\omega_{n'm}\tau\}\right)\right]d\tau\right\}w_{nn'},$$

for $n \neq n'$.

In contrast with Eq. (7.194), the right-hand side of this equation includes both a real and an imaginary part, and hence it may be represented as

$$\dot{w}_{nn'} = -(1/T_{nn'} + i\Delta_{nn'})w_{nn'}, \tag{7.201}$$

where both factors $1/T_{nn'}$ and $\Delta_{nn'}$ are real. As Eq. (7.201) shows, the second term in the right-hand side of this equation causes slow oscillations of the matrix elements $w_{nn'}$, which, after returning to the Schrödinger picture, add just small corrections[70] to the unperturbed frequencies (7.186) of their oscillations, and are not important for most applications. More important is the first term, proportional to

---

[69] Sometimes Eq. (7.200) (in any of its numerous alternative forms) is called the *Redfield equation*, after the 1965 work by A Redfield. Note, however, that several other authors, notably including (in alphabetical order) H Haken, W Lamb, M Lax, W Louisell, and M Scully, also made key contributions into the very fast development of the density-matrix approach to open quantum systems in the mid-1960s.

[70] Such corrections are sometimes called the *Lamb shift*, because such an effect was first observed experimentally in 1947 by W Lamb and R Retherford, as a minor, ~1 GHz shift between energy levels of $2s$ and $2p$ states of hydrogen, due to the electric-dipole coupling of hydrogen atoms to the free-space electromagnetic environment. (These energies are equal not only in the non-relativistic theory (section 3.6), but also in the relativistic, Dirac theory (section 9.7), if the electromagnetic environment is ignored.) The explanation of the shift, by H Bethe, also in 1947, essentially launched the whole field of quantum electrodynamics—to be briefly discussed in chapter 9.

$$\frac{1}{T_{nn'}} = \int_0^\infty \left[\frac{1}{\hbar^2} K_F(\tau) \right.$$

$$\times \left(\sum_{m \neq n} |x_{nm}|^2 \cos \omega_{nm}\tau + \sum_{m \neq n'} |x_{n'm}|^2 \cos \omega_{n'm}\tau + (x_{nn} - x_{n'n'})^2 \right) \qquad (7.202)$$

$$\left. - \frac{1}{2\hbar} G(\tau) \left(\sum_{m \neq n} |x_{nm}|^2 \sin \omega_{nm}\tau + \sum_{m \neq n'} |x_{n'm}|^2 \sin \omega_{n'm}\tau \right) \right] d\tau, \quad \text{for } n \neq n',$$

because it describes the effect completely absent without the environment: an exponential decay of the off-diagonal matrix elements, i.e. the dephasing. Comparing the first two terms of Eq. (7.202) with Eq. (7.195), we see that the dephasing rates may be described by a very simple formula:

$$\frac{1}{T_{nn'}} = \frac{1}{2} \left(\sum_{m \neq n} \Gamma_{n \to m} + \sum_{m \neq n'} \Gamma_{n' \to m} \right) + \frac{\pi}{\hbar^2} (x_{nn} - x_{n'n'})^2 S_F(0)$$

$$\equiv \frac{1}{2} \left(\sum_{m \neq n} \Gamma_{n \to m} + \sum_{m \neq n'} \Gamma_{n' \to m} \right) + \frac{k_B T}{\hbar^2} \eta (x_{nn} - x_{n'n'})^2, \quad \text{for } n \neq n',$$

$$(7.203)$$

where the low-frequency drag coefficient η is again the defined as $\lim_{\omega \to 0} \chi''(\omega)/\omega$—see Eq. (7.138).

This result shows that two effects yield independent contributions into the dephasing. The first of them may be interpreted as a result of 'virtual' transitions of the system, from the levels n and n' of our interest, to other energy levels m; according to Eq. (7.195), this contribution is proportional to the strength of coupling to environment at relatively high frequencies ω_{nm} and $\omega_{n'm}$. (If the energy quanta $\hbar\omega$ of these frequencies are much larger than the thermal fluctuation scale $k_B T$, only the lower levels, with $E_m < \max[E_n, E_{n'}]$ are important.) In contrast, the second contribution is due to low-frequency, essentially classical fluctuations of the environment, and hence to the low-frequency dissipative susceptibility. In the Ohmic dissipation case, when the ratio $\eta \equiv \chi''(\omega)/\omega$ is frequency-independent, both contributions are of the same order, but their exact relation depends on the relation between the matrix elements $x_{nn'}$ of a particular system.

For example, returning for a minute to the two-level system discussed in section 7.3, described by our current theory with the replacement $\hat{x} \to \hat{\sigma}_z$, the high-frequency contributions to dephasing vanish because of the absence of transitions between the energy levels, while the low-frequency contribution yields

$$\frac{1}{T_2} \equiv \frac{1}{T_{12}} = \frac{k_B T}{\hbar^2} \eta (x_{nn} - x_{n'n'})^2 \to \frac{k_B T}{\hbar^2} \eta [(\sigma_z)_{11} - (\sigma_z)_{22}]^2 = \frac{4 k_B T}{\hbar^2} \eta, \qquad (7.204)$$

thus exactly reproducing the result (7.142) of the Heisenberg–Langevin approach[71]. Note also that the expression for T_2 is very close in structure to Eq. (7.199) for T_1 (in the high-temperature limit). However, for the simple interaction model (7.70) that was explored in section 7.3, the off-diagonal elements of the operator $\hat{x} = \hat{\sigma}_z$ in the stationary-state z-basis vanish, so that $T_1 \to \infty$, while T_2 says finite. The physics of this result is very clear, for example, from the two-well implementation of the model (see figure 7.4 and its discussion): it is suitable for the case of a very high energy the barrier between the wells, which inhibits tunneling, and hence any change of the well occupancies. However, T_1 may become finite, and comparable with T_2, if tunneling between the wells is substantial[72].

Because of the reason explained above, the derivation of Eqs. (7.200)–(7.204) is not valid for systems with equidistant energy spectra—for example, the harmonic oscillator. For this particular, but very important system, with its simple matrix elements $x_{nn'}$, given by Eqs. (5.92), it is longish but straightforward to repeat the above calculations, starting from (7.183), to obtain an equation similar in structure to Eq. (7.200), but with two other terms, proportional to $w_{n\pm1,n'\pm1}$, on its right-hand side. Neglecting the minor Lamb–shift term, the equation reads

$$\dot{w}_{nn'} = -\delta\left\{\begin{array}{l} [(n_e + 1)(n + n') + n_e(n + n' + 2)]w_{nn'} \\ - 2(n_e + 1)[(n + 1)(n' + 1)]^{1/2}w_{n+1,n'+1} \\ - 2n_e(nn')^{1/2}w_{n-1,n'-1} \end{array}\right\}. \qquad (7.205)$$

Here δ is the effective damping coefficient[73],

$$\delta \equiv \frac{x_0^2}{2\hbar}\text{Im}\,\chi(\omega_0) \equiv \frac{\text{Im}\,\chi(\omega_0)}{2m\omega_0}, \qquad (7.206)$$

equal to just $\eta/2m$ for the Ohmic dissipation, while n_e is the equilibrium number of oscillator's excitations, given by Eq. (7.26b), with the environment's temperature T. (I am using this new notation because in dynamics, the instant expectation value $\langle n \rangle$ may be time-dependent, and is generally different from its equilibrium value n_e.)

(Actually, the derivation of Eq. (7.205) might be started at a somewhat earlier point, from the Markov approximation applied to Eq. (7.181), expressing the

---

[71] The first form of Eq. (7.203), as well as the analysis of section 7.3, imply that low-frequency fluctuations of any other origin, not taken into account in their own current analysis (say, an unintentional noise from experimental equipment), may also cause dephasing; such 'technical fluctuations' are indeed a very serious challenge for experimental implementation of coherent qubit systems—see section 8.5 below.

[72] As was discussed in section 5.1, the tunneling may be described by using, instead of Eq. (7.70), the full two-level Hamiltonian (5.3). Let me leave for the reader's exercise to spell out the equations for the time evolution of the density matrix elements of this system, and of the expectation values of the Pauli operators, for this case.

[73] This coefficient participates prominently in the classical theory of damped oscillations (see, e.g. *Part CM* section 5.1), in particular defining the oscillator's Q-factor as $Q = \omega_0/2\delta$, and the decay time of the amplitude A and the energy E of free oscillations: $A(t) = A(0)\exp\{-\delta t\}$, $E(t) = E(0)\exp\{-2\delta t\}$.

coordinate operator via the creation–annihilation operators (5.65). This procedure gives the result in the operator (basis-independent) form[74]:

$$\dot{\hat{w}} = -\delta[(n_e + 1)(\{\hat{a}^\dagger\hat{a}, \hat{w}\} - 2\hat{a}\hat{w}\hat{a}^\dagger) + n_e(\{\hat{a}\hat{a}^\dagger, \hat{w}\} - 2\hat{a}^\dagger\hat{w}\hat{a})]. \qquad (7.207)$$

In the Fock state basis, this equation immediately reduces to Eq. (7.205). However, Eq. (7.207) may be more useful for some applications.)

Returning to Eq. (7.205), we see that it relates only the elements $w_{nn'}$ located at the same distance $(n - n')$ from the principal diagonal of the density matrix. This means, in particular, that the dynamics of the diagonal elements w_{nn} of the matrix, i.e. the Fock state probabilities W_n, is independent of the off-diagonal elements, and may be represented in the form (7.194), truncated to the transitions between the adjacent energy levels only ($n' = n \pm 1$):

$$\dot{W}_n = (\Gamma_{n+1 \to n} W_{n+1} - \Gamma_{n \to n+1} W_n) + (\Gamma_{n-1 \to n} W_{n-1} - \Gamma_{n \to n-1} W_n) \qquad (7.208)$$

with the rates

$$\begin{aligned}
&\Gamma_{n+1 \to n} = 2\delta\,(n + 1)(n_e + 1), \qquad &\Gamma_{n \to n+1} = 2\delta\,(n + 1)\,n_e, \\
&\Gamma_{n-1 \to n} = 2\delta\,n\,n_e, \qquad &\Gamma_{n \to n-1} = 2\delta\,n\,(n_e + 1).
\end{aligned} \qquad (7.209)$$

Since according to the definition of n_e, given by Eq. (7.26b),

$$n_e = \frac{1}{\exp\{\hbar\omega_0/k_B T\} - 1}, \qquad \text{so that}$$

$$n_e + 1 = \frac{\exp\{\hbar\omega_0/k_B T\}}{\exp\{\hbar\omega_0/k_B T\} - 1} \equiv -\frac{1}{\exp\{-\hbar\omega_0/k_B T\} - 1}, \qquad (7.210)$$

taking into account Eqs. (5.92), (7.186), (7.206), and the asymmetry of the function $\chi''(\omega)$, we see that these rates are again described by Eq. (7.196), despite the fact that the last formula was derived for non-equidistant energy spectra.

Hence the only substantial new feature of the master equation for the harmonic oscillator, is that the decay of the off-diagonal elements of its density matrix is scaled by the same parameter (2δ) as that of the decay of its diagonal elements, i.e. there is no radical difference between the dephasing and energy-relaxation times T_2 and T_1. This fact may be interpreted as the result of the independence of the energy level distances, $\hbar\omega_0$, of the fluctuations $F(t)$ exerted on the oscillator by the environment, so that their low-frequency density, $S_F(0)$, does not contribute into the dephasing.

---

[74] Sometimes Eq. (7.207) is called the *Lindblad equation*, but I believe this terminology is inappropriate. Though its structure indeed falls into the general category of equations, suggested by G Lindblad in 1976 for the density operators in the Markov approximation, whose diagonalized form in the interaction picture is

$$\dot{\hat{w}} = \sum_j \gamma_j (2\hat{L}_j \hat{w} \hat{L}_j^\dagger - \{\hat{L}_j \hat{L}_j^\dagger, \hat{w}\}),$$

Eq. (7.207) was derived much earlier (by L Landau in 1927 for zero temperature, and by M Lax in 1960 for an arbitrary temperature), and in contrast to the general Lindblad equation, spells out the participating operators and coefficients γ_j for a particular physical system—the harmonic oscillator.

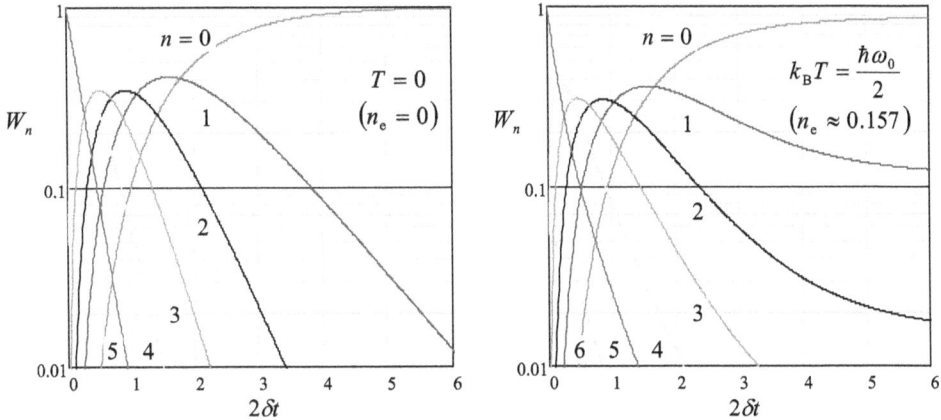

Figure 7.8. Relaxation of a harmonic oscillator, initially in its 5th Fock state, at: (a) $T = 0$, and (b) $T > 0$. Note that in the latter case, even the energy levels with $n > 5$ get populated, due the their thermal excitation.

(This fact formally follows also from Eq. (7.203) as well, taking into account that for the oscillator, $x_{nn} = x_{n'n'} = 0$.)

The simple equidistant structure of the oscillator's spectrum makes it possible to readily solve the system of Eqs. (7.208), with $n = 0, 1, 2, \ldots$, for some important cases. In particular, if the initial state of the oscillator is a classical mixture, with no off-diagonal elements, its further relaxation proceeds as such a mixture: $w_{nn'}(t) = 0$ for all $n' \neq n$.[75] In particular, it is straightforward to use Eq. (7.208) to verify that if the initial classical mixture obeys the Gibbs distribution (7.25), but with a temperature T_i different from that of the environment (T_e), then the relaxation process is reduced to a simple exponential transient of the effective temperature from T_i to T_e:

$$W_n(t) = \exp\left\{-n\frac{\hbar\omega_0}{k_B T_{ef}(t)}\right\}\left(1 - \exp\left\{-\frac{\hbar\omega_0}{k_B T_{ef}(t)}\right\}\right), \tag{7.211}$$

with $T_{ef}(t) = T_i e^{-2\delta t} + T_e(1 - e^{-2\delta t})$,

with the corresponding evolution of the expectation value of the energy E—cf Eq. (7.26b):

$$\langle E \rangle = \frac{\hbar\omega_0}{2} + \hbar\omega_0\langle n \rangle, \qquad \langle n \rangle = \frac{1}{\exp\{\hbar\omega_0/k_B T_{ef}(t)\} - 1} \to_{t\to\infty} n_e. \tag{7.212}$$

However, if the initial state of the oscillator is different (say, corresponds to some upper Fock state), the relaxation process described by Eqs. (7.208) and (7.209) is more complex—see, e.g. figure 7.8. At low temperatures (figure 7.8a), it may be interpreted as a gradual 'roll' of the probability distribution down the energy staircase, with a gradually decreasing velocity $dn/dt \propto n$. However, at substantial

---

[75] Note, however, that this is not true for many applications, in which a damped oscillator is also under the effect of an external time-dependent field, which should be described by additional, typically off-diagonal terms on the right-hand side of Eqs. (7.205).

temperatures, with $k_B T \sim \hbar\omega_0$ (figure 7.8b), this 'roll-down' is saturated when the level occupancies $W_n(t)$ approach their equilibrium values (7.25).[76]

The analysis of this process may be simplified in the case when $W(n, t) \equiv W_n(t)$ is a smooth function of the energy level number n, limited to high levels: $n \gg 1$. In this limit, we may use the Taylor expansion of this function (written for the points $\Delta n = \pm 1$), truncated to three leading terms:

$$W_{n\pm1}(t) \equiv W(n \pm 1, t) \approx W(n, t) \pm \frac{\partial W(n, t)}{\partial n} + \frac{1}{2}\frac{\partial^2 W(n, t)}{\partial n^2}. \qquad (7.213)$$

Plugging this expression into Eqs. (7.208) and (7.209), we get for the function $W(n, t)$, a partial differential equation, which may be recast in the following form:

$$\frac{\partial W}{\partial t} = -\frac{\partial}{\partial n}[f(n)\ W] + \frac{\partial^2}{\partial n^2}[d(n)\ W], \qquad (7.214)$$

$$\text{with} \quad f(n) \equiv 2\delta(n_e - n), \quad d(n) \equiv 2\delta(n_e + \tfrac{1}{2})\, n.$$

Since at $n \gg 1$, the oscillator's energy E is close to $\hbar\omega_0 n$, this *energy diffusion equation* (sometimes incorrectly called the Fokker–Planck equation—see below) essentially describes the time evolution of the continuous probability density $w(E, t)$, which may be defined as $w(E, t) \equiv W(E/\hbar\omega_0, t)/\hbar\omega_0$.[77]

This continuous approximation naturally reminds us of the need to discuss dissipative systems with a *continuous* spectrum. Unfortunately, for such systems the few (relatively) simple results that may be obtained from the basic equation (7.181) are essentially classical in nature, and will be discussed in detail in *Part SM* of this series. Here, I will give only a simple illustration. Let us consider a 1D particle that interacts weakly with a thermally-equilibrium environment, but otherwise is free to move along the x-axis. As we know from chapters 2 and 5, in this case the most convenient basis is that of the momentum eigenstates p. In the momentum representation, the density matrix is just the c-number function $w(p, p')$, defined by Eq. (7.54), which was already discussed in brief in section 7.2. On the other hand, the coordinate operator, which participates in the right-hand side of Eq. (7.181), has the form given by the first of Eqs. (4.269),

$$\hat{x} = i\hbar\frac{\partial}{\partial p}, \qquad (7.215)$$

dual to the coordinate-representation formula (4.268). As we already know, such operators are local—see, e.g. Eq. (4.244). Due to this locality, the whole right-hand side of Eq. (7.181) is local as well, and hence (within the framework of our perturbative treatment) the interaction with environment affects essentially only

---

[76] The reader may like to have a look at the results of nice measurements of such functions $W_n(t)$ in microwave oscillators, performed using their coupling with Josephson-junction circuits [5] and with Rydberg atoms [6].

[77] In the classical limit $n_e \gg 1$, this equation is analytically solvable for any initial conditions—see, e.g. the paper by B Zeldovich *et al* [7], which also gives some more intricate solutions of Eqs. (7.208) and (7.209). Note, however, that most important properties of the damped harmonic oscillator (including its relaxation dynamics) may be analyzed more simply using the Heisenberg–Langevin approach discussed in the previous section.

the diagonal values $w(p, p)$ of the density matrix, i.e. the momentum probability density $w(p)$.

Let us find the equation governing the evolution of this function in time in the Markov approximation, when the time scale of the density matrix evolution is much longer than the correlation time τ_c of the environment, i.e. the time scale of the functions $K_F(\tau)$ and $G(\tau)$. In this approximation, we may take the matrix elements out of the first integral of Eq. (7.181),

$$
\begin{aligned}
&-\frac{1}{\hbar^2} \int_{-\infty}^{t} K_F(t - t') \, dt' [\hat{x}(t), [\hat{x}(t'), \hat{w}(t')]] \\
&\approx -\frac{1}{\hbar^2} \int_{0}^{\infty} K_F(\tau) \, d\tau [\hat{x}, [\hat{x}, \hat{w}]] \\
&= -\frac{\pi}{\hbar^2} S_F(0)[\hat{x}, [\hat{x}, \hat{w}]] = -\frac{k_B T}{\hbar^2} \eta \, [\hat{x}, [\hat{x}, \hat{w}]],
\end{aligned}
\tag{7.216}
$$

and calculate the last double commutator in the Schrödinger picture. This may be done either using an explicit expression for the matrix elements of the coordinate operator, or in a simpler way, using the same trick as at the derivation of the Ehrenfest theorem in section 5.2. Namely, expanding an arbitrary function $f(p)$ into the Taylor series in p,

$$
f(p) = \sum_{k=0}^{\infty} \frac{1}{k!} \frac{\partial^k f}{\partial p^k} p^k,
\tag{7.217}
$$

and using Eq. (7.215), we can prove the following simple commutation relation:

$$
\begin{aligned}
[\hat{x}, f] &= \sum_{k=0}^{\infty} \frac{1}{k!} \frac{\partial^k f}{\partial p^k} [\hat{x}, p^k] = \sum_{k=0}^{\infty} \frac{1}{k!} \frac{\partial^k f}{\partial p^k} (i\hbar k p^{k-1}) \\
&= i\hbar \sum_{k=1}^{\infty} \frac{1}{(k-1)!} \frac{\partial^{k-1}}{\partial p^{k-1}} \left(\frac{\partial f}{\partial p} \right) p^{k-1} = i\hbar \frac{\partial f}{\partial p}.
\end{aligned}
\tag{7.218}
$$

Now applying this result sequentially, first to w and then to the resulting commutator, we get

$$
[\hat{x}, [\hat{x}, w]] = \left[\hat{x}, i\hbar \frac{\partial w}{\partial p} \right] = i\hbar \frac{\partial}{\partial p} \left(i\hbar \frac{\partial w}{\partial p} \right) = -\hbar^2 \frac{\partial^2 w}{\partial p^2}.
\tag{7.219}
$$

It may look like the second integral in Eq. (7.181) might be simplified similarly. However, it vanishes at $p' \to p$, and $t' \to t$, so that in order to calculate the first nonvanishing contribution from that integral for $p = p'$, we have to take into account the small difference $\tau \equiv t - t' \sim \tau_c$ between the arguments of the coordinate operators under that integral. This may be done using Eq. (7.169) with the free-particle's Hamiltonian consisting of the kinetic-energy contribution alone:

$$
\hat{x}(t') - \hat{x}(t) \approx -\tau \dot{\hat{x}} = -\tau \frac{1}{i\hbar} [\hat{x}, \hat{H}_s] = -\tau \frac{1}{i\hbar} \left[\hat{x}, \frac{\hat{p}^2}{2m} \right] = -\tau \frac{\hat{p}}{m},
\tag{7.220}
$$

where the exact argument of the operator on the right-hand side is already unimportant, and may be taken for t. As a result, we may use the last of Eqs. (7.136) to reduce the second term on the right-hand side of Eq. (7.181) to

$$-\frac{i}{2\hbar}\int_{-\infty}^{t}G(t-t')[\hat{x}(t),\{\hat{x}(t'),\hat{w}(t')\}]dt' \approx \frac{i}{2\hbar}\int_{0}^{\infty}G(\tau)\tau d\tau \times \left[\hat{x},\left\{\frac{\hat{p}}{m},\hat{w}\right\}\right]$$

$$= \frac{\eta}{2i\hbar}\left[\hat{x},\left\{\frac{\hat{p}}{m},\hat{w}\right\}\right]. \tag{7.221}$$

In the momentum representation, the momentum operator and the density matrix w are just c-numbers and commute, so that applying Eq. (7.218) to the product pw, we get

$$\left[\hat{x},\left\{\frac{\hat{p}}{m},\hat{w}\right\}\right] = \left[\hat{x},2\frac{p}{m}w\right] = 2i\hbar\frac{\partial}{\partial p}\left(\frac{p}{m}w\right), \tag{7.222}$$

and may finally reduce the integro-differential equation (7.181) to a partial differential equation:

$$\frac{\partial w}{\partial t} = -\frac{\partial}{\partial p}(Fw) + \eta\, k_B T \frac{\partial^2 w}{\partial p^2}, \qquad \text{with} \quad F \equiv -\eta\frac{p}{m}. \tag{7.223}$$

This is the 1D form of the famous *Fokker–Planck equation* describing the classical statistics of motion of a particle (in our particular case, of a free particle) in an environment providing a linear drag characterized by the coefficient η; it belongs to the same drift–diffusion type as Eq. (7.214). The first, *drift* term on its right-hand side describes the particle's deceleration due to the friction force (7.137), $F = -\eta p/m = -\eta v$, provided by the environment. The second, *diffusion* term on the right-hand side of Eq. (7.223) describes the effect of fluctuations: the particle's momentum' random walk about its average (drift-affected, and hence time-dependent) value. The walk obeys the law similar to Eq. (7.85), but with the *momentum-space* diffusion coefficient

$$D_p = \eta k_B T. \tag{7.224}$$

This is the reciprocal-space version of the fundamental Einstein relation between the dissipation (friction) and fluctuations, in this classical limit represented by their thermal energy scale $k_B T$.[78]

Just for the reader's reference, let me note that the Fokker–Planck equation (7.223) may be readily generalized to the 3D motion of a particle under the effect of

---

[78] Note that Eq. (7.224), as well as the original Einstein's relation between the diffusion coefficient D in the direct space and temperature, may be derived much more simply by other means—for example, from the Nyquist formula (7.139). These issues are discussed in detail in *Part SM* chapter 5.

an additional external force[79], and in this more general form is the basis for many important applications; however, due to its classical character, its discussion is also left for *Part SM* of this series[80].

To summarize our discussion of the two alternative approaches to the analysis of quantum systems interacting with a thermally-equilibrium environment, described in the last three sections, let me emphasize that they give different descriptions of the same phenomena, and are characterized by the same two functions $G(\tau)$ and $K_F(\tau)$. Namely, in the Heisenberg–Langevin approach we describe the system by operators that change (fluctuate) in time, even in the thermal equilibrium, while in the density-matrix approach the system is described by non-fluctuating probability functions, such as $W_n(t)$ or $w(p, t)$, which are stationary in equilibrium. In the cases when a problem may be solved to the end by either method (for example, for a harmonic oscillator), they give identical results for all observables.

7.7 Problems

Problem 7.1. Calculate the density matrix of a two-level system described by the Hamiltonian matrix

$$\mathbf{H} = \mathbf{c} \cdot \boldsymbol{\sigma} \equiv c_x \sigma_x + c_y \sigma_y + c_z \sigma_z,$$

where σ_k are the Pauli matrices, and c_j are c-numbers, in thermodynamic equilibrium at temperature T.

Problem 7.2. In the usual z-basis, spell out the density matrix of a spin-½ with gyromagnetic ratio γ:

  (i) in the pure state with the spin definitely directed along the z-axis,
  (ii) in the pure state with the spin definitely directed along the x-axis,
  (iii) in the thermal equilibrium at temperature T, in a magnetic field directed along the z-axis, and
  (iv) in the thermal equilibrium at temperature T, in a magnetic field directed along the x-axis.

Problem 7.3. Calculate the Wigner function of a harmonic oscillator in:

  (i) at the thermodynamic equilibrium at temperature T,
  (ii) in the ground state, and
  (iii) in the Glauber state with dimensionless complex amplitude α.

Discuss the relation between the first of the results and the Gibbs distribution.

---

[79] Moreover, Eq. (7.223) may be generalized to the motion of a *quantum* particle in an additional periodic potential $U(\mathbf{r})$. In this case, due to the band structure of the energy spectrum (which was discussed in sections 2.7 and 3.4), the coupling to the environment produces not only a continuous drift–diffusion of the probability density in the space of the quasi-momentum $\hbar\mathbf{q}$, but also quantum transitions between discrete energies of different bands at the same $\hbar\mathbf{q}$—see, e.g. [8].

[80] See *Part SM* sections 5.6–5.7. For a more detailed discussion of *quantum* effects in dissipative systems with continuous spectra see, e.g. either [9] or [10].

Problem 7.4. Calculate the Wigner functions of a harmonic oscillator, with mass m and frequency ω_0, in its first excited stationary state ($n = 1$).

Problem 7.5.* A harmonic oscillator is weakly coupled to an Ohmic environment.

(i) Use the rotating-wave approximation to write the reduced equations of motion for the Heisenberg operators of the complex amplitude of oscillations.
(ii) Calculate the expectation values of the correlators of the fluctuation force operators, participating in these equations, and express them via the average number $\langle n \rangle$ of thermally-induced excitations in equilibrium, given by the second of Eqs. (7.26b).

Problem 7.6. Calculate the average potential energy of long-range electrostatic interaction between two similar isotropic, 3D harmonic oscillators, each with the electric dipole moment $\mathbf{d} = q\mathbf{s}$, where $\mathbf{s}$ is the oscillator's displacement from its equilibrium position, at arbitrary temperature T.

Problem 7.7. A semi-infinite string with mass μ per unit length is attached to a wall, and stretched with a constant force (tension) $\mathcal{T}$. Calculate the spectral density of the transverse force exerted on the wall, in thermal equilibrium.

Problem 7.8.* Calculate the low-frequency spectral density of small fluctuations of the voltage V across a Josephson junction, shunted with an Ohmic conductor, and biased with a dc external current $\bar{I} > I_c$.

Hint: You may use Eqs. (1.73)–(1.74) to describe the junction's dynamics, and assume that the shunting conductor remains in thermal equilibrium.

Problem 7.9. Prove that in the interaction picture of quantum dynamics, the expectation value of an arbitrary observable A may be indeed calculated using Eq. (7.167).

Problem 7.10. Show that the quantum-mechanical Golden Rule (6.149) and the master equation (7.196) give the same results for the rate of spontaneous quantum transitions $n' \rightarrow n$ in a system with discrete energy spectrum, weakly coupled to a low-temperature heat bath ($k_B T \ll \hbar\omega_{nn'}$).

Hint: Establish a relation between the function $\chi''(\omega_{nn'})$, which participates in Eq. (7.196), and the density of states ρ_n, which participates in the Golden Rule formula, by considering a particular case of sinusoidal classical oscillations in the system of interest.

Problem 7.11. For a harmonic oscillator with weak Ohmic dissipation, use Eqs. (7.208)–(7.209) to find the evolution of the expectation value $\langle E \rangle$ of oscillator's energy in time at arbitrary initial state, and compare the result with that following from the Heisenberg–Langevin approach.

Problem 7.12. Derive Eq. (7.219) in an alternative way, using an expression dual to Eq. (4.244).

Problem 7.13. A particle in a system of two coupled potential wells (see, e.g. figure 7.4 in the lecture notes) is weakly coupled to an Ohmic environment.

 (i) Derive equations describing time evolution of the density matrix elements.
(ii) Solve these equations in the low-temperature limit, when the energy level splitting is much larger than $k_B T$, to calculate the time evolution of the probability of finding the particle in one of the wells, after it had been placed there at $t = 0$.

Problem 7.14.* A spin-½ with gyromagnetic ratio γ is placed into magnetic field $\mathscr{B}(t) = \mathscr{B}_0 + \tilde{\mathscr{B}}(t)$ with an arbitrary but relatively small time-dependent component, and is also weakly coupled to a dissipative environment. Derive differential equations describing the time evolution of the expectation values of spin's Cartesian components, at arbitrary temperature.

References

[1] Bimbard E 2014 *Phys. Rev. Lett.* **112** 033601
[2] Callen H and Welton T 1951 *Phys. Rev.* **83** 34
[3] Koch R *et al* 1982 *Phys. Lev. B* **26** 74
[4] Blum K 2012 *Density Matrix Theory and Applications* 3rd ed (Springer)
[5] Wang H *et al* 2008 *Phys. Rev. Lett.* **101** 240401
[6] Brune M *et al* 2008 *Phys. Rev. Lett.* **101** 240402
[7] Zeldovich B *et al* 1969 *Sov. Phys. JETP* **28** 308
[8] Likharev K and Zorin A 1985 *J. Low Temp. Phys.* **59** 347
[9] Weiss U 1999 *Quantum Dissipative Systems* 2nd ed (World Scientific)
[10] Breuer H-P and Petruccione F 2007 *The Theory of Open Quantum Systems* (Oxford University Press)

IOP Publishing

Quantum Mechanics
Lecture notes
Konstantin K Likharev

Chapter 8

Multiparticle systems

This chapter provides a brief introduction to quantum mechanics of systems of similar particles, with special attention on the case when they are indistinguishable. For such systems, theory predicts (and experiment confirms) very specific effects, even in the case of negligible explicit ('direct') interaction between the particles. These effects notably include the Bose–Einstein condensation of bosons, and the exchange interaction of fermions.

8.1 Distinguishable and indistinguishable particles

The importance of quantum systems of many similar particles is probably self-evident; just the very fact that most atoms include several/many electrons is sufficient to attract our attention. There are also important systems where the number of electrons is much higher than in one atom; for example, a cubic centimeter of a typical metal houses $\sim 10^{23}$ conduction electrons that cannot be attributed to particular atoms, and have to considered as common parts of the system as the whole. Though quantum mechanics offers virtually no exact analytical results for systems of substantially interacting particles[1], it reveals very important new quantum effects even in the simplest case when particles do not interact, and least explicitly (*directly*).

If non-interacting particles are either different from each other by their nature, or physically similar but still *distinguishable* because of other reasons, everything is simple—at least, conceptually. Then, as was already discussed in section 6.7, a

---

[1] As was emphasized in section 7.3, for such systems of similar particles the powerful methods discussed in the last chapter, based on the separation of the whole Universe into the 'system of our interest' and the 'environment', typically do not work well—mostly because the quantum state of the 'particle of interest' may be substantially correlated (in particular, entangled) with those of similar particles of its 'environment'—see below.

system of two particles, 1 and 2, each in a pure quantum state, may be described by a ket-vector being a direct product,

$$|\alpha\rangle = |\beta\rangle_1 \otimes |\beta'\rangle_2, \tag{8.1a}$$

of the single-particle ket-vectors, describing their states β and β' defined in different Hilbert spaces. (Below, I will frequently use, for the direct product, the following convenient shorthand:

$$|\alpha\rangle = |\beta\beta'\rangle, \tag{8.1b}$$

in which the state symbol's position within the vector codes the particle's number.) Hence the *permuted state*

$$\hat{P}|\beta\beta'\rangle \equiv |\beta'\beta\rangle \equiv |\beta'\rangle_1 \otimes |\beta\rangle_2, \tag{8.2}$$

where $\hat{P}$ is the *permutation operator* (defined by this equality), is clearly different from the initial one.

Such operator may be also used for states of systems of *identical particles*. This term may be used to describe:

(i) the 'really elementary' particles like electrons, which (at least at this stage of development of physics) are considered as structure-less entities, and hence are all identical;

(ii) any objects (e.g. hadrons or mesons) that may be considered as a system of 'more elementary' particles (e.g. quarks and gluons), but still are reliably placed in the same quantum state—most simply, though not necessarily, to the ground state[2].

It is important to note that identical particles still may be *distinguishable*—say by their clear spatial separation. Such systems of similar but distinguishable particles (or subsystems) are broadly discussed nowadays, for example in the context of quantum computing and encryption—see section 8.5 below. This is why it is insufficient to use the term 'identical particles' if we want to say that they are genuinely *indistinguishable*, so below I will use the latter term, despite it being rather unpleasant grammatically.

It turns out that for a quantitative description of systems of indistinguishable particles we need to use, instead of direct products of the type (8.1), linear combinations of products such products, for example of $|\beta\beta'\rangle$ and $|\beta'\beta\rangle$.[3] To see

---

[2] Note that from this point of view, even complex atoms or molecules, in the same quantum state, may be considered on the same footing as the 'really elementary' particles. For example, the already mentioned recent spectacular interference experiments by R Lopes *et al*, which require particle identity, were carried out with couples of ^{4}He atoms in the same internal state.

[3] A very legitimate question is why, in this situation, we need to introduce the particles' numbers to start with. A partial answer is that in this approach, it is much simpler to derive (or guess) the system Hamiltonians from the correspondence principle—see, e.g. Eq. (8.27) below. Later in this chapter, we will discuss an alternative approach (the so-called 'second quantization'), in which particle numbering is avoided. While this approach is more logical, writing adequate Hamiltonians (which, in particular, would avoid spurious self-interaction of the particles) in it is much more challenging—see section 8.3 below.

this, let us discuss properties of the permutation operator defined by Eq. (8.2). Consider an observable A, and a system of eigenstates of its operator:

$$\hat{A}|a_j\rangle = A_j|a_j\rangle. \tag{8.3}$$

If the particles are indistinguishable, the observable's expectation value should not be affected by their permutation. Hence the operators $\hat{A}$ and $\hat{\mathcal{P}}$ have to commute, and share their eigenstates. This is why eigenstates of the operator $\hat{\mathcal{P}}$ are so important: in particular, they include the eigenstates of the Hamiltonian, i.e. the stationary states of a system of indistinguishable particles.

Let us have a look at the action of the permutation operator squared, on an elementary ket-vector product:

$$\hat{\mathcal{P}}^2|\beta\beta'\rangle = \hat{\mathcal{P}}(\hat{\mathcal{P}}|\beta\beta'\rangle) = \hat{\mathcal{P}}|\beta'\beta\rangle = |\beta\beta'\rangle, \tag{8.4}$$

i.e. $\hat{\mathcal{P}}^2$ brings the state back to its original form. Since any pure state of a two-particle system may be represented as a linear combination of such products, this result does not depend on the state, and may be represented as the following operator relation:

$$\hat{\mathcal{P}}^2 = \hat{I}. \tag{8.5}$$

Now let us find the possible eigenvalues $\mathcal{P}_j$ of the permutation operator. Acting by both sides of Eq. (8.5) on any of eigenstates $|\alpha_j\rangle$ of the permutation operator, we get a very simple equation for its eigenvalues:

$$\mathcal{P}_j^2 = 1, \tag{8.6}$$

with two possible solutions:

$$\mathcal{P}_j = \pm 1. \tag{8.7}$$

Let us find the eigenstates of the permutation operator in the simplest case when each of the component particles can be only in one of two single-particle states—say, β and β'. Evidently, none of the simple products $|\beta\beta'\rangle$ and $|\beta'\beta\rangle$, taken alone, does qualify for the eigenstate—unless the states β and β' are identical. For this reason let us try their linear combination

$$|\alpha_j\rangle = a|\beta\beta'\rangle + b|\beta'\beta\rangle, \tag{8.8}$$

so that

$$\hat{\mathcal{P}}|\alpha_j\rangle = \mathcal{P}_j|\alpha_j\rangle = a|\beta'\beta\rangle + b|\beta\beta'\rangle. \tag{8.9}$$

For the case $\mathcal{P}_j = +1$ we have to require the states (8.8) and (8.9) to be the same, so that $a = b$, giving the so-called *symmetric eigenstate*[4]

---

[4] As in many situations we have met earlier, the kets given by Eqs. (8.10) and (8.11) may be multiplied by $\exp\{i\varphi\}$ with an arbitrary real phase φ. However, until we discuss coherent superpositions of various states α, there is no good motivation for taking the phase different from 0; that would only clutter the notation.

$$|\alpha_+\rangle = \frac{1}{\sqrt{2}}(|\beta\beta'\rangle + |\beta'\beta\rangle), \tag{8.10}$$

where the front coefficient guarantees the orthonormality of the two-particle state, provided that the single-particle states are orthonormal. Similarly, for $\mathcal{P}_j = -1$ we get $a = -b$, i.e. an *antisymmetric eigenstate*

$$|\alpha_-\rangle = \frac{1}{\sqrt{2}}(|\beta\beta'\rangle - |\beta'\beta\rangle). \tag{8.11}$$

These are the simplest (two-particle, two-state) examples of *entangled states*, defined as multiparticle system's states whose vectors cannot be factored into a direct product (8.1) of single-particle vectors.

So far, our math does not preclude either sign of $\mathcal{P}_j$, in particular the possibility that the sign depends on the state (i.e. on the index j). Here, however, comes in a crucial *experimental* fact: all indistinguishable particles fall into two groups[5]:

(i) *bosons*, particles with integer spin s, for whose states $\mathcal{P}_j = +1$, and
(ii) *fermions*, particles with half-integer spin, with $\mathcal{P}_j = -1$.

In the non-relativistic theory we are discussing now, this key fact should be considered as an experimental one. (The relativistic quantum theory, whose elements will be discussed in chapter 9, offers a proof that the half-integer-spin particles cannot be bosons and the integer-spin ones cannot be fermions.) However, our discussion of spin in section 5.7 enables the following plausible *interpretation* of the fermion–boson difference. In the free space, the permutation of particles 1 and 2 may be viewed as a result of this pair's rotation by angle $\pm\pi$ about a certain axis. As we have seen in section 5.7, at the rotation by such an angle, the state vector $|\beta\rangle$ of a particle with a quantum number m_s (which ranges from $-s$ to $+s$, and hence may take only integer values for integer s, and only half-integer values for half-integer s) changes by the factor $\exp\{\pm i\pi m_s\}$, so that the state product $|\beta\beta'\rangle$ has to change by $\exp\{\pm i2\pi m_s\}$, i.e. by the factor $+1$ for any integer s, and by the factor (-1) for any half-integer s.

The most impressive corollaries of Eqs. (8.10) and (8.11) are for the case when the partial states of the two particles are the same: $\beta = \beta'$. The corresponding Bose state α_+ is possible; in particular, at sufficiently low temperatures, a set of non-interacting Bose particles condenses on the ground state of each of them—the so-called *Bose–Einstein condensate* ('BEC').[6] Perhaps the most fascinating feature of a Bose–Einstein condensate is that the dynamics of its observables is governed by laws of

---

[5] Sometimes this fact is described as having two different 'statistics': the *Bose–Einstein statistics* of bosons, and *Fermi–Dirac statistics* of fermions, because their statistical distributions in thermal equilibrium are indeed different—see, e.g. *Part SM* section 2.8. However, this difference is actually deeper: we are dealing with *two different quantum mechanics*.
[6] For a quantitative discussion of the Bose–Einstein condensation see, e.g. *Part SM* section 3.4. Examples of such condensates include *superfluids* like helium, Cooper-pair condensates in superconductors, and BECs of weakly interacting atoms.

quantum mechanics, while (for nearly all purposes) they may be treated as *c*-numbers—see, e.g. Eqs. (1.73) and (1.74).[7]

On the other hand, if we take $\beta = \beta'$ in Eq. (8.11), we see that state α_- becomes the null-state, i.e. *cannot exist* at all. This is the mathematical expression of the *Pauli exclusion principle*[8]: two indistinguishable fermions cannot be in the same quantum state. (As will be discussed below, this is true for systems with more than two fermions as well.) Probably, the key importance of this principle is self-evident: if it was not valid for electrons (that are fermions), all electrons of each atom would condense on its ground ($1s$-like) level, and all the usual chemistry (and biochemistry, and biology, including dear us!) would not exist. The Pauli principle makes fermions implicitly interacting even if they do not interact *directly*, i.e. in the usual sense of this word.

8.2 Singlets, triplets, and the exchange interaction

Now let us discuss possible approaches to quantitative analyses of identical particles, starting from a simple case of two spin-½ particles (say, electrons), whose interaction with each other and the external world does not involve spin. The description of such a system may be based on factorable states with ket-vectors

$$|\alpha_-\rangle = |o_{12}\rangle \otimes |s_{12}\rangle, \tag{8.12}$$

with the orbital function $|o_{12}\rangle$ and the spin function $|s_{12}\rangle$ (that depends on the state of both spins of the pair) belonging to different Hilbert spaces. It is frequently convenient to use the coordinate representation of such a state, sometimes called the *spinor*:

$$\langle \mathbf{r}_1, \mathbf{r}_2 | \alpha_- \rangle = \langle \mathbf{r}_1, \mathbf{r}_2 | o_{12} \rangle \otimes |s_{12}\rangle \equiv \psi(\mathbf{r}_1, \mathbf{r}_2)|s_{12}\rangle. \tag{8.13}$$

Since the spin-½ particles are fermions, the particle permutation has to change the sign:

$$\hat{\mathscr{P}}\psi(\mathbf{r}_1, \mathbf{r}_2)|s_{12}\rangle \equiv \psi(\mathbf{r}_2, \mathbf{r}_1)|s_{21}\rangle = -\psi(\mathbf{r}_1, \mathbf{r}_2)|s_{12}\rangle, \tag{8.14}$$

of either the orbital factor of the spinor, or its spin factor.

In particular, in the case of a symmetric orbital factor,

$$\psi(\mathbf{r}_2, \mathbf{r}_1) = \psi(\mathbf{r}_1, \mathbf{r}_2), \tag{8.15}$$

the spin factor has to obey the relation

$$|s_{21}\rangle = -|s_{12}\rangle. \tag{8.16}$$

---

[7] For example, for the Bose–Einstein condensate of $N \gg 1$ particles, the Heisenberg uncertainty relation may be reduced to $\delta N \delta \varphi > 1$, where φ is the condensate wavefunction's phase, so that it may have $\delta N/\langle N \rangle \ll 1$ and $\delta \varphi \ll 1$ simultaneously.

[8] It was formulated by W Pauli in 1925, on the basis of less general rules suggested by G Lewis (1916), I Langmuir (1919), N Bohr (1922), and E Stoner (1924) for the explanation of experimental spectroscopic data.

Let us use the ordinary z-basis (where z, in the absence of external magnetic field, is an arbitrary spatial axis) for both spins. In this basis, the ket-vector of any two-spin state may be represented as a linear combination of the four following basis vectors:

$$|\uparrow\uparrow\rangle, \quad |\downarrow\downarrow\rangle, \quad |\uparrow\downarrow\rangle, \quad \text{and } |\downarrow\uparrow\rangle. \tag{8.17}$$

The first two kets evidently do not satisfy Eq. (8.16), and cannot participate in the state. Applying to the remaining kets the same argumentation as has resulted in Eq. (8.11), we get

$$|s_{12}\rangle = |s_-\rangle \equiv \frac{1}{\sqrt{2}}(|\uparrow\downarrow\rangle - |\downarrow\uparrow\rangle). \tag{8.18}$$

Such an orbital-symmetric and spin-asymmetric state is called the *singlet*.

The origin of this term becomes clear from the analysis of the opposite (orbital-asymmetric and spin-symmetric) case:

$$\psi(\mathbf{r}_2, \mathbf{r}_1) = -\psi(\mathbf{r}_1, \mathbf{r}_2), \quad |s_{12}\rangle = |s_{21}\rangle. \tag{8.19}$$

For the composition of such a symmetric spin state, the first two kets of Eq. (8.17) are completely acceptable (with arbitrary weights), and so is an entangled spin state that is a symmetric combination of the two last kets, similar to Eq. (8.10):

$$|s_+\rangle \equiv \frac{1}{\sqrt{2}}(|\uparrow\downarrow\rangle + |\downarrow\uparrow\rangle), \tag{8.20}$$

so that the general spin state is a *triplet*:

$$|s_{12}\rangle = c_+|\uparrow\uparrow\rangle + c_-|\downarrow\downarrow\rangle + c_0\frac{1}{\sqrt{2}}(|\uparrow\downarrow\rangle + |\downarrow\uparrow\rangle). \tag{8.21}$$

Note that such a state, with values of the coefficients c (satisfying the normalization condition), corresponds to the same orbital wavefunction and hence the same energy. However, each of these three states has a specific value of the z-component of the net spin—evidently equal to, respectively, $+\hbar$, $-\hbar$, and 0. Because of this, even a small external magnetic field lifts their degeneracy, splitting the energy level in three; hence the term 'triplet'.

In the particular case when the particles do not interact at all, for example

$$\hat{H} = \hat{h}_1 + \hat{h}_2, \quad \hat{h}_k = \frac{\hat{p}_k^2}{2m} + \hat{u}(\mathbf{r}_k), \quad \text{with } k = 1, 2, \tag{8.22}$$

the two-particle Schrödinger equation for the symmetrical orbital wavefunction (8.15) is obviously satisfied by the direct products,

$$\psi(\mathbf{r}_1, \mathbf{r}_2) = \psi_n(\mathbf{r}_1)\psi_{n'}(\mathbf{r}_2), \tag{8.23}$$

of single-particle eigenfunctions, with arbitrary sets n, n' of quantum numbers. For the particular but very important case $n = n'$, this means that the eigenenergy of the (only acceptable) singlet state,

$$\frac{1}{\sqrt{2}}(|\uparrow\downarrow\rangle - |\downarrow\uparrow\rangle)\psi_n(\mathbf{r}_1)\psi_n(\mathbf{r}_2), \tag{8.24}$$

is just $2\varepsilon_n$, where ε_n is the single-particle energy level[9]. In particular, for the ground state of the system, such singlet spin state gives the lowest energy $E_g = 2\varepsilon_g$, while any triplet spin state (8.19) would require one of the particles to be in a different orbital state, i.e. in a state of higher energy, so that the total energy of the system would be also higher.

Now moving to the systems in which two indistinguishable spin-½ particles do interact, let us consider, as their simplest but important[10] example, the lower energy states of a neutral atom[11] of helium—more exactly, ^{4}He. Such an atom consists of a nucleus with two protons and two neutrons, with the total electric charge $q = +2e$, and two electrons 'rotating' about the nucleus. Neglecting the small relativistic effects that were discussed in section 6.3, the Hamiltonian describing the electron motion may be expressed as

$$\hat{H} = \hat{h}_1 + \hat{h}_2 + \hat{U}_{int}, \quad \hat{h}_k = \frac{\hat{p}_k^2}{2m} - \frac{2e^2}{4\pi\varepsilon_0 r_k}, \quad \hat{U}_{int} = \frac{e^2}{4\pi\varepsilon_0|\mathbf{r}_1 - \mathbf{r}_2|}. \tag{8.25}$$

As most problems of multiparticle quantum mechanics, the eigenvalue/eigenstate problem for this Hamiltonian does not have an exact analytical solution, so let us carry out its approximate analysis considering the electron–electron interaction U_{int} as a perturbation. As was discussed in chapter 6, we have to start with the '0th-order' approximation in which the perturbation is ignored, so that the Hamiltonian is reduced to the sum (8.22). In this approximation, the ground state of the atom is the singlet (8.24), with the orbital factor

$$\psi_g(\mathbf{r}_1, \mathbf{r}_2) = \psi_{100}(\mathbf{r}_1)\psi_{100}(\mathbf{r}_2), \tag{8.26}$$

and the energy $2\varepsilon_g$. Here each factor $\psi_{100}(\mathbf{r})$ is the single-particle wavefunction of the ground (1s) state of the hydrogen-like atom with $Z = 2$, with quantum numbers $n = 1$, $l = 0$, $m = 0$. According to Eqs. (3.174) and (3.208),

$$\psi_{100}(\mathbf{r}) = Y_0^0(\theta, \varphi)\,\mathcal{R}_{1,0}(r) = \frac{1}{\sqrt{4\pi}}\frac{2}{r_0^{3/2}}e^{-r/r_0}, \quad \text{with } r_0 = \frac{r_B}{Z} = \frac{r_B}{2}, \tag{8.27}$$

so that according to Eqs. (3.191) and (3.201), in this approximation the total ground state energy is

$$E_g^{(0)} = 2\varepsilon_g^{(0)} = 2\left(-\frac{\varepsilon_0}{2n^2}\right)_{n=1,Z=2} = 2\left(-\frac{Z^2 E_H}{2}\right)_{Z=2} = -4E_H \approx -109 \text{ eV}. \tag{8.28}$$

---

[9] In this chapter, I try to use lower-case letters for all single-particle observables (in particular, ε for their energies), in order to distinguish them as clearly as possible from system's observables (including the total energy E of the system), typeset in capital letters.

[10] Indeed, helium makes up more than 20% of all 'ordinary' matter of our Universe.

[11] Evidently, the positive ion He^{+1} of this atom, with just one electron, is fully described by the hydrogen-like atom theory with $Z = 2$, whose ground-state energy, according to Eq. (3.191), is $-Z^2 E_H/2 = -2E_H \approx -55.4$ eV.

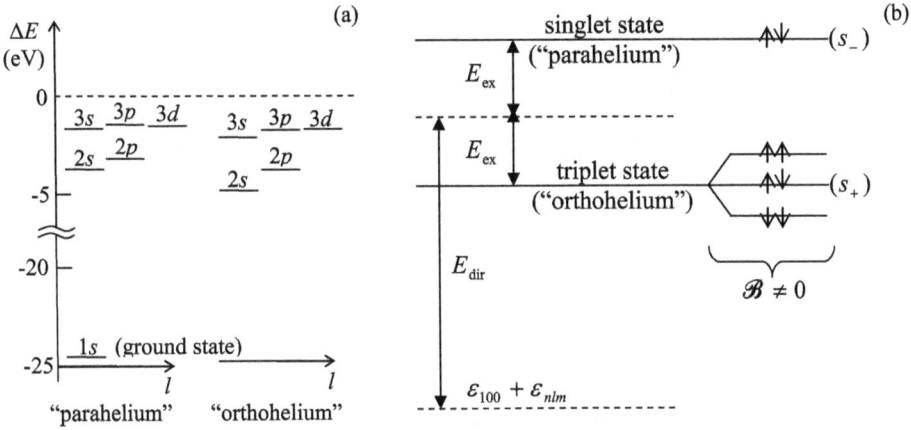

Figure 8.1. The lower energy levels of a helium atom: (a) experimental data and (b) a schematic structure of an excited state in the first order of the perturbation theory. On panel (a), all energies are referred to that $(-2E_H \approx -55.4$ eV) of the ground state of the positive ion He^{+1}, so that their magnitudes are the (readily measurable) energies of atom's ionization starting from the corresponding bound state. Note that the 'spin direction' nomenclature on panel (b) is rather crude: it does not reflect the difference between the entangled states s_+ and s_-.

This is still somewhat far (though not terribly far!) from the experimental value $E_g \approx -78.8$ eV—see the bottom level in figure 8.1a.

Making a small (but useful) detour from our main topic, let us note that we can get a much better agreement with experiment by calculating the electron interaction energy in the 1st order of the perturbation theory. Indeed, in application to our system, Eq. (6.14) reads

$$E_g^{(1)} = \langle g| \tilde{U}_{\text{int}}|g\rangle = \int d^3r_1 \int d^3r_2 \psi_g^*(\mathbf{r}_1, \mathbf{r}_2) U_{\text{int}}(\mathbf{r}_1, \mathbf{r}_2)\psi_g(\mathbf{r}_1, \mathbf{r}_2). \qquad (8.29)$$

Plugging in Eqs. (8.25)–(8.27), we get

$$E_g^{(1)} = \left(\frac{1}{4\pi}\frac{4}{r_0^3}\right)^2 \int d^3r_1 \int d^3r_2 \frac{e^2}{4\pi\varepsilon_0|\mathbf{r}_1 - \mathbf{r}_2|} \exp\left\{-\frac{2(r_1 + r_2)}{r_0}\right\}. \qquad (8.30)$$

As may be readily evaluated analytically (this exercise is left for the reader), this expression equals $(5/4)E_H$, so that the corrected ground state energy,

$$E_g \approx E_g^{(0)} + E_g^{(1)} = (-4 + 5/4)E_H = -74.8\,\text{eV}, \qquad (8.31)$$

is much closer to experiment.

There is still room for ready improvement, using the variational method discussed in section 2.9. For our particular case of a ^{4}He atom, we may try to use, as the trial state, the wavefunction given by Eqs. (8.26) and (8.27), but with the atomic number

Z considered as an adjustable parameter $Z_{ef} < Z = 2$ rather than a fixed number. The physics behind this approach is that the electric charge density $\rho(\mathbf{r}) = -e|\psi(\mathbf{r})|^2$ of each electron forms a negatively charged 'cloud' that reduces the effective charge of the nucleus, as seen by another electron, to $Z_{ef}e^2$, with some $Z_{ef} < 2$. As a result, the single-particle wavefunction spreads further in space (with the scale $r_0 = r_B/Z_{ef} > r_B/Z$), while keeping its functional form (8.27) nearly intact. Since the kinetic energy T in system's Hamiltonian (8.25) is proportional to $r_0^{-2} \propto Z_{ef}^2$, while the potential energy is proportional to $r_0^{-1} \propto Z_{ef}^1$, we can write

$$E_g(Z_{ef}) = \left(\frac{Z_{ef}}{2}\right)^2 \langle T_g \rangle_{Z=2} + \frac{Z_{ef}}{2}\langle U_g \rangle_{Z=2}. \tag{8.32}$$

Now we can use the fact that according to Eq. (3.212), for any stationary state of a hydrogen-like atom (just as for the classical circular motion in the Coulomb potential), $\langle U \rangle = 2E$, and hence $\langle T \rangle = E - \langle U \rangle = -E$. Using Eq. (8.30), and adding the correction $U_g^{(1)} = -(5/4)E_H$, calculated above, to the potential energy, we get

$$E_g(Z_{ef}) = \left[4\left(\frac{Z_{ef}}{2}\right)^2 + \left(-8 + \frac{5}{4}\right)\frac{Z_{ef}}{2}\right]E_H. \tag{8.33}$$

This expression allows an elementary calculation of the optimal value of Z_{ef}, and the corresponding minimum of the function $E_g(Z_{ef})$:

$$(Z_{ef})_{opt} = 2\left(1 - \frac{5}{32}\right) = 1.6875, \quad (E_g)_{min} \approx -2.85E_H \approx -77.5\,\text{eV}. \tag{8.34}$$

Given the trial state's crudeness, this number is in a surprisingly good agreement with experimental value cited above, with a difference of the order of 1%.

Now let us return to the basic topic of this section—the effects of particle (in this case, electron) indistinguishability. As we have just seen, the ground level energy of the helium atom is not affected directly by this fact, but the situation is different for its excited states—even the lowest ones. The reasonably good convergence of the perturbation theory, which we have seen for the ground state, tells us that we can base our analysis of wavefunctions (ψ_e) of the lowest excited state orbitals, on products like $\psi_{100}(\mathbf{r}_k)\psi_{nlm}(\mathbf{r}_{k'})$, with $n > 1$. However, in order to satisfy the fermion permutation rule, $\mathcal{P}_j = -1$, we have to take the orbital factor of the state in the either symmetric or asymmetric form:

$$\psi_e(\mathbf{r}_1, \mathbf{r}_2) = \frac{1}{\sqrt{2}}[\psi_{100}(\mathbf{r}_1)\psi_{nlm}(\mathbf{r}_2) \pm \psi_{nlm}(\mathbf{r}_1)\psi_{100}(\mathbf{r}_2)], \tag{8.35}$$

with the proper total permutation asymmetry provided by the corresponding spin factor (8.18) or Eq. (8.21), so that the upper/lower sign in Eq. (8.35) corresponds to the singlet/triplet spin state. Let us calculate the expectation values of the total energy of the system in the first order of the perturbation theory. Plugging Eq. (8.35) into the 0th-order expression

$$\langle E_e \rangle^{(0)} = \int d^3 r_1 \int d^3 r_2 \, \psi_e^*(\mathbf{r}_1, \mathbf{r}_2)(\hat{h}_1 + \hat{h}_2)\psi_e(\mathbf{r}_1, \mathbf{r}_2), \tag{8.36}$$

we get two groups of similar terms that differ only by the particle index. We can merge the terms of each pair by changing the notation as $(\mathbf{r}_1 \to \mathbf{r}, \mathbf{r}_2 \to \mathbf{r}')$ in one of them, and $(\mathbf{r}_1 \to \mathbf{r}', \mathbf{r}_2 \to \mathbf{r})$ in the other term. Using Eq. (8.25), and the mutual orthogonality of wavefunctions $\psi_{100}(\mathbf{r})$ and $\psi_{nlm}(\mathbf{r})$, we get the following result:

$$\langle E_e \rangle^{(0)} = \int \psi_{100}^*(\mathbf{r})\left(-\frac{\hbar^2 \nabla_\mathbf{r}^2}{2m} - \frac{2e^2}{4\pi\varepsilon_0 r}\right)\psi_{100}(\mathbf{r})d^3 r$$

$$+ \int \psi_{nlm}^*(\mathbf{r}')\left(-\frac{\hbar^2 \nabla_{\mathbf{r}'}^2}{2m} - \frac{2e^2}{4\pi\varepsilon_0 r'}\right)\psi_{nlm}(\mathbf{r}')d^3 r' \equiv \varepsilon_{100} + \varepsilon_{nlm}. \tag{8.37}$$

It may be interpreted as the sum of eigenenergies of two separate single particles, one in the ground state 100, and another in the excited state nlm—despite the fact that actually the electron states are entangled. Thus, in the 0th order of the perturbation theory, the electron entanglement does not affect their energy.

However, the potential energy of the system also includes the interaction term U_{int}, which does not allow such separation. Indeed, in the 1st approximation of the perturbation theory, the total energy E_e of the system may be expressed as $\varepsilon_{100} + \varepsilon_{nlm} + E_{\text{int}}^{(1)}$, with

$$E_{\text{int}}^{(1)} = \langle U_{\text{int}} \rangle = \int d^3 r_1 \int d^3 r_2 \, \psi_e^*(\mathbf{r}_1, \mathbf{r}_2) U_{\text{int}}(\mathbf{r}_1, \mathbf{r}_2) \, \psi_e(\mathbf{r}_1, \mathbf{r}_2), \tag{8.38}$$

Plugging Eq. (8.35) into this result, using the symmetry of the function U_{int} with respect to the particle number permutation, and the same particle coordinate re-numbering as above, we get

$$E_{\text{int}}^{(1)} = E_{\text{dir}} \pm E_{\text{ex}}, \tag{8.39}$$

with the following, deceivingly similar expressions for the two terms:

$$E_{\text{dir}} \equiv \int d^3 r \int d^3 r' \psi_{100}^*(\mathbf{r})\psi_{nlm}^*(\mathbf{r}') U_{\text{int}}(\mathbf{r}, \mathbf{r}')\psi_{100}(\mathbf{r})\psi_{nlm}(\mathbf{r}'), \tag{8.40}$$

$$E_{\text{ex}} \equiv \int d^3 r \int d^3 r' \psi_{100}^*(\mathbf{r})\psi_{nlm}^*(\mathbf{r}') U_{\text{int}}(\mathbf{r}, \mathbf{r}')\psi_{nlm}(\mathbf{r})\psi_{100}(\mathbf{r}'). \tag{8.41}$$

Since the single-particle orbitals can be always made real, both components are positive (or at least non-negative). However, their physics is completely different. The integral (8.40), called the *direct interaction energy*, allows a simple semi-classical interpretation as the Coulomb energy of interacting electrons, each distributed in space with the electric charge density $\rho(\mathbf{r}) = -e\psi^*(\mathbf{r})\psi(\mathbf{r})$:[12]

$$E_{\text{dir}} = \int d^3r \int d^3r' \frac{\rho_{100}(\mathbf{r})\rho_{nlm}(\mathbf{r}')}{4\pi\varepsilon_0|\mathbf{r} - \mathbf{r}'|} \equiv \int \rho_{100}(\mathbf{r})\phi_{nlm}(\mathbf{r})d^3r$$
$$\equiv \int \rho_{nlm}(\mathbf{r})\phi_{100}(\mathbf{r})d^3r, \tag{8.42}$$

where $\phi(\mathbf{r})$ are the electrostatic potentials created by the electrons' 'electric charge clouds':[13]

$$\phi_{100}(\mathbf{r}) = \frac{1}{4\pi\varepsilon_0}\int d^3r' \frac{\rho_{100}(\mathbf{r}')}{|\mathbf{r} - \mathbf{r}'|}, \quad \phi_{nlm}(\mathbf{r}) = \frac{1}{4\pi\varepsilon_0}\int d^3r' \frac{\rho_{nlm}(\mathbf{r}')}{|\mathbf{r} - \mathbf{r}'|}. \tag{8.43}$$

However, the integral (8.41), called the *exchange interaction energy*, evades a classical interpretation, and (as is clear from its derivation) is the direct corollary of electrons' indistinguishability. The magnitude of E_{ex} is also very much different from E_{dir}, because the function under the integral (8.41) disappears in the regions where single-particle wavefunctions do not overlap. This is in a full agreement with the discussion in section 8.1: if two particles are identical but well separated, i.e. their wavefunctions do not overlap, the exchange interaction disappears, i.e. measurable effects of particle indistinguishability vanish.

Figure 8.1b shows the structure of an excited energy level, with certain quantum numbers $n > 1$, l, and m, given by Eqs. (8.39)–(8.41). The upper, so-called *parahelium*[14] level, with the energy

$$E_{\text{para}} = (\varepsilon_{100} + \varepsilon_{nlm}) + E_{\text{dir}} + E_{\text{ex}} > \varepsilon_{100} + \varepsilon_{nlm}, \tag{8.44}$$

corresponds to the symmetric orbital state and hence to the singlet spin state (8.18), while the lower, *orthohelium* level, with

$$E_{\text{orth}} = (\varepsilon_{100} + \varepsilon_{nlm}) + E_{\text{dir}} - E_{\text{ex}} < E_{\text{para}}, \tag{8.45}$$

corresponds to the degenerate triplet spin state (8.21). This degeneracy may be lifted by an external magnetic field, whose effect of the electron spins[15] is described by the following evident generalization of the Pauli Hamiltonian (4.163),

---

[12] See, e.g. *Part EM* section 1.3, in particular Eq. (1.54).

[13] Note that the result for E_{dir} correctly reflects the basic fact that a charged particle does not interacts with itself, even if its wavefunction is quantum-mechanically spread over a finite space volume. Unfortunately, this is not true for some other approximate theories of multiparticle systems—see section 8.4 below.

[14] This traditional terminology reflects the historic fact that the observation of two different hydrogen-like spectra, corresponding to opposite signs in Eq. (8.39), was first taken as an evidence for two different species of ^{4}He, which were called, respectively, the 'orthohelium' and the 'parahelium'.

[15] As we know from section 6.4, the field also affects the orbital motion of the electrons, so that the simple analysis based on Eq. (8.46) is strictly valid only for the s excited state ($l = 0$, and hence $m = 0$). However, orbital effects of a very weak magnetic field do not affect the triplet level splitting we are analyzing now.

$$\hat{H}_{\text{field}} = -\gamma \hat{\mathbf{s}}_1 \cdot \mathscr{B} - \gamma \hat{\mathbf{s}}_2 \cdot \mathscr{B} \equiv -\gamma \hat{\mathbf{S}} \cdot \mathscr{B}, \quad \text{with } \gamma = \gamma_e \equiv -\frac{e}{m_e} \equiv -\frac{2}{\hbar}\mu_B, \quad (8.46)$$

where

$$\hat{\mathbf{S}} \equiv \hat{\mathbf{s}}_1 + \hat{\mathbf{s}}_2, \qquad\qquad (8.47)$$

is the operator of the (vector) sum of the system of two spins[16].

In order to analyze this effect, we need first to make one more detour, to address the general issue of *spin addition*. The main rule[17] here is that in a full analogy with the net spin of a single particle, defined by Eq. (5.170), the net spin operator (8.47) of *any* system of two spins, and its component $\hat{S}_z$ along an arbitrary axis, obey the same commutation relations (5.168) as the component operators, and hence have properties similar to those expressed by Eqs. (5.169) and (5.175):

$$\hat{S}^2|S, M_S\rangle = \hbar^2 S(S+1)|S, M_S\rangle,$$
$$\hat{S}_z|S, M_S\rangle = \hbar M_S|S, M_S\rangle, \quad \text{with } -S \leqslant M_S \leqslant +S, \qquad (8.48)$$

where the ket vectors correspond to the coupled basis of joint eigenstates of the operators of S^2 and S_z (but not necessarily all component operators—see again the Venn shown in figure 5.12 and its discussion, with the replacements $\mathbf{S}, \mathbf{L} \to \mathbf{s}_{1,2}$ and $\mathbf{J} \to \mathbf{S}$). Repeating the discussion of section 5.7 with these replacements, we see that in both coupled and uncoupled bases, the net magnetic number M_S is simply expressed via those of the components

$$M_S = (m_s)_1 + (m_s)_2. \qquad (8.49)$$

However, the net spin quantum number S (in contrast to the Nature-given spins $s_{1,2}$ of its elementary components) is not quite certain, and we may immediately say only that it has to obey the following analog of the relation $|l - s| \leqslant j \leqslant l + s$, discussed in section 5.7:

$$|s_1 - s_2| \leqslant S \leqslant s_1 + s_2. \qquad (8.50)$$

What exactly S is (within these limits), depends on the spin state of the system.

For the simplest case of two spin-½ components, with each $s = ½$ and $m_s = \pm½$, Eq. (8.49) gives three possible values of M_S, equal to 0 and ± 1, while Eq. (8.50) limits the possible values of S to just either 0 or 1. Using the last of Eqs. (8.48), we see that the possible combinations of the quantum numbers are

---

[16] Note that similarly to Eqs. (8.22) and (8.25), here the uppercase notation of the component spins is replaced with their lowercase notation, to avoid any possibility of their confusion with the total spin of the system.

[17] Since we already know that the spin of a particle is physically nothing more than a (specific) part of its angular momentum, the similarity of the properties (8.48) of the sum (8.47) of spins of different particles to those of the sum (5.170) of different spin components of the same particle is very natural, but still has to be considered as a new law of Nature—confirmed by a vast body of experimental data.

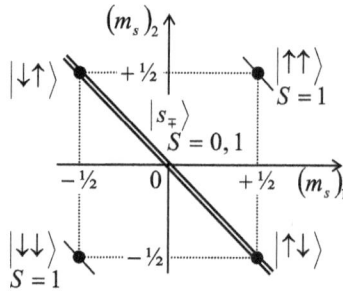

Figure 8.2. The 'rectangular diagram' showing the relation between the uncoupled-representation states (dots) and the coupled-representation states (straight lines) of a system of two spins-½—cf. figure 5.14.

$$\begin{cases} S = 0, \\ M_S = 0, \end{cases} \text{and} \begin{cases} S = 1, \\ M_S = 0, \pm 1. \end{cases} \tag{8.51}$$

It is virtually evident that the singlet spin state s_- belongs to the first class, while the simple (separable) triplet states $\uparrow\uparrow$ and $\downarrow\downarrow$ belong to the second class, with $M_S = +1$ and $M_S = -1$, respectively. However, for the entangled triplet state s_+, evidently with $M_S = 0$, the value of S is less obvious. Perhaps the easiest way to recover it[18] is to use the 'rectangular diagram', similar to that shown in figure 5.14, but redrawn for our case, i.e. with the replacements $m_l \to (m_s)_1 = \pm\frac{1}{2}$, $m_s \to (m_s)_2 = \pm\frac{1}{2}$—see figure 8.2.

Just as at the addition of angular momenta of a single particle, the top-right and bottom-left corners of this diagram correspond to the factorable triplet states $\uparrow\uparrow$ and $\downarrow\downarrow$, which participate in both the uncoupled-representation and coupled-representation bases, and have the largest value of S, i.e. 1. However, the entangled states $s_\pm$, which are linear combinations of the uncoupled-representation states $\uparrow\downarrow$ and $\downarrow\uparrow$, cannot have the same value of S, so that for the triplet state s_+, S has to take the value different from that (0) of the singlet state, i.e. 1. With that, the first of Eqs. (8.48) gives the following expectation values for the square of the net spin operator:

$$\langle S^2 \rangle = \begin{cases} 2\hbar^2, & \text{for each triplet state,} \\ 0, & \text{for the singlet state.} \end{cases} \tag{8.52}$$

Note that for the entangled triplet state s_+, whose ket-vector (8.20) is a linear superposition of two kets of states with *opposite* spins, this result is highly counter-intuitive, and shows how careful we should be interpreting quantum entangled states. (As will be discussed in chapter 10, the entanglement brings even more surprises for quantum measurements.)

Now returning for a moment to the particular issue of the magnetic field effect on the spins of ^{4}He atom's electrons, directing the axis z along the field, we may reduce Eq. (8.46) to

---

[18] Another, somewhat longer but perhaps more prudent way is to directly calculate the expectation values of $\hat{S}^2$ for the states $s_\pm$, and then find S by comparing the results with the first of Eqs. (8.48); it is highly recommended to the reader as a useful exercise.

$$\hat{H}_{\text{field}} = -\gamma_{e}\hat{S}_{z}\mathscr{B} \equiv 2\mu_{B}\mathscr{B}\frac{\hat{S}_{z}}{\hbar}. \tag{8.53}$$

Since all three triplet states (8.21) are eigenstates, in particular, of the operator $\hat{S}_{z}$, and hence of this Hamiltonian, we may use the second of Eqs. (8.48) to calculate their energy change simply as

$$\Delta E_{\text{field}} = 2\mu_{B}\mathscr{B}M_{S} = 2\mu_{B}\mathscr{B} \times \begin{cases} +1, & \text{for the factorable triplet state } \uparrow\uparrow, \\ 0, & \text{for the entangled triplet state } s_{+}, \\ -1, & \text{for the factorable triplet state } \downarrow\downarrow. \end{cases} \tag{8.54}$$

This splitting of the 'orthohelium' level is schematically shown in figure 8.1b.[19]

8.3 Multiparticle systems

Leaving several other problems on two-particle systems for the reader's exercise, let me proceed to the discussion of systems with $N > 2$ indistinguishable particles, whose list notably includes atoms, molecules, and condensed-matter systems. In this case, Eq. (8.7) for fermions is generalized as

$$\hat{\mathscr{P}}_{kk'}|\alpha_{-}\rangle = -|\alpha_{-}\rangle, \quad \text{for all } k, k' = 1, 2, \dots, N, \tag{8.55}$$

where the operator $\hat{\mathscr{P}}_{kk'}$ permutes particles with numbers k and k'. As a result, for systems with non-directly-interacting fermions, the Pauli principle forbids any state in which *any* two particles have similar single-particle wavefunctions. Nevertheless, it permits two fermions to have similar *orbital* wavefunctions, provided that their spins are in the singlet state (8.18), because this satisfies the permutation requirement (8.55). This fact has the paramount importance for the ground state of the systems whose Hamiltonians do not depend on spin, because it allows the fermions to be in their orbital single-particle ground states, with two electrons of the spin singlet sharing the same orbital state. Hence, for the limited (but very important!) goal of finding ground-state energies of multi-fermion systems with negligible direct interaction, we may ignore the actual singlet spin structure, and reduce the Pauli exclusion principle to the simple picture of single-particle orbital energy levels, each 'occupied' with two fermions.

As a very simple example, let us find the ground energy of five fermions, confined in a hard-wall, cubic-shaped 3D volume of side a, ignoring their direct interaction. From section 1.7, we know the single-particle energy spectrum of the system:

---

[19] It is interesting that another very important two-electron system, the hydrogen (H$_2$) molecule, which was briefly discussed in section 2.6, also has two similarly named forms, *parahydrogen* and *orthohydrogen*. However, their difference is due to two possible (respectively, singlet and triplet) states of the system of two spins of the hydrogen *nuclei* (protons), which are also spin-½ particles. The resulting energy of the parahydrogen is lower than that of the orthohydrogen by only ~45 meV per molecule—the difference comparable with $k_{B}T$ at room temperature (~26 meV). As a result, at the ambient conditions, the equilibrium ratio of these two *spin isomers* is close to 3:1. Curiously, the theoretical prediction of this minor effect by W Heisenberg (together with F Hund) in 1927 was cited in his 1932 Nobel Prize award as the most noteworthy application of quantum theory.

$$\varepsilon_{n_x,n_y,n_z} = \varepsilon_0\left(n_x^2 + n_y^2 + n_z^2\right), \quad \text{with} \quad \varepsilon_0 \equiv \frac{\pi^2\hbar^2}{2ma^2}, \quad \text{and } n_x, n_y, n_z = 1, 2, \ldots \quad (8.56)$$

so that the lowest-energy states are:

- one ground state with $\{n_x, n_y, n_z\} = \{1,1,1\}$, and energy $\varepsilon_{111} = (1^2 + 1^2 + 1^2)\varepsilon_0 = 3\varepsilon_0$, and
- three excited states, with $\{n_x, n_y, n_z\}$ equal to $\{2,1,1\}$, $\{1,2,1\}$, and $\{1,1,2\}$, with equal energies $\varepsilon_{211} = \varepsilon_{121} = \varepsilon_{112} = (2^2 + 1^2 + 1^2)\varepsilon_0 = 6\varepsilon_0$.

According to the above simple formulation of the Pauli principle, each of these orbital energy levels can accommodate up to two fermions. Hence the lowest-energy (ground) state of the five-fermion system is achieved by placing two of them on the ground level $\varepsilon_{111} = 3\varepsilon_0$, and the remaining three particles, in any degenerate 'excited' states of energy $6\varepsilon_0$, so that the ground-state energy of the system is

$$E_g = 2 \times 3\varepsilon_0 + 3 \times 6\varepsilon_0 \equiv 24\varepsilon_0 \equiv \frac{12\pi^2\hbar^2}{ma^2}. \quad (8.57)$$

Moreover, in many cases a relatively weak interaction between fermions does not blow up such a simple quantum state classification scheme qualitatively, and the Pauli principle allows tracing the order of single-particle state filling. This is exactly the simple approach that has been used at our discussion of atoms in section 3.7. Unfortunately, it does not allow for a more specific characterization of the ground states of most atoms, in particular the evaluation of the corresponding values of the quantum numbers S, L, and J that characterize the net angular momenta of the atom, and hence its response to external magnetic field. These numbers are defined by relations similar to Eqs. (8.48), for the vector-operators of total angular momenta:

$$\hat{\mathbf{S}} \equiv \sum_{k=1}^{N}\hat{\mathbf{s}}_k, \quad \hat{\mathbf{L}} \equiv \sum_{k=1}^{N}\hat{\mathbf{l}}_k, \quad \hat{\mathbf{J}} \equiv \sum_{k=1}^{N}\hat{\mathbf{j}}_k; \quad (8.58)$$

note that these definitions are consistent with Eq. (5.170) applied both to the angular momenta $\mathbf{s}_k$, $\mathbf{l}_k$, and $\mathbf{j}_k$ of each particle, and to the full vectors $\mathbf{S}$, $\mathbf{L}$, and $\mathbf{J}$. When the numbers S, L, and J for a state are known, they are traditionally recorded in the form of the so-called *Russell–Saunders symbols*[20]:

$$^{2S+1}\mathscr{L}_J, \quad (8.59)$$

where S and J are the corresponding values of these quantum numbers, while $\mathscr{L}$ is a capital *letter*, encoding the quantum number L via the same spectroscopic notation as for single particles (see section 3.6): $\mathscr{L} = S$ for $L = 0$, $\mathscr{L} = P$ for $L = 1$, $\mathscr{L} = D$ for $L = 2$, etc. (The reason why the front superscript of the Russel–Saunders symbol lists $2S + 1$ rather than S, is that according to the last of Eqs. (8.48), it shows the number

---

[20] Named after H Russell and F Saunders, whose pioneering (circa 1925) processing of experimental spectral-line data has established the very idea of vector addition of electron spins, described by the first of Eqs. (8.58).

of possible values of the quantum number M_S, which characterizes the state's spin degeneracy, and is called its *multiplicity*.)

For example, for the simplest, hydrogen atom ($Z = 1$), with its single electron in the ground $1s$ state, $L = l = 0$, $S = s = \frac{1}{2}$, and $J = S = \frac{1}{2}$, so that its Russell–Saunders symbol is $2S_{1/2}$. Next, the discussion of the helium atom ($Z = 2$) in the previous section has shown that in its ground state $L = 0$ (because of the $1s$ orbital state of both electrons), and $S = 0$ (because of the singlet spin state), so that the total angular momentum also vanishes: $J = 0$. As a result, the Russell–Saunders symbol is $^1S_0$. The structure of the next atom, lithium ($Z = 3$) is also easy to predict, because, as was discussed in section 3.7, its ground-state electron configuration is $1s^22s^1$, i.e. includes two electrons in the 'helium shell', i.e. on the $1s$ orbitals (now we know that they are actually in a singlet spin state), and one electron in the $2s$ state, of much higher energy, also with zero orbital moment, $l = 0$. As a result, the total L in this state is evidently equal to 0, and S is equal to $\frac{1}{2}$, so that $J = \frac{1}{2}$, meaning that the Russell-Saunders symbol of lithium is $^2P_{1/2}$. Even the next atom, beryllium ($Z = 4$), with the ground-state configuration $1s^22s^2$, is readily predictable, because none of its electrons has orbital momentum, giving $L = 0$. Also, each electron pair is in the singlet spin state, i.e. we have $S = 0$, so that $J = 0$—the quantum number set described by the Russell-Saunders symbol $^1S_0$—just as for helium.

However, for the next, boron atom ($Z = 5$), with its ground-state electron configuration $1s^22s^22p^1$ (see, e.g. figure 3.24), there is no obvious way to predict the result. Indeed, this atom has two pairs of electrons, with opposite spins, on its two lowest s-orbitals, giving zero contributions to the net S, L, and J. Hence these total quantum numbers may be only contributed by the last, fifth electron with $s = \frac{1}{2}$ and $l = 1$, giving $S = \frac{1}{2}$, $L = 1$. As was discussed in section 5.7 for the single-particle case, the vector addition of the angular momenta $\mathbf{S}$ and $\mathbf{L}$ enables two values rather than one of the quantum number J: either $L + S = 3/2$, or $L - S = \frac{1}{2}$. Experiment shows that the difference between the energies of these two states is very small ($\sim$2 meV), so that at room temperature they are both occupied, with the genuine ground state having $J = \frac{1}{2}$, so that its Russell–Saunders symbol is $^2P_{1/2}$.

Such energy differences, which become larger for heavier atoms, are determined both by the Coulomb and spin–orbit[21] interactions between the electrons. Their quantitative analysis is rather involved (see below), but the results tend to follow simple phenomenological *Hund rules*, with the following hierarchy:

Rule 1. For a given electron configuration, the ground state has the *largest* possible S, and hence the largest multiplicity.

Rule 2. For a given S, the ground state has the *largest* possible L.

Rule 3. For given S and L, J has its *smallest* possible value, $|L - S|$, if the given sub-shell $\{n, l\}$ is filled not more than by half, while in the opposite case, J has its *largest* possible value, $L + S$.

---

[21] In light atoms, the spin–orbit interaction is so weak that it may be reasonably well described as interaction of the total momenta $\mathbf{L}$ and $\mathbf{S}$ of the system—the so-called LS (or 'Russell–Saunders') *coupling*. On the other hand, in very heavy atoms, the interaction is effectively between the net momenta $\mathbf{j}_k = \mathbf{l}_k + \mathbf{s}_k$ of the individual electrons—the so-called *jj coupling*. This is the reason why in such atoms the Hund's Rule 3 may be violated.

Let us see how these rules work for the boron atom. For it, the Hund Rules 1 and 2 are satisfied automatically, while the sub-shell $\{n = 2, l = 1\}$, which can house up to $(2l + 1)s = 6$ electrons, is filled less than by half with just one $2p$ electron. As a result, the Hund Rule 3 predicts the ground state's value $J = \frac{1}{2}$, in agreement with experiment. Generally, for lighter atoms the Hund rules are well obeyed. However, the lower down the Hund rule hierarchy, the less 'powerful' the rules are, i.e. in more heavier atoms they are violated.

Now let us discuss possible approaches to a quantitative theory of multiparticle systems—not only atoms. As was discussed in section 8.1, if fermions do not interact directly, the stationary states of the system have to be the antisymmetric eigenstates of the permutation operator, i.e. satisfy Eq. (8.55). In order to understand how such states may be formed from the single-electron ones, let us return for a minute to the case of two electrons, and rewrite Eq. (8.11) in the following compact form:

$$
\begin{array}{cc}
\text{state 1} & \text{state 2} \\
\downarrow & \downarrow
\end{array}
\tag{8.60a}
$$

$$
|\alpha_-\rangle \equiv \frac{1}{\sqrt{2}}\big(|\beta\rangle \otimes |\beta'\rangle - |\beta'\rangle \otimes |\beta\rangle\big) \equiv \frac{1}{\sqrt{2}}\begin{Vmatrix} |\beta\rangle & |\beta'\rangle \\ |\beta\rangle & |\beta'\rangle \end{Vmatrix}
\begin{array}{l} \leftarrow \text{particle number 1,} \\ \leftarrow \text{particle number 2,} \end{array}
$$

where the direct product signs are just implied. In this way, the Pauli principle is mapped on the well-known property of matrix determinants: if any of two columns of a matrix coincide, its determinant vanishes. This *Slater determinant* approach[22] may be readily generalized to N fermions in N (not necessarily the lowest) single-particle states β, β', β'', etc:

$$
|\alpha_-\rangle = \frac{1}{(N!)^{1/2}} \underbrace{\begin{Vmatrix} |\beta\rangle & |\beta'\rangle & |\beta''\rangle & \cdots \\ |\beta\rangle & |\beta'\rangle & |\beta''\rangle & \cdots \\ |\beta\rangle & |\beta'\rangle & |\beta''\rangle & \cdots \\ \cdots & \cdots & \cdots & \cdots \end{Vmatrix}}_{N} \left.\begin{array}{l} \\ \\ \\ \end{array}\right\}N \quad \begin{array}{c} \text{particle} \\ \text{list} \\ \downarrow \end{array}
\tag{8.60b}
$$

Even though the Slater determinant form is extremely nice and compact (in comparison with direct writing of a sum of $N!$ products, each of N ket factors), there are two major problems with using it for practical calculations:

(i) For the calculation of any bra–ket product (say, within the perturbation theory) we still need to spell out each bra- and ket-vector as a sum of component terms. Even for a limited number of electrons (say $N \sim 10^2$ in a typical atom), the number $N! \sim 10^{160}$ of terms in such a sum is impracticably large for any analytical or numerical calculation.

(ii) In the case of interacting fermions, the Slater determinant does not describe the eigenvectors of the system; rather the stationary state is a *superposition* of such

---

[22] It was suggested in 1929 by J Slater.

basis functions, i.e. the Slater determinants—each for a specific selection of N states from the general set of single-particle states—that is generally larger than N.

For atoms and simple molecules, whose filled-shell electrons may be excluded from an explicit analysis (by describing their effects, approximately, with effective *pseudo-potentials*), the effective number N may be reduced to a smaller number N_{ef} of the order of 10, so that $N_{ef}! < 10^6$, and the Slater determinants may be used for numerical calculations—for example, in the Hartree–Fock theory—see the next section. However, for condensed-matter systems, such as metals and semiconductors, with the number of free electrons is of the order of 10^{23} per cm^3, this approach is generally unacceptable, though with some smart tricks (such as using crystal periodicity) it may be still used for some approximate (also numerical) calculations.

These challenges make the development of a more general theory that would not use particle numbers (which are superficial for indistinguishable particles to start with) a must for getting any final analytical results for multiparticle systems. The most effective formalism for this purpose, that avoids particle numbering at all, is called the *second quantization*[23]. Actually, we have already discussed a particular version of this formalism, for the case of 1D harmonic oscillator, in section 5.4. As a reminder, after the definition (5.65) of the 'creation' and 'annihilation' operators via those of the particle's coordinate and momentum, we have derived their key properties (5.89),

$$\hat{a}|n\rangle = n^{1/2}|n-1\rangle, \quad \hat{a}^\dagger|n\rangle = (n+1)^{1/2}|n+1\rangle, \qquad (8.61)$$

where n were the stationary (Fock) states of the oscillator. This property allows an interpretation of the operators' actions as the creation/annihilation of a single *excitation* with the energy $\hbar\omega_0$—thus justifying the operator names. In the next chapter, we will show that such an excitation of an electromagnetic field mode may be interpreted as a massless *boson* with $s = 1$, called the *photon*.

In order to generalize this approach to *arbitrary* bosons, not appealing to a specific system, we may use relations similar to Eq. (8.61) to *define* the creation and annihilation operators. The definitions look simple in the language of the so-called *Dirac states*, described by ket-vectors

$$|N_1, N_2, ...N_j, ...\rangle, \qquad (8.62)$$

where N_j is the state *occupancy*, i.e. the number of bosons in the single-particle state j. Let me emphasize that here the indices 1, 2, ...j,... number single-particle *states* (including their spin parts) rather than *particles*. Thus the very notion of an individual particle's number is completely (and for indistinguishable particles, very relevantly) absent from this formalism. Generally, the set of single-particle

---

[23] It was invented (first for photons and then for arbitrary bosons) by P Dirac in 1927, and then modified in 1928 for fermions by E Wigner and P Jordan. Note that the term 'second quantization' is rather misleading for the non-relativistic applications we are discussing here, but finds certain justification in the quantum field theory.

states participating in the Dirac state may be selected in an arbitrary way, provided
that it is full and orthonormal in the sense

$$\left\langle N_1', N_2'..., N_{j'}', ...\middle| N_1, N_2..., N_j, ...\right\rangle = \delta_{N_1 N_1'}\delta_{N_2 N_2'}...\delta_{N_j N_j'}..., \tag{8.63}$$

though for systems of non- (or weakly) interacting bosons, using the stationary states
of individual particles in the system under analysis is almost always the best choice.

Now we can define the *particle annihilation operator* as follows:

$$\hat{a}_j|N_1, N_2, ...N_j, ...\rangle \equiv N_j^{1/2}|N_1, N_2, ...N_j - 1, ...\rangle. \tag{8.64}$$

Note that the pre-ket coefficient, similar to that in the first of Eqs. (8.61), guarantees
that an attempt to annihilate a particle in an unpopulated state gives the non-
existing ('null') state:

$$\hat{a}_j|N_1, N_2, ...0_j, ...\rangle = 0, \tag{8.65}$$

where the symbol 0_j means zero occupancy of the jth state. According to Eq. (8.63),
an alternative way to write Eq. (8.64) is

$$\left\langle N_1', N_2', ... , N_j', ...\middle|\hat{a}_j\middle|.N_1, N_2, .., N_j, ...\right\rangle = N_j^{1/2}\delta_{N_1 N_1'}\delta_{N_2 N_2'}...\delta_{N_j', N_j-1}... \tag{8.66}$$

According to the general Eq. (4.65), the matrix element of the Hermitian conjugate
operator $\hat{a}_j^\dagger$ is

$$\left\langle N_1', N_2', ... , N_j', ...\middle| \hat{a}_j^\dagger\middle|N_1, N_2, ...N_j, ...\right\rangle$$

$$= \left\langle N_1, N_2, ... , N_j, ...\middle|\hat{a}_j\middle|N_1', N_2', ..., N_j', ...\right\rangle^*$$

$$= \left\langle N_1, N_2, ... , N_j, ...\middle|\left(N_j'\right)^{1/2}\middle|N_1', N_2', ... , N_j' - 1, ...\right\rangle \tag{8.67}$$

$$= \left(N_j'\right)^{1/2}\delta_{N_1 N_1'}\delta_{N_2 N_2'}...\delta_{N_j, N_j'-1}...$$

$$= (N_j + 1)^{1/2}\,\delta_{N_1 N_1'}\delta_{N_2 N_2'}...\delta_{N_j+1, N_j'}...,$$

meaning that

$$\hat{a}_j^\dagger|N_1, N_2, ... , N_j, ...\rangle = (N_j + 1)^{1/2}|N_1, N_2, ... , N_j + 1, ...\rangle, \tag{8.68}$$

in the total compliance with the second of Eqs. (8.61). In particular, this *particle
creation operator* allows the description of the generation of a single particle from
the *vacuum* (not null!) *state* $|0, 0, ...\rangle$:

$$\hat{a}_j^\dagger|0, 0, ... , 0_j, ... , 0\rangle = |0, 0, ... , 1_j, ...0\rangle, \tag{8.69}$$

and hence a product of such operators may create, from the vacuum, a multiparticle state with an arbitrary set of occupancies[24]:

$$\underbrace{\hat{a}_1^\dagger \hat{a}_1^\dagger ... \hat{a}_1^\dagger}_{N_1 \text{ times}} \underbrace{\hat{a}_2^\dagger \hat{a}_2^\dagger ... \hat{a}_2^\dagger}_{N_2 \text{ times}} ... |0, 0, ...\rangle = (N_1! N_2! ...)^{1/2} |N_1, N_2, ...\rangle. \tag{8.70}$$

Next, combining Eqs. (8.64) and (8.68), we get

$$\hat{a}_j^\dagger \hat{a}_j |N_1, N_2, ... N_j, ...\rangle = N_j |N_1, N_2, ... , N_j, ...\rangle, \tag{8.71}$$

so that, just as for the particular case of harmonic oscillator excitations, the operator

$$\hat{N}_j \equiv \hat{a}_j^\dagger \hat{a}_j \tag{8.72}$$

'counts' the number of particles in the jth single-particle state, while preserving the whole multiparticle state. Acting on a state by the creation–annihilation operators in the reverse order, we get

$$\hat{a}_j \hat{a}_j^\dagger |N_1, N_2, ... , N_j, ...\rangle = (N_j + 1)|N_1, N_2, ... , N_j, ...\rangle. \tag{8.73}$$

Eqs. (8.71) and (8.73) show that for *any* state of a multiparticle system (which always may be represented as a linear superposition of Dirac states with all possible sets of numbers N_j), we may write

$$\hat{a}_j \hat{a}_j^\dagger - \hat{a}_j^\dagger \hat{a}_j \equiv [\hat{a}_j, \hat{a}_j^\dagger] = \hat{I}, \tag{8.74}$$

again in agreement with what we had for the 1D oscillator—cf. Eq. (5.68). According to Eqs. (8.63), (8.64) and (8.68), the creation and annihilation operators corresponding to different single-particle states do commute, so that Eq. (8.74) may be generalized as

$$[\hat{a}_j, \hat{a}_{j'}^\dagger] = \hat{I}\delta_{jj'}, \tag{8.75}$$

while the similar operators commute, regardless of which states do they act upon:

$$\left[\hat{a}_j^\dagger, \hat{a}_{j'}^\dagger\right] = [\hat{a}_j, \hat{a}_{j'}] = \hat{0}. \tag{8.76}$$

As was mentioned earlier, a major challenge in the Dirac approach is to rewrite the Hamiltonian of a multiparticle system, that naturally carries particle numbers k (see, e.g. Eq. (8.22) for $k = 1, 2$), in the second quantization language, in which there are not these numbers. Let us start with *single-particle* components of such Hamiltonians, i.e. operators of the type

$$\hat{F} = \sum_{k=1}^{N} \hat{f}_k. \tag{8.77}$$

---

[24] The resulting Dirac state is *not* an eigenstate of every multiparticle Hamiltonian. However, we will see below that for a set of non-interacting particles it *is* a stationary state, so that the full set of such states may be used as a good basis in perturbation theories of systems of weakly interacting particles.

where all N operators $\hat{f}_k$ are similar, besides that each of them acts on one specific (kth) particle, and N is the total number of particles in the system, which is evidently equal to the sum of single-particle state occupancies:

$$N = \sum_j N_j. \tag{8.78}$$

The most important examples of such operators are the kinetic energy of N similar single particles, and their potential energy in an external field:

$$\hat{T} = \sum_{k=1}^{N} \frac{\hat{p}_k^2}{2m}, \qquad \hat{U} = \sum_{k=1}^{N} \hat{u}(\mathbf{r}_k). \tag{8.79}$$

For bosons, instead of the Slater determinant (8.60), we have to write a similar expression, but without the sign alternation at permutations:

$$|N_1, \ldots N_j, \ldots\rangle = \left(\frac{N_1! \ldots N_j! \ldots}{N!}\right)^{1/2} \sum_P \left|\underbrace{\ldots \beta\beta'\beta'' \ldots}_{N \text{ operands}}\right\rangle, \tag{8.80}$$

sometimes called the *permanent*. Note again that the left-hand side of this relation is written in the Dirac notation (that does not use particle numbering), while on its right-hand side, just in relations of sections 8.1 and 8.2, the particle numbers are coded with the positions of the single-particle states inside the ket-vectors, and the sum is over all different permutations of the states in the ket—cf. Eq. (8.10). (According to the basic combinatorics[25], there are $N!/(N_1! \ldots N_j! \ldots)$ such permutations, so that the front coefficient in Eq. (8.80) ensures the normalization of the Dirac state, provided that the single-particle states β, β', …are normalized.) Let us use Eq. (8.80) to spell out the following matrix element for a system with $(N-1)$ particles:

$$\langle \ldots N_j, \ldots N_{j'} - 1, \ldots | \hat{F} | \ldots N_j - 1, \ldots N_{j'}, \ldots \rangle$$

$$= \frac{N_1! \ldots (N_j - 1)! \ldots (N_{j'} - 1)! \ldots}{(N-1)!} (N_j N_{j'})^{1/2}$$

$$\times \sum_{P\langle N-1|} \sum_{P|N-1\rangle} \langle \ldots \beta\beta'\beta'' \ldots | \sum_{k=1}^{N-1} \hat{f}_k | \ldots \beta\beta'\beta'' \ldots \rangle, \tag{8.81}$$

where all non-specified occupation numbers in the corresponding positions of the bra- and ket-vectors are equal to each other. Each single-particle operator $\hat{f}_k$, participating in the operator sum, acts on the bra- and ket-vectors of states only in one (kth) position, giving the result, which does not depend on the position number:

---

[25] See, e.g. Eq. (A.6).

$$\langle \beta_j |_{\text{in } k^{\text{th}} \text{ position}} \, \hat{f}_k \, |\beta_{j'}\rangle_{\text{in } k^{\text{th}} \text{ position}} = \langle \beta_j| \hat{f} |\beta_{j'}\rangle \equiv f_{jj'}. \qquad (8.82)$$

Since in both permutation sets participating in Eq. (8.81), with $(N-1)$ state vectors each, all positions are equivalent, we can fix the position (say, take the first one) and replace the sum over k with the multiplication by of the bracket by $(N-1)$. The fraction of permutations with the necessary bra-vector (with number j) in that position is $N_j/(N-1)$, while that with the necessary ket-vector (with number j') in the same position in $N_{j'}/(N-1)$. As a result, the permutation sum in Eq. (8.81) reduces to

$$(N-1)\frac{N_j}{N-1}\frac{N_{j'}}{N-1}f_{jj'} \sum_{P\langle N-2|}\sum_{P|N-2\rangle} \langle ...\beta\beta'\beta''...|...\beta\beta'\beta''...\rangle, \qquad (8.83)$$

where our specific position k is now excluded from both the bra- and ket-vector permutations. Each of these permutations now includes only $(N_j - 1)$ states j and $(N_{j'} - 1)$ states j', so that, using the state orthonormality, we finally arrive at a very simple result:

$$\begin{aligned} &\langle ...N_j, \,...N_{j'} - 1, \,...|\hat{F}|...N_j - 1, \,...N_{j'}, \,...\rangle \\ &= \frac{N_1!...(N_j - 1)!...(N_{j'} - 1)!...}{(N-1)!}(N_jN_{j'})^{1/2}(N-1) \\ &\times \frac{N_j}{N-1}\frac{N_{j'}}{N-1}f_{jj'}\frac{(N-2)!}{N_1!...(N_j - 1)!...(N_{j'} - 1)!...} \equiv (N_jN_{j'})^{1/2}f_{jj'}. \end{aligned} \qquad (8.84)$$

On the other hand, let us calculate matrix elements of the following operator:

$$\sum_{j,j'} f_{jj'} \, \hat{a}_j^\dagger \hat{a}_{j'}. \qquad (8.85)$$

A direct application of Eqs. (8.64) and (8.68) shows that the only nonvanishing of the elements are

$$\langle ...N_j, \,...N_{j'} - 1, \,...|f_{jj'}\hat{a}_j^\dagger \hat{a}_{j'}|...N_j - 1, \,... , \,N_{j'}, \,...\rangle = (N_jN_{j'})^{1/2}f_{jj'}. \qquad (8.86)$$

But this is exactly the last form of Eq. (8.84), so that in the basis of Dirac states, the operator (8.77) may be represented as

$$\hat{F} = \sum_{j,j'} f_{jj'} \, \hat{a}_j^\dagger \hat{a}_{j'}. \qquad (8.87)$$

This beautifully simple equation is the key formula of the second quantization theory, and is essentially the Dirac-language analog of Eq. (4.59) of the single-particle quantum mechanics. Each term of the sum (8.87) may be described by a very simple mnemonic rule: for each pair of single-particle states j and j', kill a particle in the state j', create one in the state j, and weigh the result with the corresponding single-particle matrix element. One of corollaries of Eq. (8.87) is that the expectation value of an operator whose eigenstates coincide with the Dirac states, is

$$\langle F \rangle \equiv \langle ...N_j, \, ...|\hat{F}|...N_j, \, ...\rangle \; = \; \sum_j f_{jj} N_j, \tag{8.88}$$

with an evident physical interpretation as the sum of single-particle expectation values over all states, weighed by the occupancy of each state.

Proceeding to *fermions*, which have to obey the Pauli principle, we immediately notice that any occupation number N_j may only take two values, 0 or 1. In order to account for that, and also make the key relation (8.87) valid for fermions as well, the creation–annihilation operators are now defined by the following relations:

$$\hat{a}_j|N_1, \, N_2, \, ... \, , \, 0_j, \, ...\rangle = 0,$$
$$\hat{a}_j|N_1, \, N_2, \, ... \, , \, 1_j, \, ...\rangle = (-1)^{\Sigma(1, j-1)}|N_1, \, N_2, \, ... \, , \, 0_j, \, ...\rangle, \tag{8.89}$$

$$\hat{a}_j^{\dagger}|N_1, \, N_2, \, ... \, , \, 0_j, \, ...\rangle = (-1)^{\Sigma(1, j-1)}|N_1, \, N_2, \, ... \, , \, 1_j, \, ...\rangle,$$
$$\hat{a}_j^{\dagger}|N_1, \, N_2, \, ... \, , \, 1_j, \, ...\rangle = 0, \tag{8.90}$$

where the symbol $\Sigma(J, J')$ means the sum of all occupancy numbers in the states with numbers from J to J', including the border points:

$$\Sigma(J, \, J') \equiv \sum_{j=J}^{J'} N_j, \tag{8.91}$$

so that the sum participating in Eqs. (8.89) and (8.90) is the total occupancy of all states with the numbers below j. (The states are supposed to be numbered in a fixed albeit arbitrary order.) As a result, these relations may be conveniently summarized in the following verbal form: if an operator replaces the jth state's occupancy with the opposite one (either 1 with 0, or vice versa), it also changes the sign before the result if (and only if) the total number of particles in the states with $j' < j$ is odd.

Let us use this (perhaps somewhat counter-intuitive) sign alternation rule to spell out the ket-vector $|11\rangle$ of a completely filled two-state system, formed from the vacuum state $|00\rangle$ in two different ways. If we start from creating a fermion in the state 1, we get

$$\hat{a}_1^{\dagger}|0, \, 0\rangle = (-1)^0|1, \, 0\rangle \equiv |1, \, 0\rangle,$$
$$\hat{a}_2^{\dagger}\hat{a}_1^{\dagger}|0, \, 0\rangle = \hat{a}_2^{\dagger}|1, \, 0\rangle = (-1)^1|1, \, 1\rangle \equiv -|1, \, 1\rangle, \tag{8.92a}$$

while if the operator order is different, the result is

$$\hat{a}_2^{\dagger}|0, \, 0\rangle = (-1)^0|0, \, 1\rangle \equiv |0, \, 1\rangle,$$
$$\hat{a}_1^{\dagger}\hat{a}_2^{\dagger}|0, \, 0\rangle = \hat{a}_1^{\dagger}|0, \, 1\rangle = (-1)^0|1, \, 1\rangle \equiv |1, \, 1\rangle, \tag{8.92b}$$

so that

$$\left(\hat{a}_1^\dagger \hat{a}_2^\dagger + \hat{a}_2^\dagger \hat{a}_1^\dagger\right)|0,\,0\rangle = 0. \tag{8.93}$$

Since the action of any of these operator products on any initial state rather than the vacuum one also gives the null ket, we can write the following operator equality:

$$\hat{a}_1^\dagger \hat{a}_2^\dagger + \hat{a}_2^\dagger \hat{a}_1^\dagger \equiv \left\{\hat{a}_1^\dagger,\, \hat{a}_2^\dagger\right\} = \hat{0}. \tag{8.94}$$

It is straightforward to check that this result is valid for the Dirac vector of an arbitrary length, and does not depend on the occupancy of other states, so that we can always write

$$\left\{\hat{a}_j^\dagger,\, \hat{a}_{j'}^\dagger\right\} = \{\hat{a}_j,\, \hat{a}_{j'}\} = \hat{0}; \tag{8.95}$$

these equalities hold for $j = j'$ as well. On the other hand, an absolutely similar calculation shows that the mixed creation–annihilation commutators do depend on whether the states are different or not[26]:

$$\{\hat{a}_j,\, \hat{a}_{j'}^\dagger\} = \hat{I}\delta_{jj'}. \tag{8.96}$$

These equations look very much like Eqs. (8.75) and (8.76) for bosons, 'only' with the replacement of commutators with anticommutators. Since the core laws of quantum mechanics, including the operator compatibility (section 4.5) and the Heisenberg equation (4.199) of operator evolution in time, involve commutators rather than anticommutators, one might think that all the behavior of bosonic and fermionic multiparticle systems should be dramatically different. However, the difference is not as huge as one could expect; indeed, a straightforward check shows that the sign factors in Eqs. (8.89) and (8.90) just compensate those in the Slater determinant, and thus make the key relation (8.87) valid for the fermions as well. (Indeed, this is the very goal of the introduction of these factors.)

To illustrate this fact on the simplest example, let us examine what the second quantization formalism says about the dynamics of non-interacting particles in the system whose single-particle properties we have discussed repeatedly, namely two nearly-similar potential wells, coupled by tunneling through the separating potential barrier—see, e.g. figures 2.21 or 7.4. If the coupling is so small that the states localized in the wells are only weakly perturbed, then in the basis of these states, the single-particle Hamiltonian of the system may be represented by the 2×2 matrix (5.3). With the energy reference selected at the middle between the energies of unperturbed states, the coefficient b vanishes, this matrix is reduced to

$$\mathsf{h} = \mathbf{c} \cdot \boldsymbol{\sigma} \equiv \begin{pmatrix} c_z & c_- \\ c_+ & -c_z \end{pmatrix}, \quad \text{with } c_\pm \equiv c_x \pm ic_y, \tag{8.97}$$

---

[26] A by-product of this calculation is a proof that the operator defined by Eq. (8.72) counts the number of particles N_j (now equal to either 1 or 0), just at it does for bosons.

and its eigenvalues to

$$\varepsilon_\pm = \pm c, \quad c \equiv |\mathbf{c}| \equiv \left(c_x^2 + c_y^2 + c_z^2\right)^{1/2}. \tag{8.98}$$

Now following the recipe (8.87), we can use Eq. (8.97) to represent the Hamiltonian of the whole system of particles in terms of the creation–annihilation operators:

$$\hat{H} = c_z \hat{a}_1^\dagger \hat{a}_1 + c_- \hat{a}_1^\dagger \hat{a}_2 + c_+ \hat{a}_2^\dagger \hat{a}_1 - c_z \hat{a}_2^\dagger \hat{a}_2, \tag{8.99}$$

where $\hat{a}_{1,2}^\dagger$ and $\hat{a}_{1,2}$ are the operators of creation and annihilation of a particle in the corresponding *potential well*. (Again, in the second quantization approach the *particles* are not numbered at all!) As Eq. (8.72) shows, the first and the last terms of the right-hand side of Eq. (8.99) describe the particle energies $\varepsilon_{1,2} = \pm c_z$ in uncoupled wells,

$$c_z \hat{a}_1^\dagger \hat{a}_1 = c_z \hat{N}_1 \equiv \varepsilon_1 \hat{N}_1, \quad -c_z \hat{a}_2^\dagger \hat{a}_2 = -c_z \hat{N}_2 \equiv \varepsilon_2 \hat{N}_2, \tag{8.100}$$

while the sum of the middle two terms is the second-quantization description of tunneling between the wells.

Now we can use the general Eq. (4.199) of the Heisenberg picture to spell out the equations of motion of the creation–annihilation operators. For example,

$$i\hbar \dot{\hat{a}}_1 = [\hat{a}_1, \hat{H}] = c_z[\hat{a}_1, \hat{a}_1^\dagger \hat{a}_1] + c_-[\hat{a}_1, \hat{a}_1^\dagger \hat{a}_2] + c_+[\hat{a}_1, \hat{a}_2^\dagger \hat{a}_1] - c_z[\hat{a}_1, \hat{a}_2^\dagger \hat{a}_2]. \tag{8.101}$$

Since the Bose and Fermi operators satisfy different commutation relations, one could expect the right hand part of this equation to be different for bosons and fermions. However, it is not so. Indeed, all commutators on the right-hand side of Eq. (8.101) have the following form:

$$[\hat{a}_j, \hat{a}_{j'}^\dagger \hat{a}_{j''}] \equiv \hat{a}_j \hat{a}_{j'}^\dagger \hat{a}_{j''} - \hat{a}_{j'}^\dagger \hat{a}_{j''} \hat{a}_j. \tag{8.102}$$

According to Eqs. (8.74) and (8.94), the first pair product of the operators may be recast as

$$\hat{a}_j \hat{a}_{j'}^\dagger = \hat{I}\delta_{jj'} \pm \hat{a}_{j'}^\dagger \hat{a}_j, \tag{8.103}$$

where the upper sign pertains to bosons and the lower one to fermions, while according to Eqs. (8.76) and (8.95), the very last pair product is

$$\hat{a}_{j''} \hat{a}_j = \pm \hat{a}_j \hat{a}_{j''}, \tag{8.104}$$

with the same sign convention. Plugging these expressions into Eq. (8.102), we see that regardless of the particle type, we arrive at a universal (and generally very useful) commutation relation

$$[\hat{a}_j, \hat{a}_{j'}^\dagger \hat{a}_{j''}] = \hat{a}_{j''} \delta_{jj'}, \tag{8.105}$$

valid for both bosons and fermions. As a result, the Heisenberg equation of motion for the operator $\hat{a}_1$, and the equation for $\hat{a}_2$ (which may be obtained absolutely similarly), are also universal[27]:

$$i\hbar\dot{\hat{a}}_1 = c_z\hat{a}_1 + c_-\hat{a}_2,$$
$$i\hbar\dot{\hat{a}}_2 = c_+\hat{a}_1 - c_z\hat{a}_2.$$

(8.106)

This is a system of two coupled, linear differential equations, which is similar to the equations for the c-number probability amplitudes of single-particle wavefunctions of a two-level system—see, e.g. Eq. (2.201) and the model solution of problem 4.25. Their general solution is a linear superposition

$$\hat{a}_{1,2}(t) = \sum_{\pm}\hat{a}_{1,2}^{(\pm)}\exp\{\lambda_{\pm}t\}.$$

(8.107)

As usual, in order to find the exponents $\lambda_{\pm}$, it is sufficient to plug in a particular solution $\hat{a}_{1,2}(t) = \hat{a}_{1,2}\exp\{\lambda t\}$ into Eq. (8.106) and require that the determinant of the resulting homogeneous, linear system for the 'coefficients' (actually, time-independent operators) $\hat{a}_{1,2}$ equals zero. This gives us the following characteristic equation

$$\begin{vmatrix} c_z - i\hbar\lambda & c_- \\ c_+ & -c_z - i\hbar\lambda \end{vmatrix} = 0,$$

(8.108)

with two roots $\lambda_{\pm} = \pm i\Omega/2$, where $\Omega \equiv 2c\hbar$—cf. Eq. (5.20). Now plugging each of the roots, one by one, into the system of equations for $\hat{a}_{1,2}$, we can find these operators, and hence the general solution of system (8.98) for arbitrary initial conditions.

Let us consider the simple case $c_y = c_z = 0$ (meaning in particular that the wells are exactly aligned, see figure 2.21), so that $\hbar\Omega/2 \equiv c = c_x$; then the solution of Eq. (8.106) is

$$\hat{a}_1(t) = \hat{a}_1(0)\cos\frac{\Omega t}{2} - i\hat{a}_2(0)\sin\frac{\Omega t}{2},$$
$$\hat{a}_2(t) = -i\hat{a}_1(0)\sin\frac{\Omega t}{2} + \hat{a}_2(0)\cos\frac{\Omega t}{2}.$$

(8.109)

Multiplying the first of these relations by its Hermitian conjugate, and ensemble-averaging the result, we get

$$\langle N_1 \rangle \equiv \langle \hat{a}_1^\dagger(t)\hat{a}_1(t) \rangle = \langle \hat{a}_1^\dagger(0)\hat{a}_1(0) \rangle\cos^2\frac{\Omega t}{2} + \langle \hat{a}_2^\dagger(0)\hat{a}_2(0) \rangle\sin^2\frac{\Omega t}{2}$$
$$- i\langle \hat{a}_1^\dagger(0)\hat{a}_2(0) + \hat{a}_2^\dagger(0)\hat{a}_1(0) \rangle\sin\frac{\Omega t}{2}\cos\frac{\Omega t}{2}.$$

(8.110)

---

[27] Equations of motion for the creation operators $\hat{a}_{1,2}^\dagger$ are just the Hermitian-conjugates of Eqs. (8.106), and do not add any new information about system's dynamics.

Let the initial state of the system be a Dirac state, i.e. have a definite number of particles in each well; in this case only the two first terms on the right hand side of Eq. (8.110) are different from zero, giving[28]:

$$\langle N_1 \rangle = N_1(0)\cos^2\frac{\Omega t}{2} + N_2(0)\sin^2\frac{\Omega t}{2}. \tag{8.111}$$

For one particle, initially placed in either well, this gives us our old result (2.181) describing quantum oscillations of the particle between two wells with the frequency Ω. However, Eq. (8.111) is valid for any set of initial occupancies; let us use this fact. For example, starting from two particles, with initially one particle in each well, we get $\langle N_1 \rangle = 1$, regardless of time. So, the occupancies do not oscillate, and no experiment may detect the quantum oscillations, though their frequency Ω is still formally present in the time evolution equations. This fact may be interpreted as the simultaneous quantum oscillations of two particles between the wells, exactly in anti-phase. For bosons, we can go to even larger occupancies by preparing the system, for example, in the state with $N_1(0) = N$, $N_2(0) = 0$. The result (8.111) says that in this case we see that the quantum oscillation amplitude increases N-fold; this is a particular manifestation of the general fact that bosons can be (and evolve in time) in the same quantum state. On the other hand, for fermions we cannot increase the initial occupancies beyond 1, so that the largest oscillation amplitude we can get is if we initially fill just one well.

The Dirac approach may be readily generalized to more complex systems. For example, Eq. (8.99) implies that an arbitrary system of potential wells with weak tunneling coupling between the adjacent wells may be described by the Hamiltonian

$$\hat{H} = \sum_j \varepsilon_j a_j^\dagger \hat{a}_j + \sum_{\langle j,j' \rangle} \delta_{jj'} a_j^\dagger \hat{a}_{j'} + \text{h.c.}, \tag{8.112}$$

where the symbol $\langle j,j' \rangle$ means that the second sum is restricted to pairs of next-neighbor wells—see, e.g. Eq. (2.203) and its discussion. Note that this Hamiltonian is still a quadratic form of the creation–annihilation operators, so the Heisenberg-picture equations of motion of these operators are still linear, and its exact solutions, though possibly cumbersome, may be studied in detail. Due to this fact, the Hamiltonian (8.112) is widely used for the study of some phenomena, for example the very interesting *Anderson localization* effects, in which a random distribution of the localized-site energies ε_j prevents tunneling particles, within a certain energy range, from spreading to unlimited distances[29].

8.4 Perturbative approaches

The situation becomes much more difficult if we need to account for direct interactions between the particles. Let us assume that the interaction may be reduced to that between their pairs (as is the case at their Coulomb interaction

---

[28] For the second well's occupancy, the result is complementary, $N_2(t) = N_1(0)\sin^2\Omega t + N_2(0)\cos^2\Omega t$, giving in particular a good sanity check: $N_1(t) + N_2(t) = N_1(0) + N_2(0) = \text{const}$.

[29] For a review of the 1D version of this problem, see, e.g. [1].

and most other interactions[30]), so that it may be described by the following 'pair-interaction' Hamiltonian

$$\hat{U}_{\text{int}} = \frac{1}{2} \sum_{\substack{k,k'=1 \\ k \neq k'}}^{N} \hat{u}_{\text{int}}(\mathbf{r}_k, \mathbf{r}_{k'}), \tag{8.113}$$

with the front factor ½ compensating the double-counting of each particle pair. The translation of this operator to the second-quantization form may be done absolutely similarly to the derivation of Eq. (8.87), and gives a similar (though naturally more involved) result

$$\hat{U}_{\text{int}} = \frac{1}{2} \sum_{j,j',l,l'} u_{jj'll'} \hat{a}_j^\dagger \hat{a}_{j'}^\dagger \hat{a}_{l'} \hat{a}_l, \tag{8.114}$$

where the two-particle matrix elements are defined similarly to Eq. (8.82):

$$u_{jj'll'} \equiv \langle \beta_j \beta_{j'} | \hat{u}_{\text{int}} | \beta_l \beta_{l'} \rangle. \tag{8.115}$$

The only new feature of Eq. (8.114) is a specific order of the indices of the creation operators. Note the mnemonic rule of writing this expression, similar to that for Eq. (8.87): each term corresponds to moving a pair of particles from states l and l' to states j' and j (in this order!) factored with the corresponding two-particle matrix element (8.115).

However, with the account of such term, the resulting Heisenberg equations of time evolution of the creation/annihilation are nonlinear, so that solving them and calculating observables from the results is usually impossible, at least analytically. The only case when some general results may be obtained is the *weak interaction* limit. In this case the unperturbed Hamiltonian contains only single-particle terms such as (8.79), and we can always (at least conceptually) find such a basis of orthonormal single-particle states β_j in which that Hamiltonian is diagonal in the Dirac representation:

$$\hat{H}^{(0)} = \sum_j \varepsilon_j^{(0)} \hat{a}_j^\dagger \hat{a}_j. \tag{8.116}$$

Now we can use Eq. (6.14), in this basis, to calculate the interaction energy as a first-order perturbation:

$$\begin{aligned} E_{\text{int}}^{(1)} &= \langle N_1, N_2, \ldots | \hat{U}_{\text{int}} | N_1, N_2, \ldots \rangle \\ &= \frac{1}{2} \langle N_1, N_2, \ldots | \sum_{j,j',l,l'} u_{jj'll'} \hat{a}_j^\dagger \hat{a}_{j'}^\dagger \hat{a}_{l'} \hat{a}_l | N_1, N_2, \ldots \rangle \\ &= \frac{1}{2} \sum_{j,j',l,l'} u_{jj'll'} \langle N_1, N_2, \ldots | \hat{a}_j^\dagger \hat{a}_{j'}^\dagger \hat{a}_{l'} \hat{a}_l | N_1, N_2, \ldots \rangle. \end{aligned} \tag{8.117}$$

---

[30] A simple but important example from the condensed matter theory is the so-called *Hubbard model*, in which there may be only two particles on each of localized sites, which strongly interact, with negligible interaction of the particles on different sites—though the next-neighbor sites are still connected by tunneling—as in Eq. (8.112).

Since, according to Eq. (8.63), the Dirac states with different occupancies are orthogonal, the last bracket is different from zero only for three particular subsets of its indices:

(i) $j \neq j'$, $l = j$, and $l' = j'$. In this case the four-operator product in Eq. (8.117) is equal to $\hat{a}_j^\dagger \hat{a}_{j'}^\dagger \hat{a}_{j'} \hat{a}_j$, and applying the commutation rules twice, we can bring it to the so-called *normal ordering*, with each creation operator standing to the right of the corresponding annihilation operator, thus forming the particle number operator (8.72):

$$\hat{a}_j^\dagger \hat{a}_{j'}^\dagger \hat{a}_{j'} \hat{a}_j = \pm \hat{a}_j^\dagger \hat{a}_{j'}^\dagger \hat{a}_j \hat{a}_{j'} = \pm \hat{a}_j^\dagger (\pm \hat{a}_j \hat{a}_{j'}^\dagger) \hat{a}_{j'} = \hat{a}_j^\dagger \hat{a}_j \hat{a}_{j'}^\dagger \hat{a}_{j'} = \hat{N}_j \hat{N}_{j'}, \tag{8.118}$$

with the similar sign of the final result for bosons and fermions.

(ii) $j \neq j'$, $l = j'$, and $l' = j$. In this case the four-operator product is equal to $\hat{a}_j^\dagger \hat{a}_{j'}^\dagger \hat{a}_j \hat{a}_{j'}$, and bringing it to the form $\hat{N}_j \hat{N}_{j'}$ requires only one commutation:

$$\hat{a}_j^\dagger \hat{a}_{j'}^\dagger \hat{a}_j \hat{a}_{j'} = \hat{a}_j^\dagger (\pm \hat{a}_j \hat{a}_{j'}^\dagger) \hat{a}_{j'} = \pm \hat{a}_j^\dagger \hat{a}_j \hat{a}_{j'}^\dagger \hat{a}_{j'} = \pm \hat{N}_j \hat{N}_{j'}, \tag{8.119}$$

with the upper sign for bosons and the lower sign for fermions.

(iii) All indices equal to each other, giving $\hat{a}_j^\dagger \hat{a}_j^\dagger \hat{a}_l \hat{a}_l = \hat{a}_j^\dagger \hat{a}_j^\dagger \hat{a}_j \hat{a}_j$. For fermions, such operator (that 'tries' to create or to kill two particles in a row, in the same state) immediately gives the null vector. In the case of bosons, we may use Eq. (8.74) to commute the internal pair of operators, getting

$$\hat{a}_j^\dagger \hat{a}_j^\dagger \hat{a}_j \hat{a}_j = \hat{a}_j^\dagger (\hat{a}_j \hat{a}_j^\dagger - \hat{I}) \hat{a}_j = \hat{N}_j (\hat{N}_j - \hat{I}). \tag{8.120}$$

Note, however, that this expression formally covers the fermion case as well (always giving zero). As a result, Eq. (8.117) may be rewritten in the following universal form:

$$E_{\text{int}}^{(1)} = \frac{1}{2} \sum_{\substack{j,j' \\ j \neq j'}} N_j N_{j'} (u_{jj'jj'} \pm u_{jj'j'j}) + \frac{1}{2} \sum_j N_j (N_j - 1) \, u_{jjjj}. \tag{8.121}$$

The corollaries of this important result are very different for bosons and fermions. In the former case, the last term usually dominates, because the matrix elements (8.115) are typically the largest when all basis functions coincide. Note that this term allows a very simple interpretation: the number of the diagonal matrix elements it sums up for each state (j) is just the number of interacting particle pairs residing in that state.

In contrast, for fermions the last term is zero, and the interaction energy is the difference of the two terms inside the first parentheses. In order to spell them out, let us consider the case when there is no direct spin–orbit interaction. Then the vectors

$|\beta\rangle_j$ of the single-particle state basis may be represented as direct products $|o_j\rangle \otimes |m_j\rangle$ of their orbital and spin-orientation parts. (Here, for brevity, I am using m instead of m_s.) For spin-½ particles, including electrons, these orientations m_j may equal only either +½ or −½; in this case the spin part of the first matrix element, $u_{jj'jj'}$, equals

$$\langle m| \otimes \langle m'\|m\rangle \otimes |m'\rangle, \tag{8.122}$$

where, as in the general Eq. (8.115), the position of a particular vector in a direct product codes the particle's number. Since the spins of different particles are defined in different Hilbert spaces, we may move their vectors around to get

$$\langle m| \otimes \langle m'\|m\rangle \otimes |m'\rangle = (\langle m|m\rangle)_1 \times (\langle m'|m'\rangle)_2 = 1, \tag{8.123}$$

for any pair of j and j'. On the other hand, the second matrix element, $u_{jj'j'j}$, is proportional to

$$\langle m| \otimes \langle m'\|m'\rangle \otimes |m\rangle = (\langle m|m'\rangle)_1 \times (\langle m'|m\rangle)_2 = \delta_{mm'}. \tag{8.124}$$

In this case, it is convenient to rewrite Eq. (8.121) in the coordinate representation, using single-particle wavefunctions called *spin–orbitals*

$$\psi_j(\mathbf{r}) \equiv \langle \mathbf{r}|\beta_j\rangle = (\langle \mathbf{r}|o\rangle \otimes |m\rangle)_j. \tag{8.125}$$

They differ from the spatial parts of the usual orbital wavefunctions of the type (4.233) only in that their index j should be understood as the set of the orbital-state and the spin-orientation indices[31]. Also, due to the Pauli-principle restriction of numbers N_j to either 0 or 1, Eq. (8.121) may be also rewritten without the explicit occupancy numbers, with the understanding that the summation is extended only over the pairs of occupied states. As a result, it becomes

$$E_{\text{int}}^{(1)} = \frac{1}{2} \sum_{\substack{j,j' \\ j\neq j'}} \int d^3r \int d^3r' \left[\begin{array}{l} \psi_j^*(\mathbf{r})\, \psi_{j'}^*(\mathbf{r}')\, u_{\text{int}}(\mathbf{r},\mathbf{r}')\, \psi_j(\mathbf{r})\, \psi_{j'}(\mathbf{r}') \\ - \psi_j^*(\mathbf{r})\, \psi_{j'}^*(\mathbf{r}')\, u_{\text{int}}(\mathbf{r},\mathbf{r}')\, \psi_{j'}(\mathbf{r})\, \psi_j(\mathbf{r}') \end{array} \right]. \tag{8.126}$$

In particular, for a system of two electrons, we may limit the summation to just two states ($j, j' = 1, 2$). As a result, we return to Eqs. (8.39)–(8.41), with the bottom (minus) sign in Eq. (8.39), corresponding to the triplet spin states. Hence, Eq. (8.126) may be considered as the generalization of the direct and exchange interaction balance picture to an arbitrary number of orbitals and an arbitrary total number N of electrons. Note, however, that this equation cannot correctly describe the energy of the singlet spin state, corresponding to the plus sign in Eq. (8.39), and also of the

---

[31] The spin–orbitals (8.125) are also close to spinors (8.13), besides that the former definition takes into account that the spin s of a single particle is fixed, so that the spin–orbital may be indexed by the spin's orientation $m \equiv m_s$ only. Also, if an orbital index is used, it should be clearly distinguished from j, i.e. the set of the orbital and spin indices. This is why I believe that the frequently met notation of spin-orbitals as $\psi_{j,s}(\mathbf{r})$ may lead to confusion.

entangled triplet state[32]. The reason is that the description of entangled spin states, given in particular by Eqs. (8.18) and (8.20), requires *linear superpositions* of different Dirac states. (A proof of this fact is left for the reader's exercise.)

Now comes a very important fact: the *approximate* result (8.126), added to the sum of unperturbed energies $\varepsilon_j^{(0)}$, equals the sum of *exact* eigenenergies of the so-called *Hartree–Fock equation*[33]:

$$
\left(-\frac{\hbar^2}{2m}\nabla^2 + u(\mathbf{r}) \right) \psi_j(\mathbf{r})
$$
$$
+ \sum_{j' \neq j} \int [\psi_{j'}^*(\mathbf{r}')u_{\text{int}}(\mathbf{r},\mathbf{r}')\psi_j(\mathbf{r})\psi_{j'}(\mathbf{r}') - \psi_{j'}^*(\mathbf{r}')u_{\text{int}}(\mathbf{r},\mathbf{r}')\psi_{j'}(\mathbf{r})\psi_j(\mathbf{r})]d^3r' = \varepsilon_j\psi_j(\mathbf{r}),
$$

(8.127)

where $u(\mathbf{r})$ is the external-field potential acting on each particle separately—see the second of Eqs. (8.79). An advantage of this equation in comparison with Eq. (8.126) is that it allows the (approximate) calculation of not only the energy spectrum of the system, but also the corresponding spin–orbitals, taking into account their electron–electron interaction. Of course Eq. (8.127) is an *integro-differential* rather than just differential equation. There are, however, efficient methods of numerical solution of such equations, typically based on iterative methods. One more important trick is the exclusion of the filled internal electron shells (see section 3.7) from the explicit calculations, because the shell states are virtually unperturbed by the valence electron effects involved in typical atomic phenomena and chemical reactions. In this approach, the Coulomb field of the shells, described by fixed, pre-calculated and tabulated *pseudo-potentials*, is added to that of the nuclei. This approach dramatically cuts the computing resources necessary for systems of relatively heavy atoms, enabling a pretty accurate simulation of electronic and chemical properties of rather complex molecules, with thousands of electrons[34]. As a result, the Hartree–Fock approximation has become the de-facto baseline of all so-called *ab initio* ('first-principle') calculations in the most important field of quantum chemistry[35].

In departures from this baseline, there are two opposite trends. For larger accuracy (and typically smaller systems), several 'post-Hartree–Fock methods', notably including the *configuration interaction* method[36], that are more complex but may provide higher accuracy, have been developed.

---

[32] Indeed, due to the condition $j' \neq j$, and Eq. (8.124), the calculated negative exchange interaction is limited to electron state pairs with the same spin direction—such as the factorable triplet states (↑↑ and ↓↓) of a two-electron system, in which the contribution of E_{ex}, given by Eq. (8.41), to the total energy is also negative.

[33] This equation was suggested in 1929 by D Hartree for the direct interaction, and extended to the exchange interaction by V Fock in 1930. In order to verify its equivalence to Eq. (8.126), it is sufficient to multiply all terms of Eq. (8.127) by $\psi_j^*(\mathbf{r})$, integrate them over all $\mathbf{r}$ space (so that the right-hand side would give ε_j), and then sum these single-particle energies over all occupied states j.

[34] For condensed-matter systems, this and other computational methods are applied to single elementary spatial cells, with a limited number of electrons in them, using cyclic boundary conditions.

[35] See, e.g. [2].

[36] That method, in particular, allows the calculation of proper linear superpositions of the Dirac states (such as the entangled states for $N = 2$, discussed above) which are missing in the generic Hartree–Fock approach—see, e.g. the just-cited monograph by Szabo and Ostlund.

There is also a strong opposite trend of extending *ab initio* methods to larger systems, while sacrificing the results' accuracy and reliability. The ultimate case of this trend is applicable when the single-particle wavefunction overlaps are small and hence the exchange interaction is negligible, the last term in the square brackets of Eq. (8.127) may be ignored, the term $\psi_j(\mathbf{r})$ may be taken out of the integral, and it is reduced to a differential equation, which is formally just the Schrödinger equation for a single particle in the following self-consistent effective potential:

$$u_{\text{ef}}(\mathbf{r}) = u(\mathbf{r}) + u_{\text{dir}}(\mathbf{r}), \quad u_{\text{dir}}(\mathbf{r}) = \sum_{j' \neq j} \int \psi_{j'}^*(\mathbf{r}')u_{\text{int}}(\mathbf{r}, \mathbf{r}')\psi_{j'}(\mathbf{r}')d^3r'. \tag{8.128}$$

This is the so-called *Hartree approximation*—that gives reasonable results for some systems[37], especially those with low electron density. However, in dense electron systems (such as typical atoms, molecules, and condensed matter) the exchange interaction, described by the second term in the square brackets of Eqs. (8.126) and (8.127), may be as high as ~30% of the direct interaction, and frequently cannot be ignored.

The tendency of taking this interaction in the simplest possible form is currently dominated by the *Density Functional Theory*[38], universally known by its acronym DFT. In this approach, the equation solved for each eigenfunction $\psi_j(\mathbf{r})$ is a differential, Schrödinger-like *Kohn–Sham equation*

$$\left[-\frac{\hbar^2}{2m}\nabla^2 + u(\mathbf{r}) + u_{\text{dir}}^{\text{KS}}(\mathbf{r}) + u_{\text{xc}}(\mathbf{r})\right]\psi_j(\mathbf{r}) = \varepsilon_j\psi_j(\mathbf{r}), \tag{8.129}$$

where

$$u_{\text{dir}}^{\text{KS}}(\mathbf{r}) = -e\phi(\mathbf{r}), \quad \phi(\mathbf{r}) = \frac{1}{4\pi\varepsilon_0}\int d^3r' \frac{\rho(\mathbf{r}')}{|\mathbf{r} - \mathbf{r}'|}, \quad \rho(\mathbf{r}) = -en(\mathbf{r}), \tag{8.130}$$

and $n(\mathbf{r})$ is the total electron density in a particular point, calculated as

$$n(\mathbf{r}) \equiv \sum_j \psi_j^*(\mathbf{r})\psi_j(\mathbf{r}). \tag{8.131}$$

The most important feature of the Kohn–Sham Hamiltonian is the simplified description of the exchange and correlation effects by the effective *exchange-correlation* potential $u_{\text{xc}}(\mathbf{r})$. This potential is calculated in various approximations, most valid only in the limit when the number of electrons in the system is very high. The simplest of them (proposed by Kohn *et al* in the 1960s) is the *Local Density Approximation* (LDA) in which the effective exchange potential at each point is a

---

[37] An extreme example the Hartree approximation is the *Thomas–Fermi model* of heavy atoms (with $Z \gg 1$), in which atomic electrons, at each distance r from the nucleus, are treated as an ideal, uniform Fermi gas, with a certain density $n(r)$ corresponding to the local value $u_{\text{ef}}(r)$, but a common value of their highest full single-particle energy, $\varepsilon = 0$, to ensure the equilibrium. (The analysis of this model is left for the reader's exercise.)
[38] It had been developed by W Kohn and his associates in 1965–66, and eventually (in 1998) was marked with a Nobel Prize in Chemistry.

function only of the electron density (8.131) at the same point, taken from the theory of a *uniform* gas of free electrons[39]. However, for many tasks of quantum chemistry, the accuracy given by the LDA is insufficient, because inside molecules the density n typically changes very fast. As a result, the DFT has become widely accepted in this field only after the introduction, in the 1980s, of more accurate, though more cumbersome models for $u_{xc}(\mathbf{r})$, notably the so-called *Generalized Gradient Approximations* (GGAs).

Due to its relative simplicity, the DFT enables the calculation, with the same computing resources and reasonable precision, some properties of much larger systems than the methods based on the Hartree–Fock theory. As a result, is has become a very popular tool of *ab initio* calculations[40]. Please note, however, that despite this undisputable success, this approach has its problems. From my personal point of view, the most offensive of them is the implicit assumption of the unphysical Coulomb interaction of an electron with itself (by dropping, on the way from Eq. (8.128) to Eq. (8.130), the condition $j' \neq j$ at the calculation of u_{dir}^{KS}). As a result of these issues, for a reasonable description of some effects, the available DFT packages are either inapplicable at all or require substantial artificial tinkering[41]. Unfortunately, because of lack of time, for details I have to refer the reader to specialized literature[42].

8.5 Quantum computation and cryptography

Now I have to review the emerging fields of *quantum computation* and *encryption*[43]. These fields are currently the subject of a very intensive research effort, which has already brought (besides much hype) some results of general importance. My coverage, by necessity short, will focus on these results, referring the reader interested in details to special literature[44]. Because of the very active stage of the fields, I will also provide, in the last part of the section, quite a few references to recent publications, making its style closer to a brief research review than to a part of a textbook.

Presently, most work on quantum computation and encryption is based on systems of spatially-separated (and hence *distinguishable*) two-level systems—in this context, commonly called *qubits*[45]. Due to this distinguishability, the issues that were

---

[39] Just for the reader's reference: for a uniform, degenerate Fermi-gas of electrons (with the Fermi energy $\varepsilon_F \gg k_B T$), the most important, exchange part u_x of u_{xc} may be calculated analytically: $u_x = -(3/4\pi)e^2 k_F/4\pi\varepsilon_0$, where the Fermi momentum $k_F = (2m_e\varepsilon_F)^{1/2}/\hbar$ is defined by the electron density: $n = 2(4\pi/3)k_F^3/(2\pi)^3 \equiv k_F^3/3\pi^2$.

[40] This popularity is enhanced by the availability of several advanced DFT software packages, some of them (such as SIESTA, see https://departments.icmab.es/leem/siesta/) in public domain.

[41] As just a few examples, see [3–5].

[42] See, e.g. either the monograph by [6], or the later textbook [7]. For a popular recent review, and references to more recent work in this still-developing field, see [8].

[43] Since these fields are much related, they are often referred to under the common title of 'quantum information science', though this term is somewhat misleading, obscuring the physical aspects of the field.

[44] Despite the recent flood of new books on the field, one of its first surveys, [9], is perhaps still the best one.

[45] In some texts, the term qubit (or 'Qbit', or 'Q-bit') is used instead for the *information contents* of a two-level system—very much like the classical bit of information (in this context, frequently called 'Cbit' or 'C-bit') describes the information contents of a classical bistable system—see, e.g. *Part SM* section 2.2.

the focus of the first sections of this chapter, including the second quantization approach, are irrelevant here. On the other hand, systems of qubits have some interesting properties that have not been discussed in this course yet.

First of all, a system of $N \gg 1$ qubits may contain much more information than the same number of N classical bits. Indeed, according to the discussions in chapter 4 and section 5.1, an arbitrary pure state of a single qubit may be represented by its ket vector (4.37)—see also Eq. (5.1):

$$|\alpha\rangle_{N=1} = \alpha_1|u_1\rangle + \alpha_2|u_2\rangle, \tag{8.132}$$

where $\{u_j\}$ is any orthonormal two-state basis. It is natural and common to employ, as u_j, the eigenstates a_j of the observable A that is eventually measured in the particular physical implementation of the qubit—say, a certain Cartesian component of spin-½. It is also common to write the kets of these base states as $|0\rangle$ and $|1\rangle$,[46] so that Eq. (8.132) takes the form

$$|\alpha\rangle_{N=1} = a_0|0\rangle + a_1|1\rangle \equiv \sum_{j=0,1} a_j|j\rangle. \tag{8.133}$$

(Here, and in the balance of this section, the letter j is used to denote an integer equal to either 0 or 1.) According to this relation, any state α of a qubit is completely defined by two complex c-numbers a_j, i.e. by four real numbers. Moreover, due to the normalization condition $|a_1|^2 + |a_2|^2 = 1$, we need just three independent real numbers—say, the Bloch sphere coordinates θ and φ (see figure 5.3), plus the common phase γ, which becomes important only when we consider coherent states of a several-qubit system.

This is a good time to note that a qubit is very different from any classical bistable system used to store single bits of information—such as two possible voltage states of the usual SRAM cell (a positive-feedback loop of two transistor-based inverters). Namely, the stationary states of a classical bistable system, due to its nonlinearity, are stable with respect to small perturbations, so that they may be rather robust with respect to unintentional interaction with its environment. In contrast, the qubit's state may be readily disturbed (i.e. its representation point on the Bloch sphere shifted) by even minor perturbations, because it does not have such internal state stabilization mechanism[47]. Due to this reason, qubit-based systems are rather vulnerable to environment-induced drifts, including the dephasing and relaxation discussed in the previous chapter, creating major experimental challenges—see below.

Now, if we have a system of 2 qubits, the vector (4.37) of its arbitrary pure state may be represented as a sum of $2^2 = 4$ terms[48],

---

[46] In this notation, at the Bloch sphere representation (figure 5.3), the North Pole state (that is traditionally denoted as ↑ in quantum mechanics) is taken for 0, and the South Pole state ↓ for 1, so that in Eq. (8.133), $a_0 = \cos(\theta/2)$, $a_1 = \sin(\theta/2)\exp\{i\varphi\}$.

[47] In this aspect as well, the information processing systems based on qubits are closer to classical analog computers (which were popular once, but are now virtually abandoned) rather than classical digital ones.

[48] Here and in most instances below I use the same shorthand notation as was used in the beginning of this chapter—cf. Eq. (8.1). In this short form, qubit's number is coded by the order of its state index inside the single ket-vector, while in the long form, such as in Eq. (8.137), it is coded by the order of single-qubit vectors.

$$|\alpha\rangle_{N=2} = a_{00}|00\rangle + a_{01}|01\rangle + a_{10}|10\rangle + a_{11}|11\rangle \equiv \sum_{j_1,j_2=0,1} a_{j_1 j_2}|j_1 j_2\rangle, \qquad (8.134)$$

with four complex coefficients, i.e. $4 \times 2 = 8$ real numbers, subject to just one normalization condition, which follows from the requirement $\langle\alpha|\alpha\rangle = 1$:

$$\sum_{j_{1,2}=0,1} |a_{j_1 j_2}|^2 = 1. \qquad (8.135)$$

The evident generalization of Eqs. (8.133) and (8.134) to an arbitrary pure state of an N-qubit system is given by a sum of 2^N terms:

$$|\alpha\rangle_N = \sum_{j_1,j_2,..j_N=0,1} a_{j_1 j_2 ... j_N}|j_1 j_2 ... j_N\rangle, \qquad (8.136)$$

including all possible combinations of 0s and 1s inside the ket, so that the state is fully described by 2^N complex numbers, i.e. $2 \cdot 2^N = 2^{N+1}$ real numbers, with only one constraint, similar to Eq. (8.135), imposed by the normalization condition. Let me emphasize that this exponential growth of the information contents would not be possible without the qubit state entanglement. Indeed, in the particular case when qubit states are unentangled (factorable),

$$|\alpha\rangle_N = |\alpha_1\rangle|\alpha_2\rangle...|\alpha_N\rangle, \qquad (8.137)$$

where each $|\alpha_n\rangle$ is described by an equality similar to Eq. (8.133) with its individual expansion coefficients, the system state description requires only $3N - 1$ real numbers—e.g. N sets $\{\theta, \varphi, \gamma\}$ less one common phase.

However, it would be wrong to project this exponential growth of information contents directly on the capabilities of quantum computation, because this process has to include the output information readout, i.e. qubit state measurements. Due to the fundamental intrinsic uncertainty of quantum systems, the measurement of a single qubit even in a pure state (8.133) generally may give either of two results, with probabilities $W_0 = |a_0|^2$ and $W_1 = |a_1|^2$. In order to comply with the general notion of computation, any quantum computer has to provide certain (or virtually certain) results, and hence the probabilities W_j have to be very close to either 0 or 1, so that before the measurement, each measured qubit has to be in a basis state—either 0 or 1. This means that the computational system with N output qubits, just before the final readout, has to be in one of the factorable states

$$|\alpha\rangle_N = |j_1\rangle|j_2\rangle...|j_N\rangle \equiv |j_1 j_2 ... j_N\rangle, \qquad (8.138)$$

which is a very small subset even of the set of all unentangled states (8.137), and whose maximum information contents is just N classical bits.

Now the reader may start thinking that this constraint strips quantum computations of any advantages over their classical counterparts, but this view is also

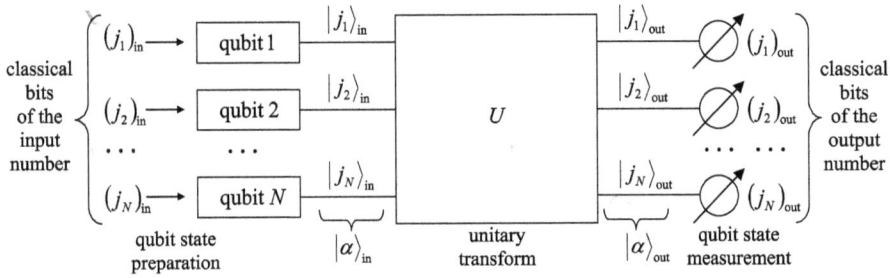

Figure 8.3. The baseline scheme of quantum computation.

superficial. In order to show that, let us consider the scheme of the most actively explored type of quantum computation, shown in figure 8.3.[49]

Here each horizontal line (sometimes called a 'wire'[50]) corresponds to a single qubit, tracing its time evolution in the same direction as at the usual time function plots: from left to right. This means that the left column $|\alpha\rangle_{in}$ of ket-vectors describes the initial state of the qubits[51], while the right column $|\alpha\rangle_{out}$ describes their final (pre-measurement) state. The box labeled U represents the qubit evolution in time due to their specially arranged interactions between each other and/or external drive 'forces'. Besides these forces, during this evolution the system is supposed to be ideally isolated from the dephasing and energy-dissipating environment, so that the evolution may be described by a unitary operator defined in the 2^N-dimensional Hilbert space of N qubits:

$$|\alpha\rangle_{out} = \hat{U}|\alpha\rangle_{in}. \tag{8.139}$$

With the condition that the input and output states have the simple form (8.138), this equality reads

$$|(j_1)_{out}(j_2)_{out}\ldots(j_N)_{out}\rangle = \hat{U}|(j_1)_{in}(j_2)_{in}\ldots(j_N)_{in}\rangle. \tag{8.140}$$

---

[49] Numerous modifications of this 'baseline' scheme have been suggested, for example with the number of output qubits different from that of input qubits, etc. Some other options are discussed at the end of this section.

[50] The notion of 'wires' stems from the similarity between these diagrams and the drawings used to describe classical computation circuits (see, e.g. figure 8.4a below); in the classical case the lines may be indeed understood as physical wires connecting physical devices: logic gates and/or memory cells. In this context, note that classical computer components also have nonvanishing time delays, so that even in this case the left-to-right device ordering is useful to indicate the timing of (and frequently the causal relation between) the signals.

[51] As follows from our discussions in chapter 7, the preparation of a pure state (8.133) is (conceptually) straightforward. Placing a qubit into a weak contact with an environment of temperature $T \ll \Delta/k_B$, where Δ is the difference between energies of the eigenstates 0 and 1, we may achieve its relaxation into the lowest-energy state. Then, if the qubit must be set into a different pure state, it may be driven there by the application of a pulse of a proper external classical 'force'. For example, if an actual spin-½ is used as qubits, a pulse of magnetic field with proper direction and duration may be applied to arrange its torque-induced precession to the required Bloch sphere point—see figure 5.3c. In most qubit systems, using a proper part of the Rabi oscillation period (see section 6.5) is more practicable for this purpose.

The art of quantum computer design consists of selecting such unitary operators $\hat{U}$ that would:

- satisfy Eq. (8.140),
- be physically implementable, and
- enable substantial performance advantages of the quantum computation over its classical counterparts with similar functionality, at least for some digital functions (algorithms).

I will have time/space to demonstrate the possibility of such advantages on just one, perhaps the simplest example—the so-called *Deutsch problem*[52]. Let us consider the family of single-bit classical Boolean functions $j_{out} = f(j_{in})$. Since both j are Boolean variables, i.e. may take only values 0 and 1, there are evidently only four such functions[53]:

f	$f(0)$	$f(1)$	class	F	$f(1)$-$f(0)$
f_1	0	0	constant	0	0
f_2	0	1	balanced	1	1
f_3	1	0	balanced	1	-1
f_4	1	1	constant	0	0

(8.141)

Of them, the functions f_1 and f_4, whose values are independent of their arguments, are called *constants*, while the functions f_2 (called 'YES' or 'IDENTITY') and f_3 ('NOT' or 'INVERSION') are called *balanced*. The Deutsch problem is to determine the class of a single-bit function, implemented with a 'black box', as being either constant or balanced, using just one experiment.

Classically, this is clearly impossible, and the simplest way to perform the function's classification involves two similar black boxes f—see figure 8.4a.[54] This solution uses the so-called *exclusive-OR* (for short, XOR) *gate* whose output is described by the following function F of its two Boolean arguments j_1 and j_2:[55]

$$F(j_1, j_2) = j_1 \oplus j_2 \equiv \begin{cases} 0, & \text{if } j_1 = j_2, \\ 1, & \text{if } j_1 \neq j_2. \end{cases}$$

(8.142)

In the particular circuit shown in figure 8.4a, the gate produces the following output:

$$F = f(0) \oplus f(1),$$

(8.143)

---

[52] It is named after D Deutsch, whose 1985 paper (motivated by an inspirational but not very specific publication by R Feynman in 1982) launched the whole field of quantum computation.

[53] The function F will be defined imminently—see Eq. (8.142).

[54] Alternatively, we may perform two sequential experiments on the same black box f, first recording and then recalling the first experiment's result. However, the Deutsch problem calls for a single experiment.

[55] The XOR sign $\oplus$ should not be confused with the sign $\otimes$ of the direct product of state vectors (which in this section is just implied).

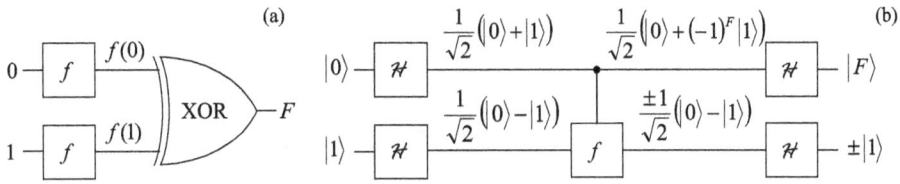

Figure 8.4. The simplest (a) classical and (b) quantum ways to classify a single-bit Boolean function f.

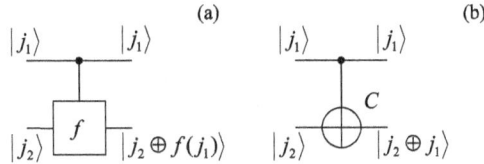

Figure 8.5. Two-qubit quantum gates: (a) a two-qubit function f and (b) its particular case C (CNOT), and their actions on a basis state.

which is equal to 1 if $f(0) \neq f(1)$, i.e. if the function f is balanced, and to 0 in the opposite case—see the 5th column in Eq. (8.141).

On the other hand, as will be proved below, all four functions f may be implemented quantum-mechanically, for example as a unitary transform of two input qubits, acting as follows on each basis component $|j_1 j_2\rangle \equiv |j_1\rangle|j_2\rangle$ of the general input state (8.134):

$$\hat{f}|j_1\rangle|j_2\rangle = |j_1\rangle|j_2 \oplus f(j_1)\rangle, \qquad (8.144)$$

where f is any of the classical Boolean functions listed in the table of Eq. (8.141)—see figure 8.5a.

In the particular case when f in Eq. (8.144) is just the YES function: $f(j) = f_2(j) = j$, this 'circuit' is reduced to the so-called *CNOT gate*, a key ingredient of many other quantum computation schemes, performing the following two-qubit transform:

$$\hat{C}|j_1 j_2\rangle = |j_1\rangle|j_2 \oplus j_1\rangle. \qquad (8.145a)$$

Let us use Eq. (8.142) to spell out this function for all four possible input qubit combinations:

$$\hat{C}|00\rangle = |00\rangle, \quad \hat{C}|01\rangle = |01\rangle, \quad \hat{C}|10\rangle = |11\rangle, \quad \hat{C}|11\rangle = |10\rangle. \qquad (8.145b)$$

In plain English, this means that acting on a basis state $j_1 j_2$, the CNOT gate leaves the state of the first, *source* qubit (shown by the upper lines in figure 8.5) intact, but flips the state of the second, *target* qubit if the first one is in the basis state 1. In even simpler words, the state j_1 of the source qubit controls the NOT function acting on the target qubit—hence the gate's name CNOT (the semi-acronym of 'Controlled NOT').

For the quantum function (8.144), with an arbitrary and unknown f, the Deutsch problem may be solved within the general scheme shown in figure 8.3, with the particular structure of the unitary-transform box U spelled out in figure 8.4b, which involves just one implementation of the function f. Here the singe-qubit quantum gate $\mathcal{H}$ performs the so-called *Hadamard* (or 'Walsh–Hadamard') *transform*[56], whose operator is defined by the following actions on the qubit's basis states:

$$\hat{\mathcal{H}}|0\rangle = \frac{1}{\sqrt{2}}(|0\rangle + |1\rangle), \quad \hat{\mathcal{H}}|1\rangle = \frac{1}{\sqrt{2}}(|0\rangle - |1\rangle), \tag{8.146}$$

—see also the two leftmost state label columns in figure 8.4b.[57] Since its quantum-mechanical operator has to be linear (to be physically realistic), it needs to perform the action (8.146) on the basis states even when they are parts of an arbitrary linear superposition—as they are, for example, for the two right Hadamard gates in figure 8.4b. For example, as immediately follows from Eqs. (8.146) and the operator's linearity,

$$\hat{\mathcal{H}}(\hat{\mathcal{H}}|0\rangle) = \hat{\mathcal{H}}\left(\frac{1}{\sqrt{2}}(|0\rangle + |1\rangle)\right) = \frac{1}{\sqrt{2}}\hat{\mathcal{H}}(|0\rangle + \hat{\mathcal{H}}|1\rangle)$$

$$= \frac{1}{\sqrt{2}}\left(\frac{1}{\sqrt{2}}(|0\rangle + |1\rangle) + \frac{1}{\sqrt{2}}(|0\rangle - |1\rangle)\right) = |0\rangle, \tag{8.147a}$$

Absolutely similarly, we may get[58]

$$\hat{\mathcal{H}}(\hat{\mathcal{H}}|1\rangle) = |1\rangle. \tag{8.147b}$$

Now let us carry out a sequential analysis of the 'circuit' shown in figure 8.4b. Since the input states of the gate f in this particular circuit are described by Eqs. (8.146), its output state's ket is

$$\hat{f}(\hat{\mathcal{H}}|0\rangle\,\hat{\mathcal{H}}|1\rangle) = \hat{f}\left(\frac{1}{\sqrt{2}}(|0\rangle + |1\rangle)\frac{1}{\sqrt{2}}(|0\rangle - |1\rangle)\right)$$

$$= \frac{1}{2}(\hat{f}|00\rangle - \hat{f}|01\rangle + \hat{f}|10\rangle - \hat{f}|11\rangle). \tag{8.148}$$

Now we may apply Eq. (8.144) to each basis ket to get:

---

[56] In order to exclude any chance of confusion between the Hadamard transform's operator $\hat{\mathcal{H}}$ and the Hamiltonian operator $\hat{H}$, they are typeset using different fonts.

[57] Note that according to Eq. (8.146), the operator $\hat{\mathcal{H}}$ does *not* belong to the class $\hat{U}$ described by Eq. (8.140)—while the whole 'circuit' shown in figure 8.4b, does—see below.

[58] Since the states 0 and 1 form a full basis of a single qubit, both Eqs. (8.147) may be summarized as an operator equality: $\hat{\mathcal{H}}^2 = \hat{I}$. It is also easy to check that the Hadamard transform of an arbitrary state may be represented on the Bloch sphere (figure 5.3) as a π-rotation about the axis that bisects the angle between x and z.

$$\hat{f}\,|00\rangle - \hat{f}\,|01\rangle + \hat{f}\,|10\rangle - \hat{f}\,|11\rangle$$
$$\equiv \hat{f}\,|0\rangle|0\rangle - \hat{f}\,|0\rangle|1\rangle + \hat{f}\,|1\rangle|0\rangle - \hat{f}\,|1\rangle|1\rangle$$
$$= |0\rangle|0 \oplus f(0)\rangle - |0\rangle|1 \oplus f(0)\rangle + |1\rangle|0 \oplus f(1)\rangle - |1\rangle|1 \oplus f(1)\rangle$$
$$\equiv |0\rangle(|0 \oplus f(0)\rangle - |1 \oplus f(0)\rangle) + |1\rangle(|0 \oplus f(1)\rangle - |1 \oplus f(1)\rangle).$$

(8.149)

Note that the expression in the first parentheses, characterizing the state of the target qubit, is equal to $(|0\rangle - |1\rangle) \equiv (-1)^0 (|0\rangle - |1\rangle)$ if $f(0) = 0$ (and hence $0 \oplus f(0) = 0$ and $1 \oplus f(0) = 1$), and to $(|1\rangle - |0\rangle) \equiv (-1)^1(|0\rangle - |1\rangle)$ in the opposite case $f(0) = 1$, so that both cases may be described in one shot by rewriting the parentheses as $(-1)^{f(0)}(|0\rangle - |1\rangle)$. The second parentheses is absolutely similarly controlled by the value of $f(1)$, so that the outputs of the gate f are unentangled:

$$\hat{f}(\hat{H}|0\rangle\hat{H}|1\rangle) = \frac{1}{2}((-1)^{f(0)}|0\rangle + (-1)^{f(1)}|1\rangle)(|0\rangle - |1\rangle)$$
$$= \pm \frac{1}{\sqrt{2}}(|0\rangle + (-1)^F|1\rangle)\frac{1}{\sqrt{2}}(|0\rangle - |1\rangle),$$

(8.150)

where the last step has used the fact that the classical Boolean function F, defined by Eq. (8.142), equals $\pm[f(1) - f(0)]$—please compare the last two columns in Eq. (8.141). The front sign $\pm$ in Eq. (8.150) may be prescribed to any of the component ket-vectors—for example to that of the target qubit, as shown by the third column of state labels in figure 8.4b.

This intermediate result is already rather remarkable. Indeed, it shows that, despite the impression one could get from figure 8.5, the gates f and C, being 'controlled' by the source qubit, may change that qubit's state as well! This fact (partly reflected by the vertical direction of the control lines in figures 8.4 and 8.5, symbolizing the same stage of system's time evolution) shows how careful one should be interpreting quantum-computational 'circuits', thriving on qubits' entanglement, because the 'signals' on different sections of a 'wire' may differ—see figure 8.4b again.

At the last stage of the circuit shown in figure 8.4b, the qubit components of the state (8.150) are fed into one more pair of Hadamard gates, whose outputs therefore are

$$\hat{H}\left(\frac{1}{\sqrt{2}}(|0\rangle + (-1)^F|1\rangle)\right) = \frac{1}{\sqrt{2}}(\hat{H}|0\rangle + (-1)^F \hat{H}|1\rangle), \quad \text{and}$$

(8.151)

$$\hat{H}\left(\pm\frac{1}{\sqrt{2}}(|0\rangle - |1\rangle)\right) = \pm\frac{1}{\sqrt{2}}(\hat{H}|1\rangle - \hat{H}|0\rangle).$$

Now using Eqs. (8.146) again, we see that the output state ket-vectors of the source and target qubits are, respectively,

$$\frac{1 + (-1)^F}{2}|0\rangle + \frac{1 - (-1)^F}{2}|1\rangle, \quad \text{and} \quad \pm |1\rangle.$$

(8.152)

Since, according to Eq. (8.142), the Boolean function F may take only values 0 or 1, the final state of the source qubit is always one of its basis states j, namely the one with $j = F$. Its measurement tells us whether the function f, participating in Eq. (8.144), is constant or balanced—see Eq. (8.141) again[59].

Thus, the quantum circuit shown in figure 8.4b indeed solves the Deutsch problem in one shot. Reviewing our analysis, we may see that this is possible because the unitary transform performed by the quantum gate f is applied to the entangled states (8.146) rather than to the basis states. Due to this trick, the quantum state components depending on $f(0)$ and $f(1)$ are processed simultaneously, in parallel. This *quantum parallelism* may be extended to circuits with many ($N \gg 1$) qubits and, for some tasks, provide a dramatic performance increase—for example, reducing the necessary circuit component number from $O(2^N)$ to $O(N^p)$, where p is a finite (and not very big) number.

However, this efficiency comes at a high price. Indeed, let us discuss the possible physical implementation of quantum gates, starting from the Hadamard gate, which performs a single-qubit transform—see Eq. (8.146). With the linearity requirement, its action on the arbitrary state (8.133) should be

$$\hat{H}|\alpha\rangle = a_0 \hat{H}|0\rangle + a_1 \hat{H}|1\rangle = a_0\frac{1}{\sqrt{2}}(|0\rangle + |1\rangle) + a_1\frac{1}{\sqrt{2}}(|0\rangle - |1\rangle)$$
$$= \frac{1}{\sqrt{2}}(a_0 + a_1)|0\rangle + \frac{1}{\sqrt{2}}(a_0 - a_1)|1\rangle, \tag{8.153}$$

meaning that the state probability amplitudes in the end ($t = \mathcal{T}$) and beginning ($t = 0$) of the qubit evolution in time have to be related as

$$a_0(\mathcal{T}) = \frac{a_0(0) + a_1(0)}{\sqrt{2}}, \quad a_1(\mathcal{T}) = \frac{a_0(0) - a_1(0)}{\sqrt{2}}. \tag{8.154}$$

This task may be again performed using the Rabi oscillations, which were discussed in section 6.5, i.e. by applying to the qubit (a two-level system), for a limited time period $\mathcal{T}$, a weak sinusoidal external signal of frequency ω equal to the intrinsic quantum oscillation frequency $\omega_{nn'}$ defined by Eq. (6.85). A perturbative analysis of the Rabi oscillations was carried out in section 6.5, even for nonvanishing (though small) detuning $\Delta = \omega - \omega_{nn}$, but only for the particular initial conditions when at $t = 0$ the system was in one on the basis states (there labeled as n'), i.e. another state (there labeled n) was empty. For our current purposes we need to find the amplitudes $a_{0,1}(t)$ for arbitrary initial conditions $a_{0,1}(0)$, subject only to the time-independent normalization condition $|a_0|^2 + |a_1|^2 = 1$. For the case of exact tuning,

---

[59] Note that the last Hadamard transform of the target qubit (i.e. the Hadamard gate shown in the lower right corner of figure 8.4b) is not necessary for the Deutsch problem's solution—though it should be included if we want the whole circuit to satisfy the general condition (8.140).

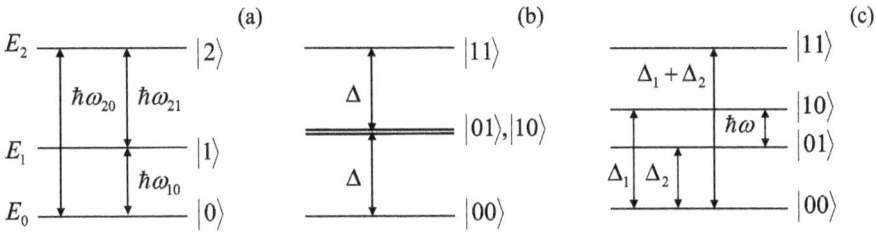

Figure 8.6. Energy-level schemes used for unitary transformations of (a) single qubits and (b, c) two-qubit systems.

$\Delta = 0$, the solution of the system (6.94) is elementary[60], and gives the following solution[61]:

$$a_0(t) = a_0(0)\cos \Omega t - ia_1(0)e^{i\varphi} \sin \Omega t,$$
$$a_1(t) = a_1(0)\cos \Omega t - ia_0(0)e^{-i\varphi} \sin \Omega t, \tag{8.155}$$

where Ω is the Rabi oscillation frequency (6.99), in the exact-tuning case proportional to the amplitude $|A|$ of the external ac drive $A = |A|\exp\{i\varphi\}$—see Eq. (6.86). Comparing these expressions with Eqs. (8.154), we see that for $t = \mathcal{T} = \pi/4\Omega$ and $\varphi = \pi/2$ they 'almost' coincide, besides the opposite sign of $a_1(\mathcal{T})$. Conceptually the simplest way to correct this deficiency is to follow the ac '$\pi/4$-pulse', just discussed, by a short dc 'π-pulse' of the duration $\mathcal{T} = \pi/\delta$, which temporarily creates a small additional energy difference δ between the basis states 0 and 1. According to the basic Eq. (1.62), such a difference creates an additional phase difference $\mathcal{T} = \delta/\hbar$ between the states, equal to π for the 'π-pulse'.

Another way (that may be also useful for two-qubit operations) is to use another, auxiliary energy level E_2 whose distances from the basic levels E_1 and E_0 are significantly different from the difference $(E_1 - E_0)$—see figure 8.6a. In this case, the weak external ac field tuned to any of three potential quantum transition frequencies $\omega_{nn'} \equiv (E_n - E_{n'})/\hbar$ initiates such transitions between the corresponding states only, with a negligible perturbation of the third state. (Such transitions may be again described by Eqs. (8.155), with the appropriate index changes.) For the Hadamard transform implementation, it is sufficient to apply (after the already discussed $\pi/4$-pulse of frequency ω_{10}, and with the initially empty level E_2), an additional π-pulse of frequency ω_{20}, with any phase φ. Indeed, according to the first of Eqs. (8.155), with the due replacement $a_1(0) \rightarrow a_2(0) = 0$, such a pulse flips the sign of the amplitude $a_0(t)$, while the amplitude $a_1(t)$, not involved in this additional transition, remains unchanged.

Now let me describe the conceptually simplest (though, for some qubit types, not the most practically convenient) scheme for the implementation of the CNOT gate, whose action is described by a linear unitary operator satisfying Eq. (8.145). For

---

[60] An alternative way to analyze the qubit evolution is to use the Bloch equation (5.21), with an appropriate function $\Omega(t)$ describing the control field.

[61] To comply with our current notation, the coefficients $a_{n'}$ and a_n of section 6.5 are replaced with a_0 and a_1.

that, evidently, the involved qubits have to interact for some time $\mathcal{T}$. As was repeatedly discussed in the two last chapters, in most cases such interaction of two subsystems is factorable—see Eq. (6.145). For qubits, i.e. two-level systems, each of the component operators may be represented by a 2×2 matrix in the basis of the states 0 and 1. According to Eq. (4.106), such a matrix may be always expressed as a linear combination $(b\mathrm{I} + \mathbf{c} \cdot \boldsymbol{\sigma})$, where b and three Cartesian components of the vector $\mathbf{c}$ are c-numbers. Let us consider the simplest form of such factorable interaction Hamiltonian:

$$\hat{H}_{\mathrm{int}}(t) = \begin{cases} \kappa\hat{\sigma}_z^{(1)}\hat{\sigma}_z^{(2)}, & \text{for } 0 < t < \mathcal{T}, \\ 0, & \text{otherwise}, \end{cases} \tag{8.156}$$

where the upper index is the qubit number, and κ is a c-number constant[62] According to Eq. (4.175), by the end of the interaction period, this Hamiltonian produces the following unitary transform:

$$\hat{U}_{\mathrm{int}} = \exp\left\{-\frac{i}{\hbar}\hat{H}_{\mathrm{int}}\mathcal{T}\right\} \equiv \exp\left\{-\frac{i}{\hbar}\kappa\hat{\sigma}_z^{(1)}\hat{\sigma}_z^{(2)}\mathcal{T}\right\}. \tag{8.157}$$

Since in the basis of unperturbed two-bit basis states $|j_1 j_2\rangle$, the product operator $\hat{\sigma}_z^{(1)}\hat{\sigma}_z^{(2)}$ is diagonal, so is the unitary operator (8.157), with the following action on these states:

$$\hat{U}_{\mathrm{int}}|j_1 j_2\rangle = \exp\left\{i\theta\sigma_z^{(1)}\sigma_z^{(2)}\right\}|j_1 j_2\rangle, \tag{8.158}$$

where $\theta = -\kappa\mathcal{T}/\hbar$, and σ_z are the eigenvalues of the Pauli matrix σ_z for the basis states of the corresponding qubit: $\sigma_z = +1$ for $|j\rangle = |0\rangle$, and $\sigma_z = -1$ for $|j\rangle = |1\rangle$. Let me, for clarity, spell out Eq. (8.158) for the particular case $\theta = -\pi/4$ (corresponding to the qubit coupling time $\mathcal{T} = \pi\hbar/4\kappa$):

$$\begin{aligned} \hat{U}_{\mathrm{int}}|00\rangle &= e^{-i\pi/4}|00\rangle, \quad \hat{U}_{\mathrm{int}}|01\rangle = e^{i\pi/4}|01\rangle, \\ \hat{U}_{\mathrm{int}}|10\rangle &= e^{i\pi/4}|10\rangle, \quad \hat{U}_{\mathrm{int}}|11\rangle = e^{-i\pi/4}|11\rangle. \end{aligned} \tag{8.159}$$

In order to compensate the undesirable parts of this joint phase shift of the basis states, let us now apply similar individual 'rotations' of each qubit by angle $\theta' = +\pi/4$,

---

[62] The assumption of simultaneous time independence of the basis state vectors and the interaction operator (within the time interval $0 < t < \mathcal{T}$) is possible only if the basis state energy difference Δ of both qubits is exactly the same. For this case, the simple physical explanation of the time evolution (8.156) follows from figure 8.6b, c, which shows the spectrum of the total energy $E = E_1 + E_2$ of the two-bit system. In the absence of interaction (figure 8.6b), the energies of two basis states, $|01\rangle$ and $|10\rangle$, are equal, enabling even a weak qubit interaction to cause their substantial evolution in time—see section 6.7. If the qubit energies are different (figure 8.6c), the interaction may still be reduced, in the rotating-wave approximation, to Eq. (8.156), by compensating the energy difference $(\Delta_1 - \Delta_2)$ with an external ac signal of frequency $\omega = (\Delta_1 - \Delta_2)/\hbar$—see section 6.5.

using the following product of two independent operators, plus (just for the result clarity) a common, and hence inconsequential, phase shift $\theta' = -\pi/4$:[63]

$$\hat{U}_{\text{com}} = \exp\left\{ i\theta' \left(\hat{\sigma}_z^{(1)} + \hat{\sigma}_z^{(2)} \right) + i\theta'' \right\} \equiv \exp\left\{ i\frac{\pi}{4}\hat{\sigma}_z^{(1)} \right\} \exp\left\{ i\frac{\pi}{4}\hat{\sigma}_z^{(2)} \right\} e^{-i\pi/4}. \quad (8.160)$$

Since this operator is also diagonal in the $|j_1 j_2\rangle$ basis, it is easy to calculate the change of the basis states by the total unitary operator $\hat{U}_{\text{tot}} \equiv \hat{U}_{\text{com}}\hat{U}_{\text{int}}$:

$$\begin{aligned} \hat{U}_{\text{tot}}|00\rangle = |00\rangle, \quad & \hat{U}_{\text{tot}}|01\rangle = |01\rangle, \\ \hat{U}_{\text{tot}}|10\rangle = |10\rangle, \quad & \hat{U}_{\text{tot}}|11\rangle = -|11\rangle. \end{aligned} \quad (8.161)$$

This result already shows the main 'miracle action' of two-qubit gates, such as the one shown in figure 8.4b: the source qubit is left intact (only if it is in a basis state!), while the state of the target qubit is altered. True, this change (of the sign) is still different from the CNOT operator's action (8.145), but may be readily used for its implementation by sandwiching of the transform U_{tot} between two Hadamard transforms of the target qubit alone:

$$\hat{C} = \frac{1}{2}\hat{\mathscr{H}}^{(2)}\hat{U}_{\text{tot}}\hat{\mathscr{H}}^{(2)}. \quad (8.162)$$

So, we have spent quite a bit of time on the discussion of the CNOT gate[64], and now I can reward the reader for his/her effort with a bit of good news: it has been proved that an *arbitrary* unitary transform that satisfies Eq. (8.140), i.e. may be used within the general scheme outlined in figure 8.3, may be decomposed into a set of CNOT gates, possibly augmented with simpler single-qubit gates—for example, the Hadamard gate plus the $\pi/2$ rotation discussed above[65]. Unfortunately, I have no time for a detailed discussion of more complex circuits[66]. The most famous of them is the scheme for integer number factoring, suggested in 1994 by P Shor[67]. Due to its potential practical importance for breaking broadly used communication encryption

---

[63] It Eq. (4.175) shows, each of component unitary transforms $\exp\{i\theta'\hat{\sigma}_z\}$ may be created by applying to each qubit, for a time period $\mathscr{T} = \hbar\theta'/\kappa'$, a constant external field described by Hamiltonian $\hat{H} = -\kappa'\hat{\sigma}_z$. We already know that for a charged, spin-½ particle, this Hamiltonian may be created by applying z-oriented external constant magnetic field—see Eq. (4.163). For most other physical implementations of qubits, the organization of such Hamiltonian is also straightforward—see, e.g. figure 7.4 and its discussion.

[64] As was discussed above, this gate is identical to the two-qubit gate shown in figure 8.5a for $f = f_3$, i.e. $f(j) = j$. The implementation of the gate of f for 3 other possible functions f requires straightforward modifications, whose analysis is left for reader's exercise.

[65] This fundamental importance of the CNOT gate was perhaps a major reason why D Wineland, the leader of the NIST group that had demonstrated its first experimental implementation in 1995 (following the theoretical suggestion by J Cirac and P Zoller), was awarded the 2012 Nobel Prize in Physics—shared with S Haroche, the leader of another group working towards quantum computation.

[66] For that, the reader may be referred to either the monographs by Nielsen–Chuang and Reiffel–Polak, cited above, or to a shorter (but much more formal) textbook [10].

[67] A clear description of this algorithm may be found in several accessible sources, including *Wikipedia*—see the article *Shor's Algorithm*.

schemes such as the RSA code[68], this opportunity has incited a huge wave of enthusiasm, and triggered experimental efforts to implement quantum gates and circuits using a broad variety of two-level quantum systems. By now, the following experimental options have given most significant results[69]:

(i) *Trapped ions.* The first experimental demonstrations of quantum state manipulation (including the already mentioned first CNOT gate) have been carried out using deeply cooled atoms in optical traps, similar to those used in frequency and time standards. Their total spins are natural qubits, whose states may be manipulated using the Rabi transfers excited by suitably tuned lasers. The spin interactions with the environment may be very weak, resulting in large dephasing times T_2—up to a few seconds. Since the distances between ions in the traps are relatively large (of the order of a micron), their direct spin–spin interaction is even weaker, but the ions may be made effectively interacting either via their mechanical oscillations about the potential minima of the trapping field, or via photons in external electromagnetic resonators ('cavities')[70]. Perhaps the main challenge of using this approach for quantum computation is a poor 'scalability', i.e. the enormous experimental difficulty of creating large, ordered systems of individually addressable qubits. So far, only a-few-qubit systems have been demonstrated[71].

(ii) *Nuclear spins* are also typically very weakly connected to environment, with dephasing times T_2 exceeding 10 s in some cases. Their eigenenergies E_0 and E_1 may be split by external dc magnetic fields (typically, of the order of 10 T), while the interstate Rabi transfers may be readily achieved by using the nuclear magnetic resonance, i.e. the application of external ac fields with frequencies $\omega = (E_1 - E_0)/\hbar$—typically, of a few hundred MHz. The challenges of this option include the weakness of spin–spin interactions (typically mediated through molecular electrons), resulting in a very slow spin evolution, whose time scale $\hbar/\kappa$ may become comparable with T_2, and also very small level separations $E_1 - E_0$, corresponding to a few K, i.e. much smaller than the room temperature, creating a challenge of qubit state preparation[72]. Despite these challenges, the nuclear spin option was used for the first implementation of the Shor algorithm for factoring of a small number ($15 = 5 \times 3$) as early as in 2001[73]. However, the extension of this success to larger systems, beyond the set of spins inside one molecule, is extremely challenging.

---

[68] Named after R Rivest, A Shamir, and L Adleman, the authors of the first open publication of the code in 1977, but actually invented earlier (in 1973) by C Cocks.
[69] For a discussion of other possible implementations (such as quantum dots and dopants in crystals) see, e.g. [11], and references therein.
[70] A brief discussion of such interactions (so-called Cavity QED) will be given in section 9.4 below.
[71] See, e.g. [12]. Note also the related work on arrays of trapped, optically-coupled neutral atoms—see, e.g. [13] and references therein.
[72] This challenge may be partly mitigated using ingenious spin manipulation techniques such as *refocusing*—see, e.g. either section 7.7 in Nielsen and Chuang, or the J Keeler's monograph cited in the end of section 6.5.
[73] [14].

(iii) *Josephson-junction devices.* Much better scalability may be achieved with solid state devices, especially using superconductor integrated circuits including weak contacts—Josephson junctions (see their brief discussion in section 1.6). The qubits of this type all based on the fact that the energy U of such a junction is a highly nonlinear function of the Josephson phase difference φ—see section 1.6. Indeed, combining Eqs. (1.73) and (1.74), we can readily calculate $U(\varphi)$ as the work $\mathscr{W}$ of an external circuit increasing the phase from, say, zero to some value φ:

$$
U(\varphi) - U(0) \;=\; \int_{\varphi'=0}^{\varphi'=\varphi} d\mathscr{W} = \int_{\varphi'=0}^{\varphi'=\varphi} IV dt = \frac{2eI_c}{\hbar} \int_{\varphi'=0}^{\varphi'=\varphi} \sin \varphi' \frac{d\varphi'}{dt} dt
$$

$$
=\; \frac{2eI_c}{\hbar}(1 - \cos \varphi).
$$

(8.163)

There are several options of using this nonlinearity for creating qubits[74]; currently the leading option, called the *phase qubit*, is using two lowest eigenstates localized in one of the potential wells of the periodic potential (8.163). A major problem of such qubits is that at the very bottom of this well the potential $U(\varphi)$ is almost quadratic, so that the energy levels are nearly equidistant—cf. Eqs. (2.262), (6.16), and (6.23). This is even more true for the so-called 'transmons' (and 'Xmons', and 'Gatemons', and several other similar devices[75])—the currently used phase qubits versions, where a Josephson junction is made a part of an external electromagnetic oscillator, making its relative net nonlineartity (anharmonism) even smaller. As a result, the external rf drive of frequency $\omega = (E_1 - E_0)/\hbar$, used to arrange the state transforms described by Eq. (8.155), may induce simultaneous undesirable transitions to (and between) higher energy levels. This effect may be mitigated by a reduction of the ac drive amplitude, but at a price of the proportional increase of the operation time. (I am leaving a quantitative estimate of this increase for the reader's exercise.)

Since the coupling of Josephson-junction qubits may be most readily controlled (and, very importantly, kept stable if so desired), they have been used to demonstrate the largest prototype quantum computing systems to date, despite quite modest dephasing times T_2—for purely integrated circuits, in the tens of microseconds at best, even at the operation temperatures in tens of mK. By the time of this writing (mid-2018), several groups have announced chips with more than 10

---

[74] The 'most quantum' option in this technology is to use Josephson junctions very weakly coupled to their dissipative environment (so that the effective resistance shunting the junction is much higher than the quantum resistance unit $R_Q \equiv (\pi/2)\,\hbar/e^2 \sim 10^4\ \Omega$). In this case, the Josephson phase variable φ behaves as a coordinate of a 1D quantum particle, moving in the 2π-periodic potential (8.163), forming the energy band structure $E(q)$ similar to those discussed in section 2.7. Both theory and experiment show that in this case, the quantum states in adjacent Brillouin zones differ by the charge of one Cooper pair $2e$. (This is exactly the effect responsible for the Bloch oscillations of frequency (2.252).) These two states may be used as the basis states of a *charge qubit*. Unfortunately, such a qubit is rather sensitive to random charged impurities in the junction's vicinity, causing uncontrollable changes of it parameters, so that currently, to the best of my knowledge, this option is not actively pursued.

[75] For a recent review of these devices see, e.g. [15], and references therein.

qubits, but to the best of my knowledge, only their smaller subsets could be used for high-fidelity quantum operations[76].

(iv) *Optical systems*, attractive because of their inherently enormous bandwidth, pose a special challenge for quantum computation: due to the virtual linearity of most electromagnetic media at reasonable light power, the implementation of qubits (i.e. two-level systems), and interaction Hamiltonians such as the one given by Eq. (8.156), is problematic. In 2001, a very smart way around this hurdle was invented[77]. In this *KLM scheme* (also called the 'linear optical quantum comput-ing'), nonlinear elements are not needed at all, and quantum gates may be composed just of linear devices (such as optical waveguides, mirrors and beam splitters), plus single-photon sources and detectors. However, estimates show that this approach requires a much larger number of physical components than those using nonlinear quantum systems such as usual qubits[78], so that right now it is not very popular.

So, despite more than two decades of large-scale efforts, the progress of the quantum computing development has been rather modest. The main culprit here is the unintentional coupling of qubits to environment, leading most importantly to their state dephasing, and eventually to errors. Let me discuss this major issue in detail.

Of course, some error probability exists in classical digital logic gates and memory cells as well[79]. However, in this case, there is no conceptual problem with the device state measurement, so that the error may be detected and corrected in many ways; perhaps the simplest one is the so-called *majority voting*. For that, the input bit set is reproduced in several (say, three) copies and sent to three similar devices whose outputs are measured and compared. If the outputs differ, at least one of the devices has made at error. This error may be not only *detected*, but also *corrected* by taking the two coinciding outputs for the correct one. If the probability of a single device error is $W \ll 1$, the probability of error of one device pair is close to W^2, and that of two pairs (and hence of the whole majority voting scheme) is close to W^3. Since for the currently dominating CMOS integrated circuits, W is extremely small ($<10^{-5}$ even for relatively complex logic blocks), even such a simple error correction circuit creates a dramatic fidelity improvement—at the cost of higher circuit complexity and consumed power.

For quantum computation, the general idea of using several devices (say, qubits) for coding the same information remains valid; however, there are two major complications, both due to the analog nature of qubit states. First, as we know from chapter 7, the dephasing effect of environment may be described as a slow random drift of the probability amplitudes a_j, leading to the deviation of the output state α_{fin} from the required form (8.140), and hence to a nonvanishing probability of wrong qubit state readout—see figure 8.3. Hence the quantum error correction has to

---

[76] See, e.g. [16] and references therein.
[77] [17].
[78] See, e.g. [18].
[79] In modern integrated circuits, such 'soft' (runtime) errors are created mostly by the high-energy neutron component of cosmic rays, and also by the α-particles emitted by radioactive impurities in silicon chips and their packaging.

protect the result not against possible random state flips $0 \leftrightarrow 1$, as in the classical digital computer, but against these 'creeping' analog errors.

Second, the qubit state is impossible to copy exactly (*clone*) without disturbing it, as follows from the following simple calculation[80]. Cloning some state α of one qubit to another qubit that is initially in an independent state (say, the basis state 0), without any change of α, means the following transformation of the two-qubit ket: $|\alpha 0\rangle \rightarrow |\alpha\alpha\rangle$. If we want such a transform to be performed by a real quantum system whose operation is described by a unitary operator $\hat{u}$, and to be correct for an arbitrary state α, it has to work not only for both basis states of the qubit:

$$\hat{u}|00\rangle = |00\rangle, \quad \hat{u}|10\rangle = |11\rangle, \tag{8.164}$$

but also for their arbitrary linear combination (8.133). Since the operator $\hat{u}$ has to be linear, we may use that relation, and then Eq. (8.164) to write

$$\begin{aligned}\hat{u}|\alpha 0\rangle &\equiv \hat{u}(a_0|0\rangle + a_1|1\rangle)|0\rangle \equiv a_0\hat{u}|00\rangle + a_1\hat{u}|10\rangle \\ &= a_0|00\rangle + a_1|11\rangle.\end{aligned} \tag{8.165}$$

On the other hand, the desired result of the state cloning is

$$\begin{aligned}|\alpha\alpha\rangle &= (a_0|0\rangle + a_1|1\rangle)(a_0|0\rangle + a_1|1\rangle) \\ &\equiv a_0^2|00\rangle + a_0 a_1(|10\rangle + |01\rangle) + a_1^2|11\rangle,\end{aligned} \tag{8.166}$$

i.e. is evidently different, so that, for an arbitrary state α, and an arbitrary unitary operator $\hat{u}$,

$$\hat{u}|\alpha 0\rangle \neq |\alpha\alpha\rangle, \tag{8.167}$$

meaning that the qubit state cloning is indeed impossible[81].

This problem may be partly circumvented—for example, in the way shown in figure 8.7a. Here the CNOT gate, whose action is described by Eq. (8.145), entangles an arbitrary input state (8.133) of the source qubit with a basis initial state of an

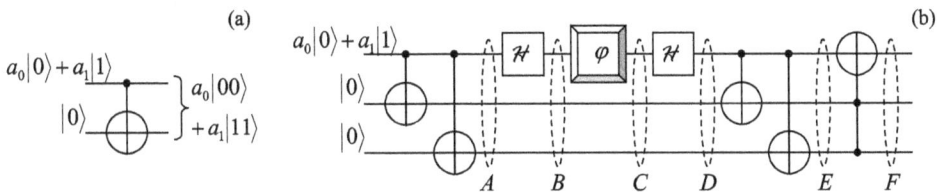

Figure 8.7. (a) Quasi-cloning, and (b) detection and correction of dephasing errors in a single qubit.

---

[80] Amazingly, this simple *no-cloning theorem* was discovered as late as in 1982 (to the best of my knowledge, independently by W Wooters and W Zurek, and by D Dieks), in the context of work toward quantum cryptography—see below.

[81] Note that this does not mean that the two (or several) qubits cannot be put into the same, arbitrary quantum state—theoretically, with arbitrary precision. Indeed, they may be first set into their lowest-energy stationary states, and then driven into the same arbitrary state (8.133) by exerting on them similar classical external fields. So, the no-cloning theorem pertains only to qubits in *unknown* states α—but this is exactly what we need for error correction—see below.

ancillary target qubit—frequently called the *ancilla*. Using Eq. (8.145), we may readily calculate the output two-qubit state's vector:

$$|\alpha\rangle_{N=2} = \hat{C}(a_0|0\rangle + a_1|1\rangle)|0\rangle \equiv a_0\hat{C}|00\rangle + a_1\hat{C}|10\rangle$$
$$= a_0|00\rangle + a_1|11\rangle. \qquad (8.168)$$

We see that this circuit does perform the operation (8.165), i.e. gives the initial source qubit's probability amplitudes a_0 and a_1 equally to two qubits, i.e. duplicates the input information. However, in contrast with the 'genuine' cloning, it changes the state of the source qubit as well, making it entangled with the target (ancilla) qubit. Such 'quasi-cloning' is the key element of most suggested quantum error correction techniques.

Consider, for example, the three-qubit 'circuit' shown in figure 8.7b, which uses two ancilla qubits (see two lower lines). At its first two stages, the double application of the quasi-cloning produces an intermediate state A with the following ket-vector:

$$|A\rangle = a_0|000\rangle + a_1|111\rangle, \qquad (8.169)$$

which is an evident generalization of Eq. (8.168).[82] Next, subjecting the source qubit to the Hadamard transform (8.146), we get the three-qubit state B represented by the vector

$$|B\rangle = a_0\frac{1}{\sqrt{2}}(|0\rangle + |1\rangle)|00\rangle + a_1\frac{1}{\sqrt{2}}(|0\rangle - |1\rangle)|11\rangle. \qquad (8.170)$$

Now let us assume that at this stage, the source qubit comes into a contact with a dephasing environment (in figure 8.7b, symbolized by the single-qubit 'gate' φ). As we know from section 7.3, its effect (besides some inconsequential shift of the *common* phase) may be described by a random *mutual* phase shift of the basis states[83]:

$$|0\rangle \rightarrow e^{i\varphi}|0\rangle, \quad |1\rangle \rightarrow e^{-i\varphi}|1\rangle. \qquad (8.171)$$

As a result, for the intermediate state C (see figure 8.7b) we may write

$$|C\rangle = a_0\frac{1}{\sqrt{2}}(e^{i\varphi}|0\rangle + e^{-i\varphi}|1\rangle)|00\rangle$$
$$+ a_1\frac{1}{\sqrt{2}}(e^{i\varphi}|0\rangle - e^{-i\varphi}|1\rangle)|11\rangle. \qquad (8.172)$$

At this stage, in this simple theoretical model, the coupling with environment is completely stopped (ahhh, if this could be possible! we might have quantum

---

[82] Such a state is also the 3 qubit example of the so-called *Greeenberger–Horne–Zeilinger* (GHZ) *states*, which are frequently called the 'most entangled' states of a system of $N > 2$ qubits.

[83] For example, in the Hilbert space of this qubit, the model Hamiltonian (7.70), which was explored in section 7.3, is diagonal in the z-basis of states 0 and 1, so that the unitary transform it provides is also diagonal, giving phase shifts described by Eq. (8.171). Let me emphasize again that Eq. (8.171) is strictly valid only if the interaction with environment is a pure dephasing, i.e. does not include the energy relaxation of the qubit or its thermal activation to the higher eigenstate; however, it is a reasonable description of errors at $T_2 \ll T_1$.

computers by now :-), and the source qubit is fed into one more Hadamard gate. Using Eqs. (8.146) again, for the state D after this gate we get

$$|D\rangle = a_0(\cos\varphi|0\rangle + i\sin\varphi|1\rangle)|00\rangle$$
$$+ a_1(i\sin\varphi|0\rangle + \cos\varphi|1\rangle)|11\rangle. \qquad (8.173)$$

Now the qubits are passed through the second, similar pair of CNOT gates—see figure 8.7b. Using Eq. (8.145), for the resulting state E we readily get the following expression:

$$|E\rangle = a_0\cos\varphi|000\rangle + a_0 i\sin\varphi|111\rangle + a_1 i\sin\varphi|011\rangle$$
$$+ a_1\cos\varphi|100\rangle, \qquad (8.174a)$$

whose right-hand side may by evidently grouped as

$$|E\rangle = (a_0|0\rangle + a_1|1\rangle)\cos\varphi|00\rangle$$
$$+ (a_1|0\rangle + a_0|1\rangle)i\sin\varphi|11\rangle. \qquad (8.174b)$$

This is already a rather remarkable result. It shows that if we measured the ancilla qubits at the stage E, and both results corresponded to states 0, we might be 100% sure that the source qubit (which is not affected by these measurements!) is in its initial state even after the interaction with environment. The only result of an increase of this unintentional interaction (as quantified by the magnitude of the random phase shift φ) is the growth of the probability,

$$W = \sin^2\varphi, \qquad (8.175)$$

of getting the opposite result, which signals a dephasing-induced error in the source qubit. Such implicit measurement, without disturbing the source qubit, is called the *quantum error detection*. An even more impressive result may be achieved by the last component of the circuit, the so-called *Toffoli* (or 'CCNOT') *gate*, denoted by the rightmost symbol in figure 8.7b. This 3 qubit gate is conceptually similar to the CNOT gate discussed above, besides that it flips the basis state of its target qubit only if *both* its source qubits are in the state 1. (In the circuit shown in figure 8.7b, the former role is played by our source qubit, while the latter role, by the two ancilla qubits.) According to its definition, the Toffoli gate has no effect on the first parentheses in Eq. (8.174b), but flips the source qubit's states in the second parentheses, so that for the output 3 qubit state F we get

$$|F\rangle = (a_0|0\rangle + a_1|1\rangle)\cos\varphi|00\rangle + (a_0|0\rangle + a_1|1\rangle)i\sin\varphi|11\rangle. \qquad (8.176a)$$

Obviously, this result may be factored as

$$|F\rangle = (a_0|0\rangle + a_1|1\rangle)(\cos\varphi|00\rangle + i\sin\varphi|11\rangle), \qquad (8.176b)$$

showing that now the source qubit is again fully unentangled from the ancilla qubits. Moreover, calculating the norm squared of the second operand, we get

$$(\cos\varphi\langle00| - i\sin\varphi\langle11|)(\cos\varphi|00\rangle + i\sin\varphi|11\rangle) = \cos^2\varphi + \sin^2\varphi = 1, \qquad (8.177)$$

so that the final state of the source qubit *always*, *exactly* coincides with its initial state. This is the famous miracle of *quantum state correction*, taking place 'automatically'—without any qubit measurements, and for any random phase shift φ.

The circuit shown in figure 8.7b may be further improved by adding Hadamard gate pairs, similar to that used for the source qubit, to the ancilla qubits as well. It is straightforward to show that if the dephasing is small in the sense that the W given by Eq. (8.175) is much less than 1, this modified circuit may provide a substantial error probability reduction (to $\sim W^2$) even if the ancilla qubits are also subjected to a similar dephasing and the source qubits, at the same stage—i.e. between two Hadamard gates. Such perfect automatic correction of *any* error (not only an inner dephasing of a qubit and its relaxation/excitation, but also the mutual dephasing between qubits) of *any* used qubit needs even more parallelism. The first circuit of that kind, based on 9 parallel qubits, which is a natural generalization of the circuit discussed above, had been invented in 1995 by the same P Shor. Later, 5qubit circuits enabling similar error correction were suggested. (The further parallelism reduction has been proved impossible.)

However, all these results assume that the error correction circuits as such are perfect, i.e. completely isolated from the environment. In the real world this cannot be done. Now the key question is what maximum level W_{max} of the error probability in each gate (including those in the used error correction scheme) can be automatically corrected, and how many qubits with $W < W_{\text{max}}$ would be required to implement quantum computers producing important results—first of all, factoring of large numbers[84]. To the best of my knowledge, estimates of these two related numbers have been made only for some very specific approaches, and they are rather pessimistic. For example, using the so-called *surface codes*, which employ many physical qubits for coding an informational one, and hence increase its fidelity, W_{min} may be increased to a few times 10^{-3}, but then we would need $\sim 10^8$ physical qubits for the Shor's algorithm implementation[85]. This is very far from what currently looks doable.

Because of this hard situation, the current development of quantum computing is focused on finding at least *some* problems that could be within the reach of either the existing systems, or their immediate extensions, and simultaneously would present some practical interest—a typical example of a technology in search for applications. Currently, to my knowledge, all suggested problems of this kind address properties of some simple quantum systems—such as the molecular hydrogen[86] or the deuteron (the deuterium's nucleus, i.e. the proton–neutron system)[87]. In the simplest option of this approach, the interaction between the qubits of a system is

---

[84] In order to compete with the existing classical factoring algorithms, such numbers should have at least 10^3 bits.
[85] [19].
[86] [20].
[87] [21].

organized so that the system's Hamiltonian is similar to that of the quantum system of interest[88].

A similar work direction (for which 'quantum system modeling' would be a more appropriate name than 'quantum computation') is pursued by the teams using schemes different from that shown in figure 8.3. Of those, the most developed is the so-called *adiabatic quantum computation*[89], which drops the hardest requirement of negligible interaction with the environment. In this approach, the qubit system is first prepared in a certain initial state, and then is allowed to evolve on its own, with no effort to couple-uncouple qubits by external control signals during the evolution[90]. Due to the interaction with the environment, in particular the dephasing and the energy dissipation it imposes, the system eventually relaxes to a final incoherent state, which is then measured. (This recalls the scheme shown in figure 8.3, with the important difference that the transform U should not necessarily be unitary.) From numerous runs of such an experiment, the outcome statistics may be revealed. Thus, at this approach the interaction with the environment is allowed to play a certain role in the system evolution, though every effort is made to reduce it, thus slowing down the relaxation process—hence the word 'adiabatic' in the name of this approach. This slowness allows the system to exhibit some quantum properties, in particular quantum tunneling[91] *through* the energy barriers separating close energy minima in the multi-dimensional space of states. This tunneling may create a substantial difference of the finite state statistics from that in purely classical systems, where such barriers may be overcome only by thermally-activated jumps *over* them[92].

Due to technical difficulties of the organization and precise control of long-range interaction in multi-qubit systems, the adiabatic quantum computing demonstrations so far have been limited to a few simple arrays described by the so-called *extended quantum Ising* ('spin-glass') *model*

$$\hat{H} = -J \sum_{\{j,j'\}} \hat{\sigma}_z^{(j)} \hat{\sigma}_z^{(j')} - \sum_j h_j \hat{\sigma}_z^{(j)}, \tag{8.178}$$

where the curly brackets denote the summation over pairs of close (though not necessarily closest) neighbors. Though the Hamiltonian (8.178) is the traditional playground of phase transitions theory (see, e.g. *Part SM* chapter 4), to the best of my knowledge there are not many practically important tasks that could be achieved by studying the statistics of its solutions. Moreover, even for this limited task, the

---

[88] By the moment of this writing (mid-2018), even for such specially-tailored problems, the performance of existing quantum computing systems has been still below that of classical computers—see, e.g. [22].

[89] Note that the qualifier 'quantum' is important in this term, to distinguish this research direction from the *classical* adiabatic (or 'reversible') computation—see, e.g. *Part SM* section 2.3 and references therein.

[90] Recently, some hybrids of this approach with the 'usual' scheme of quantum computation have been demonstrated, in particular, using some control of inter-bit coupling during the relaxation process—see, e.g. [23].

[91] As a reminder, this process was repeatedly discussed in this course, starting from section 2.3.

[92] A quantitative discussion of such jumps may be found in *Part SM* section 5.6.

speed of the largest experimental adiabatic quantum 'computers', with several hundreds of Josephson-junction qubits[93] is still comparable with that of classical, off-the-shelf semiconductor processors (with the dollar cost lower by many orders of magnitude), and no dramatic change of this comparison is predicted for realistic larger systems.

To summarize the current situation with the quantum computation development, it faces a very hard challenge of mitigating the effects of unintentional coupling with the environment. This problem is exacerbated by the lack of algorithms, beyond the Shor's number factoring, that would give quantum computation a substantial advantage over the classical competition in solving real-world problems, and hence a potential customer base much broader that the communication encryption community, that would provide the field with the necessary long-term motivation and resources. So far, the leading experts in this field abstain from predictions on when the quantum computation may become a self-supporting commercial technology[94].

There seem to be better prospects for another application of entangled qubit systems, namely to telecommunication cryptography[95]. The goal here is to replace the currently dominating classical encryption, based on the public-key RSA code mentioned above, that may be broken by factoring very large numbers, with a quantum encryption system that would be fundamentally unbreakable. The basis of this opportunity are the measurement postulate and the no-cloning theorem: if a message is carried over by a qubit, it is impossible for an eavesdropper (in cryptography, traditionally called *Eve*) to either measure or copy it faithfully, without also disturbing its state. However, as we have seen from the discussion of figure 8.7a, state *quasi*-cloning using entangled qubits is possible, so that the issue is far from being simple, especially if we want to use a publicly distributed quantum key, in some sense similar to the classical public key used at the RSA encryption.

Unfortunately, I would not have time/space to discuss various options for quantum encryption, but cannot help demonstrating how counter-intuitive they may be, on the famous example of the so-called *quantum teleportation* (figure 8.8).[96] Suppose that some party A (in cryptography, traditionally called *Alice*) wants to send to party B (*Bob*) the full information about the pure quantum state α of a qubit, unknown to either party. Instead of sending her qubit directly to Bob, Alice asks *him*

---

[93] See, e.g. [24]. Similar demonstrations with trapped-ion systems so far have been on a smaller scale, with a few tens of qubits—see, e.g. [25].

[94] See, e.g. [26].

[95] This field was pioneered in the 1970s by S Wisener. Its important theoretical aspect (which I, unfortunately, also will not be able to cover) is the distinguishability of different but close quantum states—for example, of an original qubit set, and that slightly corrupted by noise. A good introduction to this topic may be found, for example, in chapter 9 of the monograph by Nielsen and Chuang, cited above.

[96] This procedure had been first suggested in 1993 by the same C Bennett, and then repeatedly demonstrated experimentally—see, e.g. [27], and literature therein.

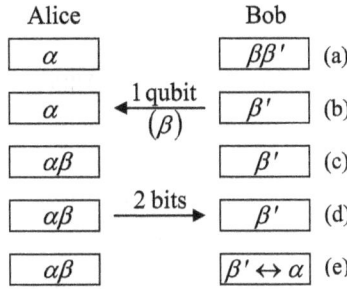

Figure 8.8. Sequential stages of a 'quantum teleportation' procedure: (a) the initial state with entangled qubits β and β', (b) back transfer of the qubit β, (c) measurement of the pair $\alpha\beta$, (d) forward transfer of 2 classical bits with the measurement results, and (e) the final state, with the state of the qubit β' mirroring the initial state of the qubit α.

to send *her* one qubit (β) of a pair of other qubits, prepared in a certain entangled state, for example in the singlet state described by Eq. (8.11); in our current notation

$$|\beta\beta'\rangle = \frac{1}{\sqrt{2}}(|01\rangle - |10\rangle).\tag{8.179}$$

The initial state of the whole 3 qubit system may be represented in the form

$$
\begin{aligned}
|\alpha\beta\beta'\rangle &= (a_0|0\rangle + a_1|1\rangle)|\beta\beta'\rangle \\
&= \frac{a_0}{\sqrt{2}}|001\rangle - \frac{a_0}{\sqrt{2}}|010\rangle + \frac{a_1}{\sqrt{2}}|010\rangle - \frac{a_1}{\sqrt{2}}|111\rangle,
\end{aligned}\tag{8.180a}
$$

which may be equivalently rewritten as the following linear superposition,

$$
\begin{aligned}
|\alpha\beta\beta'\rangle &= \frac{1}{2}|\alpha\beta\rangle_s^+ (-a_1|0\rangle + a_0|1\rangle) + \frac{1}{2}|\alpha\beta\rangle_s^- (a_1|0\rangle + a_0|1\rangle) \\
&+ \frac{1}{2}|\alpha\beta\rangle_e^+ (-a_0|0\rangle + a_1|1\rangle) + \frac{1}{2}|\alpha\beta\rangle_e^- (-a_0|0\rangle - a_1|1\rangle),
\end{aligned}\tag{8.180b}
$$

of the following four states of the qubit pair $\alpha\beta$:

$$|\alpha\beta\rangle_s^\pm \equiv \frac{1}{\sqrt{2}}(|00\rangle \pm |11\rangle), \quad |\alpha\beta\rangle_e^\pm \equiv \frac{1}{\sqrt{2}}(|01\rangle \pm |10\rangle).\tag{8.181}$$

After having received the qubit β from Bob, Alice measures which of these four states the pair $\alpha\beta$ has. This may be achieved, for example, by measurement of one observable represented by the operator $\hat{\sigma}_z^{(\alpha)}\hat{\sigma}_z^{(\beta)}$ and another one corresponding to $\hat{\sigma}_x^{(\alpha)}\hat{\sigma}_x^{(\beta)}$—cf. Eq. (8.156). (Since all four states (8.181) are eigenstates of both these operators, these two measurements do not affect each other and may be performed in any order.) The measured eigenvalue of the former operator enables distinguishing the couples of states (8.181) with different values of the lower index, while the latter measurement distinguishes the states with different upper indices.

Then Alice reports the measurement result (which may be coded with just 2 classical bits) to Bob over a classical communication channel. Since the measurement places the pair $\alpha\beta$ *definitely* into the corresponding state, the remaining Bob's bit β' is now *definitely* in the unentangled single-qubit state that is represented by the corresponding parentheses in Eq. (8.180b). Note that each of these parentheses contains both coefficients $a_{0,1}$, i.e. the whole information about the initial state that the qubit α had initially. If Bob likes, he may now use appropriate single-qubit operations, similar to those discussed earlier in this section, to move his qubit β' into the state *exactly* similar to the initial state of qubit α. (This fact does not violate the no-cloning theorem (8.167), because the measurement has already changed the state of α.) This is of course a 'teleportation' only in a very special sense of this term, but a good example of the importance of qubit entanglement's preservation at their spatial transfer[97].

Returning for just a minute to quantum cryptography: since its most common quantum key distribution protocols[98] require just a few simple quantum gates, whose experimental implementation is not a large technological challenge, the main focus of the current effort is on decreasing the single-photon dephasing in long electromagnetic-wave transmission channels[99], with sufficiently high qubit transfer fidelity. The recent progress was rather impressive, with demonstrated transfer of entangled qubits over landlines longer than 100 km,[100] and over at least one satellite-based line longer than 1000 km,[101] and also the whole quantum key distribution over a comparable distance, though as yet at a very low rate[102]. Let me hope that if not the author of these notes, then their readers will see this technology used in practical secure telecommunication systems.

8.6 Problems

Problem 8.1. Prove that Eq. (8.30) indeed yields $E_g^{(1)} = (5/4)E_H$.

Problem 8.2. For a diluted gas of helium atoms in their ground state, with n atoms per unit volume, calculate its:

(i) electric susceptibility χ_e, and
(ii) magnetic susceptibility χ_m,

and compare the results.

Hint: You may use the model solution of problems 6.8 and 6.14, and the results of the variational description of the helium atom's ground state in section 8.2.

---

[97] For this course, this is also a good primer for the forthcoming discussion of the *EPR paradox* and the *Bell's inequalities* in chapter 10.

[98] Two of them are the BB84 suggested in 1984 by C Bennett and G Brassard, and the EPRBE suggested in 1991 by A Ekert. For details, see, e.g. either section 12.6 in the repeatedly cited monograph by Nielsen and Chuang, or the review [28].

[99] For their quantitative discussion see, e.g. *Part EM* section 7.8.

[100] See, e.g. [29], and references therein.

[101] [30].

[102] [31].

Problem 8.3. Calculate the expectation values of the following observables: $\mathbf{s}_1 \cdot \mathbf{s}_2$, $S^2 \equiv (\mathbf{s}_1 + \mathbf{s}_2)^2$ and $S_z \equiv s_{1z} + s_{2z}$, for the singlet and triplet states of the system of two spins-½, defined by Eqs. (8.18) and (8.21), directly, without using the general rule (8.48) of spin addition. Compare the results with those for the system of two classical vectors of magnitude $\hbar/2$ each.

Problem 8.4. Discuss the factors $\pm 1/\sqrt{2}$ that participate in Eqs. (8.18) and (8.20) for the entangled states of the system of two spins-½, in terms of Clebsh–Gordan coefficients similar to those discussed in section 5.7.

Problem 8.5.* Use the perturbation theory to calculate the contribution into the so-called *hyperfine splitting* of the ground energy of the hydrogen atom[103], due to the interaction between spins of the nucleus (proton) and the electron.

Hint: The proton's magnetic moment operator is described by the same Eq. (4.115) as the electron, but with a positive gyromagnetic factor $\gamma_p = g_p e/2m_p \approx 2.675 \times 10^8$ $\text{s}^{-1}\,\text{T}^{-1}$, whose magnitude is much smaller than that of the electron ($|\gamma_e| \approx 1.761 \times 10^{11}\,\text{s}^{-1}\,\text{T}^{-1}$), due to the much higher mass, $m_p \approx 1.673 \times 10^{-27}\,\text{kg} \approx 1{,}835\,m_e$. (The g-factor of the proton is also different, $g_p \approx 5.586$.[104])

Problem 8.6. In the simple case of just two similar spin-interacting particles, distinguishable by their spatial location, the famous *Heisenberg model* of ferromagnetism[105] is reduced to the following Hamiltonian:

$$\hat{H} = -J\,\hat{\mathbf{s}}_1 \cdot \hat{\mathbf{s}}_2 - \gamma \mathscr{B} \cdot (\hat{\mathbf{s}}_1 + \hat{\mathbf{s}}_2),$$

where J is the spin interaction constant, γ is the gyromagnetic ratio of each particle, and $\mathscr{B}$ is the external magnetic field. Find the stationary states and eigenenergies of this system for spin-½ particles.

Problem 8.7. Two particles, both with spin-½, but different gyromagnetic ratios γ_1 and γ_2, are placed into external magnetic field $\mathscr{B}$. In addition, their spins interact as in the Heisenberg model:

$$\hat{H}_{\text{int}} = -J\,\hat{\mathbf{s}}_1 \cdot \hat{\mathbf{s}}_2.$$

Find the eigenstates and eigenenergies of the system[106].

---

[103] This effect was discovered experimentally by A Michelson in 1881, and explained theoretically by W Pauli in 1924.

[104] The anomalously large value of the proton's g-factor results from the composite quark–gluon structure of this particle. (An exact calculation of g_p remains a challenge for quantum chromodynamics.)

[105] It was suggested in 1926, independently by W Heisenberg and P Dirac. A discussion of temperature effects on this and other similar systems (especially the Ising model of ferromagnetism) may be found in *Part SM* chapter 4.

[106] For similar particles (in particular, with $\gamma_1 = \gamma_2$) the problem is evidently reduced to the previous one.

Problem 8.8. Two similar spin-½ particles, with the gyromagnetic ratio γ, localized at two points separated by distance a, interact via the field of their magnetic dipole moments. Calculate the spin eigenstates and eigenvalues of the system.

Problem 8.9. Consider the permutation of two identical particles, each of spin s. How many different symmetric and antisymmetric spin states can the system have?

Problem 8.10. For a system of two identical particles with $s = 1$:

(i) List all possible spin states in the uncoupled-representation basis.
(ii) List all possible pairs $\{S, M_S\}$ of the quantum numbers describing the states of the coupled-representation basis—see Eq. (8.48).
(iii) Which of the $\{S, M_S\}$ pairs describe the states symmetric, and which the states antisymmetric, with respect to the particle permutation?

Problem 8.11. Represent the operators of the total kinetic energy and the total orbital angular momentum of a system of two particles, with masses m_1 and m_2, as combinations of terms describing the center-of-mass motion and the relative motion. Use the results to calculate the energy spectrum of the so-called *positronium*—a metastable 'atom'[107] consisting of one electron and its positively charged antiparticle, the positron.

Problem 8.12. Two particles with similar masses m and charges q are free to move along a round, plane ring of radius R. In the limit of strong Coulomb interaction of the particles, find the lowest eigenenergies of the system, and sketch the system of its energy levels. Discuss possible effects of particle indistinguishability.

Problem 8.13. Low-energy spectra of many diatomic molecules may be well described modeling the molecule as a system of two particles connected with a light and elastic, but very stiff spring. Calculate the energy spectrum of a molecule in this approximation. Discuss possible effects of nuclear spins on the spectra of so-called *homonuclear* molecules, formed by two similar atoms.

Problem 8.14. Two indistinguishable spin-½ particles are attracting each other at contact:

$$U(x_1, x_2) = -w\delta(x_1 - x_2), \quad \text{with } w > 0,$$

but are otherwise free to move along the x-axis. Find the energy and the wavefunction of the ground state of the system.

Problem 8.15. Calculate the energy spectrum of the system of two identical spin-½ particles, moving along the x-axis, which is described by the following Hamiltonian:

---

[107] Its lifetime (either 0.124 ns or 138 ns, depending on the parallel or antiparallel configuration of the components spins), is limited by the weak interaction of its components, which causes their annihilation with the emission of several gamma-ray photons.

$$\hat{H} = \frac{\hat{p}_1^2}{2m_0} + \frac{\hat{p}_2^2}{2m_0} + \frac{m_0\omega_0^2}{2}\left(x_1^2 + x_2^2 + \varepsilon x_1 x_2\right),$$

and the degeneracy of each energy level.

Problem 8.16.* Two indistinguishable spin-½ particles are confined to move around a circle of radius R, and interact only at a very short arc distance $l = R\varphi \equiv R(\varphi_1 - \varphi_2)$ between them, so that the interaction potential U may be well approximated with a delta-function of φ. Find the ground state and its energy, for the following two cases:

(i) the 'orbital' (spin-independent) repulsion: $\hat{U} = \mathcal{W}\delta(\varphi)$,
(ii) the spin–spin interaction: $\hat{U} = -\mathcal{W}\hat{\mathbf{s}}_1 \cdot \hat{\mathbf{s}}_2 \delta(\varphi)$,

both with constant $\mathcal{W} > 0$. Analyze the trends of your results in the limits $\mathcal{W} \to 0$ and $\mathcal{W} \to \infty$.

Problem 8.17. Two particles of mass M, separated by two much lighter particles of mass $m \ll M$, are placed on a ring of radius R—see figure below. The particles strongly repulse at contact, but otherwise each of them is free to move along the ring. Calculate the lower part of the energy spectrum of the system.

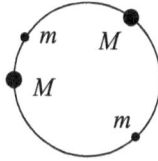

Problem 8.18. N indistinguishable spin-½ particles move in a spherically-symmetric quadratic potential $U(\mathbf{r}) = m\omega_0^2 r^2/2$. Neglecting the direct interaction of the particles, find the ground-state energy of the system.

Problem 8.19. Use the Hund rules to find the values of the quantum numbers L, S, and J in the ground states of the atoms of carbon and nitrogen. Write down the Russell–Saunders symbols for these states.

Problem 8.20. $N \gg 1$ indistinguishable, non-interacting quantum particles are placed in a hard-wall, rectangular box with sides a_x, a_y, and a_z. Calculate the ground-state energy of the system, and the average forces it exerts on each face of the box. Can we characterize the forces by certain pressure $\mathcal{P}$?

Hint: Consider separately the cases of bosons and fermions.

Problem 8.21.* Explore the *Thomas–Fermi model*[108] of a heavy atom, with the nuclear charge $Q = Ze \gg e$, in which the interaction between electrons is limited to their contribution to the common electrostatic potential $\phi(\mathbf{r})$. In particular, derive

---

[108] It was suggested in 1927, independently, by L Thomas and E Fermi.

the ordinary differential equation obeyed by the radial distribution of the potential, and use it to estimate the effective radius of the atom.

*Problem 8.22.** Use the Thomas–Fermi model, explored in the previous problem, to calculate the total binding energy of a heavy atom. Compare the result with that for the simpler model, in which the Coulomb electron–electron interaction is completely ignored.

Problem 8.23. A system of three similar but distinguishable spin-½ particles is described by the Heisenberg Hamiltonian (cf. problems 8.6 and 8.7):

$$\hat{H} = -J\,(\hat{\mathbf{s}}_1 \cdot \hat{\mathbf{s}}_2 + \hat{\mathbf{s}}_2 \cdot \hat{\mathbf{s}}_3 + \hat{\mathbf{s}}_3 \cdot \hat{\mathbf{s}}_1),$$

where J is the spin interaction constant. Find the stationary states and eigenenergies of this system, and give an interpretation of your results.

Problem 8.24. For a system of three distinguishable spins-½, find the common eigenstates and eigenvalues of the operators $\hat{S}_z$ and $\hat{S}^2$, where

$$\hat{\mathbf{S}} \equiv \hat{\mathbf{s}}_1 + \hat{\mathbf{s}}_2 + \hat{\mathbf{s}}_3$$

is the vector operator of the total spin of the system. Do the corresponding quantum numbers S and M_S obey Eqs. (8.48)?

Problem 8.25. Explore basic properties of the Heisenberg model (which was the subject of problems 8.6, 8.7, and 8.23), for a 1D chain of N spins-½:

$$\hat{H} = -J\sum_{\{j,j'\}}\hat{\mathbf{s}}_j \cdot \hat{\mathbf{s}}_{j'} - \gamma\mathscr{B} \cdot \sum_j \hat{\mathbf{s}}_j, \quad \text{with } J > 0,$$

where the summation is over all N spins, with the symbol $\{j,j'\}$ meaning that the first sum is only over the adjacent spin pairs. In particular, find the ground state of the system and its lowest excited states in the absence of external magnetic field $\mathscr{B}$, and also the dependence of their energies on the field.

Hint: For the sake of simplicity, you may assume that the first sum includes the term $\hat{\mathbf{s}}_N \cdot \hat{\mathbf{s}}_1$ as well. (Physically, this means that the chain is bent into a closed loop[109].)

Problem 8.26. Compose the simplest model Hamiltonians, in terms of the second quantization formalism, for systems of indistinguishable particles moving in the following systems:

(i) two weakly coupled potential wells, with on-site particle-pair interactions (giving additional energy J per each pair of particles in the same potential well), and

---

[109] Note that for dissipative spin systems, differences between low-energy excitations of open-end and closed-end 1D chains may be substantial even in the limit $N \to \infty$—see, e.g. *Part SM* section 4.5. However, for our Hamiltonian (and hence dissipation-free) system, the differences are relatively small.

(ii) a periodic 1D potential, with the same particle-pair interactions, in the tight-binding limit.

Problem 8.27. For each of the Hamiltonians composed in the previous problem, derive the Heisenberg equations of motion for particle creation operators, for

(i) bosons, and
(ii) fermions.

Problem 8.28. Express the ket-vectors of all possible Dirac states for the system of three indistinguishable

(i) bosons, and
(ii) fermions,

via those of their single-particle states β, β', and β'' they occupy.

Problem 8.29. Explain why the general perturbative result (8.126), when applied to the ^{4}He atom, gives the correct[110] expression (8.29) for the ground singlet state, and correct Eqs. (8.39)–(8.42) (with the minus sign in the first of these relations) for the excited triplet states, but cannot describe these results, with the plus sign in Eq. (8.39), for the excited singlet state.

Problem 8.30. For a system of two distinct qubits (i.e. two-level systems), introduce a reasonable uncoupled-representation z-basis, and find in this basis the 4×4 matrix of the operator that swaps their states.

Problem 8.31. Find a time-independent Hamiltonian that may cause the qubit evolution described by Eqs. (8.155). Discuss the relation between your result and the time-dependent Hamiltonian (6.86).

References

[1] Pendry J 1994 *Adv. Phys.* **43** 461
[2] Szabo A and Ostlund N 1996 *Modern Quantum Chemistry* Revised edn (Dover)
[3] Simonian N *et al* 2013 *J. Appl. Phys.* **113** 044504
[4] Medvedev M *et al* 2017 *Science* **335** 49
[5] Hutama A *et al* 2017 *J. Phys. Chem. C* **121** 14888
[6] Parr R and Yang W 1994 *Density-Functional Theory of Atoms and Molecules* (Oxford University Press)
[7] Steckel J and Sholl D 2009 *Density Functional Theory: Practical Introduction* (Wiley)
[8] Zangwill A 2015 *Phys. Today* **68** 34
[9] Nielsen M and Chuang I 2000 *Quantum Computation and Quantum Information* (Cambridge University Press)
[10] Mermin N 2007 *Quantum Computer Science* (Cambridge University Press)

---

[110] Correct in the sense of the first order of the perturbation theory.

[11] Ladd T *et al* 2010 *Nature* **464** 45

[12] Debnath S *et al* 2016 *Nature* **536** 63

[13] Perczel J *et al* 2017 *Phys. Rev. Lett.* **119** 023603

[14] Lanyon B *et al* 2001 *Phys. Rev. Lett.* **99** 250505

[15] Wendin G 2017 *Progr. Rep. Phys* **80** 106001

[16] Song C *et al* 2017 *Phys. Rev. Lett.* **119** 180511

[17] Knill E *et al* 2001 *Nature* **409** 46

[18] Li Y *et al* 2015 *Phys. Rev. X* **5** 041007

[19] Fowler A *et al* 2012 *Phys. Rev. A* **86** 032324

[20] O'Malley P *et al* 2016 *Phys. Rev. X* **6** 031007

[21] Dumitrescu E *et al* 2018 *Phys. Lett. Lett* **120** 210501

[22] Chen J *et al* 2018 arXiv: 1805.01450v2

[23] Barends R *et al* 2016 *Nature* **534** 222

[24] Harris R *et al* 2018 *Science* **361** 162

[25] Zhang J *et al* 2017 *Nature* **551** 601

[26] National Academies of Sciences, Engineering, and Medicine 2019 *Quantum Computing: Progress and Prospects* (Washington, DC: The National Academies Press)

[27] Steffen L *et al* 2013 *Nature* **500** 319

[28] Gizin N *et al* 2002 *Rev. Mod. Phys.* **74** 145

[29] Herbst T *et al* 2015 *Proc. Nath. Acad. Sci.* **112** 14202

[30] Yin J *et al* 2017 *Science* **356** 1140

[31] Yin H-L *et al* 2016 *Phys. Rev. Lett.* **117** 190501

IOP Publishing

Quantum Mechanics

Lecture notes

Konstantin K Likharev

Chapter 9

Introduction to relativistic quantum mechanics

This chapter gives a brief introduction to relativistic quantum mechanics. It starts with a discussion of the basic elements of the quantum theory of the electromagnetic field (usually called the quantum electrodynamics, QED), including the quantization scheme, photon statistics, radiative atomic transitions, the spontaneous and stimulated radiation, and the so-called cavity QED. We will see, in particular, that the QED may be considered as the relativistic quantum theory of quasi-particles with zero rest mass—photons. The second part of the chapter is a brief review of the relativistic quantum theory of particles with non-zero rest mass, including the Dirac theory of spin-½ particles. These theories mark the point of entry into a more complete relativistic quantum theory—the quantum field theory—which is beyond the scope of this course[1].

9.1 Electromagnetic field quantization

Classical physics gives us[2] the general relativistic relation between the momentum $\mathbf{p}$ and energy E of a free particle with rest mass m, which may be simplified in two limits—non-relativistic and ultra-relativistic:

$$E = [(pc)^2 + (mc^2)^2]^{1/2} \rightarrow \begin{cases} mc^2 + p^2/2m, & \text{for } p \ll mc, \\ pc, & \text{for } p \gg mc. \end{cases} \qquad (9.1)$$

In both limits, the transfer from classical to quantum mechanics is easier than in the arbitrary case. Since all the previous part of this course was committed to the first, non-relativistic limit, I will now jump to a brief discussion of the ultra-relativistic

---

[1] Note that some material of this chapter is frequently taught as a part of the quantum field theory. I will focus on the most important results that may be obtained without starting the heavy engines of that theory.
[2] See, e.g. *Part EM* chapter 9.

limit $p \gg mc$, for a particular but very important system—the electromagnetic field. Since the excitations of this field, called *photons*, are currently believed to have zero rest mass m,[3] the ultra-relativistic relation $E = pc$ is valid for any photon energy E, and the quantization scheme is rather straightforward.

As usual, the quantization has to be based on the classical theory of the system, in this case the Maxwell equations. As the simplest case, let us consider the electromagnetic field inside a free-space volume limited by ideal walls, which reflect incident waves perfectly[4]. Inside the volume, the Maxwell equations may be reduced to a simple wave equation[5] for the electric field

$$\nabla^2 \mathscr{E} - \frac{1}{c^2} \frac{\partial^2 \mathscr{E}}{\partial t^2} = 0, \tag{9.2}$$

and an absolutely similar equation for the magnetic field $\mathscr{B}$. We may look for the general solution of Eq. (9.2) in the variable-separating form

$$\mathscr{E}(\mathbf{r}, t) = \sum_j p_j(t) \mathbf{e}_j(\mathbf{r}). \tag{9.3}$$

Physically, each term of this sum is a standing wave whose spatial distribution and polarization ('*mode*') is described by the vector function $\mathbf{e}_j(\mathbf{r})$, and the temporal dynamics, by the function $p_j(t)$. Plugging an arbitrary term of this sum into Eq. (9.2), and separating the variables exactly as we did, for example, in the Schrödinger equation in section 1.5, we get

$$\frac{\nabla^2 \mathbf{e}_j}{\mathbf{e}_j} = \frac{1}{c^2} \frac{\ddot{p}_j}{p_j} = \text{const} \equiv -k_j^2, \tag{9.4}$$

so that the spatial distribution of the mode satisfies the 3D Helmholtz equation:

$$\nabla^2 \mathbf{e}_j + k_j^2 \mathbf{e}_j = 0. \tag{9.5}$$

The set of solutions of this equation, with appropriate boundary conditions, determines the set of the functions $\mathbf{e}_j$, and simultaneously the spectrum of wave number moduli k_j. The latter values determine the mode eigenfrequencies, following from Eq. (9.4):

$$\ddot{p}_j + \omega_j^2 p_j = 0, \quad \text{with } \omega_j \equiv k_j c. \tag{9.6}$$

There is a big philosophical difference between the quantum-mechanical approach to Eqs. (9.5) and (9.6), despite their single origin (9.4). The first

---

[3] By now this fact has been verified experimentally with an accuracy of at least $\sim 10^{-22} \, m_e$—see [1].

[4] In the case of finite energy absorption in the walls, or in the wave propagation media (say, described by complex constants ε and μ), the system is not energy-conserving, i.e. interacts with the dissipative environment. Specific cases of such interaction will be considered in sections 9.2 and 9.3 below.

[5] See, e.g. *Part EM* Eq. (7.3), for the particular case $\varepsilon = \varepsilon_0$, $\mu = \mu_0$, so that $v^2 \equiv 1/\varepsilon\mu = 1/\varepsilon_0\mu_0 \equiv c^2$.

(Helmholtz) equation may be rather difficult to solve in realistic geometries[6], but it remains intact in the basic quantum electrodynamics, with the scalar components of vector functions $e_j(\mathbf{r})$ still treated (at each point $\mathbf{r}$) as c-numbers. In contrast, the classical Eq. (9.6) is readily solvable (giving sinusoidal oscillations with frequency ω_j), but this is exactly where we can make a transfer to quantum mechanics, because we already know how to quantize a mechanical 1D harmonic oscillator, which in classics obeys the same equation.

As usual, we need to start with the appropriate Hamiltonian corresponding to the classical Hamiltonian function H of the proper set of generalized coordinates and momenta. The electromagnetic field's Hamiltonian function (which in this case coincides with the field's energy) is[7]

$$H = \int d^3r \left(\frac{\varepsilon_0 \mathscr{E}^2}{2} + \frac{\mathscr{B}^2}{2\mu_0} \right). \tag{9.7}$$

Let us represent the magnetic field in a form similar to Eq. (9.3),[8]

$$\mathscr{B}(\mathbf{r}, t) = -\sum_j \omega_j q_j(t) \mathbf{b}_j(\mathbf{r}). \tag{9.8}$$

Since, according to the Maxwell equations, in our case the magnetic field satisfies the equation similar to Eq. (9.2), the time-dependent amplitude q_j of each of its modes obeys the equation similar to Eq. (9.6), i.e. in the classical theory also changes in time sinusoidally, with the same frequency ω_j. Plugging Eqs. (9.3) and (9.8) into Eq. (9.7), we may recast it as

$$H = \sum_j \left[\frac{p_j^2}{2} \int \varepsilon_0 e_j^2(\mathbf{r}) d^3r + \frac{\omega_j^2 q_j^2}{2} \int \frac{1}{\mu_0} b_j^2(\mathbf{r}) d^3r \right]. \tag{9.9}$$

Since the distribution of constant factors between two multiplication operands in each term of Eq. (9.3) is arbitrary, we may fix it by requiring the first integral in Eq. (9.9) to equal 1. It is straightforward to check that according to the Maxwell equations, which give a specific relation between vectors $\mathscr{E}$ and $\mathscr{B}$,[9] this normalization makes the second integral in Eq. (9.9) equal 1 as well, and Eq. (9.9) becomes

$$H = \sum_j H_j, \qquad H_j = \frac{p_j^2}{2} + \frac{\omega_j^2 q_j^2}{2}. \tag{9.10}$$

---

[6] See, e.g. various problems discussed in *Part EM* chapter 7, especially in section 7.9.

[7] See, e.g. *Part EM* section 9.8, in particular, Eq. (9.225). Here I am using SI units, with $\varepsilon_0\mu_0 \equiv c^{-2}$; in the Gaussian units, the coefficients ε_0 and μ_0 disappear, but there is an additional common factor $1/4\pi$ in the equation for energy. However, if we modify the normalization conditions (see below) accordingly, all the subsequent results, starting from Eq. (9.10), look similar in any system of units.

[8] Here I am using letter q_j, instead of x_j, for the generalized coordinate of the field oscillator, in order to emphasize the difference between the former variable, and one of the Cartesian coordinates, i.e. one of arguments of the c-number functions $\mathbf{e}$ and $\mathbf{b}$.

[9] See, e.g. *Part EM* Eq. (7.6).

Now we can carry out the standard quantization procedure, namely declare H_j, p_j, and q_j the quantum-mechanical operators related exactly as in Eq. (9.10),

$$\hat{H}_j = \frac{\hat{p}_j^2}{2} + \frac{\omega_j^2 \hat{q}_j^2}{2}. \tag{9.11}$$

We see that this Hamiltonian coincides with that of a 1D harmonic oscillator with the mass m_j formally equal to 1,[10] and the eigenfrequency equal to ω_j. Next, in order to use Eq. (9.11) in the general Eq. (4.199) for the time evolution of Heisenberg-picture operators $\hat{p}_j$ and $\hat{q}_j$, we need to know the commutation relation between these operators. For that, returning to the classical case, let us calculate the Poisson bracket (4.204) for functions $A = q_{j'}$ and $B = p_{j''}$:

$$\{q_{j'}, p_{j''}\} \equiv \sum_j \left(\frac{\partial q_{j'}}{\partial p_j} \frac{\partial p_{j''}}{\partial q_j} - \frac{\partial q_{j'}}{\partial q_j} \frac{\partial p_{j''}}{\partial p_j} \right). \tag{9.12a}$$

Since in the classical Hamiltonian mechanics, all generalized coordinates q_j and the corresponding generalized momenta p_j have to be considered as independent arguments of H, only one term (with $j = j' = j''$) in only one of the sums (9.12) (namely, with $j' = j''$), gives a nonvanishing value (-1), so that

$$\{q_{j'}, p_{j''}\} = -\delta_{j'j''}. \tag{9.12b}$$

Hence, according to the general quantization rule (4.205), the commutation relation of the operators corresponding to $q_{j'}$ and $p_{j''}$ is

$$[\hat{q}_{j'}, \hat{p}_{j''}] = i\hbar\delta_{j'j''}, \tag{9.13}$$

i.e. is exactly the same as for the usual Cartesian components of the radius-vector and momentum of a mechanical particle—see Eq. (2.14).

As the reader already knows, Eqs. (9.11) and (9.13) open for us several alternative ways to proceed:

(i) Use the Schrödinger-picture wave mechanics based on wavefunctions $\Psi_j(q_j, t)$. As we know from section 2.9, this way is inconvenient for most tasks, because eigenfunctions of the harmonic oscillator are rather clumsy.

(ii) A substantially better way (for the harmonic oscillator case) is to write the equations of time evolution of the operators $\hat{q}_j(t)$ and $\hat{p}_j(t)$ in the Heisenberg-picture of quantum dynamics.

(iii) An even more convenient approach is to use equations similar to Eqs. (5.65) to decompose the Heisenberg operators $\hat{q}_j(t)$ and $\hat{p}_j(t)$ into the creation–annihilation operators $\hat{a}_j^\dagger(t)$ and $\hat{a}_j(t)$, and work with these operators.

---

[10] With different normalizations of the functions $\mathbf{e}_j(\mathbf{r})$ and $\mathbf{b}_j(\mathbf{r})$, we could readily arrange any value of m_j, and the choice corresponding to $m_j = 1$ is the best one just for the notation simplicity.

In this chapter, I will mostly use the last route. Replacing m with $m_j \equiv 1$, and ω_0 with ω_j, the last forms of Eqs. (5.65) become

$$\hat{a}_j = \left(\frac{\omega_j}{2\hbar}\right)^{1/2}\left(\hat{q}_j + i\frac{\hat{p}_j}{\omega_j}\right), \qquad \hat{a}_j^\dagger = \left(\frac{\omega_j}{2\hbar}\right)^{1/2}\left(\hat{q}_j - i\frac{\hat{p}_j}{\omega_j}\right), \tag{9.14}$$

and due to Eq. (9.13), the creation–annihilation operators obey the commutation similar to Eq. (5.68),

$$\left[\hat{a}_j, \hat{a}_{j'}^\dagger\right] = \hat{I}\delta_{jj'}. \tag{9.15}$$

As a result, according to Eqs. (9.3) and (9.8), the quantum-mechanical operators corresponding to the electric and magnetic fields are the sums over all *field oscillators*:

$$\hat{\mathscr{E}}(\mathbf{r}, t) = i\sum_j \left(\frac{\hbar\omega_j}{2}\right)^{1/2} \mathbf{e}_j(\mathbf{r})\,(\hat{a}_j^\dagger - \hat{a}_j), \tag{9.16a}$$

$$\hat{\mathscr{B}}(\mathbf{r}, t) = \sum_j \left(\frac{\hbar\omega_j}{2}\right)^{1/2} \mathbf{b}_j(\mathbf{r})\,(\hat{a}_j^\dagger + \hat{a}_j), \tag{9.16b}$$

and Eq. (9.11) for the jth mode's Hamiltonian becomes

$$\hat{H}_j = \hbar\omega_j\left(\hat{a}_j^\dagger\hat{a}_j + \frac{1}{2}\hat{I}\right) = \hbar\omega_j\left(\hat{n}_j + \frac{1}{2}\hat{I}\right), \quad \text{with } \hat{n}_j \equiv \hat{a}_j^\dagger\hat{a}_j, \tag{9.17}$$

absolutely similar to Eq. (5.72) for a mechanical oscillator.

Now comes a very important conceptual step. From section 5.4 we know that the eigenstates (Fock states) n_j of the Hamiltonian (9.17) have energies

$$E_j = \hbar\omega_j\left(n_j + \frac{1}{2}\right), \quad n_j = 0, 1, 2, \ldots \tag{9.18}$$

and, according to Eq. (5.89), the operators $\hat{a}_j^\dagger$ and $\hat{a}_j$ act on the eigenkets of these partial states as

$$\hat{a}_j|n_j\rangle = (n_j)^{1/2}|n_j - 1\rangle, \quad \hat{a}_j^\dagger|n_j\rangle = (n_j + 1)^{1/2}|n_j + 1\rangle, \tag{9.19}$$

regardless of the quantum states of other modes. These rules coincide with the definitions (8.64) and (8.68) of bosonic creation–annihilation operators, and hence their action may be considered as the creation/annihilation of certain bosons. Such a 'particle' (actually, an *excitation*, with energy $\hbar\omega_j$, of an electromagnetic field oscillator) is exactly what is, strictly speaking, called a *photon*. Note immediately that according to Eq. (9.16), such an excitation does not change the spatial distribution of the jth mode of the field. So, such a 'global' photon is an excitation created simultaneously at all points of the field confinement region.

If this picture is too contrary to the intuitive image of a particle, please recall that in chapter 2, we discussed a similar situation with the eigenstates of the non-relativistic Schrödinger equation of a free non-relativistic particle: they represent sinusoidal de Broglie waves existing simultaneously in all points of the particle confinement region. The (partial :-) reconciliation with the classical picture of a moving particle might be obtained by using the linear superposition principle to assemble a quasi-localized wave packet, as a group of sinusoidal waves with close wave numbers. Very similarly, we may form a similar wave packet using a linear superposition of the 'global' photons with close values of $\mathbf{k}_j$ (and hence ω_j), to form a quasi-localized photon. An additional simplification here is that the dispersion relation for electromagnetic waves (at least in free space) is linear:

$$\frac{\partial \omega_j}{\partial k_j} = c = \text{const}, \quad \text{i.e.} \quad \frac{\partial^2 \omega_j}{\partial k_j^2} = 0, \tag{9.20}$$

so that, according to Eq. (2.39a), the electromagnetic wave packets (i.e. space-localized photons) do not spread out during their propagation. Note also that due to the fundamental classical relations $\mathbf{p} = \mathbf{n}E/c$ for the linear momentum of the traveling electromagnetic wave packet of energy E, propagating along the direction $\mathbf{n} \equiv \mathbf{k}/k$, and $\mathbf{L} = \pm \mathbf{n}E/\omega_j$ for its angular momentum[11], such a photon may be prescribed the linear momentum $\mathbf{p} = \mathbf{n}\hbar\omega_j/c \equiv \hbar\mathbf{k}$ and the angular momentum $\mathbf{L} = \pm \mathbf{n}\hbar$, with the sign depending on the direction of its circular polarization ('helicity').

This electromagnetic field quantization scheme should look very straightforward, but it raises an important conceptual issue of the ground-state energy. Indeed, Eq. (9.18) implies that the total ground-state (i.e. the lowest) energy of the field is

$$E_\text{g} = \sum_j \left(E_\text{g}\right)_j = \sum_j \frac{\hbar\omega_j}{2}. \tag{9.21}$$

Since for any realistic model of the field-confining volume, either infinite or not, the density of electromagnetic field modes only grows with frequency[12], this sum diverges on the upper limit, leading to infinite ground-state energy per unit volume. This infinite-energy paradox cannot be dismissed by declaring the zero-point energy of field oscillators unobservable, because this would contradict numerous experimental observations—historically, starting perhaps from the famous *Casimir effect*[13]. The conceptually simplest implementation of this effect involves two parallel, well conducting plates of area A, separated by a vacuum gap of thickness $t \ll A^{1/2}$ (figure 9.1).

---

[11] See, e.g. *Part EM* sections 7.7 and 9.8.
[12] See, e.g. Eq. (1.1), which is similar to Eq. (1.90) for the de Broglie waves, derived in section 1.7.
[13] This effect was predicted in 1948 by H Casimir and D Polder, and confirmed semi-quantitatively in experiments by M Sparnaay [2]. After this, and several other experiments, a decisive error bar reduction (to about ~5%), providing a quantitative confirmation of the Casimir formula (9.23), was achieved by S Lamoreaux [3] and by U Mohideen and A Roy [4]. Note also that there are other experimental confirmations of the reality of the ground-state electromagnetic field, including, for example, the experiments by R Koch *et al* already discussed in section 7.5, and the recent spectacular direct observations by C Riek *et al* [5].

Figure 9.1. The simplest geometry of the Casimir effect manifestation.

Rather counter-intuitively, the plates attract each other with a force F, proportional to the area A and rapidly increasing with the decrease of t, even in the absence of any explicit electromagnetic field sources. The effect's explanation is that the energy of each the electromagnetic field mode, including its ground-state energy, exerts average pressure,

$$\langle \mathcal{P}_j \rangle = -\frac{\partial E_j}{\partial V}, \tag{9.22}$$

on the walls constraining it to volume V. While the field's pressure on the external surfaces on the plates is due to the contributions (9.22) of all free-space modes, with arbitrary values of k_z (the z-component of the wave vector $\mathbf{k}_j$), in the gap between the plates the spectrum of k_z is limited to the multiples of π/t, so that the pressure on the internal surfaces is lower. This is why the net force exerted on the plates may be calculated as the sum of the contributions (9.22) from all 'missing' low-frequency modes in the gap, with the minus sign. In the simplest model when the plates are made of an ideal conductor, which provides boundary conditions $\mathcal{E}_n = \mathcal{B}_\tau = 0$ on their surfaces[14], such a calculation is rather straightforward (and is hence left for the reader's exercise), and its result is

$$F = -\frac{\pi^2 A \hbar c}{240 t^4}. \tag{9.23}$$

Note that for this calculation, the high-frequency divergence of Eq. (9.21) at high frequencies is not important, because it participates in the forces exerted on all surfaces of each plate, and hence cancels out from the net pressure. In this way, the Casimir effect not only gives a confirmation of Eq. (9.21), but also teaches us an important lesson how to deal with the divergences of such sums at $\omega_j \to \infty$. The lesson is: just get accustomed to the idea that the divergence exists, and ignore this

---

[14] For realistic conductors, the reduction of t below $\sim 1\ \mu m$ causes significant deviations from this simple model, and hence from Eq. (9.23). The reason is that for gaps so narrow, the depth of field penetration into the metal (see, e.g. *Part EM* section 6.2), at the important frequencies $\omega \sim c/t$, becomes comparable with t, and an adequate theory of the Casimir effect has to involve a certain model of the penetration. (It is curious that in-depth analyses of this problem, pioneered in 1956 by E Lifshitz, have revealed a deep relation between the Casimir effect and the London dispersion force which was the subject of problems 3.16, 5.15, and 6.18—for a review see, e.g. either [6], or [7].) Recent experiments in the 100 nm—2 μm range of t, with an accuracy better than 1%, have allowed not only to observe the effects of field penetration on the Casimir force, but even to make a selection between some approximate models of the penetration—see [8].

fact while you can, i.e. if the final result you are interested in is finite. However, for some more complex problems of quantum electrodynamics (and quantum theory of any other fields), this simplest approach becomes impossible, and then more complex, *renormalization techniques* become necessary. For their study, I have to refer the reader to a quantum field theory course—see the literature cited in the end of this chapter.

9.2 Photon absorption and counting

As a matter of principle, the Casimir effect may be used to measure quantum effects in not only the free-space electromagnetic field, but also that the field arriving from active sources—lasers, etc. However, usually such studies may be done by simpler detectors, in which the absorption of a photon by a single atom leads to its ionization. This ionization, i.e. the emission of a free electron, triggers an avalanche reaction (e.g. an electric discharge in a Geiger-type counter), which may be readily registered using appropriate electronic circuitry. In order to discuss the statistics of such *photon counts*, it is sufficient to consider the field's interaction with just one, 'trigger' atom.

Here we are essentially dealing with an open, irreversible system, with the trigger atom, with the continuous spectrum of its final, ionized states, paying the role of an environment for the quantized electromagnetic field. Such systems were discussed in detail in chapter 7; however, for our current particular problem that heavy machinery is not necessary. Indeed, in section 6.6 we have discussed a simpler approach to the analysis of such problems, based on the Golden Rule of quantum mechanics—see figure 6.12 and Eq. (6.149). In our current case, we may associate the system a in this scheme with the electromagnetic field, and system b with the trigger atom. The atom's size is typically much smaller that the radiation wavelength $\lambda_j = 2\pi/k_j$, so that the field–atom interaction may be adequately described in the electric dipole approximation,

$$\hat{H}_{\text{int}} = -\hat{\mathscr{E}} \cdot \hat{\mathbf{d}}, \tag{9.24}$$

where $\hat{\mathbf{d}}$ is the dipole moment's operator[15]. Hence we may associate this operator with the operand $\hat{B}$ in Eq. (6.145), while the electric field operator $\hat{\mathscr{E}}$ is associated with the operand $\hat{A}$. Let us assume, for simplicity, that our field consists of only one mode $\mathbf{e}_j(\mathbf{r})$ of frequency ω.[16] Then we can keep only one term in the sum (9.16a), and drop the index j, so that Eq. (6.149) for the transition from a certain discrete initial state to the continuum of final states may be rewritten as

---

[15] As a reminder: Eq. (9.24), with the single-particle expression $\mathbf{d} = q\mathbf{r}$, has already been used several times in this course—see, e.g. Eq. (6.29). However, now we have to account for the quantum nature of the electromagnetic field $\mathscr{E}$, so in Eq. (9.24) it is represented by the (vector) operator (9.16a), rather than a c-number vector.

[16] In a multimode field with no inter-mode coherence, the total counting rate may be calculated as the sum of the partial rates of each mode—as will be done below for a certain case.

$$\Gamma = \frac{2\pi}{\hbar} \, |\langle \text{fin} | \hat{\mathscr{E}}(\mathbf{r}, t) | \text{ini} \rangle|^2 |\langle \text{fin} | \hat{\mathbf{d}}(t) \cdot \mathbf{n}_e | \text{ini} \rangle|^2 \rho_a$$

$$= \frac{2\pi}{\hbar} \frac{\hbar\omega}{2} \, |\langle \text{fin} | [\hat{a}^\dagger(t) - \hat{a}(t)]e(\mathbf{r}) | \text{ini} \rangle|^2 |\langle \text{fin} | \, \hat{\mathbf{d}}(t) \cdot \mathbf{n}_e | \text{ini} \rangle|^2 \rho_a, \tag{9.25}$$

where $\mathbf{n}_e \equiv \mathbf{e}(\mathbf{r})/e(\mathbf{r})$ is the local direction of the vector $\mathbf{e}(\mathbf{r})$, and the density ρ_a of the continuous atomic states should be calculated at its final energy $E_{\text{fin}} = E_{\text{ini}} + \hbar\omega$.

As a reminder, in the Heisenberg picture of quantum dynamics, the initial and final states are time-independent, while the creation–annihilation operators are functions of time. In the Golden Rule formula (9.25), as in any perturbative result, this time dependence has to be calculated ignoring the perturbation—in this case the field–atom interaction. For the field's creation–annihilation operators, this dependence coincides with that of the usual 1D oscillator—see Eq. (5.141), in which ω_0 should be, in our current notation, replaced with ω:

$$\hat{a}(t) = \hat{a}(0)e^{-i\omega t}, \quad \hat{a}^\dagger(t) = \hat{a}^\dagger(0)e^{+i\omega t}. \tag{9.26}$$

Hence Eq. (9.25) becomes

$$\Gamma = \pi\omega \, |\langle \text{fin} | [\hat{a}^\dagger(0)e^{i\omega t} - \hat{a}(0)e^{-i\omega t}]e(\mathbf{r}) | \text{ini} \rangle|^2$$

$$\times |\langle \text{fin} | \, \hat{\mathbf{d}}(t) \cdot \mathbf{n}_e | \text{ini} \rangle|^2 \rho_a. \tag{9.27a}$$

Now let us multiply the first bracket by $\exp\{i\omega t\}$, and the second one by $\exp\{-i\omega t\}$:

$$\Gamma = \pi\omega \, |\langle \text{fin} | [\hat{a}^\dagger(0)e^{2i\omega t} - \hat{a}(0)] \, e(\mathbf{r}) | \text{ini} \rangle|^2$$

$$\times |\langle \text{fin} | \, \hat{\mathbf{d}}(t) \cdot \mathbf{n}_e e^{-i\omega t} | \text{ini} \rangle|^2 \rho_a. \tag{9.27b}$$

The motivation for this, mathematically trivial, step is that at resonant photon absorption, only the *annihilation* operator gives a significant time-averaged contribution to the first bracket matrix element. (As a reminder, the quantum-mechanical Golden Rule for time-dependent perturbations is a result of averaging over a time interval much larger than $1/\omega$—see section 6.6.) Similarly, according to Eq. (4.199), the Heisenberg operator of the dipole moment, corresponding to the *increase* of atom's energy by $\hbar\omega$, has only the Fourier components that differ from ω only by $\sim\Gamma \ll \omega$, so that its time dependence virtually compensates the additional factor in the second bracket of Eq. (9.27b), and this bracket is also frequency-independent and has a substantial time average. Hence, in the first bracket we may neglect the fast-oscillating term, whose average over time interval $\sim 1/\Gamma$ is very close to zero[17].

Now let us assume that we use the same detector, characterized by the same matrix element of the quantum transition, i.e. the same second bracket in Eq. (9.27), and the same final state density ρ_a, for measurement of various electromagnetic fields

---

[17] This is essentially the same rotating wave approximation (RWA) which was already used in section 6.5 and beyond—see, e.g. the transition from Eq. (6.90) to the first of Eqs. (6.94).

—or just of the same field at different points **r**. Then we are only interested in the behavior of the first, field-related bracket, and may write

$$\Gamma \propto |\langle \text{fin}|\hat{a}e(\mathbf{r})|\text{ini}\rangle|^2 \equiv \langle \text{fin}|\hat{a}e(\mathbf{r})|\text{ini}\rangle\langle \text{fin}|\hat{a}e(\mathbf{r})|\text{ini}\rangle^*$$
$$\equiv \langle \text{ini}|\hat{a}^\dagger e^*(\mathbf{r})|\text{fin}\rangle\langle \text{fin}|\hat{a}e(\mathbf{r})|\text{ini}\rangle, \tag{9.28}$$

where the creation–annihilation operators are assumed to be taken at $t = 0$, i.e. in the Schrödinger picture, and the initial and final states are those of the field alone. As we know, any 1D harmonic oscillator (and hence the electromagnetic field oscillator) has infinitely many equidistant levels, so even if it initially was in a certain quantum state, it may undergo several coherent transitions to different final Fock states. If we want to calculate the *total* photon-absorption rate, we should sum the transition rates into all final states. Then, since the vectors of these states form a full and orthonormal set in the Hilbert space of the oscillator, we may use the closure relation (4.44) to get

$$\Gamma \propto \sum_{\text{fin}} \langle \text{ini}|\hat{a}^\dagger e^*(\mathbf{r})|\text{fin}\rangle\langle \text{fin}|\hat{a}e(\mathbf{r})|\text{ini}\rangle = \langle \text{ini}|\hat{a}^\dagger\hat{a}|\text{ini}\rangle e^*(\mathbf{r})e(\mathbf{r})$$
$$= \langle n\rangle_{\text{ini}} |e(\mathbf{r})|^2, \tag{9.29}$$

where, for a given field mode, $\langle n\rangle_{\text{ini}}$ is the expectation value of the operator $\hat{n} \equiv \hat{a}^\dagger\hat{a}$ for the initial state of the electromagnetic field, not affected by the detector.

Let us apply Eq. (9.29) to several possible quantum states of the mode.

(i) First, as a sanity check, the ground initial state ($n = 0$) gives no photon absorption at all. The interpretation is easy: the ground state field, cannot emit a photon that would ionize an atom in the counter. Again, this does not mean that the ground-state 'motion' is not observable (if you still think so, please review the Casimir effect discussion in the last section), just that it cannot ionize the trigger atom—because it does not have any *spare* energy for doing that.

(ii) All other coherent states (Fock, Glauber, squeezed, etc) of the field oscillator give the same counting rate, provided that their $\langle n\rangle$ is the same. This result may be less evident if we apply Eq. (9.29) to an interference of two light beams from the same source—say, in the double-slit or the Bragg-scattering configurations. In this case we may represent the spatial distribution of the field as a sum

$$e(\mathbf{r}) = e_1(\mathbf{r}) + e_2(\mathbf{r}). \tag{9.30}$$

Here each term describes one possible wave path, so that the operator product in Eq. (9.29) may be a rapidly changing function of the detector position. For this configuration, our result (9.29) means that the interference pattern (and its contrast) are independent of the particular state of the electromagnetic field's mode.

(iii) Surprisingly, the last statement is also valid for a classical mixture of the different eigenstates of the same field mode, for example for its thermal-equilibrium state. Indeed, in this case we need to average Eq. (9.29) over the corresponding classical ensemble, but it would only result in a different meaning of averaging n in that equation; the field part describing the interference pattern is not affected.

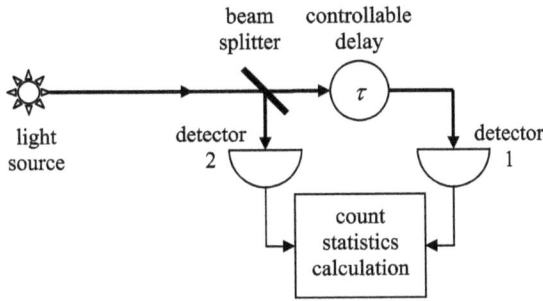

Figure 9.2. Photon counting correlation measurements. (The intensities of the split beams should be comparable, but not necessarily equal.)

The last result may look a bit counter-intuitive, because common sense tells us that the stochasticity associated with thermal equilibrium has to suppress the interference pattern contrast. These expectations are (partly) justified, because a typical thermal source of radiation produces many field modes j, rather than one mode we have analyzed. These modes may have different wave numbers k_j and hence different field distribution functions $e_j(\mathbf{r})$, resulting in shifted interference patterns. Their summation would indeed smear the interference, suppressing its contrast.

So the use of a single photon detector is not a suitable way to distinguish different quantum states of an electromagnetic field mode. This task, however, may be achieved using the photon counting correlation technique shown in figure 9.2.[18] In this experiment, the counter rate correlation may be characterized by the so-called *second-order correlation function* of the counting rates,

$$g^{(2)}(\tau) \equiv \frac{\langle \Gamma_1(t)\Gamma_2(t-\tau)\rangle}{\langle \Gamma_1(t)\rangle\langle \Gamma_2(t)\rangle}, \qquad (9.31)$$

where the averaging may be carried out either over many similar experiments, or over time t, due to the ergodicity of the system (with a stationary field source). Using the normalized correlation function (9.31) is very convenient, because the characteristics of both detectors and the beam splitter (e.g. a semi-transparent mirror, see figure 9.2) drop our from this fraction.

Very unexpectedly in the mid-1950s, Hanbury Brown and Twiss discovered that the correlation function depends on time delay τ in the way shown schematically by the solid line in figure 9.3. It is evident from Eq. (9.31) that if the counting events are completely independent, $g^{(2)}(\tau)$ should be equal 1—which is always the case in the limit $\tau \to \infty$. Hence, the observed behavior at $\tau \to 0$ corresponds to a *positive* correlation of detector counts at small time delays, i.e. to a *higher* probability of the

---

[18] It was pioneered as early as in the mid-1950s (i.e. before the advent of lasers!), by R Hanbury Brown and R Twiss. Their first experiments were also remarkable for the rather unusual light source—the star Sirius! (It was a part of an attempt to improve astrophysics interferometry techniques.)

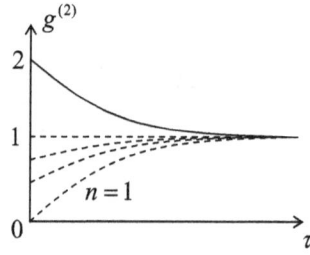

Figure 9.3. The photon bunching (solid line) and antibunching for various n (dashed lines). The lines approach level $g^{(2)} = 1$ at $\tau \to \infty$ (on the time scale depending on the light source).

nearly-simultaneous arrival of photons to both counters. This counter-intuitive effect is called the *photon bunching*.

Let us use our simple single-mode model to analyze this experiment. Now the elementary quantum process, characterized by the numerator of Eq. (9.31), is the correlated, simultaneous ionization of two trigger atoms, at two spatial–temporal points $\{\mathbf{r}_1, t\}$ and $(\mathbf{r}_2, t - \tau)$, by the same field mode, so that we need to make the following replacement in the first of Eqs. (9.25):

$$\hat{\mathscr{E}}(\mathbf{r}, t) \to \text{const} \times \hat{\mathscr{E}}(\mathbf{r}_1, t)\hat{\mathscr{E}}(\mathbf{r}_2, t - \tau). \tag{9.32}$$

Repeating all the manipulations done above for the single-counter case, we get

$$\langle \Gamma_1(t)\Gamma_2(t - \tau) \rangle \propto \langle \text{ini}|\hat{a}(t)^\dagger \hat{a}(t - \tau)^\dagger \hat{a}(t - \tau)\hat{a}(t)|\text{ini} \rangle \, e^*(\mathbf{r}_1)e^*(\mathbf{r}_2)e(\mathbf{r}_1)e(\mathbf{r}_2). \tag{9.33}$$

Plugging this expression, as well as Eq. (9.29) for single-counter rates, into Eq. (9.31), we see that the field distribution factors (as well as the detector-specific brackets and the density of states ρ_a) cancel, giving a very simple final expression:

$$g^{(2)}(\tau) = \frac{\langle \hat{a}^\dagger(t)\hat{a}^\dagger(t - \tau)\hat{a}(t - \tau)\hat{a}(t)\rangle}{\langle \hat{a}^\dagger(t)\hat{a}(t)\rangle^2}, \tag{9.34}$$

where the averaging should be carried out, as before, over the initial state of the field.

Still, the calculation of this expression for arbitrary τ may be quite complex, because the relaxation of the correlation function to the asymptotic value $g^{(2)}(\infty)$ in many cases is due to the interaction of the light source with the environment, and hence requires the open-system techniques that were discussed in chapter 7. However, the zero-delay value $g^{(2)}(0)$ may be calculated in a straightforward way, because the time arguments of all operators are equal, so that we may write

$$g^{(2)}(0) = \frac{\langle \hat{a}^\dagger \hat{a}^\dagger \hat{a}\hat{a}\rangle}{\langle \hat{a}^\dagger \hat{a}\rangle^2}. \tag{9.35}$$

Let us evaluate this ratio for the simplest states of the field.

(i) *The nth Fock state.* In this case, it is convenient to act with the annihilation operators upon the ket-vectors, and by the creation operators, upon the bra-vectors, using Eqs. (9.19):

$$g^{(2)}(0) = \frac{\langle n | \, \hat{a}^\dagger \hat{a}^\dagger \hat{a}\hat{a} \, | n \rangle}{\langle n | \, \hat{a}^\dagger \hat{a} \, | n \rangle^2} = \frac{\langle n - 2 \, | \, [n(n - 1)]^{1/2}[n(n - 1)]^{1/2} \, | n - 2 \rangle}{\langle n - 1 | n^{1/2} n^{1/2} \, | n - 1 \rangle^2}$$
$$= \frac{n(n - 1)}{n^2} = 1 - \frac{1}{n}. \tag{9.36}$$

We see that the correlation function at small delays is suppressed rather than enhanced—see the dashed lines in figure 9.3. This *photon antibunching* effect has a very simple handwaving explanation: a single photon emitted by the wave source may be absorbed by just one of the detectors. For the initial state $n = 1$, this is the only option, and it is very natural that Eq. (9.36) predicts no simultaneous counts at $\tau = 0$. Despite this theoretical simplicity, reliable observations of the antibunching have not been carried out until 1977,[19] due to the experimental difficulty of driving electromagnetic field oscillators into their Fock states—see section 9.4 below.

(ii) *The Glauber state* α. A similar procedure, but now using Eq. (5.124) and its Hermitian conjugate, $\langle \alpha | \hat{a}^\dagger = \langle \alpha | \alpha^*$, yields

$$g^{(2)}(0) = \frac{\langle \alpha | \, \hat{a}^\dagger \hat{a}^\dagger \hat{a}\hat{a} \, | \alpha \rangle}{\langle \alpha | \, \hat{a}^\dagger \hat{a} \, | \alpha \rangle^2} = \frac{\alpha^* \alpha^* \alpha \alpha}{(\alpha^* \alpha)^2} = 1, \tag{9.37}$$

for any parameter α. We see that the result is a very different result from the Fock states, unless in the latter case $n \to \infty$. (We know that the Fock and Glauber properties should also coincide for the ground state, but at that state the correlation function's value is uncertain, because there are no photon counts at all.)

(iii) *Classical mixture*. From chapter 7, we know that such statistical ensembles cannot be described by single state vectors, and require the density matrix w for their description. Here, we may combine Eqs. (9.35) and (7.5) to write

$$g^{(2)}(0) = \frac{\mathrm{Tr}(\hat{w}\hat{a}^\dagger \hat{a}^\dagger \hat{a}\hat{a})}{[\mathrm{Tr}(\hat{w}\hat{a}^\dagger \hat{a})]^2}. \tag{9.38}$$

Spelling out this expression is easy for the field in thermal equilibrium at some temperature T, because then the density matrix is diagonal in the basis of Fock states n—see Eqs. (7.24):

$$w_{nn'} = W_n \delta_{nn'}, \quad W_n = \frac{1}{Z} \exp\left\{ -\frac{E_n}{k_B T} \right\} \equiv \frac{\lambda^n}{\sum\limits_{n=0}^{\infty} \lambda^n}, \quad \text{where} \quad \lambda \equiv \exp\left\{ -\frac{\hbar\omega}{k_B T} \right\}. \tag{9.39}$$

So, for the operators in the numerator and denominator of Eq. (9.38) we also need just the diagonal terms of the operator products, which have already been calculated—see Eq. (9.36). As a result, we get

---

[19] By H Kimble [9]. For a detailed review of phonon antibunching, see, e.g. [10].

$$g^{(2)}(0) = \frac{\sum\limits_{n=0}^{\infty} W_n n(n-1)}{\left(\sum\limits_{n=0}^{\infty} W_n n\right)^2} = \frac{\sum\limits_{n=0}^{\infty} \lambda^n n(n-1) \times \sum\limits_{n=0}^{\infty} \lambda^n}{\left(\sum\limits_{n=0}^{\infty} \lambda^n n\right)^2}. \tag{9.40}$$

One of the three series involved in this expression is just the geometric progression,

$$\sum_{n=0}^{\infty} \lambda^n = \frac{1}{1-\lambda}, \tag{9.41}$$

and the remaining two may be readily calculated by its differentiation over the parameter λ:

$$\sum_{n=0}^{\infty} \lambda^n n = \lambda \sum_{n=0}^{\infty} \lambda^{n-1} n = \lambda \frac{d}{d\lambda} \sum_{n=0}^{\infty} \lambda^n = \lambda \frac{d}{d\lambda} \frac{1}{1-\lambda} = \frac{\lambda}{(1-\lambda)^2},$$

$$\sum_{n=0}^{\infty} \lambda^n n(n-1) = \lambda^2 \sum_{n=0}^{\infty} \lambda^{n-2} n(n-1) = \lambda^2 \frac{d^2}{d^2\lambda} \left(\sum_{n=0}^{\infty} \lambda^n\right) \tag{9.42}$$

$$= \lambda^2 \frac{d^2}{d\lambda^2} \frac{1}{1-\lambda} = \frac{2\lambda^2}{(1-\lambda)^3},$$

and for the correlation function we get an extremely simple result independent of the parameter λ and hence of temperature:

$$g^{(2)}(0) = \frac{[2\lambda^2/(1-\lambda)^3]\,[1/(1-\lambda)]}{[\lambda/(1-\lambda)^2]^2} \equiv 2. \tag{9.43}$$

This is the exactly the photon bunching effect first observed by Hanbury Brown and Twiss (figure 9.3). We see that in contrast to antibunching, this is an essentially classical (statistical) effect. Indeed, Eq. (9.43) allows a purely classical derivation. In the classical theory, the counting rate (of a single counter) is proportional to the wave intensity I, so that Eq. (9.31) with $\tau = 0$ is reduced to

$$g^{(2)}(0) = \frac{\langle I^2 \rangle}{\langle I \rangle^2}, \quad \text{with } I \propto \overline{E^2(t)} \propto E_\omega E_\omega^*. \tag{9.44}$$

For a sinusoidal field, the intensity is constant, and $g^{(2)}(0) = 1$. (This is also evident from Eq. (9.37), because the classical state may be considered as a Glauber state with $\alpha \to \infty$.) On the other hand, if the intensity fluctuates (either in time, or from one experiment to another), the averages in Eq. (9.44) should be calculated as

$$\langle I^k \rangle = \int_0^\infty w(I)\, I^k dI, \quad \text{with } \int_0^\infty w(I) dI = 1, \quad \text{and } k = 1, 2, \tag{9.45}$$

where $w(I)$ is the probability density. For the classical (Boltzmann) statistics, the probability is an exponential function of the electromagnetic field energy, and hence its intensity:

$$w(I) = Ce^{-\beta I}, \quad \text{where } \beta \propto 1/k_B T, \tag{9.46}$$

so that Eqs. (9.45) yield:

$$\int_0^\infty C \exp\{-\beta I\} dI \equiv C/\beta = 1, \quad \text{and hence } C = \beta,$$

$$\langle I^k \rangle = \int_0^\infty w(I) I^k dI = C \int_0^\infty \exp\{-\beta I\} I^k dI = \frac{1}{\beta^k} \int_0^\infty \exp\{-\xi\} \xi^k d\xi \tag{9.47}$$

$$= \begin{cases} 1/\beta, & \text{for } k = 1, \\ 2/\beta^2, & \text{for } k = 2. \end{cases}$$

Plugging these results into Eq. (9.44), we get $g^{(2)}(0) = 2$, in complete agreement with Eq. (9.43).[20]

9.3 Photon emission: spontaneous and stimulated

In our simple model of photon counting, considered in the last section, the trigger atom in the counter *absorbed* a photon. Now let us have a look at the opposite process of *spontaneous emission* of photons by an atom in an excited state, still using the same electric-dipole approximation (9.24) for the atom-to-field interaction. For this, we may still use the Golden Rule for the model depicted in figure 6.12, but now the roles have changed: we have to associate the operator $\hat{A}$ with the electric dipole moment of the atom, while the operator $\hat{B}$, with the electric field, and the continuous spectrum of the system b represents the plurality of the electromagnetic field modes into which the spontaneous radiation may happen. Since now the transition *increases* the energy of the electromagnetic field, after the multiplication of the field bracket in Eq. (9.27a) by $\exp\{i\omega t\}$, we may keep only the photon *creation* operator whose time evolution compensates this fast 'rotation'. As a result, the Golden Rule takes the following form:

$$\Gamma_s = \pi\omega |\langle\text{fin}|\hat{a}^\dagger|0\rangle|^2 |\langle\text{fin}| \hat{\mathbf{d}} \cdot \mathbf{e}(\mathbf{r})|\text{ini}\rangle|^2 \rho_f, \tag{9.48}$$

where all operators and states are again time-independent (i.e. taken in the Schrödinger picture), and ρ_f is now the density of final states of the electromagnetic field—which in this problem plays the role of the atom's environment. Here the electromagnetic field oscillator has been assumed to be initially in the ground state—the assumption that will be altered later in this section.

This relation, together with Eq. (9.19), shows that in order for the field's matrix element to be different from zero, the final state of the field has to be the first excited Fock state, $n = 1$. (By the way, *this* is exactly the most practicable way of generating an excited Fock state of a field oscillator.) With that, Eq. (9.48) yields

---

[20] For some field states, including the squeezed ground states ζ discussed in the end of section 5.5, values $g^{(2)}(0)$ may be even higher than 2—the so-called *super-bunching*. Analyses of two cases of such super-bunching are offered for the reader's exercise—see the problem list at the chapter's end.

$$\Gamma_s = \pi\omega \, |\langle \text{fin}|\hat{\mathbf{d}} \cdot \mathbf{e}(\mathbf{r})|\text{ini}\rangle|^2 \rho_{\text{fin}} \equiv \pi\omega \, |\langle \text{fin}|\hat{d}e_d(\mathbf{r})|\text{ini}\rangle|^2 \rho_{\text{fin}}, \tag{9.49}$$

where the density ρ_{fin} of excited electromagnetic field states should be calculated at the energy $E = \hbar\omega$, and e_d is the component of the vector $\mathbf{e}(\mathbf{r})$ along the electric dipole's direction[21]. The formula for the density was our first step in this course—see Eq. (1.1).[22] From it, we get

$$\rho_{\text{fin}} \equiv \frac{dN}{dE} = V\frac{\omega^2}{\pi^2 \hbar c^3}, \tag{9.50}$$

where the bounding volume V should be large enough to ensure the spectrum's virtual continuity: $V \gg \lambda^3 = (2\pi c/\omega)^3$. Because of that, in the normalization condition, used in section 1 to simplify Eq. (9.9), we may consider $e^2(\mathbf{r})$ constant. Let us represent this square as a sum of squares of the three Cartesian components of the vector $\mathbf{e}(\mathbf{r})$: one of those, e_d, aligned with the dipole direction; due to the space isotropy we may write

$$e^2 \equiv e_d^2 + e_{\perp 1}^2 + e_{\perp 2}^2 = 3e_d^2. \tag{9.51}$$

As a result, the normalization condition yields

$$e_d^2 = \frac{1}{3\varepsilon_0 V}. \tag{9.52}$$

and Eq. (9.49) gives the famous (and very important) formula[23]

$$\begin{aligned}
\Gamma_s &= \frac{1}{4\pi\varepsilon_0} \frac{4\omega^3}{3\hbar c^3} \, |\langle \text{fin}|\hat{\mathbf{d}}|\text{ini}\rangle|^2 \\
&\equiv \frac{1}{4\pi\varepsilon_0} \frac{4\omega^3}{3\hbar c^3} \, \langle \text{fin}|\hat{\mathbf{d}}|\text{ini}\rangle \cdot \langle \text{ini}|\hat{\mathbf{d}}|\text{fin}\rangle^*.
\end{aligned} \tag{9.53}$$

Leaving a comparison of this formula with the classical theory of radiation[24], and the exact evaluation of Γ_s for a particular transition in the hydrogen atom, for reader's exercises, let me just estimate its order of magnitude. Assuming that $d \sim er_B \equiv e\hbar^2/m_e(e^2/4\pi\varepsilon_0)$ and $\hbar\omega \sim E_H \equiv m_e(e^2/4\pi\varepsilon_0)^2/\hbar^2$, and taking into account the definition (6.62) of the fine structure constant $\alpha \approx 1/137$, we get

---

[21] Here the sum over all electromagnetic field modes j has been smuggled back. Since in the quasistationary approximation, $k_j a \ll 1$, which is necessary for the interaction representation by Eq. (9.24), the matrix elements in Eq. (9.49) are virtually independent on k_j, the summation is reduced to the calculation of the total ρ_f for all modes.

[22] Note the essential dependence of Eq. (9.50), and hence of Eq. (9.53) on the field geometry; all following formulas of this section are strictly valid for the free 3D space only. If the same atom is placed into a high-Q resonant cavity (see, e.g. *Part EM* 7.9), the rate of its photon emission is strongly suppressed at frequencies between the cavity resonances (where $\rho_f \to 0$)—see, e.g. the review [11]. On the other hand, the emission is strongly (by a factor $\sim (\lambda^3/V)Q$, where V is cavity's volume) enhanced at resonance frequencies—the so-called *Purcell effect*, discovered by E Purcell already in the 1940s. For a brief discussion of this and other quantum electrodynamic effects in cavities, see the next section.

[23] An equivalent expression was first obtained (from more formal arguments) in 1930 by V Weisskopf and E Wigner, so that the whole calculation is sometimes referred to as the *Weisskopf–Wigner theory*.

[24] See, e.g. *Part EM* section 8.2, in particular Eq. (8.29).

$$\frac{\Gamma}{\omega} \sim \left(\frac{e^2}{4\pi\varepsilon_0\hbar c}\right)^3 \equiv \alpha^3 \sim 3 \times 10^{-7}. \tag{9.54}$$

This estimate shows that the emission lines at atomic transitions are typically very sharp. With the present-day availability of high-speed electronics, it also makes sense to evaluate the time scale $\tau = 1/\Gamma$ of the typical quantum transition: for a typical optical frequency $\omega \sim 3 \times 10^{15}$ s^{-1}, it is close to 1 ns. This is exactly the time constant that determines the time-delay dependence of the photon counting statistics of the spontaneously emitted radiation—see figure 9.3. Colloquially, this is the temporal scale of the photon spontaneously emitted by an atom[25].

Note, however, that the above estimate of τ is only valid for a transition with a non-vanishing electric-dipole matrix element. If it equals zero, i.e. the transition does not satisfy the *selection rules*[26],—say, due to the initial and final state symmetry—it is 'forbidden'. The 'forbidden' transition may still take place due to a different, smaller interaction (say, via a magnetic dipole field of the atom, or its quadrupole electric field[27]), but takes much longer. In some cases the increase of τ is rather dramatic—sometimes to hours! Such long-lasting radiation is called the *luminescence*—or the *fluorescence* if the initial atom's excitation was due to an external radiation of a higher frequency, followed first by non-radiative transitions down the energy level ladder.

Now let us consider a more general case when the electromagnetic field mode of frequency ω is initially in an arbitrary Fock state n, and from it may either get energy $\hbar\omega$ from the atomic system (*photon emission*) or, vice versa, give such energy back to the atom (*photon absorption*). For the photon emission rate, an evident generalization of Eq. (9.48) gives

$$\frac{\Gamma_e}{\Gamma_s} \equiv \frac{\Gamma_{n\to\text{fin}}}{\Gamma_{0\to1}} = \frac{|\langle\text{fin}|\hat{a}^\dagger|n\rangle|^2}{|\langle1|\hat{a}^\dagger|0\rangle|^2}, \tag{9.55}$$

where both brackets should be calculated in the Schrödinger picture, and Γ_s is the spontaneous emission rate (9.48) of the same atomic system. According to the second of Eqs. (9.19), at the photon emission, the final field state has to be the Fock state with $n' = n + 1$, and Eq. (9.55) yields

$$\Gamma_e = (n + 1)\,\Gamma_s. \tag{9.56}$$

Thus the initial field increases the photon emission rate; this effect is called the *stimulated emission of radiation*. Note that the spontaneous emission may be

---

[25] The scale $c\tau$ of the spatial extension of the corresponding wave packet is surprisingly macroscopic—in the range of a few millimeters. Such 'human' size of the spontaneously emitted photons makes the usual optical table, with its 1 cm-scale components, the key instrument for many optical experiments—see, e.g. figure 9.2.

[26] As was already discussed in section 5.6, for a single spinless particle moving in a spherically-symmetric potential (e.g. a hydrogen-like atom), the orbital selection rules are simple: the only allowed electric-dipole transitions are those with $\Delta l \equiv l_{\text{fin}} - l_{\text{ini}} = \pm 1$ and $\Delta m \equiv m_{\text{fin}} - m_{\text{ini}} = 0$ or ± 1. The simplest example of the transition that does *not* satisfy this rule, i.e. is 'forbidden', is that between the s-states ($l = 0$) with $n = 2$ and $n = 1$; because of that, the lifetime of the lowest excited s-state of a hydrogen atom is as long as ~ 0.15 s.

[27] See, e.g. *Part EM* section 8.9.

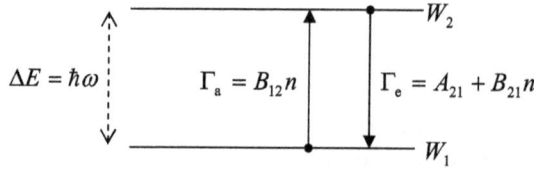

Figure 9.4. The Einstein coefficients on the atomic quantum transition diagram—cf. figure 7.6.

considered as a particular case of the stimulated emission for $n = 0$, and hence interpreted as the emission stimulated by zero-point fluctuations of the electromagnetic field.

On the other hand, in accordance with the arguments of section 9.2,[28] for the description of radiation *absorption*, the photon creation operator has to be replaced with the annihilation operator, giving the rate ratio

$$\frac{\Gamma_a}{\Gamma_s} = \frac{|\langle \text{fin}| \, \hat{a} \, |n\rangle|^2}{|\langle 1| \, \hat{a}^\dagger \, |0\rangle|^2}. \tag{9.57}$$

According to this relation and the first of Eqs. (9.19), the final state of the field at the photon absorption is the Fock state with $n' = n - 1$, and Eq. (9.57) yields

$$\Gamma_a = n\Gamma_s. \tag{9.58}$$

The results (9.56) and (9.58) are usually formulated in terms of between the *Einstein coefficients A* and *B* defined in the way shown in figure 9.4, where the two energy levels are those of the atom, Γ_a is the rate of energy absorption from the electromagnetic field, and Γ_e is that of the energy emission into the field. In this notation, Eqs. (9.56) and (9.58) yield[29]

$$A_{21} = B_{21} = B_{12}, \tag{9.59}$$

because each of these coefficients equals the spontaneous emission rate Γ_s.

I cannot resist the temptation to use this point for a small detour—an alternative derivation of the Bose–Einstein statistics for photons. Indeed, in the thermodynamic equilibrium, the average probability flows between the levels 1 and 2 (see figure 9.4 again) should be equal[30]:

$$W_2 \langle \Gamma_e \rangle = W_1 \langle \Gamma_a \rangle, \tag{9.60}$$

where W_1 and W_2 are the probabilities for the atomic system to occupy the corresponding levels, so that Eqs. (9.56) and (9.58) yield

---

[28] Note, however, a major difference between the rate Γ discussed in section 9.2, and Γ_a in Eq. (9.57). In our current case, the atomic transition is still between two *discrete* energy levels (see figure 9.4 below), so that the rate Γ_a is proportional to ρ_f, the density of final states of the electromagnetic *field*, i.e. the same density as in Eq. (9.48) and beyond, while the rate (9.27) is proportional to ρ_a, the density of final (ionized) states of the 'trigger' *atom*, more exactly of its released electron.

[29] This relation was conjectured, from very general arguments, by A Einstein as early as in 1916.

[30] This is just a particular embodiment of the detailed balance equation (7.198).

$$W_2 \Gamma_s \langle 1 + n \rangle = W_1 \Gamma_s \langle n \rangle, \qquad \text{i.e. } \frac{W_2}{W_1} = \frac{\langle n \rangle}{\langle n \rangle + 1}. \tag{9.61}$$

But, on the other hand, for the atomic subsystem, only weakly coupled to its electromagnetic environment, we ought to have the Gibbs distribution of these probabilities:

$$\frac{W_2}{W_1} = \frac{\exp\{-E_2/k_B T\}}{\exp\{-E_1/k_B T\}} = \exp\left\{-\frac{\Delta E}{k_B T}\right\} = \exp\left\{-\frac{\hbar\omega}{k_B T}\right\}. \tag{9.62}$$

Requiring Eqs. (9.61) and (9.62) to give the same result for the probability ratio, we get the Bose–Einstein distribution for the electromagnetic field in thermal equilibrium:

$$\langle n \rangle = \frac{1}{\exp\{\hbar\omega/k_B T\} - 1} \tag{9.63}$$

—the same as obtained in section 7.1 by other means—see Eq. (7.26b).

Now returning to the discussion of Eqs. (9.56) and (9.58), their very important implication is the possibility to achieve the stimulated emission of coherent radiation using the level *occupancy inversion*. Indeed, if the ratio W_2/W_1 is larger than that given by Eq. (9.62), the net power flow from the atomic system into the electromagnetic field,

$$\text{power} = \hbar\omega \times \Gamma_s[W_2(\langle n \rangle + 1) - W_1\langle n \rangle], \tag{9.64}$$

may be positive. The necessary inversion may be produced using several ways, notably by intensive quantum transitions to level 2 from an even higher energy level (which, in turn, is populated, e.g. by absorption of an external radiation, usually called *pumping*, at a higher frequency.)

A less obvious, but crucial feature of the stimulated emission is spelled out by Eq. (9.55): as was mentioned above, it shows that the final state of the field after the absorption of energy $\hbar\omega$ from the atom is a pure (coherent) Fock state $(n + 1)$. Colloquially, one may say that the new, $(n + 1)$st photon emitted from the atom is automatically in phase with the n photons that had been in the field mode initially, i.e. joins them coherently[31]. The idea of stimulated emission of coherent radiation using population inversion[32] was first implemented in the early 1950s in the microwave range (*masers*) and in 1960 in the optical range (*lasers*). Nowadays, lasers are ubiquitous components of almost all high-tech systems, and constitute one of the cornerstones of our technological civilization.

A quantitative discussion of laser operation is well beyond the framework of this course, and I have to refer the reader to special literature[33], but would like to briefly mention only two key points:

---

[31] It is straightforward to show that this fact is also true if the field is initially in the Glauber state—which is more typical for modes in practical lasers.
[32] This idea may be traced back at least to an obscure 1939 publication by V Fabrikant.
[33] I can recommend, for example, [12], and a less technical text by A Yariv [13].

(i) In a typical laser, each generated electromagnetic field mode is in the Glauber (rather than the Fock) state, so that Eqs. (9.56) and (9.58) are applicable only for the n averaged over the Fock-state decomposition of the Glauber state—see Eq. (5.134).

(ii) Since in a typical laser $\langle n \rangle \gg 1$, its operation may be well described using quasi-classical theories that use Eq. (9.64) to describe the electromagnetic energy balance (with the addition of a term describing the energy loss due to field absorption in external components of the laser, including the useful load), plus the equation describing the balance of the occupancies $W_{1,2}$ due to all inter-level transitions—similar to Eq. (9.60), but including also the contribution(s) from the particular population inversion mechanism used in the laser. At this approach, the role of quantum mechanics in laser science is essentially reduced to the calculation of the parameter Γ_s for the particular system.

This role becomes more prominent when one needs to describe *fluctuations* of the laser field. Here two approaches are possible, following the two options discussed in chapter 7. If the fluctuations are relatively small, one can linearize the Heisenberg equations of motion of the field oscillator operators near their stationary-lasing 'values', with the Langevin 'forces' (also time-dependent operators) describing the fluctuation sources, and use these Heisenberg–Langevin equations to calculate the radiation fluctuations, just as was described in section 7.5. On the other hand, near the lasing threshold the field fluctuations are relatively large, smearing the phase transition between the no-lasing and lasing states. Here the linearization is not an option, but one can use the density-matrix approach described in section 7.6, for the fluctuation analysis[34]. Note that while the laser radiation fluctuations may look like a peripheral issue, pioneering research in that field has led to the development of the general theory of open quantum systems, which was discussed in chapter 7.

9.4 Cavity QED

Now I have to mention, at least in passing, the field of *cavity quantum electro-dynamics* (usually called *cavity QED* for short)—the art and science of creating and using entanglement between quantum states of an atomic system (either an atom, or an ion, or a molecule, etc) and the electromagnetic field in a macroscopic volume called the *resonant cavity* (or just 'resonator', or just 'cavity'). This field is very popular nowadays, especially in the context of the quantum computation and communication research discussed in section 8.5.[35]

The discussion in the previous section was based on the implicit assumption that the energy spectrum of the electromagnetic field interacting with an atomic subsystem is essentially continuous, so that its final state is spread among many field modes, effectively loosing its coherence with the quantum state of the atomic

---

[34] This path has been developed (also in the mid-1960s), by several researchers, notably including M Sully and W Lamb—see, e.g. [14].

[35] This popularity was demonstrated, for example, by the award of the 2012 Nobel Prize in Physics to cavity QED experimentalists S Haroche and D Wineland.

subsystem. This assumption has justified using the quantum-mechanical Golden Rule for calculation of the spontaneous and stimulated transition rates. However, the assumption becomes invalid if the electromagnetic field is contained inside a relatively small volume, with its linear size comparable with the radiation wavelength. If the walls of such a cavity mostly reflect, rather than absorb, radiation, then the 0th approximation the power dissipation may be disregarded, and the particular solutions $e_j(\mathbf{r})$ of the Helmholtz equation (9.5) correspond to discrete, well separated mode wave numbers k_j and hence well separated eigenfrequencies ω_j.[36] Due to the energy conservation, an atomic transition corresponding to energy $\Delta E = |E_{\text{ini}} - E_{\text{fin}}|$ may be effective only if the corresponding quantum oscillation frequency $\Omega \equiv \Delta E/\hbar$.[37] As a result of such resonant interaction, the quantum states of the atomic system and the resonant electromagnetic mode may become entangled.

A very popular approximation for the qualitative description of this effect is the so-called *Rabi model*[38], in which the atom is treated as a two-level system[39] interacting with a single electromagnetic field mode of the resonant cavity. As the reader knows well from chapters 4–6 (see in particular section 5.1), any two-level system may be described, just as a spin-½, with the Hamiltonian $b\hat{I} + \mathbf{c} \cdot \hat{\boldsymbol{\sigma}}$. Since we may always select the energy origin and the state basis in that $\mathbf{b} = 0$ and $\mathbf{c} = c\mathbf{n}_z$, the Hamiltonian of the atomic subsystem may be taken in the diagonal form

$$\hat{H}_{\text{a}} = c\hat{\sigma}_z \equiv \frac{\hbar\Omega}{2}\hat{\sigma}_z, \qquad (9.65)$$

where $\hbar\Omega \equiv 2c$ is the difference between the energy levels in the absence of interaction with the field. Next, according to Eq. (9.17), ignoring the constant ground-state energy $\hbar\omega/2$ (which may be added to the energy in the end—if necessary), the contribution of a single field mode of frequency ω to the total Hamiltonian of the system is

$$\hat{H}_{\text{f}} = \hbar\omega\hat{a}^\dagger\hat{a}. \qquad (9.66)$$

Finally, according to Eq. (9.16a), the electric field of the mode may be represented as

$$\hat{\mathscr{E}}(\mathbf{r},\, t) = \frac{1}{i}\left(\frac{\hbar\omega}{2}\right)^{1/2} \mathbf{e}(\mathbf{r})(\hat{a} - \hat{a}^\dagger), \qquad (9.67)$$

---

[36] The calculation of such modes and corresponding frequencies for several simple cavity geometries was the subject of *Part EM* section 7.8 of this series.

[37] Conversely, if Ω is far from any ω_j, the interaction is suppressed; in particular, the spontaneous emission rate may be much lower than that given by Eq. (9.53)—so that this result is not as fundamental as it may look.

[38] After the pioneering work by I Rabi in 1936–37.

[39] As was shown in section 6.5, this model is justified, e.g. if transitions between all other energy level pairs have considerably different frequencies.

so that in the electric-dipole approximation (9.24), the cavity–atom interaction may be represented as a product of the field by some (say, y-) Cartesian component[40] of the Pauli spin-½ operator:

$$\hat{H}_{\text{int}} = \text{const} \times \hat{\sigma}_y \times \mathscr{E} = \text{const} \times \hat{\sigma}_y \times \left(\frac{\hbar\omega}{2}\right)^{1/2} \frac{1}{i}(\hat{a} - \hat{a}^\dagger)$$

$$= i\hbar\kappa\hat{\sigma}_y(\hat{a} - \hat{a}^\dagger), \tag{9.68}$$

where κ is a coupling constant (with the dimension of frequency). The sum of these three terms,

$$\hat{H} \equiv \hat{H}_a + \hat{H}_f + \hat{H}_{\text{int}} = \frac{\hbar\Omega}{2}\hat{\sigma}_z + \hbar\omega\hat{a}^\dagger\hat{a} + i\hbar\kappa\hat{\sigma}_y(\hat{a} - \hat{a}^\dagger). \tag{9.69}$$

giving a very reasonable description of the system, is called the *Rabi Hamiltonian*. Despite its apparent simplicity, using this Hamiltonian for calculations is not that straightforward[41]. Only in the case when the electromagnetic field is large and hence may be treated classically, the results following from Eq. (9.69) are reduced to Eqs. (6.94) describing, in particular, the Rabi oscillations discussed in section 6.3.

The situation becomes simpler in the most important case when the frequencies Ω and ω are very close, enabling an effective interaction between the cavity field and the atom even if the coupling constant κ is relatively small. Indeed, if both the κ and the so-called *detuning* (defined similarly to the parameter Δ used in section 6.5),

$$\xi \equiv \Omega - \omega, \tag{9.70}$$

are much smaller than $\Omega \approx \omega$, the Rabi Hamiltonian may be simplified using the rotating-wave approximation, already used several times in this course. For this, it is convenient to use the *spin ladder operators*, defined absolutely similarly for those of the orbital angular momentum—see Eqs. (5.153):

$$\hat{\sigma}_\pm \equiv \hat{\sigma}_x \pm i\hat{\sigma}_y, \text{ so that } \hat{\sigma}_y = \frac{\hat{\sigma}_+ - \hat{\sigma}_-}{2i}. \tag{9.71}$$

From Eq. (4.105), it is very easy to find the matrices of these operators in the standard z-basis,

$$\sigma_+ = \begin{pmatrix} 0 & 2 \\ 0 & 0 \end{pmatrix}, \qquad \sigma_- = \begin{pmatrix} 0 & 0 \\ 2 & 0 \end{pmatrix}, \tag{9.72}$$

and their commutation rules—which turn out to be naturally similar to Eqs. (5.154):

$$[\hat{\sigma}_+, \hat{\sigma}_-] = 4\hat{\sigma}_z, \qquad [\hat{\sigma}_z, \hat{\sigma}_\pm] = \pm 2\hat{\sigma}_\pm. \tag{9.73}$$

---

[40] The exact component is not important, while the intermediate formulas simplify if it is proportional to either pure σ_x or pure σ_y.

[41] For example, an exact quasi-analytical expression for its eigenenergies (as zeros of a Taylor series in the parameter κ, with coefficients determined by a recurrence relation) was found only recently—see [15].

In this notation, the Rabi Hamiltonian becomes

$$\hat{H} = \frac{\hbar\Omega}{2}\hat{\sigma}_z + \hbar\omega\hat{a}^\dagger\hat{a} + \frac{\hbar\kappa}{2}(\hat{\sigma}_+ - \hat{\sigma}_-)(\hat{a} - \hat{a}^\dagger), \tag{9.74}$$

and it is straightforward to use Eqs. (4.199) and (9.73) to derive the Heisenberg-picture equations of motion for the involved operators. (Doing this, we have to remember that operators of the 'spin' subsystem, on one hand, and of the field mode, on the other hand, are defined in different Hilbert spaces and hence commute—at least at coinciding time moments.) The result (so far, exact!) is

$$\dot{\hat{a}} = -i\omega\hat{a} + \frac{i\kappa}{2}(\hat{\sigma}_+ - \hat{\sigma}_-), \quad \dot{\hat{a}}^\dagger = i\omega\hat{a}^\dagger - \frac{i\kappa}{2}(\hat{\sigma}_+ - \hat{\sigma}_-),$$

$$\dot{\hat{\sigma}}_\pm = \pm i\Omega\hat{\sigma}_\pm + i2\kappa(\hat{a} - \hat{a}^\dagger)\hat{\sigma}_z, \quad \dot{\hat{\sigma}}_z = i\kappa(\hat{a}^\dagger - \hat{a})(\hat{\sigma}_+ + \hat{\sigma}_-). \tag{9.75}$$

At negligible coupling, $\kappa \to 0$, these equations have simple solutions,

$$\hat{a}(t) \propto e^{-i\omega t}, \quad \hat{a}^\dagger(t) \propto e^{i\omega t}, \quad \hat{\sigma}_\pm(t) \propto e^{\pm i\Omega t}, \quad \hat{\sigma}_z(t) \approx \text{const}, \tag{9.76}$$

and the small terms proportional to κ on the right-hand sides of Eqs. (9.75) cannot affect these time evolution laws dramatically even if κ is not exactly zero. Of those terms, ones with frequencies close to the 'basic' frequency of each variable would act in resonance and hence may have a substantial impact on the system's dynamics, while non-resonant terms may be ignored. In this rotating-wave approximation, Eqs. (9.75) are reduced to a much simpler system of equations:

$$\dot{\hat{a}} = -i\omega\hat{a} - \frac{i\kappa}{2}\hat{\sigma}_-, \quad \dot{\hat{a}}^\dagger = i\omega\hat{a}^\dagger + \frac{i\kappa}{2}\hat{\sigma}_+,$$

$$\dot{\hat{\sigma}}_+ = i\Omega\hat{\sigma}_+ + i2\kappa\hat{a}^\dagger\hat{\sigma}_z, \quad \dot{\hat{\sigma}}_- = -i\Omega\hat{\sigma}_- - i2\kappa\hat{a}\hat{\sigma}_z, \quad \dot{\hat{\sigma}}_z = i\kappa(\hat{a}^\dagger\hat{\sigma}_- - \hat{a}\hat{\sigma}_+). \tag{9.77}$$

Alternatively, these equations of motion may be obtained *exactly* from the Rabi Hamiltonian (9.74), preliminarily cleared of the terms proportional to $\hat{\sigma}_+\hat{a}^\dagger$ and $\hat{\sigma}_-\hat{a}$, that oscillate fast and hence self-average to virtually zero:

$$\hat{H} = \frac{\hbar\Omega}{2}\hat{\sigma}_z + \hbar\omega\hat{a}^\dagger\hat{a} + \frac{\hbar\kappa}{2}(\hat{\sigma}_+\hat{a} + \hat{\sigma}_-\hat{a}^\dagger), \quad \text{at } \kappa, |\xi| \ll \omega, \Omega. \tag{9.78}$$

This is the famous *Janes–Cummings Hamiltonian*[42], which is the basic model used in the cavity QED and its applications[43]. In order to find its eigenstates and eigenenergies, let us note that at negligible interaction ($\kappa \to 0$), the spectrum of the total energy E of the system, which in this limit is the sum of two independent contributions from the atomic and cavity-field subsystems,

---

[42] It was first proposed and analyzed in 1963 by two engineers, E Janes and F Cummings, and it took the physics community a while to recognize and acknowledge the fundamental importance of that work (published in *Proceedings of IEEE*).

[43] For most applications, the baseline Hamiltonian (9.78) has to be augmented by additional term(s) describing, for example, the incoming radiation and/or the coupling to environment, for example due to the electromagnetic energy loss in a finite-Q-factor cavity—see Eq. (7.68).

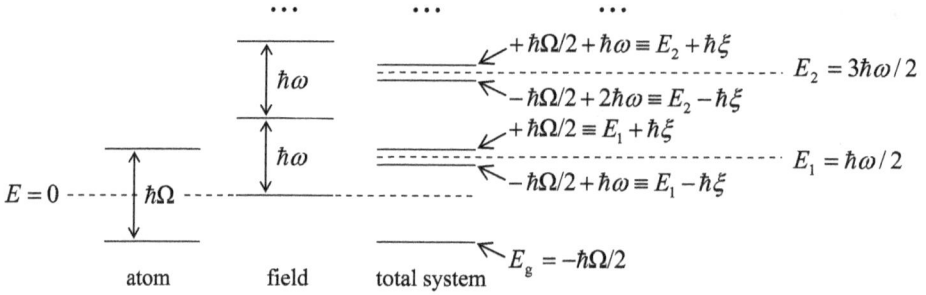

Figure 9.5. The energy spectrum (9.79) of the Janes–Cummings Hamiltonian in the limit $\kappa \ll |\xi|$. Note again that the energy is referred to the ground-state energy $\hbar\omega/2$ of the cavity field.

$$E\,|_{\kappa=0} = \pm\frac{\hbar\Omega}{2} + \hbar\omega n \equiv E_n \pm \frac{\hbar\xi}{2}, \qquad \text{with } n = 1,\, 2,\, ...,\qquad (9.79)$$

consists[44] of close level pairs (figure 9.5) centered to values

$$E_n \equiv \hbar\omega\left(n - \frac{1}{2}\right). \qquad (9.80)$$

(At the exact resonance $\omega = \Omega$, i.e. at $\xi = 0$, each pair merges into one double-degenerate level E_n.) Since at $\kappa \to 0$ the two subsystems do not interact, the eigenstates corresponding to the sublevels of the nth pair may be represented by direct products of their independent state vectors:

$$|+\rangle \equiv |\uparrow\rangle \otimes |n - 1\rangle \quad \text{and} \quad |-\rangle \equiv |\downarrow\rangle \otimes |n\rangle, \qquad (9.81)$$

where the first ket of each product represents the state of the two-level (spin-½-like) atomic subsystem, and the second ket, that of the field oscillator.

As we know from chapter 6, even weak interaction may lead to strong coherent mixing[45] of quantum states with close energies (in this case, the two states (9.81) within each pair with the same n), while their mixing with the states with farther energies is still negligible. Hence, at $0 < \kappa$, $|\xi| \ll \omega \approx \Omega$, a good approximation of an eigenstate with $E \approx E_n$ is given by a linear superposition of the states (9.81):

$$|a_n\rangle = c_+\,|+\rangle + c_-\,|-\rangle \equiv c_+\,|\uparrow\rangle \otimes |n - 1\rangle + c_-\,|\downarrow\rangle \otimes |n\rangle, \qquad (9.82)$$

with certain c-number coefficients $c_\pm$. This relation describes the entanglement of the atomic eigenstates $\uparrow$ and $\downarrow$ with the Fock states number n and $n - 1$ of the field mode. Let me leave the (straightforward) calculation of the coefficients $(c_\pm)^\pm$ for each of two entangled states (for each n) for the reader's exercise. (The result for the corresponding two eigenenergies $(E_n)^\pm$ may be again represented by the same anticrossing diagram as shown in figures 2.29 and 5.1, now with the detuning ξ as the argument.) This calculation shows, in particular, that at $\xi = 0$ (i.e. at $\omega = \Omega$), $|c_+| = |c_-| = 1/\sqrt{2}$ for both states of the pair, in a clear analogy with the entangled

---

[44] Only the ground state level $E_g = -\hbar\Omega/2$ is non-degenerate—see figure 9.5.

[45] In some fields, especially chemistry, such mixing is frequently called *hybridization*.

singlet and triplet states (8.18) and (8.21). This fact may be interpreted as a (coherent!) equal sharing of an energy quantum $\hbar\omega = \hbar\Omega$ by the atom and the cavity field at the exact resonance.

A (hopefully, self-evident) by-product of the calculation of $c_\pm$ is the fact that the dynamics of the state a_n described by Eq. (9.82) is similar to that of the generic two-level system that was repeatedly discussed in this course—first time in section 2.6 and then in chapters 4–6. In particular, if the composite system had been initially prepared to be in one component state, for example $|\uparrow\rangle \otimes |0\rangle$ (i.e. the atom excited, while the cavity in its ground state), and then allowed to evolve on its own, after some time interval $\Delta t \sim 1/\kappa$ it may be found definitely in the counterpart state $|\downarrow\rangle \otimes |1\rangle$, including the first excited Fock state $n = 1$ of the field mode. If the process is allowed to continue, after the equal time interval Δt, the system returns to the initial state $|\uparrow\rangle \otimes |0\rangle$, etc[46]. This most striking prediction of the Janes–Cummings model was directly observed, by G Rempe *et al*, only in 1987, although less directly this model was repeatedly confirmed by numerous experiments carried out in the 1960s and 1970s.

Unfortunately, my time/space allowance for the cavity QED is over, and for further reading I have to refer the reader to special literature[47].

9.5 The Klein–Gordon and relativistic Schrödinger equations

Now let me switch gears and discuss the basics of the relativistic quantum mechanics of particles with a nonvanishing rest mass m. In the ultra-relativistic limit $pc \gg mc^2$ the quantization scheme of such particles may be essentially the same as for electromagnetic waves, but for the intermediate energy range, $pc \sim mc^2$, a more general approach is necessary. Historically, the first attempts[48] to extend the non-relativistic wave mechanics into the relativistic energy range were based on performing the same transitions from classical observables to their quantum-mechanical operators as in the non-relativistic limit:

$$\mathbf{p} \to \hat{\mathbf{p}} = -i\hbar\nabla, \qquad E \to \hat{H} = i\hbar\frac{\partial}{\partial t}. \qquad (9.83)$$

The substitution of these operators, acting on the Schrödinger-picture wavefunction $\Psi(\mathbf{r},t)$, into the classical relation (9.1) between the energy E and momentum $\mathbf{p}$ (for of a free particle) leads to the following formulas:

---

[46] This quantized version of the Rabi oscillations can only persist in time if the inevitable electromagnetic energy losses (not described by the basic Janes–Cummings Hamiltonian) are somehow replenished—for example, by a passing a beam of particles, externally excited into the higher-energy state $\uparrow$, through the cavity. If the losses become higher, the dissipation suppresses quantum coherence, in our case the coherence between two components of each pair (9.82), as was discussed in chapter 7. As a result, the transition from the higher-energy atomic state $\uparrow$ to the lower-energy state $\downarrow$, giving energy $\hbar\omega$ to the cavity ($n - 1 \to n$), which is then rapidly drained into the environment, becomes incoherent, so that the system's dynamics is reduced to the Purcell effect, already mentioned in section 9.3. A quantitative analysis of this effect is left for the reader's exercise.

[47] I can recommend, for example, either [16], or [17].

[48] This approach was suggested in 1926–27, i.e. virtually simultaneously, by (at least) V Fock, E Schrödinger, O Klein and W Gordon, J Kudar, T de Donder and F -H van der Dungen, and L de Broglie.

Table 9.1. Deriving the Klein–Gordon equation for a free relativistic particle[49].

	Non-relativistic limit	Relativistic case
Classical mechanics	$E = \dfrac{1}{2m}p^2$	$E^2 = c^2 p^2 + (mc^2)^2$
Wave mechanics	$i\hbar\dfrac{\partial}{\partial t}\Psi = \dfrac{1}{2m}(-i\hbar\nabla)^2\Psi$	$\left(i\hbar\dfrac{\partial}{\partial t}\right)^2\Psi = c^2(-i\hbar\nabla)^2\Psi + (mc^2)^2\Psi$

The resulting equation for the non-relativistic limit, in the left-bottom cell of the table 9.1, is just the usual Schrödinger equation (1.28) for a free particle. Its relativistic generalization, in the right-bottom cell, usually rewritten as

$$\left(\frac{1}{c^2}\frac{\partial^2}{\partial t^2} - \nabla^2\right)\Psi + \mu^2\Psi = 0, \qquad \text{with } \mu \equiv \frac{mc}{\hbar}, \tag{9.84}$$

is called the *Klein–Gordon* (or sometimes 'Klein–Gordon-Fock') *equation*. The fundamental solutions of this equation are the same plane, monochromatic waves

$$\Psi(r, t) \propto \exp\{i(k \cdot r - \omega t)\} \tag{9.85}$$

as in the non-relativistic case. Indeed, such waves are eigenstates of the operators (9.83), with eigenvalues, respectively,

$$\mathbf{p} = \hbar\mathbf{k}, \qquad \text{and } E = \hbar\omega, \tag{9.86}$$

so that their substitution into Eq. (9.84) immediately returns us to Eq. (9.1) with the replacements (9.86):

$$E_{\pm} = \hbar\omega_{\pm} = \pm[(\hbar c k)^2 + (mc^2)^2]^{1/2}. \tag{9.87}$$

Though one may say that this dispersion relation is just a simple combination of the classical relation (9.1) and the same basic quantum-mechanical relations (9.86) as in non-relativistic limit, it attracts our attention to the fact that the energy $\hbar\omega$ as a function of the momentum $\hbar\mathbf{k}$ has two branches, with $E_-(\mathbf{p}) = -E_+(\mathbf{p})$—see figure 9.6a. Historically, this fact has played a very important role for spurring the fundamental idea of *particle–antiparticle pairs*. In this idea (very similar to the concept of electrons and holes in semiconductors, which was discussed in section 2.8), what we call the vacuum actually corresponds to all states of the lower branch, with energies $E_-(\mathbf{p}) < 0$, being completely filled, while the states on the upper branch, with energies $E_+(\mathbf{p}) > 0$, being empty. Then an externally supplied energy,

$$\Delta E = E_+ - E_- \equiv E_+ + (-E_-) \geqslant 2mc^2 > 0, \tag{9.88}$$

may bring the system from the lower branch to the upper one (figure 9.6b). The resulting excited state is interpreted as a combination of a particle (formally, of the

---

[49] Note that in the left (non-relativistic) column of table 9.1, the energy is referred to the rest energy mc^2, while in the right (relativistic) column, it is referred to zero—see Eq. (9.1).

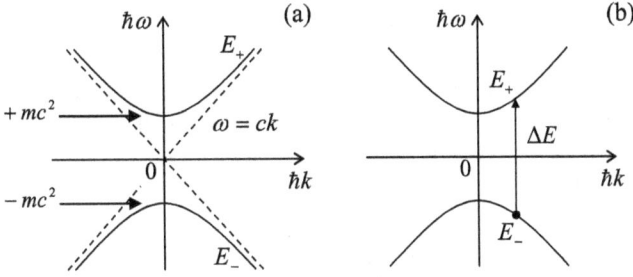

Figure 9.6. (a) The free-particle dispersion relation resulting from the Klein–Gordon and Dirac equations, and (b) the scheme of creation of a particle–antiparticle pair from the vacuum.

infinite spatial extension) with the energy E_+ and the momentum $\mathbf{p}$, and a 'hole' (antiparticle) of the *positive* energy $(-E_-)$ and the momentum $-\mathbf{p}$. This idea[50] has led to a search for, and discovery of the positron: the electron's antiparticle with charge $q = +e$, in 1932, and later of the antiproton and other antiparticles.

Free particles of a finite spatial extension may be described, in this approach, just in the non-relativistic Schrödinger equation, by wave packets, i.e. linear super-positions of the de Broglie waves (9.85) with close wave vectors $\mathbf{k}$, and the corresponding values of ω given by Eq. (9.87), with the positive sign for the 'usual' particles, and negative sign for antiparticles—see figure 9.6a above. Note that in order to form, from a particle's wave packet, a similar wave packet for the antiparticle, with the same phase and group velocities (2.33a) in each direction, we need to change the sign not only before ω, but also before $\mathbf{k}$, i.e. to replace all component wavefunctions (9.85), and hence the full wavefunction, with their complex conjugates.

Of more formal properties of Eq. (9.84), it is easy to prove that its solutions satisfy the same continuity equation (1.52), with the probability current density $\mathbf{j}$ still given by Eq. (1.47), but a different expression for the probability density w—which becomes very similar to that for $\mathbf{j}$:

$$w = \frac{i\hbar}{2mc^2}\left(\Psi^*\frac{\partial\Psi}{\partial t} - \text{c.c.}\right), \qquad \mathbf{j} = \frac{i\hbar}{2m}(\Psi\nabla\Psi^* - \text{c.c.}) \qquad (9.89)$$

—very much in the spirit of relativity theory, treating space and time on equal footing. (In the non-relativistic limit $p/mc \to 0$, Eq. (9.84) allows a reduction of this expression for w the non-relativistic Eq. (1.22): $w \to \Psi\Psi^*$.)

The Klein–Gordon equation may be readily generalized[51] to describe a single particle moving in external fields; for example, the electromagnetic field effects on a particle with charge q may be described by the same replacement as in the non-relativistic limit (see section 3.1):

$$\hat{\mathbf{p}} \to \hat{\mathbf{P}} - q\mathbf{A}(\mathbf{r}, t), \qquad \hat{H} \to \hat{H} - q\phi(\mathbf{r}, t), \qquad (9.90)$$

---

[50] Due to the same P A M Dirac!
[51] After such generalization, it is usually called the *relativistic Schrödinger equation*.

where $\hat{\mathbf{P}} = -i\hbar\nabla$ is the canonical momentum operator (3.25), and the vector- and scalar potentials, $\mathbf{A}$ and ϕ, should be treated appropriately—either as c-number functions if the electromagnetic field quantization is not important for the particular problem, or as operators (see sections 9.1–9.4 above) if it is.

However, the practical value of the relativistic Schrödinger equation is rather limited, because of two main reasons. First of all, it does not give the correct description of particles with spin. For example, for the hydrogen-like atom, i.e. the motion of an electron with the electric charge $-e$, in the Coulomb central field of an immobile nucleus with charge $+Ze$, the equation may be readily solved exactly[52] and yields the following spectrum of (doubly-degenerate) energy levels:

$$E = mc^2\left(1 + \frac{Z^2\alpha^2}{\lambda^2}\right)^{-1/2}, \quad \text{with } \lambda \equiv n + [(l + \frac{1}{2})^2 - Z^2\alpha^2]^{1/2} - (l + \frac{1}{2}), \quad (9.91)$$

where $n = 1, 2,...$ and $l = 0, 1,..., n - 1$ are the same quantum numbers as in the non-relativistic theory (see section 3.6), and $\alpha \approx 1/137$ is the fine structure constant (6.62). The three leading terms of the Taylor expansion of this result in the small parameter $Z\alpha$ are as follows:

$$E \approx mc^2\left[1 - \frac{Z^2\alpha^2}{2n^2} - \frac{Z^4\alpha^4}{2n^4}\left(\frac{n}{l + \frac{1}{2}} - \frac{3}{4}\right)\right]. \quad (9.92)$$

The first of these terms is just the rest energy of the particle. The second term,

$$E_n = -mc^2\frac{Z^2\alpha^2}{2n^2} \equiv -\frac{mZ^2e^4}{(4\pi\varepsilon_0)^2\hbar^2}\frac{1}{2n^2} \equiv -\frac{E_0}{2n^2}, \quad \text{with } E_0 = Z^2E_{\text{H}}, \quad (9.93)$$

reproduces the non-relativistic Bohr's formula (3.201). Finally, the third term,

$$-mc^2\frac{Z^4\alpha^4}{2n^4}\left(\frac{n}{l + \frac{1}{2}} - \frac{3}{4}\right) \equiv -\frac{2E_n^2}{mc^2}\left(\frac{n}{l + \frac{1}{2}} - \frac{3}{4}\right), \quad (9.94)$$

is just the perturbative kinetic-relativistic contribution (6.51) to the fine structure of the Bohr levels (9.93). However, as we already know from section 6.3, for a spin-$\frac{1}{2}$ particle such as the electron, the spin–orbit interaction (6.55) gives an additional contribution to the fine structure, of the same order, so that the net result, confirmed by experiment, is given by Eq. (6.60), i.e. is different from Eq. (9.94). This is very natural, because the relativistic Schrödinger equation does not have the very notion of spin.

Second, even for massive spinless particles (such as the Z^0 bosons), for which this equation is believed to be valid, the most important problems are related to particle interactions at high energies of the order of $\Delta E \sim 2mc^2$ and beyond—see Eq. (9.88). Due to the possibility of creation and annihilation of particle–antiparticle pairs at such energies, the number of particles participating in such interactions is typically

---

[52] This task is left for the reader's exercise.

considerable (and variable), and its adequate description of the system is given not by the relativistic Schrödinger equation (which is formulated in single-particle terms), but by the *quantum field theory*—to which I will devote only a few sentences at the very end of this chapter.

9.6 Dirac's theory

The real breakthrough toward the quantum relativistic theory of electrons (and any spin-½ fermions) was achieved in 1928 by P A M Dirac. For that time, the structure of his theory was highly nontrivial. Namely, while formally preserving, in the coordinate representation, the same Schrödinger-picture equation of quantum dynamics as in the non-relativistic quantum mechanics[53],

$$i\hbar \frac{\partial \Psi}{\partial t} = \hat{H}\Psi, \tag{9.95}$$

it postulates that the wavefunction Ψ it describes is not a scalar complex function of time and coordinates, but a *four-component* column-vector (sometimes called the *bispinor*) of such functions, its Hermitian-conjugate bispinor $\Psi^\dagger$ being a four-component row-vector of their complex conjugates:

$$\Psi = \begin{pmatrix} \Psi_1(\mathbf{r}, t) \\ \Psi_2(\mathbf{r}, t) \\ \Psi_3(\mathbf{r}, t) \\ \Psi_4(\mathbf{r}, t) \end{pmatrix}, \qquad \Psi^\dagger = \left(\Psi_1^*(\mathbf{r}, t), \quad \Psi_2^*(\mathbf{r}, t), \quad \Psi_3^*(\mathbf{r}, t), \quad \Psi_4^*(\mathbf{r}, t) \right), \tag{9.96}$$

and that the Hamiltonian participating in Eq. (9.95) is a 4×4 matrix defined in the Hilbert space of bispinors Ψ. For a free particle, the postulated Hamiltonian looks amazingly simple[54]:

---

[53] After the 'naturally-relativistic' form of the Klein–Gordon equation (9.84), this apparent return to the non-relativistic Schrödinger equation may look very counter-intuitive. However, it becomes a bit less surprising taking into account the fact (whose proof is left for the reader's exercise) that Eq. (9.84) may be also recast into the form (9.95) for a *two-component* column-vector (spinor) Ψ, with a Hamiltonian which may be represented by a 2×2 matrix - and hence expressed via the Pauli matrices (4.105) and the identity matrix I—see Eq. (5.3).

[54] Moreover, if the time derivative participating in Eq. (9.95), and the three coordinate derivatives participating (via the momentum operator) in Eq. (9.97), are merged into one four-vector operator $\partial/\partial x_k \equiv \{\nabla, \partial/\partial(ct)\}$, the Dirac equation (9.95) may be rewritten in an even simpler, manifestly Lorentz-invariant four-vector form (with the implicit summation over the repeated index $k = 1, \ldots, 4$—see, e.g. *Part EM* section 9.4):

$$\left(\hat{\gamma}_k \frac{\partial}{\partial x_k} + \mu \right) \Psi = 0, \quad \text{where } \hat{\boldsymbol{\gamma}} \equiv \{\hat{\gamma}_1, \hat{\gamma}_2, \hat{\gamma}_3\} = \begin{pmatrix} 0 & -i\hat{\boldsymbol{\sigma}} \\ i\hat{\boldsymbol{\sigma}} & 0 \end{pmatrix}, \quad \hat{\gamma}_4 = \hat{\beta},$$

where $\mu \equiv mc/\hbar$—just as in Eq. (9.84). Note also that, very counter-intuitively, the Dirac Hamiltonian (9.97) is *linear* in the momentum, while the non-relativistic Hamiltonian of a particle, as well as the relativistic Schrödinger equation, are *quadratic* in **p**. In my humble opinion, the Dirac theory (including the concept of antiparticles it has inspired) may compete for the title of the most revolutionary theoretical idea in physics of all times, despite such heavy contenders as Newton's laws, Maxwell's equations, Gibbs' statistical distribution, Bohr's theory of the hydrogen atom, and Einstein's general relativity.

$$\hat{H} = c\hat{\boldsymbol{\alpha}} \cdot \hat{\mathbf{p}} + \hat{\beta}\, mc^2. \tag{9.97}$$

where $\hat{\mathbf{p}} = -i\hbar\nabla$ is the same 3D vector operator of momentum as in the non-relativistic case, while the operators $\hat{\boldsymbol{\alpha}}$ and $\hat{\beta}$ may be represented in the following shorthand 2×2 form:

$$\hat{\boldsymbol{\alpha}} \equiv \begin{pmatrix} \hat{0} & \hat{\boldsymbol{\sigma}} \\ \hat{\boldsymbol{\sigma}} & \hat{0} \end{pmatrix}, \qquad \hat{\beta} \equiv \begin{pmatrix} \hat{I} & \hat{0} \\ \hat{0} & -\hat{I} \end{pmatrix}. \tag{9.98a}$$

The operator $\hat{\boldsymbol{\alpha}}$, composed of the Pauli vector operators $\hat{\boldsymbol{\sigma}}$, is also a vector in the usual 3D space, so that each of its three Cartesian components is a 4×4 matrix. The particular form of the 2×2 matrices corresponding to the operators $\hat{\boldsymbol{\sigma}}$ and $\hat{I}$ in Eq. (9.98a) depends on the basis selected for representation of the spin states; for example, in the standard z-basis, in which the Cartesian components of $\hat{\boldsymbol{\sigma}}$ are represented by the Pauli matrices (4.105), the 4×4 matrix form of Eq. (9.98a) is

$$\alpha_x = \begin{pmatrix} 0 & 0 & 0 & 1 \\ 0 & 0 & 1 & 0 \\ 0 & 1 & 0 & 0 \\ 1 & 0 & 0 & 0 \end{pmatrix}, \qquad \alpha_y = \begin{pmatrix} 0 & 0 & 0 & -i \\ 0 & 0 & i & 0 \\ 0 & -i & 0 & 0 \\ i & 0 & 0 & 0 \end{pmatrix},$$

$$\alpha_z = \begin{pmatrix} 0 & 0 & 1 & 0 \\ 0 & 0 & 0 & -1 \\ 1 & 0 & 0 & 0 \\ 0 & -1 & 0 & 0 \end{pmatrix}, \qquad \beta = \begin{pmatrix} 1 & 0 & 0 & 0 \\ 0 & 1 & 0 & 0 \\ 0 & 0 & -1 & 0 \\ 0 & 0 & 0 & -1 \end{pmatrix}. \tag{9.98b}$$

It is straightforward to use Eqs. (9.98) to verify that the matrices α_x, α_y, α_z and β satisfy the following relations:

$$\alpha_x^2 = \alpha_y^2 = \alpha_z^2 = \beta^2 = 1, \tag{9.99}$$

$$\alpha_x\alpha_y + \alpha_y\alpha_x = \alpha_y\alpha_z + \alpha_z\alpha_y = \alpha_z\alpha_x + \alpha_x\alpha_z = \alpha_x\beta + \beta\alpha_x$$
$$= \alpha_y\beta + \beta\alpha_y = \alpha_z\beta + \beta\alpha_z = 0, \tag{9.100}$$

i.e. anticommute. Using these commutation relations, and acting essentially as in section 1.4, it is straightforward to show that any solution to the Dirac equation obeys the probability conservation law, i.e. the continuity equation (1.52), with the probability density,

$$w = \Psi^\dagger\Psi, \tag{9.101}$$

and the probability current,

$$\mathbf{j} = \Psi^\dagger c\hat{\boldsymbol{\alpha}}\,\Psi, \tag{9.102}$$

looking *almost* as in the non-relativistic theory—cf. Eqs. (1.22) and (1.47). Note, however, the Hermitian conjugation used in these formulas instead of the complex conjugation, in order to form the scalars w, j_x, j_y, and j_z from the four-component vectors (9.96).

This close similarity is extended to the fundamental, plane-wave solutions of the Dirac equations is free space. Indeed, plugging such solution, in the form

$$\Psi = u e^{i(\mathbf{k}\cdot\mathbf{r}-\omega t)} = \begin{pmatrix} u_1 \\ u_2 \\ u_3 \\ u_4 \end{pmatrix} e^{i(\mathbf{k}\cdot\mathbf{r}-\omega t)}, \tag{9.103}$$

into Eqs. (9.95) and (9.97), we see that they are indeed satisfied, provided that a system of four coupled, linear algebraic equations for four complex c-number amplitudes $u_{1,2,3,4}$ is satisfied. The condition of its consistency yields the same dispersion relation (9.87), i.e. the same two-branch diagram shown in figure 9.6, as follows from the Klein–Gordon equation. The difference is that plugging each value of ω, given by Eq. (9.87), back into the system of the linear equations for four amplitudes u, we get two solutions for their vector $\mathbf{u} \equiv (u_1, u_2, u_3, u_4)$ for each of the two energy branches. In the standard spin z-basis, they may be represented as follows:

$$\text{for } E = E_+ > 0: \quad \mathbf{u}_{+\uparrow} = c_{+\uparrow} \begin{pmatrix} 1 \\ 0 \\ \dfrac{cp_z}{E_+ + mc^2} \\ \dfrac{c(p_x + ip_y)}{E_+ + mc^2} \end{pmatrix}, \quad \mathbf{u}_{+\downarrow} = c_{+\downarrow} \begin{pmatrix} 0 \\ 1 \\ \dfrac{c(p_x - ip_y)}{E_+ + mc^2} \\ \dfrac{-cp_z}{E_+ + mc^2} \end{pmatrix}, \tag{9.104a}$$

$$\text{for } E = E_- < 0: \quad \mathbf{u}_{-\uparrow} = c_{-\uparrow} \begin{pmatrix} \dfrac{cp_z}{E_- - mc^2} \\ \dfrac{c(p_x + ip_y)}{E_- - mc^2} \\ 1 \\ 0 \end{pmatrix}, \quad \mathbf{u}_{-\downarrow} = c_{-\downarrow} \begin{pmatrix} \dfrac{c(p_x - ip_y)}{E_- - mc^2} \\ \dfrac{-cp_z}{E_- - mc^2} \\ 0 \\ 1 \end{pmatrix}, \tag{9.104b}$$

where $c_\pm$ are normalization coefficients.

The simplest interpretation of these solutions is that Eq. (9.103), with the vectors $\mathbf{u}_+$ given by Eq. (9.104a), represents a spin-½ particle (say, an electron), while with the vectors $\mathbf{u}_-$ given by Eq. (9.104b), it represents an antiparticle (a positron), and the two solutions for each particle, indexed with opposite arrows, correspond to two possible directions of the spin-½, $\sigma_z = \pm 1$, i.e. $S_z = \pm\hbar/2$. This interpretation is indeed solid in the non-relativistic limit, when two last components of the vector (9.104a), and two first components of the vector (9.104b) are negligibly small:

$$\mathbf{u}_{+\uparrow} \to \begin{pmatrix} 1 \\ 0 \\ 0 \\ 0 \end{pmatrix}, \quad \mathbf{u}_{+\downarrow} \to \begin{pmatrix} 0 \\ 1 \\ 0 \\ 0 \end{pmatrix}, \quad \mathbf{u}_{-\uparrow} \to \begin{pmatrix} 0 \\ 0 \\ 1 \\ 0 \end{pmatrix}, \quad \mathbf{u}_{-\downarrow} \to \begin{pmatrix} 0 \\ 0 \\ 0 \\ 1 \end{pmatrix}, \quad \text{at } \frac{p_k}{mc} \to 0. \tag{9.105}$$

In order to show this, let us use the Dirac equation to calculate the Heisenberg-picture law of time evolution of the operators of the Cartesian component of the orbital angular momentum $\mathbf{L} \equiv \mathbf{r} \times \mathbf{p}$, for example of $L_x = y p_z - z p_y$, taking into account that the Dirac operators (9.98a) commute with those of $\mathbf{r}$ and $\mathbf{p}$, and also the Heisenberg commutation relations (2.14):

$$i\hbar \frac{\partial \hat{L}_x}{\partial t} = [\hat{L}_x, \hat{H}] = c\hat{\boldsymbol{\alpha}} \cdot \left[(\hat{y}\hat{p}_z - \hat{z}\hat{p}_y), \hat{\mathbf{p}} \right] = -i\hbar c(\hat{\alpha}_z \hat{p}_y - \hat{\alpha}_y \hat{p}_z), \qquad (9.106)$$

with similar relations for two other Cartesian components. Since the right-hand side of these equations is different from zero, the orbital momentum is generally *not* conserved—even for a free particle! Let us, however, consider the following vector operator,

$$\hat{\mathbf{S}} \equiv \frac{\hbar}{2} \begin{pmatrix} \hat{\boldsymbol{\sigma}} & \hat{0} \\ \hat{0} & \hat{\boldsymbol{\sigma}} \end{pmatrix}. \qquad (9.107a)$$

According to Eqs. (4.105), its Cartesian components, in the z-basis, are represented by 4×4 matrices

$$S_x = \frac{\hbar}{2} \begin{pmatrix} 0 & 1 & 0 & 0 \\ 1 & 0 & 0 & 0 \\ 0 & 0 & 0 & 1 \\ 0 & 0 & 1 & 0 \end{pmatrix}, \quad S_y = \frac{\hbar}{2} \begin{pmatrix} 0 & -i & 0 & 0 \\ i & 0 & 0 & 0 \\ 0 & 0 & 0 & -i \\ 0 & 0 & i & 0 \end{pmatrix}, \quad S_z = \frac{\hbar}{2} \begin{pmatrix} 1 & 0 & 0 & 0 \\ 0 & -1 & 0 & 0 \\ 0 & 0 & 1 & 0 \\ 0 & 0 & 0 & -1 \end{pmatrix}. \qquad (9.107b)$$

Let us calculate the Heisenberg-picture law of time evolution of these components, for example

$$i\hbar \frac{\partial \hat{S}_x}{\partial t} = [\hat{S}_x, \hat{H}] = c[\hat{S}_x, (\hat{\alpha}_x \hat{p}_x + \hat{\alpha}_y \hat{p}_y + \hat{\alpha}_z \hat{p}_z)]. \qquad (9.108)$$

A direct calculation of the commutators of the matrices (9.98) and (9.107) yields

$$[\hat{S}_x, \hat{\alpha}_x] = 0, \quad [\hat{S}_x, \hat{\alpha}_y] = i\hbar \hat{\alpha}_z, \quad [\hat{S}_x, \hat{\alpha}_z] = -i\hbar \hat{\alpha}_y, \qquad (9.109)$$

so that we finally get

$$i\hbar \frac{\partial \hat{S}_x}{\partial t} = i\hbar c(\hat{\alpha}_z \hat{p}_y - \hat{\alpha}_y \hat{p}_z), \qquad (9.110)$$

with similar expressions for other two components of the operator. Comparing this result with Eq. (9.106), we see that any Cartesian component of the operator defined similarly to Eq. (5.170),

$$\hat{\mathbf{J}} \equiv \hat{\mathbf{L}} + \hat{\mathbf{S}}, \qquad (9.111)$$

is an integral of motion[55], so that this operator may be interpreted as the one representing the total angular momentum of the particle. Hence, the operator

---

[55] It is straightforward to show that this result remains valid for a particle in the field of any central potential $U(\mathbf{r})$.

(9.107) *may* be interpreted as the spin operator of a spin-½ particle (e.g. electron). As follows from the last of Eq. (9.107*b*), the columns (9.105) represent the eigenkets of the *z*-component of that operator, with eigenstates $S_z = \pm\hbar/2$, depending on the arrow index. So, the Dirac theory provides a justification for spin-½—or, somewhat more humbly, replaces the Pauli Hamiltonian postulate (4.163) with that of a simpler (and hence more plausible), Lorentz-invariant Hamiltonian (9.97).

Note, however, that this simple interpretation, fully separating a particle from its antiparticle, is not valid for the exact solutions (9.103)–(9.104), so that generally the eigenstates of the Dirac Hamiltonian are certain linear (coherent) superpositions of the components describing the particle and its antiparticle—each with both directions of spin. This fact leads to several interesting effects, including the so-called *Klien paradox* at reflection of a relativistic electron from a potential barrier[56]. It is curious that some of these effects may be reproduced in such non-relativistic systems as particles moving in a 2D honeycomb lattice (e.g. of the graphene), since they feature a locally linear dispersion relation—see Eq. (3.122).[57]

9.7 Low-energy limit

The generalization of Dirac's theory to the case of a (spin-½) particle with an electric charge q, moving in a classically-described electromagnetic field, may be obtained using the same replacement (9.90). As a result, Eq. (9.95) becomes

$$[c\hat{\boldsymbol{\alpha}} \cdot (-i\hbar\nabla - q\mathbf{A}) + mc^2\hat{\beta} + (q\phi - \hat{H})]\,\Psi = 0, \tag{9.112}$$

where the Hamiltonian operator $\hat{H}$ is understood in the sense of Eq. (9.95), i.e. as the partial time derivative with the multiplier $i\hbar$. Let us prepare this equation for a low-energy approximation by acting on its left-hand side by a similar square bracket (also an operator!), but with the opposite sign before the last parentheses. Using Eqs. (9.99) and (9.100), and the fact that space- and time-independent operators $\hat{\boldsymbol{\alpha}}$ and $\hat{\beta}$ commute with the spin-independent, c-number functions $\mathbf{A}(\mathbf{r}, t)$ and $\phi(\mathbf{r}, t)$, as well as with the Hamiltonian operator $i\hbar\partial/\partial t$, the result is

$$\{c^2[\hat{\boldsymbol{\alpha}} \cdot (-i\hbar\nabla - q\mathbf{A})]^2 + (mc^2)^2$$
$$- c[\hat{\boldsymbol{\alpha}} \cdot (-i\hbar\nabla - q\mathbf{A}), (q\phi - \hat{H})] - (q\phi - \hat{H})^2\}\,\Psi = 0. \tag{9.113}$$

A direct calculation of the first square bracket, using Eqs. (9.98) and (9.107), yields

$$[\hat{\boldsymbol{\alpha}} \cdot (-i\hbar\nabla - q\mathbf{A})]^2 \equiv (-i\hbar\nabla - q\mathbf{A})^2 - 2q\hat{\mathbf{S}} \cdot \nabla \times \mathbf{A}. \tag{9.114}$$

But the last vector product on the right-hand side is just the magnetic field—see, e.g. Eqs. (3.21):

$$\mathscr{B} = \nabla \times \mathbf{A}. \tag{9.115}$$

---

[56] See, e.g. [18].
[57] For a review see, e.g. [19].

Similarly, we may use the first of Eqs. (3.21), for the electric field,

$$\mathscr{E} = -\nabla\phi - \frac{\partial \mathbf{A}}{\partial t}, \tag{9.116}$$

to simplify the commutator participating in Eq. (9.113):

$$[\hat{\boldsymbol{\alpha}} \cdot (-i\hbar\nabla - q\mathbf{A}), (q\phi - \hat{H})] \equiv -q\hat{\boldsymbol{\alpha}} \cdot [\hat{H}, \mathbf{A}] - i\hbar q\hat{\boldsymbol{\alpha}} \cdot [\nabla, \phi]$$
$$\equiv -i\hbar q\frac{\partial \mathbf{A}}{\partial t} - i\hbar\hat{\boldsymbol{\alpha}} \cdot \nabla\phi \equiv i\hbar q\hat{\boldsymbol{\alpha}} \cdot \mathscr{E}. \tag{9.117}$$

As a result, Eq. (9.113) becomes

$$\{c^2(-i\hbar\nabla - q\mathbf{A})^2 + (q\phi - \hat{H})^2 - (mc^2)^2 - 2qc^2\hat{\mathbf{S}} \cdot \mathscr{B} + i\hbar cq\hat{\boldsymbol{\alpha}} \cdot \mathscr{E}\}\Psi$$
$$= 0. \tag{9.118}$$

So far, this is an exact result, equivalent to Eq. (9.112), but it is more convenient for an analysis of the low-energy limit, in which not only the energy offset $E - mc^2$ (which is just the energy used in the non-relativistic mechanics), but also the electrostatic energy of the particle, $|q\langle\phi\rangle|$, are much smaller than the rest energy mc^2. In this limit, the second and third terms of Eq. (9.118) almost cancel, and introducing the offset Hamiltonian

$$\hat{\tilde{H}} \equiv \hat{H} - mc^2\hat{I}. \tag{9.119}$$

we may approximate their difference, up to the first nonvanishing term, as

$$(q\phi\hat{I} - \hat{H})^2 - (mc^2)^2\hat{I} \equiv (q\phi\,\hat{I} - mc^2\hat{I} - \hat{\tilde{H}})^2 - (mc^2)^2\hat{I}$$
$$\approx 2mc^2(\hat{\tilde{H}} - q\phi\,\hat{I}). \tag{9.120}$$

As a result, after the division of all terms by $2mc^2$, Eq. (9.118) may be approximated as

$$\hat{\tilde{H}}\Psi = \left[\frac{1}{2m}(-i\hbar\nabla - q\mathbf{A})^2 + q\phi - \frac{q}{m}\hat{\mathbf{S}} \cdot \mathscr{B} + \frac{i\hbar q}{2mc}\hat{\boldsymbol{\alpha}} \cdot \mathscr{E}\right]\Psi. \tag{9.121}$$

Let us discuss this important result. The first two terms in the square brackets give the non-relativistic Hamiltonian (3.26), which was extensively used in chapter 3 for the discussion of charged particle motion. Note again that the contribution of the vector-potential $\mathbf{A}$ into that Hamiltonian is essentially *relativistic*, in the following sense: when used for the description of magnetic interaction of two charged particles, due to their orbital motion with speed $v \ll c$, the magnetic interaction is a factor of $(v/c)^2$ smaller than the electrostatic interaction of the particles[58]. The reason why we did discuss the effects of $\mathbf{A}$ in chapter 3 was that is was used there to describe *external* magnetic fields, keeping our analysis valid even for the cases when that field is strong

---

[58] This difference may be traced by classical means—see, e.g. *Part EM* section 5.1.

by being produced by relativistic effects—such as aligned spins of a permanent magnet.

The next, third term in the square brackets of Eq. (9.121) is also familiar to the reader: this is the Pauli Hamiltonian—see Eqs. (4.3), (4.5), and (4.163). When justifying this form of interaction in chapter 4, I referred mostly to results of Stern–Gerlach-type experiments, but it is extremely pleasing that this result[59] follows from such a fundamental relativistic treatment as Dirac's theory. As we already know from the discussion of the Zeeman effect in section 6.4, the effects of magnetic field on the orbital motion of an electron (described by orbital angular momentum $\mathbf{L}$) and its spin $\mathbf{S}$ are of the same order, i.e. represent an essentially relativistic effect.

Finally, the last term in the square brackets of Eq. (9.121) is also not quite new for us: in particular it describes the spin–orbit interaction. Indeed, in the case of a classical, spherical-symmetric electric field $\mathscr{E}$ corresponding to the potential $\phi(r) = U(r)/q$, this term may be reduced to Eq. (6.56):

$$\hat{H}_{\text{so}} = \frac{1}{2m^2c^2}\hat{\mathbf{S}} \cdot \hat{\mathbf{L}}\frac{1}{r}\frac{dU}{dr} \equiv -\frac{q}{2m^2c^2}\hat{\mathbf{S}} \cdot \hat{\mathbf{L}}\frac{1}{r}\mathscr{E}. \qquad (9.122)$$

The proof of this correspondence requires a bit of additional work, to which we will now proceed[60]. In Eq. (9.121), the term responsible for the spin–orbit interaction acts on four-component wavefunctions, while the Hamiltonian (9.122) is supposed to act on non-relativistic state vectors with an account of spin, whose coordinate representation may be given by two-component spinors[61]:

$$\psi = \begin{pmatrix} \psi_\uparrow \\ \psi_\downarrow \end{pmatrix}. \qquad (9.123)$$

The simplest way to prove the identity of these two expressions is not to use Eq. (9.121) directly, but to return to the Dirac equation (9.112), for the particular case of motion in a static electric field and no magnetic field, when Dirac's Hamiltonian is reduced to

$$\hat{H} = c\hat{\boldsymbol{\alpha}} \cdot \hat{\mathbf{p}} + \hat{\beta}\,mc^2 + U(\mathbf{r}), \qquad \text{with } U = q\phi. \qquad (9.124)$$

---

[59] Note that in this result, the g-factor is still equal to exactly 2—see Eq. (4.115) and its discussion in section 4.4. In order to describe the small deviation of g_e from 2, the electromagnetic field should be quantized (just as was discussed in sections 9.1–9.4 of this chapter), and its potentials $\mathbf{A}$ and ϕ, participating in Eq. (9.121), should be treated as operators—rather than as c-number functions as was assumed above.

[60] The only facts immediately evident from Eq. (9.121) are that the term we are discussing is proportional to the electric field, as required by Eq. (9.122), and that it is of the proper order of magnitude. Indeed, Eqs. (9.101) and (9.102) imply that in the Dirac theory, $c\hat{\boldsymbol{\alpha}}$ plays the role of the velocity operator, so that the expectation values of the term are of the order of $\hbar q v \mathscr{E}/2mc^2$. Since the expectation values of the operators participating in the Hamiltonian (9.122) scale as $S \sim \hbar/2$ and $L \sim mvr$, the spin–orbit interaction energy has the same order of magnitude.

[61] In this course, the notion of spinor (popular in some textbooks) was not used much; it was introduced earlier only for two-particle states—see Eq. (8.13). For a single particle, this definition is reduced to $\psi(\mathbf{r})|s\rangle$, whose representation in a particular spin-½ basis is the column (9.123). Note that such spinors may be used as a basis for expansion of the spin-orbitals $\psi_j(\mathbf{r})$ defined by Eq. (8.125), where the index j is used for numbering both the direction of spin (i.e. the particular component of the spinor's column) and the orbital eigenfunction.

Since this Hamiltonian is time-independent, we may look for its four-component eigenfunctions in the form

$$\Psi(\mathbf{r}, t) = \begin{pmatrix} \psi_+(\mathbf{r}) \\ \psi_-(\mathbf{r}) \end{pmatrix} \exp\left\{-i\frac{E}{\hbar}t\right\}, \tag{9.125}$$

where each of $\psi_\pm$ is a two-component column of the type (9.123), representing two spin states of the particle (index +) and antiparticle (index −). Plugging Eq. (9.125) into Eq. (9.95) with the Hamiltonian (9.124), and using Eq. (9.98a), we get the following system of two linear equations:

$$
\begin{aligned}
[E - mc^2 - U(\mathbf{r})]\,\psi_+ - c\hat{\boldsymbol{\sigma}} \cdot \hat{\mathbf{p}}\psi_- &= 0, \\
[E + mc^2 - U(\mathbf{r})]\,\psi_- - c\hat{\boldsymbol{\sigma}} \cdot \hat{\mathbf{p}}\psi_+ &= 0.
\end{aligned}
\tag{9.126}
$$

Expressing ψ_- from the latter equation, and plugging the result into the former one, we get the following single equation for the particle's spinor:

$$\left[E - mc^2 - U(\mathbf{r}) - c^2\hat{\boldsymbol{\sigma}} \cdot \hat{\mathbf{p}}\frac{1}{E + mc^2 - U(\mathbf{r})}\hat{\boldsymbol{\sigma}} \cdot \hat{\mathbf{p}} \right]\psi_+ = 0. \tag{9.127}$$

So far, this is an exact equation for eigenstates and eigenvalues of the Hamiltonian (9.124). It may be substantially simplified in the low-energy limit when both the potential energy[62] and the non-relativistic eigenenergy

$$\tilde{E} \equiv E - mc^2 \tag{9.128}$$

are much lower than mc^2. Indeed, in this case the expression in denominator of the last term in the brackets of Eq. (9.127) is close to $2mc^2$. Since $\boldsymbol{\sigma}^2 = 1$, with that replacement, Eq. (9.127) is reduced to the non-relativistic Schrödinger equation, similar for both spin components of ψ_+, and hence giving spin-degenerate energy levels. In order to recover small relativistic and spin–orbit effects, we need a slightly more accurate approximation:

$$
\begin{aligned}
\frac{1}{E + mc^2 - U(\mathbf{r})} &\equiv \frac{1}{2mc^2 + \tilde{E} - U(\mathbf{r})} \equiv \frac{1}{2mc^2}\left[1 + \frac{\tilde{E} - U(\mathbf{r})}{2mc^2}\right]^{-1} \\
&\approx \frac{1}{2mc^2}\left[1 - \frac{\tilde{E} - U(\mathbf{r})}{2mc^2}\right],
\end{aligned}
\tag{9.129}
$$

in which Eq. (9.127) is reduced to

$$\left[\tilde{E} - U(\mathbf{r}) - \frac{\hat{p}^2}{2m} + \hat{\boldsymbol{\sigma}} \cdot \hat{\mathbf{p}}\frac{\tilde{E} - U(\mathbf{r})}{(2mc^2)^2}\hat{\boldsymbol{\sigma}} \cdot \hat{\mathbf{p}} \right]\psi_+ = 0. \tag{9.130}$$

---

[62] Strictly speaking, this requirement is imposed on the expectation values of $U(\mathbf{r})$ in the eigenstates to be found.

As follows from Eqs. (5.33) and (5.34), the operators of the momentum and of a function of coordinates commute as

$$[\hat{\mathbf{p}}, U(\mathbf{r})] = -i\hbar\nabla U, \tag{9.131}$$

so that the last term in the square brackets of Eq. (9.130) may be rewritten as

$$\hat{\boldsymbol{\sigma}} \cdot \hat{\mathbf{p}}\frac{\tilde{E}-U(\mathbf{r})}{(2mc)^2}\hat{\boldsymbol{\sigma}} \cdot \hat{\mathbf{p}} \equiv \frac{\tilde{E}-U(\mathbf{r})}{(2mc)^2}\hat{p}^2 - \frac{i\hbar}{(2mc)^2}(\hat{\boldsymbol{\sigma}} \cdot \nabla U)(\hat{\boldsymbol{\sigma}} \cdot \hat{\mathbf{p}}). \tag{9.132}$$

Since in the low-energy limit, both terms on the right-hand side of this relation are much smaller than the three leading terms of Eq. (9.130), we may replace the first term's numerator with its non-relativistic value $\hat{p}^2/2m$. With this replacement, the term coincides with the first relativistic correction to the kinetic energy operator—see Eq. (6.47). The second term, proportional to the electric field $\mathscr{E} = -\nabla\phi = -\nabla U/q$, may be transformed further on, using a readily verifiable identity

$$(\hat{\boldsymbol{\sigma}} \cdot \nabla U)(\hat{\boldsymbol{\sigma}} \cdot \hat{\mathbf{p}}) \equiv (\nabla U) \cdot \hat{\mathbf{p}} + i\hat{\boldsymbol{\sigma}} \cdot [(\nabla U) \times \hat{\mathbf{p}}]. \tag{9.133}$$

Of the two terms on the right-hand side of this relation, only the second one depends on spin[63], giving the following spin–orbital interaction contribution to the Hamiltonian,

$$\hat{H}_{\text{so}} = \frac{\hbar}{(2mc)^2}\hat{\boldsymbol{\sigma}} \cdot [(\nabla U) \times \hat{\mathbf{p}}] \equiv \frac{q}{2m^2c^2}\hat{\mathbf{S}} \cdot [(\nabla\phi) \times \hat{\mathbf{p}}]. \tag{9.134}$$

For a central electric field, with potential $\phi(r)$, its gradient has only the radial component: $\nabla\phi = (d\phi/dr)\mathbf{r}/r = -\mathscr{E}\mathbf{r}/r$, and with the angular momentum definition $\hat{\mathbf{L}} \equiv \mathbf{r} \times \hat{\mathbf{p}}$, Eq. (9.134) is (finally!) reduced to Eq. (9.122).

As was shown in section 6.3, the perturbative treatment of Eq. (9.122), together with the kinetic-relativistic correction (6.47), in the hydrogen-like atom problem, leads to the fine structure of each Bohr level E_n, given by Eq. (6.60):

$$\Delta E_{\text{fine}} = -\frac{2E_n}{mc^2}\left(3 - \frac{4n}{j + \frac{1}{2}}\right). \tag{9.135}$$

This result receives a confirmation from the surprising fact that for the hydrogen-like atom/ion problem, the Dirac equation may be solved *exactly*—without any assumptions. I would not have time/space to reproduce the solution[64], and will only list the final result for the energy spectrum:

$$\frac{E}{mc^2} = \left\{1 + \frac{Z^2\alpha^2}{[n + \{(j + \frac{1}{2})^2 - Z^2\alpha^2\}^{1/2} - (j + \frac{1}{2})]^2}\right\}^{-1/2}. \tag{9.136}$$

---

[63] The first term gives a small, spin-independent shift of the energy spectrum, which is very difficult to verify experimentally.

[64] Good descriptions of the solution are available in many textbooks (the older the better :-), see for example section 53 in [20].

Here $n = 1, 2, \ldots$ is the same principal quantum number as in Bohr's theory, while j is the quantum number specifying the eigenvalues (5.175) of the total angular momentum's square, in our case of a spin-½ particle taking half-integer values: $j = l \pm \frac{1}{2} = 1/2, 3/2, 5/2, \ldots$—see Eq. (5.189). Such a set of quantum numbers is rather natural, because due to the spin–orbit interaction, the orbital and spin angular momenta are not conserved, while their vector sum, $\mathbf{J} = \mathbf{L} + \mathbf{S}$, is—in the absence of an external magnetic field. Each energy level (9.136) is doubly-degenerate, with two eigenstates representing two directions of the spin, i.e. two values of $l = j \mp \frac{1}{2}$, at fixed j.

Since according to Eq. (1.13) for E_{H}, the square of the fine-structure constant $\alpha \equiv e^2/4\pi\varepsilon_0\hbar c$ may be represented as the ratio E_{H}/mc^2, the low-energy limit ($E - mc^2 \sim E_{\mathrm{H}} \ll mc^2$) may be pursued by expanding Eq. (9.136) into the Taylor series in $(Z\alpha)^2 \ll 1$. The result,

$$E \approx mc^2 \left[1 - \frac{Z^2\alpha^2}{2n^2} - \frac{Z^4\alpha^4}{2n^4} \left(\frac{n}{j + \frac{1}{2}} - \frac{3}{4} \right) \right], \tag{9.137}$$

has the same structure, and allows the same interpretation as Eq. (9.92), but with the last term coinciding with Eq. (6.60)—and with experimental results. Historically, this correct description of the fine structure of hydrogen atomic levels provided a decisive proof of Dirac's theory.

However, even such an impressive theory does not have too many direct applications. The main reason for that was already discussed in brief at the end of section 9.5: due to the possibility of creation and annihilation of particle–antiparticle pairs by an energy influx higher than $2mc^2$, the number of particles participating in high-energy interactions is not fixed. An adequate general description of such a situation is given by the quantum field theory, in which the particle's wavefunction is treated as a field to be quantized, using so-called *field operators* $\hat{\Psi}(\mathbf{r}, t)$—very much similar to the electromagnetic field operators (9.16). The Dirac equation follows from the quantum field theory in the single-particle approximation.

As was mentioned above on several occasions, the quantum field theory is beyond the scope of the time/space limits of this course, and I have to stop here, referring the interested reader to one of several excellent textbooks on this discipline[65]. However, I would strongly encourage the student going in this direction to start by playing with the field operators on his/her own, taking clues from Eqs. (9.16), but replacing the creation/annihilations operators $\hat{a}_j^{\dagger}$ and $\hat{a}_j$ of the field oscillators with those of the general second quantization formalism outlined in section 8.3.

---

[65] For a gradual introduction see, e.g. either [21] or [22]; on the other hand, [23] and [24], among others, offer a steeper learning curve.

9.8 Problems

Problem 9.1. Prove the Casimir formula, given by Eq. (9.23), by calculating the net force $F = \mathcal{P}A$ exerted by the electromagnetic field, in its ground state, on two perfectly conducting parallel plates of area A, separated by a vacuum gap of width $t \ll A^{1/2}$.

Hint: Calculate the field energy in the gap volume with and without the account of the plate effect, and then apply the Euler–Maclaurin formula[66] to the difference of these two results.

Problem 9.2. Electromagnetic radiation by some single-mode quantum sources may have such a high degree of coherence that it is possible to observe the interference of waves from two independent sources with virtually the same frequency, incident on one detector.

(i) Generalize Eq. (9.29) to this case.
(ii) Use this generalized expression to show that incident waves in different Fock states do not create an interference pattern.

Problem 9.3. Calculate the zero-delay value $g^{(2)}(0)$ of the second-order correlation function of a single-mode electromagnetic field in the so-called *Schrödinger-cat state*[67]: a coherent superposition of two Glauber states, with equal amplitudes, equal but sign-opposite parameters α, and a certain phase shift between them.

Problem 9.4. Calculate the zero-delay value $g^{(2)}(0)$ of the second-order correlation function of a single-mode electromagnetic field in the squeezed ground state ζ defined by Eq. (5.142).

Problem 9.5. Calculate the rate of spontaneous photon emission (into unrestricted free space) by a hydrogen atom, initially in the $2p$ state ($n = 2$, $l = 1$) with $m = 0$. Would the result be different for $m = \pm 1$? for the $2s$ state ($n = 2$, $l = 0$, $m = 0$)? Discuss the relation between these quantum-mechanical results and those given by the classical theory of radiation, using the simplest classical model of the atom.

Problem 9.6. An electron has been placed on the lowest excited level of a spherically-symmetric, quadratic potential well $U(\mathbf{r}) = m_e \omega^2 r^2 / 2$. Calculate the rate of its relaxation to the ground state, with the emission of a photon (into unrestricted free space). Compare the rate with that for a similar transition of the hydrogen atom, for the case when the radiation frequencies of these two systems are equal.

Problem 9.7. Derive an analog of Eq. (9.53) for the spontaneous photon emission into the free space, due to a change of the *magnetic* dipole moment $\mathbf{m}$ of a small-size system.

---

[66] See, e.g. Eq. (A.15a).
[67] Its name stems from the well-known *Schrödinger cat paradox*, which is (very briefly) discussed in section 10.1.

Problem 9.8. A spin-½ particle, with gyromagnetic ratio γ, is in its orbital ground state in dc magnetic field $\mathscr{B}_0$. Calculate the rate of its spontaneous transition from the higher to the lower energy level, with the emission of a photon into the free space. Evaluate the rate for an electron in a field of 10 T, and discuss the implications of this result for laboratory experiments with electron spins.

Problem 9.9. Calculate the rate of spontaneous transitions between the two sublevels of the ground state of a hydrogen atom, formed as a result of its hyperfine splitting. Discuss the implications of the result for the width of the 21-cm spectral line of hydrogen.

Problem 9.10. Find the eigenstates and eigenvalues of the Janes–Cummings Hamiltonian (9.78), and discuss their behavior near the resonance point $\omega = \Omega$.

Problem 9.11. Analyze the Purcell effect, mentioned in sections 9.3 and 9.4, qualitatively; in particular, calculate the so-called *Purcell factor* F_P, defined as the ratio of the spontaneous emission rates Γ_s of an atom to a resonant cavity (tuned exactly to the quantum transition frequency) and that to the free space.

Problem 9.12. Prove that the Klein–Gordon equation (9.84) may be rewritten in the form similar to the non-relativistic Schrödinger equation (1.25), but for a two-component wavefunction, with the Hamiltonian represented (in the usual z-basis) by the following 2×2-matrix:

$$\mathrm{H} = -(\sigma_z + i\sigma_y)\frac{\hbar^2}{2m}\nabla^2 + mc^2\sigma_z.$$

Use your solution to discuss the physical meaning of the wavefunction's components.

Problem 9.13. Calculate and discuss the energy spectrum of a relativistic, spinless, charged particle placed into an external uniform, time-independent magnetic field $\mathscr{B}$. Use the result to formulate the condition of validity of the non-relativistic theory in this situation.

Problem 9.14. Prove Eq. (9.91) for the energy spectrum of a hydrogen-line atom, starting from the relativistic Schrödinger equation.

Hint: A mathematical analysis of Eq. (3.193) shows that its eigenvalues are given by Eq. (3.201), $\varepsilon_n = -\frac{1}{2}n^2$, with $n = l + 1 + n_r$, where $n_r = 0, 1, 2,...$, even if the parameter l is not integer.

Problem 9.15. Derive the general expression for the differential cross-section of the elastic scattering of a spinless relativistic particle by a static potential $U(\mathbf{r})$, in the Born approximation, and formulate the conditions of its validity. Use these results to calculate the differential cross-section of scattering of a particle with the electric charge $-e$ by the Coulomb electrostatic potential $\phi(\mathbf{r}) = Ze/4\pi\varepsilon_0 r$.

Problem 9.16. Starting from Eqs. (9.95)–(9.98), prove that the probability density w given by Eq. (9.101) and the probability current density $\mathbf{j}$ defined by Eq. (9.102) do indeed satisfy the continuity equation (1.52): $\partial w/\partial t + \nabla\cdot\mathbf{j} = 0$.

Problem 9.17. Calculate the commutator of the operator $\hat{L}^2$ and the Dirac's Hamiltonian of a free particle. Compare the result with that for the non-relativistic Hamiltonian, and interpret the difference.

Problem 9.18. Calculate the commutators of the operators $\hat{S}^2$ and $\hat{J}^2$ with the Dirac's Hamiltonian (9.97), and give an interpretation of the results.

Problem 9.19. In the Heisenberg picture of quantum dynamics, derive an equation describing time evolution of free electron's velocity in the Dirac theory. Solve the equation for the simplest state, with definite energy and momentum, and discuss the solution.

Problem 9.20. Calculate the eigenstates and eigenenergies of a relativistic spin-½ particle with charge q, placed into a uniform, time-independent external magnetic field $\mathscr{B}$. Compare the calculated energy spectrum with those following from the non-relativistic theory and the relativistic Schrödinger equation.

Problem 9.21.* Following the discussion in the very end of section 9.7, introduce quantum field operators $\hat{\psi}$ that would be related to the usual wavefunctions ψ just as the electromagnetic field operators (9.16) are related to the classical electromagnetic fields, and explore the basic properties of these operators. (For this preliminary study, consider the fixed-time situation.)

References

[1] Eidelman S *et al* 2004 *Phys. Lett.* B **592** 1
[2] Sparnaay M 1957 *Nature* **180** 334
[3] Lamoreaux S 1997 *Phys. Rev. Lett.* **78** 5
[4] Mohideen U and Roy A 1998 *Phys. Rev. Lett.* **81** 004549
[5] Riek C *et al* 2015 *Science* **350** 420
[6] Dzhyaloshinskii I *et al* 1961 *Sov. Phys. Uspekhi* **4** 153
[7] Milton K 2001 *The Casimir Effect* (World Scientific)
[8] Garcia-Sanchez D *et al* 2012 *Phys. Rev. Lett.* **109** 027202
[9] Kimble H *et al* 1977 *Phys. Rev. Lett.* **39** 691
[10] Paul H 1982 *Rev. Mod. Phys.* **54** 1061
[11] Haroche S and Klepner D 1989 *Phys. Today* **42** 24
[12] Milloni P and Eberly J 2010 *Laser Physics* 2nd edn (Wiley)
[13] Yariv A 1989 *Quantum Electronics* 3rd edn (Wiley)
[14] Sargent M III, Scully M and Lamb W Jr 1977 *Laser Physics* (Westview)
[15] Braak D 2011 *Phys. Rev. Lett.* **107** 100401
[16] Gerry C and Knight P 2005 *Introductory Quantum Optics* (Cambridge University Press)
[17] Agarwal G 2012 *Quantum Optics* (Cambridge University Press)
[18] Calogeracos A and Dombey N 1999 *Contemp. Phys.* **40** 313

[19] Robinson T 2012 *Am. J. Phys.* **80** 141
[20] Schiff L 1968 *Quantum Mechanics* 3rd edn (McGraw-Hill)
[21] Brown L 1994 *Quantum Field Theory* (Cambridge University Press)
[22] Klauber R 2013 *Student Friendly Quantum Field Theory* (Sandtrove)
[23] Srednicki M 2007 *Quantum Field Theory* (Cambridge University Press)
[24] Zee A 2010 *Quantum Field Theory in a Nutshell* 2nd edn (Princeton University Press)

IOP Publishing

Quantum Mechanics
Lecture notes
Konstantin K Likharev

Chapter 10

Making sense of quantum mechanics

This (rather brief) chapter addresses the conceptually important issues of quantum measurements and quantum state interpretation. Please note that some of these issues are still subjects of debate[1]—fortunately not affecting practical results of quantum mechanics, discussed in the previous chapters.

10.1 Quantum measurements

Now we have got a sufficient background for a (by necessity, very brief) discussion of *quantum measurements*[2]. Let me start with reminding the reader the only postulate of the quantum theory that relates it to experiment—so far, meaning a *perfect* measurement. In the simplest case when the system is in a coherent (pure) state, its ket-vector may be represented as a linear superposition

$$|\alpha\rangle = \sum_j \alpha_j |a_j\rangle, \tag{10.1}$$

where a_j are the eigenstates of the operator of an observable A, related to its eigenvalues A_j by Eq. (4.68):

$$\hat{A}\,|a_j\rangle = A_j|a_j\rangle. \tag{10.2}$$

In such state, the outcome of each particular measurement of the observable A may be uncertain, but is restricted to the set of eigenvalues A_j, with the jth outcome probability equal to

---

[1] For an excellent review of these controversies, as presented in a few leading textbooks, I highly recommend J Bell's paper in the review collection by A Miller [1].

[2] 'Quantum measurements' is a very unfortunate term; it would be more sensible to speak about 'measurements of observables in quantum mechanical systems'. However, the former term is so common and compact that I will use it—albeit rather reluctantly.

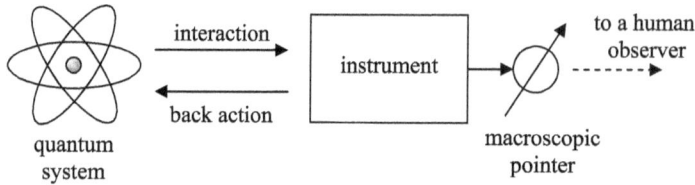

Figure 10.1. The general scheme of a quantum measurement.

$$W_j = |\alpha_j|^2. \tag{10.3}$$

As was discussed in chapter 7, the state of the system (or rather of the statistical ensemble of similar systems we are using for experiments) may be not coherent, and hence even more uncertain than the state described by Eq. (10.1). Hence, the measurement postulate means that even if the system is in the least uncertain state, the measurement outcomes are *still* probabilistic[3].

If we believe that such measurements may be always made perfectly, and do not worry too much how exactly, we are subscribing to the *mathematical* notion of measurement, that was, rather reluctantly, used in these notes—up to this point. However, actual (*physical*) quantum measurements are always imperfect, first of all because they face a huge gap between the energy-time scale $\hbar \sim 10^{-34}$ J s of the quantum phenomena in 'microscopic' quantum systems, such as atoms, and the 'macroscopic' scale of direct human perception, so that the role of the instruments bridging this gap (figure 10.1), is highly nontrivial. These instruments are *physical devices*, which also must obey the laws of physics, and for a physicist, it is rather important to understand the basic laws of their operation, as their contribution (if any) to the origin of the measurement postulate.

Besides the famous Bohr–Einstein discussion in the mid-1930s, which will be discussed in section 10.3, the founding fathers of quantum mechanics have not paid much attention to these issues, apparently because of the following reason. At that time it looked like the experimental instruments (at least the best of them :-) were doing exactly what the measurement postulate was telling them. For example, the z-oriented Stern–Gerlach experiment (figure 4.1) turns two complex coefficients $\alpha_\uparrow$ and $\alpha_\downarrow$, describing the spin state of the incoming electrons, into a set of particle-counter clicks, with the rates proportional to, respectively, $|\alpha_\uparrow|^2$ and $|\alpha_\downarrow|^2$. The crude internal nature of these instruments makes more detailed questions unnatural. For example, the electron counting with a Geiger counter involves an effective disappearance of one observed electron in a zillion-particle electric discharge avalanche it has triggered. A century ago, it seemed much more important to extend the newly born quantum mechanics to more complex systems (such as atomic nuclei, etc) than to think about the physics of such instruments.

However, since that time the experimental techniques, notably including high-vacuum and low temperature systems, micro- and nano-fabrication, and low-noise

---

[3] The outcomes become definite only in the trivial case when the system is definitely in one of the eigenstates a_j, say a_0; then $\alpha_j = \delta_{j,0}\exp\{i\varphi\}$, and $W_j = \delta_{j,0}$.

electronics, have much improved. In particular, we now may observe quantum-mechanical behavior of more and more macroscopic objects—such as the micro-mechanical oscillators mentioned in section 2.9. Moreover, some of 'macroscopic quantum systems' (in particular, special systems of Josephson junctions, see below) have properties enabling their use as essential parts of measurement setups. Such developments are making the line separating the 'micro' and 'macro' worlds more and more fine, so that more inquisitive inquiries into the physical nature of quantum measurements are not so hopeless now. In my scheme of things[4], these inquiries may be grouped as follows:

(i) Does a quantum measurement involve any laws besides those of quantum mechanics? In particular, should it necessarily involve a human/intelligent observer? (The last question is not as laughable as it may look—see below.)

(ii) What is the state of the measured system *just after* a *single-shot* measurement—meaning a measurement process limited to a time interval much shorter than the time scale of a measured system's evolution? (This question is a necessary part of the discussion of *repeated measurements* and their ultimate form—a *continuous monitoring* of a certain observable.)

(iii) If a measurement of an observable A has produced a certain outcome A_j, what statements may be made about the state of the system *just before* the measurement? (This question is most closely related to various interpretations of quantum mechanics.)

Let me discuss these issues in the listed order. First of all, I am happy to report that there is a virtual consensus of physicists on some aspects of these issues. According to this consensus, any reasonable quantum measurement needs to result in a certain, distinguishable state of a *macroscopic* output component of the measurement instrument—see figure 10.1. (Traditionally, its component is called a *pointer*, though its role may be played by a printer or a plotter, an electronic circuit sending out the result as a number, etc.) This requirement implies that the measurement process should have the following features:

- provide a large 'signal gain', i.e. some means of mapping the quantum state with its $\hbar$-scale of action (i.e. of the energy-by-time product) onto a macro-scopic position of the pointer with a much larger action scale, and
- if we want to approach the fundamental limit of uncertainty, given by Eq. (10.3), the instrument should introduce as few additional fluctuations ('noise') as permitted by the laws of physics.

Both these requirements *are* fulfilled in a good Stern–Gerlach experiment—see figure 4.1 again. Indeed, the magnetic field gradient, splitting the electron beam, turns the miniscule (microscopic) energy difference (4.167) between two spin-polarized states into a macroscopic difference between the final positions of two

---

[4] Again, this list, and some other issues discussed in the balance of this section, are still controversial.

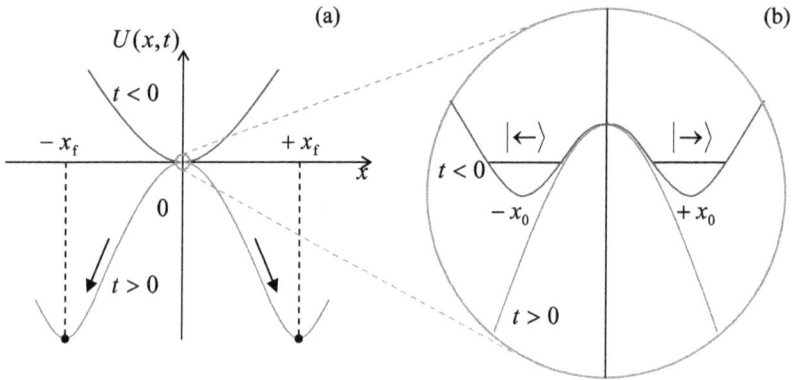

Figure 10.2. The potential inversion, as viewed on the (a) 'macroscopic' and (b) 'microscopic' scales of the generalized coordinate x.

sub-beams, where their detectors may be located. However, the internal physics of the particle detectors (say, Geiger counters) at this measurement is rather complex, and would not allow us to discuss some aspects of the measurement, in particular to answer the two inquiries we are working on.

For this reason let me describe the scheme of a virtually similar 'single-shot' measurement of a two-level quantum system, which shares the simplicity, high gain, and low internal noise of the Stern–Gerlach apparatus, but has an advantage that at its certain implementations[5], the measurement process allows a thorough, quantitative theoretical description. Let us measure a coherent state of a particle trapped in a double-well potential (figure 10.2), where x is some continuous generalized coordinate—not necessarily a linear displacement. Let the system be in a pure quantum state, with the energy close to the well's bottom. Then, as we know from the discussion of such systems in sections 2.6 and 5.1, the state may be described by the ket-vector similar to that of spin-½:

$$|\alpha\rangle = \alpha_\rightarrow |\rightarrow\rangle + \alpha_\leftarrow |\leftarrow\rangle, \tag{10.4}$$

where the component states $\rightarrow$ and $\leftarrow$ are described by wavefunctions localized near the potential well bottoms at $x \sim \pm x_0$—see the blue line in figure 10.2b. Our goal is to measure in what well the particle resides at a certain time instant, say $t = 0$. For that, let us rapidly change, at that moment, the potential profile of the system, so that at $t > 0$, and near the origin, it may be well approximated by an inverted parabola—see the red line in figure 10.2b:

---

[5] The scheme may be implemented, for example, using a simple Josephson-junction circuit called the *balanced comparator*—see, e.g. [2] and references therein. Experiments have demonstrated that this system may have a measurement variance dominated by the theoretically expected quantum-mechanical uncertainty, at practicable experimental conditions (at temperatures below ~ 1 K). A conceptual advantage of this system is that it is based on externally-shunted Josephson junctions, i.e. devices whose quantum-mechanical model, including its part describing the coupling to environment, is in quantitative agreement with experiment—see, e.g. [3]. Colloquially, the balanced comparator is a high-gain instrument with a 'well-documented Hamiltonian', eliminating the need for speculations about the environmental effects. In particular, the dephasing process in it, and its time T_2, are well described by Eqs. (7.89) and (7.142), with the coefficient η equal to the Ohmic conductances G of the shunts.

$$U(x) \approx -\frac{m\lambda^2}{2}x^2, \qquad \text{for } t > 0, \qquad |x| \ll x_f. \qquad (10.5)$$

It is straightforward to verify that the Heisenberg equation of motion in such inverted potential describes an exponential growth of the operator $\hat{x}$ in time (proportional to $\exp\{\lambda t\}$) and hence a similar, proportional growth of the expectation value $\langle x \rangle$ and its rms uncertainty δx.[6] At this 'inflation' stage, the coherence between the two component states $\rightarrow$ and $\leftarrow$ is still preserved, i.e. the time evolution of the system is, in principle, reversible.

Now let the system be weakly coupled, also at $t > 0$, to a dissipative (e.g. Ohmic) environment. As we know from chapter 7, such coupling ensures the state's dephasing on some time scale T_2. If

$$x_0 \ll x_0 \exp\{\lambda T_2\}, \quad x_f, \qquad (10.6)$$

then the process, after the potential inversion, consists of two stages, well separated in time:

- the already discussed 'inflation' stage, preserving the component state's coherence;
- the dephasing stage, at which the coherence of the component states $\rightarrow$ and $\leftarrow$ is gradually suppressed, as described by Eq. (7.89), i.e. the density matrix of the system is reduced to the diagonal form describing a classical mixture of the probability packets with the probabilities (10.3) equal to, respectively, $W_{\rightarrow} = |\alpha_{\rightarrow}|^2$ and $W_{\leftarrow} = |\alpha_{\leftarrow}|^2 \equiv 1 - |\alpha_{\rightarrow}|^2$.

Besides dephasing, the environment gives the motion a certain kinematic friction, with the drag coefficient η (7.141), so that the system eventually settles to rest at one of the macroscopically separated minima $x = \pm x_f$ of the inverted potential (figure 10.2a), thus ensuring a high 'signal gain' $x_f/x_0 \gg 1$. (The time order of these two processes, dephasing or settling, is not important.) As a result, the final probability density distribution $w(x)$ along the x-axis has two narrow, well separated peaks. But this is just the situation that was discussed in section 2.5—see, in particular, figure 2.17. Since that discussion is very important, let me repeat, or rather re-phrase it. The final state of the system is a classical mixture of two well-separated states, with the respective probabilities $W_{\leftarrow}$ and $W_{\rightarrow}$, whose sum equals 1. Now let us use some detector to test whether the system is in one of these states—say the right one. (If x_f is sufficiently large, the noise contribution of this detector into the measurement uncertainty is negligible[7], and its physics is unimportant.) If the system has been

---

[6] Somewhat counter-intuitively, the latter growth plays a positive role for measurement's fidelity. Indeed, it does not affect the intrinsic 'signal-to-noise ratio' $\delta x/\langle x \rangle$, while making the intrinsic (say, quantum-mechanical) uncertainty much larger than the possible noise contribution by the later measurement stage(s).

[7] At the balanced-comparator implementation mentioned above, the final state detection may be readily performed using a 'SQUID' magnetometer based on the same Josephson junction technology—see, e.g. *Part EM* section 6.5. In this case, the distance between the potential minima $\pm x_f$ is close to one superconducting flux quantum (3.38), while the additional uncertainty induced by the SQUID may be as low as a few millionths of that amount.

found at this location (again, the probability of this outcome is $W_\to = |\alpha_\to|^2$), the probability to find it at the counterpart, left location at a consequent detection turns to zero.

This probability 'reduction' is a purely classical (or if you like, mathematical) effect of the statistical ensemble's re-definition: $W_\leftarrow$ equals zero not in the initial ensemble of all similar experiments (where it equals $|\alpha_\leftarrow|^2$), but only in the re-defined ensemble of experiments in that the system had been found at the right location. Of course, which ensemble to use, i.e. what probabilities to register/publish is a purely accounting decision, which should be made by a human (or otherwise intelligent :-) observer. If we are only interested in an objective recording of results of a pre-fixed sequence of experiments (i.e. the members of a pre-defined, fixed statistical ensemble), there is no need to include such an observer in any discussion. In any case, this detection/registration process, very common in classical statistics, leaves no space for any mysterious 'wave packet reduction'—understood as a hypothetical process that would not obey the regular laws of quantum mechanical evolution.

The ensemble re-definition at measurement is in the core of several paradoxes, of which the so-called *quantum Zeno paradox* is perhaps the most spectacular[8]. Let us return to a two-level system with the unperturbed Hamiltonian given by Eq. (4.166), the quantum oscillation period $2\pi/\Omega$ much longer than the single-shot measurement time, in which the system initially (at $t = 0$) is definitely in one of the partial quantum states—for example, a certain potential well of the double-well potential. Then, as we know from sections 2.6 and 4.6, the probability to find the system in this initial state at time $t > 0$ is

$$W(t) = \cos^2(\Omega t/2). \tag{10.7}$$

If the time is small enough ($t = dt \ll 1/\Omega$), we may use the Taylor expansion to write

$$W(dt) \approx 1 - \frac{\Omega^2 dt^2}{4}. \tag{10.8}$$

Now, let us use some 'good' measurement scheme (say, the potential inversion discussed above) to measure whether the system is still in this partial state. If it is (as Eq. (10.8) shows, the probability of such an outcome is nearly 100%), then the system, after the measurement, is in the same partial state. Let us allow it to evolve again, with the same Hamiltonian. Then the evolution of W will follow the same law

---

[8] This name, coined by E Sudarshan and B Mishra in 1997 (though the paradox had been discussed in detail by A Turing in 1954) is due to its apparent similarity to the classical paradoxes by ancient Greek philosopher Zeno of Elea. By the way, just for fun, let us have a look at what happens when Mother Nature is discussed by people to do not understand math and physics. The most famous of the classical Zeno paradoxes is the case of *Achilles and Tortoise*: a fast runner Achilles can apparently never overtake a slower Tortoise, because (in Aristotle's words) 'the pursuer must first reach the point whence the pursued started, so that the slower must always hold a lead'. For a physicist, the paradox has a trivial, obvious resolution, but here is what a philosopher writes about it—not in some year BC, but in the 2010 AD: 'Given the history of 'final resolutions', from Aristotle onwards, it's probably foolhardy to think we've reached the end.' For me personally, this is a sad symbol of modern philosophy.

as in Eq. (10.7), but with the initial value given by Eq. (10.8) Thus, when the system is measured again at time $2dt$, the probability to find it in the same partial state is

$$W(2dt) \approx W(dt)\left(1 - \frac{\Omega^2 dt^2}{4}\right) = \left(1 - \frac{\Omega^2 dt^2}{4}\right)^2. \qquad (10.9)$$

After repeating this cycle N times (with the total time $t = Ndt$ still much less than $N^{1/2}/\Omega$), the probability that the system is still in its initial state is

$$W(Ndt) \equiv W(t) \approx \left(1 - \frac{\Omega^2 dt^2}{4}\right)^N = \left(1 - \frac{\Omega^2 t^2}{4N^2}\right)^N \approx 1 - \frac{\Omega^2 t^2}{4N}. \qquad (10.10)$$

Comparing this result with Eq. (10.7), we see that the process of system's transfer to the opposite partial state has been slowed down rather dramatically, and in the limit $N \to \infty$ (at fixed t), its evolution is virtually stopped by the measurement process. There is of course nothing mysterious here; the evolution slowdown is due to the statistical ensemble's re-definition. Indeed, the slowdown is true only for the ensemble of experiments in which the system has been found in the initial state at each moment nt.

This may be the only acceptable occasion for me to mention, very briefly, the more famous—or rather infamous—*Schrödinger cat paradox*, so much overplayed in popular publications[9]. For this thought experiment, there is no need to discuss the (rather complicated :-) physics of the cat. As soon as the charged particle, produced at the radioactive decay, reaches the Geiger counter, the initial coherent super-position of the two possible quantum states ('the decay has happened'/'the decay has not happened') of the system is rapidly dephased, i.e. reduced to their classical mixture, leading, correspondingly, to the classical mixture of the final macroscopic states 'cat dead'/'cat alive'. So, despite attempts by numerous authors, typically without proper physics background, to present this situation as a mystery whose discussion needs an involvement of professional philosophers, hopefully by this point the reader knows enough about dephasing to ignore all this babble.

10.2 QND measurements

I hope that the above discussion has sufficiently illuminated the issues of the group (i), so let me proceed to the question group (ii), in particular to the general issue of the *back action* of the instrument upon the system under measurement—symbolized with the back arrow in figure 10.1. In instruments like the Geiger counter, such back action is very large: the instrument essentially destroys ('demolishes') the initial state of the system under measurement. Even the 'cleaner' potential-inversion measure-ment, shown in figure 10.2, fully destroys the initial coherence of the system, i.e. perturbs it rather substantially.

---

[9] S Hawking has been quoted to say, 'When I hear about the Schrödinger cat, I reach for my gun.' The only good aspect of this popularity is that the formulation of this paradox should be so well known to the reader that I do not need to waste time/space for repeating it.

However, in the 1970s it was understood that this is not really necessary. For example, in section 7.3, we have already discussed an example of a two-level system coupled with its environment and described by the Hamiltonian (7.68)–(7.70):

$$\hat{H} = \hat{H}_s + \hat{H}_{\text{int}} + \hat{H}_e\{\lambda\}, \quad \text{with} \quad \hat{H}_s = a\hat{\sigma}_z, \quad \hat{H}_{\text{int}} = -f\{\lambda\}\hat{\sigma}_z, \qquad (10.11)$$

so that

$$[\hat{H}_s, \hat{H}_{\text{int}}] \propto [\hat{\sigma}_z, \hat{\sigma}_z] = 0. \qquad (10.12)$$

Comparing this equality with Eq. (4.199), applied to the explicitly-time-independent Hamiltonian $\hat{H}_s$,

$$i\hbar\dot{\hat{H}}_s = [\hat{H}_s, \hat{H}] \equiv [\hat{H}_s, (\hat{H}_s + \hat{H}_{\text{int}} + \hat{H}_e\{\lambda\})] = [\hat{H}_s, \hat{H}_{\text{int}}], \qquad (10.13)$$

we see that in the Heisenberg picture, the Hamiltonian operator (and hence the energy) of the system of our interest does not change in time. On the other hand, if the 'environment' in this discussion is the instrument used for the measurement (see figure 10.1 again), the interaction can change its state, so it may be used to measure the system's energy—or another observable whose operator commutes with the interaction Hamiltonian. Such a trick is called the *quantum non-demolition* (QND), or sometimes 'back-action-evading' measurements[10]. Due to the lack of back action of the instrument on the corresponding variable, such measurements allow its continuous monitoring. Let me present a fine example of an actual measurement of this kind—see figure 10.3.[11]

In this experiment, a single electron is captured in a *Penning trap*—a combination of a (virtually) uniform magnetic field $\mathscr{B}$ and a quadrupole electric field[12]. Such an electric field stabilizes the cyclotron orbits but does not have any noticeable effect on electron motion in the plane perpendicular to the magnetic field, and hence on its Landau level energies—see Eq. (3.50):

$$E_n = \hbar\omega_c\left(n + \frac{1}{2}\right), \quad \text{with } \omega_c = \frac{e\mathscr{B}}{m_e}. \qquad (10.14)$$

(In the cited work, with $\mathscr{B} \approx 5.3\text{T}$, the cyclic frequency $\omega_c/2\pi$ was about 147 GHz, so that the Landau level splitting $\hbar\omega_c$ was close to 10^{-22} J, i.e. corresponded to temperature ~ 10 K, while the physical temperature of the system might be reduced well below that, down to ~ 80 mK.) Now note that the analogy between a particle on a Landau level and a harmonic oscillator goes beyond the energy spectrum (10.14). Indeed, since the Hamiltonian of a 2D particle in a perpendicular magnetic field may be reduced to Eq. (3.47), similar to that of a 1D oscillator, we may repeat all procedures of section 5.4 and rewrite this Hamiltonian in terms of the creation–annihilation operators—see Eq. (5.72):

---

[10] For a detailed discussion of this field see, e.g. [4]; for an earlier review, see [5].
[11] [6]
[12] It is similar to the 2D system discussed in *Part EM* section 2.7, but with additional rotation about one of the axes.

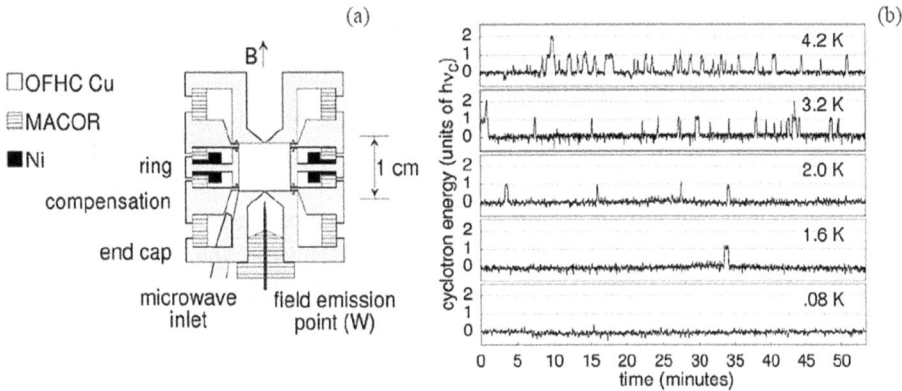

Figure 10.3. QND measurements of single electron's energy by Peil and Gabrielse: (a) the experimental setup core, and (b) a record of the thermal excitation and spontaneous relaxation of Fock states. Reprinted with permission from [6]. Copyright 1999 by the American Physical Society.

$$\hat{H}_s = \hbar\omega_c\left(\hat{a}^\dagger\hat{a} + \frac{1}{2}\right). \tag{10.15}$$

In the Peil and Gabrielse experiment, the trapped electron had one more degree of freedom—along the magnetic field. The electric field of the Penning trap created a soft confining potential along this direction (vertical in figure 10.3a; I will take it for the z-axis), so that small electron oscillations along that axis could be well described as those of a 1D harmonic oscillator of much lower eigenfrequency, in that particular experiment with $\omega_z/2\pi \approx 64$ MHz. This frequency could be measured very accurately (with error ~ 1 Hz) by sensitive electronics whose electric field does affect the z-motion of the electron, but not its motion in the perpendicular plane. In an exactly uniform magnetic field, the two modes of electron motion would be completely uncoupled. However, the experimental setup included two special superconducting rings made of niobium (see figure 10.3a), which slightly distorted the magnetic field and created an interaction between the modes, which might be well approximated by the Hamiltonian[13]

$$\hat{H}_{\text{int}} = \text{const} \times \left(\hat{a}^\dagger\hat{a} + \frac{1}{2}\right)\hat{z}^2, \tag{10.16}$$

so that the main condition (10.12) of a QND measurement was very closely satisfied. At the same time, the coupling (10.16) ensured that a change of the Landau level number n by 1 changed the z-oscillation eigenfrequency by ~ 12.4 Hz. Since this shift was substantially larger than electronics' noise, spontaneous changes of n (due to an uncontrolled coupling of the electron to environment) could be readily observed—moreover, continuously monitored—see figure 10.3b. The record shows spontaneous

---

[13] Here I have simplified the real situation a bit. Actually, in this experiment there was an electron spin's contribution to the interaction Hamiltonian as well, but since the used high magnetic field polarized the spins quite reliably, their only effect was a constant shift of the frequency ω_z, which is not important for our discussion.

excitations of higher Landau levels, with their sequential relaxation, just as described by Eqs. (7.208)–(7.210). The detailed data statistics analysis showed that there was virtually no effect of the measuring instrument on these processes—at least on the scale of minutes, i.e. as many as $\sim 10^{13}$ cyclotron orbit periods[14].

It is important, however, to note that any measurement—QND or not—cannot avoid the uncertainty relations between incompatible variables; in the particular case described above, a permanent monitoring of the Landau state number n does not allow the simultaneous exact monitoring of its quantum phase (which may be defined exactly as in the harmonic oscillator). In this context, it is natural to wonder whether the QND measurement concept may be extended from quadratic-form variables like the energy to the 'usual' observables such as coordinates and momenta. whose uncertainties are bound by the ordinary Heisenberg relation (1.35). The answer is YES, but the required methods are a bit more tricky.

For example, let us place an electrically charged particle into a uniform electric field $\mathscr{E} = \mathbf{n}_x\mathscr{E}(t)$ of the instrument, so that their interaction Hamiltonian is

$$\hat{H}_{int} = -q\mathscr{E}(t)\hat{x}. \tag{10.17}$$

Such interaction may certainly pass information on the time evolution of the coordinate x to the instrument. However, in this case Eq. (10.12) is *not* satisfied—at least for the kinetic-energy part of system's Hamiltonian; as a result the interaction distorts the time evolution of the particle's momentum. Indeed, writing the Heisenberg equation (4.199) for the x-component of the momentum, we get

$$\dot{\hat{p}} - \dot{\hat{p}}|_{\mathscr{E}=0} = q\mathscr{E}(t). \tag{10.18}$$

Integrating Eq. (5.139) for the coordinate operator evolution[15], we get the expression

$$\hat{x}(t) = \hat{x}(t_0) + \frac{1}{m}\int_{t_0}^t \hat{p}(t')dt', \tag{10.19}$$

which shows that the perturbations (10.18) of the momentum would eventually find their way to the coordinate evolution, not allowing its unperturbed sequential measurements.

However, for such an important particular system as a harmonic oscillator, the following trick is possible. For this system, Eqs. (5.139) and (10.18) may be readily combined to give a second-order differential equation for the coordinate operator, that is absolutely similar to the classical equation of motion of the system, and has a similar solution[16]:

---

[14] See also the conceptually similar experiments, performed by different means: [7].
[15] This simple relation is limited to 1D systems with Hamiltonians of the type (1.41), but by now the reader certainly knows enough to understand that this discussion may be readily generalized to many other systems.
[16] Note in particular that the function $\sin\omega_0\tau$ (with $\tau \equiv t - t'$) under the integral, divided by ω_0, is nothing more than the temporal Green's function $G(\tau)$ of a loss-free harmonic oscillator—see, e.g. *Part CM* section 5.1.

$$\hat{x}(t) = \hat{x}(t)\Big|_{\mathscr{E}=0} + \frac{q}{m\omega_0} \int_{-\infty}^{t} \mathscr{E}(t')\sin \omega_0(t - t')\, dt'. \tag{10.20}$$

This formula confirms that generally the external field $\mathscr{E}(t)$ (in our case, the sensing field of the measurement instrument) affects the time evolution law—of course. However, Eq. (10.20) shows that if the field is applied only at moments t'_n separated by intervals $\mathscr{T}/2$, where $\mathscr{T} \equiv 2\pi/\omega_0$ is the oscillation period, its effect on the coordinate vanishes at similarly spaced observation instants $t_n = t_{n'} + (m + 1/2)\mathscr{T}$. This is the idea of *stroboscopic* QND measurements. Of course, according to Eq. (10.18), even such measurement strongly perturbs the oscillator momentum, so that even if the values x_n are measured with high accuracy, the Heisenberg's uncertainty relation is not violated.

A direct implementation of the stroboscopic measurements is technically complicated, but this initial idea has opened a way to more practicable solutions. For example, it is straightforward to use the Heisenberg equations of motion to show that if the coupling of two harmonic oscillators, with coordinates x and X, and unperturbed eigenfrequencies ω and Ω, is modulated in time as

$$\hat{H}_{\text{int}} \propto \hat{x}\hat{X} \cos \omega t \cos \Omega t, \tag{10.21}$$

then the process in one of the oscillators (say, that with frequency Ω) does not affect dynamics of one of the *quadrature components* of the counterpart oscillator, defined by relations[17]

$$\hat{x}_1 \equiv \hat{x} \cos \omega t - \frac{\hat{p}}{m\omega} \sin \omega t, \qquad \hat{x}_2 \equiv \hat{x} \sin \omega t + \frac{\hat{p}}{m\omega} \cos \omega t, \tag{10.22}$$

while this component's motion does affect the dynamics of one of quadrature components of the counterpart oscillator. (For the counterpart couple of quadrature components, the information transfer goes in the opposite direction.) This scheme has been successfully used for QND measurements[18].

Please note that the last two QND measurement examples are based on the idea of external periodic *modulation* of a certain parameter in time—either in the short-pulse form or the sinusoidal form. So, the reader should not be surprised if the only role of a QND measurement is a sensitive measurement of a weak *classical force* acting on a quantum *probe system*[19], i.e. a 1D oscillator of eigenfrequency ω_0, it may be implemented much more simply—just by modulating an oscillator's parameter

---

[17] The physical sense of these relations should be clear from figure 5.8: they define a system of coordinates rotating clockwise with the angular velocity equal to ω, so that the point representing unperturbed classical oscillations with that frequency is at rest in this rotating frame. (The 'probability cloud' representing a Glauber state is also stationary in the coordinates $[x_1, x_2]$.) The reader familiar with the classical theory oscillations may notice that the observables x_1 and x_2 so defined are just the *Poincaré plane* coordinates ('RWA variables')—see, e.g. *Part CM* sections 5.3–5.6, and especially figure 5.9, where these coordinates are denoted as u and v.

[18] The first, initially imperfect QND experiments were reported by M Levenson *et al* [8], and other groups soon after this, using nonlinear interactions of optical waves. Later, the results were much improved—see, e.g. [9] and references therein. Recently, such experiments were extended to mechanical systems—see, e.g. [10].

[19] As it is, for example, for gravitational wave detectors.

with a frequency $\omega \approx 2\omega_0$. From the classical dynamics, we know that if the depth of such modulation exceeds a certain threshold value, it results in the excitation of the so-called degenerate parametric oscillations with frequency $\omega/2$, and one of two opposite phases[20]. In the language of Eq. (10.22), the parametric excitation means an exponential growth of one of the quadrature components (with its sign depending on initial conditions), while the counterpart component is suppressed. Close to, but below the excitation threshold, the parameter modulation boosts all fluctuations of the almost-excited component, including its quantum-mechanical uncertainty, and suppresses (*squeezes*) those of the counterpart component. The result is a squeezed state, already discussed in section 5.5 of this course (see in particular Eqs. (5.143) and figure 5.8), which allows one to notice the effect of an external force on the oscillator on the backdrop of a quantum uncertainty much smaller than the standard quantum limit (5.99).

In electrical engineering, this fact may be conveniently formulated in terms of *noise parameter* Θ_N of a *linear amplifier*—essentially the tool for a continuous monitoring of an input 'signal'—e.g. a microwave or optical waveform[21]. Namely, Θ_N of 'usual' (say, transistor or maser) amplifiers which are equally sensitive to both quadrature components of the signal, Θ_N has the minimum value $\hbar\omega/2$, due to the quantum uncertainty pertinent to the quantum state of the amplifier itself (which therefore plays the role of its 'quantum noise')[22]. On the other hand, a degenerate parametric amplifier, sensitive to just one quadrature component, may have Θ_N well below $\hbar\omega/2$, due to the squeezing of its ground state[23].

Let me note that the parameter-modulation schemes of the QND measurements are not limited to harmonic oscillators, and may be applied to other important quantum systems, notably including two-level (i.e. spin-½-like) systems[24]. Such measurements may be an important tool for the further progress of quantum computation and cryptography[25].

Finally, let me mention that composite systems consisting of a quantum subsystem, and a classical subsystem performing its continuous weakly-perturbing measurement and using its results for providing a specially crafted feedback, may have some curious properties, in particular mock a quantum system completely detached from the environment[26].

10.3 Hidden variables and local reality

Now we are ready to proceed to the discussion of the last, hardest group (iii) of the questions posed in the previous section, namely on the state of a quantum system *just*

---

[20] See, e.g. *Part CM* section 5.5, and also figure 5.8 and its discussion in section 5.6.

[21] For a quantitative definition of the latter parameter, suitable for the quantum sensitivity range ($\Theta_N \sim \hbar\omega$) as well, see, e.g. [11]. In the classical noise limit ($\Theta_N \gg \hbar\omega$), it coincides with $k_B T_N$, where T_N is a more popular measure of electronics noise, called the *noise temperature*.

[22] This fact was recognized very early—see, e.g. [12].

[23] See, e.g. the spectacular experiments [13].

[24] See, e.g. [14].

[25] See, e.g. [15].

[26] See, e.g. the monograph by H Wiseman and G Milburn [16], and the recent experiments by R Vijay *et al* [17].

before its measurement. After a very important but inconclusive discussion of this issue by A Einstein and his collaborators on one side, and N Bohr on the other side, in the mid-1930s, such discussions resumed in the 1950s.[27] They led to a key contribution by J Bell in the early 1960s, and an important experimental work on verifying Bell's inequalities, but besides that work, the recent progress has been marginal, and opinions of even prominent physicists on the problem are still very much different.

The central controversial issue may be formulated as follows: what had been the 'real' state of a quantum-mechanical system just before a virtually-perfect single-shot measurement was performed on it, and gave a certain outcome? In order to be specific, let us focus again on the example of Stern–Gerlach measurements of spin-½ particles—because of their conceptual simplicity[28]. For a single-component system (in this case a single spin-½) the answer to the posed question may look evident. For example, if the spin is in a coherent (least-uncertain) state α, i.e. its ket-vector may be expressed in the form similar to Eq. (10.4),

$$|\alpha\rangle = \alpha_\uparrow |\uparrow\rangle + \alpha_\downarrow |\downarrow\rangle, \tag{10.23}$$

where, as usual, $\uparrow$ and $\downarrow$ denote the states with the corresponding definite spin orientations along the z-axis, then the probabilities of the corresponding outcomes of the z-oriented Stern–Gerlach experiment are $W_\uparrow = |\alpha_\uparrow|^2$ and $W_\downarrow = |\alpha_\downarrow|^2$. Then it looks natural to suggest that if a particular experiment gave the outcome corresponding to the state $\uparrow$, the spin had been in that state just before the experiment. For a classical system such an answer would be certainly correct, and the fact that the probability $W_\uparrow = |\alpha_\uparrow|^2$, defined for the statistical ensemble of *all* experiments (regardless of their outcome), may be less than 1, would merely reflect our ignorance about the *real state* of this particular system before the measurement—which just *reveals* this *situation*.

However, as was first argued in the famous *EPR paper* published in 1935 by A Einstein, B Podolsky, and N Rosen, such an answer becomes impossible in the case of an entangled quantum system, if only one of its components is measured with an instrument. The original EPR paper discussed thought experiments with a pair of 1D particles prepared in a quantum state in that both the *sum* of their momenta and the *difference* of their coordinates have definite values: $p_1 + p_2 = 0$, $x_1 - x_2 = a$.[29] However, usually this discussion is recast into an equivalent Stern–Gerlach experiment shown in figure 10.4a.[30] A source emits rare pairs of spin-½ particles, propagating in opposite directions, with exactly zero net spin of the pair, but otherwise in random spin states. After the spatial separation of the particles has

---

[27] See, e.g. the review collection [18].

[28] As was discussed in section 1, the Stern–Gerlach-type experiments may be readily made virtually perfect, provided that we do not care about the evolution of the system *during* the single-shot measurement.

[29] This is possible, because the corresponding operators commute: $[\hat{p}_1 + \hat{p}_2, \hat{x}_1 - \hat{x}_2] = [\hat{p}_1, \hat{x}_1] - [\hat{p}_2, \hat{x}_2] = 0$.

[30] A more convenient, frequently used experimental technique for entangled state generation is the parametric excitation (also called the *four-wave mixing*, FWM) of optical photon pairs—see, e.g. the publications cited in the end of this section.

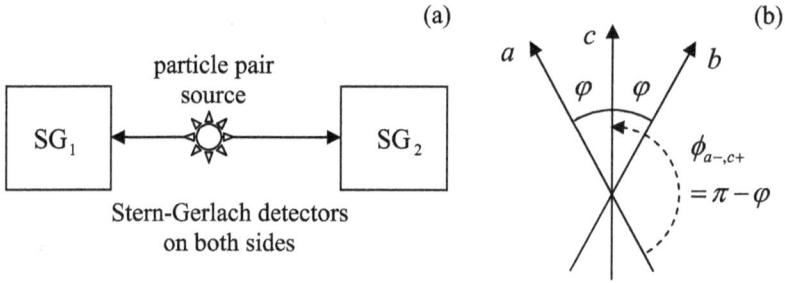

Figure 10.4. (a) General scheme of two-particle Stern–Gerlach experiments, and (b) the orientation of the detectors, assumed at the deviation of Bell's inequality (10.36).

become sufficiently large (see below), the spin state of each of them is measured with a Stern–Gerlach detector, with one of them (figure 10.1, SG_1) somewhat closer to the particle source, so it makes the measurement first, at a time $t_1 < t_2$.

First, let the detectors be oriented along the same direction, say axis z. Evidently, the probability of each detector to give any of values $S_z = \pm\hbar/2$ is 50%. However, if the first detector had given the result $S_z = -\hbar/2$, then even before the second detector's measurement, we know that the latter will give the result $S_z = +\hbar/2$ with the 100% probability. So far, this situation still allows for a classical interpretation, just as for the single-particle measurements: we may fancy that the second particle really has a definite spin before the measurement, and the first measurement has just removed our ignorance about that reality. In other words, the change of the probability of the outcome $S_z = +\hbar/2$ at the second detection from 50% to 100% is due to the statistical ensemble re-definition: the 50% probability of this detection belongs to the ensemble of all experiments, while the 100% probability belongs to the sub-ensemble of experiments with the $S_z = -\hbar/2$ outcome of the first experiment.

However, let the source generate the spin pairs in the entangled, singlet state (8.18),

$$|s_{12}\rangle = \frac{1}{\sqrt{2}}(|\uparrow\downarrow\rangle - |\downarrow\uparrow\rangle), \tag{10.24}$$

that certainly satisfies the above assumptions: the probability of each S_z value of any particle is 50%, the sum of both S_z is definitely zero, and if the first detector's result is $S_z = -\hbar/2$, then the state of the remaining particle is $\uparrow$, with zero uncertainty. Now let us use Eqs. (4.123) to represent the same state (10.24) in a different form:

$$|s_{12}\rangle = \frac{1}{\sqrt{2}}\left[\frac{1}{\sqrt{2}}(|\rightarrow\rangle + |\leftarrow\rangle)\frac{1}{\sqrt{2}}(|\rightarrow\rangle - |\leftarrow\rangle)\right.$$
$$\left. - \frac{1}{\sqrt{2}}(|\rightarrow\rangle - |\leftarrow\rangle)\frac{1}{\sqrt{2}}(|\rightarrow\rangle + |\leftarrow\rangle)\right]. \tag{10.25}$$

Opening the parentheses (carefully, without swapping the ket-vector order, which encodes the particle numbers!), we get an expression similar to Eq. (10.24), but now for the x-basis:

$$|s_{12}\rangle = \frac{1}{\sqrt{2}}(|\rightarrow \ \leftarrow\rangle - |\leftarrow \ \rightarrow\rangle). \tag{10.26}$$

Hence if we use the first detector (closest to the particle source) to measure S_x rather than S_z, then after it had given a certain result (say, $S_x = -\hbar/2$), we know for sure, before the second particle spin's measurement, that its S_x component definitely equals $+\hbar/2$.

So, depending on the experiment performed on the first particle, the second particle, before its measurement, may be in one of two states—either with a definite component S_z or with a definite component S_x, in each case with zero uncertainty. Evidently, this situation cannot be interpreted in classical terms if the particles (or instruments) do not interact during the measurements. A Einstein was deeply unhappy with such a situation, because it did not satisfy what, in his view, was the general requirement to any theory, which nowadays is called the *local reality*. His definition of this requirement was as follows: 'The real factual situation of system 2 is independent of what is done with system 1 that is spatially separated from the former'. (Here the term 'spatially separated' is not defined, but from the context it is clear that Einstein meant the detector separation by a *superluminal interval*, i.e. by distance

$$|\mathbf{r}_1 - \mathbf{r}_2| > c|t_1 - t_2|, \tag{10.27}$$

where the measurement time difference on the right-hand side includes the measurement duration.) In Einstein's view, since quantum mechanics did not satisfy the local reality condition, it could not be considered a complete theory of Nature.

This situation naturally raises the question whether *something* (usually called *hidden variables*) may be added to the quantum-mechanical description in order to enable it to satisfy the local reality requirement. The first definite statement in this regards was J von Neumann's 'proof'[31] (first famous, then infamous) that such variables cannot be introduced; for a while his work satisfied the quantum mechanics practitioners, who apparently did not pay much attention[32]. A major new contribution to the problem was made only in the 1960s by J Bell[33]. First of all, he found an elementary (in his words, 'foolish') error in von Neumann's logic, which voids his 'proof'. Second, he demonstrated that Einstein's local reality condition is *incompatible* with conclusions of quantum mechanics—that had been, by that time, confirmed by too many experiments to be seriously questioned.

---

[31] In his very early book [19].
[32] Perhaps, it would not satisfy A Einstein, but reportedly he did not know about the von Neumann's publication before signing the EPR paper.
[33] See, e. g. either [20], or [21].

Let me describe a particular version of Bell's proof (suggested by E Wigner), using the same EPR pair experiment (figure 10.4a), in that each SG detector may be oriented in any of three directions: a, b, or c—see figure 10.4b. As we already know from chapter 4, if a fully-polarized beam of spin-½ particles is passed through a Stern–Gerlach apparatus forming angle ϕ with the polarization axis, the probabilities of two alternative outcomes of the experiment are

$$W(\phi_+) = \cos^2 \frac{\phi}{2}, \quad W(\phi_-) = \sin^2 \frac{\phi}{2}. \tag{10.28}$$

Let us use this formula to calculate all joint probabilities of measurement outcomes, starting from the detectors 1 and 2 oriented, respectively, in the directions a and c. Since the angle between the *negative* direction of the a-axis and the *positive* direction of the c-axis is $\phi_{a+,c-} = \pi - \varphi$ (see the dashed arrow in figure 10.4b), we get

$$W(a_+, c_+) \equiv W(a_+)W(c_+|a_+) = \frac{1}{2} \cos^2 \frac{\phi_{a-,\,c+}}{2}$$
$$= \frac{1}{2} \cos^2 \frac{\pi - \varphi}{2} \equiv \frac{1}{2} \sin^2 \frac{\varphi}{2}, \tag{10.29}$$

where $W(x|y)$ is the conditional probability of the outcome x if the outcome y has certainly happened. (The first equality in Eq. (10.29) is the well-known identity of the probability theory.) Absolutely similarly,

$$W(c_+, b_+) \equiv W(c_+)W(b_+|c_+) = \frac{1}{2} \sin^2 \frac{\varphi}{2}, \tag{10.30}$$

$$W(a_+, b_+) \equiv W(a_+)W(b_+|a_+) = \frac{1}{2} \cos^2 \frac{\pi - 2\varphi}{2} \equiv \frac{1}{2} \sin^2 \varphi. \tag{10.31}$$

Now note that for any angle φ smaller than $\pi/2$ (as in the case shown in figure 10.4b), the trigonometry gives

$$\frac{1}{2} \sin^2 \varphi \geqslant \frac{1}{2} \sin^2 \frac{\varphi}{2} + \frac{1}{2} \sin^2 \frac{\varphi}{2} \equiv \sin^2 \frac{\varphi}{2}. \tag{10.32}$$

(For example, for $\varphi \to 0$ the left-hand side of this inequality tends to $\varphi^2/2$, and the right-hand side, to $\varphi^2/4$.) Hence the quantum-mechanical result gives, in particular,

$$W(a_+, b_+) \geqslant W(a_+, c_+) + W(c_+, b_+), \quad \text{for } |\varphi| \leqslant \pi/2. \tag{10.33}$$

On the other hand, we may compose another inequality for the same probabilities without calculating them from any particular theory, but using the local reality assumption. Let us list all possible outcomes of detector measurements, taking into account the zero net spin of the system:

$W(a_+,b_+)$ $W(a_+,c_+)$ $W(c_+,b_+)$

Detector 1 results	Detector 2 results	Probability
a_+, b_+, c_+	a_-, b_-, c_-	W_1
a_+, b_+, c_-	a_-, b_-, c_+	W_2
a_+, b_-, c_+	a_-, b_+, c_-	W_3
a_+, b_-, c_-	a_-, b_+, c_+	W_4
a_-, b_+, c_+	a_+, b_-, c_-	W_5
a_-, b_+, c_-	a_+, b_-, c_+	W_6
a_-, b_-, c_+	a_+, b_+, c_-	W_7
a_-, b_-, c_-	a_+, b_+, c_+	W_8

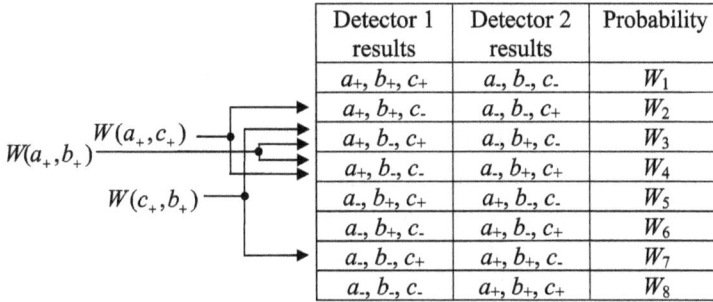

From the local reality point of view, these measurement options are independent, so we may write:

$$W(a_+, c_+) = W_2 + W_4, \quad W(c_+, b_+) = W_3 + W_7,$$
$$W(a_+, b_+) = W_3 + W_4. \tag{10.34}$$

On the other hand, since no probability may be negative (by its very definition), we may always write

$$W_3 + W_4 \leqslant (W_2 + W_4) + (W_3 + W_7). \tag{10.35}$$

Plugging into this inequality the values of these two parentheses, given by Eq. (10.34), we get

$$W(a_+, b_+) \leqslant W(a_+, c_+) + W(c_+, b_+). \tag{10.36}$$

This is (one of several possible forms of) *Bell's inequality*, which has to be satisfied by *any* local-reality theory; it directly contradicts the quantum-mechanical result (10.33).

Though experimental tests of the Bell's inequalities had been started in the late 1960s, the interpretation of first results was vulnerable to two criticisms:

(i) The detectors were not fast enough and not far enough to have the relation (10.27) satisfied. This is why, as a matter of principle, there was a chance that information on one measurement outcome had been transferred (by some, mostly implausible) means to particles before the second measurement—the so-called *locality loophole*.

(ii) The particle/photon detection efficiencies were too low to have sufficiently small error bars for both parts of the inequality—the *detection loophole*.

Gradually, these loopholes have been closed[34]. As expected, substantial violations of the Bell inequalities (10.36) (or their equivalent forms) have been proved,

---

[34] Important milestones on that way were the experiments [22] and [23]. Detailed reviews of the experimental situation were given, for example, in [24] and [25]; see also the later paper [26]. Presently, a high-fidelity demonstration of the Bell inequality violation has become a standard test in virtually each experiment with entangled qubits used for quantum encryption research—see section 8.5, in particular the paper by J Lin cited there.

essentially rejecting any possibility to reconcile quantum mechanics with Einstein's local reality requirement.

10.4 Interpretations of quantum mechanics

The fact that quantum mechanics is incompatible with local reality, makes its reconciliation with our (classically-bred) 'common sense' rather challenging. Here is a brief list of the major interpretations of quantum mechanics, that try to provide at least a partial reconciliation of this kind:

(i) The so-called *Copenhagen interpretation*—to which most physicists (myself including) adhere. This 'interpretation' does not really interpret anything; it just states the intrinsic stochasticity of measurement results in quantum mechanics, essentially saying: 'Do not worry; this is just how it is; live with it'. My only remark on this school of thought is as follows: while this interpretation implies statistical ensembles (otherwise, how would you define the probability?—see section 1.3), its most frequently stated formulations do not put a sufficient emphasis on their role, in particular on the ensemble re-definition as the only possible point of a human observer's involvement in the measurement process—see section 10.1 above. Perhaps the most impressive objection to the Copenhagen interpretation is attributed to A Einstein: 'God does not play dice.' OK, when Einstein speaks, we all should listen, but perhaps when God speaks (through experimental results), we have to pay even more attention.

(ii) *Non-local reality*. Since the dismissal of von Neumann's 'proof' by J Bell, to the best of my knowledge, there has been *no* proof that hidden parameters *could not* be introduced, provided that they do not imply the local reality. Of constructive approaches, perhaps the most notable contribution was made by D Bohm[35], who developed the initial L de Broglie's interpretation of the wavefunction as a 'pilot wave', making it quantitative. In the wave-mechanics version of this concept, the wavefunction governed by the Schrödinger equation just guides a 'real', point-like classical particle whose coordinates serve as hidden variables. However, this concept does not satisfy the notion of local reality. For example, the measurement of the particle's coordinate at a certain point r_1 has to *instantly* change the wavefunction everywhere, including the points r_2 in the superluminal range (10.27). After A Einstein's criticism, D Bohm essentially abandoned his theory[36]. I also think that admitting the 'spooky action on distance' would be too much of a sacrifice for regaining the classical determinism.

(iii) The *many-world interpretation* introduced in 1957 by H Everitt, and popularized in the 1960s and 1970s by B de Witt. In this interpretation, *all* possible measurement outcomes *do* happen, splitting the Universe into the corresponding

---

[35] [27].
[36] See, e.g. section 22.19 of his (generally good) textbook [28].

number of 'parallel' universes (or rather 'multiverses'), so that from one of them, other multiverses and hence other outcomes cannot be observed. Let me leave to the reader an estimate of the rate at which the parallel multiverses have to be constantly generated (say, per second), taking into account that such generation should take place not only in explicit lab experiments, but in any quasi-measurement irreversible process such as a fission of any atomic nucleus or an absorption/emission of a photon, everywhere in each multiverse—whether its result is recorded or not. N van Kampen has called the many-world interpretation a 'mind-boggling fantasy'[37]. Even the main proponent of this interpretation, de Witt, has confessed: 'The idea is not easy to reconcile with common sense.' I agree.

(iv) *Quantum logic.* In desperation, some physicists turned philosophers have decided to dismiss the formal logic we are using—in science and elsewhere. From what (admittedly, very little) I have read about this school of thought, it seems that from its point of view the statements like 'the SG detector has found the spin to be directed along the magnetic field' should not necessarily be either true or false. OK, if we dismiss the formal logic, I do not know how we can use any scientific theory to make any predictions—until the quantum logic experts tell us what to replace it with. To the best of my knowledge, so far they have not done that. I personally trust the opinion by J Bell, who certainly thought more about these issues: 'It is my impression that the whole vast subject of Quantum Logic has arisen [...] from the misuse of a word.'

As far as I know, neither of these interpretations has yet provided a suggestion as to how it might be tested experimentally to exclude other ones. On the positive side, there is a consensus that quantum mechanics makes correct (if sometimes proba-balistic) predictions, which do not contradict any reliable experimental results we are aware of. Maybe, this is not that bad for a scientific theory[38].

References

[1] Miller A (ed) 1989 *Sixty-Two Years of Uncertainty* (Plenum)
[2] Walls T *et al* 2007 *IEEE Trans. on Appl. Supercond.* **17** 136
[3] Schwartz D *et al* 1985 *Phys. Rev. Lett.* **55** 1547
[4] Braginsky V and Khalili F 1992 *Quantum Measurement* ed K Thorne (Cambridge University Press)
[5] Braginsky V *et al* 1980 *Science* **209** 547
[6] Peil S and Gabrielse G 1999 *Phys. Rev. Lett.* **83** 1287
[7] Nogues G *et al* 1999 *Nature* **400** 239

---

[37] [29]. By the way, I highly recommend the very reasonable summary of the quantum measurement issues, given in this paper, though believe that the quantitative theory of dephasing, discussed in chapter 7 of this course, might be used to bring additional clarity in some of its statements.

[38] For the reader who is not satisfied with this 'positivistic' approach, and wants to improve the situation, my earnest advice is to start not from square one, but from reading what other (including some very clever!) people thought about it. The review collection by J Wheeler and W Zurek, cited above, may be a good starting point.

[8] Levenson M *et al* 1986 *Phys. Rev. Lett.* **57** 2473

[9] Grangier P *et al* 1998 *Nature* **396** 537

[10] Lecocq F *et al* 2015 *Phys. Rev. X* **5** 041037

[11] Devyatov I *et al* 1986 *J. Appl. Phys.* **60** 1808

[12] Haus H and Mullen J 1962 *Phys. Rev.* **128** 2407

[13] Yurke B *et al* 1988 *Phys. Rev. Lett.* **60** 764

[14] Averin D 2002 *Phys. Rev. Lett.* **88** 207901

[15] Jaeger G 2006 *Quantum Information: An Overview* (Springer)

[16] Wiseman H and Milburn G 2009 *Quantum Measurement and Control* (Cambridge University Press)

[17] Vijay R *et al* 2012 *Nature* **490** 77

[18] Wheeler J and Zurek W (ed) 1983 *Quantum Theory and Measurement* (Princeton University Press)

[19] von Neumann J 1932 *Mathematische Grundlagen der Quantenmechanik [Mathematical Foundations of Quantum Mechanics]* (Springer) (The first English translation was published only in 1955)

[20] Bell J 1966 *Rev. Mod. Phys.* **38** 447

[21] Bell J 1982 *Found. Phys.* **12** 158

[22] Aspect A *et al* 1982 *Phys. Rev. Lett.* **49** 91

[23] Rowe M *et al* 2001 *Nature* **409** 791

[24] Genovese M 2005 *Phys. Rep.* **413** 319

[25] Aspect A 2015 *Physics* **8** 123

[26] Handsteiner J *et al* 2017 *Phys. Rev. Lett.* **118** 060401

[27] Bohm D 1952 *Phys. Rev.* **85**(165) 180

[28] Bohm D 1979 *Quantum Theory* (Dover)

[29] van Kampen N 1988 *Physica A* **153** 97

IOP Publishing

Quantum Mechanics
Lecture notes
Konstantin K Likharev

Appendix A

Selected mathematical formulas

This appendix lists selected mathematical formulas that are used in this lecture course series, but not always remembered by students (and some instructors :-).

A.1 Constants

- Euclidean circle's *length-to-diameter ratio*:

$$\pi = 3.141\ 592\ 653\ ...; \qquad \pi^{1/2} \approx 1.77. \tag{A.1}$$

- *Natural logarithm base*:

$$e \equiv \lim_{n\to\infty}\left(1 + \frac{1}{n}\right)^{n} = 2.718\ 281\ 828\ ...; \tag{A.2a}$$

from that value, the logarithm base conversion factors are as follows ($\xi > 0$):

$$\frac{\ln \xi}{\log_{10}\xi} = \ln 10 \approx 2.303, \qquad \frac{\log_{10}\xi}{\ln \xi} = \frac{1}{\ln 10} \approx 0.434. \tag{A.2b}$$

- The *Euler* (or 'Euler–Mascheroni') *constant*:

$$\gamma \equiv \lim_{n\to\infty}\left(1 + \frac{1}{2} + \frac{1}{3} + ...\ \frac{1}{n} - \ln n\right) = 0.577\ 156\ 649\ 0\ ...; \tag{A.3}$$

$$e^{\gamma} \approx 1.781.$$

A.2 Combinatorics, sums, and series

(i) *Combinatorics*

- The number of different *permutations*, i.e. *ordered* sequences of k elements selected from a set of n distinct elements ($n \geqslant k$), is

$$^{n}P_{k} \equiv n \cdot (n - 1) \cdots (n - k + 1) = \frac{n!}{(n - k)!}; \tag{A.4a}$$

in particular, the number of different permutations of *all* elements of the set $(n = k)$ is

$$^{k}P_k = k \cdot (k - 1) \cdots 2 \cdot 1 = k!. \tag{A.4b}$$

- The number of different *combinations*, i.e. *unordered* sequences of k elements from a set of $n \geqslant k$ distinct elements, is equal to the binomial coefficient

$$^{n}C_k \equiv \binom{n}{k} \equiv \frac{^{n}P_k}{^{k}P_k} = \frac{n!}{k!(n - k)!}. \tag{A.5}$$

In an alternative, very popular 'ball/box language', $^{n}C_k$ is the number of different ways to put in a box, in an arbitrary order, k balls selected from n distinct balls.

- A generalization of the binomial coefficient notion is the multinomial coefficient,

$$^{n}C_{k_1,k_2,\ldots k_l} \equiv \frac{n!}{k_1!k_2!\ldots k_l!}, \quad \text{with } n = \sum_{j=1}^{l} k_j, \tag{A.6}$$

which, in the standard mathematical language, is a number of different permutations in a multiset of l distinct element types from an n-element set which contains k_j ($j = 1, 2,\ldots l$) elements of each type. In the 'ball/box language', the coefficient (A.6) is the number of different ways to distribute n distinct balls between l distinct boxes, each time keeping the number (k_j) of balls in the jth box fixed, but ignoring their order inside the box. The binomial coefficient $^{n}C_k$ (A.5) is a particular case of the multinomial coefficient (A.6) for $l = 2$ - counting the explicit box for the first one, and the remaining space for the second box, so that if $k_1 \equiv k$, then $k_2 = n - k$.

- One more important combinatorial quantity is the number $M_n^{(k)}$ of ways to place n *indistinguishable* balls into k distinct boxes. It may be readily calculated from Eq. (A.5) as the number of different ways to select $(k - 1)$ partitions between the boxes in an imagined linear row of $(k - 1 + n)$ 'objects' (balls in the boxes *and* partitions between them):

$$M_n^{(k)} = {}^{n-1+k}C_{k-1} \equiv \frac{(k - 1 + n)!}{(k - 1)!n!}. \tag{A.7}$$

(ii) *Sums and series*
- *Arithmetic progression*:

$$r + 2r + \cdots + nr \equiv \sum_{k=1}^{n} kr = \frac{n(r + nr)}{2}; \tag{A.8a}$$

in particular, at $r = 1$ it is reduced to the sum of n first natural numbers:

$$1 + 2 + \cdots + n \equiv \sum_{k=1}^{n} k = \frac{n(n+1)}{2}. \tag{A.8b}$$

- Sums of squares and cubes of n first natural numbers:

$$1^2 + 2^2 + \cdots + n^2 \equiv \sum_{k=1}^{n} k^2 = \frac{n(n+1)(2n+1)}{6}; \tag{A.9a}$$

$$1^3 + 2^3 + \cdots + n^3 \equiv \sum_{k=1}^{n} k^3 = \frac{n^2(n+1)^2}{4}. \tag{A.9b}$$

- The *Riemann zeta function*:

$$\zeta(s) \equiv 1 + \frac{1}{2^s} + \frac{1}{3^s} + \cdots \equiv \sum_{k=1}^{\infty} \frac{1}{k^s}; \tag{A.10a}$$

the particular values frequently met in applications are

$$\zeta\left(\frac{3}{2}\right) \approx 2.612, \quad \zeta(2) = \frac{\pi^2}{6}, \quad \zeta\left(\frac{5}{2}\right) \approx 1.341,$$
$$\zeta(3) \approx 1.202, \quad \zeta(4) = \frac{\pi^4}{90}, \quad \zeta(5) \approx 1.037. \tag{A.10b}$$

- Finite geometric progression (for real $\lambda \neq 1$):

$$1 + \lambda + \lambda^2 + \cdots + \lambda^{n-1} \equiv \sum_{k=0}^{n-1} \lambda^k = \frac{1-\lambda^n}{1-\lambda}; \tag{A.11a}$$

in particular, if $\lambda^2 < 1$, the progression has a finite limit at $n \to \infty$ (called the *geometric series*):

$$\lim_{n\to\infty} \sum_{k=0}^{n-1} \lambda^k = \sum_{k=0}^{\infty} \lambda^k = \frac{1}{1-\lambda}. \tag{A.11b}$$

- *Binomial sum* (or the 'binomial theorem'):

$$(1+a)^n = \sum_{k=0}^{n} {}^nC_k a^k, \tag{A.12}$$

where ${}^nC_k$ are the binomial coefficients defined by Eq. (A.5).

- The *Stirling formula*:

$$\lim_{n\to\infty} \ln(n!) = n(\ln n - 1) + \frac{1}{2}\ln(2\pi n) + \frac{1}{12n} - \frac{1}{360n^3} + \ldots; \tag{A.13}$$

for most applications in physics, the first term[1] is sufficient.

- The *Taylor* (or 'Taylor–Maclaurin') *series*: for any infinitely differentiable function $f(\xi)$:

$$\lim_{\tilde{\xi} \to 0} f(\xi + \tilde{\xi}) = f(\xi) + \frac{df}{d\xi}(\xi)\ \tilde{\xi} + \frac{1}{2!}\frac{d^2 f}{d\xi^2}(\xi)\ \tilde{\xi}^2 + \cdots$$

$$= \sum_{k=0}^{\infty} \frac{1}{k!}\frac{d^k f}{d\xi^k}(\xi)\ \tilde{\xi}^k; \tag{A.14a}$$

note that for many functions this series converges only within a limited, sometimes small range of deviations $\tilde{\xi}$. For a function of several arguments, $f(\xi_1, \xi_2, \ldots, \xi_N)$, the first terms of the Taylor series are

$$\lim_{\tilde{\xi}_k \to 0} f(\xi_1 + \tilde{\xi}_1,\ \xi_2 + \tilde{\xi}_2,\ \cdots) = f(\xi_1,\ \xi_2,\ \cdots)$$

$$+ \sum_{k=1}^{N} \frac{\partial f}{\partial \xi_k}(\xi_1,\ \xi_2,\ \cdots)\ \tilde{\xi}_k$$

$$+ \frac{1}{2!} \sum_{k,k'=1}^{N} \frac{\partial^2 f}{\partial_k \xi\, \partial \xi_{k'}} \tilde{\xi}_k \tilde{\xi}_{k'} + \cdots \tag{A.14b}$$

- The *Euler–Maclaurin formula*, valid for any infinitely differentiable function $f(\xi)$:

$$\sum_{k=1}^{n} f(k) = \int_0^n f(\xi)d\xi + \frac{1}{2}[f(n) - f(0)] + \frac{1}{6} \cdot \frac{1}{2!}\left[\frac{df}{d\xi}(n) - \frac{df}{d\xi}(0)\right]$$

$$- \frac{1}{30} \cdot \frac{1}{4!}\left[\frac{d^3 f}{d\xi^3}(n) - \frac{d^3 f}{d\xi^3}(0)\right] \tag{A.15a}$$

$$+ \frac{1}{42} \cdot \frac{1}{6!}\left[\frac{d^5 f}{d\xi^5}(n) - \frac{d^5 f}{d\xi^5}(0)\right] + \cdots;$$

the coefficients participating in this formula are the so-called *Bernoulli numbers*[2]:

$$B_1 = \frac{1}{2}, \quad B_2 = \frac{1}{6}, \quad B_3 = 0, \quad B_4 = \frac{1}{30}, \quad B_5 = 0,$$

$$B_6 = \frac{1}{42}, \quad B_7 = 0, \quad B_8 = \frac{1}{30}, \quad \cdots \tag{A.15b}$$

---

[1] Actually, this leading term was derived by A de Moivre in 1733, before J Stirling's work.
[2] Note that definitions of B_k (or rather their signs and indices) vary even among the most popular handbooks.

A.3 Basic trigonometric functions

- Trigonometric functions of the sum and the difference of two arguments[3]:

$$\cos(a \pm b) = \cos a \cos b \mp \sin a \sin b, \tag{A.16a}$$

$$\sin(a \pm b) = \sin a \cos b \pm \cos a \sin b. \tag{A.16b}$$

- Sums of two functions of arbitrary arguments:

$$\cos a + \cos b = 2 \cos \frac{a+b}{2} \cos \frac{b-a}{2}, \tag{A.17a}$$

$$\cos a - \cos b = 2 \sin \frac{a+b}{2} \sin \frac{b-a}{2}, \tag{A.17b}$$

$$\sin a \pm \sin b = 2 \sin \frac{a \pm b}{2} \cos \frac{\pm b - a}{2}. \tag{A.17c}$$

- Trigonometric function products:

$$2 \cos a \cos b = \cos(a+b) + \cos(a-b), \tag{A.18a}$$

$$2 \sin a \cos b = \sin(a+b) + \sin(a-b), \tag{A.18b}$$

$$2 \sin a \sin b = \cos(a-b) - \cos(a+b); \tag{A.18c}$$

For the particular case of equal arguments, $b = a$, these three formulas yield the following expressions for the squares of trigonometric functions, and their product:

$$\cos^2 a = \frac{1}{2}(1 + \cos 2a), \quad \sin a \cos a = \frac{1}{2} \sin 2a,$$

$$\sin^2 a = \frac{1}{2}(1 - \cos 2a). \tag{A.18d}$$

- Cubes of trigonometric functions:

$$\cos^3 a = \frac{3}{4} \cos a + \frac{1}{4} \cos 3a, \quad \sin^3 a = \frac{3}{4} \sin a - \frac{1}{4} \sin 3a. \tag{A.19}$$

- Trigonometric functions of a complex argument:

$$\sin(a + ib) = \sin a \cosh b + i \cos a \sinh b,$$
$$\cos(a + ib) = \cos a \cosh b - i \sin a \sinh b. \tag{A.20}$$

---

[3] I am confident that the reader is quite capable of deriving the relations (A.16) by representing the exponent in the elementary relation $e^{i(a \pm b)} = e^{ia}e^{\pm ib}$ as a sum of its real and imaginary parts, Eqs. (A.18) directly from Eqs. (A.16), and Eqs. (A.17) from Eqs. (A.18) by variable replacement; however, I am still providing these formulas to save his or her time. (Quite a few formulas below are included because of the same reason.)

- Sums of trigonometric functions of n equidistant arguments:

$$\sum_{k=1}^{n} \begin{Bmatrix} \sin \\ \cos \end{Bmatrix} k\xi = \begin{Bmatrix} \sin \\ \cos \end{Bmatrix} \left(\frac{n+1}{2} \xi \right) \sin\left(\frac{n}{2}\xi \right) \Big/ \sin\left(\frac{\xi}{2} \right). \qquad (A.21)$$

A.4 General differentiation

- Full differential of a product of two functions:

$$d(fg) = (df)g + f(dg). \qquad (A.22)$$

- Full differential of a function of several independent arguments, $f(\xi_1, \xi_2, \ldots, \xi_n)$:

$$df = \sum_{k=1}^{n} \frac{\partial f}{\partial \xi_k} d\xi_k. \qquad (A.23)$$

- Curvature of the Cartesian plot of a 1D function $f(\xi)$:

$$\kappa \equiv \frac{1}{R} = \frac{|d^2f/d\xi^2|}{[1 + (df/d\xi)^2]^{3/2}}. \qquad (A.24)$$

A.5 General integration

- Integration *by parts* - immediately follows from Eq. (A.22):

$$\int_{g(A)}^{g(B)} f \ dg = fg \Big|_A^B - \int_{f(A)}^{f(B)} g \ df. \qquad (A.25)$$

- Numerical (approximate) integration of 1D functions: the simplest *trapezoidal rule*,

$$\int_a^b f(\xi)d\xi \approx h\left[f\left(a + \frac{h}{2}\right) + f\left(a + \frac{3h}{2}\right) + \cdots + f\left(b - \frac{h}{2}\right) \right]$$
$$= h\sum_{n=1}^{N} f\left(a - \frac{h}{2} + nh\right), \quad h \equiv \frac{b-a}{N}. \qquad (A.26)$$

has relatively low accuracy (error of the order of $(h^3/12)d^2f/d\xi^2$ per step), so that the following *Simpson formula*,

$$\int_a^b f(\xi)d\xi \approx \frac{h}{3}[f(a) + 4f(a + h) + 2f(a + 2h) + \cdots + 4f(b - h) + f(b)],$$
$$h \equiv \frac{b-a}{2N}, \qquad (A.27)$$

whose error per step scales as $(h^5/180)d^4 f/d\xi^4$, is used much more frequently[4].

A.6 A few 1D integrals[5]

(i) *Indefinite integrals*:
- Integrals with $(1 + \xi^2)^{1/2}$:

$$\int (1 + \xi^2)^{1/2} d\xi = \frac{\xi}{2}(1 + \xi^2)^{1/2} + \frac{1}{2}\ln|\xi + (1 + \xi^2)^{1/2}|, \qquad (A.28)$$

$$\int \frac{d\xi}{(1 + \xi^2)^{1/2}} = \ln|\xi + (1 + \xi^2)^{1/2}|, \qquad (A.29a)$$

$$\int \frac{d\xi}{(1 + \xi^2)^{3/2}} = \frac{\xi}{(1 + \xi^2)^{1/2}}. \qquad (A.29b)$$

- Miscellaneous indefinite integrals:

$$\int \frac{d\xi}{\xi(\xi^2 + 2a\xi - 1)^{1/2}} = \arccos\frac{a\xi - 1}{|\xi|(a^2 + 1)^{1/2}}, \qquad (A.30a)$$

$$\int \frac{(\sin \xi - \xi \cos \xi)^2}{\xi^5} d\xi = \frac{2\xi \sin 2\xi + \cos 2\xi - 2\xi^2 - 1}{8\xi^4}, \qquad (A.30b)$$

$$\int \frac{d\xi}{a + b \cos \xi} = \frac{2}{(a^2 - b^2)^{1/2}}\tan^{-1}\left[\frac{(a - b)}{(a^2 - b^2)^{1/2}}\tan\frac{\xi}{2}\right], \qquad (A.30c)$$

$$\text{for } a^2 > b^2.$$

$$\int \frac{d\xi}{1 + \xi^2} = \tan^{-1}\xi. \qquad (A.30d)$$

(ii) *Semi-definite integrals*:
- Integrals with $1/(e^\xi \pm 1)$:

$$\int_a^\infty \frac{d\xi}{e^\xi + 1} = \ln(1 + e^{-a}), \qquad (A.31a)$$

---

[4] Higher-order formulas (e.g. the *Bode rule*), and other guidance including ready-for-use codes for computer calculations may be found, for example, in the popular reference texts by W H Press *et al* [1]. In addition, some advanced codes are used as subroutines in the software packages listed in the same section. In some cases, the Euler–Maclaurin formula (A.15) may also be useful for numerical integration.

[5] A powerful (and free) interactive online tool for working out indefinite 1D integrals is available at http://integrals.wolfram.com/index.jsp.

$$\int_{a>0}^{\infty} \frac{d\xi}{e^{\xi} - 1} = \ln \frac{1}{1 - e^{-a}}. \tag{A.31b}$$

(iii) *Definite integrals*:
- Integrals with $1/(1 + \xi^2)$:[6]

$$\int_0^{\infty} \frac{d\xi}{1 + \xi^2} = \frac{\pi}{2}, \tag{A.32a}$$

$$\int_0^{\infty} \frac{d\xi}{(1 + \xi^2)^{3/2}} = 1; \tag{A.32b}$$

more generally,

$$\int_0^{\infty} \frac{d\xi}{(1 + \xi^2)^n} = \frac{\pi}{2} \frac{(2n - 3)!!}{(2n - 2)!!} \equiv \frac{\pi}{2} \frac{1 \cdot 3 \cdot 5 \dots (2n - 3)}{2 \cdot 4 \cdot 6 \dots (2n - 2)}, \tag{A.32c}$$
$$\text{for } n = 2, 3, \dots$$

- Integrals with $(1 - \xi^{2n})^{1/2}$:

$$\int_0^1 \frac{d\xi}{(1 - \xi^{2n})^{1/2}} = \frac{\pi^{1/2}}{2n} \Gamma\left(\frac{1}{2n}\right) \Big/ \Gamma\left(\frac{n + 1}{2n}\right), \tag{A.33a}$$

$$\int_0^1 (1 - \xi^{2n})^{1/2} d\xi = \frac{\pi^{1/2}}{4n} \Gamma\left(\frac{1}{2n}\right) \Big/ \Gamma\left(\frac{3n + 1}{2n}\right), \tag{A.33b}$$

where $\Gamma(s)$ is the *gamma-function*, which is most often defined (for $\mathrm{Re}\, s > 0$) by the following integral:

$$\int_0^{\infty} \xi^{s-1} e^{-\xi}\, d\xi = \Gamma(s). \tag{A.34a}$$

The key property of this function is the recurrence relation, valid for any $s \neq 0, -1, -2, \dots$:

$$\Gamma(s + 1) = s\Gamma(s). \tag{A.34b}$$

Since, according to Eq. (A.34a), $\Gamma(1) = 1$, Eq. (A.34b) for non-negative integers takes the form

$$\Gamma(n + 1) = n!, \quad \text{for } n = 0, 1, 2, \cdots \tag{A.34c}$$

---

[6] Eq. (A.32a) follows immediately from Eq. (A.30d), and Eq. (A.32b) from Eq. (A.29b)—a couple more examples of the (intentional) redundancy in this list.

(where $0! \equiv 1$). Because of this, for integer $s = n + 1 \geqslant 1$, Eq. (A.34a) is reduced to

$$\int_0^\infty \xi^n e^{-\xi} d\xi = n!. \tag{A.34d}$$

Other frequently met values of the gamma-function are those for positive semi-integer arguments:

$$\Gamma\left(\frac{1}{2}\right) = \pi^{1/2}, \quad \Gamma\left(\frac{3}{2}\right) = \frac{1}{2}\pi^{1/2}, \quad \Gamma\left(\frac{5}{2}\right) = \frac{1}{2} \cdot \frac{3}{2}\pi^{1/2},$$

$$\Gamma\left(\frac{7}{2}\right) = \frac{1}{2} \cdot \frac{3}{2} \cdot \frac{5}{2}\pi^{1/2}, \quad \ldots. \tag{A.34e}$$

- Integrals with $1/(e^\xi \pm 1)$:

$$\int_0^\infty \frac{\xi^{s-1}d\xi}{e^\xi + 1} = (1 - 2^{1-s})\,\Gamma(s)\zeta(s), \quad \text{for } s > 0, \tag{A.35a}$$

$$\int_0^\infty \frac{\xi^{s-1}d\xi}{e^\xi - 1} = \Gamma(s)\zeta(s), \quad \text{for } s > 1, \tag{A.35b}$$

where $\zeta(s)$ is the Riemann zeta-function—see Eq. (A.10). Particular cases: for $s = 2n$,

$$\int_0^\infty \frac{\xi^{2n-1}d\xi}{e^\xi + 1} = \frac{2^{2n-1} - 1}{2n}\pi^{2n}B_{2n}, \tag{A.35c}$$

$$\int_0^\infty \frac{\xi^{2n-1}d\xi}{e^\xi - 1} = \frac{(2\pi)^{2n}}{4n}B_{2n}. \tag{A.35d}$$

where B_n are the Bernoulli numbers—see Eq. (A.15). For the particular case $s = 1$ (when Eq. (A.35a) yields uncertainty),

$$\int_0^\infty \frac{d\xi}{e^\xi + 1} = \ln 2. \tag{A.35e}$$

- Integrals with $\exp\{-\xi^2\}$:

$$\int_0^\infty \xi^s e^{-\xi^2} d\xi = \frac{1}{2}\Gamma\left(\frac{s+1}{2}\right), \quad \text{for } s > -1; \tag{A.36a}$$

for applications the most important particular values of s are 0 and 2:

$$\int_0^\infty e^{-\xi^2} d\xi = \frac{1}{2}\Gamma\left(\frac{1}{2}\right) = \frac{\pi^{1/2}}{2}, \tag{A.36b}$$

$$\int_0^\infty \xi^2 e^{-\xi^2} d\xi = \frac{1}{2}\Gamma\left(\frac{3}{2}\right) = \frac{\pi^{1/2}}{4}, \tag{A.36c}$$

although we will also run into the cases $s = 4$ and $s = 6$:

$$\int_0^\infty \xi^4 e^{-\xi^2} d\xi = \frac{1}{2}\Gamma\left(\frac{5}{2}\right) = \frac{3\pi^{1/2}}{8},$$

$$\int_0^\infty \xi^6 e^{-\xi^2} d\xi = \frac{1}{2}\Gamma\left(\frac{7}{2}\right) = \frac{15\pi^{1/2}}{16}; \tag{A.36d}$$

for odd integer values $s = 2n + 1$ (with $n = 0, 1, 2,...$), Eq. (A.36a) takes a simpler form:

$$\int_0^\infty \xi^{2n+1} e^{-\xi^2} d\xi = \frac{1}{2}\Gamma(n+1) = \frac{n!}{2}. \tag{A.36e}$$

- Integrals with cosine and sine functions:

$$\int_0^\infty \cos(\xi^2)\, d\xi = \int_0^\infty \sin(\xi^2)\, d\xi = \left(\frac{\pi}{8}\right)^{1/2}. \tag{A.37}$$

$$\int_0^\infty \frac{\cos \xi}{a^2 + \xi^2} d\xi = \frac{\pi}{2a} e^{-a}. \tag{A.38}$$

$$\int_0^\infty \left(\frac{\sin \xi}{\xi}\right)^2 d\xi = \frac{\pi}{2}. \tag{A.39}$$

- Integrals with logarithms:

$$\int_0^1 \ln\frac{a + (1 - \xi^2)^{1/2}}{a - (1 - \xi^2)^{1/2}} d\xi = \pi[a - (a^2 - 1)^{1/2}], \quad \text{for } a \geqslant 1. \tag{A.40}$$

$$\int_0^1 \ln\frac{1 + (1 - \xi)^{1/2}}{\xi^{1/2}} d\xi = 1. \tag{A.41}$$

- Integral representations of the Bessel functions of integer order:

$$J_n(\alpha) = \frac{1}{2\pi}\int_{-\pi}^{+\pi} e^{i(\alpha \sin \xi - n\xi)} d\xi,$$

$$\text{so that } e^{i\alpha \sin \xi} = \sum_{k=-\infty}^{\infty} J_k(\alpha) e^{ik\xi}; \tag{A.42a}$$

$$I_n(\alpha) = \frac{1}{\pi}\int_0^\pi e^{\alpha \cos \xi} \cos n\xi \; d\xi. \tag{A.42b}$$

A.7 3D vector products

(i) *Definitions*:
- *Scalar ('dot-') product*:

$$\mathbf{a} \cdot \mathbf{b} = \sum_{j=1}^{3} a_j b_j, \tag{A.43}$$

where a_j and b_j are vector components in any orthogonal coordinate system. In particular, the vector squared (the same as the norm squared):

$$a^2 \equiv \mathbf{a} \cdot \mathbf{a} = \sum_{j=1}^{3} a_j^2 \equiv \| \mathbf{a} \|^2. \tag{A.44}$$

- *Vector ('cross-') product*:

$$\mathbf{a} \times \mathbf{b} \equiv \mathbf{n}_1(a_2 b_3 - a_3 b_2) + \mathbf{n}_2(a_3 b_1 - a_1 b_3) + \mathbf{n}_3(a_1 b_2 - a_2 b_1)$$

$$= \begin{vmatrix} \mathbf{n}_1 & \mathbf{n}_2 & \mathbf{n}_3 \\ a_1 & a_2 & a_3 \\ b_1 & b_2 & b_3 \end{vmatrix}, \tag{A.45}$$

where $\{\mathbf{n}_j\}$ is the set of mutually perpendicular unit vectors[7] along the corresponding coordinate system axes[8]. In particular, Eq. (A.45) yields

$$\mathbf{a} \times \mathbf{a} = 0. \tag{A.46}$$

(ii) *Corollaries* (readily verified by Cartesian components):
- Double vector product (the so-called *bac minus cab* rule):

$$\mathbf{a} \times (\mathbf{b} \times \mathbf{c}) = \mathbf{b}(\mathbf{a} \cdot \mathbf{c}) - \mathbf{c}(\mathbf{a} \cdot \mathbf{b}). \tag{A.47}$$

- Mixed scalar–vector product (the *operand rotation rule*):

$$\mathbf{a} \cdot (\mathbf{b} \times \mathbf{c}) = \mathbf{b} \cdot (\mathbf{c} \times \mathbf{a}) = \mathbf{c} \cdot (\mathbf{a} \times \mathbf{b}). \tag{A.48}$$

- Scalar product of vector products:

$$(\mathbf{a} \times \mathbf{b}) \cdot (\mathbf{c} \times \mathbf{d}) = (\mathbf{a} \cdot \mathbf{c})(\mathbf{b} \cdot \mathbf{d}) - (\mathbf{a} \cdot \mathbf{d})(\mathbf{b} \cdot \mathbf{c}); \tag{A.49a}$$

---

[7] Other popular notations for this vector set are $\{\mathbf{e}_j\}$ and $\{\hat{\mathbf{r}}_j\}$.

[8] It is easy to use Eq. (A.45) to check that the direction of the product vector corresponds to the well-known 'right-hand rule' and to the even more convenient *corkscrew rule*: if we rotate a corkscrew's handle from the first operand toward the second one, its axis moves in the direction of the product.

in the particular case of two similar operands (say, $\mathbf{a} = \mathbf{c}$ and $\mathbf{b} = \mathbf{d}$), the last formula is reduced to

$$(\mathbf{a} \times \mathbf{b})^2 = (ab)^2 - (\mathbf{a} \cdot \mathbf{b})^2. \tag{A.49b}$$

A.8 Differentiation in 3D Cartesian coordinates

- Definition of the *del* (or 'nabla') vector-operator ∇:[9]

$$\nabla \equiv \sum_{j=1}^{3} \mathbf{n}_j \frac{\partial}{\partial r_j}, \tag{A.50}$$

where r_j is a set of linear and orthogonal (*Cartesian*) coordinates along directions $\mathbf{n}_j$. In accordance with this definition, the operator ∇ acting on a *scalar* function of coordinates, $f(\mathbf{r})$,[10] gives its gradient, i.e. a new *vector*:

$$\nabla f \equiv \sum_{j=1}^{3} \mathbf{n}_j \frac{\partial f}{\partial r_j} \equiv \mathbf{grad}\, f. \tag{A.51}$$

- The *scalar product* of del by a *vector* function of coordinates (a *vector field*),

$$\mathbf{f}(\mathbf{r}) \equiv \sum_{j=1}^{3} \mathbf{n}_j f_j(\mathbf{r}), \tag{A.52}$$

compiled formally following Eq. (A.43), is a *scalar* function—the *divergence* of the initial function:

$$\nabla \cdot \mathbf{f} \equiv \sum_{j=1}^{3} \frac{\partial f_j}{\partial r_j} \equiv \mathrm{div}\, \mathbf{f}, \tag{A.53}$$

while the *vector product* of ∇ and $\mathbf{f}$, formed in a formal accordance with Eq. (A.45), is a new vector - the *curl* (in European tradition, called rotor and denoted **rot**) of $\mathbf{f}$:

$$\nabla \times \mathbf{f} \equiv \begin{vmatrix} \mathbf{n}_1 & \mathbf{n}_2 & \mathbf{n}_3 \\ \frac{\partial}{\partial r_1} & \frac{\partial}{\partial r_2} & \frac{\partial}{\partial r_3} \\ f_1 & f_2 & f_3 \end{vmatrix} = \mathbf{n}_1\left(\frac{\partial f_3}{\partial r_2} - \frac{\partial f_2}{\partial r_3}\right) + \mathbf{n}_2\left(\frac{\partial f_1}{\partial r_3} - \frac{\partial f_3}{\partial r_1}\right)$$
$$+ \mathbf{n}_3\left(\frac{\partial f_2}{\partial r_1} - \frac{\partial f_1}{\partial r_2}\right) \equiv \mathbf{curl}\, \mathbf{f}. \tag{A.54}$$

---

[9] One can run into the following notation: $\nabla \equiv \partial/\partial \mathbf{r}$, which is convenient is some cases, but may be misleading in quite a few others, so it will be not used in these notes.
[10] In this, and four next sections, all scalar and vector functions are assumed to be differentiable.

- One more frequently met 'product' is $(\mathbf{f}\cdot\nabla)\mathbf{g}$, where $\mathbf{f}$ and $\mathbf{g}$ are two arbitrary vector functions of $\mathbf{r}$. This product should be also understood in the sense implied by Eq. (A.43), i.e. as a vector whose jth Cartesian component is

$$[(\mathbf{f}\cdot\nabla)\,\mathbf{g}]_j = \sum_{j'=1}^{3} f_{j'}\frac{\partial g_j}{\partial r_{j'}}. \tag{A.55}$$

A.9 The Laplace operator $\nabla^2 \equiv \nabla\cdot\nabla$

- Expression in Cartesian coordinates—in the formal accordance with Eq. (A.44):

$$\nabla^2 = \sum_{j=1}^{3}\frac{\partial^2}{\partial r_j^2}. \tag{A.56}$$

- According to its definition, the Laplace operator acting on a *scalar* function of coordinates gives a new scalar function:

$$\nabla^2 f \equiv \nabla\cdot(\nabla f) = \operatorname{div}(\mathbf{grad}\,f) = \sum_{j=1}^{3}\frac{\partial^2 f}{\partial r_j^2}. \tag{A.57}$$

- On the other hand, acting on a *vector* function (A.52), the operator ∇^2 returns another *vector*:

$$\nabla^2\mathbf{f} = \sum_{j=1}^{3}\mathbf{n}_j\nabla^2 f_j. \tag{A.58}$$

Note that Eqs. (A.56)–(A.58) are only valid in Cartesian (i.e. orthogonal and linear) coordinates, but generally not in other (even orthogonal) coordinates—see, e.g. Eqs. (A.61), (A.64), (A.67) and (A.70) below.

A.10 Operators ∇ and ∇^2 in the most important systems of orthogonal coordinates[11]

(i) *Cylindrical*[12] *coordinates* $\{\rho, \varphi, z\}$ (see figure below) may be defined by their relations with the Cartesian coordinates:

$$r_1 = \rho\cos\varphi,$$
$$r_2 = \rho\sin\varphi, \tag{A.59}$$
$$r_3 = z.$$

[11] Some other orthogonal curvilinear coordinate systems are discussed in *Part EM*, section 2.3.
[12] In the 2D geometry with fixed coordinate z, these coordinates are called *polar*.

- Gradient of a scalar function:

$$\nabla f = \mathbf{n}_\rho \frac{\partial f}{\partial \rho} + \mathbf{n}_\varphi \frac{1}{\rho} \frac{\partial f}{\partial \varphi} + \mathbf{n}_z \frac{\partial f}{\partial z}. \qquad (A.60)$$

- The Laplace operator of a scalar function:

$$\nabla^2 f = \frac{1}{\rho} \frac{\partial}{\partial \rho} \left(\rho \frac{\partial f}{\partial \rho} \right) + \frac{1}{\rho^2} \frac{\partial^2 f}{\partial \varphi^2} + \frac{\partial^2 f}{\partial z^2}, \qquad (A.61)$$

- Divergence of a vector function of coordinates ($\mathbf{f} = \mathbf{n}_\rho f_\rho + \mathbf{n}_\varphi f_\varphi + \mathbf{n}_z f_z$):

$$\nabla \cdot \mathbf{f} = \frac{1}{\rho} \frac{\partial(\rho f_\rho)}{\partial \rho} + \frac{1}{\rho} \frac{\partial f_\varphi}{\partial \varphi} + \frac{\partial f_z}{\partial z}. \qquad (A.62)$$

- Curl of a vector function:

$$\nabla \times \mathbf{f} = \mathbf{n}_\rho \left(\frac{1}{\rho} \frac{\partial f_z}{\partial \varphi} - \frac{\partial f_\varphi}{\partial z} \right) + \mathbf{n}_\varphi \left(\frac{\partial f_\rho}{\partial z} - \frac{\partial f_z}{\partial \rho} \right) + \mathbf{n}_z \frac{1}{\rho} \left(\frac{\partial(\rho f_\varphi)}{\partial \rho} - \frac{\partial f_\rho}{\partial \varphi} \right). \qquad (A.63)$$

- The Laplace operator of a vector function:

$$\nabla^2 \mathbf{f} = \mathbf{n}_\rho \left(\nabla^2 f_\rho - \frac{1}{\rho^2} f_\rho - \frac{2}{\rho^2} \frac{\partial f_\varphi}{\partial \varphi} \right) + \mathbf{n}_\varphi \left(\nabla^2 f_\varphi - \frac{1}{\rho^2} f_\varphi + \frac{2}{\rho^2} \frac{\partial f_\rho}{\partial \varphi} \right) + \mathbf{n}_z \nabla^2 f_z. \quad (A.64)$$

(ii) *Spherical coordinates* $\{r, \theta, \varphi\}$ (see figure below) may be defined as:

$$\begin{aligned} r_1 &= r \sin \theta \cos \varphi, \\ r_2 &= r \sin \theta \sin \varphi, \\ r_3 &= r \cos \theta. \end{aligned} \qquad (A.65)$$

- Gradient of a scalar function:

$$\nabla f = \mathbf{n}_r \frac{\partial f}{\partial r} + \mathbf{n}_\theta \frac{1}{r} \frac{\partial f}{\partial \theta} + \mathbf{n}_\varphi \frac{1}{r \sin \theta} \frac{\partial f}{\partial \varphi}. \qquad (A.66)$$

- The Laplace operator of a scalar function:

$$\nabla^2 f = \frac{1}{r^2} \frac{\partial}{\partial r} \left(r^2 \frac{\partial f}{\partial r} \right) + \frac{1}{r^2 \sin \theta} \frac{\partial}{\partial \theta} \left(\sin \theta \frac{\partial f}{\partial \theta} \right) + \frac{1}{(r \sin \theta)^2} \frac{\partial^2 f}{\partial \varphi^2}. \qquad (A.67)$$

- Divergence of a vector function $\mathbf{f} = \mathbf{n}_r f_r + \mathbf{n}_\theta f_\theta + \mathbf{n}_\varphi f_\varphi$:

$$\nabla \cdot \mathbf{f} = \frac{1}{r^2} \frac{\partial(r^2 f_r)}{\partial r} + \frac{1}{r \sin \theta} \frac{\partial(f_\theta \sin \theta)}{\partial \theta} + \frac{1}{r \sin \theta} \frac{\partial f_\varphi}{\partial \varphi}. \tag{A.68}$$

- Curl of a similar vector function:

$$\nabla \times \mathbf{f} = \mathbf{n}_r \frac{1}{r \sin \theta} \left(\frac{\partial(f_\varphi \sin \theta)}{\partial \theta} - \frac{\partial f_\theta}{\partial \varphi} \right) + \mathbf{n}_\theta \frac{1}{r} \left(\frac{1}{\sin \theta} \frac{\partial f_r}{\partial \varphi} - \frac{\partial(r f_\varphi)}{\partial r} \right)$$
$$+ \mathbf{n}_\varphi \frac{1}{r} \left(\frac{\partial(r f_\theta)}{\partial r} - \frac{\partial f_r}{\partial \theta} \right). \tag{A.69}$$

- The Laplace operator of a vector function:

$$\nabla^2 \mathbf{f} = \mathbf{n}_r \left(\nabla^2 f_r - \frac{2}{r^2} f_r - \frac{2}{r^2 \sin \theta} \frac{\partial}{\partial \theta}(f_\theta \sin \theta) - \frac{2}{r^2 \sin \theta} \frac{\partial f_\varphi}{\partial \varphi} \right)$$
$$+ \mathbf{n}_\theta \left(\nabla^2 f_\theta - \frac{1}{r^2 \sin^2 \theta} f_\theta + \frac{2}{r^2} \frac{\partial f_r}{\partial \theta} - \frac{2 \cos \theta}{r^2 \sin^2 \theta} \frac{\partial f_\varphi}{\partial \varphi} \right) \tag{A.70}$$
$$+ \mathbf{n}_\varphi \left(\nabla^2 f_\varphi - \frac{1}{r^2 \sin^2 \theta} f_\varphi + \frac{2}{r^2 \sin \theta} \frac{\partial f_r}{\partial \varphi} + \frac{2 \cos \theta}{r^2 \sin^2 \theta} \frac{\partial f_\theta}{\partial \varphi} \right).$$

A.11 Products involving ∇

(i) *Useful zeros*:

- For any scalar function $f(\mathbf{r})$,

$$\nabla \times (\nabla f) \equiv \mathbf{curl}(\mathbf{grad}\, f) = 0. \tag{A.71}$$

- For any vector function $\mathbf{f}(\mathbf{r})$,

$$\nabla \cdot (\nabla \times \mathbf{f}) \equiv \text{div}(\mathbf{curl}\, f) = 0. \tag{A.72}$$

(ii) The *Laplace operator* expressed via the curl of a curl:

$$\nabla^2 \mathbf{f} = \nabla(\nabla \cdot \mathbf{f}) - \nabla \times (\nabla \times \mathbf{f}). \tag{A.73}$$

(iii) Spatial differentiation of a product of a *scalar* function by a *vector* function:x

- The scalar 3D generalization of Eq. (A.22) is

$$\nabla \cdot (f\, \mathbf{g}) = (\nabla f) \cdot \mathbf{g} + f(\nabla \cdot \mathbf{g}). \tag{A.74a}$$

- Its vector generalization is similar:

$$\nabla \times (f\, \mathbf{g}) = (\nabla f) \times \mathbf{g} + f(\nabla \times \mathbf{g}).$$ (A.74b)

(iv) Spatial differentiation of products of *two vector* functions:

$$\nabla \times (\mathbf{f} \times \mathbf{g}) = \mathbf{f}(\nabla \cdot \mathbf{g}) - (\mathbf{f} \cdot \nabla)\mathbf{g} - (\nabla \cdot \mathbf{f})\mathbf{g} + (\mathbf{g} \cdot \nabla)\mathbf{f},$$ (A.75)

$$\nabla(\mathbf{f} \cdot \mathbf{g}) = (\mathbf{f} \cdot \nabla)\mathbf{g} + (\mathbf{g} \cdot \nabla)\mathbf{f} + \mathbf{f} \times (\nabla \times \mathbf{g}) + \mathbf{g} \times (\nabla \times \mathbf{f}),$$ (A.76)

$$\nabla \cdot (\mathbf{f} \times \mathbf{g}) = \mathbf{g} \cdot (\nabla \times \mathbf{f}) - \mathbf{f} \cdot (\nabla \times \mathbf{g}).$$ (A.77)

A.12 Integro-differential relations

(i) For an *arbitrary surface S* limited by closed contour *C*:

- The *Stokes theorem*, valid for any differentiable vector field $\mathbf{f(r)}$:

$$\int_S (\nabla \times \mathbf{f}) \cdot d^2\mathbf{r} \equiv \int_S (\nabla \times \mathbf{f})_n d^2r = \oint_C \mathbf{f} \cdot d\mathbf{r} \equiv \oint_C f_\tau dr,$$ (A.78)

where $d^2\mathbf{r} \equiv \mathbf{n}d^2r$ is the elementary area vector (normal to the surface), and $d\mathbf{r}$ is the elementary contour length vector (tangential to the contour line).

(ii) For an *arbitrary volume V* limited by closed surface *S*:

- *Divergence* (or 'Gauss') *theorem*, valid for any differentiable vector field $\mathbf{f(r)}$:

$$\int_V (\nabla \cdot \mathbf{f})\, d^3r = \oint_S \mathbf{f} \cdot d^2\mathbf{r} \equiv \oint_S f_n d^2r.$$ (A.79)

- *Green's theorem*, valid for two differentiable scalar functions $f(\mathbf{r})$ and $g(\mathbf{r})$:

$$\int_V (f\, \nabla^2 g - g\nabla^2 f)\, d^3r = \oint_S (f\, \nabla g - g\nabla f)_n d^2r.$$ (A.80)

- An identity valid for any two scalar functions f and g, and a vector field $\mathbf{j}$ with $\nabla \cdot \mathbf{j} = 0$ (all differentiable):

$$\int_V [f(\mathbf{j} \cdot \nabla g) + g(\mathbf{j} \cdot \nabla f)]\, d^3r = \oint_S fg j_n d^2r.$$ (A.81)

A.13 The Kronecker delta and Levi-Civita permutation symbols

- The *Kronecker delta symbol* (defined for integer indices):

$$\delta_{jj'} \equiv \begin{cases} 1, & \text{if } j' = j, \\ 0, & \text{otherwise.} \end{cases}$$ (A.82)

- The *Levi-Civita permutation symbol* (most frequently used for 3 integer indices, each taking one of values 1, 2, or 3):

$$\varepsilon_{jj'j''} \equiv \begin{cases} +1, & \text{if the indices follow in the 'correct' ('even')} \\ & \quad \text{order: } 1 \to 2 \to 3 \to 1 \to 2 \dots, \\ -1, & \text{if the indices follow in the 'incorrect' ('odd')} \\ & \quad \text{order: } 1 \to 3 \to 2 \to 1 \to 3 \dots, \\ 0, & \text{if any two indices coincide.} \end{cases} \tag{A.83}$$

- Relation between the Levi-Civita and the Kronecker delta products:

$$\varepsilon_{jj'j''}\varepsilon_{kk'k''} = \sum_{l,l',l''=1}^{3} \begin{vmatrix} \delta_{jl} & \delta_{jl'} & \delta_{jl''} \\ \delta_{j'l} & \delta_{j'l'} & \delta_{j'l''} \\ \delta_{j''l} & \delta_{j''l'} & \delta_{j''l''} \end{vmatrix}; \tag{A.84a}$$

summation of this relation, written for 3 different values of $j = k$, over these values yields the so-called *contracted epsilon identity*:

$$\sum_{j=1}^{3}\varepsilon_{jj'j''}\varepsilon_{jk'k''} = \delta_{j'k'}\delta_{j''k''} - \delta_{j'k''}\delta_{j''k'}. \tag{A.84b}$$

A.14 Dirac's delta-function, sign function, and theta-function

- Definition of 1D *delta-function* (for real $a < b$):

$$\int_{a}^{b} f(\xi)\delta(\xi)d\xi = \begin{cases} f(0), & \text{if } a < 0 < b, \\ 0, & \text{otherwise,} \end{cases} \tag{A.85}$$

where $f(\xi)$ is any function continuous near $\xi = 0$. In particular (if $f(\xi) = 1$ near $\xi = 0$), the definition yields

$$\int_{a}^{b} \delta(\xi)d\xi = \begin{cases} 1, & \text{if } a < 0 < b, \\ 0, & \text{otherwise.} \end{cases} \tag{A.86}$$

- Relation to the *theta-function* $\theta(\xi)$ and *sign function* $\text{sgn}(\xi)$

$$\delta(\xi) = \frac{d}{d\xi}\theta(\zeta) = \frac{1}{2}\frac{d}{d\xi}\text{sgn}(\xi), \tag{A.87a}$$

where

$$\theta(\xi) \equiv \frac{\text{sgn}(\xi) + 1}{2} = \begin{cases} 0, & \text{if } \xi < 0, \\ 1, & \text{if } \xi > 1, \end{cases}$$

$$\text{sgn}(\xi) \equiv \frac{\xi}{|\xi|} = \begin{cases} -1, & \text{if } \xi < 0, \\ +1, & \text{if } \xi > 1. \end{cases} \tag{A.87b}$$

- An important integral[13]:

$$\int_{-\infty}^{+\infty} e^{is\,\xi} ds = 2\pi\delta(\xi).\qquad (A.88)$$

- 3D generalization of the delta-function of the radius-vector (the 2D generalization is similar):

$$\int_V f(\mathbf{r})\delta(\mathbf{r})d^3r = \begin{cases} f(0), & \text{if } 0 \in V, \\ 0, & \text{otherwise;} \end{cases} \qquad (A.89)$$

it may be represented as a product of 1D delta-functions of Cartesian coordinates:

$$\delta(\mathbf{r}) = \delta(r_1)\delta(r_2)\delta(r_3).\qquad (A.90)$$

A.15 The Cauchy theorem and integral

Let a complex function $f(z)$ be analytic within a part of the complex plane z, that is limited by a closed contour C and includes point z'. Then

$$\oint_C f(z)dz = 0,\qquad (A.91)$$

$$\oint_C f(z)\frac{dz}{z - z'} = 2\pi i f(z')\qquad (A.92)$$

The first of these relations is usually called the *Cauchy integral theorem* (or the 'Cauchy–Goursat theorem'), and the second one—the *Cauchy integral* (or the 'Cauchy integral formula').

A.16 Literature

(i) Properties of some *special functions* are briefly discussed at the relevant points of the lecture notes; in the alphabetical order:
- Airy functions: *Part QM* section 2.4;
- Bessel functions: *Part EM* section 2.7;
- Fresnel integrals: *Part EM* section 8.6;
- Hermite polynomials: *Part QM* section 2.9;
- Laguerre polynomials (both simple and associated): *Part QM* section 3.7;

---

[13] The coefficient in this relation may be readily recalled by considering its left-hand part as the Fourier-integral representation of function $f(s) \equiv 1$, and applying Eq. (A.85) to the reciprocal Fourier transform

$$f(s) \equiv 1 = \frac{1}{2\pi}\int_{-\infty}^{+\infty} e^{-is\xi}[2\pi\delta(\xi)]d\xi.$$

- Legendre polynomials, associated Legendre functions: *Part EM* section 2.8, and *Part QM* section 3.6;
- Spherical harmonics: *Part QM* section 3.6;
- Spherical Bessel functions: *Part QM* sections 3.6 and 3.8.

(ii) For *more formulas*, and their discussion, I can recommend the following handbooks[14]:
- *Handbook of Mathematical Formulas* [2];
- *Tables of Integrals, Series, and Products* [3];
- *Mathematical Handbook for Scientists and Engineers* [4];
- *Integrals and Series* volumes 1 and 2 [5];
- A popular textbook *Mathematical Methods for Physicists* [6] may be also used as a formula manual.

Many formulas are also available from the symbolic calculation modules of the commercially available software packages listed in section (iv) below.

(iii) Probably the most popular collection of *numerical calculation codes* are the twin manuals by W Press *et al* [1]:
- *Numerical Recipes in Fortran 77*;
- *Numerical Recipes* [in C++—KKL].

My lecture notes include very brief introductions to numerical methods of differential equation solution:
- ordinary differential equations: *Part CM*, section 5.7;
- partial differential equations: *Part CM* section 8.5 and *Part EM* section 2.11, which include references to literature for further reading.

(iv) The following are the most popular *software packages* for numerical and symbolic calculations, all with plotting capabilities (in the alphabetical order):
- Maple (www.maplesoft.com/products/maple/);
- MathCAD (www.ptc.com/engineering-math-software/mathcad/);
- Mathematica (www.wolfram.com/mathematica/);
- MATLAB (www.mathworks.com/products/matlab.html).

References

[1] Press W *et al* 1992 *Numerical Recipes in Fortran 77* 2nd edn (Cambridge: Cambridge University Press)
Press W *et al* 2007 *Numerical Recipes* 3rd edn (Cambridge: Cambridge University Press)
[2] Abramowitz M and Stegun I (eds) 1965 *Handbook of Mathematical Formulas* (New York: Dover), and numerous later printings. An updated version of this collection is now available online at http://dlmf.nist.gov/.

---

[14] On a personal note, perhaps 90% of all formula needs throughout my research career were satisfied by a tiny, wonderfully compiled old book [7], used copies of which, rather amazingly, are still available on the Web.

[3] Gradshteyn I and Ryzhik I 1980 *Tables of Integrals, Series, and Products* 5th edn (New York: Academic)

[4] Korn G and Korn T 2000 *Mathematical Handbook for Scientists and Engineers* 2nd edn (New York: Academic)

[5] Prudnikov A *et al* 1986 *Integrals and Series* vol 1 (Boca Raton, FL: CRC Press)
Prudnikov A *et al* 1986 *Integrals and Series* vol 2 (Boca Raton, FL: CRC Press)

[6] Arfken G *et al* 2012 *Mathematical Methods for Physicists* 7th edn (New York: Academic)

[7] Dwight H 1961 *Tables of Integrals and Other Mathematical Formulas* 4th edn (London: Macmillan)

IOP Publishing

Quantum Mechanics
Lecture notes
Konstantin K Likharev

Appendix B

Selected physical constants

The listed numerical values of the constants are from the most recent (2014) International CODATA recommendation (see, e.g. http://physics.nist.gov/cuu/Constants/index.html), besides a newer result for k_B—see [1]. Please note the recently announced (but, by this volume's press time, not yet official) adjustment of the SI values - see, e.g. https://www.nist.gov/si-redefinition/meet-constants. In particular, the Planck constant will also get a definite value (within the interval specified in table B.1), enabling a new, fundamental standard of the kilogram.

Table B.1.

Symbol	Quantity	SI value and unit	Gaussian value and unit	Relative rms uncertainty
c	speed of light in free space	$2.99\ 792\ 458 \times 10^8$ m s^{-1}	$2.99\ 792\ 458 \times 10^{10}$ cm s^{-1}	0 (defined value)
G	gravitation constant	6.6741×10^{-11} m^3 kg^{-1} s^{-2}	6.6741×10^{-8} cm^3 g^{-1} s^{-2}	$\sim 5 \times 10^{-5}$
$\hbar$	Planck constant	$1.05\ 457\ 180 \times 10^{-34}$ J s	$1.05\ 457\ 180 \times 10^{-27}$ erg s	$\sim 2 \times 10^{-8}$
e	elementary electric charge	$1.6\ 021\ 762 \times 10^{-19}$ C	$4.803\ 203 \times 10^{-10}$ statcoulomb	$\sim 6 \times 10^{-9}$
m_e	electron's rest mass	$0.91\ 093\ 835 \times 10^{-30}$ kg	$0.91\ 093\ 835 \times 10^{-27}$ g	$\sim 1 \times 10^{-8}$
m_p	proton's rest mass	$1.67\ 262\ 190 \times 10^{-27}$ kg	$1.67\ 262\ 190 \times 10^{-24}$ g	$\sim 1 \times 10^{-8}$
μ_0	magnetic constant	$4\pi \times 10^{-7}$ N A^{-2}	–	0 (defined value)
ε_0	electric constant	$8.854\ 187\ 817 \times 10^{-12}$ F m^{-1}	–	0 (defined value)
k_B	Boltzmann constant	$1.380\ 649 \times 10^{-23}$ J K^{-1}	$1.3\ 806\ 490 \times 10^{-16}$ erg K^{-1}	$\sim 2 \times 10^{-6}$

Comments:

1. The fixed value of c was defined by an international convention in 1983, in order to extend the official definition of the second (as 'the duration of 9 192 631 770 periods of the radiation corresponding to the transition between the two hyperfine levels of the ground state of the cesium-133 atom') to that of the meter. The values are back-compatible with the legacy definitions of the meter (initially, as 1/40 000 000th of the Earth's meridian length) and the second (for a long time, as $1/(24 \times 60 \times 60) = 1/86\ 400$th of the Earth's rotation period), within the experimental errors of those measures.

2. ε_0 and μ_0 are not really the fundamental constants; in the SI system of units one of them (say, μ_0) is selected arbitrarily[1], while the other one is defined via the relation $\varepsilon_0 \mu_0 = 1/c^2$.

3. The Boltzmann constant k_B is also not quite fundamental, because its only role is to comply with the independent definition of the kelvin (K), as the temperature unit in which the triple point of water is exactly 273.16 K. If temperature is expressed in energy units $k_B T$ (as is done, for example, in *Part SM* of this series), this constant disappears altogether.

4. The dimensionless *fine structure* ('Sommerfeld's') *constant* α is numerically the same in any system of units:

$$\alpha \equiv \begin{cases} e^2/4\pi\varepsilon_0\hbar c & \text{in SI units} \\ e^2/\hbar c & \text{in Gaussian units} \end{cases} \approx 7.297\ 352\ 566 \times 10^{-3}$$

$$\approx \frac{1}{137.035\ 999\ 14},$$

and is known with a much smaller relative rms uncertainty (currently, $\sim 3 \times 10^{-10}$) than those of the component constants.

References

[1] Gaiser C *et al* 2017 *Metrologia* **54** 280
[2] Newell D 2014 *Phys. Today* **67** 35–41

---

[1] Note that the selected value of μ_0 may be changed (a bit) in a few years—see, e.g., [2].

IOP Publishing

Quantum Mechanics
Lecture notes
Konstantin K Likharev

Bibliography

This section presents a partial list of textbooks and monographs used in the work on the EAP series[1,2].

Part CM: Classical Mechanics

Fetter A L and Walecka J D 2003 *Theoretical Mechanics of Particles and Continua* (New York: Dover)

Goldstein H, Poole C and Safko J 2002 *Classical Mechanics* 3rd edn (Reading, MA: Addison Wesley)

Granger R A 1995 *Fluid Mechanics* (New York: Dover)

José J V and Saletan E J 1998 *Classical Dynamics* (Cambridge: Cambridge University Press)

Landau L D and Lifshitz E M 1976 *Mechanics* 3rd edn (Oxford: Butterworth-Heinemann)

Landau L D and Lifshitz E M 1986 *Theory of Elasticity* (Oxford: Butterworth-Heinemann)

Landau L D and Lifshitz E M 1987 *Fluid Mechanics* 2nd edn (Oxford: Butterworth-Heinemann)

Schuster H G 1995 *Deterministic Chaos* 3rd edn (New York: Wiley)

Sommerfeld A 1964 *Mechanics* (New York: Academic)

Sommerfeld A 1964 *Mechanics of Deformable Bodies* (New York: Academic)

Part EM: Classical Electrodynamics

Batygin V V and Toptygin I N 1978 *Problems in Electrodynamics* 2nd edn (New York: Academic)

Griffiths D J 2007 *Introduction to Electrodynamics* 3rd edn (Englewood Cliffs, NJ: Prentice-Hall)

Jackson J D 1999 *Classical Electrodynamics* 3rd edn (New York: Wiley)

Landau L D and Lifshitz E M 1984 *Electrodynamics of Continuous Media* 2nd edn (Auckland: Reed)

Landau L D and Lifshitz E M 1975 *The Classical Theory of Fields* 4th edn (Oxford: Pergamon)

---

[1] The list does not include the sources (mostly, recent original publications) cited in the lecture notes and problem solutions, and the mathematics textbooks and handbooks listed in section A.16.

[2] Recently several high-quality teaching materials on advanced physics became available online, including R. Fitzpatrick's text on *Classical Electromagnetism* (farside.ph.utexas.edu/teaching/jk1/Electromagnetism.pdf), B Simons' 'lecture shrunks' on *Advanced Quantum Mechanics* (www.tcm.phy.cam.ac.uk/~bds10/aqp.html), and D Tong's lecture notes on several advanced topics (www.damtp.cam.ac.uk/user/tong/teaching.html).

Panofsky W K H and Phillips M 1990 *Classical Electricity and Magnetism* 2nd edn (New York: Dover)

Stratton J A 2007 *Electromagnetic Theory* (New York: Wiley)

Tamm I E 1979 *Fundamentals of the Theory of Electricity* (Paris: Mir)

Zangwill A 2013 *Modern Electrodynamics* (Cambridge: Cambridge University Press)

Part QM: Quantum Mechanics

Abers E S 2004 *Quantum Mechanics* (London: Pearson)

Auletta G, Fortunato M and Parisi G 2009 *Quantum Mechanics* (Cambridge: Cambridge University Press)

Capri A Z 2002 *Nonrelativistic Quantum Mechanics* 3rd edn (Singapore: World Scientific)

Cohen-Tannoudji C, Diu B and Laloë F 2005 *Quantum Mechanics* (New York: Wiley)

Constantinescu F, Magyari E and Spiers J A 1971 *Problems in Quantum Mechanics* (Amsterdam: Elsevier)

Galitski V *et al* 2013 *Exploring Quantum Mechanics* (Oxford: Oxford University Press)

Gottfried K and Yan T-M 2004 *Quantum Mechanics: Fundamentals* 2nd edn (Berlin: Springer)

Griffith D 2005 *Quantum Mechanics* 2nd edn (Englewood Cliffs, NJ: Prentice Hall)

Landau L D and Lifshitz E M 1977 *Quantum Mechanics (Nonrelativistic Theory)* 3rd edn (Oxford: Pergamon)

Messiah A 1999 *Quantum Mechanics* (New York: Dover)

Merzbacher E 1998 *Quantum Mechanics* 3rd edn (New York: Wiley)

Miller D A B 2008 *Quantum Mechanics for Scientists and Engineers* (Cambridge: Cambridge University Press)

Sakurai J J 1994 *Modern Quantum Mechanics* (Reading, MA: Addison-Wesley)

Schiff L I 1968 *Quantum Mechanics* 3rd edn (New York: McGraw-Hill)

Shankar R 1980 *Principles of Quantum Mechanics* 2nd edn (Berlin: Springer)

Schwabl F 2002 *Quantum Mechanics* 3rd edn (Berlin: Springer)

Part SM: Statistical Mechanics

Feynman R P 1998 *Statistical Mechanics* 2nd edn (Boulder, CO: Westview)

Huang K 1987 *Statistical Mechanics* 2nd edn (New York: Wiley)

Kubo R 1965 *Statistical Mechanics* (Amsterdam: Elsevier)

Landau L D and Lifshitz E M 1980 *Statistical Physics, Part 1* 3rd edn (Oxford: Pergamon)

Lifshitz E M and Pitaevskii L P 1981 *Physical Kinetics* (Oxford: Pergamon)

Pathria R K and Beale P D 2011 *Statistical Mechanics* 3rd edn (Amsterdam: Elsevier)

Pierce J R 1980 *An Introduction to Information Theory* 2nd edn (New York: Dover)

Plishke M and Bergersen B 2006 *Equilibrium Statistical Physics* 3rd edn (Singapore: World Scientific)

Schwabl F 2000 *Statistical Mechanics* (Berlin: Springer)

Yeomans J M 1992 *Statistical Mechanics of Phase Transitions* (Oxford: Oxford University Press)

Multidisciplinary/specialty

Ashcroft W N and Mermin N D 1976 *Solid State Physics* (Philadelphia, PA: Saunders)

Blum K 1981 *Density Matrix and Applications* (New York: Plenum)

Breuer H-P and Petruccione E 2002 *The Theory of Open Quantum Systems* (Oxford: Oxford University Press)

Cahn S B and Nadgorny B E 1994 *A Guide to Physics Problems, Part 1* (New York: Plenum)

Cahn S B, Mahan G D and Nadgorny B E 1997 *A Guide to Physics Problems, Part 2* (New York: Plenum)

Cronin J A, Greenberg D F and Telegdi V L 1967 *University of Chicago Graduate Problems in Physics* (Reading, MA: Addison Wesley)

Hook J R and Hall H E 1991 *Solid State Physics* 2nd edn (New York: Wiley)

Joos G 1986 *Theoretical Physics* (New York: Dover)

Kaye G W C and Laby T H 1986 *Tables of Physical and Chemical Constants* 15th edn (London: Longmans Green)

Kompaneyets A S 2012 *Theoretical Physics* 2nd edn (New York: Dover)

Lax M 1968 *Fluctuations and Coherent Phenomena* (London: Gordon and Breach)

Lifshitz E M and Pitaevskii L P 1980 *Statistical Physics, Part 2* (Oxford: Pergamon)

Newbury N *et al* 1991 *Princeton Problems in Physics with Solutions* (Princeton, NJ: Princeton University Press)

Pauling L 1988 *General Chemistry* 3rd edn (New York: Dover)

Tinkham M 1996 *Introduction to Superconductivity* 2nd edn (New York: McGraw-Hill)

Walecka J D 2008 *Introduction to Modern Physics* (Singapore: World Scientific)

Ziman J M 1979 *Principles of the Theory of Solids* 2nd edn (Cambridge: Cambridge University Press)